STATICS AND STRENGTH OF MATERIALS

Robert L. Mott, P.E.
University of Dayton

Prentice Hall

Boston   Columbus   Indianapolis   New York   San Francisco   Upper Saddle River
Amsterdam   Cape Town   Dubai   London   Madrid   Milan   Munich   Paris   Montreal   Toronto
Delhi   Mexico City   Sao Paulo   Sydney   Hong Kong   Seoul   Singapore   Taipei   Tokyo

Vice President and Executive Editor:
Vernon R. Anthony
Acquisitions Editor: David Ploskonka
Editorial Assistant: Nancy Kesterson
Director of Marketing: David Gesell
Senior Marketing Coordinator: Alicia
Wozniak
Project Manager: Maren L. Miller
Managing Editor: Alexandrina Benedicto Wolf
Senior Managing Editor: JoEllen Gohr
Senior Operations Supervisor:
Pat Tonneman

Operations Specialist: Laura Weaver
Senior Art Director: Diane Y. Ernsberger
Text and Cover Designer: Ali Mohrman
Cover Image: Photodisc
AV Project Manager: Janet Portisch
Full-Service Project Management:
Ravi Bhatt/Aptara®, Inc.
Composition: Aptara®, Inc.
Printer/Binder: Edwards Brothers
Cover Printer: LeHigh Phoenix
Text Font: Minion

Credits and acknowledgments borrowed from other sources and reproduced, with permission, in this textbook appear on appropriate page within text.

Library of Congress Cataloging-in-Publication Data

Mott, Robert L.
    Statics and strength of materials/Robert L. Mott. — 1st ed.
      p. cm.
    Includes index.
    ISBN-13: 978-0-13-515982-8
    ISBN-10: 0-13-515982-2
  1. Statics.   2. Strains and stresses.   3. Strength of materials.   I. Title.

  TA351.M68 2010
  620.1'12—dc22   2009007393

Prentice Hall
is an imprint of

10 9 8 7 6 5 4 3 2 1
ISBN-13: 978-0-13-515982-8
ISBN-10:   0-13-515982-2

PROPERTIES OF AREAS

Circle

$$A = \frac{\pi D^2}{4} = \pi R^2 \qquad r = \frac{D}{4} = \frac{R}{2}$$

$$I = \frac{\pi D^4}{64} \qquad J = \frac{\pi D^4}{32}$$

$$S = \frac{\pi D^3}{32} \qquad Z_p = \frac{\pi D^3}{16}$$

$$\text{Circumference} = \pi D = 2\pi R$$

Hollow circle (tube)

$$A = \frac{\pi(D^2 - d^2)}{4} \qquad r = \frac{\sqrt{D^2 + d^2}}{4}$$

$$I = \frac{\pi(D^4 - d^4)}{64} \qquad J = \frac{\pi(D^4 - d^4)}{32}$$

$$S = \frac{\pi(D^4 - d^4)}{32D} \qquad Z_p = \frac{\pi(D^4 - d^4)}{16D}$$

Square

$$A = s^2$$

$$I_x = \frac{s^4}{12} \qquad\qquad r_x = \frac{s}{\sqrt{12}}$$

$$S_x = \frac{s^3}{6}$$

Rectangle

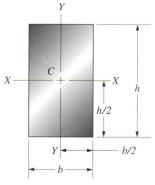

$$A = bh$$

$$I_x = \frac{bh^3}{12} \qquad S_x = \frac{bh^2}{6} \qquad r_x = \frac{h}{\sqrt{12}}$$

$$I_y = \frac{hb^3}{12} \qquad S_y = \frac{hb^2}{6} \qquad r_y = \frac{b}{\sqrt{12}}$$

Triangle

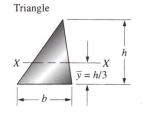

$$A = \frac{bh}{2}$$

$$I_x = \frac{bh^3}{36} \qquad\qquad r_x = \frac{h}{\sqrt{18}}$$

$$S_x = \frac{bh^2}{24}$$

Semicircle

$$A = \frac{\pi D^2}{8}$$

$$I_x = 0.00686D^4 \qquad S_x = 0.0238D^3 \qquad r_x = 0.132D$$

$$I_y = 0.0245D^4 \qquad S_y = 0.0491D^3 \qquad r_y = D/4 = R/2$$

To my wife, Marge;
our children, Lynné, Robert Jr., and Stephen;
and my mother and father.

CONTENTS

Preface   vii

Acknowledgments   x

1 Basic Concepts in Statics and Strength of Materials 1

1–1   The Big Picture—Uses of Statics and Strength of Materials   1

1–2   Basic Concepts of Forces, Moments, and Torques   3

1–3   The Concept of Load Path and Its Relation to Statics and Strength of Materials   8

1–4   Approaches to Problem Solving   15

1–5   Calculations   16

1–6   Unit Systems   16

1–7   Relationship Among Mass, Force, and Weight   17

1–8   Sizes of Standard Shapes   19

2 Forces and Free Body Diagrams 30

2–1   The Big Picture   30

2–2   How Forces from Real Systems Are Represented for Analysis   31

2–3   Components of Forces and Resultants   36

2–4   Forces at Supports   43

2–5   Creating Free Body Diagrams   44

2–6   Building Codes and Bridge Standards   48

3 Static Equilibrium 64

3–1   The Big Picture   64

3–2   Analysis of Static Equilibrium   65

3–3   The Concept of Mechanical Advantage   77

3–4   Pulley Systems   77

4 Forces in Frames, Structures, and Trusses 93

4–1   The Big Picture   93

4–2   Analysis of Pin-Connected Frames and Mechanisms   96

4–3   Analysis of Trusses   107

4–4   Equilibrium in Three-Dimensional Frames and Structures   123

5 Frictional Forces 146

5–1   The Big Picture   146

5–2   Coefficient of Friction   147

5–3   Analysis of Frictional Forces   150

6 Centroids and Moments of Inertia of Areas 169

6–1   The Big Picture—Centroids and Moments of Inertia of Areas   169

6–2   The Concept of Centroid—Simple Shapes   171

6–3   Centroid of Complex Shapes   171

6–4   The Concept of Moment of Inertia of an Area   174

6–5   Moment of Inertia for Composite Shapes Whose Parts Have the Same Centroidal Axis   175

6–6   Moment of Inertia for Composite Shapes—General Case—Use of the Parallel Axis Theorem   176

6–7   Mathematical Definition of Moment of Inertia   178

6–8   Composite Sections Made from Commercially Available Shapes   179

6–9   Moment of Inertia for Shapes with All Rectangular Parts   182

6–10   Radius of Gyration   182

6–11   Section Modulus   186

7 Basic Concepts in Strength of Materials 195

7–1   The Big Picture—Basic Concepts in Strength of Materials   195

7–2   Objective of This Book—To Ensure Safety   196

7–3   The Concept of Stress   200

7–4   Direct Normal Stress   201

7–5   Stress Elements for Direct Normal Stresses   203

7–6   The Concept of Strain   204

7–7   Direct Shear Stress   204

7–8   Stress Elements for Shear Stresses   209

7–9   Experimental and Computational Stress Analysis   209

8 Design Properties of Materials 221

8–1   The Big Picture—Design Properties of Materials 221

8–2   Design Properties of Materials 222

8–3   Steel 233

8–4   Cast Iron 237

8–5   Aluminum 239

8–6   Copper, Brass, and Bronze 240

8–7   Zinc, Magnesium, Titanium, and Nickel-Based Alloys 240

8–8   Nonmetals in Engineering Design 241

8–9   Wood 241

8–10   Concrete 242

8–11   Plastics 244

8–12   Composites 245

8–13   Materials Selection 255

9 Direct Stress, Deformation, and Design 262

9–1   The Big Picture—Direct Stresses Deformation, and Design 262

9–2   Design of Members Under Direct Tension or Compression 264

9–3   Design Normal Stresses 265

9–4   Design Factor 265

9–5   Design Approaches and Guidelines for Design Factors 267

9–6   Methods of Computing Design Stress 270

9–7   Elastic Deformation in Tension and Compression Members 273

9–8   Deformation Due to Temperature Changes 278

9–9   Thermal Stress 281

9–10   Members Made of More Than One Material 283

9–11   Stress Concentration Factors for Direct Axial Stresses 285

9–12   Bearing Stress 287

9–13   Design Bearing Stress 290

9–14   Design Shear Stress 296

10 Torsional Shear Stress and Torsional Deformation 318

10–1   The Big Picture—Torsional Shear Stress and Torsional Deformation 318

10-2   Torque, Power, and Rotational Speed 321

10-3   Torsional Shear Stress in Members with Circular Cross Sections 323

10-4   Development of the Torsional Shear Stress Formula 325

10-5   Polar Moment of Inertia for Solid Circular Bars 326

10-6   Torsional Shear Stress and Polar Moment of Inertia for Hollow Circular Bars 326

10-7   Design of Circular Members Under Torsion 328

10-8   Comparison of Solid and Hollow Circular Members 330

10-9   Stress Concentrations in Torsionally Loaded Members 332

10–10 Twisting—Elastic Torsional Deformation 337

10–11 Torsion in Noncircular Sections 344

11 Shearing Forces and Bending Moments in Beams 355

11–1   The Big Picture—Shearing Forces and Bending Moments in Beams 355

11–2   Beam Loading, Supports, and Types of Beams 359

11–3   Reactions at Supports 364

11–4   Shearing Forces and Bending Moments for Concentrated Loads 367

11–5   Guidelines for Drawing Beam Diagrams for Concentrated Loads 371

11–6   Shearing Forces and Bending Moments for Distributed Loads 376

11–7   General Shapes Found in Bending Moment Diagrams 381

11–8   Shearing Forces and Bending Moments for Cantilever Beams 381

11–9   Beams with Linearly Varying Distributed Loads 382

11–10 Free Body Diagrams of Parts of Structures 383

11–11 Mathematical Analysis of Beam Diagrams 387

11–12 Continuous Beams—Theorem of Three Moments 393

12 Stress Due to Bending 407

12–1   The Big Picture—Stress Due to Bending 407

12–2   The Flexure Formula 409

12–3   Conditions on the Use of the Flexure Formula 411

12–4   Stress Distribution on a Cross Section of a Beam 413

12–5   Derivation of the Flexure Formula 414

12–6   Applications—Beam Analysis 415

12–7   Applications—Beam Design and Design Stresses 417

12–8   Section Modulus and Design Procedures   419

12–9   Stress Concentrations   424

12–10   Flexural Center or Shear Center   428

12–11   Preferred Shapes for Beam Cross Sections   431

12–12   Design of Beams to Be Made from Composite Materials   434

13 Shearing Stresses in Beams   454

13–1   The Big Picture—Shearing Stresses in Beams   454

13–2   Importance of Shearing Stresses in Beams   456

13–3   The General Shear Formula   457

13–4   Distribution of Shearing Stress in Beams   462

13–5   Development of the General Shear Formula   467

13–6   Special Shear Formulas   468

13–7   Design for Shear   471

13–8   Shear Flow   472

14 Deflection of Beams   482

14–1   The Big Picture—Deflection of Beams   482

14–2   The Need for Considering Beam Deflections   485

14–3   General Principles and Definitions of Terms   487

14–4   Beam Deflections Using the Formula Method   489

14–5   Comparison of the Manner of Support for Beams   493

14–6   Superposition Using Deflection Formulas   499

14–7   Successive Integration Method   505

14–8   Moment-Area Method   513

15 Combined Stresses   537

15–1   The Big Picture—Combined Stresses   537

15–2   The Stress Element   539

15–3   Stress Distribution Created by Basic Stresses   540

15–4   Creating the Initial Stress Element   543

15–5   Combined Normal Stresses   545

15–6   Combined Normal and Shear Stresses   550

15–7   Equations for Stresses in Any Direction   554

15–8   Maximum Stresses   556

15–9   Mohr's Circle for Stress   558

15–10   Stress Condition on Selected Planes   569

15–11   Special Case in Which Both Principal Stresses Have the Same Sign   572

15–12   Use of Strain-Gage Rosettes to Determine Principal Stresses   575

16 Columns   591

16–1   The Big Picture—Columns   591

16–2   Slenderness Ratio   593

16–3   Transition Slenderness Ratio   596

16–4   The Euler Formula for Long Columns   597

16–5   The J. B. Johnson Formula for Short Columns   598

16–6   Summary—Buckling Formulas   598

16–7   Design Factors for Columns and Allowable Load   600

16–8   Summary—Method of Analyzing Columns   600

16–9   Column Analysis Spreadsheet   603

16–10   Efficient Shapes for Column Cross Sections   603

16–11   Specifications of the AISC   605

16–12   Specifications of the Aluminum Association   607

16–13   Noncentrally Loaded Columns   607

17 Pressure Vessels   618

17–1   The Big Picture—Pressure Vessels   618

17–2   Distinction Between Thin-Walled and Thick-Walled Pressure Vessels   620

17–3   Thin-Walled Spheres   620

17–4   Thin-Walled Cylinders   622

17–5   Thick-Walled Cylinders and Spheres   624

17–6   Analysis and Design Procedures for Pressure Vessels   625

17–7   Spreadsheet Aid for Analyzing Thick-Walled Spheres and Cylinders   630

17–8   Shearing Stress in Cylinders and Spheres   631

17–9   Other Design Considerations for Pressure Vessels   633

17–10   Composite Pressure Vessels   635

18 Connections   639

18–1   The Big Picture—Connections   639

18–2   Modes of Failure for Bolted Joints   641

18–3   Design of Bolted Connections   642

18–4   Riveted Joints   644

18–5   Eccentrically Loaded Riveted and Bolted Joints   646

18–6   Welded Joints with Concentric Loads   648

Appendix   655

Answers to Selected Problems   712

Index   729

PREFACE

INTRODUCTION

Statics and Strength of Materials is a new book that parallels and complements my book, *Applied Strength of Materials,* Fifth Edition, which has served a broad array of programs emphasizing the applications of engineering mechanics. *Statics and Strength of Materials* offers instructors in a variety of fields the flexibility to arrange curricula differently from the traditional three-credit course. In this time of dynamic change and continuous improvement, educators need textbooks that permit new arrangements of subject matter and new approaches to teaching.

The book is appropriate for virtually any college program wherein strength of materials, often called mechanics of materials, is a fundamental part: *mechanical, manufacturing, aeronautical, civil, construction, construction management,* and *architecture.* The emphasis on applications in all of these fields makes the book attractive for *engineering technology, applied engineering,* and *industrial technology programs,* in either *associate degree* or *bachelor degree* arrangements.

PHILOSOPHY FOR THE BOOK

Students must be competent in the topics of statics before fully comprehending the topics of strength of materials. Many instructors and curriculum designers, however, believe that the order of coverage and the integration of the principles of statics with those of strength of materials can be accomplished better within a single book, regardless of whether the subjects are taught separately or in the same course. While some have taught an integrated statics and strength of materials course for quite some time, it has become more popular in recent years.

Statics and Strength of Materials provides strong fundamentals on which students and professionals can build additional skills and knowledge in their chosen specialty. Study of the subjects of statics and strength of materials occurs early in the curricula of both two-year and four-year programs, typically in the second, third, or fourth semester or the corresponding quarters.

The book presumes that students have acquired competencies in the principles of mathematics and physical science to include at least college algebra and trigonometry (precalculus), and basic physics mechanics. Some programs do require coursework in calculus and, therefore, the book uses calculus primarily in the sections where basic concepts and fundamental equations of strength of materials are developed and they are organized so their use in the course is

optional. This approach has been used successfully in all three of the author's other books. Equilibrium analyses are performed with scalar, nonvector approaches.

The subjects of statics and strength of materials are inextricably linked. A significantly large and growing group of mechanics educators takes advantage of this linkage by combining both subjects in the same course. Sometimes this is done with the application of fewer credit hours than if the two courses were taught separately. For example, in a semester system the separate statics and strength of materials courses would typically be three credits each (six total) while the combined statics and strength of materials course may be a total of four or five credits. This book allows flexibility in curriculum planning while providing a comprehensive set of topics from which a given program can select those appropriate to its mission.

The statics and strength of materials topics are integrated to a large degree to avoid duplication and to focus on the *essential* concepts and methods. As topics in statics are presented in the text and example problems, students are informed of how the subsequent coverage of strength of materials concepts use the results of the statics analyses and references are given to later chapters. The concepts of stress, strain, and deformation are introduced conceptually early in the book so students can relate to the effects produced by the forces and other loads being analyzed using statics principles. Thus, students acquire a qualitative appreciation of why careful determination of forces, torques, and moments is essential to analysis and design. Also, the book includes a wide variety of applications in the several fields for which the study of strength of materials is needed so students gain an appreciation for related disciplines in addition to their chosen specialty.

In the strength of materials parts of the book, students continually use the techniques learned in the statics sections to analyze the behavior of load-carrying members to determine the internal stresses and deformations caused by externally applied forces, moments, and torques. Furthermore, students are introduced to the design of such members to ensure that they are safe and that they will perform satisfactorily in given applications. Included is the study of the strength and elastic properties of materials and how they are related to a load-carrying member's responses to external loads. The text and appendix content on the design properties of materials in the author's *Applied Strength of Materials* has earned high praise from users and is included in this book as well.

Units are a mixture of SI metric and U.S. Customary units, in keeping with the dual usage evident in U.S. industry and construction.

FEATURES OF THE BOOK

The Big Picture

Students should see the relevance of the material they study. They should be able to visualize where devices and systems that they are familiar with depend on the principles of strength of materials. For this reason each chapter starts with a section called *The Big Picture*. Here the basic concepts developed in the chapter are identified and students are asked to think about examples from their own experience. Sometimes they are asked to explore new things on their own to discover how a product works or how it can fail. They are coached to make observations about the behavior of common mechanical devices, vehicles, industrial machinery, consumer products, and structures. Educational philosophy indicates that students learn better and retain more when such methods are employed.

Activity-Based Learning

Activity-based learning is integrated into the popular Big Picture section, a successful feature in all of the author's books. The activities can be used independently by the students, by the instructor as a classroom demonstration, or a combination of these approaches. These activities allow the instructor and the students to extend the Big Picture dialog into hands-on experiences that give an enhanced appreciation and greater physical feel for the phenomena involved. Activities can help students from different disciplines work together and learn from each other. The activities are generally simple and can be completed in a short amount of time with inexpensive materials and quick setups. The emphasis is on qualitative appreciation of the physical phenomena with a modest amount of measurement involved. Educational research has shown that students learn better when they are personally involved in activities as opposed to listening to lectures. Furthermore, retention of abilities learned is improved along with greater ability to transfer learning to new and different applications.

Problem-Solving Techniques

Students must be able to solve real problems, complete the necessary calculations, manipulate units in equations, seek appropriate data, and make good design decisions. The example problems in this book are designed to help students master these processes. In addition, students must learn to communicate the results of their work to others in the field. One important means of communication is the presentation of the problem solutions in an orderly, well-documented manner using established methods. Example problems are set off with a distinctive graphic design and type font. Readers are guided in the process of formulating an approach to problem solving that includes

  a. Statement of the objective of the problem
  b. Summary of the given information
  c. Definition of the analysis technique to be used

  d. Detailed development of the results with all of the equations used and unit manipulations
  e. At times, comments on the solution to remind the reader of the important concepts involved and to judge the appropriateness of the solution
  f. At times, comments present alternate approaches or improvements to the machine element or structural member being analyzed or designed

The reader's thought process is carried beyond the requested answer into a critical review of the result. With this process, designers gain good habits of organization when solving their own problems.

Design Approaches

This text provides more extensive guidelines for design of mechanical devices and structural members than in most books on this subject. The design approaches are based on my other book, *Machine Elements in Mechanical Design.* References are also included to design methodology espoused by the civil engineering and construction industries through the American Society of Civil Engineers, the American Institute of Steel Construction, and related organizations. Design details are typically presented in follow-up courses in these fields.

Learning about design in addition to analysis increases the usefulness of the book to students and professional users. There are some students who will not go on to a following course that emphasizes design. They should get some introduction to the principles of design in the introductory course in strength of materials. For those who do proceed to a design course, they should enter that course with a higher level of capability.

Design Properties of Materials

Chapter 8 includes extensive information and discussion on the proper application of engineering materials of many types, both metallic and nonmetallic. There is an extensive introduction to the nature of composite materials given along with commentary throughout the book on the application of composites to various kinds of load-carrying members. Information about the advantages of composites relative to traditional structural materials such as metals, wood, concrete, and plastics is given. The reader is encouraged to seek more education and experience to learn the unique analysis and design techniques required for the proper application of composite materials. Such materials are, in fact, tailored to a specific application, and general tables of material properties are not readily available.

End-of-Chapter Problems

There is an extensive set of problems for student practice at the end of each chapter. The problems are typically organized around the main topics in the chapter. In general, they are presented in a graded manner with simpler problems followed by more comprehensive problems. There are many additional problems at the end of most chapters for practice, review, and design.

Extensive Appendix

To complement the use of design approaches, the Appendix provides additional information on material properties, geometry of common areas and commercially available structural shapes, stress concentration factors, formulas for beam deflection, conversion factors, and many others. This allows for a wider variety of problems in the book and for creating tests and projects. It adds to the realism of the book and gives the student practice in looking for the necessary information to solve a problem or to complete a design.

This book includes a significant amount of data in SI metric units. All commercially available section property data for structural shapes include separate tables of SI data in addition to U.S. Customary unit data. The SI data are taken from the latest versions of publications by the American Institute of Steel Construction (AISC). The SI data tables and the U.S. data tables are coordinated so students and instructors can quickly compare the designations and specific data from the two systems. Problems stated in SI metric data should be solved using the SI property data; instructors can develop their own quiz and exam problems completely in the SI system.

LEARNING AIDS FOR PROBLEM SOLVING AND DESIGN

Many chapters include *computer assignments* along with suggestions for the use of spreadsheets, computer programs, computer algebra software, and graphing calculators pertinent to strength of materials. Such electronic aids, when used to supplement the basic understanding of the principles presented in the book, lead to a deeper appreciation of those principles and their application to simple and complex problems. Examples of spreadsheets are given in the chapters on column analysis and pressure vessels.

Lists of *Internet sites* that are pertinent to the topics covered in the chapter are included at the end of most chapters. The sites direct users to a wealth of additional resources to supplement the text. Some sites are for vendors of commercially available products or materials, while others provide additional examples of analyses done in the book or more in-depth discussion of certain topics. Users are encouraged to perform their own searches to explore other aids to understanding of strength of materials.

The accompanying *CD-ROM* contains useful software to help students understand several important principles of strength of materials. An extensive set of learning aids reinforces the book's presentation of key topics, with dynamic, colorful illustrations and examples worked out in a step-by-step manner. A notation is given in the margin of the sections of the book where these learning aids are pertinent.

A *powerful beam analysis software package* is included that is useful for five chapters of the book dealing with shearing force and bending moments in beams, centroids and moments of inertia, stresses due to bending, shearing stresses in beams, statically indeterminate beams, and beam deflections (Chapters 11–15).

Instructors and students are cautioned about proper use of the beam analysis software:

> *Users of computer software must have a solid understanding of the principles of analysis and design to ensure that analysis results and design decisions are based on reliable foundations. We recommend that the beam analysis software be used only after mastering pertinent principles on which the software is based using careful study and manual techniques.*

The advantage of using the software after appropriate prior study is that many additional and more complex practice problems can be solved in the allocated time to give students more in-depth understanding of the behavior of beams for a variety of beam geometries, beam materials, loading patterns, and support conditions. They can pursue several alternative designs and work toward more optimum results.

Physical models have been constructed to aid students' visualization of design details for objects and structures for selected example problems or practice problems for student solution in both the statics and the strength of materials parts of the book. Photographs of the physical models are included in the book. Also, colorful three-dimensional computer-aided design drawings of the objects are in the CD-ROM that contains the learning modules and the beam analysis software. Extensive feedback from students who used some of the physical models in classes attests to their effectiveness in improving understanding of design details and the analyses required to determine forces, stress, and deformations.

An online Instructor's Manual is also available to instructors. To access supplementary materials online, instructors need to request an instructor access code. Go to www.pearsonhighered.com/irc, where you can register for an instructor access code. Within 48 hours after registering, you will receive a confirming e-mail, including an instructor access code. Once you have received your code, go to the site and log on for full instructions on downloading the materials you wish to use.

Visual Attractiveness

The use of a second color makes the book visually appealing; illustrations, graphs, and tables are easier to use and interpret. Many illustrations employ three-dimensional graphics, a high level of realism, and effective use of shading.

ACKNOWLEDGMENTS

I appreciate the feedback provided by both students and instructors who have used the earlier editions of my applied strength of materials book. I am also grateful to my colleagues at the University of Dayton. I would like to thank the participants of a focus group that provided input for this book: Janice Chambers, Portland Community College; Janak Dave, University of Cincinnati; David Dvorak, University of Maine; Frank Gourley, West Virginia University Institute of Technology; and Jack Zecher, Indiana University–Purdue University at Indianapolis (IUPUI). I would also like to thank my reviewers. Anthony L. Breizendine, University of North Carolina, Charlotte; Charles Drake, Ferris State University; Gary S. Drigel, Miami University; Jason Durfee, Eastern Washington University; Mohammed F. Fahmy, University of Northern Iowa; R. L. Alan Jordan, Purdue University; Tang-Hung Nguyen, California State University, Long Beach; Basavapatna S. Sridhara, Middle Tennessee State University; and Arnold Wallace, Jr., Camden County College, reviewed the text, and William A. Wood, Youngstown State University, and Jack Zecher, IUPUI, reviewed the software to accompany the text. Thank you all for your valuable input.

Special recognition is given to Professor Jack Zecher, Indiana University–Purdue University at Indianapolis (IUPUI), for his expert development of the software provided on the CD that accompanies this book. The interactive learning modules and the beam analysis software blend well with the text and offer students and instructors valuable supplemental resources.

Professor Gary Drigel, Miami University of Ohio, also deserves special recognition for producing the physical models to aid students' visualization of design details for objects and structures for selected example problems or practice problems for student solution in both the statics and the strength of materials parts of the book. He also created the colorful three-dimensional computer-aided design drawings of the objects in the CD-ROM that contains the learning modules and the beam analysis software. His careful surveying of student learning and student reactions to the use of such physical models in the classroom validates their effectiveness.

Robert L. Mott
University of Dayton

CHAPTER ONE

BASIC CONCEPTS IN STATICS AND STRENGTH OF MATERIALS

MAJOR TOPICS

1–1 The Big Picture—Uses of Statics and Strength of Materials
1–2 Basic Concepts of Forces, Moments, and Torques
1–3 The Concept of Load Path and Its Relation to Statics and Strength of Materials
1–4 Approaches to Problem Solving
1–5 Calculations
1–6 Unit Systems
1–7 Relationship Among Mass, Force, and Weight
1–8 Sizes of Standard Shapes

1–1 THE BIG PICTURE—USES OF STATICS AND STRENGTH OF MATERIALS

The principles of *statics and strength of materials* are used in any case in which forces are exerted on members that must be designed to be safe under the influence of those forces. Consider the examples shown in Figure 1–1:

- Part (a) shows a new home being constructed with a framework made from wood. Note the vertical studs for the walls, the horizontal beams and base plates, the lintels over the openings for windows and doors, and the sloped rafters and gables of the roof structure. Each of these components must be designed to carry forces exerted by other structural members, wind, snow, and the weight of people, furniture, appliances, furnace, air conditioner, and other equipment inside the home.

- Part (b) shows a sleek bicycle. The need for careful analysis of forces exists for every component—the main frame, front and rear forks, seat and its support system, handlebars, crank and pedals, wheel rims, spokes, axels, brake system, and shifting mechanism.

- Part (c) shows the engine for a hydrogen-powered car, containing virtually hundreds of mechanical components that experience forces and other types of loading, such as pressures, torques, and moments. Consider the pistons, connecting rods, valves, valve actuators, springs, crank, oil pump, driveshaft, transmission components, and numerous others. Some of these components and subsystems are discussed in more detail later in this chapter.

- Part (d) shows construction equipment on a job site. The track-mounted excavator digs the earth and transfers the heavy load from its bucket to a large dump truck for movement to another location. Look at the booms extending from the cab of the excavator, the hydraulic fluid power cylinders that move them through the graceful motions from the earth to the truck, the bucket itself, the support system for the cab that allows it to rotate while carrying heavy loads, the drive system that moves the track system, and many structural elements of the machine. For the truck, consider the frame, drive system, its engine, the steering system, brakes, suspension, load hopper, and the dumping mechanism.

- Part (e) shows a huge passenger aircraft designed to carry aloft several hundred people, baggage, freight, and the weight of the massive airframe, engines, landing gear, control systems, and fuel. Designers must analyze forces on the structural members; actuators for controls on the wings, tail, and horizontal stabilizers; engine components that develop the power required for takeoff, cruising at high altitudes, and safe landings; doors, hatches, seats, tray tables, overhead storage compartments, and cockpit systems.

Other examples can be drawn from many fields such as the structures of buildings, the frames of machines, mechanical devices, components of power transmission systems, test and measurement equipment, manufacturing machinery, as well as the manufacturing processes, construction equipment, and the structures and components of land, sea, air, and space vehicles. A few examples are discussed in more detail later in this chapter to give you a feel for the importance of the study of this field and the relationships between statics and strength of materials. Consider the following basic definitions of the terms you will use while studying this book.

(a)

(d)

(b)

(e)

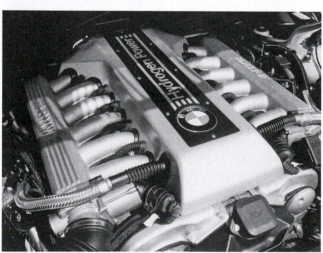

(c)

FIGURE 1–1   Examples of mechanical and structural systems in which the principles of statics and strength of materials are used.

Source: (a) The Stock Connection; (b) Doring Kindersley Media Library; (c) Photo Researchers, Inc.; (d) The Stock Connection; (e) Doring Kindersley Media Library.

1–2 BASIC CONCEPTS OF FORCES, MOMENTS, AND TORQUES

Forces

Let's start with the definition of a *force*.

> *A force is a pulling, pushing, or cutting effort exerted on one body by another that tends to move it or change its shape.*

The following paragraphs define several types of forces and describe the kind of failure mode produced by each. In the first part of this book, *Statics,* you will learn how to analyze many cases in which such forces act, calculating how they act together and how they must be balanced to ensure that a structure or a machine element is in stable equilibrium. In general, the term *statics* refers to the analysis of forces that are exerted on load-carrying members that are stationary or, if they are moving, without regard for the dynamic effects of the accelerations of the members themselves.

In Part 2 of this book, *Strength of Materials,* you will learn how to analyze the stresses and deformations caused by forces, to design load-carrying members to be safe under expected loads, and to verify that they will perform to specifications.

Tensile Forces   Pulling on a rope attached to a heavy machine produces a *tensile force* in the rope that tends to stretch it. The rope will break in tension if the pulling force is higher than the *tensile strength* of the rope itself. You can demonstrate the effect of tensile forces by pulling on a relatively light string fastened to a rigid support as shown in Figure 1–2(a). Hanging a heavy load from a string can also be used to

demonstrate tensile forces. What other examples of members subjected to tensile forces can you find? You will study direct tensile stresses in Chapters 7–9 in Part 2 of this book.

Compressive Forces   You have likely experienced pushing on a piece of clay to change its shape by crushing it. Any load that tends to crush an object is called a *compressive force.* Figure 1–2(b) shows samples of different shapes and materials that could be crushed by placing them between the jaws of a vise. Try this yourself and observe how much force is required to produce a noticeable change in the shape of the sample or to break it completely. An example of a structural member subjected to compressive forces and stresses is a short concrete cylinder supporting a load at the base of a column. Find examples of such bases around buildings you visit today. The concrete cylinder will break if the compressive force is higher than the *compressive strength* of the concrete. You will study direct compressive stresses in Chapters 7–9. When the sample has a relatively small cross section, be sure it is short so that it fails by crushing. If a compressive force is applied to a long, slender member, it would fail by column buckling as shown in Figure 1–2(c). You can demonstrate that with a simple ruler having a thin cross section. Part (d) of the figure shows several long, slender rods that would fail by column buckling if loaded in compression. You will study columns in Chapter 16.

Shearing Forces   Scissors or shears are devices used to cut paper, cloth, or sheets of metal by applying two forces simultaneously in opposite directions to produce the cutting action. Such forces are called *shearing forces* and they produce *shearing stress* in the materials. Another important example is the process of cutting shapes from flat

(a) Tensile force applied to a string

FIGURE 1–2   Examples of simple setups to demonstrate a variety of load-carrying members and kinds of stresses produced by different load and support conditions.

FIGURE 1–2 (*Continued*)

(b) Vise for applying compressive forces with an array of sample specimens

(c) Applying a compressive force to a long, thin, slender ruler illustrating column buckling

(d) Long, slender rods that would fail by buckling under compressive forces

(e) Handheld paper punch designed to apply shearing forces to cut holes in paper

(f) Sheet metal parts with holes and slots punched out to illustrate punching shear

FIGURE 1–2   (*Continued*)

(g) Clevis and pin illustrating shearing forces on the pin   (h) Simply supported beams subjected to bending moments

sheets of metal, plastic, paper, card stock, or other materials. Figure 1–2(e) shows the familiar handheld paper punch. Part (f) of the figure shows several objects made from sheet metal with holes, slots, or curved edges punched out. Several of the objects were also cut out of larger sheets. These are examples of actually making the material fail in shear to produce a desirable part. At other times, you may *not* want the material to be cut through. Consider the pin joint shown in Figure 1–2(g). One part, called a *clevis,* is held to a rigid support. A cylindrical pin then connects that part to a link that can swivel around the pin. When a tensile force is applied to the link, it produces shearing forces that tend to cut the pin. In Chapter 9, you will study shearing stresses caused by such shearing forces and design the pin to be safe under expected loads.

Moments

When a force is applied that causes bending or rotation of a member, we say that it produces a *moment* with respect to some point of interest. The fundamental definition of a moment follows:

> *A moment is the product of a force times the perpendicular distance from the line of action of the force to a point of interest.*

An important example of a load-carrying member subjected to moments is illustrated in Figure 1–2(h). Three relatively long, thin, flat members (such as the ruler in the back) are supported near their ends. A load, in this case a D-cell battery, is placed on the ruler such that its weight applies a force to the ruler. What happens to the ruler? It bends downward because of the *bending moments* developed in the ruler under the influence of the downward-acting weight of the battery and the upward-acting forces

provided by the two supports. When loaded by a force acting perpendicular to its long axis, the ruler is called a *beam.* You will learn more about bending moments and beams in Chapters 3 and 11–14.

Figure 1–3 shows another important application of moments. A woman is pushing a refrigerator across a floor. Pushing to the left is required to move the refrigerator in the desired direction. The floor provides the vertical force to support the weight. As the refrigerator tends to move to the left, frictional forces between the feet and the floor oppose that tendency. But what else could happen? Applying the force at a distance h above the floor causes a counterclockwise tipping moment of

$$M_1 = F \cdot h \quad \text{Counterclockwise}$$

Depending on the placement of the applied force and the amount of frictional force provided at the feet, the refrigerator could either slide across the floor as shown in (*b*) or tip on the feet at the left side as shown in (*c*). Note that the clockwise moment caused by the weight of the refrigerator acting downward through its center of gravity opposes the moment produced by the applied force. The opposing moment is

$$M_2 = W \cdot (w/2) \quad \text{Clockwise}$$

where w is the width of the refrigerator.

Try this for yourself with any object having a rectangular shape that is relatively taller than it is wide (e.g., a cereal box). Pushing on the side of a cereal box at a point near its bottom causes the box to move as desired. Now pushing progressively farther up on the box will cause it to tilt with respect to one of the bottom edges and fall over. This activity demonstrates that applying a simple pushing force has the effect of tending to rotate an object.

(a) Woman pushing a refrigerator

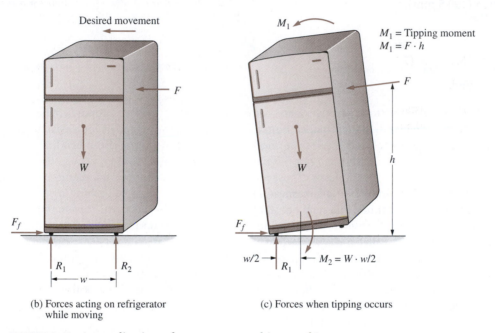

(b) Forces acting on refrigerator
    while moving

(c) Forces when tipping occurs

FIGURE 1–3   An application of moments—pushing a refrigerator.

That tendency occurs because the product of the force times the distance from its line of action to the edge of the box about which the box tends to rotate is an applied *moment*.

Let's do a numerical example. Considering the refrigerator, assume that the woman pushes horizontally with a force of 320 N (72 lb), applying the force 1.50 m (59 in) above the floor. The moment with respect to the feet of the refrigerator would be

$$\text{Moment} = M = \text{Force} \times \text{distance}$$
$$= (320 \text{ N}) \times (1.50 \text{ m}) = 480 \text{ N} \cdot \text{m}$$

Using U.S. units, we find

$$\text{Moment} = M = (72 \text{ lb}) \times (59 \text{ in}) = 4248 \text{ lb} \cdot \text{in}$$

That moment is a measure of the tendency to overturn the refrigerator. Whether it tilts or slides along the floor depends on several other factors, such as the placement of the center of gravity of the refrigerator, the magnitude of the applied force, and the amount of frictional force between the feet of the refrigerator and the floor. You will study this case in more detail in Chapter 5.

Torques

A torque is similar to a moment in that it is created by a force acting at a certain distance from a point of reference. However, we use the term *torque* to apply to cases in which actual rotation occurs or there is a tendency to twist a member about its long axis.

Figure 1–4 shows hands grasping a flexible cylinder by its ends and twisting it by rotating the hands in opposite directions. We say that the person is applying a torque to the cylinder. Note how the cylinder deforms as it twists.

Figure 1–5 shows another familiar example. A socket wrench is attached to a bolt head while a force F is applied near the end of the wrench handle at a distance d from

Original grid

(a)

Deformed grid

(b)

FIGURE 1–4   Twisting of a cylindrical rod.

the axis of the bolt. If the force is 15.5 lb (68.9 N) and the distance is 7.50 in (190.5 mm), we then say that the applied torque is

$$\text{Torque} = T = \text{Force} \times \text{distance} = F \times d$$
$$= (15.5 \text{ lb}) \times (7.50 \text{ in}) = 116.3 \text{ lb} \cdot \text{in}$$

Using SI units, we find,

$$\text{Torque} = T = \text{Force} \times \text{distance} = F \times d$$
$$= (68.9 \text{ N}) \times (190.5 \text{ mm}) = 13\ 125 \text{ N} \cdot \text{mm}$$

A more convenient unit is the N · m. Using unit conversion, we find

$$T = (13\ 125 \text{ N} \cdot \text{mm}) \times (1.0 \text{ m})/(1000 \text{ mm}) = 13.125 \text{ N} \cdot \text{m}$$

Expressing this value with typical three-place precision, we could report the result as 13.1 N · m.

Note that the calculations and unit manipulations for torques are virtually the same as for moments. The difference is the intended outcome. Moments tend to bend beams

or tilt solid objects. Torques tend to rotate members, usually circular rods or shafts, about their long axes.

Among the important applications of torques are power transmission shafts, torsion bars, and certain links in mechanisms. Figure 1–6 shows an engine and transmission powering a driveshaft for a boat. The propeller driveshaft, the engine crank, and several other shafts internal to the transmission and the engine are subjected to applied torques that result in rotational motion and/or *torsional shearing stresses* in the members. You will study torsional shearing stresses in Chapter 10.

Couples

A special case that can be added to the discussion of either moments or torques is the concept of a *couple*, defined as

> *A couple is a pair of equal forces acting in opposite directions and aligned parallel to each other.*

Care is required to compute the magnitude of a couple:

> *The moment or torque created by a couple is the product of one applied force times the perpendicular distance between the two forces.*

A simple example, called a T-wrench, is shown in Figure 1–7. You may be familiar with this if you work on cars, trucks, motorcycles, or bicycles as they are sometimes used for tightening lug nuts on wheels or assembling axles with multiple parts. An advantage of a T-wrench is that the applied forces are balanced and a pure torque is applied to the bolt or nut to drive it home. Compare this with the standard wrench shown in Figure 1–5, for which you would have to apply an opposing force in the opposite direction to the applied force to keep the wrench positioned properly over the bolt or nut. Typically, you would apply the opposing force at the socket and in the same plane with the applied force. If you do not, you would apply a significant bending moment to the bolt or create a moment that tends to disengage the socket from the bolt head or nut.

Calculations are simple. If you apply two equal forces of 150 N to the ends of the T-wrench shown in Figure 1–7, the torque applied to the bolt is

F

d

Socket wrench

Socket

$\text{Torque} = T = F \times d$

Extension shaft

Bolt

T

FIGURE 1–5   Wrench applying a torque to a bolt.

$$T = 150 \text{ N} \times 350 \text{ mm} \times (1 \text{ m}/1000 \text{ mm}) = 52.5 \text{ N} \cdot \text{m}$$

FIGURE 1–6   Drive system for a boat.

FIGURE 1–7   T-wrench illustrating a couple.

More will be said about couples in Chapter 3 where they will be used in more general applications of the principle of static equilibrium.

1–3 THE CONCEPT OF LOAD PATH AND ITS RELATION TO STATICS AND STRENGTH OF MATERIALS

We discuss here several examples in more detail to explain how the application of forces, moments, and torques must be analyzed to determine how they are passed along a *load path* and how they produce stresses and deformations that must be proven to be safe and acceptable to the intended use of the system. Examples are drawn from typical structures, mechanical devices, and power transmission systems with which you should be familiar. Only qualitative analyses are illustrated here to familiarize you with the concepts of load paths. Later chapters include quantitative examples and practice problems.

Structures

When you look at a platform supporting a machine above you (see Figure 1–8), it is obviously important that the designer of the platform considered all of the loads applied to it and that the design of the structure and its component members assures you that it will be safe under the influence of those loads. Otherwise the platform will collapse and the machine will fall causing injury or even death to those under it and costly damage to the platform and the machine.

Designers of such structures must apply the principles of statics to determine all the applied loads and how they will be supported. In the case of the platform shown in Figure 1–8, the loads are

- the weight of the machine
- the weight of people who install, operate, and service the machine
- the weight of materials that the machine processes
- the weight of accessory equipment that may be located on the platform such as storage racks and bins
- the weight of the components of the platform itself

Application of the principles of statics enables the designer to compute the magnitude and direction of the forces that act on each load-carrying member of the structure.

The designer decides where the machine is to be located and where the support points are to be placed on the platform. The load path is then determined to describe how the forces due to the weight of the machine are transferred in turn from those support points to the ultimate support in the earth or to some other known stable foundation. Figure 1–8 illustrates the following design considerations:

1. The weight of the machine is supported on four legs and feet of the machine structure. The designer must know how the weight of the machine is distributed. Is it uniform throughout so that each foot carries one-fourth of the total weight? Or does each foot carry a different load? To answer these questions, the designer must know where the *center of gravity*, sometimes called the *center of mass*, of the machine is located in relation to the supporting feet.

FIGURE 1–8 Platform supporting a machine.

2. The forces are transmitted from the feet of the machine through support pads that are in turn mounted to the floor system of the platform. The feet of the machine and all parts of the support pads must be able to carry the loads without being crushed. The tendency to be crushed is called *compressive stress* and the designer must ensure that the magnitude of the stresses in the support pads is safe.

3. Notice that the pads are wide to distribute the load over a fairly large area of the floor of the platform so it does not indent the floor material. This requires the analysis of the *bearing stress* between mating surfaces.

4. The floor system transfers the forces from the machine and all other loads (such as people, process materials, and other equipment) to rails along the four edges of the platform. What could be used for the floor system? Perhaps heavy wood boards could span the space between the rails (see Appendix A–4 for standard sizes). Steel or aluminum channels (see Appendixes A–6 and A–10) could be laid side by side with their flat sides up providing the flat surface of the floor. Hollow structural sections (HSS shapes) like those shown in Appendix A–9 could be used. The boards, channels, or HSS shapes in such a design would act as beams, carrying the downward forces from the support pads to the rails. They must be strong enough not to break under the loads, but they also must not deflect too much so as to hold the machine steady. Are there other designs you can think of for the floor system?

5. The side rails also act as beams, carrying the loads to the vertical columns at the end of each rail. Note that the loads transferred to the rails from the floor system include all forces on the platform. The rails must be strong and rigid under the applied loads. In Figure 1–7, the rails shown are called "I-beams" with the name coming from the I-shape of the cross section. This shape is an excellent choice for the rails because it provides a strong, stiff beam with efficient use of material. Commercially available I-beam shapes are shown in Appendixes A–7, A–8, and A–11.

6. The force on the top of the columns pushes down, tending to crush the material from which the column is made. Thus, it experiences a compressive stress. If the column is relatively long and slender, it could *buckle*, which means it could become severely deformed by bowing out between its ends and eventually collapsing suddenly—a very dangerous mode of failure. The columns appear to be square HSS sections, although I-beam shapes or hollow circular pipes or tubes (Appendixes A–12 and A–13) could also be used.

7. At the base of each column is a support system, such as a concrete cylinder, to transfer the load to the floor of

the building or directly to a foundation supported in the earth. All parts of the support system must be designed to ensure that the *compressive strength* and the *bearing strength* of each material carrying the load are adequate.

The description of the platform includes many situations in which the principles of statics and strength of materials must be applied carefully to ensure safety and proper function. Load paths in a typical wood frame house are discussed next. Many other kinds of structures could have been chosen for the example: commercial or industrial buildings;

machine frames; or structures of automobiles, trucks, aircraft, or spacecraft.

Wood Frame House

Figure 1–9 illustrates nine stages of building a wood frame house with two stories and a basement. Studying the details can help you to follow the load paths through the structure. This, in turn, will help you complete the necessary analyses to ensure that each component of the home is safe and performs to required standards. Later chapters discuss the external loads experienced by such structures, the analysis of

(a) Design plans for the house

(b) Foundation–concrete walls and floor

(c) Beams added to support first floor

(d) First floor–joists and end framing

(e) First-floor walls with door and window frames

(f) Beams added to support second floor

FIGURE 1–9 Stages of building a wood frame house.

Source: Jordan Long, Author, Nascor International Ltd., Calgary, Alberta, Canada.

FIGURE 1–9   (*Continued*)

(g) Second floor–joists and end framing

(h) Second-floor walls

(i) Roof trusses and related framing

forces, and the stress analysis of load-carrying members. Follow along with the description of the steps:

a. **Plan Generation:** The architect and structural engineer generate a plan and detailed drawings showing the arrangement of components from the foundation through the roof.

b. **Foundation:** Contractors clear the land where the house will be located and excavate as needed to accommodate the foundation for the building. Typically, soil conditions permit the use of a wide concrete footer that rests directly on the earth or on a gravel bed spread over the earth. The *bearing strength* of earth and other materials are listed in Chapter 9. Concrete walls that rest on the footers are then poured and cured. Their functions are to support the loads of the structure above them and to resist the forces exerted by the earth on the sides of the walls. The basement floor would also be poured.

c. **First-Floor Posts and Beams:** Most homes require beams in the interior spaces of the basement to support the floor above. Beam analysis and design are discussed in Chapters 11–14. The beams are supported on vertical columns or posts. See Chapter 16 for column analysis and design. The posts are set on pads that perform a

similar function as the footer for the walls: the transfer of loads from the posts to the earth.

d. **First-Floor Assembly:** The support system for the first floor of the home is typically made from joists that act as beams spanning distances between two beams or from a basement wall to a beam. End framing is added as needed on the outside perimeter. A subfloor is then attached to the joists with nails or screws to ensure that the floor system acts as a single unit. Finished flooring is then attached to the subfloor. Floor loading is discussed in Chapter 2. The forces and stresses in the fasteners are discussed in Chapter 13.

e. **First-Floor Walls:** Exterior walls, interior load-bearing walls, and partitions are then installed on the floor to provide the desired living spaces. Walls are typically made from studs using standard-sized boards as listed in Appendix A–4, with sheeting applied to the outside and interior wall materials (e.g., gypsum wallboard or plaster). The combination provides a stiff, strong structure to resist the loads provided by the weight of the upper part of the house and wind loads. Framing is included for windows and doors.

f. **Upper-Floor Beams:** Beams may be needed to support the upper-floor structure, load-bearing walls, and special equipment such as bathtubs or air conditioners.

(a) Complete assembly

(b) Free body diagram of upper jaw piece

FIGURE 1–10   Pliers.

g. **Upper-Floor Assembly:** Joists, end framing, subfloor, and finished flooring are added.

h. **Upper-Floor Walls:** Continuations of the exterior walls and interior partitions for the upper-floor rooms are then added.

i. **Roof System:** Typically, the roof is sloped to promote runoff of rain and most of the snow in cold climates. Roof loads are discussed in Chapter 2. The roof is supported on rafters or trusses that rest on the exterior walls of the house. The analysis of trusses is discussed in Chapter 4. The primary stresses in truss members are either direct tension or direct compression and are discussed in Chapters 7 and 9.

Mechanical Devices

From a different field, a second example of how you can use the principles of statics and strength of materials is shown in Figure 1–10. A simple pair of pliers is used to grasp an object like the head of a bolt with a large force while exerting a moderate force on the handles.

Note in Figure 1–10 that two different drawings are used to show the forces of interest on the parts of the pliers. Part (a) shows the complete assembled pair of pliers consisting of the upper and lower jaws with their corresponding handles and the pin connecting the two that allows the jaws to be rotated by the application of forces on the handles. This figure shows the forces on the handles and the resulting forces acting on the jaws as they clamp on the head of a bolt. One of the jaw pieces is shown in part (b), isolated from the other parts. Here we show the force on the handle, the force on the jaw, and a third force at the place where the pin contacts the hole. A *free body diagram* of a load-carrying member shows all of the applied forces acting on it. Free body diagrams are essential in performing force analyses and more will be said about them in Chapter 2.

Three important features of the force analysis of the pliers are discussed here.

1. Statics can be used to determine the relationship between the clamping force at the jaws and the applied forces on the handles.

2. Another important force is at the pin joint between the two moving parts. The pin acts as a pivot or fulcrum and the force on the pin tends to cut it by shearing it in two. If you were the designer of the pliers, you would need to know how to compute these forces and to ensure that the pin does not fail in shear.

3. The handles must be carefully designed to be safe against the applied loads that produce moments that tend to bend them.

Power Transmission Systems

When a motor or an engine drives a vehicle or a machine, power is transmitted by shafts, through couplings, and by chain drives, belt drives, or gears. Because these machine elements are not stationary, a complete analysis requires principles of dynamics, which are not covered in this book. However, when operating at a constant speed, the same relationships that we will develop for statics can be used to compute the forces acting on various parts of the power transmission system. Then the principles of strength of materials can be used to analyze the stresses produced in the members to ensure safety and long life.

Consider the drive system for a boat shown in Figure 1–11. The engine and transmission are mounted inside the boat and the driveshaft and propeller extend into the water

FIGURE 1–11   Drive system for a boat.

Spark plug

Exhaust valve

Combusting
gases

Net force
on piston

Cylinder

Piston   Pin

Connecting
rod

Total force on
crank pin

Crank

Component force
that produces torque

Torque on
crank

FIGURE 1–12   Engine components.

to provide the thrust to move the boat. The following list points out several ways in which the analytical techniques you will learn in this book can be used to design components of such systems.

1. Inside the internal combustion engine are numerous components that are subjected to forces. Some of the forces produce moments or torques that cause the drive elements to rotate and, therefore, to transmit power. Let's begin with the piston, connecting rod, and crankshaft system (Figure 1–12). A large force is developed on the head of the piston by the pressure in the cylinder created when the fuel/air mixture is exploded by the spark plug. The force is passed through the piston pin to the connecting rod and then through the rod to the journal on the crank. At this

point, a component of the force on the journal causes torque to be developed on the crankshaft. Repeated actions of the piston, or the combined actions of several pistons, create the desired continuous rotation of the crankshaft.

2. Other mechanisms in the engine require a similar analysis. The camshaft is often driven by a chain. Figure 1–13 shows how the intake and exhaust valves are actuated by the camshaft through linkages made up of pushrods, rocker arms, and valve stems. The forces on all of these members and the stresses created by the forces must be analyzed. As the cam rotates, it pushes upward on the lifter. The pushrod transmits this load to one end of the rocker arm. The rocker arm pivots on a pin and pushes down on the valve, moving

Rocker arm
Spring
L_i
Valve
Pushrod
Piston
Lifter
L_o L_i
Cam

(a) Valve closed   (b) Valve open

FIGURE 1–13   Valve actuation system in an engine.

it off its seat at the top of the combustion chamber. As the cam continues to rotate, the pushrod moves down allowing the valve to close, assisted by the upward spring force.

3. The output end of the crankshaft of the engine is connected to the input of the transmission that typically contains several gears that change speed or direction of the rotation of the various shafts. Eventually, the power is delivered to the driveshaft that actually rotates the propeller. Each shaft must be able to transmit the torque developed in it safely. This requires the analysis of *torsional shear stress* (Chapter 10).

4. Some of the gears in the transmission are connected to the shaft through keys and keyseats similar to the arrange-

ment sketched in Figure 1–14. The action of the hub of the gear creates a force on the side of the key that tends to cut or shear it across its horizontal cross section. You will study shearing stresses in Chapter 9.

5. The power is transmitted from the driveshaft to the hub of the propeller through a pin, sometimes called a *shear pin*. Figure 1–15 shows the arrangement. Under normal driving conditions, the torque is transmitted reliably from the driveshaft through the pin to the hub of the propeller. But the pin is designed to break through by shearing if the propeller hits an obstruction such as a rock or submerged log. This saves the expensive propeller and allows the boater to return to the water after the simple replacement of the inexpensive pin.

You are encouraged to look for other machines, devices, and structures that illustrate the importance of being able to analyze the forces and stresses on load-carrying members. Look at the structure of houses, commercial buildings, kitchen appliances, lawn and garden equipment, woodworking tools, cars, trucks, tractors, construction equipment, agricultural implements, production equipment in a factory, measuring devices, boats, airplanes, spacecraft, office equipment, furniture, sports equipment, toys, mountain bikes, snowboards, surfboards, housings for mobile phones, music systems, or instruments. Try to identify the kinds of forces that are exerted on the components of the device and the kinds of failure modes that could occur. Preventing such failures is the prime goal of knowing statics and strength of materials.

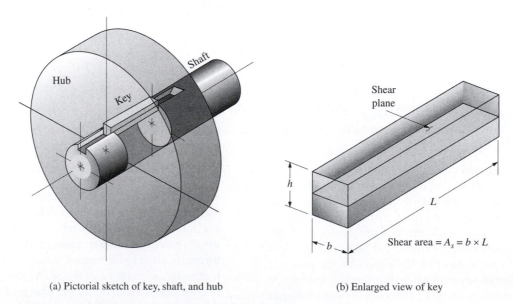

Hub
Shaft
Key
Shear plane
h
b
L
Shear area = $A_s = b \times L$

(a) Pictorial sketch of key, shaft, and hub   (b) Enlarged view of key

FIGURE 1–14   Key connecting a hub to a shaft.

(a) Cross section through propeller hub and shaft   (b) Propeller drive pin

FIGURE 1–15   Shear pin in a propeller driveshaft.

1–4 APPROACHES TO PROBLEM SOLVING

The application of the principles of statics and strength of materials involves several kinds of personal actions on your part. In your job, you must be able to recognize situations in which these principles apply and be able to visualize how loads are generated and how the load path through the structure or machine affects the design of each component. Examples were given in the previous section using the platform, the pliers, and the engine/transmission system. When analyzing or designing a given component of a machine or system, you must understand the overall objective of the analysis and plan how the analysis is to be completed. You will need to ask and answer such questions as the following:

- What forces are applied to this load-carrying member?
- How are those forces reacted at points of support to keep the member in equilibrium?
- What are the effects of the forces? Do they tend to stretch the member in tension? Do they tend to compress it? Will any members be in shear? Does bending occur? Is any member being twisted? Is there some combination of these effects?
- Given the knowledge of the general nature of the effects on the member, what type of analysis must be performed?
- What are the pertinent data to be used?
- In what unit system are the data presented and what units are appropriate for the result?

Then you can complete the analysis and determine the desired result. You should critically evaluate the result for reasonableness and check your work. Make sure your work is clear and well communicated so a colleague, supervisor, or customer can follow what you have done and use your results with confidence. Summarize the results and highlight key final values of desired quantities. If the result is a design for the member, be sure that design specifications are complete and convenient to implement. This will be more important in the study of strength of materials in the latter part of this book than in the early part that deals with statics.

The following organization is recommended for your problem solutions:

Problem Statement	Typically given in the book. On the job, you must create the problem statement describing clearly and completely the information available and the necessary analysis or design.
Solution Objective:	State clearly what is to be determined.
Given:	Summarize data needed to solve the problem.
Analysis:	Decide on the strategy to be used in solving the problem.
	State the formulas that govern the calculations.
	Solve algebraically for the desired result.
	State any assumptions made.
Results:	Complete required calculations by entering data and manipulating units. Summarize and highlight key results.
Comments:	Are the results reasonable? Is the design acceptable? How could improvements be made? What alternative analysis techniques could have been used? What generalizations can be made from the results?

Example problems in this book use this approach and you are advised to follow it or use something similar.

1–5 CALCULATIONS

Most exercises you will do while using this book require calculations. You should have a scientific calculator available that can do basic arithmetic operations, trigonometric functions, exponentiation, square roots, and reciprocals. It should handle scientific notation for quantities in addition to the normal decimal form. Greater speed and more capability can be obtained with calculators that have several memories for storing data, programmability, graphing capability, and the ability to solve systems of equations.

Access to a computer and the ability to use a technical programming language, such as Visual BASIC, C++, FORTRAN, PASCAL, or others, is helpful to do repetitive analyses, perform parametric analyses, or compute results for an entire cycle of loading in a machine. Similar analyses can be completed using spreadsheets or mathematical analysis software such as MathCAD, MATLAB, TK Solver, or Mathematica.

When performing calculations, it is important to achieve a reasonable level of precision and to report results that are meaningful. In scientific work and precise analysis, the concept of *significant figures* is used to judge the precision of a given calculation. The method used in this book is more pragmatic, but it does not relieve you of the responsibility of evaluating the reasonableness of every calculation.

In this book, when making calculations and for reporting results:

1. **Given data are taken to be precise.** This applies regardless of the number of significant digits given. For example, a diameter for a shaft may be reported as 150 mm. *This should be considered to be an exact number.* In scientific work, one would interpret the diameter to be anywhere between 145 mm and 155 mm. In practical terms, if the shaft is a part of a machine, it is likely to be produced to within 0.1 mm of the stated dimension. Therefore, you must know more about the actual application to judge the degree of precision.

2. **Avoid premature rounding of data.** While in the process of making calculations in a given problem, you should carry intermediate results in calculator memory whenever possible to avoid premature rounding that may produce compounded inaccuracies. If necessary, write down intermediate results and reenter them later, carrying at least four significant digits into subsequent calculations.

3. **Report at least three computed digits of accuracy for results.** For example, if your calculator reads 3489.3926 for the value of the force in pounds that a given load-carrying member can safely support, you may report it as 3489 pounds or even 3490 pounds. It is unlikely that the accuracy of the given data or the judgments that were made for the actual strength of a material or other quantities were more precise.

4. **Use scientific notation when appropriate.** Typically, very small or very large numbers should be reported in scientific notation with at least three computed digits of accuracy. For example, if the change in length of an aluminum bar when heated shows on your calculator to be 0.00029714 mm, it should be reported as 2.97×10^{-4} mm if the desired result was a length in mm. A value for a moment that is computed to be 241 593 922 lb · in should be reported as 2.42×10^8 lb · in. Other approaches to reporting units with prefixes are discussed in the following section.

1–6 UNIT SYSTEMS

Computations required in the application of statics and strength of materials involve the manipulation of several sets of units in equations. For numerical accuracy it is of great importance to ensure that consistent units are used in the equations. Throughout this book units will be carried with the applicable numbers.

We will use two different unit systems in this book:

- The *U.S. Customary System* uses
  - pounds for force
  - inches or feet for length
- The *International System of Units*, called SI, uses
  - newtons for force
  - meters for length

This approach considers both the long tradition in the United States with the U.S. Customary System and the widespread worldwide adoption of the International System of Units. It is very likely that you will be required to work in both systems and to contribute to the ongoing dialog about the pace of moving toward the uniform use of the SI system everywhere. References 1–4 provide extensive guidance on the proper use of SI units in technical work including mechanical design, manufacturing, architecture, and building construction.

Because of the present transition in the United States from the traditional U.S. Customary units to metric units, both are used in this book. It is expected that persons entering or continuing an industrial career within the next several years will be faced with having to be familiar with both systems. On the one hand, many new products such as automobiles and business machines are being built using metric dimensions. Thus, components and manufacturing equipment will be specified in those units. However, the transition is not being made uniformly in all areas. Designers will continue to deal with such items as structural steel, aluminum, and wood whose properties and dimensions are given in English units in standard references. Also, designers, sales and service people, and those in manufacturing must work with equipment that has already been installed and that was made to U.S. Customary System dimensions. Therefore, it seems logical that persons now working in industry should be capable of working and thinking in both systems.

TABLE 1–1	Basic Quantities in the SI Metric Unit System	
Quantity	**SI unit**	**Other metric units**
Length	meter (m)	millimeter (mm)
Time	second (s)	minute (min), hour (h)
Force	newton (N)	$kg \cdot m/s^2$
Mass	kilogram (kg)	$N \cdot s^2/m$
Temperature	kelvin (K)	degrees Celsius (°C)
Angle	radian (rad)	degree (deg)

TABLE 1–3	Prefixes for SI Units	
Prefix	**SI symbol**	**Factor**
giga	G	$10^9 = 1\ 000\ 000\ 000$
mega	M	$10^6 = 1\ 000\ 000$
kilo	k	$10^3 = 1\ 000$
milli	m	$10^{-3} = 0.001$
micro	μ	$10^{-6} = 0.000\ 001$

strength of materials are listed in Table 1–3. Table 1–4 shows how computed results should be converted to the use of the standard prefixes for units.

The formal name for the U.S. Customary System is the English Gravitational Unit System (EGU). The metric system, which has been adopted internationally, is called Système International d'Unités, the International System of Units (SI). See Reference 4.

In most cases, the problems in this book are worked either in the U.S. Customary System or the metric system rather than mixing units. In problems for which data are given in both unit systems, it is most desirable to change all data to the same system before completing the problem solution. Appendix A–26 gives factors for use in performing conversions. Internet sites 2–7 listed at the end of this chapter link to unit conversion calculators that are available online. Most of the simpler calculators can be used online without cost, whereas the more versatile and complete calculators may be downloaded for a fee.

The basic quantities for any unit system are length, time, force, mass, temperature, and angle. Table 1–1 lists these quantities in SI units and Table 1–2 lists the quantities in U.S. Customary units.

Prefixes for SI Units

In SI units, prefixes should be used to indicate orders of magnitude, thus eliminating digits and providing a convenient substitute for writing powers of 10, as generally preferred for computation. Prefixes representing steps of 1000 are recommended. Those usually encountered in statics and

1–7 RELATIONSHIP AMONG MASS, FORCE, AND WEIGHT

Force and mass are separate and distinct quantities. Weight is a special kind of force.

Mass refers to the amount of the substance in a body.

Force is a push or pull effort exerted on a body either by an external source or by gravity.

Weight is the force of gravitational pull on a body.

Mass, force, and weight are related by Newton's law:

$$\text{Force} = \text{Mass} \times \text{Acceleration}$$

We often use the symbols F for force, m for mass, and a for acceleration. Thus,

$$F = m \times a \quad \text{or} \quad m = F/a$$

When the pull of gravity is involved in the calculation of the weight of a mass, a takes the value of g, the acceleration due to gravity. Then, using W for weight,

Weight–Mass Relationship ▼

$$W = m \times g \quad \text{or} \quad m = W/g \qquad \textbf{(1–1)}$$

We will use the following values for g:

SI units: $g = 9.81 \text{ m/s}^2$ U.S. units: $g = 32.2 \text{ ft/s}^2$

Units for Mass, Force, and Weight

Tables 1–1 and 1–2 show the preferred units and some other convenient units for mass and force in both the SI and U.S.

TABLE 1–2	Basic Quantities in the U.S. Customary System	
Quantity	**U.S. Customary unit**	**Other U.S. units**
Length	foot (ft)	inch (in)
Time	second (s)	minute (min), hour (h)
Force	pound (lb)	kip*
Mass	slug	$lb \cdot s^2/ft$
Temperature	degrees Fahrenheit (°F)	
Angle	degree (deg)	radian (rad)

*1.0 kip = 1000 pounds. The name is derived from the term *kilopound*.

TABLE 1–4	Proper Method of Reporting Computed Quantities	
Computed result	**Reported result**
0.005 48 m	5.48×10^{-3} m, or 5.48 mm
12 750 N	12.75×10^3 N, or 12.75 kN
34 500 kg	34.5×10^3 kg, or 34.5 Mg (megagrams)

unit systems. The units for force are also used as the units for weight.

The newton (N) in the SI unit system is named in honor of Sir Isaac Newton, and it represents the amount of force required to give a mass of 1.0 kg an acceleration of 1.0 m/s². Equivalent units for the newton can be derived as follows using Newton's law with units only:

$$F = m \times a = \text{kg} \cdot \text{m/s}^2 = \text{newton}$$

In the U.S. Customary System, the unit for force is defined to be the pound, whereas the unit of mass (slug) is derived from Newton's law as follows:

$$m = \frac{F}{a} = \frac{\text{lb}}{\text{ft/s}^2} = \frac{\text{lb} \cdot \text{s}^2}{\text{ft}} = \text{slug}$$

The conversion of weight and mass is illustrated in the following example problems.

Example Problem 1–1 (SI system)	A hoist lifts 425 kg of concrete. Compute the weight of the concrete that is the force exerted on the hoist by the concrete.

Solution Objective: Compute the weight of a mass of concrete.

Given: $m = 425$ kg

Analysis: $W = m \times g$; $g = 9.81$ m/s²

Results: $W = 425$ kg $\times 9.81$ m/s² $= 4170$ kg $\cdot$ m/s² $= 4170$ N

Comment: Thus, 425 kg of concrete weighs 4170 N.

Example Problem 1–2 (U.S. system)	A hopper of coal weighs 8500 lb. Determine its mass.

Solution Objective: Compute the mass of a hopper of coal.

Given: $W = 8500$ lb

Analysis: $m = W/g$; $g = 32.2$ ft/s²

Results: $m = 8500$ lb/32.2 ft/s² $= 264$ lb $\cdot$ s²/ft $= 264$ slugs

Comment: Thus, 8500 lb of coal has a mass of 264 slugs.

Density and Specific Weight

To characterize the mass or weight of a material relative to its volume, we use the terms *density* and *specific weight*, defined as follows:

Density is the amount of mass per unit volume of a material.

Specific weight is the amount of weight per unit volume of a material.

We will use the Greek letter ρ (rho) as the symbol for density. For specific weight we will use γ (gamma).

The units for density and specific weight are summarized in Table 1–5. Other conventions are sometimes used, often leading to confusion. For example, in the United States, density is occasionally expressed in lb/ft³ or lb/in³. Two interpretations are used for this. One is that the term implies *weight density* with the same meaning as specific weight. Another is that the quantity *lb* is meant to be *pound mass* rather than *pound weight*, and the

two have equal numerical values when g is the standard value.

Table 1–6 lists typical densities, ρ, and specific weights, γ, of common materials that are useful for problem solving in this book. Internet site 1 at the end of this chapter links to listings of the density of a large number of bulk materials, woods, metals, and liquids in the SI unit of kg/m³. These data may be useful for problem solving or design problems. If other units are desired, the unit conversion factors in Appendix A–26 can be used. Alternatively, Internet sites 2–7 include unit conversion calculators that can be used online, most without cost.

TABLE 1–5 Density and Specific Weight Units

	U.S. customary	SI metric
Density	slugs/ft³	kg/m³
Specific weight	lb/ft³	N/m³

TABLE 1–6 Typical Densities and Specific Weights of Common Materials

Material	Density, ρ kg/m³	lb$_m$/ft³	Specific weight, γ kN/m³	lb$_f$/ft³
Gasoline	680	42.5	6.67	42.5
Hydraulic fluid	895	55.9	8.78	55.9
Water	1000	62.4	9.81	62.4
Mercury	13 540	845	131.9	845
Coal, pieces bulk	1100	68.7	10.8	68.7
Earth, moist	1440	89.9	14.1	89.9
Sand, dry	1600	100	15.7	100
Gravel, wet	2000	125	19.6	125
Balsa wood	170	10.6	1.67	10.6
Cedar, red	380	23.7	3.73	23.7
Pine wood	530	33.1	5.20	33.1
Maple wood	755	47.1	7.41	47.1
Magnesium	1830	114	18.0	114
Aluminum	2770	173	27.2	173
Steel	7680	479	75.3	479
Bronze	8800	549	86.3	549

Pressure, Load Per Unit Area, and Stress

The study of statics and strength of materials frequently involves loads that are distributed over an area. For example, in an engine or compressor, gases are compressed to high pressures that act uniformly over the area of a piston. We indicate the magnitude of pressure as the amount of force applied to a given unit of area. In the building construction field, the load on a roof structure is typically made up of several layers of material spread uniformly over the roof. Such a load is often called a *uniformly distributed load* and is expressed as a force per unit area of the roof. When we discuss stress analysis later, we define *stress* as the internal force exerted on a unit area of the material from which a member is made due to an externally applied load.

In each of the cases just described, the general dimensions for the quantity will be *force per unit area.* Typical units in the U.S. Customary System are

lb/in² or abbreviated as psi

lb/ft² or abbreviated as psf

K/in² or abbreviated as ksi (1.0 K = 1.0 kip = 1000 lb)

K/ft² or abbreviated as ksf

Typical units in the SI metric system are

N/m² called the *pascal,* abbreviated as Pa

kPa = 10³ Pa called the kilopascal

MPa = 10⁶ Pa called the megapascal

The name *pascal* (Pa) is in honor of the seventeenth-century French scientist, Blaise Pascal, who performed important work in the analysis of fluids under pressure. The magnitude

of 1.0 Pa is very small. For example, if you wanted to increase the pressure on a 1.0 m² area by 1.0 Pa, you would need only to pour a depth of slightly more than 0.1 mm of water on the area. This is a volume of only 0.10 L or 0.0264 gal. For this reason, practical levels of pressure, unit loading, or stress are in the range of several kPa (10³ Pa) or MPa (10⁶ Pa).

1–8 SIZES OF STANDARD SHAPES

Because the early chapters of this book deal mostly with the study of forces, moments, and torques on structural members and machine components, it is important for you to be familiar with the kinds of standard structural members that are typically specified along with the commonly used schemes for designating standard sizes. Appendix tables A–1 to A–13 provide a large amount of data. However, they are only samples of the huge amount of data available from the ultimate sources listed in the References and Internet sites at the end of the chapter. Included in the Appendix are tables for

- Properties of the areas of several common shapes found in structural members
- Properties of standard wood beam shapes and sizes
- Properties of standard steel structural shapes such as angles, channels, and I-beam shapes
- Properties of aluminum channels and I-beam shapes
- Properties of standard steel pipe and mechanical tubing

You will need to be familiar with these tables for some example problems and problems at the end of chapters. Dimensional data and areas of standard shapes are used in the Statics part of this book. Other properties of the areas are defined in the Strength of Materials chapters and data for them are listed in the tables.

The following sections describe the designations for standard structural shapes and how data are presented in both U.S. Customary System and SI metric units.

Standard Wood Beams

Appendix A–4 gives the dimensions and section properties for many standard sizes of wood beams. Note that the nominal size is simply the "name" of the beam, and it relates to the approximate rough size prior to finishing. Actual finished dimensions are significantly smaller than the nominal sizes. For example, a common "2 × 4" board is actually 1.5 in wide and 3.5 in high. Also note the sketch of the orientation of the beams for the standard designation of the x- and y-axes. When used as a beam in bending, the long dimension should be vertical for maximum strength and stiffness.

Steel Structural Shapes

Steel manufacturers provide a large array of standard structural shapes that are efficient in the use of material and that are convenient for specification and installation into building structures or machine frames. Included, as shown in Table 1–7, are standard angles (L-shapes), channels

TABLE 1–7 Designations for Steel and Aluminum Shapes

Name of shape	Shape	Symbol	Example designation
Angle		L	L4×3×1/2 Appendix A–5 Metric: L102×76×12.7
Channel		C	C15×50 Appendix A–6 Metric: C380×74
Wide-flange beam		W	W14×43 Appendix A–7 Metric: W360×64
American Standard beam		S	S10×35 Appendix A–8 Metric: S250×52
Structural tubing—square			HSS4×4×1/4 Appendix A–9 Metric: HSS 102×102×6.4
Structural tubing—rectangular			HSS6×4×1/4 Appendix A–9 Metric: HSS 152×102×6.4
Pipe			PIPE 4 STD 4-inch Schedule 40 Appendix A–12 Metric: Pipe 102 STD
Aluminum Association channel		C	C4×1.738 Appendix A–10 Metric: C102×2.586
Aluminum Association I-beam		I	I8×6.181 Appendix A–11 Metric: I203×9.197

(C-shapes), wide-flange beams (W-shapes), American Standard beams (S-shapes), structural tubing, and pipe called *hollow structural sections* (HSS). Note that the W-shapes and S-shapes are often referred to in general conversation as "I-beams" because the shape of the cross section looks like the capital letter I.

Appendix tables A–5 through A–9 give geometric properties of selected structural shapes that cover a fairly wide range of sizes. Note that many more sizes are available as presented in Reference 7. The appendix tables give data for the area of the cross section (A), the weight per foot of length, the location of the centroid of the cross section, the moment of inertia (I), the section modulus (S), and the radius of gyration (r). Some of these properties may be new to you at this time and they will be defined as needed later in the book. The values of I and S are important in the analysis and design of beams. For column analysis, I and r are needed.

Metrication Applied to Structural Shapes   Metrication in the construction industry is progressing even though it is expected that, in the United States, use of standard structural shapes that have been produced for many years to basic dimensions expressed in the U.S. Customary System (pound-foot-second system) will remain dominant in the near future. It is important for you to become aware of the trend toward more use of SI metric units in design and fabrication of structures while maintaining the ability to use U.S. units.

Appendix tables A–5 through A–9 include separate pages for U.S. units and SI units for L-shapes, W-shapes, S-shapes, C-shapes, and HSS shapes. In each table, the first column indicates a reference letter for each size of the given shape. This facilitates the correlation of the data for a specific shape between the U.S. and SI units. All of the tables listing SI data have been *soft-converted* from the U.S. units. This means that the long-used standard shapes made to U.S. unit conventions have not been changed. For the SI tables, the designations and the properties of the cross sections have been converted to their metric equivalent using the definitions given in Section 1–6 of this chapter and the conversion factors listed in Appendix A–26. A complete table of equivalencies between the U.S. and SI designations is included in Reference 7. The metric nomenclature for structural shapes is included in Reference 8. Some observations are discussed here.

1. Linear dimensions given in inches are typically converted to the nearest millimeter. However, small dimensions, such as the thickness of the legs of an angle (L-shape), are rounded to the nearest tenth of a mm. The factor used is 1.0 in = 25.4 mm.

2. Converted nominal sizes are typically rounded to the nearest 10 mm. It is important to use actual dimensions of critical features of shapes in calculations. Nominal sizes are merely the *names* of the shapes suggesting the approximate size.

3. The U.S. designation system lists *weight per foot* (lb/ft) for each shape. More precisely, these data are *mass per foot*,

where the units are lb_m/ft, read as *pounds mass per foot*. Common usage often does not include the subscript *m*. For uses on or near the surface of the earth, the numerical value of mass in lb_m is the same as the value for weight (the force due to gravity), often expressed in lb_f or *pounds force*.

4. Conversion of the mass per unit length from lb_m/ft to SI units of kg/m applies the factor

$$\frac{1.488 \text{ kg/m}}{lb_m/ft}$$

For example, a beam shape listed as weighing 60 lb/ft is converted to SI units as

$$60 \text{ lb/ft } [(1.488 \text{ kg/m})/(1.0 \text{ lb/ft})] = 89.28 \text{ kg/m}$$

The result is typically rounded to the nearest whole kg/m value, except for very small numbers. Thus, this value is reported as 89 kg/m in the table.

5. The weight (force) per unit length is often desired when considering the force applied by one member to another. Then the mass per unit length must be converted to force per unit length by multiplying by $g = 9.81 \text{ m/s}^2$, the acceleration due to gravity. This follows from the relationship $w = mg$ discussed in Section 1–7. For example, for the beam described in Item 4 above, the weight per unit length is

$$89.27 \text{ kg/m } [9.81 \text{ m/s}^2] = (876 \text{ kg} \cdot \text{m/s}^2)/m$$
$$= 876 \text{N/m}$$

Larger values may be listed in kN/m. See Table A–7(SI) in the Appendix.

Steel Angles (L-Shapes)

Appendix A–5 shows sketches of the typical shapes of steel angles having equal or unequal leg lengths. Called L-shapes because of the appearance of the cross section, angles are often used as tension members of trusses and towers, framing members for machine structures, lintels over windows and doors in construction, stiffeners for large plates used in housings and beams, brackets, and ledge-type supports for equipment. Some refer to these shapes as "angle iron." The standard designation takes the following form, using one example size:

$$L4 \times 3 \times 1/2$$

where L refers to the L-shape, 4 is the length of the longer leg, 3 is the length of the shorter leg, and 1/2 is the thickness of the legs. Dimensions are in inches.

Section property data (I and S) are given for the *x*-axis and the *y*-axis as shown in the drawing. These data are used in later chapters (6 and 11–16), and it is essential that values for the proper axis are used in analyzing beams and columns incorporating L-shapes. The intersection of the X- and Y-axes locates the centroid of the shape, and the listed dimensions *y* and *x* are used to define that position.

Additional data are given for the radius of gyration, *r*, of the L-shape relative to the Z-axis, defined as the axis having the smallest value of r. This is important to the analysis of column loading of the shape as described in Chapter 16. The column tends to buckle about the Z-axis that passes through the centroid and has an angle of inclination of α with respect to the Y-axis.

The corresponding SI designation for the L4×3×1/2 shape is

$$L102 \times 76 \times 12.7$$

This indicates the longer leg length of 102 mm, the shorter leg length of 76 mm, and a thickness of 12.7 mm. See Appendix A–5(SI).

American Standard Channels (C-Shapes)

See Appendix A–6 for the appearance of channels and their geometric properties. Channels are used in applications similar to those described for angles. The flat web and the two flanges provide a generally stiffer shape than angles that are more resistant to bending and twisting under load.

The sketch at the top of the table shows that channels have tapered flanges and webs with constant thickness. The slope of the flange taper is approximately 2 in in 12 in, and this makes it difficult to attach other members to the flanges. Special tapered washers are available to facilitate fastening. Note the designation of the X- and Y-axes in the sketch, defined with the web of the channel vertical, which gives it the characteristic "C" shape. This is most important when using channels as beams or columns. The X-axis is located on the horizontal axis of symmetry while the dimension x, given in the table, locates the Y-axis relative to the back of the web. The centroid is at the intersection of the X- and Y-axes.

The form of the standard designation for channels is

$$C15 \times 50$$

where   C indicates that it is a standard C-shape
15 is the nominal (and actual) depth in inches with the web vertical
50 is the weight per unit length in lb/ft

The corresponding SI designation for this shape is

$$C380 \times 74$$

This indicates the nominal depth of 380 mm for the channel and a mass per unit length of 74 kg/m. From Appendix A–6(SI) we find the weight per unit length to be 0.730 kN/m and the actual depth to be 381 mm.

Wide-Flange Shapes (W-Shapes)

Refer to Appendix A–7. This is the most common shape used for beams, as will be discussed in Chapters 11–14. W-shapes have relatively thin webs and somewhat thicker, flat flanges with constant thickness. Most of the area of the cross section is in the flanges, farthest away from the horizontal centroidal axis (X-axis), thus making the moment of inertia very high for a given amount of material. (See Chapter 6.)

Note that the properties of moment of inertia and section modulus are very much higher with respect to the X-axis than they are for the Y-axis. Therefore, W-shapes are typically used in the orientation shown in the sketch in Appendix A–7. Also, these shapes are best when used in pure bending without twisting because they are quite flexible in torsion.

The standard designation for W-shapes carries much information. Consider the following example:

$$W14 \times 43$$

where   W indicates that it is a W-shape
14 is the nominal depth in inches
43 is the weight per unit length in lb/ft

The term *depth* is the standard designation for the vertical height of the cross section when placed in the orientation shown in Appendix A–7. Note from the data in the table that the actual depth is often different from the nominal depth. For the W14×43, the actual depth is 13.7 in.

The corresponding SI designation for this shape is

$$W360 \times 64$$

This indicates the nominal depth of 360 mm for the W-beam and a mass per unit length of 64 kg/m. From Appendix A–7(SI) we find the weight per unit length to be 0.628 kN/m and the actual depth to be 348 mm.

American Standard Beams (S-Shapes)

Appendix A–8 shows the properties for S-shapes. Much of the discussion given for W-shapes applies to S-shapes as well. Note that, again, the weight per foot of length is included in the designation such as the S10×35 that weighs 35 lb/ft. For most, but not all of the S-shapes, the actual depth is the same as the nominal depth. The flanges of the S-shapes are tapered at a slope of approximately 2 in in 12 in, similar to the flanges of the C-shapes. The X- and Y-axes are defined as shown with the web vertical.

For the S10×35 shape discussed here, the corresponding SI designation is

$$S250 \times 52$$

This indicates the nominal depth of 250 mm for the S-beam and a mass per unit length of 52 kg/m. From Appendix A–8(SI) we find the weight per unit length to be 0.511 kN/m and the actual depth to be 254 mm.

Often wide-flange shapes (W-shapes) are preferred over S-shapes because of their relatively wide flanges, the constant thickness of the flanges, and generally higher section properties for a given weight and depth.

Structural Tubing (Square and Rectangular)

See Appendix A–9 for the appearance and properties for steel structural tubing, also called *hollow structural section* (HSS). These shapes are usually formed from flat sheet and welded along the length. The section properties account for the corner radii. Note the sketches showing the X- and Y-axes. The standard designation takes the form,

$$HSS6 \times 4 \times 1/4$$

where   HSS indicates that it is a standard hollow structural section

6 is the depth of the longer side in inches

4 is the width of the shorter side in inches

1/4 is the nominal wall thickness in inches

Square tubing and rectangular tubing are very useful in machine structures because they provide good section properties for members loaded as beams in bending and for torsional loading (twisting) because of the closed cross section. The flat sides often facilitate fastening of members together or the attachment of equipment to the structural members. Some frames are welded into an integral unit that functions as a stiff space-frame. Square tubing makes an efficient section for columns.

Note that the design wall thickness is somewhat smaller than the nominal. The nominal $\frac{1}{4}$-in wall has a design thickness of 0.233 in. A nominal $\frac{1}{2}$-in wall has a design thickness of 0.465 in.

Appendix A–9 lists two torsional constants, J and C, for the square and rectangular HSS shapes. These are discussed in Chapter 10 in the section called *Torsion of noncircular sections*. The term J is the polar moment of inertia related to the stiffness of the section in resisting twisting. The term C is used for computing the shearing stress in the shape when subjected to torsion. Because these shapes perform well as columns, the radius of gyration is given for each member size because it relates to the tendency for a column to buckle as described in Chapter 16.

For the HSS shape discussed previously (HSS6×4×1/4), the corresponding SI designation is:

$$HSS152 \times 102 \times 6.4$$

This indicates the depth of 152 mm, the width of 102 mm, and nominal wall thickness of 6.4 mm. From Appendix A–9(SI) we find the mass per unit length to be 23.2 kg/m and the weight per unit length to be 228 N/m. The design wall thickness is 5.92 mm.

Pipe and Round Tubing

Hollow circular sections, commonly called *pipe*, are very efficient for use as beams, torsion members, and columns. The placement of the material uniformly away from the center of the pipe enhances the moment of inertia for a given amount of material and gives the pipe uniform properties with respect to all axes through the center of the cross section. The closed cross-sectional shape gives it high strength and stiffness in torsion, as well as in bending.

Appendix A–12 gives the properties for a representative few sizes of pipe that may be used for structural applications in construction and mechanical design. Included are the designation nomenclature, actual dimensions, area, moment of inertia, section modulus, radius of gyration, and the torsional constants J and Z_p.

There are several types of round hollow members in hundreds of sizes available from which to choose as detailed in Reference 7. See also Internet sites 8–14. The following is a summary of common types.

- Construction applications are typically specified from round *hollow structural sections* (HSS) ranging in outside diameter from 1.660 in (42.2 mm) to 20.0 in (508 mm) as detailed in Reference 7. A total of 59 different outside diameters are produced, each with multiple wall thicknesses, providing a total of 139 specific sizes of round HSS sections.
  - An example designation for this type of member is HSS4.000×0.250, indicating an actual outside diameter of 4.000 in and a wall thickness of 0.250 in.
  - The SI equivalent for this member is designated as HSS101.6×6.4, indicating an actual outside diameter of 101.6 mm and a wall thickness of 6.4 mm.
- **Steel pipe used for structural application** in construction is available in three classes, *Standard Weight (Std)*, *Extra Strong (X-Strong or XS)*, and *Double-Extra Strong (XX-Strong or XXS)*. Each series has the same set of outside diameters with different increasing wall thicknesses and decreasing inside diameters for the XS and XXS sizes. Designations are related to the *nominal diameters* that range from $\frac{1}{2}$ in to 12 in (13 mm to 310 mm). Actual outside and inside diameters are significantly different from the nominal sizes as indicated in Appendix A–12, and they must be used for design and analysis.
  - An example designation for construction pipe is PIPE4STD, indicating a nominal 4-in size and standard weight. This pipe has an actual outside diameter of 4.500 in, a wall thickness of 0.237 in, and an inside diameter of 4.026 in.
  - The SI equivalent pipe is called PIPE102STD, indicating a nominal size of 102 mm, an actual outside diameter of 114.3 mm, a wall thickness of 6.02 mm, and an inside diameter of 102.3 mm.
  - Note that the complete set of construction pipe is included within the round HSS grouping described earlier. The example pipe, PIPE4STD, can also be designated as HSS4.500×0.237 for U.S. units or HSS114.3×6 for SI units.
- **Steel pipe is produced primarily for carrying fluids** and commonly distributed by plumbing supply centers are designated by different *schedules*, ranging from lightweight Schedule 5 to heavyweight Schedule 160. The most common schedules are 40, 80, and 160, and the majority of these sizes are identical to the STD, XS, and XXS classes described for construction uses. Steel pipe is available in nominal sizes from $\frac{1}{8}$-in to 42-in. In Appendix A–12, all of the listed sizes are Schedule 40. Those from $\frac{1}{2}$-in to 10-in can also be designated with the PIPE notation described previously for construction applications.
  - The pipe called PIPE4STD can also be named 4-in Schedule 40 pipe, and this form will typically be used in this book.
  - The SI equivalent designations for fluid-handling pipe have not yet been defined.

- *Mechanical tubing* is used for numerous applications in mechanical design, and its size ranges and designations are significantly different from round HSS or pipe. They are also available in many materials, such as carbon steel, alloy steel, stainless steel, brass, copper, titanium, tantalum, and aluminum. Appendix A–13 and A–13(SI) give a few examples of mechanical tubing designated by outside diameter and wall thickness with the wall thickness corresponding to a standard called *Birmingham Wall Gage* or BWG. Many more sizes and wall gages are available as seen in Internet site 16. See Internet sites 8–10 and 16–18 for other examples of commercially available steel and copper tubing.

- *Plastic pipe and tubing* are also commercially available in many materials and sizes and are useful for mechanical design. Some plastic pipe is made to the same dimensions as Schedule 40 steel pipe. Tubing is available in a wide range of diameters and wall thicknesses. See Internet sites 11–13 and 18.

Aluminum Association Standard Channels and I-Beams

Appendixes A–10 and A–11 give the dimensions and section properties of channels and I-beams developed by the Aluminum Association (Reference 1). These are extruded shapes having uniform thicknesses of the webs and flanges with generous radii where they meet. The proportions of these sections are somewhat different from those of the rolled steel sections described earlier. The extruded form offers advantages in the efficient use of material and in the joining of members. This book will use the following forms for the designation of aluminum sections:

$$C4 \times 1.738 \quad \text{or} \quad I8 \times 6.181$$

where   C or I indicates the basic section shape
4 or 8 indicates the depth of the shape when in the orientation shown
1.738 or 6.181 indicates the weight per unit length in lb/ft

Tables A–10(SI) and A–11(SI) give the soft-converted sizes for Aluminum Association channels and I-beam shapes. The format for the designation is the similar to that for U.S. units. The example for a channel, C4×1.738, becomes

$$C102 \times 2.586$$

where   102 is the depth in mm
2.586 is the mass per unit length in kg/m
From Table A–10(SI) we find the weight per unit length, 25.37 N/m

The example for an I-beam shape, I8×6.181, becomes

$$I203 \times 9.197$$

where   203 is the depth in mm
9.197 is the mass per unit length in kg/m
From Table A–10(SI) we find the weight per unit length, 90.22 N/m

REFERENCES

1. ASTM International, ASTM E621, *Standard Practice for the Use of Metric (SI) Units in Building Design and Construction,* West Conshohocken, PA, 1999.

2. ASTM International, ASTM SI-10, *American National Standard for Use of the International System of Units (SI): The Modern Metric System,* West Conshohocken, PA, 2002.

3. Taylor, Barry N., *Guide for the Use of the International System of Units (SI),* NIST Special Publication 811-1995 Edition, Gaithersburg, MD, National Institute of Standards and Technology (U.S. Government Printing Office, Washington, DC), 1995.

4. Taylor, Barry N., *The International System of Units (SI),* NIST Special Publication 330-2001 Edition, Gaithersburg, MD, National Institute of Standards and Technology (U.S. Government Printing Office, Washington, DC), 2001.

5. Aluminum Association, *Aluminum Design Manual,* Washington, DC, 2005.

6. American Institute of Steel Construction (AISC), *Designing with Structural Steel—A Guide for Architects,* 2nd ed., AISC, Chicago, IL, 2002.

7. American Institute of Steel Construction (AISC), *Steel Construction Manual,* 13th ed., AISC, Chicago, IL, 2005.

8. American Society for Testing and Materials, ASTM Standard A6/A6M-05a, *Standard Specification for General Requirements for Rolled Structural Steel Bars, Plates, Shapes, and Sheet Piling,* West Conshohocken, PA, 2005.

INTERNET SITES

1. SImetric.co.uk www.simetric.co.uk/si_materials.htm Listing of the density of a large number of bulk materials in the SI metric unit of kg/m^3. From this page, you can link to pages giving the density of metals, wood, and liquids.

2. Accelware www.accelware.com/Unit_Conversion_Tool Description of a very versatile tool for unit conversions for a large number of engineering quantities in mechanical, electrical, and chemical fields. The tool can be downloaded for a fee.

3. Knovel www.knovel.com/knovel2/unitconverter.jsp A versatile unit converter can be used on screen at no cost. A copy of the converter can be downloaded for a fee. Other Knovel products and publications can be accessed from the site.

4. Engineers Edge www.engineersedge.com/calculators.htm This site has numerous calculators that can be accessed at no cost from this page. Scrolling down the right column leads to links to unit conversion tools for several quantities. Other types of calculators for technical calculations are also listed.

5. Online Conversion www.onlineconversion.com A no-cost conversion tool for most commonly encountered units that can be used online.

6. Digital Dutch http://digitaldutch.com/unitconverter A modest-sized unit converter that can be used for free online.

7. Unit-Converter.org http://unit-converter.org/index.html A modest-sized unit converter that can be used for free online.

8. TubeNet www.tubenet.org/tubes.shtml A listing of dimensions, properties, and suppliers of pipe and tubing in steel, copper, titanium, tantalum, and other materials.

9. Stainless Tubular Products www.stainlesstubular.com A supplier of stainless steel pipe and tubing.

10. Copper Development Association www.copper.org A professional association of the copper industry; the site offers a large amount of data on sizes and physical characteristics of copper tubing. A *Copper Tube Handbook* or parts of it can be downloaded from the site.

11. Plastics Pipe Institute www.plasticpipe.org An association representing all segments of the plastics piping industry; the site includes a list of manufacturers of plastic pipe from which much data for sizes, materials, and application information can be found.

12. Charter Plastics www.charterplastics.com A supplier of polyethylene plastic pipe and tubing.

13. Expert Piping Supply www.expertpiping.com A supplier of copper, steel, PVC, CPVC, polyethylene, and polypropylene pipe in a wide range of diameters and wall thicknesses.

14. The Piping Tool Box www.piping-toolbox.com A site containing data for pipe dimensions, pipe standards, and related topics.

15. American Institute of Steel Construction (AISC) www.aisc.org A trade association that serves the structural steel design community and the construction industry. Publisher of the AISC Steel Construction Manual.

16. Webco Mechanical Tubing Products www.webcoindustries.com/tubing/mechanical Manufacturer of mechanical tubing in round, square, and rectangular forms made from a variety of carbon and alloy steels.

17. Central Steel & Wire Co. www.centralsteel.com Supplier of a variety of metals (carbon and alloy steel, stainless, aluminum, copper, brass, bronze) in many shapes (structural shapes, strip, sheets, plates, bars), tubing (round, square, rectangular), pipe, and wire.

18. Ryerson, Inc. www.ryerson.com Supplier of a wide variety of metals (carbon and alloy steel, stainless, nickel alloys, aluminum, copper, brass, bronze) in many shapes (structural shapes, strip, sheets, plates, bars), tubing (round, square, rectangular), pipe, and wire. Also supplies plastics in sheet, rod, and tubing.

PROBLEMS

Definitions and Unit Systems

For Problems 1–1 to 1–5, state the standard unit for the following basic quantities in both the U.S. Customary System and the SI metric system.

1–1. Length
1–2. Time
1–3. Force
1–4. Temperature
1–5. Angle

For Problems 1–6 to 1–10, state typically used units in the SI metric system other than the standard unit.

1–6. Length
1–7. Time
1–8. Force
1–9. Temperature
1–10. Angle

For Problems 1–11 to 1–15, state typically used units in the U.S. Customary System other than the standard unit.

1–11. Length
1–12. Time
1–13. Force
1–14. Temperature
1–15. Angle

For Problems 1–16 to 1–25, express the given computed numerical values with standard prefixes and proper units in the SI metric system, giving three computed digits of accuracy.

1–16. 0.002 493 638 m
1–17. 45 218.184 m
1–18. 74 481.529 m
1–19. 0.048 033 255 m
1–20. 2518 N
1–21. 337 810 N
1–22. 75 226 N
1–23. 5329 N
1–24. 643.48×10^4 N
1–25. 298.153×10^5 N
1–26. Define *mass*.
1–27. State the standard unit and at least one other typical unit for mass in the SI metric system.
1–28. State the standard unit and at least one other typical unit for mass in the U.S. Customary System.
1–29. Define *weight*.
1–30. State the standard unit and at least one other typical unit for weight in the SI metric system.
1–31. State the standard unit and at least one other typical unit for weight in the U.S. Customary System.
1–32. Define *density*.
1–33. State the standard unit for density in the SI metric system.
1–34. State the standard unit for density in the U.S. Customary System.
1–35. Define *specific weight*.
1–36. State the standard unit for specific weight in the SI metric system.
1–37. State the standard unit for specific weight in the U.S. Customary System.
1–38. Define *pressure*.
1–39. State at least two typical units for pressure in the SI metric system.
1–40. State at least two typical units for pressure in the U.S. Customary System.

Mass–Weight Conversions

1–41. A truck carries 1800 kg of gravel. What is the weight of the gravel in newtons?
1–42. A four-wheeled truck having a total mass of 4000 kg is sitting on a bridge. If 60% of the weight is on the rear wheels and 40% is on the front wheels, compute the force exerted on the bridge at each wheel.
1–43. A total of 6800 kg of a bulk fertilizer is stored in a flat-bottomed bin having side dimensions 5.0 m × 3.5 m. Compute the loading on the floor in newtons per square meter, or pascals.

1–44. A mass of 25 kg is suspended by a spring that has a spring scale of 4500 N/m. How much will the spring be stretched?

1–45. Measure the length, width, and thickness of this book in millimeters.

1–46. Determine your own weight in newtons and your mass in kilograms.

1–47. Express the weight found in Problem 1–41 in pounds.

1–48. Express the forces found in Problem 1–42 in pounds.

1–49. Express the loading in Problem 1–43 in pounds per square foot.

1–50. For the data in Problem 1–44, compute the weight of the mass in pounds, the spring scale in pounds per inch, and the stretch of the spring in inches.

1–51. A cast iron base for a machine weighs 2750 lb. Compute its mass in slugs.

1–52. A roll of steel hanging on a scale causes a reading of 12 800 lb. Compute its mass in slugs.

1–53. Determine your own weight in pounds and your mass in slugs.

Unit Conversions

1–54. A pressure vessel contains a gas at 1200 psi. Express the pressure in pascals.

1–55. A structural steel has an allowable stress of 21 600 psi. Express this in pascals.

1–56. The stress at which a material will break under a direct tensile load is called the *ultimate strength*. The range of ultimate strengths for aluminum alloys ranges from about 14 000 to 76 000 psi. Express this range in pascals.

1–57. An electric motor shaft rotates at 1750 rpm. Express the rotational speed in radians per second.

1–58. Express an area of 14.1 in^2 in the units of square millimeters.

1–59. An allowable deformation of a certain beam is 0.080 in. Express the deformation in millimeters.

1–60. A base for a building column measures 18.0 in by 18.0 in on a side and 12.0 in high. Compute the cross-sectional area in both square inches and square millimeters. Compute the volume in cubic inches, cubic feet, cubic millimeters, and cubic meters.

1–61. Compute the area in square inches of a rod having a diameter of 0.505 in. Then convert the result to square millimeters.

Moments, Torques, and Couples

1–62. For the cantilever beam shown in Figure P1–62, compute the moment due to the force *F* with respect to the support location *A*.

FIGURE P1–62   Cantilever beam.

1–63. If the force on the beam in Problem 1–62 acts upward instead of downward, compute the moment due to the force with respect to the support location *A*.

FIGURE P1–64   Machining fixture.

1–64. The machining fixture shown in Figure P1–64 resists three forces. Compute the moment due to each force with respect to point *A*.

1–65. Sum the moments found in Problem 1–64 vectorially to determine the net moment acting with respect to point *A*.

1–66. A basketball player dunks a shot and then hangs on the rim with his full weight of 180 lb. Compute the moment due to this force with respect to the attachment point of the basket to the backboard. See Figure P1–66.

FIGURE P1–66   Basketball backboard and hoop.

1–67. Figure P1–67 shows a traffic light suspended from an arm extending out over a street from a pole mounted on the side of the street. Compute the moment due to the traffic light with respect to the mounting point *A*.

FIGURE P1–67   Traffic light support.

1–68. The arm holding the traffic light described in Problem 1–67 is made from a steel pipe made to the metric specification PIPE 152 STD. Using the dimensions listed in Appendix A–12(SI) and the properties of steel given in Table 1–6, compute the weight of the arm. Then, considering the weight of the arm to act at its midpoint, compute the moment due to the arm with respect to the mounting point *A* of the pole.

1–69. Considering both the traffic light and the arm described in Problems 1–67 and 1–68, compute the total moment with respect to the mounting point *A* of the pole.

1–70. A home includes a balcony off an upper level bedroom as shown in Figure P1–70. The building materials for the balcony have a total weight of 480 lb. Assuming that the center of mass for the balcony is out 65% of the total length, compute the moment with respect to the mounting system at the wall of the house.

FIGURE P1–70   Balcony.

1–71. For the balcony described in Problem 1–70, consider the case when four people weighing an average of 180 lb each stand side by side at the outer railing. Compute the moment due to those people with respect to the mounting system for the balcony at the wall of the house.

1–72. For the balcony described in Problems 1–70 and 1–71, list other types of loading that may be applied to the balcony. Then describe how they would produce moments with respect to the mounting system for the balcony at the wall of the house.

1–73. For the beam shown in Figure P1–73, compute the moment due to each of the two forces with respect to point *A* at its left end.

FIGURE P1–73   Beam.

1–74. Considering the support for the beam in Figure P1–73 to be a pin that allows free rotation, describe what would happen to the beam under the influence of the two given forces.

1–75. Consider the beam in Figure P1–73. The forces F_1 and F_2 must remain in the same location and act in the same direction. The magnitude of the force F_1 must remain as shown. Determine what the magnitude of the force F_2 must be in order to have all moments with respect to point *A* to be balanced.

1–76. Consider the socket wrench shown in Figure 1–5. Compute the torque applied to the bolt when a force *F* of 145 N acts as shown at a distance *d* of 220 mm from the axis of the bolt.

1–77. Consider the socket wrench shown in Figure 1–5. Assume that the force *F* = 145 N is moved laterally from the given location to act directly on the side of the socket. Compute the torque applied to the bolt with respect to its axis. Then compute the moment on the bolt if the length of the extension shaft from the bolt to the top of the socket is 300 mm.

1–78. Consider the socket wrench shown in Figure 1–5 and the data from Problems 1–76 and 1–77. Assume that the force *F* = 145 N is placed at the end of the wrench as described in Problem 1–76. Now consider that the mechanic applies a second force of 145 N directly to the side of the socket in a direction opposite to the first force. Compute the torque applied to the bolt with respect to its axis. Then compute the net moment on the bolt if the length of the extension shaft from the bolt to the top of the socket is 300 mm.

1–79. A driver grasps the steering wheel of a car on its two sides as shown in Figure P1–79. She pulls down on the right with a force of 12.6 lb and applies an upward force of the same amount with the left hand. Compute the resulting torque on the steering column. (Note that the driver is applying a *couple* to the steering wheel)

$F = 12.6$ lb

$F = 12.6$ lb

FIGURE P1–79   Steering wheel.

1–80. See the photograph of a bicycle shown in Figure 1–1(b). The front pedal is shown to be horizontal. The length of the crank arm is 205 mm from the axis of the crankshaft to the mounting bolt of the pedal. If the rider pushes down with a force of 360 N on this pedal, compute the torque produced with respect to the axle of the crank.

1–81. See the photograph of a bicycle shown in Figure 1–1(b). Note that the front sprocket of the chain drive is mounted on the same shaft as the crank. The act of pushing down on the pedal as described in Problem 1-80 tends to pull on the chain, putting its top part in tension. The diameter of the sprocket is 230 mm and the top part of the chain is horizontal. For the torque applied to the crank found in Problem 1–80, compute the force produced in the chain.

1–82. See the photograph of a bicycle shown in Figure 1–1(b). Note that the top part of the chain is horizontal and that it transmits the chain force to the rear sprocket that has a diameter of 110 mm. Use the value of the chain force from the result of Problem 1–81. Compute the torque applied to the rear wheel.

1–83. See the photograph of a bicycle shown in Figure 1–1(b). Note the braking system near the top of the front wheel. As the rider actuates the brake lever, pads press on the side of the wheel rim creating a friction force acting tangential to the wheel. The diameter of the rim is 660 mm. If the braking force is 790 N, compute the braking torque applied to the wheel.

1–84. Consider the engine for the aircraft shown in Figure 1–1(e). The turbine engine has a drive rotor with a diameter of 1420 mm to the center of the blades. If each blade receives a net force acting tangential to the rotor of 135 N, compute the torque applied to the rotor for each blade. (The general appearance of the turbine drive rotor is similar to that shown in Figure 8–22 in the chapter on the design properties of materials.)

1–85. For the turbine rotor described in Problem 1–84, compute the total torque produced if each of the 56 blades receives the same force.

1–86. Consider the vise shown in Figure 1–2(a) and (b). Clamping force is applied through the jaws at the top. The right jaw is moved horizontally by the drive screw passing through the body on the left of the vise and supported by the movable jaw on the right. The screw is rotated by the bar-type handle on the right side. Complete the following computations that illustrate moments, torques, and couples inherent in the action of the vise.

    a. The user of the vise is able to exert a force of 60 lb perpendicular to the handle that has a length of 4.25 in

from the end to the center of the axis of the screw. What torque can be exerted on the screw?

    b. An alternate operation allows the user to move the handle to the midpoint with 2.125 in of length to either side. Then, if 60 lb is applied downward on one side while 60 lb is applied upward on the other side, what torque is exerted on the screw?

    c. The screw action exerts the horizontal clamping force at the jaws with the right, movable jaw pushing on the clamped object toward the left and up against the fixed left jaw. The reaction force back on the right jaw pushes it toward the right. The jaw is held stable by the rectangular bar across the bottom. If the clamping force is 120 lb, compute the moment due to the reaction to the clamping force if the line of action of the clamping force is 3.50 in above the midline of the rectangle.

ACTIVITIES

1. Examine the chair you are sitting in as you read this book with regard to the forces exerted on it due to your weight. Discuss how those forces are distributed from one component to the next until the force reaches the floor of the room.

2. Raise the hood of a car. Track and document the forces required to complete the process. First, from the release lever inside the car, how is the force from your hand transmitted to the latch on the front of the hood? How does that force actuate the release of the latch? Then what actions are required to disengage the safety catch before the hood can actually be raised?

3. After unlatching the hood of a car, track the force you have to exert to raise it to its fully open position. How does the magnitude of the force you have to exert change with position? This may be different depending on how an assist is provided by springs and gas-filled cylinders. Try this on different makes of car.

4. After being fully raised, how is the hood held in a stable, open position? Is a prop rod used? Or is there a mechanical spring or gas-filled cylinder to hold the hood up? Where are springs used in the latch mechanism? How are the spring forces applied to the mechanism?

5. Design one or more experiments to measure how much force is required for the various actions required to open the latches, raise the hood, and hold it open.

6. Examine a bicycle with hand-actuated brakes. How is the force from your hand transferred through the lever to the actual brake pads? Isolate each component of the brake system, determine how forces are applied to it, and how it transfers the forces to the frame of the bicycle or to another component. Make a sketch of the key components and show the forces that act on each one. Such sketches called *free body diagrams* are very useful in analyzing forces.

7. Take the emergency jack out of a car and study its construction. When it is extended and carrying the weight of the car, how is the force transferred from the car through the structural members and then to the earth? As the jack is raised, how is the force from your arm and hand transferred through the mechanism to raise the car? Why can you raise a heavy car with a relatively small force?

8. Raise a manually operated garage door. What are all of the forces that act on the door as you open it? How are springs used to reduce the amount of force you have to exert? How does the amount of force you must exert change as the door is raised? Why does it change? Do an experiment with force gages (like spring scales) to make actual measurements of force versus position of the door. Make sketches of the door, its springs, any cables attached to the door, the tracks it rides on, and how the tracks are supported on the garage structure.

9. Examine a power-operated garage door system. How does the actuator work? How does it use the rotational motion of the electric motor to cause the linear motion of the door as it opens? How are all of the components of the door opener supported during operation? Make free body diagrams of key components showing all forces.

FORCES AND FREE BODY DIAGRAMS

MAJOR TOPICS

2–1   The Big Picture

2–2   How Forces From Real Systems Are Represented for Analysis

2–3   Components of Forces and Resultants

2–4   Forces at Supports

2–5   Creating Free Body Diagrams

2–6   Building Codes and Bridge Standards

2–1 THE BIG PICTURE

Forces and their effects were important parts of the discussion in Chapter 1. As you followed the examples given for the platform, the pliers, and the various parts of the engine/transmission system for the boat drive illustrated in Section 1–1, you were guided to identify the forces that were exerted on various load-carrying members. Activities at the end of Chapter 1 advised you to do experiments to measure forces required to actuate familiar devices, such as the hood of a car, the garage door, the hand brakes on a bicycle, or the jack used to raise the car to change a tire.

Most people understand the concept of forces from their life experiences. However, most of those experiences deal with forces qualitatively rather than quantitatively. They may sense that a force is "large" or "small" based on their own ability to apply the force, but rarely do people without a technical education make actual measurements of forces required to complete their daily tasks at home, at school, or at work. Nor do they make calculations to predict how much force must be exerted. They just do it or employ some tool or machine to assist them.

After you complete your technical education, you will likely be responsible for designing a product, tool, or machine, or you may be asked to analyze how much force is exerted on critical components of a structure or a mechanism. You must be able to represent the applied forces accurately in a manner that truly represents the physical situation. Typically this requires a drawing or a sketch called a *free body diagram* that shows the object and all forces acting on it. Then you must complete the analysis of other forces caused by the applied forces.

When a force acting on an object tends to cause rotation of the object or some part of it, we say that the force produces a *moment*. Your ability to recognize moments and

to include them in your analyses of machines and structures is critical to a complete analysis. When a moment is restrained so that rotation is prohibited, it causes the member to either bend or twist. This is the typical case in the study of statics. Moments are discussed more completely in Chapter 3, where you will use both forces and moments to ensure that structures and individual members are at rest and will stay at rest. That is, the members must be in *static equilibrium*.

Moments are also very important in the study of strength of materials. Look ahead to Chapters 11–14 of this book dealing with bending stresses and deflections in beams. When the applied moment tends to twist the member, such as when a motor drives a shaft, the moment is often called a *torque*. See Chapter 10 on torsional shear stresses.

Objectives of This Chapter

This chapter will help you represent forces in standard ways and complete basic computations for their magnitudes and directions. Such skills are critical to understanding later material. After completing this chapter, you should be able to:

1. Represent forces as vectors having both magnitude and direction.

2. Recognize all of the forces that act on real objects and represent them on sketches of the objects called *free body diagrams.*

3. Break forces into perpendicular components that are useful for analysis.

4. Combine the effects of two or more forces and compute the magnitude and direction of the single force, called the *resultant,* which has the same overall effect.

Relationships of Statics to Future Study and Work

As the objectives of this chapter and the more complete study of statics were discussed previously, their relationships to the study of strength of materials were stated as well so you can see where you are headed as you complete the study of this book. It is also important to look farther ahead to future courses and to your career. One way to do that is to peruse the numerous references listed at the end of this chapter. Which are most pertinent to you depends on your discipline:

1. For *civil, construction, architecture,* and related fields, References 1–8, 11–16, and 18–21 are the most relevant. These books, codes, and standards apply the concepts of forces and moments to ensure the safe design and construction of buildings, bridges, and other structures. Of critical importance are References 3–8 and 15 because they identify the codes and standards that such structures must meet by law and standard industry practice. Internet sites 1–5 are most pertinent to these fields, some of which are the organizations that develop and distribute codes and standards. Site 5 provides an example of how structural design for wind loads should be done. Also, there is a list of software that performs computer-aided calculations of loads to ensure that designs meet the required codes and standards.

2. For *mechanical, manufacturing, industrial, aeronautical,* and related fields, References 9, 10, 17, 18 and 22 are most pertinent. They describe important parts of the fields involving mechanical design, product design, aerospace structure design, fixture design, manufacturing processes, production equipment design, and the operation and management of production systems.

Regardless of your specific discipline, the skills you will learn from this book will be essential to success in your field. Those studying this course with colleagues from other disciplines will acquire a broad view of how the principles of statics and strength of materials are used every day.

2–2 HOW FORCES FROM REAL SYSTEMS ARE REPRESENTED FOR ANALYSIS

A force has been defined as a pulling or pushing effort exerted on one body by another that tends to move it or change its shape. Pulling forces are called *tensile* whereas pushing forces are called *compressive*. Some forces tend to cut the material of the body on which they act and are called *shearing forces*.

In general, every force is a *vector*, meaning that it has both magnitude and direction. Changing either the magnitude or the direction results in a different effect on the body to which the force is applied. You can demonstrate this yourself by simple actions. For example, place a book flat on a table. Then use your hand and the muscles in your arm to push straight down on the book. You are compressing the

book but it does not tend to move unless you are strong enough to break the table!

Next, try to exert an equal *magnitude* of force on the book but change the direction gradually until you are pushing at an angle of 45° with respect to the vertical. What happens? You probably notice that at some angle the book begins to slide on the table surface. The actual angle will be different depending on the materials of the book and the tabletop, but it is likely that the book moved when the angle of the force was between 30° and 45°.

The resulting movement of the book is not surprising, but it does illustrate the importance of both magnitude and direction on the effect of a force. As the angle of inclination of the force increases, you are still pushing the book down toward the table. But at the same time, you are tending to move it horizontally. At first it does not move because there is friction between the book and the table that acts against the tendency for the book to move. More is said about friction in Chapter 5. But as the angle increases, the tendency for causing the horizontal motion increases until it is greater than the frictional force and then the book moves.

How can we represent this situation in a sketch? In Figure 2–1, parts (a) and (b) show pictorial representations of the action of pushing straight down on the book and also pushing at an angle, represented as the Greek letter θ (theta). For purposes of analysis, we can represent the force exerted by your arm as a vector, a simple arrow pointing in the direction of the force application, and show the point of the arrow applied at the middle of the area of the book on which you were pushing. See parts (c) and (d) of Figure 2–1. Of course, your hand was covering a fairly sizable area, perhaps 75 mm (3.0 in) across. But, for many analyses, it is suitable to represent the entire force as if it were acting at a point. You should be alert to situations in which it is *not* suitable to use the pointed arrow. Examples are shown later in this chapter.

Weight Forces

Another force that acts on the book is its weight because of the effect of gravity. On earth, the weight of an object always acts toward the center of the earth, straight downward. We picture the weight force acting through the *center of gravity* of the object, sometimes called the *center of mass*. More will be said in Chapter 3 about how to locate the center of gravity. For the present, you can visualize it to be at the geometric center of the object. Figure 2–2 shows how we will represent weight forces acting through the center of gravity.

When the weight of the object of interest is small in comparison to the forces applied by other means, it is acceptable to neglect the weight in the analysis. You must carefully evaluate this issue for each problem to see if weight is significant or not.

Reaction Forces and Free Body Diagrams

If, in our use of Figure 2–1, it is our objective to analyze the effects of all of the forces applied to the book, then we must also show the forces that are exerted on the bottom of the book by the table. You should be able to visualize that when

(a)

(b)

(c)

F

(d)

θ

F

FIGURE 2–1 Illustration of forces as vectors.

you push down on the book as in Figure 2–1(a), the table pushes upward in order to hold the book in position. This force is called the *reaction force* or simply *reaction,* because it only exists in reaction to the applied forces. This is shown in Figure 2–3, labeled R. Again, the force exerted by the table is actually distributed over the entire bottom of the book. But it is satisfactory for many analyses to show the total reaction force as a single vector.

Now consider again the situation in Figure 2–1(b) and (d) where force is applied to the book at an angle. You observed a tendency for the book to slide to the left as the angle of inclination of the applied force increased. But it did not move until the angle was fairly large, and we observed a frictional force acting between the book and the table. Frictional

forces always act in the direction that opposes the impending motion. So if the book tends to slide to the left, the frictional force acts toward the right. The frictional force, called F_f, is shown acting adjacent to the surface where the friction occurs. In our example, that is the bottom surface. Combining all of the discussion to this point, the situation shown in Figure 2–1(b) can be represented as a free body diagram of the book as shown in Figure 2–4.

Forces Exerted by Cables and Ropes

Pulling on an object with a cable provides an action force that is directed in line with the cable. Consider the winch system on a truck as shown in Figure 2–5. As the winch draws

W

FIGURE 2–2 Book with weight force shown.

F

W

R

FIGURE 2–3 Free body diagram of book as loaded in Figure 2–1(a).

θ

F

Frictional force = F_f

W

R

FIGURE 2–4 Free body diagram of book as loaded in Figure 2–1(b).

(a) Pictorial representation

Force on winch

F_c F_c Cable segment F_c F_c Force on crate

(b) Cable force, F_c

FIGURE 2–5 Winch system mounted on a truck pulling a crate up a ramp.

the cable in and wraps it around the drum, the cable is placed in tension and pulls on the crate, moving it up the ramp into the bed of the truck. The force applied to the crate by the cable, F_c, acts in the same direction as the alignment of the cable. Another effect of the tension in the cable is a force exerted by the cable on the winch. Any segment of the cable can be drawn as a free body with the force, F_c, acting at each end. So the cable is sometimes called a *two-force*

member. Other kinds of two-force members are described in Chapter 3.

Ropes behave similarly to cables. Either can be used with a pulley to change the direction of the force. Figure 2–6 shows a person pulling on a rope with a force, T, at an angle that passes over a pulley and exerts a vertical force on a bucket of building materials to be raised to an upper floor of a building. The pulley's rim is grooved to

Pulley

T

T

T = Tension in rope

FIGURE 2–6 Rope on a pulley.

FIGURE 2–7   Casting supported by a set of three cables.

guide the rope and it is normally assumed that the magnitude of the tension force in the rope is equal on both sides of the pulley if the pulley is free to rotate about its axle.

Sometimes two or more cables act together as shown in Figure 2–7. The vertical force in the lifting cable coming off the traveling crane leads to the two angled forces in the cables that are attached to the heavy casting. The three forces can be shown acting on the connection point as three vectors in line with the three cables. We will analyze such force systems in Chapter 3.

Distributed Forces on Surfaces

When a load covers a significant part of a surface, it may not be appropriate to represent the load as a single concentrated force. Subsequent analyses may have to take into account that the force acts over the entire area. An example is shown in Figure 2–8(a) that illustrates one design for the roof of a building. Follow the structure of the roof from the walls upward and notice that several beams support the roof load and transfer that load to the walls. Decking spans the space between beams to carry the other materials that make up the roof. Figure 2–8(b) shows corrugated metal decking onto which concrete is poured and weatherproofing materials are placed on top. In colder climates, snow will accumulate on the roof adding more load. The result is a very large load distributed uniformly over the total roof area.

Distributed loads on surfaces are expressed in the units of force per unit area, such as pounds per square inch (lb/in^2), pounds per square foot (lb/ft^2), or kilonewtons per square meter (kN/m^2). You must know the specific weight or density of the materials that make up the roof and their thickness. Then you can compute the weight per unit area of the material as it is applied to the surface.

(a) Roof construction

(b) Detail of roof materials and snow load

(c) Schematic diagram of load on roof beam per unit length

FIGURE 2–8   Uniformly distributed loads on a roof.

Example Problem 2–1

For the roof structure shown in Figure 2–8, compute the load on each square foot of area including the decking, roofing materials, and a snow depth of 42 in. Each piece of corrugated metal decking covers 4.50 ft by 12.00 ft and weighs 180 lb. Concrete, with a density of 150 lb/ft^3, covers the decking at an average depth of 3.00 in. The top coating of weatherproofing weighs 6.0 lb/ft^2. The snow density is 6.5 lb/ft^3.

Solution

Objective: Compute the load on the roof in lb/ft^2

Given: Snow: Depth = 42 in; density = 6.5 lb/ft^3

Top coat—weatherproofing: 6.0 lb/ft^2

Concrete: Depth = 3.00 in; density = 150 lb/ft^3

Decking: 4.50 ft × 12.00 ft; 180 lb total weight

Analysis: The load per unit area of each material is computed:

Results: Snow: Depth = 42 in(1 ft/12 in) = 3.50 ft

Each ft^2 of roof carries 3.50 ft^3 of snow.

$$W_s = 6.5 \text{ lb/ft}^3 \times 3.50 \text{ ft}^3 = 22.75 \text{ lb}$$

Top Coat: Given that each ft^2 weighs 6.0 lb = W_{tc}

Concrete: Depth = 3.00 in(1 ft/12 in) = 0.25 ft. Each ft^2 of roof carries 0.25 ft^3 of concrete.

$$W_c = 15 \text{ lb/ft}^3 \times 0.25 \text{ ft}^3 = 37.5 \text{ lb}$$

Decking: Total area = 4.50 ft × 12.0 ft = 54.0 ft^2/piece.

Each ft^2 of roof carries 180 lb/54.0 = 3.33 lb = W_d

Now compute the total load per ft^2, W_t:

$$W_t = W_s + W_{tc} + W_c + W_d$$
$$W_t = 22.75 + 6.0 + 37.5 + 3.33 = 69.6 \text{ lb}$$

Comments: The result shows that each square foot of roof carries 69.6 lb. It could be reported that the *rate of loading per unit area* is 69.6 lb/ft^2.

An example of distributed forces on a surface from the mechanical field is shown in Figure 2–9. A certain type of pressure transducer employs a thin, flat disc sealed into the wall of a cylinder. As gas pressure is introduced on one side of the disc, sometimes called a *diaphragm*, the disc is subjected to a uniform rate of loading in lb/in^2 (psi) or N/m^2 (Pa). The disc deforms by bowing outward and movement of the center of the disc can be related to the pressure.

Distributed Forces Per Unit Length

Looking again at Figure 2–8, it would be appropriate to determine how much load each beam carries as it spans the distance between the two walls of the building. The design or analysis of the beam, which you will learn in Chapters 11 and 12 of this book, depends on the amount of load carried by each lineal unit of length, indicated in the units of lb/in, lb/ft, K/ft, N/m, kN/m, and so forth. In Figure 2–8, for example, near the middle of the building, each 30-ft-long beam carries a portion of the total load on the roof equal to that on an area corresponding to the distance between beams, in this case 6.00 feet. This is 3.0 feet on each side of the centerline of the beam. Thus, the total area of roof carried by each beam is 30 ft × 6 ft = 180 ft^2. From Example Problem 2–1 we found that 69.6 lb was applied to each square foot of the roof. Then the total load on one beam is

$$\text{Total load} = 69.6 \text{ lb/ft}^2 \times 180 \text{ ft}^2 = 12\,530 \text{ lb}$$

Each foot of length of the beam carries a rate of loading per unit length of

$$w_b = 12\,530 \text{ lb/30 ft} = 418 \text{ lb/ft}$$

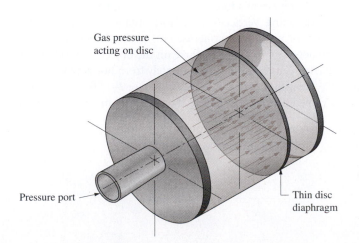

FIGURE 2–9 Pressure transducer showing gas pressure producing a uniformly distributed load on a thin disc.

Gas pressure acting on disc

Pressure port

Thin disc diaphragm

This type of load is called a *uniformly distributed load.* We typically represent this type of load in a schematic diagram like that shown in Figure 2–8(c).

Varying Distributed Forces Per Unit Length

Some distributed loads vary with position instead of being uniformly distributed as in previous examples. Figure 2–10 shows an extended roof over a loading dock on which snow has accumulated at depths varying from only a few inches at its end to several feet at the wall. We call this a *varying distributed load* and typically report it as a straight-line, sloped loading pattern with defined rates of loading at each end of the slope.

Of course, some real loads are not so neatly distributed that they can be represented as a straight line. More complex equations for loading could be used or the total load could be broken into segments in which each segment can be approximated as a straight line. An example is shown in Figure 2–11 in which gravel at varying depth is stored in a hopper. The load on one of the floor beams of the hopper might be represented as shown in Figure 2–11(a).

(a) Schematic representation of snow load on a beam

(b) Pictorial example—snow load on a projecting roof

FIGURE 2–10 Example of linearly varying distributed load on a cantilever.

(a) Schematic representation of load on floor beam due to gravel

(b) Pictorial example–gravel at varying depth in a hopper

FIGURE 2–11 Example of varying distributed load.

2–3 COMPONENTS OF FORCES AND RESULTANTS

We saw earlier that forces can be applied in many ways and in virtually any direction. The analysis of forces is sometimes easier if forces are shown acting in two perpendicular directions, such as vertically and horizontally, especially when we want to know how much effect a given force has in lifting an object or pulling it along the flat table of a machine or up a ramp. Any single force can be broken into *components* acting at the point of application of the original force and having the same effect as the original force. Sometimes we know the magnitudes and directions of two or more forces acting on an object and we would like to know the single force that would cause the same effect. That single force is called the *resultant.* This section shows how components and resultants are found.

Perpendicular Components of a Force

We can represent a force acting at some angle as a pair of two forces acting vertically and horizontally by breaking the given force into components. This process does not change the net effect of the given force on the object, but it allows us to consider vertical forces separately from horizontal forces.

Consider the crate shown in Figure 2–12 being pulled by a cable force upward and to the left at an angle of 40° to the

FIGURE 2–12 Cable force acting on a crate.

FIGURE 2–13 Resolving the cable force into horizontal and vertical components.

FIGURE 2–14 Horizontal force resolved into components parallel and perpendicular to surface of ramp.

horizontal. One effect of the cable force is the tendency to raise the crate off the floor. Another effect is to pull it to the left. We may want to analyze either situation.

The process is called *resolving a force into perpendicular components*. Figure 2–13 shows the basic elements of the process. We represent the cable force, F_c, as a vector acting at the place of attachment of the cable to the crate. Horizontal and vertical lines are drawn from the origin of the vector. Then horizontal and vertical lines are drawn through the tip of the vector until they intersect the first pair of lines. The horizontal distance from the origin to the intersection with the new vertical line represents the horizontal component of the cable force, which we will call, F_{ch}. Similarly, the vertical distance from the origin to the intersection with the new horizontal line represents the vertical component of the cable force, called F_{cv}. The whole process can be done graphically by carefully drawing the cable force to some convenient scale and at the proper angle, drawing the two pairs of vertical and horizontal lines, and measuring the final lengths of the component vectors F_{ch} and F_{cv}. Then the magnitude of the components can be computed by applying the scale factor.

Because an analytical solution is usually preferred, however, we use the functions of a right triangle to compute the forces F_{ch} and F_{cv}. Note that a right triangle is formed by the cable force, the horizontal component, and the vertical line connecting their tips. Notice also that the vertical line has a length equal to the vector F_{cv}. The magnitudes of the components are computed as follows:

$$\cos \theta = F_{ch}/F_c$$

Then

$$F_{ch} = F_c \cos \theta$$

Also

$$\sin \theta = F_{cv}/F_c$$

Then

$$F_{cv} = F_c \sin \theta$$

As a numerical example, assume that the cable force in Figure 2–12 is 850 lb and that it acts at an angle $\theta = 40°$ with respect to the horizontal. Then

$$F_{ch} = F_c \cos \theta = (850 \text{ lb}) \cos 40° = 651 \text{ lb to the left}$$
$$F_{cv} = F_c \sin \theta = (850 \text{ lb}) \sin 40° = 546 \text{ lb upward}$$

The directions of these forces can be seen in Figure 2–13. In summary, the applied force of $F_c = 850$ lb acting at an angle of 40° with the horizontal can be replaced by the two forces with the same effect on the crate, where $F_{ch} = 651$ lb acting horizontally to the left and $F_{cv} = 546$ lb acting vertically upward. The vertical component tends to lift the crate and the horizontal component tends to move it along the surface.

Sometimes directions other than horizontal and vertical are more convenient for the components. The nature of the problem often dictates the directions of the components that should be computed. For example, in Figure 2–14, the crate sits on an inclined ramp and has a horizontal force of 850 lb pushing on it. The analysis, as presented in Chapter 5, might want to predict if the applied force is adequate to push the crate up the ramp. Then components parallel to and perpendicular to the surface of the ramp would be more appropriate. We can define the x-direction to be parallel to the ramp and the y-direction to be perpendicular to the ramp as shown in the figure. The analysis for the magnitudes of the components proceeds as before:

$$F_{cx} = F_c \cos \theta = (850 \text{ lb}) \cos 40° = 651 \text{ lb}$$
$$\text{parallel to ramp surface}$$
$$F_{cy} = -F_c \sin \theta = -(850 \text{ lb}) \sin 40° = -546 \text{ lb}$$
$$\text{perpendicular to ramp surface}$$

The minus sign for the value of F_{cy} indicates that it acts in the opposite direction from the assumed positive y-direction. F_{cx} acts in the positive x-direction.

Resultant of Two Perpendicular Components

There are other times when the two perpendicular components of a force are known as the result of other analyses and it is desired to compute the single force that would have the same effect. That single force is called the *resultant* and it is computed by again using the right triangle analysis. Note that a complete analysis must determine both the *magnitude* and *direction* of the resultant. Example Problem 2–2 illustrates the procedure.

Example Problem 2–2   Figure 2–15(a) shows a platform on which a 145-lb person is standing while a horizontal force of 550 lb pulls to the right. Compute the resultant force acting on the platform.

(a) Loads applied to platform

(b) Loads acting at point of effective application

(c) Vector diagram used to determine resultant, R

FIGURE 2–15   Resultant force on a platform for Example Problem 2–2.

Solution   Objective: Compute the resultant.

Given: In Figure 2–15, $W = F_v = 145$ lb acting vertically downward.

$F_h = 550$ lb acting to the right.

Analysis: Extending the lines of action of the two applied forces until they intersect indicates the effective point of application of the resultant within the platform. Showing W and F_h acting at this point allows the determination of the resultant for which W and F_h are components. See Figure 2–15(b).

The actual analysis is shown in Figure 2–15(c). Vertical and horizontal lines are drawn through the ends of the two vectors and their intersection is found. The resultant is the vector from the origin to that intersection. The right triangle thus formed has sides equal to the two known forces, and the hypotenuse is the resultant.

Results: The magnitude of the resultant is found using the Pythagorean Theorem.

$$R = \sqrt{F_h^2 + W^2} = \sqrt{(550 \text{ lb})^2 + (145 \text{ lb})^2} = 569 \text{ lb}$$

The direction of R is

$$\tan \theta = W/F_h = 145/550 = 0.2636$$
$$\theta = \text{Tan}^{-1}(0.2636) = 14.8°$$

LESSON 2

Comments: The resultant force, R, is a vector with a magnitude of 569 lb acting downward and to the right at an angle of 14.8° below the horizontal. The effective point of application is shown in Figure 2–15 as point O.

General Case of the Resultant of Two Forces

If two forces that are not perpendicular to each other act on an object, their resultant cannot be found by the method shown in Example Problem 2–2. However, we now show a more general procedure.

Figure 2–16 shows a pertinent case. A valve body is being drilled simultaneously by two drills. The one on top

acts vertically downward and exerts a force of 550 N during the drilling operation. The second drill on the side exerts a force of 380 N in a direction 30° above the horizontal. We would like to know the resultant force on the valve body in order to design fixturing to hold it during machining. To do that, we must perform vector addition, adding F_1 and F_2 to determine the resultant, R. First we show the vector addition procedure and then an example problem illustrates its use.

$F_1 = 560$ N

$F_2 = 380$ N

30°

FIGURE 2–16   Drills acting on a valve body for Example Problem 2–3.

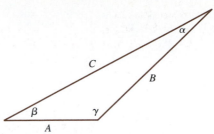

FIGURE 2–17   General notation for law of sines and law of cosines.

4. The triangle thus formed satisfies the equation for vector addition.

$$R = F_1 + F_2$$

5. If the vector triangle is drawn accurately, the length and direction could be measured. However, it is better to compute these values using the following steps.

6. Label all known lengths and angles for the given vectors on the diagram.

7. Note that the triangle will not, in general, be a right triangle. Therefore, the simple applications of sine, cosine, and tangent cannot be used. Instead, the *law of sines* and the *law of cosines* are typically needed. The general forms of these laws follow, using the general notation in Figure 2–17.

$$\frac{\sin \alpha}{A} = \frac{\sin \beta}{B} = \frac{\sin \gamma}{C} \quad \text{[Law of sines]} \quad (2\text{–}1)$$

Note that A is the side opposite angle α, B is opposite β, and C is opposite γ. You will have to arrange the notation for angles and sides for your problems in a similarly consistent manner.

$$C^2 = A^2 + B^2 - 2AB \cos \gamma \quad \text{[Law of cosines]} \quad (2\text{–}2)$$

Again, you will have to ensure that C is opposite γ.

Typically, a combination of the law of sines and the law of cosines must be used to solve for the desired vector.

Procedure for the addition of two vectors (see Figure 2–17):

1. From a point called the origin, draw the first vector in its proper direction with a length proportional to the magnitude of the force.

2. From the tip of the first vector, draw the second vector in its true direction and with a length proportional to its magnitude.

3. The resultant is the vector drawn from the origin to the end of the second vector. The figure will now be a closed triangle with the length of the final vector proportional to the magnitude of the resultant. The direction of this vector is the true direction of the resultant.

Example Problem 2–3   Figure 2–16 shows the valve body being machined by two drills simultaneously. Compute the resultant force acting on the valve body.

Solution   Objective: Compute the resultant.

Given: In Figure 2–16, $F_1 = 550$ N acting vertically downward.

$F_2 = 380$ N acting downward and to the left at an angle of 30° below the horizontal.

Analysis: Figure 2–18 shows the vector addition triangle. First, F_1 was drawn from the origin, O. Then F_2 was drawn with its tail added to the tip of F_1. The resultant is the vector from the origin to the end of the tip of F_2. We will use the law of cosines to solve for the magnitude of R. Then the law of sines will be used to find the angle of orientation of R.

Results: The law of cosines [Equation (2–2)] can be used to find the magnitudes of the resultant force in this problem. From the labeling of the parts of the triangle, we can state

$$R^2 = F_1^2 + F_2^2 - 2F_1 F_2 \cos \gamma$$

The value of γ must be determined from the geometry of the triangle. Note that $\gamma = 30° + 90° = 120°$. Then

$$R^2 = (550)^2 + (380)^2 - 2(550)(380) \cos(120°)$$
$$R = 810 \text{ N}$$

FIGURE 2–18 Vector triangle solution for resultant R for Example Problem 2–3.

The direction of R can be defined by finding the value of the angle α using the law of sines.

$$\frac{\sin \alpha}{F_2} = \frac{\sin \gamma}{R}$$

We know the values of F_2, R, and γ, so we can solve for α.

$$\alpha = \text{Sin}^{-1}[\sin \gamma (F_2/R)] = \text{Sin}^{-1}[\sin(120°)(380)/810)] = 24.0°$$

Comments: The resultant force, R, is a vector with a magnitude of 810 N acting downward and to the left at an angle of 24.0° from the vertical.

Parallelogram Rule

A slight modification of the method just described is now shown. Called the *parallelogram rule*, it is useful when the two vectors being added both act at the same point.

Example Problem 2–4

Figure 2–19 shows two people standing on the sides of an open ditch pulling on ropes to raise the cover from an access hole to an underground utility vault. What is the resultant force applied to the cover?

Solution   Objective: Compute the resultant.

Given: In Figure 2–19, F_1 = 80 lb acting upward and to the left at an angle of 60° from the horizontal. F_2 = 65 lb acting upward and to the right at an angle of 45° above the horizontal.

Analysis: Figure 2–20 shows the vector addition using the parallelogram rule. First, F_1 and F_2 were drawn from the origin, O, representing the point where the two ropes are attached to the cover. Then lines were drawn through the tips of each given vector, parallel to the other vector, creating a parallelogram in which two sides are equal to F_1 and two are equal to F_2. The resultant is the vector from the origin to the opposite corner of the parallelogram. The resultant splits the parallelogram into two parts, each of which is a triangle made up of F_1, F_2, and R. Now the law of sines and the law of cosines can be applied to either triangle to solve for the resultant using the same technique as demonstrated in Example Problem 2–3.

Results: The law of cosines [Equation (2–2)] can be written in terms of the variables in this problem to find the magnitude of the resultant.

$$R^2 = F_1'^2 + F_2^2 - 2F_1'F_2 \cos \gamma$$

FIGURE 2–19 Workers pulling on ropes to raise a cover.

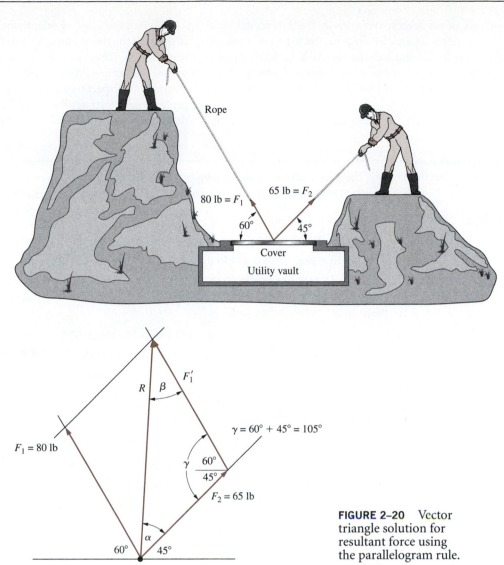

FIGURE 2–20 Vector triangle solution for resultant force using the parallelogram rule.

The value of γ must be determined from the geometry of the triangle. Note that $\gamma = 45° + 60° = 105°$.

$$R^2 = (80)^2 + (65)^2 - 2(80)(65)\cos(105°) = 115.4 \text{ lb}$$

The direction of R can be defined by finding the value of the angle α using the law of cosines.

$$\frac{\sin\alpha}{F_1'} = \frac{\sin\gamma}{R}$$

We know the values of F_1', R, and γ, so we can solve for α.

$$\alpha = \text{Sin}^{-1}[\sin\gamma(F_1'/R)] = \text{Sin}^{-1}[\sin(105°)(80)/15.4)] = 42.0°$$

Adding this to the 45° angle gives the total angle from the horizontal to R of 87.0° or 3.0° to the right of vertical.

LESSON 1

Comments: The resultant force, R, is a vector with a magnitude of 115.4 lb acting upward and to the right at an angle of 3.0° from the vertical.

Alternate Approach to Computing Resultants

So far, we have shown two methods of computing resultants of two or more forces: the *vector triangle method* and the application of the *parallelogram rule*. An alternate approach is shown here that may be preferred for some types of problems, particularly when more than two force vectors are applied at a single point.

We call this method the *summation of components method*. The process involves first computing the perpendicular components of all forces, typically in the horizontal and vertical directions. Then we compute the algebraic sum of all x-direction forces and call the result, ΣF_x. Similarly, we compute the algebraic sum of all y-direction forces and call the result, ΣF_y. The resultant force R is then computed from

$$R = \sqrt{(\Sigma F_x)^2 + (\Sigma F_y)^2} \qquad (2\text{--}3)$$

To find the angle of orientation of the resultant force, we first sketch the two vectors, ΣF_x and ΣF_y with their tails at the same origin. Then we can use a technique similar to that shown in Example Problem 2–2 to find the direction of the resultant with respect to the x- or y-axes. The following example problem demonstrates the use of this technique with the same data used for Example Problem 2–4 and shown in Figure 2–19.

Example Problem 2–5   Use the data from Example Problem 2–4 and the illustration shown in Figure 2–19 to compute the resultant of the two forces exerted on the cover using the summation of components method.

Solution   Objective: Compute the resultant.

Given: In Figure 2–19, $F_1 = 80$ lb acting upward and to the left at an angle of 60° from the horizontal.
$F_2 = 65$ lb acting upward and to the right at an angle of 45° above the horizontal.

Analysis: Figure 2–21(a) shows the two given force vectors acting at the same origin. The x- and y-components of each vector are also shown. *Notice the sign convention for directions.* Now we can compute the magnitudes of the components.

$$F_{1x} = -F_1 \cos(60°) = (-80 \text{ lb})\cos(60°) = -40.00 \text{ lb}$$
$$F_{1y} = +F_1 \sin(60°) = (+80 \text{ lb})\sin(60°) = +69.28 \text{ lb}$$
$$F_{2x} = +F_2 \cos(45°) = (+65 \text{ lb})\cos(45°) = +45.96 \text{ lb}$$
$$F_{2y} = +F_2 \sin(45°) = (+65 \text{ lb})\sin(45°) = +45.96 \text{ lb}$$

The summations ΣF_x and ΣF_y can now be computed.

$$\Sigma F_x = F_{1x} + F_{2x} = -40.00 \text{ lb} + 45.96 \text{ lb} = 5.96 \text{ lb}$$
$$\Sigma F_y = F_{1y} + F_{2y} = 69.28 \text{ lb} + 45.96 \text{ lb} = 115.24 \text{ lb}$$

These two forces are shown in Figure 2–21(b), along with their resultant, which we now compute.

$$R = \sqrt{(\Sigma F_x)^2 + (\Sigma F_y)^2} = \sqrt{(5.96)^2 + (115.24)^2} = \underline{115.4 \text{ lb}}$$

The angle or orientation of R is now found from

$$\tan\theta = (\Sigma F_y / \Sigma F_x)$$
$$\theta = \text{Tan}^{-1}(\Sigma F_y / \Sigma F_x) = \text{Tan}^{-1}(115.4/5.96) = \underline{87.0°}$$

Comments: These results match those found from Example Problem 2–4.

FIGURE 2–21   Resultant by summation of components method.

(a) Given force vectors

(b) Resultant

2–4 FORCES AT SUPPORTS

Beams in structures, links in mechanisms, and various components of assemblies of any type have joints where two members are connected to one another and places where loads are transferred by supports to the ultimate foundation of a building or the frame of a machine. We often use the term *support* for either a joint or an actual support.

The kinds of forces exerted by supports depend on the way the joined members are connected. Here we will describe several types of supports for members that can be analyzed in a single plane: *roller, pin, smooth surface, rough surface,* and *fixed*.

Roller Support

Figure 2–22(a) shows a load being supported on two cylindrical rollers. Because the load acts vertically downward due to its weight, the rollers are able to support it with upward reaction forces as shown in the free body diagram in part (b). But this arrangement is unstable in the sense that, if the load is not directly vertical or if the rollers are on an inclined plane, they would tend to roll and not resist the applied loads. Therefore, we will normally use no more than one roller as a part of a support system. The following rule will be applied in this book.

A roller used as a support for a load-carrying member can exert only a force acting perpendicular to the member.

Pin Support

Figure 2–22(c) shows a beam supported at its right end with a roller while a smooth cylindrical pin connection is used for the left end. Although the roller can resist only vertical loads, the pin connection can resist forces perpendicular to it acting in any direction. Thus, the beam is stable even when the applied force may act in any direction. The reaction at a pin support is typically shown as a set of two mutually perpendicular forces, here called R_{1x} and R_{1y}, as shown in part (d). The member is, therefore, restrained from moving in any direction at the point of support. Note, however, that the smooth pin would allow the member to rotate about the centerline of the pin, requiring the second support for complete equilibrium of the beam. In summary, then,

A pin support can resist loads in any direction but it will permit rotation of the load-carrying members about the centerline of the pin.

Smooth Surface

We represent a smooth-surface support as shown in Figure 2–22(e) with a round shape contacting a plane surface, here shown as the top of a ladder resting against a vertical wall. Furthermore, we assume that no friction occurs between the support and the surface, much like very smooth ice that would cause the round-shaped member to slide if any force acted parallel to the surface. As long as the force between the

(a) Beam on two rollers

(b) Free body diagram for (a)

(c) Beam with pinned support and one roller

(d) Free body diagram for (c)

(e) Ladder

(f) Reaction at wall

(g) Reaction at floor

FIGURE 2–22 Support types.

support and the surface acts perpendicular to the surface, the system is stable. In summary, we will use the following rule.

A smooth surface used as a support for a load-carrying member can exert only a force acting perpendicular to the member.

See Figure 2–22(f) for the method of showing the reaction. Thus, the roller and the smooth surface are assumed to provide the same kind of support. In reality, there is some small amount of friction present in either a roller or smooth surface. Our assumption is that the frictional force is very small compared with applied loads on the members and can be neglected.

Rough Surface

A support surface is called *rough* if a sizable frictional force can be developed between it and a mating member that acts parallel to the surface. In fact, we assume

A rough surface can exert a frictional force parallel to the surface along with a force perpendicular to that surface.

Figure 2–22(e) shows a rough surface at the feet of the ladder to keep it from sliding away from the wall. The analysis of frictional forces is included in Chapter 5. In summary, when a support surface is called rough, the reactions it can exert are shown in Figure 2–22(g).

Fixed Support

The most rigid type of support is one that prohibits linear motion in any direction and also prohibits rotation of the load-carrying member. Figure 2–23(a) shows a device called a jib crane whose boom is a beam with a fixed end at the left with a load applied at the other end. This type of beam is called a *cantilever*. The fixed support can be visualized as a close-fitting socket in a rigid structure into which the beam is installed and held rigidly as if by welding. In summary,

A fixed support can resist loads in any direction and also restrain the load-carrying member from rotation at the support.

The jib crane shows only the gravity load being lifted. However, the force applied to the boom may be at some angle, resulting in both horizontal and vertical components as shown in part (b) of the figure. The reaction at the fixed support will, in general, consist of vertical and horizontal forces and a moment resisting the tendency to tilt the crane.

2–5 CREATING FREE BODY DIAGRAMS

Now we will combine the principles discussed in preceding sections to create free body diagrams of various parts of a structure or a mechanical device. This is an essential skill for the study of statics and, in turn, for applying statics to strength of materials. You will first use free body diagrams in the following chapter in which you will learn and apply the concepts of *equilibrium* of load-carrying members to determine supporting forces necessary to maintain equilibrium. Only by using complete, accurate free body diagrams can you ensure that the ultimate designs of all parts of a structure or a machine are safe and that they will perform their expected function.

The concept of *static equilibrium* requires that

1. The body of interest is at rest or is moving with constant speed and direction.

2. All forces acting on the body are balanced such that there are no net forces that tend to move it. This requires that

    The sum of all forces in any direction must be zero.

3. All moments acting on the body are balanced such that the body will not tend to rotate about any point or axis. This can be stated as

    The sum of all moments about any point must be zero.

The process for preparing free body diagrams requires the following six steps:

1. Select the component of a structure for which the free body diagram will be prepared and isolate it from the assembly.

FIGURE 2–23 Cantilever beam with a fixed support.

(a) Pictorial example—jib crane

(b) Schematic representation—free body diagram

2. Show all externally applied loads that act on the body with careful attention to their location and direction.

3. Show the weight of the component acting through its center of mass and directed downward toward the center of the earth. Note that in some problems the weight of the component can be ignored because it is negligibly small compared with the externally applied loads.

4. When there are two or more components to be analyzed, draw each separately. At joints, the principle of *action–reaction* will be applied. Each *action* force exerted by one member on a mating member produces an equal and oppositely directed *reaction* force against the other. It is helpful to draw free body diagrams of joined members in the same orientation as they exist in the assembly.

5. Identify the type or types of supports necessary to maintain the component in equilibrium.

6. Show the forces and/or moments that the supports are capable of applying on the free body diagram. Refer to Figures 2–22 and 2–23 for the kinds of supports.

7. Inspect the resulting diagram to ensure that it is possible to have all forces and all moments balanced.

In the following chapter, you will learn how to analyze the free body diagram to determine the magnitude and direction of unknown forces or moments that will ensure equilibrium. Typically this involves finding the forces or moments at supports.

Later, in the chapters on strength of materials, you will complete the analysis or design of the components to ensure that it will be safe under the loads shown on the free body diagram. This requires the specification of the material from which the component will be made, the shape of its cross section, and the dimensions of all features. You can see that successful design is inherently dependent on your ability to prepare complete, accurate free body diagrams.

The following example problems illustrate the general method of preparing free body diagrams. You should work toward a general understanding of the basic principles and techniques so that you can apply them to any component regardless of the specific shape or type of loading and support. At the end of each problem we discuss the kinds of stresses that are likely to be developed in each component so you can see the reasons for completing the free body diagrams. Also we point out which later chapters cover the particular types of stresses produced.

Example Problem 2–6

Figure 2–24 shows a simple clevis joint consisting of a fixed base, a vertical pin, and a link installed over the pin. When in use, a force is applied through the hole on the right end of the link. The direction of the force is aligned with the axis of the link along a line between the pin and the hole. Prepare free body diagrams of the link, the pin, and the base.

Solution Link: Figure 2–25(a) shows the link isolated from the clevis.

External Forces:
a. The applied force *F* is shown acting to the right.
b. The link's weight is shown acting vertically downward.

FIGURE 2–24 Clevis joint.

| (a) Free body diagram of link | (b) Free body diagram of pin | (c) Free body diagram of base |

FIGURE 2–25   Free body diagrams of clevis parts.

Forces supplied by the Pin and the Base:

a. The pin must apply the force F_p acting toward the left and in line with the applied force F. This tends to pull the link in tension producing a *tensile stress* in its material. Additionally, the pin force F_p presses on the inside of the hole producing a surface-to-surface *bearing stress* that will be discussed in Chapter 9.

b. The base must provide an upward force to balance the weight. We show this as a single force R_1 acting vertically upward to the right of the pin for convenience. However, the supporting force will actually be spread over the entire area of the base as a distributed load. Typically it is not necessary to show that much detail unless the forces are so large that the base material could be crushed. The analysis of bearing stress on flat surfaces is presented in Chapter 9.

c. The combination of the downward weight and the upward supporting force is separated and these forces act in opposite directions, producing a couple that tends to rotate the link in a clockwise direction. An opposing force or couple is required. Assuming that there is a small clearance fit between the pin and the hole in the link, we can assume that the retaining ring at the top end of the pin supplies the required balancing force R_2.

d. Examining the final free body diagram, you should see that all horizontal forces and all vertical forces are now balanced. Also, the combination of W, R_2, and R_1 will ensure that all moments are balanced.

Pin: Figure 2–25(b) shows the free body diagram of the pin. The link and the base are shown lightly in phantom for reference only.

External Forces: It is necessary to implement the principle of *action–reaction* here.

a. The force F_p acting toward the left on the link acts toward the right on the pin. That is, the *action* is the pin pushing on the link, while the *reaction* is the link pushing back on the pin with an equal and oppositely directed force.

b. The retaining ring pulls upward on the pin with the force R_2.

Forces Supplied by the Base:

a. The horizontal force F_p to the right must be reacted by a force to the left provided by the base. We call this force F'_p. These two forces are actually distributed across the surface of the pin but are typically shown as single resultant forces acting at the middle of the areas. These two forces acting in opposite directions tend to cut the pin across its cross section at the interface between the link and the base. The cutting action causes a *shearing stress* in the pin that is discussed in Chapters 7 and 9.

b. The vertical upward force R_2 is reacted by the downward R'_2 force provided by the base acting on the flange at the bottom of the pin.

c. The combination of the two forces F_p and F'_p, being separated, causes a clockwise couple to act on the pin. This tends to cause the pin to be cocked in the hole resulting in a leftward force near its top and a rightward force near its bottom. If the clearance between the pin and the hole is small (a desirable design decision), the opposing couple is produced by a modest adjustment of the forces F_p on both the pin and the base. Such forces are typically neglected unless the couple is very large, the link is thin, or the clearance between the pin and the hole is large.

Base: Figure 2–25(c) shows the free body diagram of the base fixed to some rigid supporting structure. Thus, it is capable of providing reactions for both forces and moments as discussed earlier. In this case the fixed support provides a horizontal force R_{3h} acting toward the left, a vertical upward force R_{3v}, and a counterclockwise moment M_B to oppose the couple caused by the two horizontal forces.

Example Problem 2–7

Figure 2–26 shows an enclosure through which four pipes pass. The pipes are supported on a horizontal beam that in turn is supported by cables at the ends that are attached to the roof of the enclosure. Prepare the free body diagrams of the beam and the cables.

Solution Beam: Figure 2–27 shows the free body diagrams of the beam and the cables.

External Loads:
a. Each of the pipes exerts a concentrated downward force on the beam.
b. The weight of the beam acts vertically downward at its center of gravity.

Supports: The cables are capable of providing only upward supporting forces, F_{c1} and F_{c2} on the beam. The loading and support system causes the beam to be in bending. You will learn how to analyze beams in Chapters 11–14.

Cables: Each cable is subjected to a tensile force equal to the reaction forces at the support points of the beam. Chapters 7 and 9 discuss tensile stress.

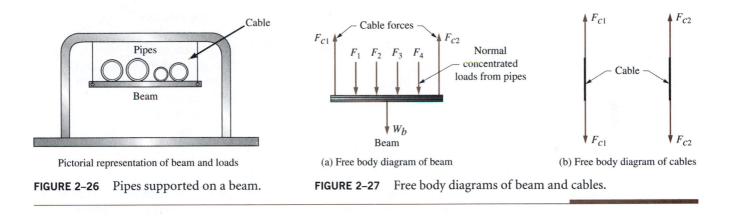

Pictorial representation of beam and loads   (a) Free body diagram of beam   (b) Free body diagram of cables

FIGURE 2–26 Pipes supported on a beam.   **FIGURE 2–27** Free body diagrams of beam and cables.

Example Problem 2–8

Figure 2–28 shows a part of a machine in which a beam is pivoted at its lower end while a hydraulic actuator actuates its upper end. Between the ends, a spring pulls down on the beam to assist in returning it to a rest position after the hydraulic force is removed. Prepare the free body diagrams of the beam, the spring, and the hydraulic actuator. Because the weight of the beam is small compared with the externally applied forces, neglect the weight in the diagram.

Solution Beam: See Figure 2–29(a).

External Loads:
a. The hydraulic actuator pushes outward, exerting a force F_1 at C perpendicular to the axis of the beam.

FIGURE 2–28 Machine part with hydraulic actuator and spring.

FIGURE 2–29 Free body diagrams of beam, spring, and hydraulic actuator.

(a) Free body diagram of beam

Schematic drawing of beam

Normal component F_{SN}

Parallel component F_{SP}

F_S

F_1

F_1

A

R_P

R_N

B

F_S

Spring

F_S

(c) Free body diagram of spring

Hydraulic actuator

F_1

(b) Free body diagram of hydraulic actuator

b. The spring pulls vertically downward on the beam at B with the force F_S. We show also the rectangular components normal (F_{SN}) and parallel (F_{SP}) to the beam.

Supporting Forces: The pin joint at the lower support at A is capable of exerting a reaction force in any direction. We show the rectangular components of this force acting normal (R_N) and parallel (R_P) to the beam.

Comments on Beam Stresses: The three normal forces F_1, F_{SN}, and R_N cause the beam to be in bending. See Chapters 11–14. The opposing forces acting along the axis of the beam, F_{SP} and R_P, exert a compressive force between the pin joints A and B. Chapters 7 and 9 discuss compressive stresses. If the beam is relatively slender and long, column buckling analysis may be necessary as discussed in Chapter 16.

Hydraulic Actuator: See Figure 2–29(b). As the actuator pushes on the beam, the beam pushes back with an equal force F_1, acting on each end of the actuator.

Spring: See Figure 2–29(c). The spring force F_S acts outward from each end of the spring, pulling it in tension.

2–6 BUILDING CODES AND BRIDGE STANDARDS

Design and construction of buildings and bridges for public use and occupancy are strictly regulated to ensure safety and ease of use. This section discusses the basics of the codes and standards applied. It is your responsibility to understand all parts of the standards, to verify that the latest version is being used, and to ensure that every structure meets the requirements. Although there has been major progress in developing and adopting common standards, it is recognized that there are local differences that should be addressed. The following is a partial list of organizations responsible for codes and standards.

■ **International Code Council (ICC):** Publisher of the *International Building Code (IBC)*, a promoter of uniform building codes and standards, and provider of

products and services for all concerned with the safety and performance of the built environment (Reference 15; Internet site 2).

■ **State building codes:** Each state in the United States establishes its own standards with attention to local climate, geological characteristics, and population issues. Many state codes are based on the IBC with special adjustments for wind, snow, and other conditions.

■ **County and municipal building codes:** Typically incorporate the state building code but may contain local requirements.

■ **American Society of Civil Engineers (ASCE):** Professional society for the civil engineering community and publisher of the standard SEI/ASCE 7-05 *Minimum Design Loads for Buildings and Other Structures*. This standard is adopted within most state building codes in the United States and is the most pertinent document for the study of statics and strength of materials and educational programs in the civil, construction, and architectural fields. Standard SEI/ASCE 37-02 is also important to ensure that a structure is safe and stable at each stage of construction (References 5 and 6; Internet site 1).

- **American Association of State Highway and Transportation Officials (AASHTO):** Developer of transportation policies and provider of technical services to support states in their efforts to efficiently and safely move people and goods. Publisher of the *AASHTO Bridge Design Specification* adopted by most states in the United States with additions to accommodate local conditions (References 3 and 4; Internet site 4).

- **National Research Council Canada—Institute for Research in Construction:** Developer and publisher of the National Building Code of Canada.

Building Codes

There are several parts of building codes dealing with various types of buildings and their geographic and governmental location, along with numerous aspects of building design that affect public safety and ease of use. The parts most pertinent to this book are those dealing with structural design and the specification of the types and magnitudes of loads for which the buildings are to be designed. Other critically important features of the codes include fire safety, ingress, egress, lighting, doors, windows, sitting, utilities, vehicle access, protection from weather, protection from damage by insects, and so forth.

The codes are separated between

- Residential (up to three stories)
- All others such as commercial, industrial, public, residential above three stories, and nonbuilding structures (towers, platforms, stadiums, etc.)

The primary types of loads considered in building codes are

- Dead loads, *D*—the weight of materials of construction incorporated into buildings. Examples are
  - Walls, floors, roofs, ceilings
  - Stairways, built-in partitions, finishes, cladding, windows, doors, and so forth
  - Weight of fixed service equipment such as cranes; plumbing stacks and risers; electrical feeders; heating, ventilating and air-conditioning systems (HVAC); and fire sprinkler systems
- Live loads, *L*—those loads produced by the use and occupancy of the building or other structures. Examples are
  - People, furniture, carpeting, decorations
  - Movable partitions, business equipment
  - Materials storage, production equipment, carts, and other interior vehicles
  - Maintenance and custodial equipment and supplies

- Roof live loads, L_r
  - Awnings and canopies
  - Maintenance workers and equipment
  - Roof gardens and landscaping
- Environmental loads
  - Wind load, *W*
  - Snow load, *S*
  - Rain load, *R*, particularly where ponding may occur
  - Flood load, F_a
  - Loads, *E*, due to earthquake and other seismic activity
  - Loads, *H*, due to lateral earth pressure, groundwater pressure, or the pressure of bulk materials
  - Forces, *T*, created by thermal expansion and contraction, shrinkage of materials, change in moisture, creep of structural materials, differential settlement of foundation elements, and so forth
  - Fluid loads, *F*, due to fluid pressure caused by fluid depth, pumps, or process equipment

Live Loads, *L* Table 2–1 lists examples of minimum design loads that are typically found in codes. They are to be used only for problem solving in this book. Actual code documents contain many more types of loads and numerous detailed discussions of special cases and annotations that qualify the conditions for using the given loads. It is the responsibility of the designer to adjust these figures if known uses of the space require higher loading than the minimums. Special considerations are required for extraordinary concentrated or distributed loads applied by machinery, vehicles, or stored materials.

Roof Live Loads, L_r See Table 2–2 for example data for minimum roof live loads. Roof live loads for sloped roofs are based on the horizontal projected area of the roof. Some parts of the code have different ratings according to the slope of the roof, typically expressed in terms of *rise per foot*. For example, a roof that slopes up 4.0 in for each 12.0 in of horizontal distance has a slope of 4/12, also noted as 1:3. Often the design snow load *S*, discussed next, is greater than the minimums shown for L_r. If so, the snow load is used.

Some of the data for roof loads vary according to the *tributary loaded area* for a given structural member. This refers to the area of the roof considered to be supported by the member. One method of computing the tributary area was shown in Example Problem 2–1. Other roof framing arrangements require different methods as described in the codes.

Snow Loads, *S* Standard ASCE 7 and Canada's National Building Code (NBC) include extensive data on snow loads to be used in different geographical locations in North America. The discussion here will be general to illustrate concepts. You must determine specific requirements for the location of the building being designed.

Snow loads vary significantly in the United States and Canada, as would be expected because of the major climatic range from south to north and the influence of the higher elevations in the Appalachian range and the various western mountain ranges. ASCE 7, Figure 7–1 shows several bands and zones giving minimum pressures (in lb/ft²) for ground snow loads in the lower 48 U.S. states. See Figure 2–30 and

TABLE 2–1   Examples of Minimum Design Live Loads from Codes

Residential buildings and apartments

Use	Live loads (lb/ft^2)	(Pa)	Use	Live loads (lb/ft^2)	(Pa)
Attics with storage	20	958	Passenger vehicle garages	50	2394
Attics without storage	10	479	Rooms other than sleeping rooms	40	1915
Decks	40	1915	Sleeping rooms	30	1436
Exterior balconies	60	2873	Stairs	40	1915
Fire escapes	40	1915			

Commercial and other nonresidential buildings and residential buildings over three stories

Use	Live loads (lb/ft^2)	(Pa)	Use	Live loads (lb/ft^2)	(Pa)
Offices—floors	50	2394	Corridors	100	4788
Computer rooms	100	4788	Library stacks	150	7182
Assembly areas and theaters			Manufacturing—heavy	250	11 970
Lobbies and ballrooms	100	4788	Manufacturing—light	125	5985
Seating—fixed seats	60	2873	Hotel rooms, classrooms	40	1915
Stages	125	5985	Stadium bleachers	100	4788
Restaurants	100	4788	Retail stores	100	4788
Hospital patient rooms	40	1915	Wholesale stores	125	5985
Hospital corridors	80	3830	Gymnasiums and balconies	100	4788
Hospital operating rooms, labs	60	2873	Elevated platforms	60	2873

Reference 5. Canada's NBC includes similar data. Then the standard relates minimum roof snow loads to the local ground snow loads. Some general observations about the magnitudes of these loads follow.

- All of Florida, most of the coastal areas around the Gulf of Mexico and the Pacific Ocean, and the regions along the Mexican border have a minimum snow load of zero.

TABLE 2–2   Roof Live Loads in lb/ft^2 (Pa)

Residential

Roof slope	Tributary loaded area for any structural member, ft^2 (m^2) 0 to 200 (0 to 18.6)	201 to 600 (18.7 to 55.8)	Over 600 (Over 55.8)
Flat or slope < 1:3	20 (958)	16 (766)	12 (575)
Rise 1:3 to < 1:1	16 (766)	14 (670)	12 (575)
Rise 1:1 and greater	12 (575)	12 (575)	12 (575)

Commercial or multistory residential

Ordinary flat, pitched, or curved roofs	20 (958)
Roofs used for promenade purposes	60 (2873)
Roofs used for roof gardens or assembly	100 (4788)

- The central plains and the Midwest consist of the following bands of snow load values:
  - 5 lb/ft^2 (239 Pa) in much of the southern parts of Georgia, Alabama, Mississippi, Louisiana, and Texas.
  - 10 lb/ft^2 (479 Pa) for most of Oklahoma, northern Mississippi and Alabama, and western Tennessee.
  - 15 lb/ft^2 (718 Pa) in a narrow band including much of western Kentucky and southern Indiana, Illinois, Missouri, and Nebraska.
  - 20 lb/ft^2 (958 Pa) for most of Ohio, Indiana, Illinois, Missouri, and Nebraska.
- The state of Michigan shows a highly varying set of bands from 20 lb/ft^2 (958 Pa) in the southeast to 60 lb/ft^2 (2873 Pa) in the northeastern part of the Lower Peninsula to 100 lb/ft^2 (4788 Pa) in the farthest northern part of the Upper Peninsula.
- The New England states also show tightly spaced bands ranging from 25 lb/ft^2 (1197 Pa) in Connecticut to 100 lb/ft^2 (4788 Pa) in northern Maine.
- In many of the mountainous regions and the areas on the eastern shores of the Great Lakes (in Michigan, Ohio, Pennsylvania, and New York) the maps show the designation "CS," indicating that *case studies* must be relied on to determine snow loads in local areas. Lake effect snow and rapid elevation changes produce highly variable snow amounts.

FIGURE 2–30   Ground snow loads from Standard ASCE-7.

Source: With permission from ASCE, Reston, VA.

- In most other parts of the United States the variations are so irregular that generalization is difficult and local codes must be consulted.
- Canadian provinces exhibit wide variations of snow loads, typically expressed as pressures in kg-force/m². The range is from a low of about 112 kg/m² to a high of about 673 kg/m². Databases are available listing design data by region or city.

Wind Loads, *W*   Standard ASCE 7 includes maps showing the variation of minimum design values of wind loads. The following gives some general observations for the United States.

- A very large percentage of the interior parts of the United States has a design wind velocity of 90 mph (40 m/s) with highly localized special wind regions ranging higher.
- Most of Washington, Oregon, and California have a design wind velocity of 85 mph (38 m/s) with highly localized special wind regions ranging higher.
- Coastal regions along the Atlantic Ocean and the Gulf of Mexico, the coastlines of Alaska, Hawaii, and the island territories (Puerto Rico, Guam, Virgin Islands, and American Samoa) have design wind velocity values ranging as high as 150 mph (67 m/s) because of hurricanes and other seaborne storms.

The actual pressures created on walls and roofs by the design wind speeds must be determined from Standard ASCE 7. For residential applications, additional data are provided for wind pressures on building components and cladding sys-

tems. It must be noted that the design pressures may be either positive (acting toward the surface) or negative (acting away from the surface), depending on the direction of the wind and the orientation of the affected surface relative to the wind. Therefore, the damage may be from either crushing or pulling away. Negative pressures on roofs are often called *uplift*. Such pressures can range from +10.0 lb/ft² to +50 lb/ft² and −10.8 lb/ft² to −54.2 lb/ft² (+479 Pa to +2394 Pa and −517 Pa to −2595 Pa).

Seismic Loads, *E*   Earthquake design loads are critical to safe design and require careful determination of the *seismic design category* for the local area. Special knowledge is also required about the reaction of various building components to the velocities and accelerations experienced by seismic activity. These topics are not included in this book.

Highway and Bridge Loads   The *AASHTO Bridge Design Specification* should be consulted to determine loads for bridge design. Included are *vehicular loads* and *lane loads*. Three frequently referenced standard vehicles are shown in Figure 2–31 and described here. Others may be used for some projects.

1. **H20 (M18) Design Truck.** Two-axle truck weighing 20 tons (18 metric tons) or 40 K (178 kN) having two axles [Figure 2–31(a)]:
   a. The forward axle carries 8000 lb (35.6 kN) under the cab.
   b. The rear axle carries 32 000 lb (142 kN) supporting the loaded bed.

FIGURE 2–31   Design truck styles.

2. **HS20 Design Truck.** Tractor-trailer (also called a *semi*) truck weighing 36 tons (32.7 metric tons) or 72 K (320 kN) and having three axles [Figure 2–31(b)]:
   a. The forward axle of the tractor carries 8000 lb (35.6 kN) under the cab.
   b. The rear axle of the tractor carries 32 000 lb (142 kN) supporting the forward part of the loaded trailer.
   c. The trailer axle carries 32 000 lb (142 kN) supporting the rear part of the loaded trailer.

3. **HS20-44 Design Truck.** Tractor-trailer truck weighing 36 tons (32.7 metric tons) having five axles [Figure 2–31(c); this design is the typical *18-wheeler*]:
   a. The forward tractor axle carries 10 000 lb (44.5 kN) typically under the engine.
   b. Each of the two tandem rear tractor axles carries 15 500 lb (68.9 kN) supporting the forward part of the loaded trailer. The axles are 4.0 ft (1.22 m) apart.
   c. Each of the two tandem trailer axles carries 15 500 lb (68.9 kN) supporting the rear part of the loaded trailer. The axles are 4.0 ft (1.22 m) apart.

Lane load is a uniform load derived from a string of closely spaced trucks moving together with a set distance between them. The current lane load is 640 lb/ft (9.34 kN/m) along the entire bridge length and spread over a 10.0 ft (3.05 m) lane width.

The widely used AASHTO HL93 design loading defines three different combinations. The first two are typically applied to single-span bridges on two supports and the third is applied to a continuous two-span bridge on three supports. See Figure 2–32.

1. HS20 truck with a uniform lane load added
2. One tandem pair of axles each carrying 25 000 lb (111 kN) at a spacing of 4.0 ft (1.22 m) plus the lane load
3. For the special case of a continuous two-span bridge with three supports, one HS20 truck on each span plus the lane load

The designer must analyze the specified load combination to determine where the concentrated loads are placed to have the worst-case influence on the bridge.

(a) HS20 truck plus lane load

(b) Tandem plus lane load

(c) Two HS20 trucks plus lane load

FIGURE 2–32   AASHTO HL93 design loadings.

The bridge design must also consider wind loads that act at varying angles of incidence. Such loads may produce moments on the deck tending to twist it or to affect the distribution of forces on supports. Winds may also cause oscillations that may become dangerously large as exhibited by the infamous Tacoma Narrows Bridge in Tacoma, Washington, in 1940. Analysis of bridge designs for dynamic behavior has become standard since that tragic event.

REFERENCES

1. Allen, E., and J. Iano, *Fundamentals of Building Construction: Materials and Methods,* 4th ed., Hoboken, NJ, John Wiley & Sons, Inc., 2004.

2. Ambrose, J., and Tripeny, P., *Simplified Design of Concrete Structures,* 8th ed., Hoboken, NJ, John Wiley & Sons, Inc., 2007.

3. American Association of State Highway and Transportation Officials (AASHTO), *AASHTO LRFD Bridge Design Specifications, Customary U.S. Units,* 4th ed., Washington, DC, AASHTO, 2007.

4. American Association of State Highway and Transportation Officials (AASHTO), *AASHTO LRFD Bridge Design Specifications, SI Units,* 4th ed., Washington, DC, AASHTO, 2007.

5. American Society of Civil Engineers, *SEI/ASCE 7-05 Standard—Minimum Design Loads for Buildings and Other Structures,* Alexandria, VA, American Society of Civil Engineers, 2006.

6. American Society of Civil Engineers, *SEI/ASCE 37-02 Standard—Design Loads on Structures During Construction,* Alexandria, VA, American Society of Civil Engineers, 2002.

7. American Wood Council, *National Design Specification® (NDS) for Wood Construction,* ANSI/AF&PA NDS-2005, Washington, DC, American Wood Council, 2005.

8. American Wood Council, *Special Design Provisions for Wind and Seismic Standard with Commentary,* ANSI/AF&PA SDPWS-2005, Washington, DC, American Wood Council, 2005.

9. Avallone, Eugene A., Theodore Baimeister, and Ali Sadegh, eds., *Marks' Standard Handbook for Mechanical Engineers,* 11th ed., New York, McGraw-Hill, 2007.

10. Backman, B. F., *Composite Structures, Design, Safety, and Innovation,* Burlington, MA, Elsevier Science and Technology Books, 2005.

11. Breyer, D. E., K. J. Fridley, D. G. Pollack, Jr., and K. E. Cobeen, *Design of Wood Structures—ASD,* 6th ed., New York, McGraw-Hill, 2007.

12. Brockenbrough, R. L., and F. S. Merritt, *Structural Steel Designer's Handbook: AISC, AASHTO, AISI, ASTM, Arema, and ASCE-07,* New York, McGraw-Hill, 2006.

13. Ching, F. D. K., and S. R. Winkel, *Building Codes Illustrated: A Guide to Understanding the 2006 International Building Code,* 2nd ed., Hoboken, NJ, John Wiley & Sons, Inc., 2007.

14. Holmes, J. D., *Wind Loading of Structures,* 2nd ed., London, Taylor & Francis, 2007.

15. International Code Council, *2006 International Building Code,* Washington, DC, International Code Council, 2006.

16. Levy, M., and M. G. Salvadori, *Why Buildings Fall Down: How Structures Fail,* New York, W. W. Norton & Co., 1994.

17. Mott, R. L., *Machine Elements in Mechanical Design,* 4th ed., Upper Saddle River, NJ, Pearson/Prentice Hall, 2004.

18. Nee, A. Y. C., A. S. Kumar, and Z. J. Tao, *An Advanced Treatise on Fixture Design and Planning,* Hackensack, NJ, World Scientific Publishing Co., 2005.

19. Salvadori, M. G., *Why Buildings Stand Up: The Strength of Architecture,* New York, W. W. Norton & Co., 1991.

20. Sandaker, B. N., *On Span and Space: Exploring Structures in Architecture,* London, Taylor & Francis, 2008.

21. Taly, Narendra, *Loads and Load Paths in Buildings: Principles of Structural Design,* Washington, DC, International Code Council, 2003.

22. Tlusty, G., *Manufacturing Processes and Equipment,* Upper Saddle River, NJ, Prentice Hall, 2000.

INTERNET SITES

1. American Society of Civil Engineers (ASCE) www.asce.org Professional society for the civil engineering community and publisher of the standard ASCE/SEI 7-05 *Minimum Design Loads for Buildings and Other Structures.*

2. International Code Council (ICC) www.iccsafe.org Publisher of the 2006 *International Building Code* and a promoter of uniform building codes, standards, products, and services for all concerned with the safety and performance of the built environment.

3. American Wood Council (AWC) www.awc.org Part of the American Forest & Paper Association and publisher of the *National Design Specification® for Wood Construction.* AWC's mission is to increase the use of wood by assuring the broad regulatory acceptance of wood products, developing design tools and guidelines for wood construction, and influencing the development of public policies affecting the use of wood products.

4. American Association of State Highway and Transportation Officials (AASHTO) www.transportation.org AASHTO advocates transportation policies and provides technical services to support states in their efforts to efficiently and safely move people and goods. Publisher of *AASHTO LRFD Bridge Design Specifications.*

5. Alpine—ITW Building Components Group, Inc. www.alpeng .com/windarticle An article titled "Designing for Wind," by Bruce Feldmann, P.E., giving an overview of the application of ASCE, IBC, and IRC standards for wind loads on structures with particular emphasis on roof trusses.

SOFTWARE

1. Mehta, K. C., McDonald, J. R., Norville, H. S., and Smith, D. A., *Wind Loads on Structures 2005,* Standards Design Group, Inc., 2005. Software to perform wind load computations called for in ASCE 7-05. Available through ASCE at www .asce.org/bookstore/book.cfm?book=7201.

2. Windloadcal.com, producer of a variety of software for calculation of wind loads, including roof, walls, structures, and signs. www.windloadcalc.com.

3. Engineering-Software.com, producer of a variety of software for civil, mechanical, manufacturing, chemical, and general engineering fields. Included is MECAWind05 for calculation of wind loads on buildings and other structures, developed by Meca Enterprises, Inc. www.mecaenterprises.com.

4. Northwest Builder's Network, distributor of services for the construction industry including books, software, and calculators; offers ConstructionCalc General loads Calculator Software for computing wind loads, snow loads, seismic forces, and dead loads for roofs, floors, and walls. www.nwbuildnet.com.

PROBLEMS

Definitions and General Principles

2–1. Describe the concept of a *free body diagram* and relate it to the study of a statics and strength of materials course.

2–2. Describe and compare the terms *applied force, reaction force,* and the *principle of action–reaction.*

2–3. Describe why a cable or a rope in tension is called a *two-force member.*

2–4. Describe the differences between a *concentrated load* and a *distributed load.*

2–5. Give three examples of a real product or structure in which concentrated loads are produced; give three examples of real products or structures in which distributed loads are produced.

2–6. Describe the difference between a *uniformly distributed load* and a *varying distributed load.*

2–7. Give three examples of real products or structures in which varying distributed loads are produced.

2–8. Describe the general method of computing perpendicular components of a force acting at any angle to a reference axis.

2–9. Describe the term *resultant* as it relates to a set of two or more forces acting on an object.

2–10. State the general *law of sines* and draw a careful sketch to illustrate terms; state the general *law of cosines* and draw a careful sketch to illustrate terms:

2–11. Describe a *roller-type support,* including a sketch of the possible forces it can exert on a load-carrying member.

2–12. Describe a *pin-type support,* including a sketch of the possible forces it can exert on a load-carrying member.

2–13. Describe a *smooth surface-type support,* including a sketch of the possible forces it can exert on a load-carrying member.

2–14. Describe a *rough surface-type support,* including a sketch of the possible forces it can exert on a load-carrying member.

2–15. Describe a *fixed-type support,* including a sketch of the possible forces it can exert on a load-carrying member.

2–16. Describe the type of information contained in building codes and how knowledge of such codes is pertinent to the study of statics and strength of materials.

2–17. Describe the term *dead load* as it pertains to building structures.

2–18. Describe the term *live load* as it pertains to building structures.

2–19. Name and describe at least three types of live loads typically applied to building structures.

2–20. Describe the role of the American Society of Civil Engineers (ASCE) as it pertains to loads on buildings.

2–21. Describe the role of the International Code Council (ICC) as it pertains to loads on buildings.

2–22. Describe the function of state and local building codes and how they are related to the ASCE and ICC.

2–23. State the full name of the AASHTO organization and describe its functions as they relate to the study of statics and strength of materials.

2–24. Describe the AASHTO standard H20 design truck.

2–25. Describe the AASHTO standard HS20 design truck.

2–26. Describe the AASHTO standard HS20-44 design truck.

2–27. Describe the AASHTO standard HL93 design loading for bridges.

2–28. Define the term *lane load* as it relates to bridge design.

Forces—Components

For Problems 2–29 to 2–38, compute the horizontal and vertical components of the given force.

2–29. Use Figure P2–29.
2–30. Use Figure P2–30.
2–31. Use Figure P2–31.
2–32. Use Figure P2–32.
2–33. Use Figure P2–33.

2–34. Use Figure P2–34.
2–35. Use Figure P2–35.
2–36. Use Figure P2–36.
2–37. Use Figure P2–37.
2–38. Use Figure P2–38.

FIGURE P2–29

FIGURE P2–30

FIGURE P2–31

FIGURE P2–32

FIGURE P2–33

FIGURE P2–34

FIGURE P2–35

FIGURE P2–36

FIGURE P2–37

FIGURE P2–38

2–39. Figure P2–39 shows a force of 1000 lb acting at an angle α above the x-axis. Compute the horizontal and vertical components of this force for values of α from 0° to 45° in increments of 5°. Also compute the ratio F_x/F_y.

FIGURE P2–39

2–40. Figure P2–40 shows a force of 10.0 kN acting at an angle α above the x-axis. Compute the horizontal and vertical components of this force for values of α from 0° to 90° in increments of 10°. Also compute the ratio F_x/F_y.

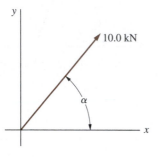

FIGURE P2–40

Distributed Loads on Surfaces

Note: Refer to Table 1–6 for densities of common materials.

2–41. A storage bin carrying pieces of bulk coal has floor dimensions of 20 ft by 12 ft. Compute the total weight of the coal if the bin is filled to a uniform depth of 8 ft.

2–42. For the coal bin described in Problem 2–41, compute the average load per unit area in lb/ft² on the floor.

2–43. Figure P2–43 shows the proposed layout of joists to support the floor of the coal bin described in Problem 2–41. Compute the distributed load in lb/ft due to the coal for a typical interior joist and for an edge beam along the short side.

2–44. The floor of the coal bin described in Problem 2–41 and shown in Figure P2–43 is to be made from 2×12 pine wood boards in 20-ft lengths laid perpendicular to the joists. Determine the weight of each board and the total weight of the flooring. Refer to Appendix A–4 for the finished dimensions of the boards. Then compute the load per unit area in lb/ft² due to the flooring.

2–45. For the wood floor described in Problem 2–44, compute the distributed load in lb/ft due to the flooring itself on a typical interior joist and on an edge beam along the short side.

2–46. For the coal bin described in Problem 2–41 and shown in Figure P2–43, consider that the coal is not of uniform depth but rather has the cross section shown in Figure P2–46 from end to end over the entire 20-ft length.
  a. Compute the total weight of the coal.
  b. Compute the distributed load due to the coal on a typical interior joist in lb/ft for the middle 6 ft where the coal has a uniform depth of 9.5 ft.
  c. Draw a sketch of the loading for a typical interior joist over its entire 12-ft length.

2–47. An aluminum bar has a rectangular cross section, 50 mm wide and 125 mm high, and a length of 1.68 m. Compute its total weight in newtons and the weight per unit length in N/m.

2–48. Repeat Problem 2–47 if the bar is magnesium.

2–49. A bronze casting is shown in Figure P2–49.
  a. Compute its total weight in pounds.
  b. Compute the force per unit area that the casting exerts on the floor in lb/ft². Consider each part, A, B, C, and D, separately.

2–50. In a factory for producing cast-iron machine bases, dry foundry sand is stored in cylindrical silos. Figure P2–50 shows the geometry of the bottom of a silo. If the sand has a uniform depth of 2.25 m,
  a. Compute the total weight of the sand.
  b. Compute the average unit load on the bottom in N/m².
  c. Then compute the load carried by each radial beam from the center to the outer rim, taken to be that within

FIGURE P2–43   Plan view of coal bin.

FIGURE P2–46 Distribution of coal for Problem 2–46.

FIGURE P2–49

40 in.

24 in.

36 in.

48 in.

72 in.

48 in.

42 in.

A   B   C   D

Floor

Dimensions in inches

FIGURE P2–50 Layout of bottom of cylindrical silo.

Example of a radial sector of the load on a beam

Segment of load on each beam

Beam (8)

2.75 m

the 45-degree segment shown by the dashed lines in the figure. Approximate the load distribution by dividing each segment into five sectors at equal radial increments from the center, computing the weight of sand in each sector, and considering that weight to act as a resultant at the middle of each sector.

Forces—Resultants

2–51. A cover for a heat treatment furnace is supported by two cables attached to a fixed point as shown in Figure P2–51. The force in cable AC is 4.92 kN and the force in cable AB is 6.28 kN. Compute the resultant force on the support using the parallelogram rule.

2–52. Repeat Problem 2–51 using the summation of components method.

2–53. Figure P2–53 shows a power boat towing two water skiers. Skier A pulls with a force of 128 lb while skier B exerts 206 lb of force as she cuts away from the path of the boat. Compute the resultant force on the boat using the parallelogram rule.

2–54. Repeat Problem 2–53 using the summation of components method.

2–55. A heavy machine base is supported on two cables from a single point A as shown in Figure P2–55. The forces in cables AB and AC are shown in the figure. Compute the resultant force on the attachment ring at A using the parallelogram method.

2–56. Repeat Problem 2–55 using the summation of components method.

2–57. Figure P2–57 shows a large marquee for a theater supported by three cables, using a cable-stayed design. The tensions in the cables are F_{AB} = 18.80 kN, F_{AC} = 15.60 kN, and F_{AD} = 13.30 kN. Use the summation of components method to compute the resultant load on the attachment point at A.

2–58. In a hot-air balloon gathering, four balloons are tethered to a single mounting point as shown in Figure P2–58. The cables pass around frictionless pulleys. The force on each balloon is different because of its size and the degree of inflation. At the given time, the forces are F_A = 2945 lb, F_B = 3572 lb, F_C = 1980 lb, and F_D = 2395 lb. Compute the resultant force on the mounting ring using the summation of components method.

FIGURE P2–51

FIGURE P2–53

FIGURE P2–55

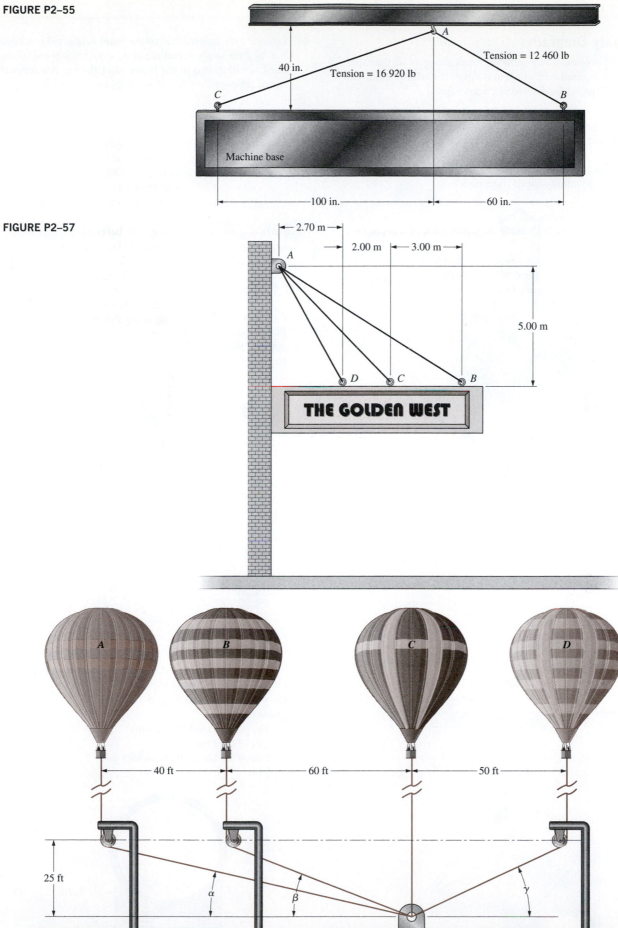

FIGURE P2–57

THE GOLDEN WEST

FIGURE P2–58

Free Body Diagrams

In this set of problems, only show the forces acting on all the members. It is not required at this time to compute the value of any forces at supports. That will be covered in a later chapter.

2–59. Figure P2–59 shows the foundation support for the column of a large machine. The steel tube exerts a force of 30 000 lb vertically downward. Draw the free body diagrams for

a. The tubular column

b. The 6.0-in square plate

c. The pyramid-shaped concrete pier

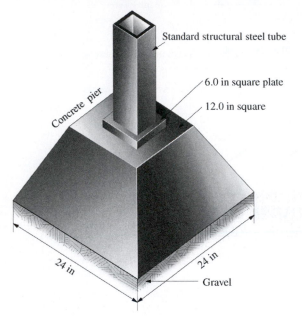

FIGURE P2–59

2–60. The shelf shown in Figure P2–60 carries a concentrated load at its middle. Draw free body diagrams of the shelf, the two support rods, and the pin-type supports.

FIGURE P2–60

2–61. Figure P2–61 shows a horizontal beam supported at a rigid wall by a pin support and at point B by a vertical rod. Draw free body diagrams of the beam CBD, the bar AB, support A, and support C.

FIGURE P2–61

2–62. Figure P2–62 shows a simple frame in which the angled bar BC supports the beam at point B. Draw the free body diagram of the beam AB, the bar BC, and the supports at A and C.

FIGURE P2–62

2–63. A large pipe is supported on a saddle that is in turn supported on two pin-type supports. The blocks that support the pins are fixed to a rigid surface. Draw free body diagrams of

a. The pipe and saddle as a unit

b. One of the pins

c. One of the blocks that carries a pin

FIGURE P2–63

FIGURE P2–64

2–64. Figure P2–64 shows a truss of the type that will be studied in more detail in a later chapter. Consider the entire truss to be one body carrying the three applied forces and supported on the pin-type supports at *A* and *B*. Draw the free body diagram of the truss and the supports.

2–65. The simple jib-crane shown in Figure P2–65 carries a load at point *D*. Draw free body diagrams of the load, the horizontal beam *DEF*, the cables *AE* and *AB*, the vertical member *AFC*, and the two supports.

FIGURE P2–65

2–66. The deck shown in Figure P2–66 will be covered by decking over all the joists. The joists rest on the 12-ft-long beams. Each beam is supported on two posts that are placed solidly in the ground and cemented in place. The deck will be designed to carry a uniformly distributed load over the entire deck surface. Draw free body diagrams of
  a. One joist in the middle portion of the deck
  b. One 12-ft-long beam along the side of the deck
  c. One of the posts

2–67. Figure P2–67 shows a system designed to use a hydraulic cylinder to press a bushing into a large casting. The two vertical members on each side can be considered to be rigid. Draw free body diagrams of
  a. The casting
  b. The bushing
  c. The hydraulic cylinder taken as an assembly
  d. The horizontal beam

2–68. Figure P2–68 is a sketch of a method for supporting a beam comprised of a fixed support at its left end (attached to a rigid structure) and a smooth simple support at the right end. Draw a free body diagram of the beam.

Side view

FIGURE P2–66

FIGURE P2–67

FIGURE P2–68

FIGURE P2–70

2–69. For the system shown in Figure P2–68, consider that the support on the right end is removed. Then draw the free body diagram of the horizontal beam.

2–70. The diving board shown in Figure P2–70 is supported by a simple strap at its left end and by a roller at point *B*. Draw a free body diagram of the board.

2–71. Consider the diving board in Figure P2–70. Replace the simple strap at the right end with a rigid clamp as shown in Figure P2–71. A heavy plate is placed on top of the board and bolted through with lag screws deeply into the lower support block. Draw the free body diagram of the board with this type of support.

FIGURE P2–71

2–72. Figure P2–72 shows a rod bent into an L-shape and welded to a rigid plate. Draw a free body diagram of the bar when the vertical force *P* is applied.

FIGURE P2–72

2–73. Figure P2–73 shows a large container suspended by three cables that are attached to two horizontal arms. The arms are bolted rigidly to the heavy vertical columns. The columns are bolted solidly to a rigid floor. Draw free body diagrams of
  a. The container
  b. The vertical cable
  c. The connector holding the three cables
  d. The angled cable on the left
  e. The left horizontal beam
  f. The left vertical column

2–74. Figure P2–74 shows a picnic table on which a person is sitting near the edge. The table is supported by a single pipe at its center and the pipe is sunk into the ground and cemented in place. Draw the free body diagrams of the table and the pipe.

2–75. The beam with an L-shaped bracket in Figure P2–75 is supported at points *A* and *C*. Only *C* can exert an axial force. Both *A* and *C* can exert forces perpendicular to the horizontal bar *ABC*. Draw the free body diagram of the bar.

Loads on Buildings and Bridges

2–76. Specify a reasonable design live load for an elevated platform in a commercial building.

2–77. Specify a reasonable design live load for a room serving as a library where books are stored.

2–78. Specify a reasonable design live load for a bedroom in a residence.

2–79. Specify a reasonable design live load for the family room in a home.

2–80. Specify a reasonable design live load for a commercial building where computer equipment is housed.

2–81. Specify a reasonable design live load for a restaurant.

2–82. Specify a reasonable design live roof load for a typical commercial building in southern Texas.

2–83. Specify a reasonable design live load for the roof of a community theater building in Florida where patrons gather during intermissions.

2–84. Specify a reasonable design ground snow load in Cincinnati, Ohio.

2–85. Specify a reasonable design ground snow load in southern Indiana.

2–86. Specify a reasonable design ground snow load for the most northern part of Michigan.

2–87. Specify a reasonable design wind velocity for most of the interior of the United States.

2–88. Describe where in the United States the design wind velocity would be less than that called for in Problem 2–87.

2–89. Describe where in the United States the design wind velocity would likely be greater than that called for in Problem 2–87.

2–90. List at least two kinds of loads included in most building codes besides live loads, snow loads, and wind loads.

FIGURE P2–73

FIGURE P2–74

FIGURE P2–75

CHAPTER THREE

STATIC EQUILIBRIUM
Balance Among Forces, Moments, Torques, Couples, Reactions, and Supports

MAJOR TOPICS
3–1 The Big Picture
3–2 Analysis of Static Equilibrium
3–3 The Concept of Mechanical Advantage
3–4 Pulley Systems

3–1 THE BIG PICTURE

The concept of *static equilibrium* is at the core of the application of statics to real structures and mechanical devices. By using this tool, you will be able to determine the forces, moments, and torques acting on a structure as a whole or on any of its individual components. In turn, you will have a skill that is essential to learn the principles of stress analysis in the later part of this book.

The fundamental principles of static equilibrium, introduced in Chapter 2 as part of the process of creating free body diagrams, are the following:

1. The sum of all forces acting on a body in any direction must be zero.

2. The sum of all moments, torques, and couples acting on a body about any point must be zero.

3. The weight of a member acts through its center of mass vertically downward toward the center of the earth.

4. Weight can be an important force in achieving total balance of forces. However, when the weight of a member is relatively light compared with the applied forces, it can be neglected.

5. A moment about a certain point is the product of a force times the perpendicular distance from the point of interest to the line of action of the force.

6. At joints, the principle of *action–reaction* is applied to track the path of loads through a structure with proper handling of both magnitudes and directions.

7. Knowledge of the forces that different kinds of supports can exert is an essential part of static equilibrium analysis. Review Section 2–5.

8. Analysis using static equilibrium principles requires a complete and accurate free body diagram of the component of interest.

Activity

Anything you can see around you that is stationary is in static equilibrium. Make a quick list of items that meet these conditions.

Here is a list that might represent things in an office or study place:

1. Computer, monitor, and keyboard
2. Printer
3. Lamp
4. Telephone
5. Chair
6. Desk
7. Tables
8. Clock
9. Books on bookshelves

What is on your list?

Now look beyond your study place to find more complex examples of things that have multiple components that must fit together as a system while maintaining static equilibrium. Here is a starter list:

1. The structure of the building you are in, including
   a. Roof and its supports that carry wind, rain, and snow loads
      i. Are trusses used?
      ii. Is the roof supported on beams spanning across supporting walls?
      iii. Are other objects supported by the roof or its supporting structures?
   b. Floors that support furniture, appliances, equipment, and people
   c. Walls that support upper floors and that resist wind and other external forces

d. Foundations that support the entire weight of the building

2. Your car or truck
   a. Frame members that provide rigidity to the structure and support functional parts of the vehicle, such as the engine, seats, and suspension elements
   b. The suspension system itself. Even though its members move during maneuvers, static equilibrium exists while it is stationary due to the geometry of struts, springs, brackets, ball joints, and other features that transfer the weight of the car to its wheels
   c. The systems that support the hood and the trunk lid of the car when they are raised
   d. Doors when they are either open or closed and supported by hinges and latches
   e. Seats
   f. Items in the instrument panel including the radio, heating, cooling, and ventilation equipment

3. Components of construction equipment (see Figure 1–1(d) and Figure 15–1)
   a. The machine itself for which its weight is distributed to wheels or tracks on which it is supported
   b. The load bed of a dump truck due to the load itself, the weight of all components, its supports mounted to the truck chassis, and the hydraulic actuator that raises the bed when dumping its load
   c. The arm that holds the shovel of an excavator due to the weight of all components, the load, and actuators that allow it to maneuver
   d. The cab of an excavator that carries the operator's station, the engine, hydraulic fluid power system, and the forces from the excavation arm while being supported on a system that allows the cab to rotate while digging and transferring a load to the dump truck
   e. For the articulated boom system mounted on a truck shown in Figure 15–1, each member must be in equilibrium after raising the bucket at the end of the boom to working height. Forces are applied due to the weight of the members, their hydraulic actuators, the bucket, the person in the bucket, and any tools on board. Support is provided by a sturdy mounting system that transfers loads to the frame of the truck while allowing the boom to rotate to reach the desired position

4. Bridges (see Figure 15–3)

Objects Moving with Constant Velocity

The principles of static equilibrium can also be applied to anything that is moving with a constant velocity because all forces and moments on it are balanced. An example is a car or truck moving on a straight, level highway at constant speed. In addition to gravity forces (weights), the car is subjected to wind resistance called *drag* and rolling friction at the interfaces between the wheels and the road. The engine and drive system provide exactly the right amount of propulsive forces that counter the friction and drag. Then

all forces and moments are balanced and the methods of static equilibrium can be applied to determine support forces on the wheels or forces on any component of the vehicle. To accelerate the vehicle or to slow it down, additional forces from the engine or braking system must be applied. Complete analysis of these actions is part of the study of *dynamics,* a course of study that typically follows *statics.*

Objectives of This Chapter

After completing this chapter, you will be able to:

1. State the principles of static equilibrium
2. Apply the principles of static equilibrium to determine reaction forces or moments on structures and mechanical devices for any component or system that is stationary or moving with constant velocity

3–2 ANALYSIS OF STATIC EQUILIBRIUM

The process of analyzing a component for static equilibrium was described in the Big Picture. We can take the first two items and state them mathematically as

- The sum of all forces acting on a body in any direction must be zero. For objects acted upon by forces in a two-dimensional plane only, we can say

$$\Sigma F_x = 0 \qquad (3\text{--}1)$$
$$\Sigma F_y = 0 \qquad (3\text{--}2)$$

- The sum of all moments, torques, and couples acting on a body about any point must be zero.

$$\Sigma M = 0 \text{ about any point} \qquad (3\text{--}3)$$

Note About Directions of Forces and Moments

It is essential that signs for forces and moments are *consistent* within any calculation. For forces, it is conventional to take rightward directed forces to be positive, corresponding to the direction of the positive *x*-direction on a typical *x–y*-axis system. Then leftward directed forces will be negative. Similarly, upward forces are taken to be positive in the direction of the positive *y*-axis. Downward forces are taken to be negative.

The weight of an object on earth is typically shown as a single vector acting vertically downward toward the center of the earth and acting through the center of mass of the object.

For moments, some technical professionals prefer to adopt a sign convention. However, in this book we will refrain from that and, instead, ensure that

Consistent signs must be used within any equation.

That is, all clockwise (CW) moments have the same sign and all counterclockwise (CCW) moments have the opposite sign. It matters not which is positive and which is negative as long as the signs are consistent. The example problems in this chapter illustrate the process.

Many applications of the principles of static equilibrium involve the calculation of reaction forces and/or moments at supports. It is important for you to try to *reason* the directions of the unknown reactions in order to maintain equilibrium and to represent them as vectors with directions indicated. For example, if a set of applied forces tends to rotate a beam in the clockwise direction with respect to a particular point, the reaction at another point must provide a force that produces a counterclockwise moment about the first point. Then, when using the forces in the equilibrium equations, give them a descriptive name and be sure to apply the proper sign. After the analysis is complete:

- *Your result will be positive if your original assumption for the direction of the unknown quantity was correct.*

- *If the result is negative, your original assumption for direction was opposite what it truly is. Adjust the free body diagram to show the actual direction.*

The Concept of Center of Mass

For a complex structure, vehicle, or machine, the total weight is sum of the weights of subsystems or components. It is convenient for analysis of static equilibrium to aggregate the total weight and place it at a point where it would have the same effect as the distributed weights acting together. If the object is supported from that point, the object would be in balance and would not tend to rotate in any direction unless acted on by an outside force.

Recall now that the weight of an object is actually the force due to gravity acting on the mass. For this reason, some call the balance point the *center of gravity*. A more accurate term that we will use in this book is the *center of mass* because it describes the point at which all the mass of the object could be concentrated and have the same effect on the system.

We can locate the center of mass of a system made from two or more components by applying the following procedure:

1. Specify a reference point from which to measure the position of the center of mass for each component.

2. Determine the weight W_i of each component. You may use a common scale. If its mass m is known, compute the weight from $W = m \cdot g$ where g is the acceleration due to gravity. This process was discussed in Chapter 1.

3. Measure the location of the center of mass of each component relative to the reference point and call that distance x_i.

4. Multiply the weight of each component by the distance to its center of mass. That is, compute $W_i \cdot x_i$ for each component.

5. Sum all of the weights of the individual components to determine the total weight, W_T.

6. Sum all of the products, $W_i \cdot x_i$. Call the sum $\Sigma(W_i \cdot x_i)$.

7. Compute the distance from the original reference point to the center of mass using the following equation:

$$x_{cm} = \Sigma(W_i \cdot x_i)/W_T \qquad (3\text{--}4)$$

FIGURE 3–1 Hopper carrying two pieces of air-conditioning equipment.

Let's illustrate this process by considering a rigid hopper carrying two parts of a roof-mounted air conditioner as shown in Figure 3–1. The hopper is to be raised to the roof by a crane and it is desired to know where to apply the support system to ensure that the hopper stays horizontal during the lift. The weight and position of the center of mass of each air conditioner part are shown. Part 1 has a mass of 475 kg and part 2 has a mass of 528 kg. The hopper itself has a mass of 34.6 kg and it is symmetrical so that its weight W_H is shown to act vertically downward at its midpoint.

1. We select the left side of the hopper as a reference point.

2. The weights are
   a. $W_1 = m_1 \cdot g = (475 \text{ kg})(9.81 \text{ m/s}^2)$
      $= 4660 \text{ kg} \cdot \text{m/s}^2 = 4660 \text{ N}$
   b. $W_2 = m_2 \cdot g = (528 \text{ kg})(9.81 \text{ m/s}^2)$
      $= 5180 \text{ kg} \cdot \text{m/s}^2 = 5180 \text{ N}$
   c. $W_H = m_H \cdot g = (34.6 \text{ kg})(9.81 \text{ m/s}^2)$
      $= 343 \text{ kg} \cdot \text{m/s}^2 = 343 \text{ N}$

3. By inspection of the figure, we can see the location of each piece:
   a. $x_1 = 1.15 \text{ m}$
   b. $x_2 = 2.20 \text{ m}$
   c. $x_H = 1.40 \text{ m}$

4. The values of $W_i \cdot x_i$ are
   a. $W_1 \cdot x_1 = (4660 \text{ N})(1.15 \text{ m}) = 5359 \text{ N} \cdot \text{m}$
   b. $W_2 \cdot x_2 = (5180 \text{ N})(2.20 \text{ m}) = 11\,396 \text{ N} \cdot \text{m}$
   c. $W_H \cdot x_H = (343 \text{ N})(1.40 \text{ m}) = 480 \text{ N} \cdot \text{m}$

5. The sum of the weights is

$$W_T = W_1 + W_2 + W_H$$
$$= 4660 \text{ N} + 5180 \text{ N} + 343 \text{ N} = 10\,274 \text{ N}$$

6. The sum of the values of $W_i \cdot x_i$ is

$$\Sigma(W_i \cdot x_i) = W_1 \cdot x_1 + W_2 \cdot x_2 + W_H \cdot x_H$$
$$= 5359\,\text{N} \cdot \text{m} + 11\,396\,\text{N} \cdot \text{m} + 480\,\text{N} \cdot \text{m}$$
$$= 17\,235\,\text{N} \cdot \text{m}$$

7. Finally, we can compute x_{cm}:

$$x_{cm} = \Sigma(W_i \cdot x_i)/W_T = (17\,235\,\text{N} \cdot \text{m})(10\,274\,\text{N})$$
$$= 1.678\,\text{m}$$

In summary, the center of mass is located 1.678 m to the right of the left side of the hopper. A support system must be applied in line with the center of mass to maintain balance of the hopper.

Examples—Applying the Principles of Static Equilibrium

Let's now apply the formulas for static equilibrium to determine the weight distribution for the motorcycle shown in Figure 3–2.

FIGURE 3–2 Motorcycle weight distribution.

Example Problem 3–1

The motorcycle shown in Figure 3–2 has a total weight of 2.16 kN. Consider the weight force W_m to act through the center of mass as shown in the figure. The rider adds her weight of 575 N shown as W_r acting on the saddle. Compute the reaction forces exerted by the road upward along the vertical centerline of each wheel.

Solution Given: Motorcycle in Figure 3–2: $W_m = 2.16$ kN $= 2160$ N; $W_r = 575$ N

Find: Reaction forces at each wheel

Analysis:
1. Apply the moment equation, $\Sigma M = 0$ twice
   a. first with respect to the point of contact of the rear wheel, A, that allows the calculation of the front reaction,
   b. then with respect to the point of contact of the front wheel, B, to find the rear reaction.
2. Apply the force equation in the vertical direction, $\Sigma F_y = 0$, to check that the results of the moment calculations were correctly done.
3. Note that there are no forces exerted in the horizontal direction.

Results: a. Sum moments about the rear wheel contact point A:

$$\Sigma M_A = 0 = \underset{\text{CW}}{W_r(354\,\text{mm})} + \underset{\text{CW}}{W_m(544\,\text{mm})} - \underset{\text{CCW}}{R_F(1524\,\text{mm})}$$

Note the signs used for the moment terms. The moments for the applied forces W_r and W_m tend to rotate the motorcycle clockwise with respect to point A at the rear wheel and we have arbitrarily assigned those moments positive (+) signs. Now note that the upward reaction R_F on the front wheel tends to rotate the motorcycle counterclockwise with respect to point A. Therefore, we have assigned that moment a negative sign (−).

Note also that the force R_R has no moment with respect to point A because its line of action passes directly through point A.

Now observe that the only unknown quantity in the moment equation is the front reaction R_F. Then we can solve algebraically for that force as follows:

$$0 = W_r(354\,\text{mm}) + W_m(544\,\text{mm}) - R_F(1524\,\text{mm})$$
$$R_F(1524\,\text{mm}) = W_r(354\,\text{mm}) + W_m(544\,\text{mm})$$
$$R_F = \frac{W_r(354\,\text{mm}) + W_m(544\,\text{mm})}{1524\,\text{mm}}$$

Substituting values for the forces gives,

$$R_F = \frac{(575\,\text{N})(354\,\text{mm}) + (2160\,\text{N})(544\,\text{mm})}{1524\,\text{mm}}$$
$$R_F = 904.6\,\text{N}\uparrow \text{ vertically upward}$$

b. Now we can apply the same technique to find the reaction on the rear wheel R_R. Less detail will be shown. We sum moments with respect to point B where the front tire meets the road.

$$\Sigma M_B = 0 = W_r(1170 \text{ mm}) + W_m(980 \text{ mm}) - R_R(1524 \text{ mm})$$

$$R_R = \frac{W_r(1170 \text{ mm}) + W_m(980 \text{ mm})}{1524 \text{ mm}}$$

$$R_R = \frac{(575 \text{ N})(1170 \text{ mm}) + (2160 \text{ N})(980 \text{ mm})}{1524 \text{ mm}}$$

$$R_R = 1830.4 \text{ N} \uparrow \text{ vertically upward}$$

As a check on the accuracy of these calculations, we can apply the sum of the forces in the vertical direction to ensure that they sum to zero, thus satisfying that principle of equilibrium.

$$\Sigma F_y = R_R + R_F - W_r - W_m = 1830.4 + 980.6 - 575 - 2160$$

$$\Sigma F_y = 0 \quad \text{[Check!]}$$

This verifies that our calculations are correct.

Summary: The reaction forces at the wheels of the motorcycle are

$$R_R = 1830.4 \text{ N} \uparrow \text{ vertically upward}$$
$$R_F = 904.6 \text{ N} \uparrow \text{ vertically upward}$$

Analysis of Reaction Forces on a Beam Carrying Concentrated Loads

Example Problem 3–2   Figure 3–3(a) shows a beam of uniform cross section carrying four pipes that are supported by two cables. The beam itself weighs 92 lb. From left to right, the forces exerted on the beam by the pipes are

$$F_1 = 630 \text{ lb}; F_2 = 1080 \text{ lb}; F_3 = 310 \text{ lb}; F_4 = 545 \text{ lb}$$

Compute the tension forces in the cables.

Solution   Given: Beam in Figure 3–3(a); beam weight = 92 lb

$$F_1 = 630 \text{ lb}; F_2 = 1080 \text{ lb}; F_3 = 310 \text{ lb}; F_4 = 545 \text{ lb}$$

Find: Tension forces in the two cables supporting the beam

Analysis: The first step is to create the free body diagram of the beam. Figure 3–3(b) shows the diagram that was developed in Example Problem 2–7 in Chapter 2.

We can now sum moments about each of the two support points to find the reaction forces in the cables.

Results: a. First sum moments about point A at the left end of the beam to find the reaction F_{c2}.

$$\Sigma M_A = 0 = F_1(a) + F_2(b) + F_3(c) + F_4(d) + W(L/2) - F_{c2}(L)$$

$$L = 90 \text{ in}$$
$$a = 15 \text{ in}$$
$$b = 35 \text{ in}$$
$$c = 55 \text{ in}$$
$$d = 70 \text{ in}$$

(a) Pictorial representation of beam and loads

(b) Free body diagram of beam

FIGURE 3–3   Beam supporting pipes for Example Problem 3–2.

All data in this equation are known except F_{c2}. Solve for that.

$$F_{c2} = \frac{[F_1(a) + F_2(b) + F_3(c) + F_4(d) + W(L/2)]}{L}$$

Using consistent units of pounds for forces and inches for distances,

$$F_{c2} = \frac{[(630)(15) + (1080)(35) + (310)(55) + (545)(70) + (92)(45)]}{90}$$

$$F_{c2} = 1184.3 \text{ lb}$$

b. Now sum moments about B to find the reaction F_{c1}.

$$\Sigma M_B = 0 = F_1(L - a) + F_2(L - b) + F_3(L - c) + F_4(L - d) + W(L/2) - F_{c1}(L)$$

Only F_{c1} is unknown so we solve for it.

$$F_{c1} = \frac{[F_1(L - a) + F_2(L - b) + F_3(L - c) + F_4(L - d) + W(L/2)]}{L}$$

$$F_{c1} = \frac{[(630)(75) + (1080)(55) + (310)(35) + (545)(20) + (92)(45)]}{90}$$

$$F_{c1} = 1472.7 \text{ lb}$$

Now check by summing forces in the vertical direction:

$$\Sigma F_y = 1184.3 - 630 - 1080 - 310 - 545 - 92 + 1472.7 = 0$$

$$\Sigma F_y = 0 \qquad [\text{Check!}]$$

LESSON 4

Summary: The force in the left cable = F_{c1} = 1472.7 lb tension

The force in the right cable = F_{c2} = 1184.3 lb tension

Analysis of Reaction Forces on a Beam Carrying Uniformly Distributed Loads

For the purposes of finding reactions, we can consider the resultant force due to distributed loads to act at the middle of the distributed load. The resultant load W is the product of the load per unit length w and the length of the beam over which the load is applied L_w. Expressing this mathematically,

Resultant force $= W = w \times L_w$ considered to act at $L_w/2$

It is important to note that this use of the resultant force is only for the purposes of finding reactions on beams. When the actual effect of the distributed load on the stresses created in the beam is analyzed, the distributed load itself is used as demonstrated in Chapters 11–14.

Example Problem 3–3   Figure 3–4(a) shows a beam of uniform cross section carrying a uniformly distributed load of $w = 418$ lb/ft due to the dead weight of roofing materials and the live load due to snow. The beam is a steel W-beam, W18×55, weighing 55 lb/ft (see Appendix A–7). Compute the reaction forces at the supports. (Note: Refer to Example Problem 2–1 and Figure 2–8 in Chapter 2 for an explanation of how the loading pattern was determined.)

Solution   Given: Beam in Figure 3–4(a)

Find: Reaction forces at the two supports for the beam

FIGURE 3–4   Beam carrying a uniformly distributed load.

Analysis: Figure 3–4(a) is the free body diagram of the beam. We first add the weight per unit length of the beam w_b to the applied distributed load w.

$$w_t = w + w_b = 418 \text{ lb/ft} + 55 \text{ lb/ft} = 473 \text{ lb/ft}$$

For this total uniformly distributed load, we can now determine the total resultant force applied to the beam.

$$\text{Resultant} = W = w_t \times L_w = (473 \text{ lb/ft})(30 \text{ ft}) = 14\ 190 \text{ lb}$$

For the purposes of finding the reactions, we can consider that this resultant force acts at the middle of the distributed load at $L/2 = 15$ ft from either support. See Figure 3–4(b).

Now we can use the equilibrium condition for moments:

$$\Sigma M = 0 \text{ about any point}$$

Let's first sum moments about the left reaction at A.

$$\Sigma M_A = 0 = W \times L/2 - R_B \times L = (14\ 190 \text{ lb})(15 \text{ ft}) - R_B(30 \text{ ft})$$

Solving for R_B gives:

$$R_B = \frac{(14\ 190 \text{ lb})(15 \text{ ft})}{30 \text{ ft}} = 7095 \text{ lb (upward)}$$

We normally would solve for the left reaction separately using the same procedure. However, in this simple case we can see by symmetry that $R_A = R_B = 7095$ lb (upward).

Reaction Forces on a Beam Carrying Partial Uniformly Distributed Loads and Concentrated Loads Together

The following example problem includes a uniformly distributed load over only a portion of the length of the beam along with two concentrated loads. Note that the right concentrated load is at the end of an overhang. Careful consideration of the signs for forces and moments is required for this kind of loading.

Example Problem 3–4   For the beam loading shown in Figure 3–5, compute the reaction forces at the two supports.

Solution   Given: Beam in Figure 3–5

Find: Reaction forces at the two supports for the beam

Analysis: Figure 3–5(a) is the free body diagram of the beam.

We first compute the resultant force due to the distributed load.

$$W = w \times L_w = (12 \text{ kN/m})(1.25 \text{ m}) = 15.0 \text{ kN}$$

This resultant acts at $L_w/2$ to the right of A, or 0.625 m.

(a) Beam is HSS 203×102×6.4 hollow structural rectangular tube   (b) Equivalent loading for finding reactions

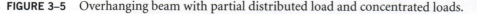

FIGURE 3–5   Overhanging beam with partial distributed load and concentrated loads.

We can now compute the resultant force due to the weight of the HSS 203×102×6.4 hollow structural tube. Appendix A–9(SI) lists the weight per unit length to be 277 N/m. Then the total weight of the tube is

$$W_t = (277\text{ N/m})(3.0\text{ m}) = 831\text{ N} = 0.831\text{ kN}$$

For finding reactions, we can consider this resultant to act at the middle of the entire length of the beam, 1.50 m from A.

Figure 3–5(b) shows the equivalent beam loading for the purposes of finding reaction forces. Dashed vectors are used for the resultants of the two distributed loads to emphasize that they are not truly concentrated loads and that the full effects of the distributed loads must be considered for stress analysis.

We can now apply the principles of static equilibrium in a manner similar to that used in the preceding example problem.

$$\Sigma M_A = 0 = W(0.625\text{ m}) + W_t(1.50\text{ m}) + F_1(1.75\text{ m}) + F_2(3.0\text{ m}) - R_B(2.25\text{ m})$$

Solving for R_B (units of kN for forces and m for distances),

$$R_B = \frac{(15.0)(0.625) + (0.831)(1.50) + (13.4)(1.75) + (8.6)(3.0)}{2.25}$$

$$R_B = 26.61\text{ kN}$$

Similarly, solving for R_A by summing moments about B,

$$\Sigma M_B = 0 = W(1.625\text{ m}) + W_t(0.75\text{ m}) + F_1(0.50\text{ m}) - F_2(0.75\text{ m}) - R_A(2.25\text{ m})$$

$$R_A = \frac{(15.0)(1.625) + (0.831)(0.75) + (13.4)(0.50) - (8.6)(0.75)}{2.25}$$

$$R_A = 11.22\text{ kN}$$

We can now sum forces in the vertical direction as a check.

$$\Sigma F_y = 15.0 + 0.831 + 13.4 + 8.6 - 26.21 - 11.22 = 0$$

$$\Sigma F_y = 0 \quad \text{[Check!]}$$

Resultant Forces for Varying Distributed Loads

Section 2–2 described cases for which distributed loads have varying rates of loading rather than being uniformly distributed (see Figures 2–10 and 2–11). We compute the resultant force for such a loading pattern by a three-step process:

1. The magnitude of the resultant force is equal to the area under the varying distributed load.

2. The line of action of the resultant force is through the centroid of the area.

3. If the varying distributed load is due to the weight of some material, the resultant force acts vertically downward.

Let's illustrate these steps by considering first the loading pattern from Figure 2–10, repeated here in Figure 3–6(a). The loading pattern is triangular approximating the accumulation of snow on a roof beam extending out from a building. The rate of loading is 200 lb/ft at the left end where the snow is deepest and zero at the right end where the snow depth is virtually zero. The length of the snow pattern is $L_w = 8.70$ ft.

Appendix A–1 gives the geometry of a triangle, summarized here:

1. The area of a triangle is $bh/2$ where $b =$ the length of the base and h is the height.

2. The centroid of a triangle is located 1/3 of the distance from any side.

For the given loading pattern, we can say the height of the triangle is a rate of loading of 200 lb/ft and the length of the base is 8.70 ft. Then the magnitude of the resultant load is

$$R = 0.5(200\text{ lb/ft})(8.70\text{ ft}) = 870\text{ lb}$$

(a) Schematic representation of snow load on a beam

(b) Resultant of varying distributed load

FIGURE 3–6   Varying distributed load.

This force acts vertically downward through the centroid of the triangle, located at

$$x = L_w/3 = (8.70\text{ ft})/3 = 2.90\text{ ft}$$

Figure 3–6(b) shows how we can use this resultant force for purposes of computing reaction forces on the cantilevered roof beams:

$$\Sigma F_y = 0 = R_A - R \qquad \text{Then } R_A = R = 870\text{ lb}$$
$$\Sigma M_A = 0 = M_A - R \cdot x \qquad \text{Then } M_A = R \cdot x =$$
$$(870\text{ lb})(2.90\text{ ft}) = 2523\text{ lb} \cdot \text{ft}$$

The same techniques can be applied to more complex loading patterns as illustrated in Figure 2–11, repeated here

$$R_1 = A_1 = (0.20\text{ kN/m})(1.50\text{ m}) = 0.300\text{ kN}$$
$$R_2 = A_2 = (0.5)(0.30\text{ kN/m})(1.50\text{ m}) = 0.225\text{ kN}$$
$$R_3 = A_3 = (0.50\text{ kN/m})(1.50\text{ m}) = 0.750\text{ kN}$$
$$R_4 = A_4 = (0.5)(0.20\text{ kN/m})(1.50\text{ m}) = 0.150\text{ kN}$$
$$R_5 = A_5 = (0.70\text{ kN/m})(1.50\text{ m}) = 1.050\text{ kN}$$

as Figure 3–7(a). The load approximates the loading due to gravel stored in a bin where the slope of the surface is relatively steep toward the ends, less steep toward the center, and nearly uniform across the middle. Here it is convenient to divide the total loading pattern into five segments of 1.50 m width. The result is four trapezoids and one rectangle. The rectangle can be analyzed like we have shown for uniformly distributed loads. The trapezoids are not so simple. However, we can separate each trapezoid into a rectangle and a triangle and then apply the methods of analysis used before to find the resultant of each part. This is shown in Figure 3–7(b) where we identify a total of nine parts and the following calculations give the values for the individual resultant forces. Note that symmetry allows some simplification.

$$R_9 = R_1 = 0.300\text{ kN}$$
$$R_8 = R_2 = 0.225\text{ kN}$$
$$R_7 = R_3 = 0.750\text{ kN}$$
$$R_6 = R_4 = 0.150\text{ kN}$$

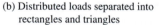

(a) Schematic representation of load on floor beam due to gravel

(b) Distributed loads separated into rectangles and triangles

(c) Resultant forces

$$R_1 = R_9 = 0.300\text{ kN}$$
$$R_2 = R_8 = 0.225\text{ kN}$$
$$R_3 = R_7 = 0.750\text{ kN}$$
$$R_4 = R_6 = 0.150\text{ kN}$$
$$R_5 = 1.050\text{ kN}$$

Total load $= W$
$$\sum_{i=1}^{9} R_i = 3.90\text{ kN}$$

By symmetry
$$R_A = R_B = W/2 = 3.90\text{ kN}/2 = 1.95\text{ kN}$$

FIGURE 3–7   Complex varying distributed load.

Figure 3–7(c) shows these eight resultant forces acting at the centroid of the individual areas from which they were calculated, all referenced from the left end of the beam. These forces can then be used to find the reactions R_A and R_B using the procedure demonstrated in Example Problem 3–2 and shown in Figure 3–7(c).

Analysis of a Structure with Forces in x- and y-directions

Previous example problems involved cases in which only vertical forces were present. Now we analyze a truss that carries forces acting in several directions.

Example Problem 3–5

Figure 3–8 shows a truss on which several forces act in many directions. Compute the reaction forces at the two supports.

Solution Given: Truss in Figure 3–8

Find: Reaction forces at the two supports for the truss

Analysis: Figure 3–8 is the free body diagram of the truss. Support A can exert only a vertical reaction while the pin-type support at F is capable of producing a reaction force in any direction. The analysis will first determine the reaction at B and then x and y components of the reaction force at A, from which the resultant force can be computed. We first compute the components of the applied forces in the x- and y- directions.

Components of forces:

Force	x-component	y-component
$F_1 = 1200\,lb$	$F_{1x} = 0$	$F_{1y} = -1200\,lb\downarrow$
$F_2 = 1500\,lb$	$F_{2x} = -F_2\cos 67° = -586\,lb \leftarrow$	$F_{2y} = -F_2\sin 67° = -1381\,lb\downarrow$
$F_3 = 860\,lb$	$F_{3x} = F_3\sin 22° = 322\,lb \rightarrow$	$F_{3y} = -F_3\cos 22° = -797\,lb\downarrow$
$F_4 = 640\,lb$	$F_{4x} = F_4\sin 38° = 394\,lb \rightarrow$	$F_{4y} = -F_4\cos 38° = -504\,lb\downarrow$
	Sum of x-components = $130\,lb \rightarrow$	Sum of y-components = $-3882\,lb\downarrow$

To find R_{Ay}: We first compute the sum of moments with respect to point F from which we will be able to find the reaction at point A, R_{Ay}. Note that the unknown force components at point F produce no moment about that point because their lines of action pass through the point. Because of the large number of forces and components, a tabular method is used.

Moments about point F (all forces in lb and distances in ft):
a. $F_{1y}(10) = (1200)(10) = 12\,000\,lb \cdot ft$ CCW
b. $F_{2y}(4) = (1381)(4) = 5524\,lb \cdot ft$ CCW
c. $F_{3y}(10) = (797)(10) = 7970\,lb \cdot ft$ CCW
d. $F_{3x}(4) = -(322)(4) = -1288\,lb \cdot ft$ CW
e. $F_{4y}(4) = (504)(4) = 2016\,lb \cdot ft$ CCW
f. $F_{4x}(4) = (394)(4) = -1576\,lb \cdot ft$ CW

Sum of moments of applied forces (a–f) = $24\,646\,lb \cdot ft$ CCW

g. $R_A(14) = $ (negative unknown value) CW

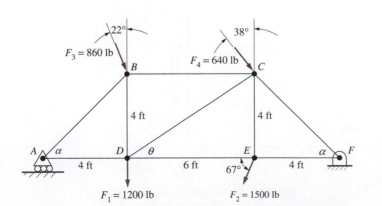

FIGURE 3–8 Forces on a truss and its joints.

Now we set up the equilibrium equation:

$$\Sigma M_F = 0 = 24\,646\ \text{lb}\cdot\text{ft} - R_{Ay}(14)$$

Solving for R_A:

$$R_{Ay} = (24\,646\ \text{lb}\cdot\text{ft})/(14\ \text{ft}) = \underline{1760.4\ \text{lb}\uparrow}$$

To find the reactions at F in the x- and y-directions, we use the sum of forces in those directions.

To find R_{Fy}: We use $\Sigma F_y = 0$.

$$\Sigma F_y = 0 = R_{Ay} - 3882\ \text{lb} + R_{Fy} = 1760.4\ \text{lb} - 3882\ \text{lb} + R_{Fy}$$

Solving for R_{Fy} gives

$$R_{Fy} = 3882\ \text{lb} - 1760.4\ \text{lb} = \underline{2121.6\ \text{lb}\uparrow}$$

To find R_{Fx}: We sum forces in the x-direction to find the last unknown force, R_{Fx}. We assume that this reaction acts to the left.

$$\Sigma F_x = 0 = 130\ \text{lb} - R_{Fx}$$
$$\underline{R_{Fx} = 130\ \text{lb} \leftarrow}$$

We note here that because the calculation for R_{Fx} produced a positive value, our assumed direction was correct.

Resultant reaction at F: We use the components of the reaction to compute the resultant.

$$R_F = \sqrt{(R_{Fx})^2 + (R_{Fy})^2} = \sqrt{(130\ \text{lb})^2 + (2121.6\ \text{lb})^2}$$
$$R_F = 2125.6\ \text{lb}$$

Figure 3–9 shows the resultant and the angle of orientation β computed next.

$$\beta = \text{Tan}^{-1}(130/2121.6) = 3.51°$$

Summary: The application of the principles of static equilibrium has enabled us to compute the reaction forces at the supports for the truss in Figure 3–8. The results are

$$R_{Ay} = \underline{1760.4\ \text{lb}\uparrow} \qquad R_{Ax} = 0$$
$$R_{Fy} = \underline{2121.6\ \text{lb}\uparrow} \qquad F_{Fx} = 130\ \text{lb}\leftarrow$$

Resultant at F: $R_F = 2125.6$ lb at 3.51° to left of positive y-axis

FIGURE 3–9 Resultant reaction force at support F for Example Problem 3–5.

Torsional (Rotational) Equilibrium Analysis

Equilibrium of rotating systems is important to designers, manufacturers, and users of a variety of devices. For example,

- Motors and mechanical power transmissions that use gears, belt drives, chain drives and similar machine elements (see Figures 2–28, 10–14, P10–67, 15–20, and Internet sites 3–5).

- Equipment used in construction, such as winches on cranes used to raise construction materials to upper floors of buildings, trucks, and track-driven machines (see Figures 1–1(d) and Internet sites 6–9).

- Components of vehicles (cars, motorcycles, bicycles, trucks, airplanes, spacecraft, ships) such as transmissions that deliver power from an engine to driveshafts and on to shafts, axles that drive wheels, propellers, fuel pumps, hydraulic systems, compressors, generators, fans, and many other devices (see Figures 1–1(b–e), 1–9 to 1–13, and 15–1, and Internet sites 10–12).

- Manufacturing equipment such as machine tools, conveyors, lift trucks, centrifuges, industrial robots, material handling systems, automation equipment, packaging systems, crushers, screw and nut drivers for assembling products with bolts and other fasteners, and devices that aid production associates by providing ergonomically safe working environments (see Figures 8–23 and 15–2, and Internet sites 13–15).

- Building systems that must be considered by architects such as elevators, escalators, door operators, laundry equipment, food mixers, waste disposers, window cleaning systems on high-rise buildings, scaffolding for painting and general maintenance, and many others (see Internet sites 16–19).

In all these cases, it is necessary to understand and to analyze the forces, torques, and moments that exist within the equipment. This book considers only the static case when the items being analyzed are stationary or moving with a constant velocity. Additional forces, moments, and torques would be required when accelerating or decelerating such systems and those cases are typically covered in a course in *dynamics*.

The following example problem demonstrates the basic principles of balancing torques and the relationships of torques and forces on various components. The problem deals with a simple hand-cranked winch that could be used in many of the work places described earlier.

Example Problem 3–6   Figure 3–10 shows a winch that is capable of lifting 1200 lb. It is actuated by the operator pushing on the handle of a crank connected to the shaft carrying a small gear called a pinion. The pinion meshes with the larger gear that is fixed to one side of the drum on which the cable is wound. Compute the following:

a. The torque on the drum shaft
b. The torque on the shaft actuated by the crank
c. The force that the operator must exert to hold a 1200-lb load on the cable

Solution   Given: Winch in Figure 3–10; Load = L = 1200 lb

Find: Force F_H required on the handle at the end of the crank

Analysis: It is necessary to prepare free body diagrams of each functional part of the winch. (Perform that process yourself before proceeding.)

Figure 3–11 shows an orthographic sketch (a) of the winch and the free body diagrams of (b) the drum with the attached large gear, and (c) the pinion with the attached crank and handle.

Dimensions in inches

FIGURE 3–10   Winch for Example Problem 3–6.

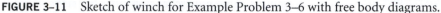

FIGURE 3–11 Sketch of winch for Example Problem 3–6 with free body diagrams.

These are adequate to determine the torques on the functional parts of the winch and to show the forces related to the torques. More complete free body diagrams would be required to design some of the individual components such as the shafts for the drum and the pinion, bearings that hold each shaft, the crank arm, and the handle. These details are not considered in this problem.

Results: *Torques and forces on the drum and large gear* [Figure 3–11(b)]

The load L acts vertically downward tangential to the surface of the drum while the force on the gear teeth F_G acts upward, tangential to the pitch diameter of the gear. The force F_G is exerted by the teeth of the pinion. The moments about the axis A of the drum must be balanced for rotational equilibrium. By summing moments about this axis, we can analyze the torque required to hold the load and the force F_G.

$$\Sigma M_A = 0 = L \times R_D - F_G \times R_G$$

All of these data are known except F_G, so we solve for that term.

$$F_G = \frac{L \times R_D}{R_G} = \frac{(1200 \text{ lb})(0.90 \text{ in})}{2.40 \text{ in}} = 450 \text{ lb}$$

The torque on the drum can be computed by considering either the load or the force on the gear teeth.

$$T_D = L \times R_D = (1200 \text{ lb})(0.90 \text{ in}) = 1080 \text{ lb} \cdot \text{in}$$
$$T_D = F_G \times R_G = (450 \text{ lb})(2.40 \text{ in}) = 1080 \text{ lb} \cdot \text{in}$$

Torque and forces on crankshaft and pinion [Figure 3–11(c)]

First we note that there is an action–reaction pair of forces at the point of contact between the teeth of the gear and the teeth of the pinion. That is,

- The pinion teeth push upward on the gear teeth as shown by the force $F_G = 450$ lb in the previous analysis.
- The gear teeth then push downward on the pinion teeth with a force F_P having an equal magnitude.

Now we can use a method similar to that used for the drum shaft to analyze the forces and torque on the crankshaft. We call the axis of this shaft B.

$$\Sigma M_B = 0 = F_P \times R_P - F_H \times R_C$$

All of these data are known except F_H, so we solve for that term.

$$F_H = \frac{F_P \times R_P}{R_C} = \frac{(450 \text{ lb})(0.60 \text{ in})}{4.00 \text{ in}} = 67.5 \text{ lb}$$

The torque on the crankshaft can be computed by considering either the force on the pinion teeth or the force on the handle.

$$T_C = F_P \times R_P = (450 \text{ lb})(0.60 \text{ in}) = 270 \text{ lb} \cdot \text{in}$$
$$T_C = F_H \times R_C = (67.5 \text{ lb})(4.00 \text{ in}) = 270 \text{ lb} \cdot \text{in}$$

Summary: Load = L = 1200 lb

Force on crank handle = F_H = 67.5 lb

Torque on drum shaft = T_D = 1080 lb·in

Torque on crankshaft = T_C = 270 lb·in

3–3 THE CONCEPT OF MECHANICAL ADVANTAGE

One reason for using a device like the winch analyzed in Example Problem 3–6 is to enable a person to lift heavy loads by exerting forces that are reasonable, safe, and easier for the person to apply. Note the result of the example problem. The operator of the winch is able to apply a force of 67.5 lb to the handle of the winch to hold a load of 1200 lb on a hook attached to the end of the cable wound around the winch drum. The term for this effect is *mechanical advantage,* defined as follows:

> Mechanical advantage *refers to the ratio of the force that can be exerted by a mechanical device to the amount of force applied to actuate the device.*

Using the data for the winch example, we can compute:

MA = Mechanical advantage

= Output force/Applied force

MA = (1200 lb)/(67.5 lb) = 17.8

The person can raise a load with only 1/17.8 (0.056) times the effort.

A related concept, called *rotational mechanical advantage,* RMA, can be defined in terms of the changes in torque that occur in rotating equipment as follows:

> *RAM = Rotational mechanical advantage*
> *= Output torque/Input torque*

Considering only the drum and the crankshaft of the winch in Example Problem 3–6, we can say that

RMA = Torque on drum shaft/Torque on crankshaft
RMA = (1080 lb·in)/(270 lb·in) = 4.0

In this example, the RMA is produced by the actuation of the larger gear by the smaller pinion. Here we can state the following generalization:

> *The rotational mechanical advantage for a pair of gears is equal to the ratio of the pitch diameters of the two gears.*

For the winch, we can demonstrate this by

RMA = D_G/D_P = (4.80 in)/(1.20 in) = 4.0

This matches the RMA calculated previously from the torque ratio.

The concept of RMA also applies to similar mechanical devices such as

- Belt drives
- Chain drives
- Friction roller drives

Relation Between Motion and Mechanical Advantage

The concept of mechanical advantage seems at first to defy common sense. Are we getting something for nothing? Or are we giving something in return for the advantage? The answer is that we are exchanging *difference in motion* for the force or torque advantage.

Consider the relationship between the large gear and the small pinion for which we computed an RMA of 4.0. The 4.0:1 size ratio of the two gears results in the need to rotate the pinion four times for each one revolution of the large gear and the drum.

For forces, we can use the idea of a *lever* to illustrate the mechanical advantage. A lever is one of the classical basic machines discovered centuries ago. The analysis of a lever follows one of the principles of static equilibrium that we have used already several times:

$$\Sigma M = 0 \text{ about any point}$$

The crank for the winch is a type of lever. The operator applies the force F_H to the end of the crank arm, 4.0 in out from the axis of the crankshaft. The pinion is mounted to the crankshaft and held in some way (pin, key, welded to the shaft, machined into the shaft) so that it rotates with the shaft. It exerts a force F_P on the gear teeth at a distance away from the centerline of the crankshaft equal to the pitch radius of the pinion, 0.60 in. Then the ratio of the moment arms of these two forces is

Moment arm for handle = 4.00 in

Moment arm for pinion = 0.60 in

Ratio = 4.00/0.60 = 6.667

This is the mechanical advantage of this part of the winch mechanism as demonstrated by the ratio of the forces:

MA = F_P/F_H = 450 lb/67.5 lb = 6.667

The motion implication here is that the length of the arc moved by the force on the handle is 6.667 times farther than the arc distance traveled by the pinion tooth.

3–4 PULLEY SYSTEMS

In Chapters 1 and 2, we discussed the use of ropes or cables passing over low-friction pulleys to lift loads or to pull from a more comfortable position. See Figure 2–6 for an example. It was noted that the tension in the rope or cable is virtually unchanged as it passes over the freely rotating pulley. If the operator stands below the pulley and pulls down on the rope (the usual case), the maximum force that

(a)   (b) Free body diagram of pulley

FIGURE 3–12   Single pulley applied with a mechanical advantage of 2.0.

can be applied using one pulley is equal to that with which the operator can pull.

Note the difference if the applied force acts in the same direction as the load moves. Figure 3–12 shows an example. The operator stands above the pulley and pulls upward to raise the load. The rope passes through the pulley and up to a point of attachment. Now, to analyze the forces involved, consider a free body diagram of only the pulley and the two lines extending upward as shown in Figure 3–12(b). We "cut" the lines and show the tension force T in each rope acting upward. The load acts vertically downward and the upward force in each cut in the rope is equal to the pull being exerted by the operator. The entire system is in equilibrium and, therefore, any part of it is also in equilibrium. Now use the equilibrium equation to sum forces in the vertical direction and set that sum equal to zero.

$$\Sigma F_v = 0 = L - 2T$$

Solving for the tension in the rope T gives

$$T = L/2$$

This shows that with the pulley system shown in Figure 3–12, the operator needs to apply only one-half the force to raise the load. Thus, the mechanical advantage is 2.0. The figure shows one way to hold the ropes in the pulley grooves while allowing the load to be attached. This design is called a *block*. The assembly of the block with its attachment device for holding the load, and the lines themselves is called a *block and tackle*.

Hoists often use a system of two or more pulleys to allow the application of modest forces to lift heavy loads. The pulley systems produce a relatively high mechanical advantage by providing two or more lines supporting the load.

Figure 3–13 shows one of the simpler types of pulley systems, using two single-groove pulleys. One pulley is attached to a rigid support and the second hangs below and supports the load. The rope passes through the top pulley, down and around the lower pulley, and then fastens to the underside of the top pulley. This design allows the operator to stand below the pulley. With a downward pull on the free end of the rope, the lower pulley rises bringing the load with it. A free body diagram of the lower pulley and the two lines around it would appear as already shown in Figure 3–12(b). Thus, this arrangement also has a mechanical advantage of 2.0.

Other arrangements of blocks using additional pulleys result in higher values of mechanical advantage. Figure 3–14 shows some possible arrangements using individual blocks having one, two, or three pulleys and exhibiting values of mechanical advantage from 1 to 7. Part (a) shows designs where the applied force is downward. In part (b) the applied force is upward and an additional unit of mechanical advantage is gained. Pay special attention to the fact that the mechanical advantage of a block and tackle system is equal to the number of lines supporting the load. You should be able to isolate the lower block including the load by cutting the connecting lines and counting the number that are cut. Then a simple application of the equilibrium of forces in the vertical direction enables the calculation of the force

	A	B	C	D	E	F
Number of upper pulleys	1	1	2	2	3	3
Number of lines	1	2	3	4	5	6
Number of lower pulleys	0	1	1	2	2	3
Mechanical advantage	1	2	3	4	5	6

(a) Block and tackle with downward applied pull on rope

	A	B	C	D	E	F
Number of upper pulleys	0	1	1	2	2	3
Number of lines	2	3	4	5	6	7
Number of lower pulleys	1	1	2	2	3	3
Mechanical advantage	2	3	4	5	6	7

(b) Block and tackle with upward applied pull on rope

FIGURE 3–13 Block and tackle.

FIGURE 3–14 Types of block and tackle arrangements.

that must be applied to raise a given load and the mechanical advantage. Figure 3–15 show a commercially available block and tackle system with a mechanical advantage of 4 like that shown in Figure 3–14(a), type C (see also Internet site 1).

The mechanical advantage discussed thus far is the theoretical maximum; the actual value will be lower for the following reasons:

1. There is friction at the axle of the pulleys, although with high-quality rolling contact bearings, this can be quite small.

2. There may be friction between the rope or cable and a stationary surface.

3. The lower blocks that move with the load have mass that must be lifted in addition to the weight of the load. However, this is typically small and should only be an issue if precise measurements are needed.

It was previously discussed that one effect of having a large mechanical advantage is that greater motion is required to lift the object a given distance. For example, to raise a 200-lb load by 2.0 ft with the hoist system shown in Figure 3–15(MA = 4), you would need to exert only 50 lb of force on the rope while you would have to pull 8.0 ft of rope through the system. You can get a feel for this effect from the interactive exercise in Internet site 2 that allows you to input an amount of weight to be lifted and select the number of lines in the block and tackle system, as shown in Figure 3–12. The software then computes the force required to accomplish the lift and you can use the cursor to "pull" on the lift rope and watch the resulting vertical lift of the load. The weight of the moving pulleys can also be input to the calculator and its weight is factored into the force required to raise the load. Entering a value of zero for the weight of the pulleys gives the theoretical force.

(a)   (b)

FIGURE 3–15 Rope hoist with a mechanical advantage of 4.

Source: Tuf-Tug, Inc., Moraine, Ohio, Photographs by Lynne' McGuire.

INTERNET SITES

1. Tuf-Tug Products www.tuf-tug.com Manufacturer of hoists, pulleys, pulley blocks, and other accessory and safety items for construction and utility work.

2. ThinkQuest http://thinkquest.org/27948/pulley.html An interactive demonstrator of pulley systems in which the mechanical advantage of the system for a selected number of pulleys is computed along with a visual display of the motion of the load versus the input motion of the applied force.

3. Baldor-Dodge-Reliance www.baldor.com A comprehensive supplier of electric motors, complete mechanical drives, and drive components.

    a. Under the pull-down menu, click on PRODUCTS to select from several types of motors or gear-driven products under the Baldor, Reliance, or Dodge brand names.

    b. On the main page, select Dodge-pt to go to the site www.dodge-pt.com for descriptions of gear-type drives, V-belt drives, conveyor pulleys, and more.

4. Emerson Power Transmission www.emerson-ept.com Manufacturer of numerous power transmission components including V-belt drives, synchronous belt drives, and roller chain drives through its Browning and Morse divisions.

5. Rexnord www.rexnord.com Manufacturer of a wide variety of power transmission products under the Rexnord (chain drives, conveyor components, couplings, bearings, and others), Falk (gear drives), and Berg (precision mechanical components) brands. Also a diversified producer for aerospace, industrial, mining, beverage and food, water treatment, and other industries.

6. Altec www.altec.com Manufacturer of truck-mounted aerial lifts and cranes for construction, utility service, tree maintenance, and other applications.

7. Link Belt www.linkbelt.com Manufacturer of construction cranes.

8. JLG Industries www.jlg.com Manufacturer of mobile aerial work platforms, scissors lifts, and load handlers for construction, maintenance, warehousing, and other industries.

9. Yale Hoists www.yalehoists.net Manufacturer of a large variety of hoists for overhead cranes using chains, wire rope, and ratchet-type actuation, either manual or power operated.

10. BorgWarner, Inc. www.bwauto.com Manufacturer of drivetrain components for automotive applications including clutches, timing chains, and all-wheel drive systems.

11. Delphi Automotive Systems https://delphi.com/shared/pdf/ppd/body/window_lift.pdf Manufacturer of numerous systems and components for automotive applications. This site features window lift mechanisms for automobiles.

12. Freightliner Trucks http://www.freightlinertrucks.com/trucks/featured-components/transmissions.aspx Manufacturer of highway trucks. This site features the models of transmissions available.

13. MAG Giddings & Lewis www.giddings.com/industries/aerospace.html Producer of highly productive machine tool systems for a variety of industries. The listed site features machines for the aerospace industry.

14. Fanuc Robotics America, Inc. www.fanucrobotics.com Producer of industrial robots and complete systems for assembly, material handling, welding, and many other applications. From the home page, click on *Robotic Applications Image Gallery* to view an extensive array of photographs of robots at work.

15. Cisco-Eagle www.cisco-eagle.com Distributor of devices for handling materials in commercial and industrial applications. Select *Products* from the home page to see a list of types of items. Some of those most relevant to the study of statics and strength of materials are racks, shelving, conveyors, lifts, pallet jacks, drum handling, carts, and hand trucks.

16. ThyssenKrupp Elevator www.thyssenkrupp.com Manufacturer of elevators, escalators, moving walkways, and passenger bridges for commercial buildings and airports. The principles of statics are needed to analyze the forces on various parts of these systems when they are stationary or moving with constant velocity.

17. Colgate Scaffolding www.colgatescaffold.com Producer of scaffolding and mobile systems used for maintenance of buildings and other commercial applications. Accessories include attachment devices to support a scaffold over the side of a building.

18. Hi-Lo Climbers http://hiloclimbers.com Manufacturer of modular platforms and powered rigging equipment to provide work space and access to high locations and the exterior panels and windows of buildings.

19. Chamberlain Group, Inc. www.liftmaster.com Manufacturer and distributor of LiftMaster operators for gates, residential garage doors, and commercial building doors.

PROBLEMS

Static Equilibrium of Equipment and Structures

Note: The given forces act through the center of mass indicated in each figure. Unless stated otherwise, the object being analyzed has a center of mass at its geometric center.

3–1. The workbench shown in Figure P3–1 holds a small drill press weighing 192 lb and a wheel grinder weighing 42 lb. The table itself weighs 132 lb. Compute the forces that the feet of the table exert on the floor. Assume that the loads act at the middle of the tabletop so that the two feet at each end share the reactions equally.

3–2. The utility cart shown in Figure P3–2 carries two castings as shown. $W_1 = 164$ lb and $W_2 = 225$ lb. The cart itself weighs 38 lb. Compute the forces on each wheel. The forces are equal on each wheel at each end.

3–3. The two-wheeled hand truck shown in Figure P3–3 carries a rectangular box weighing 280 lb whose center of gravity is at the center of the box. Compute the horizontal force that must be exerted at the top of the handle to tip the hand truck backward. The operator places a foot on the axle of the wheels to steady the load as the horizontal force is applied.

3–4. The two-wheeled hand truck shown in Figure P3–4 carries a rectangular box weighing 280 lb whose center of gravity is at the center of the box. Use the dimensions from Figure P3–3. For the position shown ($\theta = 35$ degrees), compute the vertical force that the operator must exert to

FIGURE P3–1   Table with machine tools.

FIGURE P3–2   Cart carrying two castings.   **FIGURE P3–3**   Two-wheeled hand truck.

FIGURE P3–4 Hand truck.

Dimensions in millimeters

FIGURE P3–7 Wheelbarrow.

keep the hand truck steady. Neglect the weight of the hand truck.

3–5. For the hand truck described in Problem 3–4, compute the required vertical force that the operator must exert to hold the hand truck steady for angles θ ranging from zero to 90 degrees in increments of 15 degrees. Neglect the weight of the hand truck.

3–6. For the loaded hand truck described in Problem 3–4, at what angle θ will the required vertical force on the handle be zero?

3–7. The wheelbarrow shown in Figure P3–7 carries a full load of dry sand. Determine the vertical force required at the end of the handles to hold the feet off the ground. See Table 1–6 for the properties of dry sand. Consider the sides of the hopper to be a triangle. The width of the wheelbarrow is 550 mm. Neglect the weight of the wheelbarrow.

3–8. Figure P3–8 shows a lift truck used in a warehouse that handles bags of concrete mix, each having a mass of 25.0 kg.

The pallets hold six bags per layer and there are nine layers. The center of mass is indicated in the figure. Compute the reaction forces at each wheel of the lift truck.

3–9. If the lift truck depicted in Figure P3–8 requires at least 1.20 kN of load on each rear wheel for safe traction and maneuverability, compute the number of bags of concrete that may be safely carried on the pallet. Assume that the center of mass for the entire load is at the same place as shown in the figure.

3–10. For the condition of loading on the rear wheels outlined in Problem 3–9 for the safe operation of the lift truck, consider alternate methods of stacking the bags on the pallet to permit more bags per pallet than found for Problem 3–9.

3–11. Consider the original plan in Problem 3–8 for carrying nine layers of bags of cement with six bags per layer on the pallet. Also consider the condition outlined in Problem 3–9 that each rear wheel must carry at least 1.20 kN of load. How can the lift truck be modified to satisfy both of these conditions?

FIGURE P3–8 Forklift truck.

FIGURE P3–13 Travel trailer with fifth-wheel pin.

3–12. Figure 2–31(a) shows a commercial truck with two axles, called the H20 design truck. There are two single wheels on the front axle and four wheels on the rear axle. For the purposes of this problem, ignore the notation that the total weight of the truck and its load is 40 K. Instead, consider the following:
1. The truck with no load weighs 3800 lb and its center of mass is 3.0 ft to the rear of the front axle.
2. The loading on each front wheel cannot exceed 4.0 K.
3. Each of the rear wheels can carry no more than 8.0 K.
4. When loaded, the center of mass of the load is 2.5 ft forward of the rear axle.

Compute the maximum permissible load in the truck body.

3–13. Figure P3–13 shows a travel trailer supported by its rear wheels and a landing gear assembly. The feet of the landing gear rest on a load cell that records a force of 9.76 kN. The total weight of the trailer is shown in the manual to be 22.30 kN. Determine the location of the center of mass of the trailer.

3–14. Figure P3–14 shows a pickup truck that has a total weight of 12.38 kN and the center of mass is located 1.28 m to the rear of the front axle. Compute the load on each wheel.

FIGURE P3–14 Pickup truck with fifth-wheel hitch to carry travel trailer.

3–15. The trailer in Figure P3–13 is to be mounted on the fifth-wheel fixture in the bed of the pickup truck shown in Figure P3–14. A fifth-wheel system is a means of efficiently connecting a trailer to a truck by simply backing under the trailer and allowing the pin on the front of the trailer to be inserted into the receiver hole in the fifth-wheel after retracting the trailer's landing gear. Compute the load on each of the three axles of the combined unit, using the data from Problems 3–13 and 3–14. Then compute the increase in load on the truck axles due to the added trailer load.

3–16. Weight and balance are critical to the stability and performance of aircraft. Designers and pilots constantly check to ensure that the aircraft is balanced side to side to ensure level flight. Also, front-to-rear balance is critical to ensure that the center of mass is in proper relation to the center of aerodynamic forces of lift and drag. Analyze the aircraft shown in Figure P3–16 to determine the location of the center of mass relative to its nose. Consider only the six downward loads. Then compare the position of the center of mass with the net lift force shown.

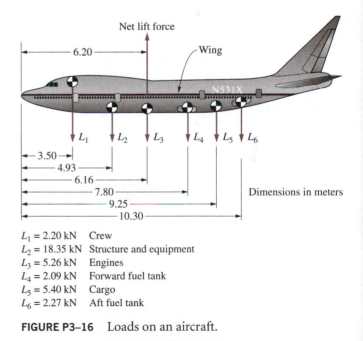

$L_1 = 2.20$ kN   Crew
$L_2 = 18.35$ kN   Structure and equipment
$L_3 = 5.26$ kN   Engines
$L_4 = 2.09$ kN   Forward fuel tank
$L_5 = 5.40$ kN   Cargo
$L_6 = 2.27$ kN   Aft fuel tank

FIGURE P3–16 Loads on an aircraft.

3–17. Changes in the location of the center of mass are also tracked by the pilot in flight, particularly as fuel is expended from the various tanks in the wings and/or the fuselage. Sometimes fuel is transferred from one tank to another to achieve a more optimum balance condition. For the aircraft in Problem 3–16, compute the location of the center of mass when the aft fuel tank in the fuselage is emptied during flight. Compare this location with the position of the net lift force.

3–18. For the aircraft described in Problem 3–16, compute the position of the center of mass when the fuel tanks are full and the cargo load is zero. Compare this location with the position of the net lift force.

Reactions on Beams—Problem Statement for Figures P3–19 to P3–56

Figures P3–19 to P3–56 show a variety of beam types and loading conditions. For the beam in each figure, compute the magnitude and direction for the reactions at the supports to maintain the beam in equilibrium. Unless stated otherwise, consider the weight of the beam to be small relative to the applied loads and, therefore, negligible.

3–19. Use Figure P3–19. **3–25.** Use Figure P3–25.
3–20. Use Figure P3–20. **3–26.** Use Figure P3–26.
3–21. Use Figure P3–21. **3–27.** Use Figure P3–27.
3–22. Use Figure P3–22. **3–28.** Use Figure P3–28.
3–23. Use Figure P3–23. **3–29.** Use Figure P3–29.
3–24. Use Figure P3–24. **3–30.** Use Figure P3–30.

FIGURE P3–19

FIGURE P3–20

FIGURE P3–21

FIGURE P3–22

FIGURE P3–23

FIGURE P3–24

FIGURE P3–25

FIGURE P3–26

FIGURE P3–27

FIGURE P3–28

FIGURE P3–29

FIGURE P3–30

3–31. Use Figure P3–31.
3–32. Use Figure P3–32.
3–33. Use Figure P3–33.
3–34. Use Figure P3–34.
3–35. Use Figure P3–35.
3–36. Use Figure P3–36.
3–37. Use Figure P3–37.

3–38. Use Figure P3–38.
3–39. Use Figure P3–39.
3–40. Use Figure P3–40.
3–41. Use Figure P3–41.
3–42. Use Figure P3–42.
3–43. Use Figure P3–43.
3–44. Use Figure P3–44.

FIGURE P3–38

FIGURE P3–31

FIGURE P3–39

FIGURE P3–32

FIGURE P3–40

FIGURE P3–33

FIGURE P3–41

FIGURE P3–34

FIGURE P3–42

FIGURE P3–35

FIGURE P3–43

FIGURE P3–36

FIGURE P3–37

FIGURE P3–44

3–45. Use Figure P3–45. **3–48.** Use Figure P3–48. **3–51.** Use Figure P3–51. **3–54.** Use Figure P3–54.
3–46. Use Figure P3–46. **3–49.** Use Figure P3–49. **3–52.** Use Figure P3–52. **3–55.** Use Figure P3–55.
3–47. Use Figure P3–47. **3–50.** Use Figure P3–50. **3–53.** Use Figure P3–53. **3–56.** Use Figure P3–56.

300 lb

5 in   15 in

8.0 lb/in

FIGURE P3–45

10 K

12 K

4 ft   4 ft

3 ft   3 ft   2.5 K/ft

FIGURE P3–46

8.6 kN   12.2 kN

2 m   2 m   2 m   9.6 kN

12 kN/m   2 m   2 m

FIGURE P3–47

220 lb

80 lb/in

18 in   4 in

FIGURE P3–48

15 K   12 K

8 ft   8 ft   4 ft

4 K/ft

FIGURE P3–49

25 K   15 K

6 ft   4 ft

3.0 K/ft

FIGURE P3–50

30 kN

2.0 m   1.0 m   20 kN

60 kN/m   2.0 m

FIGURE P3–51

1.50 kN/m

4.0 m

FIGURE P3–52

30 lb/in

18 in

FIGURE P3–53

50 lb/in

10 in   12 in

FIGURE P3–54

8.4 kN/m

2.4 m

FIGURE P3–55

1500 lb/ft

4.0 ft   2.5 ft

FIGURE P3–56

Reactions on Beams Considering Beam Weight—Problem Statement for Problems 3–57 to 3–60

For the beam in each figure, compute the magnitude and direction for the reactions at the supports to maintain the beam in equilibrium. Consider the weight of the beam. **Note:** At the discretion of the instructor, the results can be compared with the solution for the same beam loading for which the beam weight was neglected.

3–57. Use Figure P3–21. The beam is steel with a uniform rectangular cross section measuring 0.75 in × 1.25 in.

3–58. Use Figure P3–19. The beam is a steel wide-flange shape, W12×30.

3–59. Use Figure P3–22. The beam is steel with a uniform circular cross section with a diameter of 40 mm.

3–60. Use Figure P3–29. The beam is a structural steel tube, HSS203×102×12.7.

Forces at Supports for Structural Components and Mechanical Devices— Problems 3–61 to 3–68

Each figure for Problems 3–61 to 3–68 shows a structural component or mechanical device in which one or more forces are applied parallel to or at an angle from the main portion. The supports are at locations marked with an ×, which can provide reaction forces in any direction perpendicular to the axis of the beam. One of the supports has the capability of resisting horizontally directed forces. For each figure, compute the reaction forces at the supports to maintain equilibrium. Neglect the weight of the components.

3–61. Use Figure P3–61.	**3–65.** Use Figure P3–65.
3–62. Use Figure P3–62.	**3–66.** Use Figure P3–66.
3–63. Use Figure P3–63.	**3–67.** Use Figure P3–67.
3–64. Use Figure P3–64.	**3–68.** Use Figure P3–68.

FIGURE P3–61

FIGURE P3–62

FIGURE P3–63

FIGURE P3–64

FIGURE P3–65

FIGURE P3–66

FIGURE P3–67

FIGURE P3–68

Support Forces for Structures— Problems 3–69 to 3–77

Each figure for Problems 3–69 to 3–77 shows a structure such as a truss or a frame with loads applied to it. Considering the entire structure to be a rigid body, compute the forces at the supports. Unless otherwise stated, neglect the weight of the components of the structure.

3–69. Use Figure P3–69.   **3–71.** Use Figure P3–70.   **3–73.** Use Figure P3–73.   **3–76.** Use Figure P3–76.

3–70. Use Figure P3–70.   **3–72.** Use Figure P3–72.   **3–74.** Use Figure P3–74.   **3–77.** Use Figure P3–77.

                                              3–75. Use Figure P3–75.

Member specifications:

AD, DE, EF L2×2×⅛ – doubled ——— ⌐L

BD, CE, BE L2×2×⅛ – single ——— ⌐

AB, BC, CF C3×4.1 – doubled ——— ⊐⊏

FIGURE P3–69   Truss for Problem 3–69.

	Problem No.	
	3–70	3–71
L	28.9 kN	28.9 kN
W_T	3.4 kN	3.4 kN
F_W	0	11.3 kN

W_T = Weight of tower

F_W
Wind load

FIGURE P3–70   Structure for Problems 3–70 and 3–71.

FIGURE P3–72   Truss for Problem 3–72.

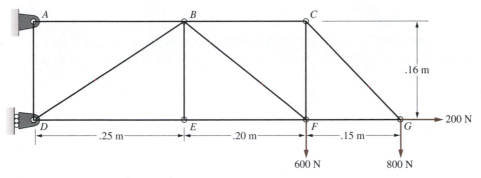

FIGURE P3–73   Truss for Problem 3–73.

FIGURE P3–74   Structure for Problem 3–74.

FIGURE P3–75   Machine lever as a simple beam carrying normal concentrated loads.

FIGURE P3–76   Machine lever as a simple beam carrying normal concentrated loads and an inclined load.

Rotational Equilibrium—
Problems 3–78 to 3–81

Each figure for Problems 3–78 to 3–81 shows a cylindrical shaft with extended arms or cranks to which external forces are applied. The forces tend to cause rotation of the shaft. Consider the entire object as a free body supported at two bearings. In each figure, Bearing A is held fixed against rotation and is capable of resisting forces in any direction. Bearing B cannot resist rotation but it can provide reaction forces directed only perpendicular to the shaft. Prepare the complete free body diagram for each shaft.

3–78. Use Figure P3–78.   **3–80.** Use Figure P3–80.

3–79. Use Figure P3–79.   **3–81.** Use Figure P3–81.

3–82. Figure P3–82 shows a lever called a *bell crank* for a machine. The applied force F_1 acts toward the left and a vertical force F_2 pulls downward. The bell crank can rotate freely around the pin-joint pivot at O. For an applied force of $F_1 = 80$ lb, compute the force F_2 and the reaction forces at the pivot to maintain equilibrium.

3–83. Figure P3–83 shows a lever called a *bell crank* for a machine. The applied force F_1 acts toward the right and a vertical

$F_{Bx} = 235$ lb
$F_{By} = 182$ lb
$F_{Cx} = 174$ lb
$F_{Cy} = 118$ lb

Dimensions in inches

FIGURE P3–77   Shaft with varying cross sections.

Shaft can rotate freely in supports *A* and *C*. All resisting torque acts to the left of *A*, provided by an adjacent element.

FIGURE P3–78   Shaft carrying a crank.

1.25 kN

Dimensions in millimeters

FIGURE P3–81

Torque is resisted at *A*

Shaft is free to rotate at *C*

Dimensions in inches

FIGURE P3–79

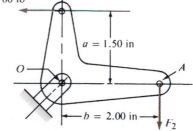

$F_1 = 80$ lb

$a = 1.50$ in

$b = 2.00$ in

F_2

FIGURE P3–82   Bell crank.

Torque is resisted at *A*

Shaft is free to rotate at *C*

200 lb

600 lb

Dimensions in inches

FIGURE P3–80

$F_1 = 450$ N

25

60

40

F_2

Dimensions in millimeters

FIGURE P3–83

force F_2 pulls downward. The bell crank can rotate freely around the pin-joint pivot at O. For an applied force of $F_1 = 450$ N, compute the force F_2 and the reaction forces at the pivot to maintain equilibrium.

3–84. Figure P3–84 shows a print head for an impact printer. The applied force F pushes the print head to the left and imprints a character onto paper through an inked ribbon. The paper is backed up by a platen. The print head is guided by four small roller guides as shown, which can resist only vertical forces. Draw the complete free body diagram of the print head.

FIGURE P3–84   Print head.

3–85. Figure P3–85 shows a shaft carrying a bevel gear and a spur gear supported on two bearings. The two gears are keyed to the shaft so that they rotate at the same speed. The forces shown on the bevel gear are the x, y, and z components of the total force exerted by a mating gear. The component in the y-direction acts tangentially on the gear causing it to rotate along with the shaft. The spur gear carries forces in only the y-and z-directions as it transmits torque to its mating gear. Because of the geometry of the spur gear teeth, the z-directed force is always 0.364 times the y-directed force. Bearing A can resist axial forces along with radially directed forces. Bearing C can resist only radially directed forces. Compute the following:

a. The forces on the spur gear
b. The reaction forces on bearing A
c. The reaction forces on bearing C

Pulley Systems

3–86. The pulley system shown in Figure 3–13 is holding a load of 120 lb attached to the lower pulley. With what force must an operator pull on the cord at the right?

3–87. The pulley system shown in Figure 3–13 is to be used to raise slowly a load of 525 N a distance of 1.75 m. With what force must an operator pull on the cord at the right and how much rope must be pulled through the system?

3–88. Describe the salient differences between the block and tackle arrangements called C in Figure 3–14 parts (a) and (b). Follow the path of the rope from the end where the applied force T is exerted through the pulleys and to the place where the other end is attached to one of the blocks. Verify the values of mechanical advantage listed in the figures.

FIGURE P3–85   Shaft with bevel gear and spur gear.

FORCES IN FRAMES, STRUCTURES, AND TRUSSES

MAJOR TOPICS

4–1 The Big Picture
4–2 Analysis of Pin-Connected Frames and Mechanisms
4–3 Analysis of Trusses
4–4 Equilibrium in Three-Dimensional Frames and Structures

4–1 THE BIG PICTURE

Previous chapters focused on the analysis of forces on complete structures and discrete objects, such as beams and certain machine components. In this chapter we consider the individual members that make up structures, frames, and trusses, recognizing that all the members must act together to provide stability, rigidity, and functionality with safety and efficiency.

It is important for structures to be constrained so that they hold their shape under all expected loads. However, it is normally desirable not to overly constrain the structure by adding elements that cause it to become statically indeterminate, that is, so it is impractical to analyze the forces in each member. At first, we consider two-dimensional structures and later we address three-dimensional situations.

Examples of the structures that you will analyze in this chapter are shown in Figure 4–1.

- Part (a) shows a type of frame for supporting a load in which all members are pin-connected. Detailed analysis of the forces is essential to completing the design of the members to ensure safety and functionality. Note that the load acts perpendicular to the end of the horizontal bar causing bending. However, the angled strut is subjected to pure compression acting along the length of the strut between two pin joints. The strut is called a *two-force member.* This kind of frame is analyzed in Section 4–2.

- Part (b) shows a simple *truss* in which all loads and reactions act at joints. The result is that all forces in the members of the truss are axial, either tension or compression. Trusses are analyzed in Section 4–3.

- Part (c) is a *rigid frame* in which the connections of the upper beam to the tops of each column are rigid and can resist moments. The discussion of rigid frames is covered in later courses.

- Part (d) shows a rectangular frame comprised of two columns pin-connected to the upper beam and covered by a sheet-like panel. When subjected to vertical and horizontal loads, the shape is maintained by the panel covering the primary horizontal and vertical members. Without the panel, the frame would not be stable and would topple toward the right. This design is called a *panel-stiffened frame,* a *shear panel,* or a *shear wall* when used in a building structure. This type of design is discussed in later courses.

Note the variety of methods of providing structural rigidity and stability to the structures shown in Figure 4–1. In the pin-connected frame of part (a) and the truss in part (b), members are connected in ways that form triangles. This approach, sometimes called *triangulation,* is highly effective and simple to construct.

Forces in Three-Dimensional Structures

The examples discussed so far have been two-dimensional, called *planar.* They are useful in cases in which the applied loads are in a single plane or when loads are balanced between two, or among three or more, planar structures that share the applied loads equally. Of course, it is necessary to brace each such structure to ensure that its shape and position are maintained as loads are applied.

However, many realistic structures are truly three-dimensional and require an analysis technique that considers forces acting in any direction in three-dimensional space to determine their magnitude and direction. The results are essential to completing the design of a safe system. Figure 4–2 shows three examples of such three-dimensional structures.

- Part (a) shows a tower held by a set of three guy wires acting at different points around the base of the tower. Wind loads and unbalanced loads from devices attached to the tower create unequal loads in the guy wires.

- The marquee in part (b) uses one strut extending from the building structure to the top of the marquee, but it is installed at an angle because of limitations for suitable attachment points. You would need to determine the force in the strut and the forces transmitted to the marquee and to the wall of the building.

(a)

(b)

Rigid joints

(c)

Cover plate

(d)

FIGURE 4–1   Types of frame construction.

(a) Antenna tower with guy wires

(b) Marquee supported by one cable

FIGURE 4–2   Examples of three-dimensional structures.

FIGURE 4–2 (*Continued*)

(c) Roof-mounted derrick crane

- Part (c) shows a roof-mounted derrick crane supporting an air-conditioning unit over the side of a building. The vertical tower is supported by two angled struts connected to the roof structure. The boom extends up at an angle to the tower and reaches out away from the façade of the building. A winch system on the top of the tower reels cable in to raise the load. When the air conditioner is above roof level, the tower swings around to allow the unit to be lowered onto the roof. The forces in the various members are functions of their placement, the angles between them, and the weight of the load.

This chapter deals first with forces in two-dimensional frames, trusses, and structures to develop concepts and procedures for analysis (Sections 4–2 and 4–3). Then the procedures are modified and extended to handle the three-dimensional cases in Section 4–4.

Activity A

Demonstrate to yourself the desirability of triangulation by making a simple triangle from three strips of heavy cardboard pinned together at their ends. The strips can be about 0.5 in (12 mm) wide and 4 to 6 in long (100 to 150 mm) and made from the backing for a pad of notepaper. Apply forces to the triangle at the joints and notice that the shape does not change provided the applied forces are not so high as to distort the strips themselves.

Now change the structure by adding a fourth strip and pin all strips together at their ends, making a four-sided figure (a quadrilateral). Hold one of the strips steady and then apply a force to any of the other joints. The shape changes because the members can rotate at the joints.

Finally, add a fifth strip across the diagonal of the quadrilateral, again using pin connections at the joints. This divides the quadrilateral into two triangles and you should show that the structure is now rigid, demonstrating the desirability of triangulation.

Activity B

Use heavy cardboard strips to fabricate the truss shape shown in Figure 4–1(b), using pin connections at each joint. Note that the truss has a series of triangles. Fix the two supports to a flat surface with pins, apply loads to any of the other joints, and notice that the truss maintains its basic shape as long as the applied forces do not cause major deformation of the strips.

However, you will likely be able to, in fact, apply sufficient force to deform the strips. Those loaded in compression will likely bend and buckle first, while those that are in tension will remain straight. Make note of which ones buckle. This illustrates the necessity to design the members to withstand all expected forces, making them sufficiently rigid and strong. The study of strength of materials in later chapters of this book will help you to develop such skill.

Activity C

Again use three heavy cardboard strips to assemble the frame shown in Figure 4–1(c). Fix the two supports to a flat surface with pins. Then connect the other members together at the top joints by gluing the strips together and letting the glue set up. Use this structure for the following exercises.

- First apply a downward force on the end of one of the side members pushing directly toward the pinned support. You should see that the force is transmitted down the member to the pin joint fixed to the surface.
- Now apply a modest force at an angle to the upper joint so that a component of the force acts horizontally. What happens? The shape of the frame will be maintained provided the applied force is small enough so as not to distort the members themselves. The basic shape is similar to the quadrilateral you made in Activity A, but the glue rigidizes the joints allowing them to transmit the applied force through the glue to the members themselves without allowing rotation. As the force is increased there will be a point at which one of the members distorts due to bending moments created. The design of the frame would have to resist the bending action.

Activity D

Again use three heavy cardboard strips to assemble the frame shown in Figure 4–1(d). Fix the two supports to a flat surface with pins. Then cut a rectangular panel from the cardboard to cover the three members and glue the assembly together. Let the glue set up and then use this structure for the following exercises.

- Apply forces in any direction to the frame and note that it remains rigid and that the applied forces cause reactions at the pinned supports.
- Note the high level of stiffness provided by the relatively thin panel.
- Apply a horizontal force to the upper left corner of the frame and then increase its magnitude slowly until the flat panel tends to deform noticeably. It may take a fairly high force depending on the thickness of the panel material. Note the nature of the deformation. It starts to buckle locally, typically along a diagonal of the panel. This is typical of *shear deformation*, *shear buckling*, or *shear wrinkling* and is the reason that such a frame is called a *shear panel*.

Activity E

Look for examples of structures that use the types of construction described in this section: pin-connected frames, trusses, diagonal bracing, rigid frames, or shear panels. Check out houses, commercial buildings, high-rise buildings, machine frames, towers, bridges, cranes, vehicle frames, the structure of toys, cabinets, desks, storage shelves, truck bodies, computer cases, bicycle and motorcycle frames, and many other types of objects. Observe whether the basic elements of the structure can be considered two-dimensional or whether a three-dimensional analysis would be required.

Objectives of This Chapter

After completing this chapter, you should be able to:

1. Analyze pin-connected frames to determine the forces acting on any member.
2. Identify two-force members in a frame, truss, or other kinds of structures.
3. Analyze trusses to determine the force in any member, including whether the forces are compression or tension.
4. Identify and describe rigid frames in which joints are fixed against rotation.
5. Identify and describe a shear panel or a panel-stiffened frame.
6. Describe how structures are made rigid using triangulation, diagonal bracing, or shear panels.
7. Analyze three-dimensional frames and structures to determine the forces in each member.

4–2 ANALYSIS OF PIN-CONNECTED FRAMES AND MECHANISMS

It is impractical to define a completely explicit method of analyzing the forces in all members of a frame or a mechanical mechanism. However, there are some general principles that can be applied, some of which were developed in earlier chapters and it would be good for you to review them.

One concept mentioned briefly in an earlier chapter is the *two-force member*. Additional attention is now given to this concept because it is critical to the procedure we describe in this section.

> *The concept of a two-force member:* By definition,
>
> *A two-force member is a component of a structure having only two connections to other members and for which forces are applied to the member only through those connections.*

For such a member, the following important observation can be made:

> *A two-force member has these properties:*
> 1. *It carries a total force that acts along the line between the two connections.*
> 2. *The force is either direct tension or direct compression.*

Applying these observations is often critical to the analysis of forces on pin-connected frames discussed in this section and to trusses discussed later.

The logic of a two-force member can be visualized by applying the principles of static equilibrium. First, look at the two members shown in Figure 4–3, one in tension and the other in compression. Part (a) shows a member in tension due to the applied forces F_{T1} and F_{T2} that are in line with the axis, called u, between the two joints at either end. If we sum forces in the direction of that axis we obtain,

$$\Sigma F_u = 0 = F_{T1} - F_{T2}$$

FIGURE 4–3   Two-force members.

Then

$$F_{T1} = F_{T2}$$

This demonstrates that the member is in direct tension. Now consider the other equation of static equilibrium, $\Sigma M = 0$. Summing moments about either connection A or B results in no applied moments at all because the lines of action of both forces are through A or B. Therefore, the member is in equilibrium when subjected to the forces as shown and it satisfies the principles of a two-force member given earlier.

A similar analysis applied to the compression member in Figure 4–3(b) would show that

$$F_{C1} = F_{C2}$$
$$\Sigma M = 0 \text{ about either point } C \text{ or } D$$

Therefore, the member is in equilibrium and is in direct compression. It satisfies the principles of a two-force member given earlier.

Now, to demonstrate that no other force system on these members satisfies the requirements for static equilibrium, consider Figure 4–4. In part (a), the two forces, F_1 and F_2 are shown acting in different directions and it is impossible to satisfy the condition that $\Sigma F = 0$ in any direction regardless of the magnitude of the forces. Therefore, this condition cannot exist for a static member. In part (b), the two forces are parallel and acting in opposite directions. You should see that if the two forces have equal magnitudes, the sum of forces in their direction of action would sum to zero, satisfying *one condition* of equilibrium. However, if we sum moments about either joint C or D, there is an unbalanced moment causing the failure to meet the other condition of

FIGURE 4–4   Improper forces applied to two-force members.

equilibrium: $\Sigma M = 0$ about any point. Therefore, this condition cannot exist either. In conclusion, we have demonstrated that the two principles of two-force members stated previously after the definition must apply.

Procedure for Analyzing Pin-Connected Frames

We will follow the general procedure listed here for analyzing pin-connected frames. You must learn to adapt this procedure for any given arrangement of members.

1. First consider the entire frame as a free body and determine the reactions at its supports. Methods you learned from Chapter 3 should be used that call for applying the equations of equilibrium:
   a. Apply the principles described in Section 2–5 to represent the possible forces that can be exerted by different types of supports: *pin, roller, smooth surface, rough surface,* and *fixed.*
   b. $\Sigma M = 0$ about any point. Sum moments about one support point to solve for forces at the other support.
   c. $\Sigma F = 0$ in any direction. Sum forces in any direction to solve for unknown forces. Often the vertical and horizontal directions are convenient. At other times it is better to consider directions perpendicular and parallel to a given member.

2. Disassemble the frame into its component parts and prepare the complete free body diagram for each. Methods you learned earlier should be used, including
   a. Apply the principle of *action–reaction* to the force systems that exist at joints.

b. Review Section 2–6 on creating free body diagrams and apply those principles to each member.

c. Observe which members of the frame perform as two-force members for which the force in the member is either direct tension or direct compression, acting along the line between the two joints of the member.

d. Where possible, use logic and reasoning to predict the directions for unknown forces.

e. Trace the path of forces from the applied loads through the connected members to the supports.

This process is sometimes called *load tracing* or *following the load path* and it was used in the example problems of Section 2–6.

f. In general, start the analysis with two-force members if possible, then proceed to other members carrying simpler force systems, and finally to the more complex members. The forces determined from the simpler members become known forces on the members to which they are connected.

g. Summarize the results and show all forces on the free body diagrams in their proper directions.

Example Problem 4–1   Figure 4–5(a) shows a pin-connected frame carrying a load having a weight of 45 kN. Determine the forces on all members and joints.

Solution   **Objective:**   Find the forces on members *ABC* and *BD* and the reactions at supports *A* and *D*.

Given: Frame in Figure 4–5(a); $F = 45$ kN

Analysis: First consider the entire structure to be a free body as shown in part (b) of the figure. Each support is pinned so two forces (*x*- and *y*-components) are to be found at *A* and *D*.

Start by summing moments about support *D*. Notice that only the applied load *F* and the horizontal reaction at *A*, R_{Ax}, have moments about *D*. This allows us to solve for that force.

$$\Sigma M_D = 0 = (45\,\text{kN})(1500\,\text{mm}) - R_{Ax}(600\,\text{mm})$$

$$R_{Ax} = (67\,500\,\text{kN} \cdot \text{mm})/(600\,\text{mm}) = 112.5\,\text{kN} \leftarrow$$

FIGURE 4–5   Pin-connected frame for Example Problem 4–1.

Now we can see that there are only two horizontal forces so summing forces in the horizontal direction allows us to determine the *x*-component of the unknown reaction at *D*.

$$(+\rightarrow)\Sigma F_h = 0 = R_{Dx} - R_{Ax}$$
$$R_{Dx} = R_{Ax} = 112.5 \text{ kN} \rightarrow$$

The vertical components of the reactions cannot yet be found. Therefore, let's turn our attention to the horizontal beam *ABC*. The free body diagram is shown in Figure 4–5(c). The components acting at support *A* are shown including the known value of $R_{Ax} = 112.5$ kN.

Now we can sum moments about *B* to find R_{Ay}. Make note that R_{Ay} must act vertically downward to counteract the moment due to the applied force *F*.

$$\Sigma M_B = 0 = (45 \text{ kN})(600 \text{ mm}) - R_{Ay}(900 \text{ mm})$$
$$R_{Ay} = (27\,000 \text{ kN} \cdot \text{mm})/(900 \text{ mm}) = 30.0 \text{ kN} \downarrow$$

This allows us to compute the total reaction at support *A* because its horizontal and vertical components are now known.

$$R_A = \sqrt{R_{Ax}^2 + R_{Ay}^2} = \sqrt{(112.5 \text{ kN})^2 + (30.0 \text{ kN})^2} = 116.4 \text{ kN}$$

The angle of orientation of R_A is

$$\alpha = \text{Tan}^{-1}(R_{Ay}/R_{Ax}) = \text{Tan}^{-1}[(30.0 \text{ kN})/(112.5 \text{ kN})] = 14.9°$$

Return your attention to point *B*. A critical part of the free body diagram is that the *direction* of the force exerted at *B* by the member *BD* is known because *BD is a two-force member* as shown in part (d) of the figure. Member *BD* is in compression so it pushes back on pin *B* with a force acting along the line between *B* and *D*. Let's call this force *BD* and its components BD_x and BD_y.

We can sum forces in the vertical direction to find BD_y and we can sum forces in the horizontal direction to find BD_x.

$$(+\uparrow)\Sigma F_v = 0 = BD_y - F - R_{Ay}; \quad BD_y = 45.0 \text{ kN} + 30.0 \text{ kN} = 75 \text{ kN} \uparrow$$
$$(+\rightarrow)\Sigma F_h = 0 = BD_x - R_{Ax}; \quad \text{then } BD_x = R_{Ax} = 112.5 \text{ kN} \rightarrow$$

We can now find the total compressive force *BD* and the angle θ between *BD* and the horizontal.

$$BD = \sqrt{BD_x^2 + BD_y^2} = \sqrt{(112.5 \text{ kN})^2 + (75.0 \text{ kN})^2} = 135.2 \text{ kN}$$
$$\tan\theta = BD_y/BD_x = (75.0)/(112.5) = 0.667$$
$$\text{Then } \theta = \text{Tan}^{-1}(0.667) = 33.7°$$

An important observation should now be made. Because *BD* is a two-force member, the line of action of the force *BD* should be along the line between the two joints *B* and *D*. Let's use the geometry of the structure in part (a) of the figure to find the angle θ.

$$\tan\theta = (600 \text{ mm})/(900 \text{ mm}) = 0.667$$
$$\text{Then } \theta = \text{Tan}^{-1}(0.667) = 33.7°$$

This is equal to the angle of the direction of the force *BD* found before. This is a check on the solution. However, there are some analyses for which the observation of the direction of a two-force member is critical to the solution of the problem.

The problem is now completed by observing that the force system acting on the strut *BD* at pin *D* is the same as that acting at *B* but acting in the opposite direction as shown in Figure 4–5(d).

Summary and Comments: Summarized here are the pertinent results. See also the drawing of the complete frame and the free body diagrams of members shown in Figure 4–5.

Reactions: $R_{Ax} = 112.5$ kN $\leftarrow$; $R_{Ay} = 30.0$ kN $\downarrow$

$R_A = 116.4$ kN at 14.9° from horizontal

$R_{Dx} = 112.5$ kN $\rightarrow$; $R_{Dy} = 75.0$ kN $\uparrow$

$R_D = 135.2$ kN at 33.7° from horizontal

It is interesting to note that R_{Ax} and R_{Dx} form a couple and that the magnitude of the forces is inversely proportional to the spacing between A and D. A closer spacing would increase the forces while a larger spacing would decrease them.

Member ABC is a beam carrying the loads shown in Figure 4–5(c). The beam is bending, which will be studied in Chapter 12. The part from A to B also sees a tensile force of 112.5 kN, resulting in combined stresses in this part, which is discussed in Chapter 15.

Member BD is in direct compression and is a two-force member. The direct compressive stress is covered in Chapters 7 and 9. If the member is relatively long and slender, it must be analyzed as a column as discussed in Chapter 16.

The pins at A, B, and D are subjected to direct shearing stress (covered in Chapter 9) where the shearing forces are equal to the resultants at those joints:

Pin A: $R_A = 116.4$ kN   Pins B and D: Force $BD = 135.2$ kN

LESSON 6

Example Problem 4–2

Figure 4–6(a) shows a model of a crane with a pin-connected frame designed to carry a load of construction materials having a weight of 800 lb. The hydraulic actuator pushes up on the horizontal beam to assist in positioning the load on the building being constructed. The hoist at the end of the beam raises the load from a delivery truck and lowers it to the floor or roof of the building. Part (b) shows a schematic representation of the crane with pertinent dimensions. Determine the forces on all members and joints, including the supports A and B. Comment on the kinds of stresses produced in the members under these forces and how the choices of shape and dimensions of the members are influenced by the magnitude and directions of the forces acting on them. Recall the discussion of the kinds of stresses in Chapter 1 and the descriptions of standard structural shapes in Section 1–8 and shown in Table 1–7.

Solution **Objective:**   Find the forces on all members and the reactions at supports A and B.

Given: Frame in Figure 4–6; $F = 800$ lb

Analysis: First, to find the reactions at the supports, consider the entire structure to be a free body as shown in Figure 4–7(a). Next prepare free body diagrams of each structural member: cable AE, vertical post $BCDE$, beam DFG, and hydraulic actuator CF.

- Note that the cable AE and the hydraulic actuator CF are two-force members.
- The cable is in tension as it pulls back on the vertical post to hold it in position [Figure 4–7(b)].
- As the actuator CF pushes up on the beam at F, the beam exerts an equal and opposite force back on F, placing the actuator in compression [Figure 4–7(c)].

FIGURE 4–6   Crane with a pin-connected frame.

Source of model: Gary Drigel.

(a) Photograph of a model of the crane

(b) Schematic drawing of the crane with dimensions

FIGURE 4–7   Free body diagrams for crane in Example Problem 4–2.

- The combination of the applied load at G, the actuator force at B, and the reaction at pin D produces bending and axial tension in the beam [Figure 4–7(d)].
- The vertical post is subjected to a complex set of forces, some acting perpendicular to its axis and some axially, making it act as a beam and a column in compression simultaneously [Figure 4–7(e)].
- The several pin joints at A, B, C, D, E, and F act as clevises. The pins are subjected to direct shearing forces while the clevis components experience tension or compression and bearing forces between the pins and the holes through which they pass.

Support Reactions: Each support is pinned so two forces (x- and y-components) are to be found at A and B. At A, we know that the cable pulls upward and to the right at the angle θ so the support must exert reaction forces downward and to the left such that their resultant is in line with the cable force and acts in the opposite direction. See in the figure the detail of the force system acting at A. At B, there must be a horizontal reaction to the right to balance the leftward reaction force at A. Similarly, there must be a vertically upward reaction force at B to counteract the downward forces of the load and the reaction at A.

At A: Refer to Figure 4–7(a). Summing moments about B allows the calculation of the reaction force A_y.

$$\Sigma M_B = 0 = (800\text{ lb})(12\text{ ft}) - A_y(7\text{ ft})$$
$$A_y = (9600\text{ lb}\cdot\text{ft})/(7\text{ ft}) = 1371\text{ lb}\downarrow$$

Now use the force parallelogram that shows the resultant A acting at the angle θ to the left of vertical along with its components A_x and A_y. We can use this to find the relationship between the two components. First we find the value of the angle θ from the geometry of the crane:

$$\tan\theta = (7\text{ ft})/(13\text{ ft}) = 0.538; \quad \text{then } \theta = \text{Tan}^{-1}(0.538) = 28.3°$$

Now we note that

$$\tan\theta = A_x/A_y$$

Because we know the value of A_y, we can solve for A_x.

$$A_x = A_y\tan\theta = (1371\text{ lb})(0.538) = 738\text{ lb}\leftarrow$$

Now we can find the resultant A.

$$\cos\theta = A_y/A$$
$$A = A_y/\sin\theta = (1371)/\cos(28.3°) = 1557\text{ lb}$$

At B: To find the horizontal component B_x, we sum forces in the x-direction.

$$(+\rightarrow)\Sigma F_x = 0 = B_x - A_x$$
$$B_x = A_x = 738\text{ lb}\rightarrow$$

Similarly, for the vertical component B_y, we sum forces in the y-direction.

$$(+\uparrow)\Sigma F_y = 0 = B_y - 800\text{ lb} - A_y$$
$$B_y = A_y + 800\text{ lb} = 1371\text{ lb} + 800\text{ lb} = 2171\text{ lb}\uparrow$$

The resultant B can now be found, including its direction.

$$B = \sqrt{B_x^2 + B_y^2} = \sqrt{(738\text{ lb})^2 + (2171\text{ lb})^2} = 2293\text{ lb}$$
$$\tan\beta = B_x/B_y = (738\text{ lb})/(2171\text{ lb}) = 0.340$$
$$\text{Then } \beta = \text{Tan}^{-1}(0.340) = 18.8° \text{ between } B \text{ and the vertical}$$

Cable AE: The axial tensile force in the cable is equal to the resultant force A.

$$AE = A = 1557\text{ lb}$$

AE acts upward at an angle of $28.3°$ to the right of vertical at the joint E.

Beam DFG: See the free body diagram in part (d). We know the direction of the force F to be in line with the actuator CF at an angle, α, below the horizontal. From the geometry of the frame we find,

$$\alpha = \text{Tan}^{-1}(5\text{ ft}/8\text{ ft}) = 32.0°$$

Forces at F: We can sum moments about D to find F_y.

$$\Sigma M_D = 0 = (800\text{ lb})(12\text{ ft}) - F_y(8\text{ ft})$$
$$F_y = (9600\text{ lb}\cdot\text{ft})/(8\text{ ft}) = 1200\text{ lb}\uparrow$$

The horizontal component of F can now be found.

$$\tan\alpha = F_y/F_x$$
$$F_x = F_y/\tan\alpha = (1200\text{ lb})/(\tan 32.0°) = 1920\text{ lb}\rightarrow$$

The resultant F can now be found.

$$\sin\alpha = F_y/F$$
$$F = F_y/\sin\alpha = (1200\text{ lb})/\sin 32.0° = 2264\text{ lb}$$

This is also the magnitude of the force on the actuator.

Forces at D: Summing forces in the horizontal direction:

$$(+\rightarrow)\Sigma F_x = 0 = F_x - D_x$$
$$D_x = F_x = 1920\,\text{lb} \leftarrow$$

Summing forces in the vertical direction:

$$(+\uparrow)\Sigma F_y = 0 = F_y - 800\,\text{lb} - D_y$$
$$D_y = F_y - 800\,\text{lb} = 1200\,\text{lb} - 800\,\text{lb} = 400\,\text{lb}\downarrow$$

The resultant at *D* can now be found.

$$D = \sqrt{D_x^2 + D_y^2} = \sqrt{(1920\,\text{lb})^2 + (400\,\text{lb})^2} = 1961\,\text{lb}$$
$$\tan\gamma = D_x/D_y = (1920\,\text{lb})/(400\,\text{lb}) = 4.80$$

Then $\gamma = \text{Tan}^{-1}(4.80) = 78.2°$.

Member CF: This is the hydraulic actuator that is subjected to an axial compressive force equal to the force on pin *F*.

$$CF = F = 2264\text{-lb direct compression}$$

Member BCDE: Even though the force system on this member seems complex, note that all have been found already. We only need to apply the principle of action–reaction to each connection point as summarized next.

At A: The force *AE* is the cable force of 1557 lb, acting down and to the left at 28.3° to the vertical. The *x*- and *y*-components are the same as those acting at *B*.

$$E_x = A_x = 738\,\text{lb} \leftarrow \qquad E_y = A_y = 1371\,\text{lb}\downarrow$$

At D: These reaction forces are the reactions to those acting on the left end of the member *DFG*. Only the directions change, acting in the opposite directions.

$$D_x = 1920\,\text{lb} \rightarrow \qquad D_y = 400\,\text{lb}\uparrow$$

At C: Here the lower part of the actuator pushes down and to the left on the vertical column.

$$C_y = 1200\,\text{lb}\downarrow \qquad C_x = 1920\,\text{lb} \leftarrow$$

At B: These components are the identical forces found in the second part of the solution in which we found the forces exerted by support *B on the structure.*

$$B_x = 738\,\text{lb} \rightarrow \qquad B_y = 2171\,\text{lb}\uparrow$$

Comments: Here we comment on the kinds of stresses produced in the members under the forces we found in the analysis. We also comment on how the choice of shape and dimensions of the members are influenced by the magnitude and directions of the forces acting on them. The discussion is qualitative at this time, but it provides connections to the study you will be doing in later chapters of this book.

Cable AE: A steel cable is an appropriate choice to carry the axial tensile force of 1557 lb. A suitable diameter for the cable depends on the strength of the steel from which the cable is made and the desired design factor. Some designers apply a factor of four to the applied load to account for accidental overloads, wear, or abuse during operation and select a cable size with a breaking strength equal to that amount.

$$\text{Breaking strength} = (4)(1557\,\text{lb}) = 6228\,\text{lb}$$

One possible choice, found from Internet site 1 at the end of this chapter, is a 1/4-in diameter, galvanized steel cable with a breaking strength of 7000 lb. Note that this seems relatively small, demonstrating that direct tensile forces are managed fairly easily.

Hydraulic Actuator CF: An actuator must be selected that is capable of exerting a pushing force on pin *F* of 2264 lb as it extends to position the beam. Clevis-type pin joints are shown at *C* and *F* to allow the actuator to pivot as the beam swings up and down. Design decisions for the actuator involve the diameter of the hollow cylindrical body into which pressurized oil is delivered by a pump at a given pressure and the diameter of the solid circular rod. The pressure acts on the face of a piston inside the cylinder, exerting a force on the piston rod that extends out of the cylinder. In addition, the rod

FIGURE 4–8 Typical hydraulic cylinder design.

must be designed to carry the compressive force, considering it to be a column subjected to possible buckling as discussed in Chapter 16. The hollow cylinder and the solid piston rod are good choices for such an application. Internet site 2 shows some examples of clevis-mounted hydraulic cylinders. Because of the large nominal length (about 9.5 ft or 2.9 m) of the cylinder in this problem, special attention to the column loading is required. Figure 4–8 shows a type of cylinder that is appropriate for this application.

Beam DFG: This member is loaded as a beam in bending with additional axial tensile force between D and G. In Chapter 11, you will learn how to compute bending moments at any point on the beam created by the forces determined in this problem and to show that the maximum moment occurs at joint F. You will discover in Chapter 12 that the I-beam shape shown in Figure 4–6 is an efficient choice in terms of the weight of the beam relative to the applied load and the resulting bending moments. The orientation shown is preferred for bending in the plane of the drawing with the web vertical and the flanges horizontal. Table 1–7 shows two types of steel I-beam shapes, the wide-flange W-shape and the American Standard S-shape. A modest selection of sizes and their properties is shown in Appendices A–7 and A–8. Aluminum I-shapes are also good choices and they are shown in Appendix A–11. Internet sites 17 and 18 in Chapter 1 are sources of alternate smaller sizes of I-shapes in steel and aluminum.

Special attention must be paid to how the actuator rod is attached to the beam at F to ensure that it is able to transfer the force without significantly weakening the beam, particularly because that is where the maximum moment occurs. The approach shown uses a filler block attached to the web of the beam with the pinhole machined through the block. The filler block reinforces the web and provides added thickness to stabilize the pin and a larger bearing area for the pin as the load is transferred from the beam to the clevis on the actuator.

Similarly, the clevis-type attachment at D at the left end must be designed. The design shown in the figure has the clevis ears attached to the web of the beam. The pin at joint D passes through the vertical post $BCDE$, described next.

The right end of the beam carries the hoist that allows the applied load to be raised and lowered. The hoist should be mounted such that it can swivel as the beam is raised and lowered and the mounting must be designed to handle the maximum expected applied load, which is 800 lb in this problem.

Vertical Post BCDE: The presence of the forces at each joint that act perpendicular to the long axis of the post indicates that it must be designed as a beam in bending as discussed in Chapter 12. You will learn in Chapter 11 to compute the resulting bending moments and to show that the maximum moment occurs at joint C. The axial compressive forces must also be considered because they cause compressive stresses and the possibility of column buckling. The highest compressive force, 2171 lb, exists in the lower part from C to B.

Although the primary strength is required in the plane of the drawing due to the bending loads, you must also anticipate moderate side loads acting perpendicular to the plane of the drawing due to unbalanced loading, wind forces, and incidental forces applied by operators or other equipment. In fact, the manner of providing stability against these side loads must be defined and is not completed in this problem.

However, the design for the post, shown to be a hollow rectangular steel tube, is a good choice. Figure 1–6 shows that such a tube is called a *hollow structural section*, HSS, and standard sizes are shown in Appendix A–9. Smaller sizes are available from suppliers such as those listed at Internet sites 17 and 18 in Chapter 1. The wider side of the rectangle is placed in the plane of the drawing to resist the bending stresses and the axial compressive stresses. The narrower width implies that side loads are smaller than the applied loads, and this must be checked. When substantial side loads may be encountered, the vertical post may be made from a square tube.

Pin Joints: The pin-type attachments at points B, C, D, and E must be carefully designed to provide for inserting the pin while also resisting the bending moments and axial forces. The design approach shown here is to install hollow cylindrical bushings into holes in the sides of the tube at the centerline and welding them in place. That provides good force transfer to the fairly thin walls of the tube while providing good stability as the members rotate relative to each other.

Example Problem 4–3   Figure 4–9 shows a scissors lift table in which a hydraulic cylinder pulls on the movable lower pivot A that rolls on the lower frame. As movement proceeds, the link BEC rotates about fixed pivot B and link AED also rotates while A moves to the right. The table thus rises while remaining level. Pivot D is fixed under the table and pivot C rolls on the lower surface of the table. Links AED and BEC are pinned together where they cross. The complete table includes two identical scissors mechanisms for which the rated capacity is 2000 lb. Therefore, consider the load on one scissors to be 1000 lb as shown. For the loading condition and position shown, compute the force required by the hydraulic cylinder to hold the load or move it slowly. Also, prepare complete free body diagrams of all components. Commercially available scissors lifts of a variety of designs and actuation methods can be viewed at Internet sites 3 and 4.

Solution   **Objective:**   Find the hydraulic cylinder force and prepare free body diagrams of all components for the scissors lift table.

Given: Scissors lift in Figure 4–9; $W = 1000$ lb placed as shown.

Analysis: **Reactions:** First, to find the reactions at the supports, consider the entire structure to be a free body and sum moments about A.

$$\Sigma M_A = 0 = (1000\ \text{lb})(14\ \text{in}) - R_{Ry}(43\ \text{in})$$
$$R_{Ry} = (14\,000\ \text{lb} \cdot \text{in})/(43\ \text{in}) = 326\ \text{lb}\uparrow$$
$$\Sigma M_B = 0 = (1000\ \text{lb})(29\ \text{in}) - R_{Ly}(43\ \text{in})$$
$$R_{Ly} = (29\,000\ \text{lb} \cdot \text{in})/(43\ \text{in}) = 674\ \text{lb}\uparrow$$

Because there are no applied loads in the horizontal direction, support reactions in the x-direction are zero.

Free Body Diagrams: Prepare *on your own* the free body diagrams of each component of the scissors lift using the principles demonstrated in Example Problems 4–1 and 4–2.

FIGURE 4–9   Scissors lift for Example Problem 4–3.

(a) Free body diagram of table

(b) Free body diagram of link *BEC*

(c) Free body diagram of link *AED*

(d) Free body diagram of hydraulic cylinder

Dimensions are in inches

FIGURE 4–10   Free body diagrams of scissors lift components.

Check your work against Figure 4–10. Now let's work through the analyses of forces, using the free body diagrams in the following order: 1—Table; 2—Link *BEC*; 3—Link *AED*; 4—Actuator.

Table: See Figure 4–10(a). The loading and support conditions are identical to those acting on the complete lift used to find reactions. Therefore,

$$C_y = R_{Ly} = 674 \text{ lb} \uparrow \qquad D_y = R_{Ry} = 326 \text{ lb} \uparrow$$

Link BEC: See Figure 4–10(b). Let's start by summing moments about E to enable calculation of the horizontal force B_x.

$$\Sigma M_E = 0 = (674 \text{ lb})(21.5 \text{ in}) + (326 \text{ lb})(21.5 \text{ in}) - B_x(16.5 \text{ in})$$
$$B_x = (21\,500 \text{ lb} \cdot \text{in})/(16.5 \text{ in}) = 1303 \text{ lb} \leftarrow$$

Now sum forces in the horizontal direction to find E_x.

$$(+\rightarrow)\Sigma F_h = 0 = E_x - B_x$$
$$E_x = B_x = 1303 \text{ lb} \rightarrow$$

Now sum forces in the vertical direction to find E_y.

$$(+\uparrow)\Sigma F_v = 0 = 326 \text{ lb} + E_y - 674 \text{ lb}$$
$$E_y = 348 \text{ lb} \uparrow$$

Link AED: See Figure 4–10(c). Several of the listed forces are transferred from earlier analyses applying the principle of action–reaction. The only unknown is the horizontal force at pivot *A*. Let's sum forces in the horizontal direction to find that.

$$(+\rightarrow)\Sigma F_h = 0 = A_x - E_x$$
$$A_x = E_x = 1303\,\text{lb}\rightarrow$$

Hydraulic Cylinder: The cylinder applies the 1303-lb force by pulling on joint *A*. Therefore, joint *A* pulls back on the cylinder rod with an equal and oppositely directed force placing the rod in tension. The reaction force to the right on the cylinder body is resisted by the way it is fastened to the base.

Comment: This completes the problem as stated. However, you can discover an interesting feature of the design of a scissors lift by examining the analysis for the horizontal force at pivot *B*. Note that its moment arm with respect to joint *E*, 16.5 inches in the configuration shown, is in the denominator of the final calculation for B_x where it was determined that the force is 1303 lb. Note also that this is the magnitude of the force required by the hydraulic cylinder to maintain the table in position or to raise it slowly. Now consider what happens when the table is lowered. *The moment arm decreases and, therefore, the force required by the cylinder will increase.* Conversely, as the table is elevated, the force required by the cylinder will decrease. A useful exercise would be to analyze the forces on the different members and the hydraulic cylinder at various elevations of the table and prepare a chart or graph of those forces. Such an analysis would be essential to the final design of the lift.

The table is loaded and supported as a beam where the applied force is reacted by the two support points at *C* and *D*. Significant stiffening should be applied to the flat plate of the tabletop to maintain a flat surface under load. The stiffening should be applied to the edges and across the center portion of the underside.

Links *BEC* and *AED* are subjected to a combination of bending stresses due to components of the forces acting perpendicular to their long axes and axial stresses due to the components acting along the axis.

The hydraulic cylinder and its piston rod are in tension under the conditions shown as it pulls roller *A* toward the right. The cylinder must exert 1303 lb of force to maintain the scissors mechanism in the position shown while holding a load of 1000 lb.

4–3 ANALYSIS OF TRUSSES

One type of truss was shown in Figure 4–1(b). Although many configurations can be created, the following general properties can be observed:

1. A truss is made from a series of pin-connected members.
2. Loads and reactions are applied only through joints.
3. Load-carrying members are all two-force members subjected to either direct tension or direct compression.
4. Tensile members can often be made from simple bars or cables as discussed in Chapters 7 and 9.
5. Compression members are typically long and slender and must be analyzed as columns as discussed in Chapter 16.
6. The resulting design of compression members will be relatively larger than the tension members and attention must be paid to the shape of the cross section to give the member rather high stiffness to resist buckling.

Why Choose a Truss?

Trusses are typically used when loads must be carried over relatively long spans between supports. Designs of trusses cause the applied loads to be distributed through members around the periphery of the structure in an efficient manner, resulting in very light structures for a given loading pattern.

Savings in material costs are often significant as compared with alternate means of carrying the loads. Also, the lighter structure transfers lower reaction forces at supports.

Where Are Trusses Typically Used?

Try to visualize where you have seen the trusses used in applications like the following:

a. **Residential and Light Commercial Roof Structures:** Look for buildings under construction that have sloped roofs and you will likely see trusses carrying the roof loads and transferring them to the walls and eventually to the foundation of the building. The trusses may be made from standard wood shapes, such as those listed in Appendix A–4, or from lightweight steel members formed as channels or angles. Figure 4–11(a) shows an example. See also Internet site 5.

b. **Large Buildings Having Wide Open Spans:** Examples are sports arenas, convention centers, industrial buildings, shopping centers, and individual "big-box" stores. Such buildings often have flat roofs with small slopes that serve to direct rainwater to the edges for disposal away from the building. The long spans produce large open interior spaces. The roofs of such buildings are typically supported by trusses as illustrated in Figure 4–11(b). See also Internet site 6.

(a)

FIGURE 4–11(a) Wood frame house with trusses making up the roof structure.

Source: Jordan Long, Author, Nascor International Ltd., Calgary, Alberta, Canada.

c. Structures Supporting Signs over Highways: Expressways, bridges, tunnels, and other roadways are equipped with large signs to give drivers adequate advice on the choice of routes and exits. Where signs are suspended over the roadway you will often see trusses used because of the large span and the desire to make the structures light but with adequate strength and rigidity. See Figure 4–11(c) for an example. See also Internet site 7.

d. Bridges: Whether used for cars and trucks, railroads, or pedestrians, bridges of relatively long spans are often made from trusses to keep the overall structure weight low. Figure 4–11(d) shows an example. See also Internet site 8.

e. Stage Lighting and Display Systems: See Figure 4–11(e). Have you seen such structures at a concert, trade show

(b)

FIGURE 4–11(b) Parallel chord roof truss for a commercial building.

Source: Commander Buildings, Inc., Monticello, IA.

(c)

FIGURE 4–11(c) Truss structure supporting direction signs over an interstate highway.

Source: AARoads (Alex Nitzman, Andy Field, Justin Cozart), San Diego, CA.

exhibition, or other special event? They are lightweight, versatile, attractive, easily handled, assembled, disassembled, and transported. See also Internet site 9.

f. Aircraft Fuselages and Aerospace Structures: Lightweight and stiff structures are essential to good design of aircraft and space vehicles. Many will incorporate trusses to achieve that goal as shown in Figure 4–11(f) and Internet site 10. Another concept called a *skin-stiffened structure* or *monocoque structure* is also used in such applications.

g. Race Cars: Some race cars are made from open-space truss structures that provide structural strength, rigidity, and light weight with a sheet-metal or fiberglass body attached to provide low drag. See Figure 4–11(g) and Internet site 11.

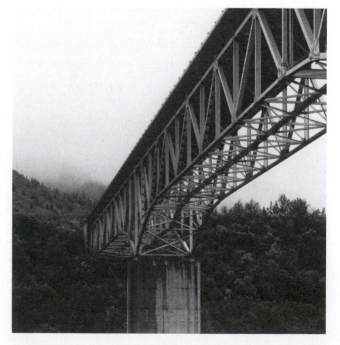

(d)

FIGURE 4–11(d) Truss bridge.

(e)

FIGURE 4–11(e) Entertainment stage structure with roof; truss tower unit.
Source: Total Structures, Inc., Ventura, CA.

(a) Truss-type fuselage frame

(b) Complete aircraft in flight

(c) Cutaway showing structure and placement of engine and other equipment

FIGURE 4–11(f) Pitts Model 12 aerobatic airplane with truss-type fuselage frame and trusses in wings.
Source: (a) and (b) Courtesy of Jim Kimball Enterprises, Inc., Zellwood, FL; (c) Tom Johnson, TJ TechArt, Inc., Portland, OR.

(g)

FIGURE 4–11(g)   Welded steel truss-type frame for a race car.

Source: Classic Craft Motorsports, Inc., Springfield, OH
Photograph by Robert L. Mott, Jr.

h. **Supports for Utilities Running over Machinery, Walkways, or Roadways:** Many are used within factories, where trucks or trains are loaded, or where large piping must extend over obstacles above ground; so look for them when you are at work, driving in an urban area, or taking a tour of an industrial facility. Figure 4–11(h) shows such a structure at an industrial site.

Types of Trusses:

Commonly seen trusses used in buildings, bridges, and machinery often employ one of the basic arrangements shown in Figure 4–12. Manufacturers in every region offer stan-

(h)

FIGURE 4–11(h)   Truss supporting industrial utility lines and piping.

dard and custom-designed trusses for architects and building contractors. (See Internet site 13.) A creative designer can adapt these basic styles to fit a particular project. Note the implementation of triangulation in all of the truss styles shown. Some observations of these styles follow:

- Typical sloped roof trusses are the *king post, queen post, Fink, Howe, fan, scissor, double W,* and *open plan.*

- The angle of the slope is specified by the architect and structural engineer to resist loads due to the deadweight of the roof structure, snow, and wind. The slope is often stated in terms of *rise over run,* such as

  i. A 5/12 slope rises 5 inches over a horizontal length of 12 inches.

  ii. If needed, the angle of inclination of the roof, θ, can be computed from basic trigonometry, illustrated here for a 5/12 slope:

  $$\theta = \text{Tan}^{-1}(\text{slope}) = \text{Tan}^{-1}(5/12) = 22.6°$$

  iii. Conventional residential roof slopes are typically in the range from 4/12 to 9/12 (18.4° to 36.9°).

  iv. Steeper roofs, up to about 12/12 (45°), are used in cold climates where heavy snow loads are expected, whereas in warmer climates lower slopes or even flat roofs are used.

- The *dual pitch* or *asymmetric* trusses can be used for architectural interest or for special interior space reasons.

- The *clerestory* design allows transparent or translucent windows to be installed in the vertical plane near the peak to allow light into the middle of the enclosed space.

- Extensions to a main building may call for single-sided trusses such as the *mono A* or *mono B* styles.

- Flat or low-sloped roofs often use the *Pratt, Warren, parallel chord,* or *low-profile designs.*

- The *vaulted parallel chord* truss is similar to the scissor truss except it offers greater strength and rigidity for heavy snow or wind loads.

- The *flat top* truss offers design flexibility and may provide space for roof-mounted heating, ventilating, and air conditioning (HVAC) equipment.

- Long spans, such as may be found in sports arenas or factories, may use more complex truss shapes such as the *bowstring* or *gambrel* styles. These are also used in highway and railroad bridges.

Fabrication of trusses from standard wood shapes typically uses metal or plywood gussets at joints to facilitate the transfer of loads among the several component members of the truss. The gussets add stiffness to the truss and they must be capable of resisting the tensile or compressive forces in the truss members.

Trusses made from steel, aluminum, and other metals are often welded assemblies. Some larger trusses may be made from shop-fabricated subassemblies that are shipped to a building site and bolted together there.

Support points for trusses must be carefully specified to ensure that loads are applied only to joints to ensure that

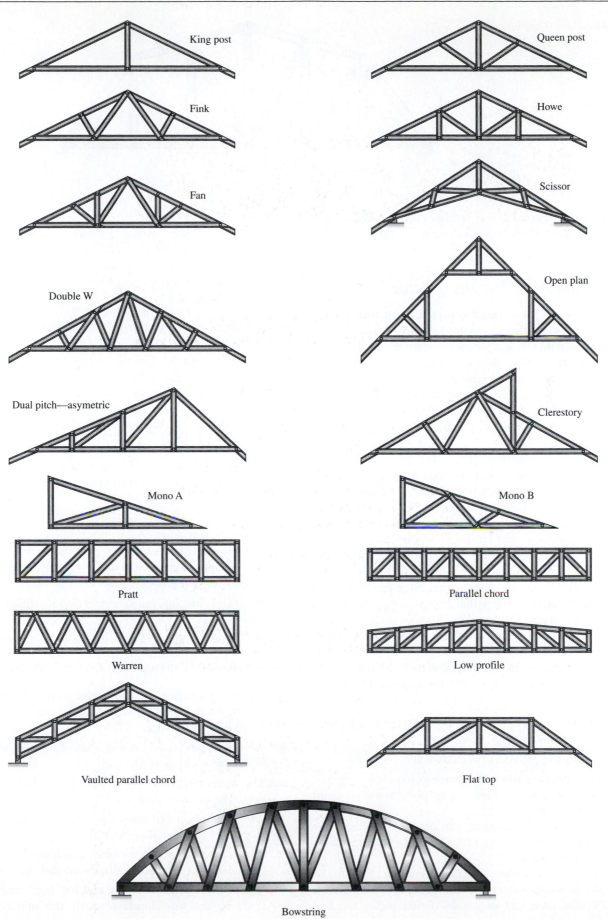

FIGURE 4–12   Types of trusses.

FIGURE 4–12 (*Continued*)

Gambrel

members are subjected to only tensile or compressive stresses. Otherwise, the analysis methods described in this section will not apply.

Arrangements of Trusses in Structures

If a single planar truss is used as part of a structure, the designer must ensure that it remains in the preferred orientation and is stable. Long, slender trusses are likely to fall over or buckle unless they are made very stiff or have braces to provide lateral stability. These comments also apply when a single truss is handled during fabrication of the structure even though it will be part of a larger structural system when finally in place. See Internet site 12 for safety guidelines.

Normally, two or more trusses are assembled together in parallel and tied together to provide the necessary stability. Look at the example shown in Figure 4–11(a) for a roof system of a wood frame home. The roof trusses are set at spacings typically approximately 24 in on center (600 mm). During construction, the trusses are set in place and temporarily braced against falling. As the sheathing is placed on the trusses, the system is tied together and remains stable under applied loads. Metal trusses are often assembled into square or triangular shapes that provide inherent stability when loaded in any direction. See Figures 4–11(c)–(h). Systems of such assemblies are often called *space frames*.

One method of providing stability to a pair of trusses can be seen in Figure 4–13 that diagrams the components of a through truss as might be used for highway or railroad bridges.

- The two main load-carrying trusses run lengthwise along the sides of the bridge.
- The main components of each truss are
  - Each primary triangular shape comprising a *vertical post*, a *diagonal*, and either a *top chord* or *bottom chord* is called a *panel*.
  - *Bottom chord* runs along the complete length of the lower part of the truss and connects to other members at panel points.
  - *Top chord* runs along the complete length of the top of the truss and connects to other members at panel points.
  - *Vertical posts* and *diagonals* connect to the top and bottom chords at panel points.

- *End posts*, when used, form the angled shape of the end panels. Some trusses have equal-length top and bottom chords and, therefore, have no end posts.
  - *Joints* or *connections* at panel points are often tied together with *gussets*, particularly for wood and structural metal shapes.

- *Floor beams* run perpendicular to the bottom chords of the trusses at major joints.
- *Stringers* connect the floor beams together and run lengthwise along the bridge.
- The bridge *deck* is laid on top of the floor beams and stringers.
- The top chords of the two trusses are tied together by *portal struts*.
- *Portal bracing* is often added under the portal struts to provide additional stiffness and to maintain the squareness of each portal.
- Cross bracing, called *top laterals*, lies in the horizontal plane of the top chords and connects to each panel point, providing stiffness to maintain the shape of the bridge against racking forces that tend to distort it. This is sometimes called *sway bracing*.
- Similar cross bracing, called *bottom laterals*, stiffens the bottom.
- *Supports* are typically placed at each end of each main truss.

Follow the load path through the members of the truss.

a. The main loading on the bridge comes from cars, trucks, trains, buses, bicycles, motorcycles, and pedestrians using the bridge.

b. Another major load is the dead load of the deck and the bridge members themselves.

c. The loads are transferred to the deck, which is supported by the floor beams and the stringers.

d. The floor beams transfer loads to the panel points along the bottom of the bridge.

e. The lower panel points distribute the loads to the bottom chords, vertical posts, diagonals, and end posts.

f. The vertical posts, diagonals, and end posts distribute loads to the top chords through the upper panel points.

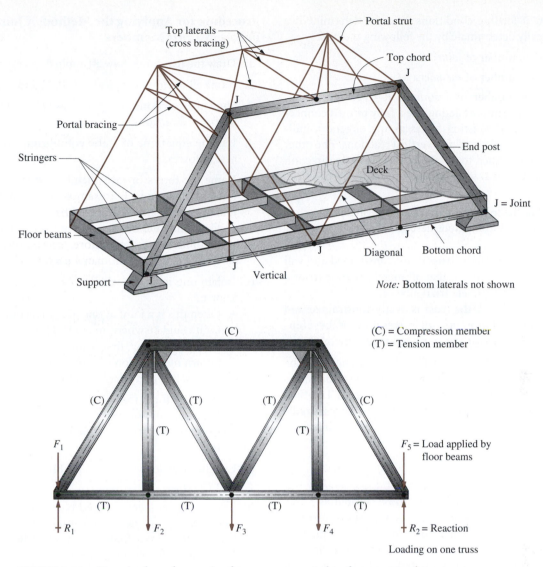

FIGURE 4–13   Terminology for a pair of trusses connected to form a complete structure.

g. Wind loads and other incidental loads may act on the bridge horizontally or in any direction, causing the tendency for racking. Such forces are resisted by the laterals and portal bracing.

h. Ultimately, the loads from the bridge are transferred through the supports to foundations and thus to the earth.

i. The supports and the foundation must provide for changes in length of the truss system due to deformations of the members under load and thermal expansion and contraction.

j. In the truss shown and with the given loading, note that the top chord and the end posts are in compression and all other members (bottom chord, verticals, and diagonals) are in tension. Other loading patterns may have different results but it is important to determine this information. The analysis methods discussed next give the tools necessary to make the calculations.

You should apply this type of thinking to the load path through other types of truss systems such as those described earlier and shown in Figure 4–11.

Methods of Analyzing Forces in Truss Members

Two primary methods of analysis are used to determine the forces in truss members, the *method of joints* and the *method of sections*. Both methods are described next and demonstrated through example problems. Either analysis method requires that the truss be *statically determinate*, indicating that there are sufficient members to maintain the shape of the structure by the inherent action of triangulation. There should be no extra members that provide redundancy because accurate computation of the forces in members cannot then be completed. However, under certain special loading situations, one or more members of a statically determinate truss may carry no loads and they are called *zero-force members*. Such members are still necessary and will carry loads under different loading conditions. You can determine if a truss is statically determinate by the following method:

1. Count the number of *joints, J*.
2. Count the number of *members, M*.
3. Count the number of *restraints, R*. Typically $R = 3$, based on one support acting like a fully pinned connection capable of resisting forces in any direction, indicated by the horizontal and vertical reactions. The other support is designed to resist loads in only one direction (typically upward) with no restraint in the perpendicular direction.
4. Verify that the following equation is satisfied:

$$M + R = 2J \qquad (4\text{--}1)$$

   a. If $M + R < 2J$, the truss is not constrained and will not be stable. Ensure that all members are part of a triangle for complete triangulation.
   b. If $M + R > 2J$, the truss is overly constrained and complete solution for all forces cannot be completed. Remove one or more redundant members.

It is critical that the force analysis indicates not only the magnitude of forces in a given member but also the direction of the forces, either tension or compression. The final design of members to resist these two kinds of forces is radically different because of the buckling type of failure of compression members.

Method of Joints

This method calls for a single joint to be isolated as a free body with all forces acting on it shown. At least one force on the joint must be known. Then you can apply the equations of equilibrium in the vertical and horizontal directions to find the unknown forces. The newly found forces are then known and can be used as inputs to the analyses of other joints to which a given member is connected. This process is repeated for other joints until the forces in all members are known. We can summarize this process into a general procedure that must be adapted to each truss design, loading pattern, and support system.

Procedure for Applying the Method of Joints to Find Forces in Truss Members

1. Draw the truss and show all applied forces.
2. Name all the joints, typically *A, B, C*, and so forth.
3. Show the reactions at supports as vectors in their most probable directions.
4. Use the equations of static equilibrium to find all reaction forces.
5. Name the forces in each member by the letters of the joints at their ends. For example, the member between joints *A* and *B* should be called *AB*.
   a. Determine the number of members, *M*, the number of joints, *J*, and the number of restraints, *R*.
   b. Ensure static determinacy such that $M + R = 2J$.
6. Isolate one joint for which there is at least one force known.
   a. Often this is a joint at one of the reactions.
   b. Or it could be where an applied force acts.
   c. If possible select first a joint having only two unknown member forces acting on it.
   d. Draw a free body diagram of the joint.
7. Resolve any force that acts at an angle into horizontal and vertical components.
   a. Compute necessary angles from the geometry of the truss.
8. Use the equations of static equilibrium to find all unknown forces.
9. Next select another joint and repeat this process. Typically it is best to move sequentially through the joints because the forces found from an earlier analysis become known forces for adjacent joints.
10. Continue the process until all forces are known. Sometimes symmetry allows the process to be shortened so that not all joints need to be analyzed.
11. Make a table summarizing the forces on all members, being careful to list both magnitude and direction, either tension (T) or compression (C).

Example Problem 4–4

Figure 4–14 shows a truss that is part of a roof structure for a commercial building. The flat part carries a platform for HVAC equipment that transmits three downward loads of 10.0 kN each through its supports to joints *B, D*, and *E*. The horizontal force, F_4, is zero for this problem. Use the method of joints to determine the forces in each member of the truss.

Solution   Given: Truss and loading as shown in Figure 4–14. $F_4 = 0$.

$$F_1 = F_2 = F_3 = 10.0 \, \text{kN}$$

Find: Forces in each member of the truss using the method of joints.

Use the procedure outlined earlier in this section.

Analysis and Results: ***Steps 1 and 2:*** Completed in the figure.

Steps 3: Vertical reactions act at points *A* and *F*. Support *A* is capable of reacting to horizontal loads. Therefore, there are three restraints; $R = 3$.

FIGURE 4–14 Truss and loading for Example Problem 4–4.

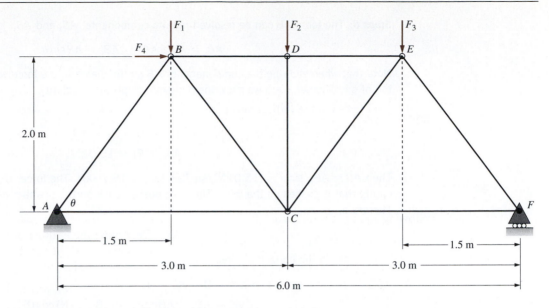

Steps 4: The total applied load is 30.0 kN downward with no horizontal forces or components. Because of symmetry,

$$R_1 = R_2 = (30.0 \text{ kN})/2 = 15.0 \text{ kN} \uparrow$$

Steps 5: There are nine members ($M = 9$) whose forces we will name:

$$AB \quad AC \quad BC \quad BD \quad CD \quad CE \quad DE \quad CF \quad EF$$

There are six joints; $J = 6$. From Step 3, we found the number of restraints, $R = 3$. Now we can check for static determinacy:

$$M + R = 9 + 3 = 12 \qquad 2J = 2(6) = 12$$

Therefore, the truss is statically determinate and we can proceed.

Steps 6: Let's start with joint A where the reaction force is known to be $R_1 = 15.0 \text{ kN} \uparrow$ and there are two unknown forces, AB and AC. Figure 4–15(a) shows the free body diagram.

Steps 7: Member AB is inclined at the angle θ from the horizontal. From the geometry of the truss,

$$\theta = \text{Tan}^{-1}(2/1.5) = 53.13°$$

(a) Free body diagram—Joint A

(b) Free body diagram—Joint B

(c) Free body diagram—Joint D

FIGURE 4–15 Free body diagrams of joints for Example Problem 4–4.

Steps 8: The force AB can be resolved into its components, AB_x and AB_y.

$$AB_x = AB \cos\theta \qquad AB_y = AB \sin\theta$$

Note that when showing the components of AB on the diagram, the force vectors AB_x and AC lay on top of one another. Here we show them slightly displaced for clarity.

Sum forces in the vertical direction:

$$+\uparrow \Sigma F_y = 0 = R_1 - AB_y$$
$$AB_y = R_1 = 15.0\,\text{kN}\downarrow$$

Then $AB = AB_y/\sin\theta = (15.0\,\text{kN})/\sin(53.13°) = 18.75\,\text{kN}$. The force AB acts toward joint A indicating that it pushes on the joint. The joint pushes back on the member placing it in compression. Therefore,

$$\underline{AB = 18.75\,\text{kN (C)}}$$

Sum forces in the horizontal direction:

$$+\rightarrow \Sigma F_x = 0 = AC - AB_x$$
$$AC = AB_x = AB\cos\theta = (18.75\,\text{kN})\cos(53.13°)$$
$$= 11.25\,\text{kN} \rightarrow$$

The force AC acts away from joint A indicating that it pulls on the joint. The joint pulls back on the member placing it in tension. Therefore,

$$\underline{AC = 11.25\,\text{kN (T)}}$$

Steps 9: Let's move to joint B next. The free body diagram is shown in Figure 4–15(b). The $F_1 = 10.0\,\text{kN}$ applied load and the compressive force $AB = 18.75\,\text{kN}$ are known and the forces BD and BC are to be found. First resolve forces into x- and y-components:

$$AB_x = 11.25\,\text{kN} \rightarrow \text{as found earlier}$$
$$AB_y = 15.0\,\text{kN}\uparrow \text{as found earlier}$$
$$BC_x = BC\cos\theta \quad BC_y = BC\sin\theta$$

Sum forces in the vertical direction:

$$+\uparrow \Sigma F_y = 0 = AB_y - 10.0\,\text{kN} - BC_y = 15.0\,\text{kN} - 10.0\,\text{kN} - BC_y$$
$$BC_y = 5.0\,\text{kN}\downarrow = BC\sin\theta$$

Then $BC = BC_y/\sin\theta = (5.0\,\text{kN})/\sin(53.13°) = \underline{6.25\,\text{kN (T)} = BC.}$

Now sum forces in the horizontal direction:

$$+\rightarrow \Sigma F_x = 0 = AB_x + BC_x - BD$$

But $BC_x = BC\cos\theta = (6.25\,\text{kN})\cos(53.13°) = 3.75\,\text{kN} \rightarrow$

Then $BD = AB_x + BC_x = 11.25\,\text{kN} + 3.75\,\text{kN} = \underline{15.0\,\text{kN (C)} = BD.}$

This completes the force analysis at joint B.

Now let's move to joint D, shown in Figure 4–15(c). All forces act in either the vertical or horizontal directions.

$$BD = 15.0\,\text{kN} \rightarrow \text{and must be balanced by } DE$$
$$F_2 = 10.0\,\text{kN}\downarrow \text{and must be balanced by } CD$$

First sum forces in the horizontal direction:

$$+\rightarrow \Sigma F_x = 0 = BD - DE$$
$$BD = DE = \underline{15.0\,\text{kN (C)}}$$

Sum forces in the vertical direction:

$$+\uparrow \Sigma F_y = 0 = CD - 10.0\,\text{kN}$$
$$CD = \underline{10.0\,\text{kN (C)}}$$

Steps 10 and 11: We can now take advantage of symmetry in both the geometry of the truss and the loading pattern. Note that all of the calculation procedures completed thus far could also have been applied starting at joint *F* and moving toward the left. Thus, we can conclude

$$AB = EF = 18.75 \text{ kN (C)}$$
$$AC = CF = 11.25 \text{ kN (T)}$$
$$BC = CE = 6.25 \text{ kN (T)}$$
$$BD = DE = 15.0 \text{ kN (C)}$$
$$CD = 10.0 \text{ kN (C)}$$

Comment: This list summarizes all member forces in the truss.

Example Problem 4–5

In Example Problem 4–4, the forces applied to joints *B*, *D*, and *E* as shown in Figure 4–14 were each equal to 10.0 kN. Examine how the forces in all members of the truss change if the applied forces were unequal:

$$F_1 = 15.0 \text{ kN} \downarrow \quad F_2 = 10.0 \text{ kN} \downarrow \quad F_3 = 5.0 \text{ kN} \downarrow \quad F_4 = 0$$

This is due to the arrangement of the equipment on the platform placing heavier pieces toward the left.

Solution Given: Truss in Figure 4–14; forces given above.

Find: Forces in all members of the truss. Compare to results from Example Problem 4–4.

Analysis and Results: The procedure for analyzing the forces in members of a truss using the method of joints will be applied as in Example Problem 4–4. Only an outline of the steps of the solution is shown here. You are advised to carry out the detailed calculations to verify the given results.

$$\text{Reactions: } R_1 = A_y = 17.5 \text{ kN} \uparrow \quad R_2 = F_y = 12.5 \text{ kN} \uparrow$$

Joint A: See the free body diagram in Figure 4–16(a).

$$AB = 21.88 \text{ kN (C)} \quad AC = 13.13 \text{ kN (T)}$$

Joint B: See the free body diagram in Figure 4–16(b).

$$BD = 15.00 \text{ kN (C)} \quad BC = 3.13 \text{ kN (T)}$$

Joint D: See the free body diagram in Figure 4–16(c).

$$DE = 15.00 \text{ kN (C)} \quad CD = 10.0 \text{ kN (C)}$$

Joint C: See the free body diagram in Figure 4–16(d).

$$CE = CF = 9.38 \text{ kN (T)}$$

Joint E: See the free body diagram in Figure 4–16(e).

$$EF = 15.63 \text{ kN (C)}$$

Notice that this completes the calculation of the forces in all members and we have not used joint *F*. We could use the analysis of joint *F* as a check on prior calculations, verifying that all equations of equilibrium are satisfied.

Comparisons of Results from Example Problems 4–4 and 4–5:

Variable	Ex. Prob. 4–4	Ex. Prob. 4–5	Difference
F_1	10.0 kN	15.0 kN	+5.0 kN
F_2	10.0 kN	10.0 kN	0
F_3	10.0 kN	5.0 kN	−5.0 kN
R_1	15.0 kN	17.5 kN	+2.5 kN
R_2	15.0 kN	12.5 kN	−2.5 kN
AB	18.75 kN (C)	21.88 kN (C)	+3.13 kN
AC	11.25 kN (T)	13.13 kN (T)	+1.88 kN
BC	6.25 kN (T)	3.13 kN (T)	−3.12 kN

FIGURE 4–16 Free body diagrams of joints for Example Problem 4–5.

BD	15.0 kN (C)	15.0 kN (C)	0
CD	10.0 kN (C)	10.0 kN (C)	0
DE	15.0 kN (C)	15.0 kN (C)	0
CE	6.25 kN (T)	9.38 kN (T)	+3.13 kN
CF	11.25 kN (T)	9.38 kN (T)	−1.87 kN
EF	18.75 kN (C)	15.63 kN (C)	−4.12 kN

Summary of Differences:

1. Three members experienced increases in force, *AB, AC, CE*.
2. Three members experienced decreases in force, *BC, CF, EF*.
3. Three members had no change in force, *BD, CD, DE*.
4. The maximum compressive force (in member *AB*) increased from 18.75 kN to 21.88 kN, a 17% increase.
5. The maximum tensile force (in member *AC*) increased from 11.25 kN to 13.13 kN, a 17% increase.

**Example Problem
4–6**

The vertical forces applied to joints *B, D*, and *E* in Example Problem 4–5 are maintained at the same level. However, in this problem, consider an additional force of $F_4 = 20.0$ kN → acting horizontally at joint *B* caused by strong winds pushing on the HVAC equipment on the platform. Examine how the forces in all members of the truss change.

Solution Given: Truss in Figure 4–14; forces applied:

$$F_1 = 15.0 \text{ kN} \downarrow \text{ at joint } B \qquad F_4 = 20.0 \text{ kN} \rightarrow \text{ at joint } B$$
$$F_2 = 10.0 \text{ kN} \downarrow \text{ at joint } D \qquad F_3 = 5.0 \text{ kN} \downarrow \text{ at joint } E$$

FIGURE 4–17   Free body diagrams of joints for Example Problem 4–6.

Find: Forces in all members of the truss.

Analysis and Results: The procedure for analyzing the forces in members of a truss using the method of joints will be applied as in Example Problems 4–4 and 4–5. Only an outline of the steps of the solution is shown here. You are advised to carry out the detailed calculations to verify the given results.

Reactions: $R_{1y} = A_y = 10.83$ kN↑ $R_{1x} = A_x = 20.0$ kN ←

Resultant $= R_1 = 22.74$ kN upward and to the left at 28.4° to horizontal

$R_{2y} = F_y = 19.17$ kN↑

Member Forces: **Joint A:** See the free body diagram in Figure 4–17(a).

$$AB = 13.54 \text{ kN (C)} \qquad AC = 28.13 \text{ kN (T)}$$

Joint B: See the free body diagram in Figure 4–17(b).

$$BD = 25.00 \text{ kN (C)} \qquad BC = 5.21 \text{ kN (C)}$$

Joint D: See the free body diagram in Figure 4–17(c).

$$DE = 25.00 \text{ kN (C)} \qquad CD = 10.0 \text{ kN (C)}$$

Joint C: See the free body diagram in Figure 4–17(d).

$$CE = 17.71 \text{ kN (T)} \qquad CF = 14.38 \text{ kN (T)}$$

Joint E: See the free body diagram in Figure 4–17(e).

$$EF = 23.96 \text{ kN (C)}$$

Comparisons of results from Example Problems 4–5 and 4–6:

Variable	Ex. Prob. 4–5	Ex. Prob. 4–6	Difference
F_1	15.0 kN↓	15.0 kN↓	0
F_2	10.0 kN↓	10.0 kN↓	0
F_3	5.0 kN↓	5.0 kN↓	0
F_4	0.0 kN	20.0 kN →	+20.0 kN
R_1	17.50 kN↑	10.83 kN↑	−6.67 kN
R_2	12.5 kN↑	19.17 kN↑	+6.67 kN
AB	21.88 kN (C)	13.42 kN (C)	−8.46 kN
AC	13.13 kN (T)	28.13 kN (T)	+15.0 kN
BC	3.13 kN (T)	5.21 kN (C)	+8.34 kN

Note: Force in BC changed from tension to compression.

BD	15.0 kN (C)	25.0 kN (C)	+10.0 kN
CD	10.0 kN (C)	10.0 kN (C)	0
DE	15.0 kN (C)	25.0 kN (C)	+10.0 kN
CE	9.38 kN (T)	17.71 kN (T)	+8.33 kN
CF	9.38 kN (T)	14.38 kN (T)	+5.00 kN
EF	15.63 kN (C)	23.96 kN (C)	+8.33 kN

Summary of Differences:

1. Seven members experienced increases in force, AC, BC, BD, DE, CE, CF, and EF.
2. One member experienced decreases in force, AB.
3. One member had no change in force, CD.
4. The maximum compressive force (in member EF) increased from 21.88 kN (in member AB) to 23.96 kN (in member EF), a 9% increase.
5. The maximum tensile force (in member AC) increased from 13.13 kN to 28.13 kN, a 114% increase.
6. The force in BC changed from 3.13 kN (T) to 5.21 kN (C).
7. These are very serious issues for the design of the truss members.

Method of Sections

This is an alternate method for determining the forces in members of a truss. Having two methods gives you flexibility for solving problems with different objectives. The method of joints allows an organized approach to finding the forces in all members of the truss and involves the sequential application of the procedure given earlier and illustrated in Example Problems 4–4, 4–5, and 4–6. But what if you wanted to compute the forces in only a few selected members? It may be tedious to analyze several joints to work to one that involves the desired members. The method of sections permits a shorter process to find the desired forces.

Procedure for Applying the Method of Sections

1. Locate the desired members within the complete truss.
2. If possible, make a selected cut through two or three members of the truss at a section that includes the targeted members. Cut no more than three members. If all desired members cannot be cut with one cut, this process may need to be applied two or more times, or it may be more convenient to use the method of joints for the remaining unknown forces.
3. Isolate either part of the truss to one side of the cut line as a free body and show all forces acting on the cut section. Show only externally applied forces and reactions along with the internal forces in each cut member. The forces in other members within the truss need not be shown.
4. If the selected section includes one or more supports, compute the reactions using the entire truss as a free body.
5. Use the equations of equilibrium ($\Sigma M = 0$, $\Sigma F_x = 0$, and $\Sigma F_y = 0$) to find the desired forces. Some general guidelines are given next:
   a. Carefully visualize from the free body diagram of the cut section the most probable direction of the unknown member forces.
   b. Make the simplest practical cut to minimize the number of equations that will need to be analyzed.
   c. When using $\Sigma M = 0$, consider taking moments at a joint through which all but one unknown force acts. Then it will be possible to compute that one force directly because the other forces produce no moment with respect to the chosen joint.
   d. Continue until all desired forces are found.

Example Problem 4-7   For the truss and loading pattern analyzed in Example Problem 4–6, use the method of sections to compute the forces in only members *CE*, *CF*, and *DE*. The truss design is shown in Figure 4–14 and the following applied force values are given:

$$F_1 = 15.0 \text{ kN} \downarrow \text{ at joint } B \qquad F_4 = 20.0 \text{ kN} \rightarrow \text{ at joint } B$$
$$F_2 = 10.0 \text{ kN} \downarrow \text{ at joint } D \qquad F_3 = 5.0 \text{ kN} \downarrow \text{ at joint } E$$

Solution   Given: Truss in Figure 4–14 and the preceding applied forces.

Find: Forces in members *CE*, *CF*, and *DE*.

Analysis and Results: Apply the procedure for applying the method of sections given previously.

Step 1: The three desired forces lie in the third panel of the truss.

Step 2: A cut to the right of joints *C* and *D* will cut all three desired forces.

Step 3: Figure 4–18(a) shows the truss with the selected cut. Let's select the part of the truss to the right of the cut as the free body. Figure 4–18(b) shows that cut section along with the force vectors *CE*, *CF*, and *DE* acting on the cut members. Prior to analysis, we show the force in the diagonal *CE* to be transposed back to the joint *E* to indicate the true lines of action of the x- and y-components.

Step 4: The reaction F_y is a part of the free body so its value must be computed. This was done in Example Problem 4–6 with the result $F_y = 19.17 \text{ kN} \uparrow$. We can now proceed with **Step 5.**

Step 5:

a.   Let's first sum moments about joint *E* because two unknown forces, *DE* and *CE*, pass through that point and produce no moment. This leaves only force *CF*. We show the vector *CF* acting toward the left in order to balance the moment due to F_y.

$$\Sigma M_E = 0 = CF(2.0 \text{ m}) - F_y(1.5 \text{ m})$$
$$CF = F_y(1.5 \text{ m})/(2.0 \text{ m}) = 19.17 \text{ kN}(0.75) = 14.375 \text{ kN} \leftarrow$$

The leftward direction of *CF* indicates that the member pulls on joint *F* and, therefore, *CF* is a tensile force. That is, $CF = 14.375 \text{ kN (T)}$.

b.   Now let's sum forces in the vertical direction because only the given force $F_3 = 5.0 \text{ kN}$, the known reaction F_y, and the unknown force CE_x act in that direction.

$$+\uparrow \Sigma F_y = 0 = F_y - 5.0 \text{ kN} - CE_y$$
$$CE_y = F_y - 5.0 \text{ kN} = 19.17 \text{ kN} - 5.0 \text{ kN} = 14.17 \text{ kN}$$

But $CE_y = CE \sin \theta$, where $\theta = 53.13°$ as found in Example Problem 4–6.

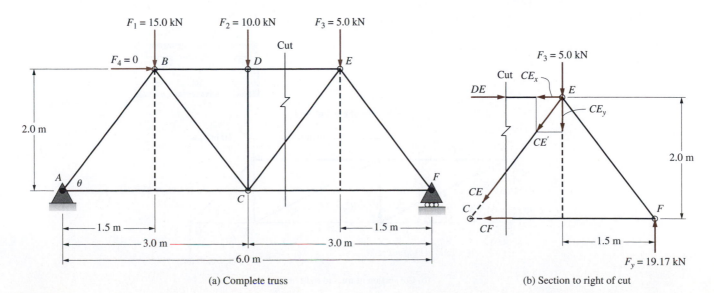

(a) Complete truss   (b) Section to right of cut

FIGURE 4–18   Truss and loading for Example Problem 4–7.

Then

$$CE = CE_y/\sin\theta = 14.17\,\text{kN}/\sin(53.13°) = \underline{17.71\,\text{kN (T)}} = CE$$

The member is in tension because the assumed direction of action pulled on joint E and a positive result confirms the correctness of the assumption.

c.   Finally, we can sum forces in the x-direction to find the last unknown force, DE.

$$+\rightarrow\Sigma F_x = 0 = DE - CE_x - CF$$
$$DE = CE_x + CF = CE\cos\theta + CF = (17.71\,\text{kN})\cos(53.13°) + 14.375\,\text{kN}$$
$$\underline{DE = 25.0\,\text{kN}\,(C)}$$

LESSON 5

The member is in compression because it pushes on joint E.

Comment: These values are identical to those found in Example Problem 4–6.

Example Problem 4–8

Figure 4–19(a) shows one of a pair of trusses used to support a 1000-lb athletic scoreboard away from the wall of an arena. The load is divided equally between the two trusses. Use the method of sections to compute the forces in members AC, BC, and BD.

Solution   Given: One truss carrying one-half a scoreboard's weight, 500 lb, suspending it 50 in from a wall as shown in Figure 4–19(a).

Find: Forces in AC, BC, BD using the method of sections.

Analysis and Results: We use the method of sections as outlined in the procedure previously.

Steps 1 and 2: The three members are in the first panel from the wall. A vertical cut is made between points B and D.

Steps 3 and 4: Let's select the right portion of the truss as a free body diagram as shown in Figure 4–19(b). Doing so does not require the calculation of the support reactions.

(a) Truss holding scoreboard

$\theta = \text{Tan}^{-1}\,(^{10}\!/_{20}) = 26.57°$

(b) Cut section of truss to right of cut

FIGURE 4–19   Truss and loading for Example Problem 4–8.

Step 5: Let's start by summing moments about joint C because only member force BD is involved.

$$\Sigma M_C = 0 = BD(10\text{ in}) - G_y(30\text{ in})$$

$$BD = (500\text{ lb})(30\text{ in})/(10\text{ in}) = \underline{1500\text{ lb}} \leftarrow (T) = BD$$

Next we can sum forces in the vertical direction to find BC.

$$+\uparrow\Sigma F_y = 0 = BC_y - G_y = BC_y - 500\text{ lb}$$

$$BC_y = 500\text{ lb} = BC\sin\theta$$

Then $BC = BC_y/\sin\theta = (500\text{ lb})/[\sin(26.57°)] = \underline{1118\text{ lb (T)}} = BC$.

Finally, we can sum forces in the horizontal direction:

$$+\Sigma F_x = AC - BD - BC\cos\theta$$

$$AC = BD + BC\cos(26.57°) = 1500\text{ lb} + (1118)\cos(26.57°)$$

$$AC = 1500\text{ lb} + 1000\text{ lb} = \underline{2500\text{ lb}} \rightarrow (C) = AC$$

Summary: The desired values of forces are

$$AC = 2500\text{ lb (C)} \qquad BC = 1118\text{ lb (T)} \qquad BD = 1500\text{ lb (T)}$$

4–4 EQUILIBRIUM IN THREE-DIMENSIONAL FRAMES AND STRUCTURES

Analysis methods to this point have been limited to frames and structures lying in a plane such that all forces act in a two-dimensional space. Now we discuss three-dimensional structures such as those shown earlier in Figure 4–2. The principles of static equilibrium remain the same:

The sum of forces in any direction must be zero.

The sum of moments taken about any axis must be zero.

The introduction of the third dimension requires additional mathematical statements to ensure equilibrium and to facilitate calculation of all unknown forces and moments.

$$\Sigma F_x = 0 \quad \Sigma F_y = 0 \quad \Sigma F_z = 0 \qquad \textbf{(4–2)}$$

$$\Sigma M_x = 0 \quad \Sigma M_y = 0 \quad \Sigma M_z = 0 \qquad \textbf{(4–3)}$$

Let's deal with the forces first. The forces, or components of forces, are oriented relative to the three mutually perpendicular axes, commonly called x, y, and z. To ensure proper management of the signs and directions for force vectors, in this book will use the *right-hand rule* for orienting the three axes.

When wrapping your right hand around the z-axis while moving your fingers from the positive x-direction toward the positive y-direction, your thumb will be pointing toward the positive z-direction.

We will typically show the three axes in the orientation shown in Figure 4–20(a) because it conforms to a common convention used in site planning for buildings and in machining parts on machining centers and other machine tools. The x-y plane is typically considered to be horizontal and the z-axis points vertically upward. The x-z plane and the y-z plane are vertical.

Other technical documents and books may use other orientations for the three axes as illustrated in Figures 4–20(b) and (c). Any consistent approach can be used provided that signs of forces are carefully accounted for.

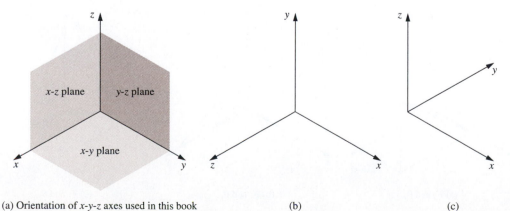

(a) Orientation of x-y-z axes used in this book   (b)   (c)

FIGURE 4–20   Orientations of x-y-z axes using right-hand rule.

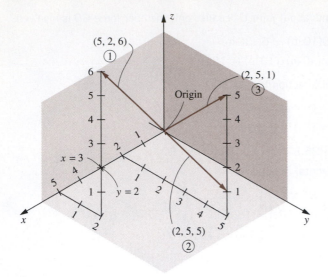

FIGURE 4–21 Representing vectors in an x-y-z space.

Components of a given force vector in the x-, y-, and z-directions require a clear visualization of the vector's relationship to the axes. When showing one vector or two or more vectors acting at a given point, it is convenient to locate the origin of the axes at the start of the vectors. Then the end of each vector can be indicated by its x-, y-, and z-coordinates as shown in Figure 4–21. The three vectors are

Vector 1: (5, 2, 6) $x = 5, y = 2, z = 6$

Vector 2: (2, 5, 1) $x = 2, y = 5, z = 1$

Vector 3: (2, 5, 5) $x = 2, y = 5, z = 5$

Of course, any of the axes can include negative values as well as positive values.

Angles of Orientation for Vectors in Three-Dimensional Space

We must define three angles to show the orientation of a vector to each axis:

θ_x = angle between the vector and the positive x-axis

θ_y = angle between the vector and the positive y-axis

θ_z = angle between the vector and the positive z-axis

These are called the *direction angles* for the vector and they are always between 0° and 180°. When the force component for a given axis is on the positive side, the angle will be acute, less than 90°. Conversely, when the force component is on the negative side, the angle will be obtuse, between 90° and 180°.

Knowing these angles allows the calculation of the components of the vector using the technique already introduced for vectors in two-dimensional space.

$$F_x = F \cos \theta_x \quad F_y = F \cos \theta_y \quad F_z = F \cos \theta_z \quad \text{(4–4)}$$

These cosine functions are called the *direction cosines* for the vector. Figure 4–22 shows the components and the direction angles for a vector having the coordinates of (4, 7, 6). When the force component is negative, the direction cosine for that force will be negative and the direction angle will be between 90° and 180°.

Often a given problem will contain data for only two of the three direction angles. The third angle can be found using the following important relationship:

$$\cos^2 \theta_x + \cos^2 \theta_y + \cos^2 \theta_z = 1 \quad \text{(4–5)}$$

This equation can be solved for any angle if the other two angles are known.

The magnitude of the resultant of the three components of a force acting in a three-dimensional space system can be computed from

$$F = \sqrt{F_x^2 + F_y^2 + F_z^2} \quad \text{(4–6)}$$

A common problem is one for which the three rectangular components of a force are known and it is desired to find their resultant and its direction angles. This can be simply done by first applying Equation (4–6) to find the resultant, F, and then computing the direction angles from the relations in Equation (4–4) as follows:

$$\theta_x = \cos^{-1}(F_x/F) \quad \theta_y = \cos^{-1}(F_y/F)$$
$$\theta_z = \cos^{-1}(F_z/F) \quad \text{(4–7)}$$

Consider the following example problem.

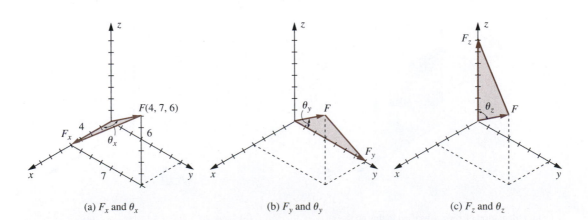

| (a) F_x and θ_x | (b) F_y and θ_y | (c) F_z and θ_z |

FIGURE 4–22 Components of a force in three-dimensional space and the direction angles of the force.

Example Problem 4–9

For the force vector *F* shown in Figure 4–22, the three rectangular components are known to be

$$F_x = 400 \text{ N} \qquad F_y = 700 \text{ N} \qquad F_z = 600 \text{ N}$$

Compute the resultant force and its direction angles.

Solution   Objective: Find the resultant, *F*, and the direction angles, θ_x, θ_y, and θ_z.

Given: Figure 4–22 and given components of the total force.

Analysis: First apply Equation (4–6) to find *F*. Then use Equations (4–4) to find the three direction angles.

Results:

$$F = \sqrt{F_x^2 + F_y^2 + F_z^2} \tag{4–6}$$

$$F = \sqrt{(400 \text{ N})^2 + (700 \text{ N})^2 + (600 \text{ N})^2} = 1005 \text{ N}$$

Now, from Equations (4–4):

$$F_x = F\cos\theta_x \rightarrow \theta_x = \cos^{-1}(F_x/F) = \cos^{-1}(400/1005)$$
$$\theta_x = 66.5°$$
$$F_y = F\cos\theta_y \rightarrow \theta_y = \cos^{-1}(F_y/F) = \cos^{-1}(700/1005)$$
$$\theta_x = 45.9°$$
$$F_z = F\cos\theta_z \rightarrow \theta_z = \cos^{-1}(F_z/F) = \cos^{-1}(600/1005)$$
$$\theta_x = 53.3°$$

Another useful method for visualizing the components of a force vector and its orientation in three-dimensional space is shown in Figure 4–23(a). Here we show a boxlike frame for which the vector *F* is the diagonal from the origin of the *x*-, *y*-, *z*-axis system to the tip of *F*. We can then see by inspection that the dimensions of the box are F_x, F_y, and F_z. For example, let

$$F_x = 30.0 \text{ lb} \qquad F_y = 40.0 \text{ lb} \qquad F_z = 60.0 \text{ lb}$$

We can then compute

$$F = \sqrt{F_x^2 + F_y^2 + F_z^2} \tag{4–6}$$

$$F = \sqrt{(30.0 \text{ lb})^2 + (40.0 \text{ lb})^2 + (60.0 \text{ lb})^2}$$
$$= 78.1 \text{ lb}$$

Now visualize a plane through the vector *F* and the vertical *z*-axis as shown shaded in blue in the figure. Where this plane intersects the horizontal *x*-*y* plane we can draw a vector F_h from the origin to the lower corner of the box. F_h is the component of the force *F* in the horizontal plane. From the dimensions of the box and using the Pythagorean Theorem, we can compute:

$$F_h = \sqrt{F_x^2 + F_y^2} = \sqrt{(30.0 \text{ lb})^2 + (40.0 \text{ lb})^2} = 50.0 \text{ lb}$$

(a) Given force vectors F_x, F_y, F_z and their resultant *F*

(b) Direct view of plane containing F_z, *F*, and F_h

(c) Direct downward view toward *x*-*y* plane

FIGURE 4–23   Force vector represented within a box.

The orientation of the vector F can be defined in terms of an angle (θ_z) from the z-axis to F in the blue-shaded plane and an angle (α) in the horizontal plane from the x-axis to F_h. The two views shown in parts (b) and (c) of the figure facilitates computation of these angles: Part (b) is the direct view of the plane containing F and the z-axis and this allows the calculation of θ_z:

$$\theta_z = \cos^{-1}(F_z/F) = \cos^{-1}[(60.0\,\text{lb})/(78.1\,\text{lb})] = 39.8°$$

Now look at part (c) of the figure to see that we can compute the angle α:

$$\alpha = \cos^{-1}(F_x/F_h) = \cos^{-1}[(30.0\,\text{lb})/(50.0\,\text{lb})] = 53.1°$$

In summary, we have completely defined the magnitude and direction of the vector F.

Moments in Three-Dimensional Space   Equations 4–3 state that the sum of the moments about any axis must equal zero for a body in static equilibrium. It is important to observe that the moments of interest are those that tend to cause rotation about one of the *primary axes, x, y, z*. In earlier work, we computed moments about a certain *point of interest*.

Now let's apply these techniques to the analysis of forces for the marquee shown earlier in Figure 4–2(b).

Example Problem 4–10   For the marquee shown in Figure 4–2(b), the following data are known:

1. The marquee has a total weight of 4.50 kN and its center of mass is at the center of the structure.
2. Figure 4–24 shows pertinent dimensions.
3. Two smooth hinges attach the upper surface of the marquee to the wall of a theater.
   a. Hinge A can resist only radial loads.
   b. Hinge B can resist radial loads and an axial load.
   c. Without the cable installed, the marquee could swing downward around the hinge pins.
4. The cable is attached at C on the wall and at D on the top surface of the marquee at the locations shown. The cable holds the marquee in a horizontal plane by exerting an upward force F_C acting in a direction from D to C.

Compute the force in the cable and in each hinge.

Solution   Objective: Find the force in the cable, F_C

Given: Figures 4–2 and 4–24; weight of marquee, W_M, is 4.50 kN acting vertically downward through the center of mass of the marquee.

Analysis: First observe the placement of the x-, y-, and z-axes in Figure 4–24.

- The origin of the x, y, z system is at hinge A.
- The x-axis is coincident with the pins in the hinges at the rear of the marquee.
- The y-axis passes through the middle of hinge A.
- The z-axis is vertical and passes through A and C.
- The center of mass of the marquee is midway between A and B and half of the width in the y-direction out from the wall.
- The cable is in direct tension and is shown as a force vector pulling upward on attachment point D on the marquee and downward on attachment point C on the wall.

We will use Equations (4–2) and (4–3) to compute the unknown forces.

Results: To aid in viewing the forces, a rectangular box is drawn in Figure 4–24 with points A, B, D, and C as corners and the cable is aligned with the diagonal from C to D. We will use the marquee itself as the free body that is held in equilibrium by the forces at A, B, and D balancing the downward weight of the marquee, W_M. The cable force at C does not enter into the equilibrium analysis.

We need to quantify the direction of the cable force and its components. Refer to part (b) of the figure showing the plane containing the z-axis and the vector, F_C, aligned with the diagonal CD. We can compute the angles α and β as follows:

$$AD = \sqrt{(5.00\,\text{m})^2 + (4.00\,\text{m})^2} = 6.403\,\text{m}$$

$$CD = \sqrt{(5.00\,\text{m})^2 + (6.403\,\text{m})^2} = 8.124\,\text{m}$$

$$\alpha = \text{Tan}^{-1}[(5.00\,\text{m})/(4.00\,\text{m})] = 51.34°$$

$$\beta = \text{Cos}^{-1}[(5.00\,\text{m})/(8.124\,\text{m})] = 52.01°$$

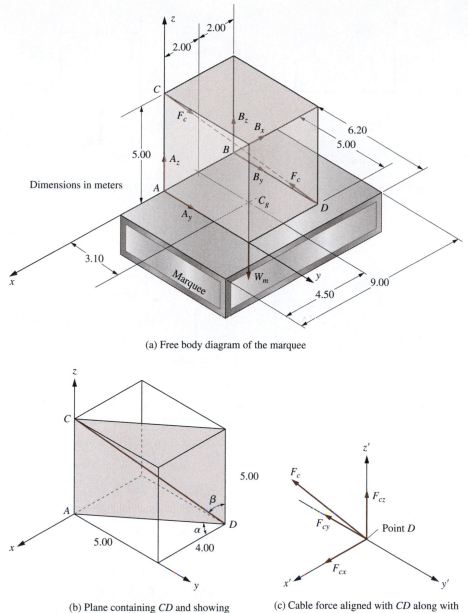

(a) Free body diagram of the marquee

(b) Plane containing *CD* and showing angles β and α

(c) Cable force aligned with *CD* along with components F_{cx}, F_{cy}, F_{cz}

FIGURE 4–24 Force systems for Example Problem 4–10.

Now we can write expressions for the components of the cable force.

$$F_{cz} = F_c \cos \beta = 0.6155\, F_c$$

$$F_{ch} = F_c \sin \beta = 0.7882\, F_c \quad \text{[Component of } F_c \text{ in horizontal plane]}$$

$$F_{cx} = F_{ch} \cos \alpha = (0.7822\, F_c)(0.6247) = 0.4924\, F_c$$

$$F_{cy} = F_{ch} \sin \alpha = -(0.7822\, F_c)(0.7809) = -0.6155\, F_c$$

Note that F_{cy} is negative because we have shown it acting in the negative *y*-direction.

The six equilibrium equations given in Equations (4–2) and (4–3) are the tools needed to solve for all the unknown forces. We write each of them here. However, in some problems, it may be evident that certain equations can be used first because they allow direct solution for key unknowns.

An aid in writing the equilibrium equations is to draw views of the object being analyzed looking directly at the *y-z*, *x-z*, and *x-y* planes as shown for this problem in Figure 4–25. In each case, one

(c) View toward x-y plane (top view)

FIGURE 4–25   Side, front, and top views of marquee—an aid to writing moment equations.

axis shows as a point and it is easier to view which forces produce moments about that axis. Note that sometimes two or more forces are aligned so care must be exercised to label all forces.

$$\Sigma F_x = 0 = F_{cx} - B_x \Rightarrow F_{cx} = B_x \tag{I}$$
$$\Sigma F_y = 0 = A_y + B_y - F_{cy} \tag{II}$$
$$\Sigma F_z = 0 = A_z + B_z - W_M + F_{cz} \tag{III}$$
$$\Sigma M_x = 0 = W_M(3.10\,\text{m}) - F_{cz}(5.00\,\text{m}) \tag{IV}$$
$$\Sigma M_y = 0 = W_M(2.00\,\text{m}) - B_z(4.00\,\text{m}) - F_{cz}(4.00\,\text{m}) \tag{V}$$
$$\Sigma M_z = 0 = B_y(4.00\,\text{m}) - F_{cy}(4.00\,\text{m}) + F_{cx}(5.00\,\text{m}) \tag{VI}$$

The Roman numerals facilitate designation of particular equations. Note in Equation (IV) that the only unknown is F_{cz} so we can solve for that value. That in turn allows computation of F_c and then F_{cx} and F_{cy}.

From Equation (IV): $0 = W_M(3.10\,\text{m}) - F_{cz}(5.00\,\text{m})$.

$$F_{cz} = W_M(3.10\,\text{m})/(5.00\,\text{m}) = (4.50\,\text{kN})(3.10\,\text{m})/(5.00\,\text{m}) = 2.79\,\text{kN}.$$

Earlier we found $F_{cz} = F_c \cos\beta = 0.6155\,F_c$.

Then $F_c = F_{cz}/0.6155 = 2.79/0.6155 = 4.533\,\text{kN}$.

Similarly $F_{cx} = 0.4924\,F_c = (0.4924)(4.533\,\text{kN}) = 2.232\,\text{kN}$

$$F_{cy} = -0.6155\,F_c = -(0.6155)(4.533\,\text{kN}) = -2.79\,\text{kN}$$

Knowing these values now permits the direct solution of the forces at the hinges at A and B. From Equation (VI) we can find B_y.

$$B_y = [F_{cy}\,4.00\,\text{m}) - F_{cx}(5.00\,\text{m})]/4.00\,\text{m}$$
$$B_y = [(2.79\,\text{kN})(4.00\,\text{m}) - (2.232\,\text{kN})(5.00\,\text{m})]/4.00\,\text{m}$$
$$B_y = 0$$

From Equation (II): $\Sigma F_y = 0 = A_y + B_y - F_{cy}$.

$$A_y = F_{cy} - B_y = 2.79 \text{ kN} - 0 = 2.79 \text{ kN}$$

From Equation (V): $\Sigma M_y = 0 = W_M(2.00\text{ m}) - B_z(4.00\text{ m}) - F_{cz}(4.00\text{ m})$.

$$B_z = [W_M(2.00\text{ m}) - F_{cz}(4.00\text{ m})]/(4.00\text{ m})$$
$$B_z = [(4.50\text{ kN})(2.00\text{ m}) - (2.79\text{ kN})(4.00\text{ m})]/(4.00\text{ m}) = -0.540 \text{ kN}$$

The negative sign here indicates that our original judgment for the direction of B_z was incorrect. It actually acts vertically downward, not upward, as shown in Figure 4–24(a). That is, $B_z = 0.540$ kN $\downarrow$.

Finally, we can use Equation (II) to determine A_z:

$$\Sigma F_z = 0 = A_z + B_z - W_M + F_{cz}$$
$$A_z = W_M - F_{cz} - B_z = 4.50\text{ kN} - 2.79\text{ kN} - (-0.540\text{ kN})$$
$$A_z = 2.25 \text{ kN} \uparrow$$

Summary: Forces in the cable: $F_c = 4.533$ kN

$$F_{cx} = 2.232 \text{ kN in positive } x\text{-direction}$$
$$F_{cy} = -2.79 \text{ kN in negative } y\text{-direction}$$
$$F_{cz} = 2.79 \text{ kN} \uparrow \text{ in positive } z\text{-direction}$$

Forces at hinge A: $A_x = 0$

$$A_y = 2.79 \text{ kN in positive } y\text{-direction}$$
$$A_z = 2.25 \text{ kN} \uparrow \text{ in positive } z\text{-direction}$$

Forces at hinge B: $B_x = -2.232$ kN in negative x-direction

$$B_y = 0$$
$$B_z = -0.540 \text{ kN} \downarrow \text{ in negative } z\text{-direction}$$

An alternate solution procedure can be used to solve Example Problem 4–10 using the concept of direction cosines described earlier. We know the direction of the cable force, F_c, to be along the diagonal CD in Figure 4–24(b). Now the dimensions of the box drawn around points A, B, D, and C can be used to compute the value of the direction cosines of F_c. We computed the length of CD to be 8.124 m in Example Problem 4–10. Let's call that value d. Then, from Figure 4–24(b), we can say that the three sides of the box have the dimensions $d_x = 4.00$ m, $d_y = 5.00$ m, and $d_z = 5.00$ m. Now from Equations (4–4), substituting d for F, we can say

$$d_x/d = \cos\theta_x \quad d_y/d = \cos\theta_y \quad d_z/d = \cos\theta_z \quad (4\text{–}7)$$

For the data in this problem,

$$\cos\theta_x = d_x/d = (4.00)(8.124) = 0.4924 = F_{cx}/F_c$$
$$\cos\theta_y = d_y/d = (5.00)(8.124) = 0.6155 = F_{cy}/F_c$$
$$\cos\theta_z = d_z/d = (5.00)(8.124) = 0.6155 = F_{cz}/F_c$$

For this problem we do not need the actual values of the direction angles, although they can be easily calculated.

The rest of the solution procedure for Example Problem 4–10 is the same as before except we use the direction cosines to relate the cable force to its components, instead of the angles α and β. In any given problem, you may choose to use either method for computing the unknown forces.

Summary of Techniques for Three-Dimensional Static Analysis

It is not feasible to define a single, general procedure for analyzing three-dimensional static equilibrium analyses. However, this section and the example problems have used several principles that can be considered for application to a given situation. You must select which principle will allow computing the results desired for a given problem. While learning, it is not unusual to use a trial-and-error technique until a suitable approach emerges.

1. If not already defined, a set of mutually perpendicular axes should be specified, typically called x, y, z, and oriented conveniently with key features of the object being analyzed.

2. The three components of a force, F, in three-dimensional space are typically called F_x, F_y, F_z. See Figures 4–20 and 4–21.

3. The six basic equations of static equilibrium for three-dimensional systems are

$$\Sigma F_x = 0 \quad \Sigma F_y = 0 \quad \Sigma F_z = 0 \quad (4\text{–}2)$$
$$\Sigma M_x = 0 \quad \Sigma M_y = 0 \quad \Sigma M_z = 0 \quad (4\text{–}3)$$

Moments are computed about the given *axis*, not a specific point.

4. The magnitudes of the components are related to the magnitude of the total force by

$$F = \sqrt{F_x^2 + F_y^2 + F_z^2} \qquad (4\text{--}6)$$

5. The components and the orientation of the total force vector can be related in terms of *direction angles* and *direction cosines,* as shown in Figure 4–22, and by the following equations:

$$F_x = F\cos\theta_x \quad F_y = F\cos\theta_y \quad F_z = F\cos\theta_z \ \ (4\text{--}4)$$

6. If the line of action of a force is known, the direction cosines can be computed by identifying a characteristic length, d, of a line along that direction and relating it to the x-, y-, z-components of that line, d_x, d_y, and d_z. The direction cosines are in turn related to the force components as given in previously item 5.

$$d_x/d = \cos\theta_x = F_x/F$$
$$d_y/d = \cos\theta_y = F_y/F$$
$$d_z/d = \cos\theta_z = F_z/F$$

7. The three direction cosines are related by

$$\cos^2\theta_x + \cos^2\theta_y + \cos^2\theta_z = 1 \qquad (4\text{--}5)$$

Thus, if any two of the values are known, the third can be easily computed.

8. Figure 4–23 shows another method of visualizing the three-dimensional components of a force.

9. When writing the equations of equilibrium (4–2) and (4–3), it is often helpful to draw projected views of the system where a two-dimensional plane is shown (x-y, x-z, or y-z) and the third axis is a point (end view). Moments about that axis are more easily visualized. See Example Problem 4–10 and Figures 4–24 and 4–25.

10. After the equilibrium equations are written, various solution methods may arise.

   a. An equation may be solved algebraically for a desired unknown.
   b. Two or more equations may have to be solved simultaneously to determine the unknowns. This may involve adding or subtracting two equations at a time to eliminate one unknown or to produce an expression of one unknown in terms of another.
   c. When three or more unknowns exist in a system of equations, determinates, matrix analysis, advanced calculators, spreadsheets, or computer-based mathematics solvers are useful.

INTERNET SITES

1. Continental Cable LLC www.gbgindustries.com/Stranded Cable.htm Manufacturer of stranded cable in galvanized or stainless steel for aerospace, automotive, and other industrial fields.

2. Eaton Fluid Power www.eaton.com/hydraulics/cylinders Manufacturer and marketer of hydraulic systems and components for construction, earthmoving, automotive, industrial machinery, agricultural equipment, and many other industries.

3. Material Handling Solutions www.e-rackonline.com/lifts/ Distributor of a variety of material handling products and systems, including scissors lift tables.

4. Genie—A Terex Company www.genieindustries.com/gs-series/ Manufacturer and marketer of a variety of lifting devices, many of which include scissors-type mechanisms, for construction, maintenance, and service applications.

5. Caterra Master Planning Ltd. www.caterra.ca/woodframe.php Designers and marketers of preengineered and factory-assembled wood trusses and building roof systems. The Web site illustrates nine key stages of constructing a wood frame home from the foundation to the roof.

6. Commander Buildings, Inc. www.commanderbuildings.com Manufacturer of a variety of preengineered building systems including steel frames, many of which employ trusses.

7. AARoads.com www.aaroads.com Developer of the AARoads.com Web site offering detailed photographs and description of interstate highways and other routes to assist travelers. Many of the photographs illustrate the use of truss structures to support highway signs.

8. Matsuo Bridge Company www.matsuo-bridge.co.jp/english/ bridges Designer, fabricator, and installer of major bridges, many of which include truss structures.

9. Total Structures, Inc. www.totalstructures.com Producer of truss systems for use in stage shows and trade shows.

10. Jim Kimball Enterprises www.jimkimballenterprises.com Manufacturer of high-performance sport and acrobatic aircraft that employ lightweight tubular truss structures for the fuselage frames.

11. Porter Race Cars www.porterracecars.com Fabricator of custom-built truss-type frames for high-performance vehicles used in off-road racing.

12. Building Component Safety Information www.sbcindustry .com/bcsi.php Guide for jobsite safety, BCSI 2008, concerning handling, installation, restraining, and bracing of metal plate connected wood trusses. Ten BCSI Summary Sheets can be viewed at the site.

13. ThomasNet www.thomasnet.com/products/roof-trusses-88320601-1.html List of manufacturers that provide roof trusses.

14. RUD Chain Company www.rud.com Manufacturer of lifting chain and accessories including swivel-type lifting eyes.

15. Edward W. Daniel Company www.ewdaniel.com Manufacturer of lifting eyes, hoist rings, turnbuckles, and related items.

16. Altec, Inc. www.altec.com Manufacturer of articulated arm bucket trucks and truck-mounted cranes for use in electrical system service, tree trimming, building maintenance, construction, and sign installation and service.

17. EnterTRAINment Junction www.entertrainmentjunction.com A multifaceted attraction featuring a 25,000-square-foot model railroad in G-scale located in a suburb of Cincinnati, Ohio. The author was one of a large team of volunteers who built the railroad.

PROBLEMS

Pin-Connected Frames

Note: The free body diagrams called for in the following problems are to include the magnitudes and directions for all forces. Unless stated otherwise, reaction forces can be expressed in terms of their *x*- and *y*-components.

4–1. For the structure shown in Figure P4–1, draw the complete free body diagrams for members *AB* and *CBD*.

FIGURE P4–1

4–2. Figure P4–2 shows a pipe rack attached to a vertical wall. Each of the two large pipes exerts a force of 3.35 kN on the rack while each small pipe exerts 1.42 kN. Draw the complete free body diagrams for members *AB*, *AC*, and *BCD*. Consider the beam *BCD* to be pin-connected at *B*. The member *AB* hangs from the pin at *B* and supports the roller at *A*, but it does not affect the forces on the beam.

FIGURE P4–2

4–3. Figure P4–3 shows a structure supporting a load at the end of a beam. Draw the complete free body diagrams for members *BC* and *ABD*.

4–4. The shelf shown in Figure P4–4 carries two pieces of equipment having weights of $W_1 = 105$ kN and $W_2 = 74$ kN. Their weights are located as shown and they are evenly

FIGURE P4–3

FIGURE P4–4 Problems 4–4 and 4–5.

balanced from side to side. Draw the complete free body diagrams for the shelf and each support rod.

4–5. For the shelf described in Problem 4–4, change the angle of inclination of the rods to 40° and compute the forces in all members. Compare the results for the forces in the rods with those of Problem 4–4.

4–6. Figure P4–6 shows a single beam carrying a hoist on a trolley. For the design shown, determine the complete free body diagrams of the beam and the support rod.

4–7. Repeat Problem 4–6 but place the trolley 2.20 m from the support. Compare the results with those of Problem 4–6.

FIGURE P4–6 Problems 4–6 and 4–7.

FIGURE P4–9

4–8. Figure P4–8 shows a frame carrying a downward load of $F_1 = 438$ lb and a horizontal load of $F_2 = 224$ lb. Draw the complete free body diagrams for all members.

FIGURE P4–8

FIGURE P4–10 Problems 4–10, 4–11, and 4–12.

4–9. The crane in Figure P4–9 carries a load that has a mass of 545 kg. Draw the complete free body diagrams of the boom *ABC* and the hydraulic cylinder *BD*.

4–10. The crane in Figure P4–10 carries a load weighing 800 lb. Draw the complete free body diagrams of the horizontal boom *ABC* and the hydraulic cylinder *BD*.

4–11. Repeat Problem 4–10 except move the attachment point *B* for the top end of the hydraulic cylinder out to 20 in.

4–12. Repeat Problem 4–10 except consider the rod of the hydraulic cylinder to be extended to raise the boom to an angle of 28° above the horizontal.

4–13. For the crane in Figure P4–13, draw the complete free body diagrams of cables *AB* and *AE*, boom *DEF*, and post *AFC*.

4–14. Figure P4–14 shows a large container being supported by a cable system and two cantilevered support arms. Compute the forces in the three sections of cable and draw the complete free body diagram of the left cantilever that is fastened rigidly to a sturdy column.

4–15. Figure P4–15 shows a casting with a mass of 940 kg supported by two rods that are in turn carried by a spreader bar and a cable assembly. The center of mass of the casting is midway between the two rods. Compute the forces in the rods, the spreader bar, and the cables.

4–16. The two lower cables in Figure P4–15 are at a lower angle than recommended for this application. Repeat Problem 4–15 but extend the vertical dimension of 250 mm as

FIGURE P4–13

FIGURE P4–14

needed to make the angle of inclination 50°. Then reanalyze the forces on the spreader bar and the cables. Compare the results to those of Problem 4–15.

4–17. Compute the forces in the three cables shown in Figure P4–17 when the load has a mass of 1500 kg.

4–18. The machine in Figure P4–18 has a mass of 752 kg. Compute the forces in all cables.

4–19. The load in Figure P4–19 has a mass of 4200 kg. Compute the forces in all cables.

4–20. The load in Figure P4–20 has a mass of 680 kg. Compute the forces in all cables.

4–21. Figure P4–21 shows a small scissors jack for automotive applications for which a power screw across the middle provides the 900-lb lifting force. At the left end, the threads

250

450

1100 mm —550

Cable from crane

Casting

FIGURE P4–15   Problems 4–15 and 4–16.

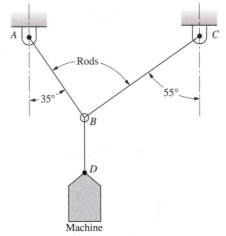

$\theta = 25°$

A

B

C

D

Load

FIGURE P4–17

A

Rods

35°

55°

B

D

Machine

C

FIGURE P4–18

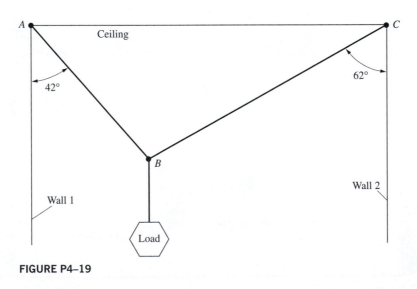

A

Ceiling

C

42°

62°

B

Wall 1

Wall 2

Load

FIGURE P4–19

of the power screw pull inward on the pin connecting the two identical scissors mechanisms. At the right end, the unthreaded part of the power screw passes through a clearance hole in the pin while a thrust bearing presses on the outer side of the pin. Compute the force developed in the power screw and the forces in the angled members of the scissors.

4–22. An aircraft carrier uses a side-mounted elevator to move aircraft to the flight deck from the hangar deck as shown in

Figure P4–22. Determine the force required in the lifting cable to raise the elevator with the plane shown in the figure. The applied forces shown are based on the total weights of the plane and the platform and are shared by two cables and two sets of rollers.

4–23. For the aircraft elevator shown in Figure P4–22, compute the forces on the rollers that ride in vertical guide rails.

4–24. For the locking pliers shown in Figure P4–24, compute the force applied to the bolt, F_b, by the jaws when the operator

FIGURE P4–20

FIGURE P4–21

applies a force to the handle of $F_h = 65$ N. Also compute the total shearing force on the pin A.

4–25. Typical household pliers are shown in Figure P4–25. Compute the force applied to the small-diameter wire being clasped. Then compute the force on the pin that connects the two parts of the pliers.

4–26. Figure P4–26(a) shows a truck-mounted crane in the rest position with key dimensions. The boom pivots about point C. It is elevated by the hydraulic cylinder AD. The load is suspended from point E by means of a cable, block, and tackle system. The boom has a mass of 748 kg considered to act at the center of mass, point CM. For this position, compute the force required by the hydraulic cylinder to lift the boom off its support. Also determine the length of the hydraulic cylinder. (See Internet site 16 for examples of commercially available truck-mounted cranes.)

4–27. Refer to Figure P4–26(b) showing the boom of the truck-mounted crane raised at an angle of 21° and carrying a load having a mass of 22.7 Mg. See Problem 4–26 for additional

data. Determine the force in the hydraulic cylinder and its total length.

4–28. Repeat Problem 4–27 when the boom has been extended to a length $CE = 19.80$ m. The center of mass of the boom has moved to 7.92 m from point C. Determine the force in the hydraulic cylinder.

4–29. Figure P4–29 shows an articulated arm-type bucket truck used to service electrical lines, trees, building façades, and signs. Hydraulic cylinder AC extends to raise the lower boom relative to the truck bed. Hydraulic cylinder FG extends to raise the upper boom relative to the lower boom. The bucket carries a service technician and equipment. Note that the bucket is separately controlled to maintain its axis vertical. (See Internet site 16 for examples of commercially available articulated arm-type bucket trucks.) The following data are known:

- Key dimensions are shown in the figure.
- The load in the bucket plus the weight of the bucket itself is 500 lb acting through the center of mass CM_B.

Force from lifting cable

F_C

Flight deck

15 ft   Rollers

Guide rails

Hangar deck

F_N

F_M

F_P

5 ft

11 ft

19 ft

29 ft

F_N = Nose wheel force = 12 400 lb
F_M = Main wheel force = 48 400 lb
F_P = Platform weight = 10 600 lb

FIGURE P4–22   Problems 4–22 and 4–23.

42 mm

138 mm

F_h

F_b

F_b

6 mm

43°

39 mm

F_h

FIGURE P4–24

F_h = 55 N

Pin

100 mm

45 mm

F_b

FIGURE P4–25   Pliers for Problem 4–25.

■ The weight of the lower boom is 835 lb and its center of mass is at CM_L.
■ The weight of the upper boom is 672 lb and its center of mass is at CM_U.

Determine the following:

a. The force exerted by hydraulic cylinder *FG* to raise the upper boom and the bucket from its initial position while the lower boom remains horizontal

b. The force exerted by hydraulic cylinder *AD* to raise the lower boom from its initial position while the upper boom remains horizontal

4–30. Figure P4–30 shows a drawing of the articulated arm system for the truck shown in Figure P4–29 with both the upper and lower booms extended. Refer to Problem 4–29 for additional data. Determine the force in both hydraulic cylinders *FG* and *AC* for the position shown.

Trusses

Note: For all truss problems compute the forces in all members unless otherwise noted in the problem statement or as indicated by the instructor. It is essential that the results show whether the forces in the members are tension (T) or compression (C) in addition to the magnitudes because the design of the members is radically different depending on the manner of loading. Consider all truss members to be pin-connected and assume that they are two-force members. Solutions can be done using the *method of joints* or the *method of sections* as assigned by the instructor. Some of each is recommended.

(a) Crane at rest

Note: Triangle ABC
fixed to Truck bed

(b) Schematic drawing of crane at 21° elevation

FIGURE P4–26   Truck-mounted crane for Problems 4–26, 4–27, and 4–28.

FIGURE P4–29   Articulated arm-type bucket truck for Problems 4–29 and 4–30.

FIGURE P4–30

4–31. Figure P4–31(a) shows a photograph of a physical model for the truss whose dimensions and loading are shown in part (b) of the figure. Additional drawings of the truss and its members are included on the CD-ROM in this book. The two loads are forces produced from a machine component supported at joints *D* and *E*.

4–32. For the same truss shown in Figure P4–31(a), solve for the forces in all members due to the loading shown in Figure P4–32.

4–33. Use Figure P4–33.

4–34. Use Figure P4–34.

4–35. Use Figure P4–34 but remove the forces at joint *D*.

4–36. Use Figure P4–34 but remove the forces at joint *D* and the horizontal force at joint *B*.

4–37. Use Figure P4–37.

4–38. Use Figure P4–38.

4–39. Use Figure P4–38 but remove the vertical force at joint *G*. Comment on the changes in member forces compared with Problem 4–38.

4–40. Use Figure P4–40. The load at joint *I* is 28.9 kN. The weight of the structure itself ($W_T = 3.40$ kN) is approximated by applying $W_{TL} = 1.58$ kN at joint *C* and $W_{TR} = 1.82$ kN at joint *D*. The vectors F_W represent wind

forces that for this problem are taken to be zero because it is a calm day.

4–41. Use Figure P4–40. The loads are the same as in Problem 4–40 except today it is windy and $F_W = 3.80$ kN at each joints *C*, *E*, and *G*.

4–42. For the truss in Figure 4–19, compute the forces in each member using the method of joints. The force exerted by the scoreboard is 500 lb.

4–43. Use Figure P4–43.

4–44. Use Figure P4–44.

4–45. Use Figure P4–45.

4–46. Use Figure P4–45 but add two forces at joint *E*: 1000 lb vertically downward and 600 lb horizontally to the left.

4–47. Figure P4–47 shows the design of one side of a deck truss bridge built by the author for the large model railroad attraction described in Internet site 17. A second identical truss and cross bracing make up the complete bridge. The bridge has six panels, each 12.0 in square. The loads on the joints in the top chord are produced by a model train running over the bridge. Crossbeams span across the two trusses mounted at the panel points. Longitudinal beams run the length of the bridge and support the track, transmitting their loads to the cross beams and thus to the trusses.

FIGURE P4–31   Truss with one pinned support and one roller support.

Source of model: Gary Drigel.

(a) Photograph of a model of the truss for Problems 4–31 and 4–32

See photograph of model in Figure P4–31(a).

FIGURE P4–32   Truss for Problem 4–32.

$F_1 = 1200$ lb $F_2 = 1500$ lb

(b) Drawing of the truss with dimensions

FIGURE P4–33   Truss for Problem 4–33.

Member specifications:

AD, DE, EF L2 × 2 × ⅛ – doubled ⎯⎯ ⅃L

BD, CE, BE L2 × 2 × ⅛ – single ⎯⎯ ⅃

AB, BC, CF C3 × 4.1 – doubled ⎯⎯ ⅃⊏

FIGURE P4–34   Truss for Problems 4–34, 4–35, and 4–36.

FIGURE P4–37   Truss for Problem 4–37.

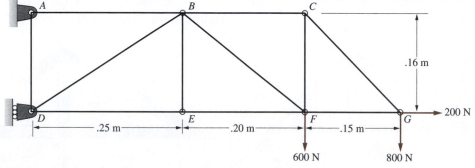

FIGURE P4–38   Truss for Problems 4–38 and 4–39.

	Problem No.	
	4-40	4-41
L	28.9 kN	28.9 kN
W_{TL}	1.58 kN	1.58 kN
W_{TR}	1.82 kN	1.82 kN
F_W	0 kN	3.80 kN

FIGURE P4–40   Truss structure for Problems 4–40 and 4–41.

FIGURE P4–43

FIGURE P4–44

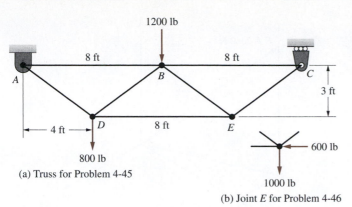

1200 lb

8 ft   8 ft

A

B

C

3 ft

4 ft

D   8 ft   E

800 lb

(a) Truss for Problem 4-45

600 lb

1000 lb

(b) Joint E for Problem 4-46

FIGURE P4–45   Truss for Problems 4–45 and 4–46.

Forces in Three-Dimensional Systems

4–48. Determine the x-, y-, and z-components of the force shown in Figure P4–48.

4–49. Determine the x-, y-, and z-components of the force shown in Figure P4–49.

4–50. Determine the x-, y-, and z-components of the force shown in Figure P4–50.

4–51. Determine the x-, y-, and z-components of the force shown in Figure P4–51.

4–52. Compute the direction cosines and the direction angles for the force shown in Figure P4–48.

4–53. Compute the direction cosines and the direction angles for the force shown in Figure P4–49.

12 lb typical

Top chord

12 in

12 in
typical

FIGURE P4–47   Deck truss bridge for a model railroad.

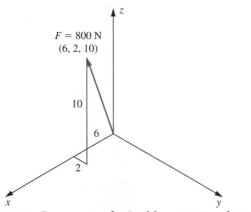

$F = 800$ N
$(6, 2, 10)$

10

6

2

x y

FIGURE P4–48   Force vector for Problems 4–48 and 4–52.

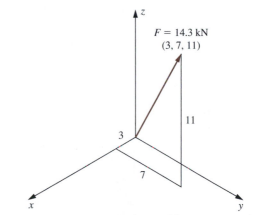

$F = 14.3$ kN
$(3, 7, 11)$

3

11

7

x y

FIGURE P4–49   Force vector for Problems 4–49 and 4–53.

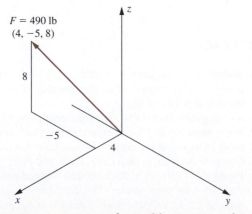

$F = 490$ lb
$(4, -5, 8)$

8

-5

4

x y

FIGURE P4–50   Force vector for Problems 4–50 and 4–54.

$F = 12\,400$ lb
$(-6, 3, 6.5)$

6.5

3

-6

x y

FIGURE P4–51   Force vector for Problems 4–51 and 4–55.

4–54. Compute the direction cosines and the direction angles for the force shown in Figure P4–50.

4–55. Compute the direction cosines and the direction angles for the force shown in Figure P4–51.

4–56. A force vector is known to have the two direction angles shown in Figure P4–56. Compute the third direction angle. Then, for $F = 116$ N, compute its x-, y-, and z-components.

FIGURE P4–58

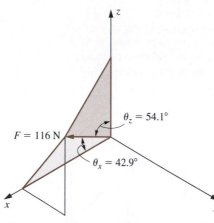

FIGURE P4–56

4–57. Figure P4–57 shows a force vector with a magnitude of 1182 lb. Its direction is denoted by the angles $\alpha = 54.0°$ and $\beta = 40.4°$.

   a. Compute the x-, y-, and z-components of the force.

   b. Compute the direction cosines and the direction angles for the force.

FIGURE P4–59

FIGURE P4–57

FIGURE P4–60

4–58. Figure P4–58 shows a force vector with a magnitude of 2360 N. Its direction is denoted by the angles $\alpha = 24.8°$ and $\beta = 43.0°$.

   a. Compute the x-, y-, and z-components of the force.

   b. Compute the direction cosines and the direction angles for the force.

4–59. A force vector is known to have the two direction angles shown in Figure P4–59. Compute the third direction angle. Then, for $F = 2352$ lb, compute its x-, y-, and z-components.

4–60. The lifting eye (eyebolt) shown in Figure P4–60 is subjected to a tensile load of 2250 lb. Compute the x-, y-, and z-components of the force acting on the eyebolt. (*Note:* This

analysis is critical because a lifting eye's load-carrying capacity is highly dependent on the direction of the load. See Internet sites 14 and 15.)

4–61. A ceiling-mounted HVAC unit is suspended by three cables as shown in Figure P4–61. Its weight is 1.85 kN and its center of mass lies at the center of the circular fan on the top. Compute the force in each cable, F_1, F_2, and F_3.

4–62. Figure P4–62 shows a platform and the equipment on it suspended by three cables. The combination has a total mass of 420 kg and its center of mass is at the point called *CM*. Compute the force in each cable, F_1, F_2, and F_3.

FIGURE P4–61

FIGURE P4–62

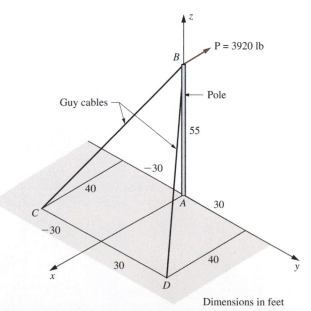

FIGURE P4–63 Strut supporting cover for railroad tank car for Problems 4–63 and 4–64.

4–63. A hatch cover on a railroad tank car is supported by a strut as shown in Figure P4–63. Compute the force on the strut if the hatch cover weighs 110 lb and its center of mass is at its center.

4–64. Using the value of the force in the strut from Problem 4–63, compute all forces acting at the hinges at points A and B. Hinge B can resist forces in the x- and z-directions only. Hinge A resists forces in the x-, y-, and z-directions.

4–65. Figure P4–65 shows a vertical pole supported at its base by a ball-and-socket joint that allows rotation in any direction but that resists forces in the x-, y-, and z-directions. The force $T = 3920$ lb represents the tension in a power line and it acts parallel to the x-axis (the ground). Find the forces in the two guy cables and in the pole.

4–66. Figure P4–66 shows an alternate layout for the pole described in Problem 4–65. Because of an obstruction in the area, mounting point D is located as shown. Find the forces in the two guy cables and in the pole. Compare the results with those of Problem 4–65.

4–67. A set of three cables supports a 1750-lb sculpture as shown in Figure P4–67. Mounting points B, C, and D all lie in the x-y plane of the ceiling of a room. The attachment to

FIGURE P4–65 Planned mounting for pole in Problem 4–65.

FIGURE P4–66   Alternate placement of Point *D* for pole from Problem 4–65.

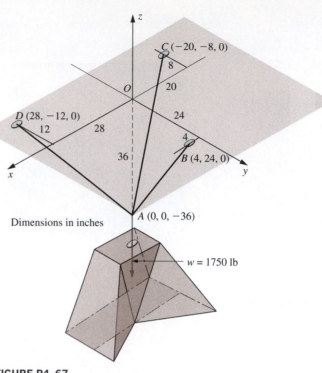

FIGURE P4–67

the sculpture is 36 in below the origin of the *x-y-z* axis system. Compute the force in each cable.

4–68. The jib-crane shown in Figure P4–68 carries a traveling hoist that moves between points *A* and *B*. The maximum load carried by the hoist is 13.3 kN. Compute the forces in the two struts *AC* and *AD* supporting the jib when the hoist is at *A*. The connection at *E* is free to rotate but can resist forces in the *x-*, *y-*, and *z*-directions.

4–69. Consider the transmission tower in Figure 4–2(a). The cable *BC* is located as shown in Figure P4–69 and carries a maximum tensile load of 66.00 kN. The lower end of the

cable at *B* is fastened by a swivel bolt to a rigid concrete pad. Determine the components of the cable force in the *x-*, *y-*, and *z*-directions and the direction cosines for the cable force.

4–70. Consider the roof-mounted derrick crane in Figure 4–2(c). Figure P4–70 shows more detail for the dimensions of the derrick crane and the placement of the supports. The pivot at *A* permits rotation about the *z*-axis while its bearing

FIGURE P4–68

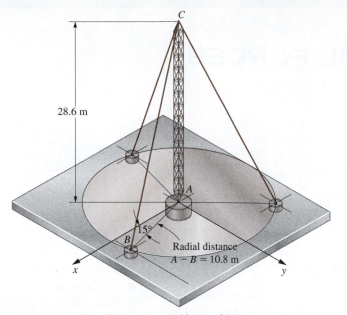

Antenna tower with guy wires

FIGURE P4–69

FIGURE P4–70   Roof-mounted derrick crane.

provides support in the x-, y-, and z-directions. The tower AB is vertical. The struts BC and BD are rigid and mounted by swivel joints at each end. The boom is hinged at its lower end and maintained at the position shown by a cable from a winch mounted at the top of the tower. For

this problem, the combination of the boom, tower, and cable can be considered a rigid body. A second winch at B raises and lowers loads. The cable force due to the HVAC unit is 4800 lb. Compute the forces on the tower and in the two struts.

FRICTIONAL FORCES

MAJOR TOPICS
5–1 The Big Picture
5–2 Coefficient of Friction
5–3 Analysis of Frictional Forces

5–1 THE BIG PICTURE

In previous chapters forces encountered were due to the weight of structural elements, externally applied loads, wind, snow, mechanical actuation, and similar conditions. When we analyzed problems using the equations of equilibrium, it was always the case that the members, their supports, and connections to other members allowed the system to withstand any force and maintain equilibrium. It is desirable to design systems that will always be fully constrained and under control, enabling them to perform as expected. We also emphasized that such analyses are necessary to be able to determine stresses and deformations that are discussed in later chapters in strength of materials.

However, an important force was neglected—*friction*.

It is quite likely that you are familiar with the basic nature of friction because it affects many aspects of your daily existence. You may also have studied the fundamental concepts in earlier science classes. Let's recall the basic definition of *friction*:

> *When a body is in contact with a surface and a net force tends to cause motion in a particular direction, a frictional force is created that acts opposite to the direction of the net force. That is, the frictional force tends to oppose impending motion.*

The field of study that deals with friction is called *tribology*. More in-depth information can be found in References 1–7 and Internet sites 1–3.

Think about your activities in the last day and identify some situations in which friction played a significant part.

- Were there cases in which you needed to use friction to accomplish a given task?
- Did friction sometimes make a job more difficult?
- Would someone have gotten injured if sufficiently high frictional forces did not exist?

Does your list include items like these?

1. Did you climb stairs or walk across a floor or up a ramp?
   - All of those simple tasks would be dangerous without friction.

2. Did you drive or ride in a car?
   - The tires would spin uncontrollably if friction between the tires and the roadway did not exist. Think of driving on ice!
   - You would have a hard time using the steering wheel unless you continuously grabbed some feature of the wheel.
   - Even if you could turn the steering wheel, the tires would not control the car in a turning maneuver without friction.
   - You would likely have slid around on the seats without friction.
   - Your brakes would not have worked to stop the car without friction.

3. Did you try to push or pull something heavy?
   - High frictional force made it more difficult.
   - Did you consider some means of reducing the friction to make it easier?
   - Are you aware of products like that featured at Internet site 8 that ease the movement of furniture and appliances?

4. Did you move a mouse with a trackball or did you operate a joystick control?
   - The trackball would not have rolled responsively without friction.
   - The joystick would not maintain a stable position without being held steady by constant force.

Objectives of This Chapter

In this chapter you will review the principles of friction and apply frictional forces as part of the static equilibrium analysis. After completing this chapter, you will be able to:

1. Define the *coefficient of friction* and recognize typical values for it in common situations.

2. Compute the frictional force acting on objects within systems you are analyzing.

3. Account for frictional forces in static equilibrium analyses.

4. Apply frictional forces to practical situations involving attempts to move objects.

5. Analyze the conditions under which an object may tend to tip instead of slide when being pushed or pulled.

6. Analyze the functions of a wedge.

7. Distinguish among static friction, kinetic friction, and rolling friction.

8. Consider the effects of lubrication as a means of reducing friction.

5–2 COEFFICIENT OF FRICTION

The basic definition of friction given in the previous section can be visualized by the simple sketch in Figure 5–1. A book is resting on the flat, horizontal surface of a desk. An applied force F acts to the left tending to push it in that direction. You know from experience that the book would not immediately move under a light force and that you would have to increase the force to some particular level to actually move the book. The reason is that a frictional force, F_f, is created between the bottom of the book and the surface of the desk. If the book does not move, then the principle of equilibrium states that the applied force and the frictional force are equal; they are balanced.

You also know that you will be able to move the book by applying a gradually increasing force. It is important to be able to predict the magnitude of the limiting force and answer these questions:

1. What factors affect the final force value that moves the book?

2. Does it matter how much the book weighs?

3. Does it matter how large it is; that is, how much surface area there is between the bottom of the book and the desktop?

4. Does it matter what material the book is made from?

5. Does it matter what the surface of the desk is made from?

These questions can be answered by considering carefully the definition of the *coefficient of friction*. Figure 5–1 is incomplete and more detail is required to fully understand the nature of frictional forces. Refer now to Figure 5–2 in which we have added the forces in the vertical direction to the horizontal forces shown in Figure 5–1, resulting in a complete free body diagram of the book. The weight of the book, W, acts vertically downward through the book's center of mass. To ensure equilibrium in the vertical direction, the desktop must push back upward with just enough force to balance the applied weight force. We call that the *normal force, N*, because it acts perpendicular, or *normal*, to the surface of the book. Furthermore, we are considering here only the *sliding situation*. Other types of friction are considered later.

The value of the frictional force in this *sliding situation* depends on the following conditions:

1. When the applied force is smaller than that required to move the book, the frictional force is equal to the applied force.

2. There is a maximum value that the frictional force can be before the book actually slides.

3. At the instant that the book tends to slide, we say that there is *impending motion*.

4. At the point of impending motion, the frictional force is maximum and has the value

$$F_{f\text{-}max} = \mu_s N \qquad (5\text{–}1)$$

5. The term μ_s (Greek mu) is the *coefficient of static friction*, now defined as

$$\mu_s = F_{f\text{-}max}/N \qquad (5\text{–}2)$$

Equations 5–1 and 5–2 are called *Coulomb's laws of friction*.

6. The value of the coefficient of static friction does, indeed, depend on the nature of the surface of the book and the surface of the desktop. It also depends on whether the surfaces are smooth or rough, dry or lubricated with water, oil, dust, grease, or any other substance between them.

7. When the book actually moves, overcoming the maximum value of the frictional force, the effective coefficient of friction typically drops. It is convenient to then define a separate term, the *coefficient of kinetic friction, μ_k*. The word *kinetic* refers to the object being in motion.

8. Some nominal values for the static coefficient of friction, μ_s, and the kinetic coefficient of friction, μ_k, for commonly encountered combinations of materials are given in Table 5–1. Be aware that variations exist based on smoothness of the surfaces, cleanliness, and the specific nature of any lubricant present. The data for coefficient of friction for lubricated surfaces are typically for light oils. For critical applications, testing using actual materials is recommended. See also Internet sites 1–3.

9. The value of the coefficient of friction does *not* depend on the size of the contacting areas for most common situations.

10. There are, however, conditions where the size of the contacting surfaces may affect the frictional force. If the area becomes quite small, approaching a knife edge, then the small object may actually "dig into" or "cut" the

FIGURE 5–1   Forces acting on a book resting on a desk.

FIGURE 5–2   Forces acting on a book resting on a desk—complete free body diagram.

TABLE 5–1 Typical Values for Coefficient of Friction for Sliding Situations

Materials	Dry/clean surfaces		Lubricated	
	μ_s	μ_k	μ_s	μ_k
Steel on steel	0.78	0.42	0.11	0.03
Steel on cast iron	—	0.23	0.18	0.13
Steel on babbitt (a bearing material)	0.42	0.17	0.35	0.14
Steel on phosphor bronze	—	0.34	—	0.17
Steel on aluminum	0.61	0.47	—	—
Nickel on nickel	0.90	0.53	0.28	0.12
Aluminum on aluminum	1.05	—	0.30	—
Cast iron on cast iron	1.10	0.15	—	0.07
Zinc on zinc	0.60	—	0.04	—
Copper on copper	1.0	—	0.08	—
Copper on steel	0.53	0.36	—	0.18
Copper on cast iron	1.05	0.29	—	—
Brake lining on cast iron	0.40	—	—	—
Graphite on steel	0.10	—	0.10	—
Glass on glass	0.94	0.40	—	—
Leather on steel	0.60	—	0.20	—
Rubber on asphalt	—	0.65	—	0.40 (water)
Rubber on concrete	—	0.73	—	0.60 (water)
Nylon and LDPE*	0.40	—	0.15 (water)	—
PVC+, Plexiglas, Polystyrene	0.50	—	—	—
Teflon (PTFE) on steel	0.04	—	0.04	0.04
Filled PTFE (15% glass)	0.12	—	—	—

Data are for example only from a variety of sources. Testing with actual materials and lubricants to be used is recommended. See References 1–7 and Internet sites 1–3.

* Low-density polyethylene
+ Polyvinyl chloride

supporting surface, causing a higher effective frictional force. The behaviors of objects on the microscale and nanoscale are also generally quite different from the more typical macroscale examples we encounter in normal work.

Other Kinds of Friction

With its coverage of *statics,* this book focuses on applications of friction when the tendency for sliding occurs, as discussed earlier. Even the mention of kinetic friction in the preceding paragraphs and Table 5–1 moves into the realm of *dynamics,* although that is a natural extension of the subject. Mentioned here, for a more complete discussion, are other kinds of friction.

Rolling Friction The ancient invention of the wheel was a major milestone in the field of mechanics. It was an important discovery that it is much easier to move an object if it is round and can be rolled along instead of having to overcome sliding frictional resistance by pushing or pulling. For example, heavier loads could be moved with significantly

less effort by placing round logs beneath them. Then the development of wheels on axles supporting platforms was put to use as carts. Eventually these carts became means of transportation. Now we see a plethora of wheels and other applications of rolling friction:

- Steel wheels on railroad locomotives rolling on steel rails allow the movement of huge payloads with relatively low amounts of energy.
- Rubber tires on automobiles, trucks, bicycles, motorcycles, material handling equipment, and taxiing aircraft allow convenient mobility for people and the transport of goods.
- Roller conveyors permit the easy movement of products and packages in manufacturing and shipping.
- Ball and roller bearings employ precision rolling elements to reduce friction for rotating shafts.
- Ball screws and other linear motion devices use rolling elements to minimize friction in machine tools and automation equipment.
- Highly engineered plastics are used for skateboard wheels and for rollers in automation equipment.

■ Rollers are often used to support doors in residential, industrial, and commercial buildings to permit easy opening by hand or with powered actuators.

The mechanism of rolling friction is quite different from the sliding situation. The inherent elasticity of the materials in contact results in small deformations of the rolling element and the surface on which it is rolling. There are obvious differences between the deformation of stiff materials, such as steel used in railroad wheels, and more flexible materials, such as rubber used in road vehicles. Often designers seek a balance between low friction and smooth, quiet motion when specifying materials for wheels and rollers.

Tests are conducted to determine the amount of force required to move the axle supporting a wheel while carrying a given load. Typically, the amount of force depends not only on the material from which the wheel is made but also on the size of the wheel, measured by its radius. The resulting equation for rolling friction is

$$F_f = f_R(N/R) \tag{5–3}$$

where

F_f = force applied at the axle required to move the wheel

f_R = coefficient of rolling friction

N = normal force between the wheel and the surface

R = radius of the wheel

TABLE 5–2 Typical Values for Coefficients of Rolling Friction

Materials	f_R	
Steel on steel	0.50 mm	0.02 in
Cast iron on cast iron	0.50 mm	0.02 in
Cast iron on wood	5.60 mm	0.22 in
Wood on steel	1.25 mm	0.05 in
Wood on wood	1.50 mm	0.06 in
Hard plastic on steel	2.00 mm	0.08 in
Pneumatic rubber tire on concrete	25.0 mm	1.00 in
Hard rubber on concrete	15.0 mm	0.60 in
Hard rubber on steel	7.50 mm	0.30 in

The resulting equation for computing f_R is

$$f_R = F_f/(R/N) \tag{5–4}$$

Table 5–2 shows typical values for f_R. Make careful note that f_f is *not* dimensionless, having the length dimension from the radius term. We will use millimeters (mm) and inches (in) because they are convenient units for measuring the radii of wheels and rollers. See Internet site 1 for more data. As with sliding friction, there are numerous variations in conditions that can affect the actual values and testing is recommended.

Example Problem 5–1

Determine the horizontal force required to move the two axles (four wheels) of a cart that carries a load having a mass of 340 kg over a floor covered by steel plate. The wheels have a diameter of 300 mm and are made from (a) hard rubber, (b) hard plastic, and (c) steel.

Solution   Given: Cart load = 340 kg

Find: Horizontal force required to move the cart against rolling friction

Analysis and Results:
1. First express the load in force units of newtons.
2. Divide by four to determine the downward force on each wheel.
3. For equilibrium, the upward normal force, N, equals the downward force.
4. Determine the coefficient of friction from Table 5–2.
5. Compute the force of rolling friction from Equation (5–3).
6. Multiply by four to determine the total force.

Step 1: $W = m \cdot g = (340\,\text{kg})(9.81\,\text{m/s}^2) = 3335\,\text{N}$

Step 2: Force on each wheel = $W/4 = (3335\,\text{N})/4 = 833.9\,\text{N}$

Step 3: Normal force between each wheel and the floor: $N = 833.9\,\text{N}$

Step 4: a. $f_R = 7.5\,\text{mm}$ for hard rubber wheels on steel floor
b. $f_R = 2.0\,\text{mm}$ for hard plastic wheels on steel floor
c. $f_R = 0.5\,\text{mm}$ for steel wheels on steel floor

Step 5: a. $F_f = f_R(N/R) = (7.5\,\text{mm})(833.9\,\text{N})/(150\,\text{mm}) = 41.7\,\text{N}$
b. $F_f = f_R(N/R) = (2.0\,\text{mm})(833.9\,\text{N})/(150\,\text{mm}) = 11.1\,\text{N}$
c. $F_f = f_R(N/R) = (0.5\,\text{mm})(833.9\,\text{N})/(150\,\text{mm}) = 2.78\,\text{N}$

Step 6: Total force for four wheels:
a. $F_T = 4.0(F_f) = 4.0(41.7\,\text{N}) = 167\,\text{N}$
b. $F_T = 4.0(F_f) = 4.0(11.1\,\text{N}) = 44.4\,\text{N}$
c. $F_T = 4.0(F_f) = 4.0(2.78\,\text{N}) = 11.1\,\text{N}$

Comment: The steel wheels have the minimum total rolling friction force. However, the noise of steel rolling on steel may be objectionable. The rubber wheels would likely be the quietest but would require more force to move the cart. The hard plastic wheels may be the optimum choice. Other factors to consider are cost and wear life.

Rotating Bearing Friction Rotating shafts often are supported by cylindrical bearings made from such materials as bronze, plastic, or babbitt (an alloy of lead, copper, and tin). Each can provide sufficient support for the shaft while offering relatively low frictional resistance and satisfactory wear life. Some such bearings are lubricated with light oils that coat the surfaces and reduce further the frictional forces. The nature of the frictional force depends on several factors:

- The load on the bearing
- The relative speed of the rotating shaft and the stationary bearing
- The size of the shaft at the bearings
- The viscosity and lubricity of the lubricant and the surfaces themselves

For relatively heavy loads operating at low speeds, the friction is likely to be of the *boundary type* where the moving and the stationary surfaces are in contact. Internet sites 4–7 show commercially available materials and bearing designs tailored for boundary-lubricated operation. See also Reference 8.

With careful design and adequate supply of lubricant with a suitable viscosity, bearings can operate with a full film of lubricant between the rotating shaft and the stationary bearing. This type of bearing is called a *full-film hydrodynamic* bearing and gives very low friction and long life. See Internet site 7 and Reference 8 for examples of commercially available hydrodynamic bearings and additional discussion of design methods.

Hydrostatic Bearings. Some very heavy, slow-moving equipment is supported on a series of pads to which high-pressure oil is supplied by hydraulic pumps. With careful design, the pressure can exert sufficient force to raise the equipment and allow it to be moved with very little force. Some large telescopes and radar antennas employ these hydrostatic bearings. See Reference 8 for additional discussion of design methods.

Air Bearings. Similar to hydrostatic bearings, a continuous supply of compressed air can be delivered to bearing surfaces and permit the devices to float gently on a film of air. Such devices can range from small discs on a physics lab air table to very large fixtures carrying locomotives and large machined parts.

Brakes, Clutches, and Belt Drives Some of the most useful applications of friction are in brakes and clutches that are used to control motion on highway vehicles, manufacturing equipment, cranes, construction machinery, and many other types of systems. Brakes slow the rotation of axles and rotors or bring them to a stop. Clutches engage and disengage a driver (such as an electric motor or engine) to a driven machine. This allows the driver to operate continuously at an efficient speed while providing intermittent motion to the driven machine. Mating friction surfaces are typically employed to actuate these devices. See Reference 8.

Belt drives are used in many vehicles and mechanical devices to transmit motion and power from a motor or engine to a driven machine, typically with a speed reduction. Flexible flat or V-shaped belts wrap around pulleys (called sheaves) requiring careful analysis of the friction between them to ensure reliable power transmission without slippage. See Reference 8.

Some of the following sections address these important applications of friction.

5–3 ANALYSIS OF FRICTIONAL FORCES

Here we show various methods of including frictional forces in the overall analysis of the forces on static structures and mechanical devices. Virtually all the techniques used in preceding chapters apply here as well: equations of static equilibrium, nature of forces at supports, free body diagrams, two-force members, solution of vector triangles, law of sines, law of cosines, and the principle of action–reaction at mating joints and surfaces. Example problems are used to illustrate the methods of including friction. It is essential that you generalize the fundamental principles and problem-solving methods so you can apply them to other problems that may look different.

Example Problem 5–2 illustrates testing to determine the coefficient of static friction.

Example Problem 5–2 A special test is set up to determine the coefficient of static friction between an aluminum block with a mass of 4.49 kg and a ground finish against a special plastic sheet material. The plastic is fixed to a rigid, flat, horizontal surface as the block is pulled with a horizontal force. A force gage is used to measure the applied force and it indicates that the maximum force exerted at the moment of the inception of motion was 9.46 N. Compute the coefficient of static friction for this pair of surfaces.

Solution Given: Mass of aluminum block = 4.49 kg

Maximum force to initiate motion of block on plastic sheet = 9.46 N

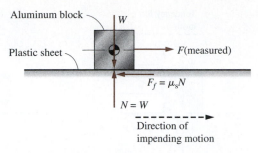

FIGURE 5–3   Basic setup for measuring static coefficient of friction for Example Problem 5–2.

Find: Coefficient of static friction for the aluminum on the plastic sheet

Analysis and Results: First draw the free body diagram of the block as shown in Figure 5–3.

The block's weight acts vertically downward through its center of mass.

The normal force, N, acts vertically upward and is equal to the weight.

Then express the weight of the aluminum in units of newtons.

$$W = m \cdot g = (4.49\,\text{kg})(9.81\,\text{m/s}^2) = 44.05\,\text{N}$$

At the point of impending motion, the friction force, F_f, equals the applied horizontal force, F. That is, $F_f = F = 9.46\,\text{N}$. Furthermore, F_f is at its maximum value.

Now set $F_f = \mu_s \cdot N = \mu_s \cdot W$

Solve for $\mu_s = F_f/W = (9.46\,\text{N})/(44.05\,\text{N}) = \underline{0.215}$ **Answer**

Possible Variation of Frictional Force

Figure 5–4 shows five possible cases for loading of an object and the resulting force system on a free body diagram of the object, including the frictional force.

a. The object sits on a flat horizontal surface while an externally applied vertical force is added. The upward normal force, N, balances the sum of W and F. Because there is no force acting in the horizontal direction that tends to slide the block on the surface, the frictional force is zero.

b. The object sits on a flat horizontal surface while a force, F, is applied to the side of the side of the object at some angle, θ, above the horizontal. Although there is a tendency to move the object to the right, no motion occurs. The frictional force, F_f, is equal to the horizontal component, F_x, and acts in the opposite direction. The normal force is

$$N = W - F_y$$

The resultant of N and F_f is called R and the angle between N and R is called α. It is important to note that *the frictional force is less than the maximum possible value.*

c. The applied force, F, has been increased above the value in (b) to the point where motion to the right is impending. At this condition, *the friction force is at its maximum value for which the coefficient of static friction is μ_s.* That is,

$$F_f = F_{f\text{-}max} = \mu_s\,N$$

Now the resultant force on the bottom of the object, R, acts at an angle α such that

$$\tan\alpha = F_f/N = \mu_s N/N = \mu_s$$

Here we define this angle to be the special value, φ_s, and we call it the *angle of friction.* That is,

$$\tan\varphi_s = \mu_s \tag{5–5}$$

and

$$\varphi_s = \text{Tan}^{-1}\mu_s \tag{5–6}$$

d. After the object begins to move, the coefficient of kinetic friction, μ_k, applies and the frictional force reduces to $F_f = \mu_k\,N$. Typically, $\mu_k < \mu_s$ and, therefore, the applied force must be decreased to maintain the velocity of the object to a set value. In this case of motion with uniform velocity, the resultant, R, on the bottom of the object acts at an angle φ_k from the normal force, where, using a logic similar to that in (c),

$$\tan\varphi_k = \mu_k \tag{5–7}$$

and

$$\varphi_k = \text{Tan}^{-1}\mu_k \tag{5–8}$$

e. If the applied force, F, is increased above the value from (d) that maintains uniform velocity of the object, the forces will no longer be uniform and the object accelerates to the right with an increasing velocity. Study of this type of motion is a part of the subject of dynamics and is not covered in this book.

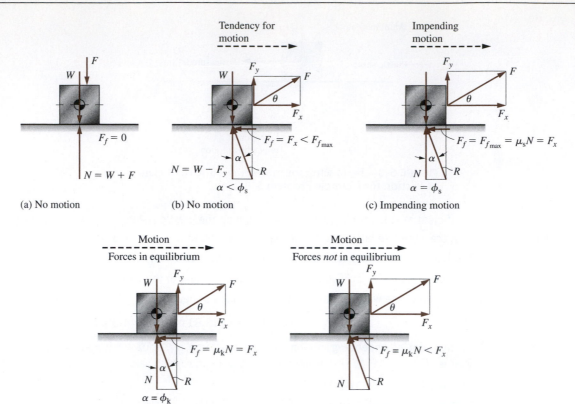

FIGURE 5–4   Possible variations of frictional force systems.

Figure 5–5 shows a graph that summarizes and illustrates the variation of the frictional force, F_f, for the cases shown in Figure 5–4 and discussed previously.

1. The frictional force starts at zero when no force is applied that tends to move the object along the supporting surface.

2. With increasing force tending to cause motion, the frictional force increases to just that value that will balance the external force. The frictional force is less than the maximum possible value.

3. When the applied force is sufficient to develop impending motion, the frictional force is equal to its maximum value, $F_f = F_{f\text{-}max} = \mu_s N$.

4. After motion starts, the frictional force decreases to the value, $F_f = \mu_k N$, called the kinetic frictional force.

5. If, after the initial acceleration of the object to a small velocity, the applied force is adjusted to balance the kinetic frictional force, the object moves with uniform velocity.

6. If the applied force is increased above the kinetic frictional force, the object will accelerate.

Testing for the Coefficient of Static Friction

Example Problem 5–2 demonstrated one method of testing for the coefficient of friction using a force gage to pull an object over a flat surface (see Figure 5–3). Figure 5–6 shows a schematic of an alternate setup that allows the measurement of the coefficient of static friction using a

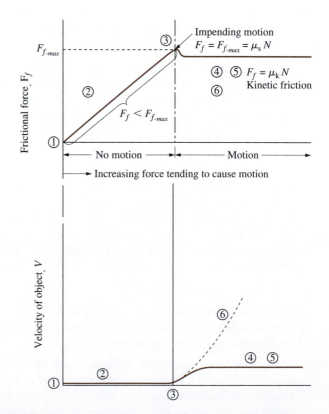

FIGURE 5–5   Friction force and velocity for the cases shown in Figure 5–4.

θ = Angle of inclination of ramp
W = Weight of block
$W_x = W\sin\theta$
$W_y = W\cos\theta$
N = Normal force on block
F_f = Maximum frictional force $= \mu_s N$
R = Resultant of N and F_f
μ_s = Coefficient of static friction

For impending motion:
$R = W$
$N = W_y$
$F_f = W_x$
$\tan\theta = \dfrac{W_x}{W_y} = \dfrac{F_f}{N} = \dfrac{\mu_s N}{N} = \mu_s$

Therefore: $\mu_s = \tan\theta$
θ = angle of static friction $= \text{Tan}^{-1}\mu_s = \phi_s$
Also: ϕ_s is called the *angle of repose*

FIGURE 5–6   Testing for μ_s with inclined ramp—angle of static friction—angle of repose

ramp with an adjustable slope. The ramp pivots about A at the lower right. While the position adjustment device on the left is moved slowly to the right, it causes the ramp's slope to increase. A protractor fixed to the frame allows measurement of the angle θ, indicating the slope of the ramp. The ramp is prepared with a surface of interest for which the value of the coefficient of static friction between it and the material on the bottom of an object of any weight is desired.

When placing the object on the ramp with a small slope, it will tend to stay in place. You should recognize that at some larger angle, the object will slide down the ramp. The goal of the test is to slowly increase the angle until motion is initiated and then to measure carefully the angle of inclination of the ramp. *The setup is shown in Figure 5–6 at the moment of impending motion of the object down the ramp.* Note carefully the force system created on the free body diagram of the object, summarized next.

1. We show the x- and y-axes to be aligned parallel and perpendicular to the ramp surface.
2. The angle of inclination of the ramp, θ, is the angle between the ramp and the horizontal.
3. The weight of the object acts vertically downward through its center of mass.
4. We resolve the weight into components in the x- and y-directions, W_x and W_y, where,

$$W_x = W\sin\theta \quad \text{and} \quad W_y = W\cos\theta$$

5. The angle θ is the angle between the vertical weight vector and the W_y vector directed perpendicular to the ramp surface. Identifying the proper placement of the angle θ is a very important observation and one that many people have difficulty with at first.
6. Now note the following relationships that describe the force system at the point of impending motion:
   a. For equilibrium: $\Sigma F_x = 0$.
   $$W_x = F_f = F_{f\text{-}max} = \mu_s N$$
   b. For equilibrium: $\Sigma F_y = 0$. $N = W_y$
   c. R is the resultant of N and $F_{f\text{-}max}$ and it is equal to W.
   d. Therefore, the angle between N and R is also equal to θ.
   e. Now, $\tan\theta = F_{f\text{-}max}/N = (\mu_s N)/N = \mu_s$.
   f. Therefore, by Equations (5–5) and (5–6), $\theta = \varphi_s$ = the angle of friction for the surfaces. *Note also that the magnitude of the weight of the object is irrelevant.*
   g. Also, $\varphi_s = \text{Tan}^{-1}\mu_s$.
   h. Thus, this setup allows the measurement of the coefficient of static friction by the simple measurement of the angle of inclination of the slope of the ramp when impending motion occurs.

Example Problem 5–3 illustrates the use of the ramp to determine the coefficient of static friction.

Example Problem
5–3

The ramp shown in Figure 5–6 is at an angle of 23.3° when the object just begins to slide down. Compute the coefficient of static friction for the pair of surfaces between the ramp and the bottom of the object.

Solution   Given: Figure 5–6 with impending motion and $\theta = 23.3°$.

Find: Coefficient of static friction for the test surfaces.

Analysis and Results: We use the analysis completed previously allowing the direct computation of the coefficient of static friction, μ_s, from

$$\text{Angle of friction} = \varphi_s = \theta = 23.3°$$

$$\mu_s = \tan \theta = \tan \varphi_s = \tan 23.3° = 0.431$$

We now summarize the principles found so far:

1. The simple ramp system shown in Figure 5–6 can be used to find the coefficient of static friction by measuring the angle at which the object begins to slide.

2. That angle, φ_s, is called the *angle of friction* and also the *angle of repose*.

3. The coefficient of static friction is found from $\mu_s = \tan \varphi_s$.

Force Analysis Including Friction

We can use the analysis methods from the ramp problem just completed to solve general problems involving horizontal or sloped surfaces as demonstrated in the following example problems.

Example Problem
5–4

Figure 5–7 shows a woman pushing a refrigerator weighing 250 lb across a kitchen floor. (a) Compute the force required to cause the refrigerator to slide if the coefficient of static friction is 0.35. (b) If the woman applies the force from (a) at a height of 56 in from the floor, will the refrigerator slide or tip?

Solution   Given: Figure 5–7; $W_R = 250$ lb; $\mu_s = 0.35$; force applied 56 in from floor

Find: a. Force required to initiate motion of the refrigerator
    b. Will refrigerator slide or tip?

Analysis and Results: a. We first develop a free body diagram of the refrigerator as shown in Figure 5–8(a) that assumes it will be in a state of impending motion to slide to the left. The weight of the refrigerator acts vertically downward through its center of mass, assumed to be at its middle. The applied force *F* acts horizontally to the left. The frictional force acts to the right, opposing the impending motion. The reaction forces on the feet of the refrigerator act vertically upward to balance the weight.

FIGURE 5–7 Woman pushing a refrigerator for Example Problem 5–4.

66 in

31 in

FIGURE 5-8 Free body diagrams for Example Problem 5-4.

Desired movement

M_t

M_t = Tipping moment
$M_t = F \cdot h$

W

F

F_f

R_1 R_2

d

(a) Forces acting on refrigerator while moving

W

h

F_f A

$M_2 = W \cdot d/2$ = Righting moment

$d/2$ R_1

(b) Forces when tipping occurs

Equilibrium analysis: Let's call the sum of the reaction forces on the four feet the *normal force, N.* Then

$$\Sigma F_y = 0 = N - W$$

Then

$$N = W = 250 \text{ lb.}$$

Assuming impending motion of the refrigerator,

$$\Sigma F_x = 0 = F_f - F$$

$$F = F_f = \mu_s N = (0.35)(250 \text{ lb}) = \textbf{87.5 lb} \qquad \textbf{Answer}$$

b. To evaluate whether the refrigerator will tip or slide with the force applied at 56 in from the floor, refer to Figure 5-8(b). If tipping occurs, the right feet will not support any load and all of the weight is borne on the left feet, called point *A.* We can then sum moments about *A.*

$$\Sigma M_A = F \cdot h - W \cdot d/2$$

Let's call the first term, $F \cdot h$, the *tipping moment* $= M_T$.

Call the second term, $W \cdot d/2$, the *righting moment*, M_R, indicating that it tends to bring the refrigerator back to its original position, that is, to *right it*. Now, if the tipping moment is greater than the righting moment, the refrigerator will tip, and vice versa. Evaluating each term separately gives

$$F \cdot h = (87.5 \text{ lb})(56 \text{ in}) = 4900 \text{ lb} \cdot \text{in} = M_T$$

$$W \cdot d/2 = (250 \text{ lb})(31 \text{ in}/2) = 3875 \text{ lb} \cdot \text{in} = M_R$$

Therefore, $M_T > M_R$, and the refrigerator will tip.   **Answer**

Comment: This type of analysis for tipping or sliding should be considered on any similar problem. The tendency to tip is greater for objects that are tall relative to their width, like the refrigerator. They, in turn, are sensitive to the position of the force tending to move them relative to the supports.

The following example problem shows an additional feature of friction problems in which the applied force acts at an angle to a ramped surface.

Example Problem 5-5   Figure 5-9 shows a part of a material handling system in an ice cream packing plant. Cartons of ice cream weighing 24.0 N are pushed up a ramp at an angle of $\theta = 22.0°$ onto a cart where the ice cream is taken to a freezer. A pneumatic cylinder provides the pushing force and its rod acts horizontally. Use a coefficient of static friction of 0.40. (a) Compute the required cylinder force for this task. (b) Will the carton slide down the ramp if not acted on by the pneumatic cylinder?

FIGURE 5–9   System for loading ice cream on a cart for Example Problem 5–5.

Solution Given: Figure 5–9; $\theta = 22°$, $W = 24.0\,N$, $\mu_s = 0.40$

Find: Force F_c required by the cylinder to initiate motion up the ramp. Determine if the carton will slide down the ramp without the cylinder force.

Analysis and Results: We first draw the free body diagram of the carton as shown in Figure 5–10(a) for impending motion up the ramp. The weight and cylinder forces are resolved into their x- and y-components parallel and perpendicular to the ramp surface. The normal force acts perpendicular to the surface and the frictional force acts downward to oppose the impending motion of the carton upward.

Components of Forces: $W_x = W\sin\theta = (24.0\,N)(\sin 22°) = 8.991\,N$

$$W_y = W\cos\theta = (24.0\,N)(\cos 22°) = 22.25\,N$$

$$F_{c-x} = F_c \cos 22° = 0.9272\,F_c$$

$$F_{c-y} = F_c \sin 22° = 0.3746\,F_c$$

Part (a): Equilibrium analysis

$$\Sigma F_y = 0 = N - W_y - F_{c-y}$$

Then

$$N = W_y + F_{c-y} = 22.25\,N + 0.3746\,F_c$$

$$\Sigma F_x = 0 = F_{c-x} - W_x - F_f$$

$$0 = 0.9272\,F_c - 8.991\,N - 0.4\,N$$

$$0 = 0.9272\,F_c - 8.991\,N - 0.4(22.25\,N + 0.3746\,F_c)$$

(a) Impending motion up the ramp with cylinder force

(b) Impending motion down the ramp with no cylinder force

FIGURE 5–10   Ramp for Example Problem 5–5.

Solving for F_c gives

$$F_c = (17.891\ N)/(0.7774) = \textbf{23.0 N} \quad \textbf{Answer}$$

Part (b): Now we draw a second free body diagram of the carton as shown in Figure 5–10(b). No force is applied by the cylinder and we check for the carton sliding down the ramp.

$$\Sigma F_y = 0 = N - W_y = N - 22.25\ N$$
$$N = 22.25\ N$$
$$\Sigma F_x = 0 = F_f - W_x = F_f - 8.991\ N$$
$$F_f = 8.991\ N \text{ required for equilibrium without sliding}$$

Maximum frictional force available if impending motion occurs:

$$F_{f\text{-}max} = \mu_s N = 0.40(22.25\ N) = 8.90\ N$$

LESSON 8

Because there is slightly higher force down the ramp than can be resisted by friction, **the carton will slide.** **Answer**

Wedges Wedges are often used to position heavy loads. Typical wedges are basically block shaped with one face machined at a small angle, say, less than 10°. Placing the angled surface under the object to be moved and then applying a force perpendicular to the desired movement causes the load to be moved. Look at the example shown in Figure 5–11 in which a beam spans between supports and carries a heavy air-conditioning unit. When initially installed, support B at the right end was lower than support A at the left end. A wedge was then used to raise B to provide a level mounting platform. The wedge is driven to the left with a hammer causing the beam to be raised.

The solution procedure for wedges shown in the following example problem illustrates the use of resultant forces, the angle of friction, and vector triangles to determine unknown quantities.

FIGURE 5–11 Air-conditioning installation for Example Problem 5–6.

Example Problem 5–6 Figure 5–11 shows a beam spanning an opening that will carry a heavy air-conditioning unit. A 6° wedge is to be inserted under the right support at B to level the beam. The air conditioner weighs 1.42 kN and its center of mass is shown. Determine the force F that must be applied to the wedge to raise the end of the beam. Use $\mu_s = 0.30$ for all surfaces.

Solution Given: Figure 5–11; $W = 1.42\ kN, \mu_s = 0.30$

Find: Force F required to raise the end of the beam with the wedge

Analysis and Results: We first determine the reaction force at support B, called R_2. Then we draw the free body diagram of the wedge and compute the force F using a vector triangle.

Reaction B on the Beam:

$$\Sigma M_A = 0 = (1.42\ kN)(1046\ mm) - R_2(1220\ mm)$$
$$R_2 = (1485.3\ kN \cdot mm)/(1220\ mm) = 1.217\ kN$$

Free Body Diagram of Wedge: See Figure 5–12(a). Note the following:

1. The actuating force to insert the wedge is F acting horizontally to the left. This force is to be found.
2. The beam reaction $R_2 = 1.217\ kN$ acts vertically downward on the sloped top of the wedge.
3. The normal force on the bottom of the wedge acts vertically upward.
4. For equilibrium in the vertical direction, $R_2 = N_B = 1.217\ kN$.

(a) Free body diagram of wedge   (b) Vector triangle for R_T, R_B, F

FIGURE 5–12 Forces on wedge for Example Problem 5–6.

5. The frictional force on the bottom of the wedge acts toward the right, opposing the motion of the wedge.

6. Because motion is impending, the resultant force on the bottom, R_B, acts at an angle equal to the friction angle, φ_s, from the normal.

$$\varphi_s = \text{Tan}^{-1} \mu_s = \text{Tan}^{-1}(0.30) = 16.7°$$

7. The frictional force on the top of the wedge acts parallel to the 6° surface and the normal force is perpendicular to that surface. Therefore, the normal force acts 6° from the vertical.

8. The resultant force on the top, R_T, acts at an angle $\varphi_s = 16.7°$ from the normal force. Therefore, R_T acts at an angle of $16.7° + 6° = 22.7°$ from the vertical.

9. The summary of the preceding items is that there is a set of three concurrent forces acting on the wedge that are in equilibrium for impending motion of the wedge. Therefore, a vector triangle can be used to solve for the unknown force, F. See Figure 5–12(b).

Solution of the Vector Triangle:

1. The vector triangle must close because the three forces are in equilibrium. Note the angles of inclination of R_B and R_T and that F acts horizontally.

2. We can show within the triangle that the vertical components of both R_B and R_T are equal to R_2 and $N_B = 1.217$ kN.

3. We can then compute R_B from

$$\cos 16.7° = R_{By}/R_B$$
$$R_B = R_{By}/\cos 16.7° = (1.217 \text{ kN})/(0.9578) = 1.2706 \text{ kN}$$

4. Now we know all the angles and the magnitude of one side of the vector triangle. Therefore, we can use the law of sines to find the unknown force, F.

$$\frac{R_B}{\sin 67.3°} = \frac{F}{\sin 39.4°}$$

$$F = R_B \left[(\sin 39.4°)/(\sin 67.3°)\right] = 1.2706 \text{ kN} \left[(0.6347)/(0.9225)\right]$$

$$\boxed{F = 0.874 \text{ kN}} \quad \textbf{Answer}$$

Comment: In summary, a horizontal force of $F = 0.874$ kN applied to the right side of the wedge will cause the wedge to move to the left and raise the end of the beam at support B, exerting an upward force of 1.217 kN on the beam.

(a) Short-shoe drum brake

(b) Band brake

(c) Bicycle brake

(d) Disc brake

FIGURE 5–13 Types of brakes.

Brakes and Clutches Here we consider important applications of friction in mechanical drives. Figure 5–13 shows sketches of brakes and clutches and indicates where the frictional forces act. In brakes, one or more fixed elements have friction material that contacts a moving part and creates a controlled frictional force at a point away from the center of rotation. The product of the frictional force and the radial distance to the center of rotation is the frictional torque that slows or stops the rotating element. In clutches, a similar arrangement causes a stationary element to be accelerated by a driver (motor or engine). Often clutches and brakes are applied together. The clutch brings an element up to speed to start a cycle. It then disengages at

the end of the cycle and a brake is engaged to quickly bring the element to a stop. In some types of equipment, this clutch-brake cycle can be quite rapid.

It should be noted that the complete analysis of the performance of clutches and brakes requires the study of dynamics because elements are accelerated or decelerated during operation. In addition, energy is either added or removed from the rotating elements and the friction elements often slide during engagement or disengagement. Specification of the friction material is critical to ensure adequate frictional force while maintaining a reasonable life for the wearing elements. See Reference 8 for a more complete analysis of these devices.

Here we consider only the force analysis at an instant or during motion at a uniform velocity. We will determine the frictional force on some part of the device or the frictional torque that tends to accelerate or decelerate an element. Actuation of clutches and brakes can be by manual lever, pneumatic cylinders, hydraulic cylinders, springs, or other means.

Figure 5–13 (a) is the *short-shoe drum brake* in which the brake shoe is brought into contact with the rim of the rotating drum by a force applied to a lever. The relative motion of the shoe and the surface of the drum creates a frictional force tangential to the drum. The product of the frictional force times the radius of the drum is the frictional torque. The free body diagrams of the lever and the drum show the result. The small shoe allows the analysis to consider that the frictional force acts at a single point.

The *band brake* shown in Figure 5–13(b) develops frictional force between the rotating drum surface and the stationary flexible band when the band is tightened by a tensile force. Here we must consider the variation of the tensile force on the two ends of the band. With the direction of drum rotation shown, pushing down on the lever with a force W creates a force P_2 in the band. The frictional force between the drum and the band causes the tension in the band to increase from the initial point of contact at A to the final point of contact at B where the tension in the band is P_2. Therefore, $P_1 > P_2$. Using calculus, it can be shown (see Reference 9) that

$$P_1/P_2 = e^{\mu\theta} \tag{5–9}$$

An alternate form of this equation is

$$\ln(P_1/P_2) = \mu\theta \tag{5–10}$$

In these equations,

P_2 = force on the right end of the band (lower value)

P_1 = force on the left end of the band (higher value)

θ = angle of contact between the band and the drum (must be in radians)

μ = coefficient of friction between the band and the drum

$\mu = \mu_s$ if the drum is at rest and a torque on the drum tends to rotate it in a state of impending motion

$\mu = \mu_k$ if the drum is rotating and the frictional torque on the drum tends to stop it

$\ln$ = natural logarithm function

e = base for the natural logarithm, approximately 2.71828

See References 8 and 9 for additional details on band brakes.

You are likely familiar with the bicycle brake shown in Figure 5–13(c). As you squeeze the brake levers on the handlebars, a cable pulls on the linkage that actuates the brake pads, moving them into contact with the rim of the wheel. The frictional force times the radius of the wheel is the stopping torque on the wheel.

Most automobiles, other vehicles, and some industrial machinery use *disc brakes* similar to the design sketched in Figure 5–13(d). Hydraulic pressure pushes the pads to make contact with the rotor attached to the wheel of the vehicle or to rotating parts of a machine. The force is distributed over a fairly large area to minimize wear of the pads and the discs. Typically a mean radius of the pad area is used to compute the braking torque.

Example Problem 5–7

Figure 5–13(a) shows a short-shoe drum brake. A is a fixed pivot and an actuation force of $W = 380 \, \text{N}$ is applied at the end of the lever. Use $\mu_k = 0.50$ for the friction pad. The following dimensions apply:

$$L = 460 \, \text{mm}, \, a = 90 \, \text{mm}, \, b = 60 \, \text{mm}, \, D_d = 300 \, \text{mm}$$

Compute the torque applied to the rotating drum.

Solution

Given: Figure 5–13(a); $W = 380 \, \text{N}$, $\mu_s = 0.50$

Find: Torque on the drum

Analysis and Results: The free body diagrams of the lever and the drum are shown in the figure. Consider first the lever and find the frictional force, F_f.

$$\Sigma M_A = 0 = W \cdot L - N \cdot a + F_f \cdot b$$

But $F_f = \mu_k \cdot N$ and $N = F_f/\mu_k$. Then

$$\Sigma M_A = 0 = W \cdot L - F_f/\mu_k \cdot a + F_f \cdot b = W \cdot L - F_f(a/\mu_k - b)$$

$$F_f = \frac{W \cdot L}{(a/\mu_k - b)} = \frac{(380 \, \text{N})(460 \, \text{mm})}{[(90 \, \text{mm}/0.50) - 60 \, \text{mm}]} = 1457 \, \text{N}$$

The frictional torque is

$$T_f = F_f \cdot (D_d/2) = (1457 \, \text{N})(150 \, \text{mm}) = 218\,500 \, \text{N} \cdot \text{mm} = \mathbf{218.5 \, N \cdot m} \quad \textbf{Answer}$$

Example Problem 5–8 Figure 5–13(b) shows a band drum brake. Compute the friction torque applied to the drum for an actuation force of $W = 380$ N applied at the end of the lever. Use $\mu_k = 0.50$ for the friction pad. The following dimensions apply:

$$L = 460 \text{ mm},\ a = 90 \text{ mm, drum diameter} = D_d = 300 \text{ mm},\ \theta = 225°$$

Solution Given: Figure 5–13(b); $W = 380$ N, $\mu_k = 0.50$

Find: Torque on the drum

Analysis and Results: The free body diagram of the lever is shown in the figure. First find the force P_2 that is the tension on the right end of the band.

$$\Sigma M_A = 0 = W \cdot L - P_2 \cdot a$$

Then $P_2 = W \cdot L/a = (380 \text{ N})(460 \text{ mm})/(90 \text{ mm}) = 1942$ N

We can now use Equation (5–9) to find P_1 on the left side of the drum.

$$P_1/P_2 = e^{\mu\theta} \qquad \text{or} \qquad P_1 = P_2 \cdot e^{\mu\theta}$$

The angle of wrap must be in radians.

$$\theta = (225°)(\pi \text{ rad})/(180°) = 3.927 \text{ rad}$$

$$P_1 = P_2 \cdot e^{\mu\theta} = (1942 \text{ N})(2.71828)^{(0.50)(3.927)} = 13\ 835 \text{ N}$$

Both P_1 and P_2 act tangential to the drum. Then the frictional torque is

$$T_f = (D_d/2)(P_1 - P_2) = (150 \text{ mm})(13\ 835 \text{ N} - 1942 \text{ N})$$

$$T_f = 1784 \text{ N} \cdot \text{m} \qquad \textbf{Answer}$$

Belt Drives and Rope Friction

Equations (5–9) and (5–10) are also useful for analyzing flat belt drives for machines and the friction of a rope around a stationary post.

Figure 5–14 shows a schematic representation of a flat belt drive. Typically the smaller pulley (also called a *sheave*) is attached to the output shaft of an electric motor, fluid power motor, or engine operating at a relatively high speed, ω_1. The belt wraps around the small sheave and the larger sheave attached to the driven machine. The result is a lower speed, ω_2, and a higher torque for the shaft carrying the larger sheave. The belt is typically made from tough rubber having a set of strong continuous fibers within it giving the belt strength and flexibility. The rubber belt acting on smooth metal sheaves depends on belt tension and friction to transmit power. The center distance between the two sheaves is tightened during installation to ensure that an appropriate level of torque can be transmitted. However, if the resistance torque from the driven machine becomes excessively high, the belt will slip. For some designs, this is a fail-safe feature. The analysis method is similar to that demonstrated for band brakes in Example Problem 5–8, except the design is based on no slippage between the belt and the sheave. Therefore, the static coefficient of friction, μ_s, is used in the analysis to ensure that the belt tension for impending slippage is adequate for the intended application. See Reference 8 for additional detail about flat belt drives along with V-belts and cogged synchronous belts.

A rope in tension passing around a stationary pipe or other cylindrical device behaves in a manner similar to the band brake. A good example is shown in Figure 5–15 where the post is called a *capstan*, used in marine applications. The rope may be wrapped more than one revolution and the total angle of contact should be included in the value of θ in Equations 5–9 and 5–10. Very large forces can be resisted by use of a suitably sized rope wrapped a few times around the capstan, allowing secure handling of large boats. Similar applications exist in construction, industry, and recreation.

FIGURE 5–14 Belt drive.

FIGURE 5–15 Capstan for rope handling on a boat.

Example Problem **5–9**	Figure 5–15 shows a capstan as used in marinas for lashing large boats to the dock. Compute the force that can be exerted on the boat for an actuation force of $F = 430$ N applied by the operator. Use $\mu_s = 0.45$ for the rope against the capstan surface. The rope is wrapped around the capstan $2\frac{1}{4}$ turns.
Solution Given:	Figure 5–15; $F = 430$ N, $\mu_s = 0.45$
Find:	Force exerted on the boat
Analysis and Results:	We can use Equation (5–9).

$$P_1/P_2 = e^{\mu\theta} \tag{5–9}$$

Recall that P_1 is the higher force, P_2 is the lower force, μ is the coefficient of friction, and θ is the angle of wrap in radians.

Let $P_2 = F = 430$ N.

$$\theta = 2.25 \text{ turns} \times (360°/\text{turn}) \times (2\pi \text{ rad}/360°) = 14.14 \text{ rad}$$

Use $\mu = \mu_s = 0.45$ because the maximum force exerted on the boat occurs when there is impending slippage of the rope on the capstan. Then

$$P_1/P_2 = e^{\mu\theta} \quad \text{or} \quad P_1 = P_2 \cdot e^{\mu\theta}$$
$$P_1 = P_2 \cdot e^{\mu\theta} = (430 \text{ N})(2.71828)^{(0.45)(14.14)} = 249\,100 \text{ N}$$
$$\boxed{P_1 = \textbf{249.1 kN applied to boat} \qquad \textbf{Answer}}$$

Comment:	The pull by the person on the dock of 430 N (97 lb) is increased by *579 times* (to 249.1 kN or 56 000 lb) with just 2.25 turns of the rope around the capstan for the data of this problem.

REFERENCES

1. Avallone, E. A., Baumeister, T., and Sadegh, A. (eds.), *Marks' Standard Handbook for Mechanical Engineers*, 11th ed., New York, McGraw-Hill, 2007.

2. Blau, Peter J., *Friction Science and Technology*, Oxford, UK, Taylor and Francis, 2008.

3. Lingaiah, K., *Machine Design Data Handbook*, New York, McGraw-Hill, 1994.

4. Persson, B. N. J., *Sliding Friction*, New York, Springer-Verlag, 2008.

5. Rabinowicz, Ernest, *Friction and Wear of Materials*, 2nd ed., New York, John Wiley & Sons, Inc., 1995.

6. Slocum, Alexander H., *Precision Machine Design*, Englewood Cliffs, NJ, Prentice Hall, 1998.

7. Smith, Robert H., *Analyzing Friction in the Design of Rubber Products and Their Paired Surfaces*, Boca Raton, FL, CRC Press —Taylor & Francis Group, 2008.

8. Mott, Robert L., *Machine Elements in Mechanical Design,* 4th ed., Upper Saddle River, NJ, Pearson-Prentice Hall, 2004.

9. Budynas, Richard, Budynas, R. G., and Nisbett, J. K., *Shigley's Mechanical Engineering Design,* 8th ed., New York, McGraw-Hill, 2006.

INTERNET SITES

1. Roymech www.roymech.co.uk A collection of useful tables relevant to mechanical design, including tribology and friction coefficients.

2. Engineers Handbook.com www.engineershandbook.com/Tables/frictioncoefficients.htm A collection of useful tables and information for engineering applications, including friction coefficients.

3. Tribology.com www.tribology-abc.com An array of useful information and data concerning the field of tribology including friction and wear.

4. Glacier Garlock Bearings www.garlockbearings.com Manufacturer of a variety of plain bearings made from a composite of PTFE plastic, porous bronze, and steel or fiberglass under brand names DU®, DX®, Gar-Fil®, and Gar-Max®.

5. Saint-Gobain Performance Plastics. www.saint-gobain.com Manufacturer of plain plastic bearings under the brand name of Rulon®, available in 15 formulations.

6. Beemer Precision, Inc. www.beemerprecision.com Manufacturer of plain bearings made from cast bronze and porous sintered bronze impregnated with oil under the Oilite® brand.

7. Dodge Bearing Products www.dodge-pt.com/products/bearing/bearing_home.html Manufacturer of mounted plain-surface sleeve bearings under the Solidlube® and Bronzoil® brands used in industrial, material handling, and bulk materials industries. A part of the Baldor Electric Company.

8. Moving Men Furniture Sliders www.asseenontv.com/prod-pages/Movingmen.html Marketing site for a line of low-friction discs used to ease the task of moving heavy furniture such as couches and appliances.

PROBLEMS

Measuring and Using Coefficient of Friction

For Problems 5–1 to 5–14, the setup involved is similar to Figure 5–3 showing an object on a flat horizontal surface being pulled by a horizontal applied force that causes impending motion of the object.

5–1. A block weighing 9.25 N requires a force of 3.80 N to initiate movement to the right. Determine the coefficient of sliding friction, μ_s.

5–2. A force of 2.64 lb is required to initiate movement of a block weighing 7.76 lb. Determine the coefficient of sliding friction, μ_s.

5–3. A force of 13.2 lb is required to initiate movement of a block weighing 23.2 lb. Determine the coefficient of sliding friction, μ_s.

5–4. A block weighing 55.7 N requires a force of 29.5 N to initiate movement to the right. Determine the coefficient of sliding friction, μ_s.

5–5. Using data from Table 5–1, determine the amount of force required to initiate sliding movement of a clean, dry, 284-lb steel casting on the steel bed of a machine tool.

5–6. For the data from Problem 5–5, compute the amount of force required to continue to slide the casting at a constant velocity.

5–7. Repeat Problem 5–5 except the casting is coated with cutting oil.

5–8. Using data from Table 5–1, determine the amount of force required to initiate sliding movement of a clean, dry, 1260-N copper casting on steel plate.

5–9. For the data from Problem 5–8, compute the amount of force required to continue to slide the casting at a constant velocity.

5–10. Repeat Problem 5–9 except the steel plate has a film of light oil on it.

5–11. A car with a mass of 1552 kg has all four wheels locked against rotation due to damage in an accident. Determine the amount of horizontal force that a tow truck must exert to pull the car along a dry concrete roadway at constant velocity.

5–12. Repeat Problem 5–11 if the roadway is wet.

5–13. Repeat Problem 5–11 if the roadway is dry asphalt.

5–14. Repeat Problem 5–12 if the roadway is wet asphalt.

For Problems 5–15 to 5–22, the setup is similar to that shown in Figure 5–6.

5–15. At an angle $\theta = 19.8°$ a block just begins to slide down the ramp. Determine the coefficient of friction between the block and the ramp surface.

5–16. At an angle $\theta = 22.8°$ a block just begins to slide down the ramp. Determine the coefficient of friction between the block and the ramp surface.

5–17. At an angle $\theta = 38.0°$ a block just begins to slide down the ramp. Determine the coefficient of friction between the block and the ramp surface.

5–18. At an angle $\theta = 2.30°$ a block just begins to slide down the ramp. Determine the coefficient of friction between the block and the ramp surface.

5–19. Using the data from Table 5–1, determine at what angle a ramp should be set to hold a steel block in impending motion down a steel ramp. Both surfaces are clean and dry.

5–20. Repeat Problem 5–19 if the surfaces are lubricated with light cutting oil from a machining process.

5–21. Using the data from Table 5–1, determine at what angle a glass plate will be in a state of impending motion downward while laying on another sheet of glass. Both glass plates are clean and dry.

5–22. Using the data from Table 5–1, determine at what angle a ramp should be set to hold a PTFE-coated casting in impending motion down a steel ramp.

Rolling Friction

Use data from Table 5–2 for coefficient of rolling friction.

5–23. Determine the force required to move one steel wheel along a steel track when it carries a load of 3550 lb. The diameter of the wheel is 36 in.

5–24. Repeat Problem 5–23 if the wheel diameter is 12 in.

5–25. A refrigerator weighing 1125 N is supported on four hard plastic rollers having a diameter of 38 mm. Determine the force required to push the refrigerator on a flat steel floor in the factory.

5–26. Using data from Problem 5–25, assume the coefficient of rolling friction for a plastic roller on a wood floor is 11.2 times higher than when rolling on steel. Determine the force required to push the refrigerator.

5–27. A car is coasting on a horizontal concrete road. The total weight of the car is 11.12 kN and the four wheels have a diameter of 620 mm. Determine the effective rolling frictional force acting to slow the car.

5–28. A scheme to move a heavy tool chest across a shop's wood floor involves placing the chest on a plywood sheet and using 1.25-in-diameter wood dowels under the plywood. If the chest weighs 438 lb and at least three dowels are under the chest at any time, compute the horizontal force required to move the chest.

5–29. Compare the rolling friction from the scheme in Problem 5–28 with the alternative of sliding the chest's steel feet across the wood floor having $\mu_s = 0.55$.

Tipping and Sliding Friction Problems

For Problems 5–30 to 5–35, use the data from Example Problem 5–4 and Figure 5–7 for the refrigerator.

5–30. Determine the maximum height from the floor at which a force from the side can be applied to the refrigerator to produce sliding motion with no tipping.

5–31. For the refrigerator, determine what the maximum coefficient of sliding friction can be if the force is to be applied at the height of 56 in with no tipping.

5–32. The refrigerator is to be pushed from the front instead of the side to place it against a wall in the kitchen. The spread between the front and rear feet is 24 inches. If the refrigerator is pushed at a height of 56 in from the floor, will it tip or slide?

5–33. Determine the maximum height from the floor at which a force from the front can be applied to the refrigerator to produce sliding motion with no tipping.

5–34. The refrigerator is to be pushed from the front at a height of 56.0 in. Determine the maximum coefficient of static friction to slide the refrigerator without tipping.

5–35. An alternate design for the refrigerator in Figure 5–7 is proposed to replace the flat feet with hard plastic cylindrical rollers having a diameter of 1.75 inches in the same locations. If the rollers have a coefficient of rolling friction, $f_R = 0.12$ in, on the kitchen floor, compute the force required to move the refrigerator into place from the front. Check whether the applied force will cause tipping if applied anywhere on the front.

General Friction Analysis Problems

5–36. For the system shown in Figure P5–36, determine the required mass M for impending motion up the ramp. Use $\mu_s = 0.40$ and $\theta = 55°$.

5–37. Repeat Problem 5–36 except find M for impending motion down the ramp.

5–38. Figure P5–38 shows a crate weighing 1200 lb being pulled with a force F while sitting on a loading dock ramp. Determine the force required for impending motion up the ramp. Use $\mu_s = 0.30$ and $\mu_k = 0.25$.

5–39. If the force F in Problem 5–38 is zero, will the crate be in equilibrium? What frictional force exists between the crate and the ramp?

5–40. Figure P5–40 shows a cylindrical drum on a ramp. Its dimensions are $D = 750$ mm and $h = 900$ mm. The center of mass is at its geometric center. The coefficient of sliding

FIGURE P5–36   Problems 5–36 and 5–37.

FIGURE P5–38   Problems 5–38 and 5–39.

FIGURE P5–40   Problems 5–40 and 5–41.

friction between the drum and the ramp is 0.60. At what angle θ will the drum begin moving down the ramp? Will it slide or will it tip?

5–41. For the drum in Problem 5–40, specify the maximum height h for the drum to ensure that it will not tip before sliding down the ramp. Use $\mu_s = 0.60$.

5–42. Figure P5–42 shows a ladder leaning against a building. Assume that the support surface at B is smooth. The feet of the ladder rest on clean dry concrete with $\mu_s = 0.80$. The mass of the ladder is 9.2 kg with its center of mass shown. If a worker with equipment has a mass of 86 kg and acts as shown, will the ladder slip?

A
←— 1.0 m —→
←——— 1.3 m ———→
←———— 1.9 m ————→

FIGURE P5–42   Problems 5–42, 5–43, and 5–44.

FIGURE P5–46   Problems 5–46 and 5–47.

FIGURE P5–48   Problems 5–48 and 5–49.

5–43. For the ladder in Problem 5–42, consider the force due to the worker to be acting downward right at B. Will the ladder slip?

5–44. For the ladder in Problem 5–42, what value of μ_s would produce a state of impending slipping when the worker is right at B?

5–45. A spiral ramp in a material handling system is made from sheet steel. The ramp delivers dry copper castings to the next processing station. The angle of incline of the ramp surface is to be 10% greater than the minimum angle for which impending motion would occur to produce reliable motion but also to control the rate of descent. Specify the angle for the spiral.

5–46. Figure P5–46 shows a warehouse worker pulling on a 75-lb carton sitting on a shelf. For $\mu_s = 0.45$, determine the angle θ at which the worker should pull with a force $F = 60$ lb to slide the carton off the shelf.

5–47. For the situation shown in Problem 5–46, repeat the calculation if the carton is on rollers with a coefficient of rolling friction $f_R = 0.07$ in. The roller diameter is 1.50 in.

5–48. Figure P5–48 shows a bar clamp for which the right jaw is fixed to the bar while the left jaw slides easily when unloaded. To clamp a workpiece, the operator slides the left jaw lightly against it and then tightens the handle by means of a screw. If $\mu_s = 0.35$, will the left jaw slide or remain in place?

5–49. For the bar clamp in Problem 5–48, at what minimum distance above the top of the bar could the clamping force be located before the left jaw would slide, making the clamp ineffective?

5–50. A crate weighing 250 lb rests on a ramp having $\mu_s = 0.30$ between the bottom of the crate and the ramp surface. Determine the force F directed parallel to the ramp surface that would be required to move the crate for angles of inclination θ from zero to 90° in increments of 10°. Plot that force versus the angle.

5–51. Figure P5–51 shows two cartons of ice cream on a ramp. Each carton weighs 45 N. Between the two cartons $\mu_s = 0.15$,

FIGURE P5–51

and between the lower carton and the ramp $\mu_s = 0.10$. What, if any, motion will occur?

5–52. Figure P5–52 shows two steel castings. *A* weighs 120 N and *B* weighs 375 N. *B* rests on an aluminum surface. Using the data in Table 5–1 for dry clean surfaces,
   a. Determine the force *F* required for impending motion of *B* to the left.
   b. Determine the tension in the bar connecting *A* to the vertical wall.

FIGURE P5–52

5–53. Figure P5–53 shows an arrangement of two machined parts. The fixed surface is steel, *B* is copper, and *A* is cast iron ($W_A = 850$ N and $W_B = 450$ N). All mating surfaces are dry and clean. Determine the maximum force *F* that can be applied as shown without causing motion.

FIGURE P5–53

5–54. Figure P5–54 shows a mechanism employing a hinged static bar resting on a cylinder. The bar is made from steel and weighs 1.25 N. The cylinder has a rubberized surface having

FIGURE P5–54

$\mu_s = 0.70$ against the steel bar. Determine the torque required to cause rotation of the cylinder, neglecting friction in the bearings supporting it. Note that this arrangement provides Coulomb friction that tends to stabilize the system under light loads.

5–55. During construction of a bridge, a large steel structural component weighing 7.8 kN be moved into final position after being set down on a concrete slab. See Figure P5–55. A wedge is placed between the left side of the component and a fixed wall having a slope of 15°. Determine the force that must be applied to the top of the wedge to move the component to the right. Use the values for $\mu_s = 0.75$ for steel on steel and $\mu_s = 0.60$ for steel on concrete.

FIGURE P5–55

5–56. Figure P5–56 shows a block resting on a wedge and against a vertical wall. For $\theta = 8°$ determine the horizontal force *F* that must be applied to the wedge to raise the block.

FIGURE P5–56 Problems 5–56, 5–57, and 5–58.

5–57. For the system shown in Figure P5–56, would the wedge move to the left with $F = 0$?

5–58. For the system shown in Figure P5–56, at what angle θ would the wedge be under impending motion to the left with $F = 0$?

Brakes, Belts, and Rope Friction

5–59. The system shown in Figure P5–59 represents a winch drum *B* with a mass of 6.12 kg suspended by a flexible cable. Consider the pulley to the right to provide negligible friction. A cylinder is fixed to one end of the winch drum

FIGURE P5–59

and it is acted on by the lever of a short-shoe brake. Determine the force F required to keep the drum from turning.

5–60. A model of a Conestoga wagon is being built for a theater production. A simple short-shoe brake is to be used as shown in Figure P5–60. The wheel has a thin steel tire shrunk fit onto a sturdy wooden wheel. Determine the torque applied to the rotating wheel when a force $F = 40$ lb is applied to the handle and the brake friction material has $\mu_k = 0.55$ against steel.

FIGURE P5–60

5–61. A band brake similar to that shown in Figure 5–13(b) must exert a braking torque of 54 lb·ft to the drum. Determine the force W that must be applied to the lever for the following dimensions: $L = 20.0$ in, $a = 3.50$ in, $r = 3.50$ in, $\mu_k = 0.75$, $\theta = 200°$.

5–62. The band brake shown in Figure P5–62 is actuated by a pneumatic cylinder providing a force of 120 lb. Determine the braking torque on the drum if $\mu_k = 0.40$.

5–63. On a construction site, an arm supporting a dispenser for concrete is extended over a slab being poured as shown in

Figure P5–63. A worker provides support for the arm by using a rope wrapped $1^3/_{16}$ turns around a stationary pipe. If the worker can exert a steady pull of 60 lb, determine the load W that can be supported at the end of the boom. Use $\mu_s = 0.35$ between the rope and the pipe.

5–64. Figure P5–64 shows a device for controlling the descent of a load from a winch drum. When a constant pull of $P = 90$ N is applied to a flat leather belt at the left, determine the maximum load W that can be supported without slipping. Use $\mu_s = 0.60$ between the belt and the 300-mm-diameter wheel.

5–65. Figure P5–65 shows a rope pulling on a 250-lb pallet of piping components in a chemical plant. The rope passes over a fixed pipe. Determine the force F required to initiate motion of the pallet.

5–66. Figure 5–14 shows a typical design for a belt drive with power being transmitted from an electric motor driving the smaller sheave ($D_1 = 6.45$ in) and the larger sheave ($D_2 = 13.9$ in) mounted to the shaft of a driven machine. The flat rubber belt has $\mu_s = 0.70$ on the sheaves. The angle of wrap on sheave 1 is 160° and on sheave 2 is 210°. If the maximum tension on the tight side of the belt is 155 lb, determine the torque applied to both sheaves.

5–67. Figure 5–14 shows a typical design for a belt drive with power being transmitted from an electric motor driving the smaller sheave ($D_1 = 140$ mm) and the larger sheave ($D_2 = 315$ mm) mounted to the shaft of a driven machine. The flat rubber belt has $\mu_s = 0.65$ on the sheaves. The angle of wrap on sheave 1 is 155° and on sheave 2 is 225°. If the maximum tension on the tight side of the belt is 680 N, determine the torque applied to both sheaves.

FIGURE P5–62

67.5°

Rope wraps one
full turn plus
given angle

22°

Fixed pipe

45°

B

A – B = 12 ft

22°

A

Concrete
dispenser

FIGURE P5–63

150 mm

90 mm

Frictionless
pulley

Wheel

Drum

30°

P

Leather belt

W

FIGURE P5–64

Wood

Steel

Fixed pipe
$\mu_s = 0.40$

$\mu_s = 0.25$

F

FIGURE P5–65

CENTROIDS AND MOMENTS OF INERTIA OF AREAS

MAJOR TOPICS

6–1 The Big Picture—Centroids and Moments of Inertia of Areas

6–2 The Concept of Centroid—Simple Shapes

6–3 Centroid of Complex Shapes

6–4 The Concept of Moment of Inertia of an Area

6–5 Moment of Inertia of Composite Shapes Whose Parts Have the Same Centroidal Axis

6–6 Moment of Inertia for Composite Shapes—General Case—Use of the Parallel Axis Theorem

6–7 Mathematical Definition of Moment of Inertia

6–8 Composite Sections Made from Commercially Available Shapes

6–9 Moment of Inertia for Shapes with All Rectangular Parts

6–10 Radius of Gyration

6–11 Section Modulus

6–1 THE BIG PICTURE— CENTROIDS AND MOMENTS OF INERTIA OF AREAS

Discussion Map

■ In Chapters 11 to 14 you will learn how to analyze and design beams, considering stresses due to bending and shear along with deflections. This requires specifying the shape of the beam's cross section and computing various properties of the area.

■ The properties of the cross-sectional area of the beam that are of interest here are the *centroid* and the *moment of inertia with respect to the centroidal axis*. The concepts and techniques presented here will enable you to solve the beam analysis problems throughout this book and in many real design situations.

Discover

Think of examples of beams that you have seen yourself, and picture the shapes of their cross sections. What shape would you see if you looked directly into the end of the beam?

Look ahead in this chapter to see the wide variety of shapes that are often specified for the cross section of beams. Figure 6–7 shows a typical "I-beam." This is one of the most efficient shapes for a beam (as you will see as you study Chapters 11–14). In this discussion, efficiency refers to a beam shape that performs well with regard to stresses and deflections while requiring a relatively small amount of material. Such a beam would usually have a lower cost and be lighter than most other shapes.

Look back to Chapter 1 where commercially available structural shapes in steel and aluminum were described. Review Section 1–8, Table 1–7, and Appendices A5 through A13. There you will see sections called *wide flange beams (W-shapes); American standard beams (S-shapes); channels (C-shapes); angles (L-shapes); square and rectangular hollow structural shapes (HSS-shapes); and circular hollow pipe and tubing.* All of these shapes are commonly used for beams, either alone or in combination.

Look ahead to Section 6–8 where we show how to analyze composite sections made from two or more kinds of shapes or add flat plates to critical sections to enhance the ability of the beam to withstand loads and limit deflections.

Have you seen any of these types of beams? Discuss this with colleagues and compare their experiences with yours.

ACTIVITY Chapter 6—Centroids and Moments of Inertia of Areas

Let's do an Internet search to find evidence of various shapes used for the cross section of beams and columns. Try the ones described here that refer to the sites listed at the end of this chapter. Each shows examples of structural products, components of machinery, or parts of consumer products that perform as beams or columns and that must be strong, stiff, and attractive. Use your own creativity to find other interesting sites.

1. Site 1 is from a company that produces a wide variety of relatively small precision cold rolled and drawn shapes. Starting with long sections of circular wire or rod, unique,

highly accurate shapes are produced by drawing or rolling that perform well as beams while meeting the needs of designers for special functionality such as guiding mating parts along a liner path. Note that some of the shapes are similar to the I-beams included in the Appendix but much smaller. Others have longitudinal ribs or curved shapes that add to the stiffness of the beams. These shapes can be applied in precision manufacturing automation equipment, business machines, consumer products, electrical distribution equipment, and a host of other devices.

2. A product for the building construction industry is illustrated in Site 2. Fabricated from a variety of sawn rectangular wood shapes, the resulting I-shape is strong and lightweight, allowing the building designer to specify floor systems that span long spaces with good strength and stiffness. An additional advantage is that the assembly and arrangement of these joists and the covering plywood sheets provide silent support as people walk across the floor.

3. Site 3 shows some of the special shapes produced by a steel fabricator for use in large structures. Note the combination of angles, channels, bar stock, and other basic forms to produce an efficient truss-like support system for use as floor joists or roof rafters. The joists are sometimes combined with steel decking and a concrete slab covering to produce a strong, stiff composite floor or roof system.

4. Special shapes produced by the extrusion process using aluminum are shown in Site 4. Molten metal is forced through dies with openings having the shapes shown. The resulting products made from these forms can be used in automotive applications, cabinetry, window and door frames, furniture, aircraft, and many other industries. Check out the examples of applications listed on this site. Then you may visit some of the sites of companies that produce the extrusions to learn more about the process and the equipment used to make them.

5. Site 5 shows shapes produced in plastics using extrusion processes similar to those described for Site 4. Nylon, ABS, polycarbonate, PVC, and several other plastics are used for applications in the automotive/transportation, architectural, building construction, aerospace, electronics, and many other industries. Also check out the design, engineering, and manufacturing processes used to produce these shapes.

6. Site 6 shows an example from the ladder industry where rolled or extruded shapes are typically used for the side frames and steps that are subjected to bending in service. Note that the side rails are special channel or I-shaped sections allowing the fastening of the steps. The steps themselves are formed, hollow shapes that provide a flat surface for good footing as the user climbs the ladder, while also offering strong, stiff support. Materials used include aluminum, fiberglass, steel, and wood. Planks used to provide a stable horizontal platform for workers are also provided by this company.

7. Have you noticed the trucks like those shown in Site 7, used by electrical utility installers, tree-trimming teams, painters, and others to raise workers to high points? The articulated booms act as beams while the hydraulic actuators allow smooth, precise positioning close to the worksite. Designers of such trucks must be able to determine the centroids and moments of inertia of these components to ensure the safety of the workers and the durability of the equipment.

8. The product featured in Site 8 provides a versatile set of structural components that can be assembled into frames tailored to a particular application. Made from roll-formed steel, the resulting channel shape provides the strength and stiffness required for the frame while also facilitating the attachment of equipment, side panels, floor plates, and other subsystems. Check out the application examples under the *Tech. Data tab.* The Introduction shows how the system works while illustrating a variety of applications. Then you can explore more specific details under the heading *Mechanical, Architectural, Industrial,* and *Electrical.* Find the detailed data in the online catalog for the variety of sizes available that list the section properties such as moment of inertia, cross-sectional area, and the weight per unit length. These shapes can be used in some of the problems in this chapter and in following chapters.

9. Site 9 shows an extruded aluminum structural framing system that can be used in similar applications as those described in Site 8. The flat sides and integral channels facilitate assembly and provide a clean, neat appearance of the finished structure. The shapes of the cross sections are optimized for load-carrying ability while being adaptable to build special machines, workstations, automation equipment, mobile units, and storage systems. Check out the online catalogs that describe technical data required for proper application including section properties of area, moment of inertia, and weight per unit length.

Inspection of these sites and those you may have found on your own demonstrate your need to develop the ability to analyze many different shapes to determine the location of their centroids and the magnitudes of their moment of inertia with respect to the centroid. Use the information and exercises in this chapter to acquire and refine your ability to do just that and be prepared to transfer this ability to the following chapters that discuss beams and columns.

Objectives of This Chapter

After completing this chapter, you should be able to:

1. Define *centroid*.

2. Locate the centroid of simple shapes by inspection.

3. Compute the location of the centroid for complex shapes by treating them as composites of two or more simple shapes.

4. Define *moment of inertia* as it applies to the cross-sectional area of beams.

5. Use formulas to compute the moment of inertia for simple shapes with respect to the centroidal axes of the area.

6. Compute the moment of inertia of complex shapes by treating them as composites of two or more simple shapes.

7. Properly use the *parallel axis theorem* in computing the moment of inertia of complex shapes.

8. Analyze composite beam shapes made from two or more standard structural shapes to determine the resulting centroid location and moment of inertia.

9. Recognize what types of shapes are efficient in terms of providing a large moment of inertia relative to the amount of area of the cross section.

6–2 THE CONCEPT OF CENTROID—SIMPLE SHAPES

The *centroid* of an area is the point about which the area could be balanced if it were supported from that point. The word is derived from the word *center,* and it can be thought of as the geometrical center of an area. For three-dimensional bodies, the term *center of gravity,* or *center of mass,* is used to define a similar point.

For simple areas, such as the circle, the square, the rectangle, and the triangle, the location of the centroid is easy to visualize. Figure 6–1 shows the locations, denoted by *C.* If these shapes were carefully made and the location for the centroid carefully found, the shapes could be balanced on a

pencil point at the centroid. Of course, a steady hand is required. How's yours?

Appendix A–1 gives more data for centroids and other properties of areas for a variety of shapes. See References 1–3 for additional shapes.

6–3 CENTROID OF COMPLEX SHAPES

Most complex shapes can be considered to be made up by combining several simple shapes. This can be used to facilitate the location of the centroid, as will be demonstrated later.

A simple rule for locating centroids can be used for some special combinations of areas:

- If the area has an axis of symmetry, the centroid will lie on that axis.
- If the area has two axes of symmetry, the centroid is at the intersection of these two axes.

Figure 6–2 shows six examples where these rules apply.

Where two axes of symmetry do not occur, the *method of composite areas* can be used to locate the centroid. For example, consider the shape shown in Figure 6–3. It has a vertical axis of symmetry but not a horizontal axis of symmetry. Such areas can be considered to be a composite of two or more simple areas for which the centroid can be found by applying the following principle:

The product of the total area times the distance to the centroid of the total area is equal to the sum of the products of the area of each component part times the distance to its centroid, with the distances measured from the same reference axis.

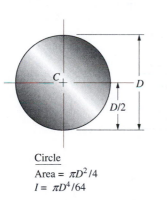

Circle
Area $= \pi D^2/4$
$I = \pi D^4/64$

Square
Area $= s^2$
$I = s^4/12$

Rectangle
Area $= bh$
$I_x = bh^3/12$
$I_y = hb^3/12$

Triangle
Area $= \frac{1}{2} bh$
$I_x = bh^3/36$

FIGURE 6–1   Properties of simple areas. The centroid is denoted as *C.*

FIGURE 6–2   Composite shapes having two axes of symmetry. The centroid is denoted as *C.*

This principle uses the concept of the *moment of an area,* that is, the product of the area times the distance from a reference axis to the centroid of the area. The principle states

> The moment of the total area with respect to a particular axis is equal to the sum of the moments of all the component parts with respect to the same axis.

This can be stated mathematically as

$$A_T \, \overline{Y} = \sum(A_i y_i) \qquad (6-1)$$

where

A_T = total area of the composite shape

$\overline{Y}$ = distance to the centroid of the composite shape measured from some reference axis

A_i = area of one component part of the shape

y_i = distance to the centroid of the component part from the reference axis

The subscript i indicates that there may be several component parts, and the product $A_i \, y_i$ for each must be formed and then summed together, as called for in Equation (6–1). Since our objective is to compute $\overline{Y}$, Equation (6–1) can be solved:

$$\overline{Y} = \frac{\sum(A_i y_i)}{A_T} \qquad (6-2)$$

A tabular form of writing the data helps keep track of the parts of the calculations called for in Equation (6–2). An example will illustrate the method.

Example Problem 6–1   Find the location of the centroid of the area shown in Figure 6–3.

Solution   Objective: Compute the location of the centroid.

Given: Shape shown in Figure 6–3.

Analysis: Because the shape has a vertical axis of symmetry, the centroid lies on that line. The vertical distance from the bottom of the shape to the centroid will be computed using Equation (6–2). The total area is divided into a rectangle (part 1) and a triangle (part 2), as shown in Figure 6–4. Each part is a simple shape for which the centroid is found using the data from Figure 6–1. The distances to the centroids from the bottom of the shape are shown in Figure 6–4 as y_1 and y_2.

Results: The following table facilitates the calculations for data required in Equation (6–2).

Part	A_i	y_i	$A_i y_i$
1	3200 mm^2	40 mm	128 000 mm^3
2	600 mm^2	90 mm	54 000 mm^3
	A_T = 3800 mm^2		$\sum(A_i y_i)$ = 182 000 mm^3

FIGURE 6–3   Shape for Example Problem 6–1.

FIGURE 6–4   Data used in Example Problem 6–1.

Now $\bar{Y}$ can be computed.

$$\bar{Y} = \frac{\sum(A_i y_i)}{A_T} = \frac{182\,000\ \text{mm}^3}{3800\ \text{mm}^2} = 47.9\ \text{mm}$$

This locates the centroid as shown in Figure 6–4.

Comment: In summary, the centroid is on the vertical axis of symmetry at a distance of 47.9 mm up from the bottom of the shape.

The composite area method works also for sections where parts are removed, as well as added. In this case the removed area is considered negative, illustrated as follows.

Example Problem 6–2   Find the location of the centroid of the area shown in Figure 6–5.

Solution   Objective: Compute the location of the centroid.

Given: Shape shown in Figure 6–5.

Analysis: Because the shape has a vertical axis of symmetry, the centroid lies on that line. The vertical distance from the bottom of the shape to the centroid will be computed using Equation (6–2). The total area is divided into three rectangles, as shown in Figure 6–6. Part 1 is the total large rectangle, 50 mm high and 40 mm wide. Part 2 is the 20 mm × 30 mm rectangle that is *removed* from the composite area and the area, A_2, will be considered to be negative. Part 3 is the 10 mm × 60 mm rectangle on top. The distances to the centroids from the bottom of the shape are shown in Figure 6–6 as y_1, y_2, and y_3.

Results: The following table facilitates the calculations for data required in Equation (6–2).

Part	A_i	y_i	$A_i y_i$
1	2000 mm²	25 mm	50 000 mm³
2	−600 mm²	15 mm	−9000 mm³
3	600 mm²	55 mm	33 000 mm³
	$A_T = 2000$ mm²		$\sum(A_i y_i) = 74\,000$ mm³

FIGURE 6–5   Shape for Example Problem 6–2.

FIGURE 6–6   Data used in Example Problem 6–2.

Then

$$\overline{Y} = \frac{\sum (A_i\, y_i)}{A_T} = \frac{74\,000 \text{ mm}^3}{2000 \text{ mm}^2} = 37.0 \text{ mm}$$

LESSON 7

Comment: In summary, the centroid is on the vertical axis of symmetry at a distance of 37.0 mm up from the bottom of the shape.

6–4 THE CONCEPT OF MOMENT OF INERTIA OF AN AREA

In the study of strength of materials, the property of *moment of inertia of an area* is an indication of the stiffness of a beam, that is, the resistance to deflection of the beam when carrying loads that tend to cause it to bend. The deflection of a beam is inversely proportional to the moment of inertia as described in Chapter 14. The use of the moment of inertia in the calculation of stress due to bending is discussed in Chapter 12 of this book. Stresses due to vertical shearing forces also depend on moment of inertia and are discussed in Chapter 13.

Some mathematicians and stress analysts use the term *second moment of area* instead of *moment of inertia*. That term is, in fact, more descriptive of the definition of this property in the following discussion. Others call this term *area moment of inertia* or *rectangular moment of inertia* to help distinguish it from *polar moment of inertia* as used in the analysis of torsional shear stress in Chapter 10. There are other uses of the term *moment of inertia* in the study of dynamics where the term refers to a property of a three-dimensional mass.

However, the term *moment of inertia* has become common usage in certain handbooks and industry publications. See References 1–3. Therefore, this book will continue to use the term *moment of inertia*. You should understand the context of a problem where this term is needed to ensure that the correct approach is used to determine its value.

Of interest is the moment of inertia of the shape of the *cross section* of the beam. For example, consider the beam shape in Figure 6–7, often referred to as an "I-beam." Another example is a hollow rectangle, as sketched in Figure 6–8.

Both examples shown in Figures 6–7 and 6–8 represent shapes that are relatively efficient in the use of material to produce large values of their moment of inertia. In most important cases in the study of strength of materials, the moment of inertia of a shape, denoted by the symbol *I*, is a function of the placement of the area with respect to the *centroidal axis* of the shape, the axis that passes through the centroid of the shape. It is most desirable, from the standpoint of the efficient use of material, to place as much of the material as far away from the centroidal axis as practical. This observation is based on the definition of moment of inertia given here.

The moment of inertia of an area with respect to a particular axis is defined as the sum of the products obtained by multiplying each infinitesimally small element of the area by the square of its distance from the axis.

Thus, you should be able to reason that if much of the area is placed far away from the centroidal axis, the moment of inertia would tend to be large.

The mathematical formula for the moment of inertia, *I*, follows from the definition. An approximate method involves the process of *summation*, indicated by Σ.

$$I = \sum y^2 (\Delta A) \qquad (6\text{–}3)$$

This would require that the total area be divided into many very small parts represented by ΔA, and that the distance *y* to the centroid of each part from the axis of interest be determined. Then the product of $y^2 (\Delta A)$ would be computed for each small part, followed by summing all such products. This is a very tedious process and, fortunately, one that is not used often.

A refinement of the summation method indicated by Equation (6–3) is the process of integration, which is the mathematical technique of summing infinitesimal quantities over an entire area. The true mathematical definition of moment of inertia requires the use of integration as follows:

$$I = \int y^2\, dA \qquad (6\text{–}4)$$

Here the term *dA* is an area of infinitesimally small size and *y*, as before, is the distance to the centroid of *dA*. We will demonstrate the development and use of Equation (6–4) in a later section. However, in many practical problems, it is not necessary to perform the integration process. We show these techniques first.

Looking closely at the definition, you should be able to see that the units for moment of inertia will be *length to the fourth power*. Examples are in^4, m^4, and mm^4.

FIGURE 6–7 Typical cross-sectional shape of an I-beam.

FIGURE 6–8 Typical cross-sectional shape of a hollow rectangular tube.

There are several methods of determining the magnitude of the moment of inertia.

1. For simple shapes it is convenient to use standard formulas that have been derived from the basic definition given earlier. Figure 6–1 shows such formulas for four shapes and Appendix A–1 gives several more. Reference 2 includes a table of formulas for I for 42 different shapes.

2. For standard commercially available shapes such as wide-flange beams (W-shapes), channels (C-shapes), angles (L-shapes), and pipe, values of moment of inertia are tabulated in published references such as Reference 1. See also Appendixes A–4 to A–13.

3. For more complex shapes for which no standard formulas are available, it is often practical to divide the shape into component parts that are themselves simple shapes. Examples are shown in Figures 6–4 to 6–8. The details of calculating the moment of inertia of such shapes, called *composite shapes,* depend on the nature of the shapes and will be demonstrated later in this chapter. Important concepts are stated here.

   a. If all component parts of a composite shape have the same centroidal axis, the total moment of inertia for the shape can be found by adding or subtracting the moments of inertia of the component parts with respect to the centroidal axis. See Section 6–6.

   b. If all component parts of a composite shape do not have the same centroidal axis, the use of a process called the parallel axis theorem is required. See Section 6–7.

4. The fundamental definition of moment of inertia, Equation (6–4), can be used if the geometry of the shape can be represented in mathematical terms that can be integrated. See Section 6–8.

5. Many computer-aided design software systems include automatic calculation of the location of the centroid and the moment of inertia of any closed shape drawn in the system.

6. For the special case of a shape that can be represented as a composite of rectangles having sides perpendicular or parallel to the centroidal axis, a special tabulation technique can be applied that is described in Section 6–10 of this chapter. This technique lends itself well to solution by using a programmable calculator, a spreadsheet, or a simple computer program.

6–5 MOMENT OF INERTIA OF COMPOSITE SHAPES WHOSE PARTS HAVE THE SAME CENTROIDAL AXIS

A composite shape is one made up of two or more parts that are themselves simple shapes for which formulas are available to calculate the moment of inertia, I. A special case is when all parts have the same centroidal axis. Then the moment of

FIGURE 6–9 Shape for Example Problem 6–3.

inertia for the composite shape is found by combining the values of I for all parts according to the following rule:

If the component parts of a composite area all have the same centroidal axis, the total moment of inertia can be found by adding or subtracting the moments of inertia of the component parts with respect to the centroidal axis. The value of I is added if the part is a positive solid area. If the part is a void, the value of I is subtracted.

Figure 6–9 shows an example of such a shape, composed of a vertical central stem, 30 mm wide and 80 mm high, and two side parts, 30 mm wide and 40 mm high. Notice that all have their own centroidal axis coincident with the centroidal axis *x-x* for the composite section. The rule just stated can then be used to compute the total value of I for the cross by adding the value of I for each of the three parts. See Example Problem 6–3.

Figure 6–10 shows an example where there is a 35-mm-diameter circular hole removed from a square whose sides measure 50 mm. The circle and the square have the same centroidal axis *x-x.* The rule can then be used to compute the value of I for the square and then to subtract the value of I for the circle to obtain the total value for I of the composite shape. See Example Problem 6–4.

FIGURE 6–10 Shape for Example Problem 6–4.

Example Problem 6–3

Compute the moment of inertia of the cross-shape shown in Figure 6–9 with respect to its centroidal axis.

Solution   Objective: Compute the centroidal moment of inertia.

Given: Shape shown in Figure 6–9.

Analysis: The centroid of the cross-shape is at the intersection of the horizontal and vertical axes of symmetry. Dividing the cross into the three parts shown in the figure results in each part having the same centroidal axis, x-x, as the entire composite section. Therefore, we can compute the value of I for each part and sum them to obtain the total value, I_T. That is,

$$I_T = I_1 + I_2 + I_3$$

Results: Referring to Figure 6–1 for the formula for I for a rectangle gives

$$I_1 = \frac{bh^3}{12} = \frac{30(80)^3}{12} = 1.28 \times 10^6 \text{ mm}^4$$

$$I_2 = I_3 = \frac{30(40)^3}{12} = 0.16 \times 10^6 \text{ mm}^4$$

Then

$$I_T = 1.28 \times 10^6 \text{ mm}^4 + 2(0.16 \times 10^6 \text{ mm}^4) = 1.60 \times 10^6 \text{ mm}^4$$

Example Problem 6–4

Compute the moment of inertia of the shape shown in Figure 6–10 with respect to its centroidal axis.

Solution   Objective: Compute the centroidal moment of inertia.

Given: Shape shown in Figure 6–10.

Analysis: The centroid of the composite shape is at the intersection of the horizontal and vertical axes of symmetry. This coincides with the centroid of both the square and the circle. The composite shape can be considered to be the square with the circle removed. Therefore, we can compute the total value of I_T by computing the value of I_1 for the square and subtracting I_2 for the circle. That is,

$$I_T = I_1 - I_2$$

Results:

$$I_1 = \frac{s^4}{12} = \frac{(50)^4}{12} = 520.8 \times 10^3 \text{ mm}^4$$

$$I_2 = \frac{\pi D^4}{64} = \frac{\pi (35)^4}{64} = 73.7 \times 10^3 \text{ mm}^4$$

For the composite section,

$$I_T = I_1 - I_2 = 447.1 \times 10^3 \text{ mm}^4$$

6–6 MOMENT OF INERTIA FOR COMPOSITE SHAPES— GENERAL CASE—USE OF THE PARALLEL AXIS THEOREM

When a composite section is composed of parts whose centroidal axes do not lie on the centroidal axis of the entire section, the process of simply summing the values of I for the parts *cannot* be used. It is necessary to employ the *parallel axis theorem*.

The general statement of the parallel axis theorem is

> The moment of inertia of a shape with respect to a certain axis is equal to the sum of the moment of inertia of the shape with respect to its own centroidal axis plus an amount called the **transfer term** computed from Ad^2, where A is the area of the shape and d is the distance from the centroid of the shape to the axis of interest.

This theorem can be applied to compute the total moment of inertia for a general composite shape by using the following procedure. In this case, the axis of interest is the centroidal axis of the composite shape that must be found using the method given in Section 6–4.

GENERAL PROCEDURE FOR COMPUTING THE MOMENT OF INERTIA FOR A COMPOSITE SHAPE

1. Divide the composite shape into component parts that are simple shapes for which formulas are available to compute the moment of inertia of the part with respect to its own centroidal axis. Identify the parts as *1, 2, 3,* and so forth.

2. Locate the distance from the centroid of each component part to some convenient reference axis, typically the bottom of the composite section. Call these distances y_1, y_2, y_3, and so forth.

3. Locate the centroid of the composite section using the method given in Section 6–4. Call the distance from the reference axis used in step 2 to the centroid, $\overline{Y}$.

4. Compute the moment of inertia of each part with respect to its own centroidal axis, calling these values, I_1, I_2, I_3, and so forth.

5. Determine the distance from the centroid of the composite shape to the centroid of each part, calling these values d_1, d_2, d_3, and so forth. Note that $d_1 = \overline{Y} - y_1$, $d_2 = \overline{Y} - y_2$, $d_3 = \overline{Y} - y_3$, and so forth. Use the absolute value of each distance.

6. Compute the *transfer term* for each part from $A_i d_i^2$, where A_i is the area of the part and d_i is the distance found in step 5.

7. Compute the total moment of inertia of the composite section with respect to its centroidal axis from

$$I_T = I_1 + A_1 d_1^2 + I_2 + A_2 d_2^2 + I_3 + A_3 d_3^2 + \cdots \quad \text{(6–5)}$$

Equation (6–5) is called the *parallel axis theorem* because it defines how to transfer the moment of inertia of an area from one axis to any parallel axis. As applied here, the two axes are the centroidal axis of the component part and the centroidal axis of the composite section. For each part of a composite section, the sum $I + Ad^2$ is a measure of its contribution to the total moment of inertia.

Implementation of the *general procedure for computing the moment of inertia for a composite shape* can be facilitated by preparing a table that is an extension of the one used in Section 6–4 to find the location of the centroid of the shape. The general design of this table follows.

Part	A_i	y_i	A_iy_i	I_i	$d_i = \overline{Y} - y_i$	$A_id_i^2$	$I_i + A_id_i^2$
1							
2							
3							
$A_T = \Sigma A_i =$		$\Sigma(A_iy_i) =$				$I_T = \Sigma(I_i + A_id_i^2) =$	

$$\text{Distance to centroid} = \overline{Y} = \frac{\Sigma(A_iy_i)}{A_T} =$$

The use of the table and the general guidelines are demonstrated in Example Problems 6–5, 6–6, and 6–7. The benefit from using this type of table becomes greater as the number of components gets greater. Also, the use of a computer spreadsheet to make the appropriate calculations is very convenient.

Example Problem 6–5 Compute the moment of inertia of the tee shape in Figure 6–11 with respect to its centroidal axis.

Solution Objective: Compute the moment of inertia.

Given: Shape shown in Figure 6–11.

Analysis: Use the General **Procedure** listed in this section. As ***Step 1,*** divide the tee shape into two parts, as shown in Figure 6–12. Part 1 is the vertical stem and part 2 is the horizontal flange.

Results: The following table summarizes the complete set of data used to compute the total moment of inertia with respect to the centroid of the tee shape. Some of the data are shown also in Figure 6–12. Comments are given here to show the manner of arriving at certain data.

Part	A_i	y_i	A_iy_i	I_i	d_i	$A_id_i^2$	$I_i + A_id_i^2$
1	4.0 in^2	2.0 in	8.0 in^3	5.333 in^4	0.75 in	2.25 in^4	7.583 in^4
2	2.0 in^2	4.25 in	8.50 in^3	0.042 in^4	1.50 in	4.50 in^4	4.542 in^4
	$A_T = 6.0$ in^2		$\Sigma(A_iy_i) = 16.5$ in^3				$I_T = 12.125$ in^4

$$\overline{Y} = \frac{\Sigma(A_iy_i)}{A_T} = \frac{16.5 \text{ in}^3}{6.0 \text{ in}^2} = 2.75 \text{ in}$$

FIGURE 6–11   Shape for Example
Problem 6–5.

FIGURE 6–12   Data used in Example
Problem 6–5.

Steps 2, 3: The first three columns of the table give data for computing the location of the centroid using the technique shown in Section 6–4. Distances, y_i, are measured upward from the bottom of the tee. The result is $\bar{Y} = 2.75$ in.

Step 4: Both parts are simple rectangles. Then the values of I are

$$I_1 = bh^3/12 = (1.0)(4.0)^3/12 = 5.333 \text{ in}^4$$
$$I_2 = bh^3/12 = (4.0)(0.5)^3/12 = 0.042 \text{ in}^4$$

Step 5: Distances, d_i, from the overall centroid to the centroid of each part,

$$d_1 = \bar{Y} - y_1 = 2.75 \text{ in} - 2.0 \text{ in} = 0.75 \text{ in}$$
$$d_2 = y_2 - \bar{Y} = 4.25 \text{ in} - 2.75 \text{ in} = 1.50 \text{ in}$$

Step 6: Transfer term for each part,

$$A_1 \, d_1^2 = (4.0 \text{ in}^2)(0.75 \text{ in})^2 = 2.25 \text{ in}^4$$
$$A_2 \, d_2^2 = (2.0 \text{ in}^2)(1.50 \text{ in})^2 = 4.50 \text{ in}^4$$

Step 7: Total moment of inertia:

$$I_T = I_1 + A_1 d_1^2 + I_2 + A_2 d_2^2$$
$$= 5.333 \text{ in}^4 + 2.25 \text{ in}^4 + 0.042 \text{ in}^4 + 4.50 \text{ in}^4$$
$$I_T = 12.125 \text{ in}^4$$

LESSON
9

Comment: Notice that the transfer terms contribute over half of the total value to the moment of inertia.

6–7 MATHEMATICAL DEFINITION OF MOMENT OF INERTIA

As stated in the Section 6–5, the moment of inertia, I, is defined as the sum of the products obtained by multiplying each element of the area by the square of its distance from the reference axis. The mathematical formula for moment of inertia follows from that definition and is now given. Note that the process of summing over an entire area is accomplished by integration.

$$I = \int y^2 dA \qquad (6\text{–}4)$$

Refer to Figure 6–13 for an illustration of the terms in this formula for the special case of a rectangle, for which we want to compute the moment of inertia with respect to its centroidal axis. The small element of area is shown as a thin strip parallel to the centroidal axis where the width of the strip is the total width of the rectangle, b, and the thickness of the strip is a small value, dy. Then the area of the element is

$$dA = b \cdot dy$$

The distance, y, is the distance from the centroidal axis to the centroid of the elemental area as shown. Substituting these values into Equation (6–4) allows the derivation of the formula for the moment of inertia of the rectangle with respect to its centroidal axis. Note that to integrate over the

FIGURE 6–13 Data used in derivation of moment of inertia for a rectangle.

6–8 COMPOSITE SECTIONS MADE FROM COMMERCIALLY AVAILABLE SHAPES

In Section 1–8 commercially available structural shapes are described for wood, steel, and aluminum. Properties of representative sizes of these shapes are listed in the following appendix tables:

> Appendix A–4 for wood beams
>
> Appendix A–5 for structural steel angles
>
> Appendix A–6 for structural steel channels
>
> Appendix A–7 for structural steel wide-flange shapes
>
> Appendix A–8 for structural steel American standard beams
>
> Appendix A–9 for structural tubing—square and rectangular
>
> Appendix A–10 for aluminum standard channels
>
> Appendix A–11 for aluminum standard I-beams
>
> Appendix A–12 for standard schedule 40 steel pipe
>
> Appendix A–13 for steel mechanical tubing

In addition to being very good by themselves for use as beams, these shapes are often combined to produce special composite shapes with enhanced properties.

When used separately, the properties for designing can be read directly from the tables for area, moment of inertia, and pertinent dimensions. When combined into composite shapes, the area and the moment of inertia of the component shapes with respect to their own centroidal axes are needed and can be read from the tables. Also, the tables give the location of the centroid for the shape, which is often needed to determine distances needed to compute the transfer-of-axis term, Ad^2, in the moment of inertia calculation. The following example problems illustrate these processes.

entire area requires the limits for the integral to be from $-h/2$ to $+h/2$.

$$I = \int_{-h/2}^{+h/2} y^2 \, dA = \int_{-h/2}^{+h/2} y^2 (b \cdot dy)$$

Because b is a constant, it can be taken outside the integral, giving

$$I = b \int_{-h/2}^{+h/2} y^2 \, dy = b \left[\frac{y^3}{3} \right]_{-h/2}^{+h/2}$$

Inserting the limits for the integral gives

$$I = b \left[\frac{h^3}{24} - \frac{(-h)^3}{24} \right] = b \left[\frac{2h^3}{24} \right] = \frac{bh^3}{12}$$

This is the formula reported in the tables. Similar procedures can be used to develop the formulas for other shapes.

Example Problem 6–6   Compute the moment of inertia of the composite I-beam shape shown in Figure 6–14 with respect to its centroidal axis. The shape is formed by welding a 0.50 in thick by 6.00 in wide plate to both the top and bottom flanges to increase the stiffness of the standard aluminum I-beam.

Solution   **Objective:**   Compute the moment of inertia.

Given: Shape shown in Figure 6–14. For the I10×10.286 beam shape:
$I = 155.79$ in⁴; $A = 8.747$ in² (from Appendix A–11)

Analysis: Use the **General Procedure** listed earlier in this chapter. As **Step 1,** divide the beam shape into three parts. Part 1 is the I-beam; part 2 is the bottom plate; part 3 is the top plate. As **Steps 2 and 3,** the centroid is coincident with the centroid of the I-beam because the composite shape is symmetrical. Thus, $\overline{Y} = 5.50$ in , or one-half of the total height of the composite shape. No separate calculation of $\overline{Y}$ is needed.

Results: The following table summarizes the complete set of data used in steps 4–7 to compute the total moment of inertia with respect to the centroid of the beam shape. Some of the data are shown also in Figure 6–14. Comments are given here to show the manner of arriving at certain data.

FIGURE 6–14 Data used in Example Problem 6–6.

Part	A_i	y_i	A_iy_i	I_i	$d_i = \bar{Y} - y_i$	$A_id_i^2$	$I_i + A_id_i^2$
1	8.747	5.50	–	155.79	0	0	155.79
2	3.00	0.25	–	0.063	5.25	82.69	82.75
3	3.00	10.75	–	0.063	5.25	82.69	82.75

$A_T = \Sigma A_i = 14.747$ in^2 $\Sigma(A_i\, y_i) = $ – – $I_T = \Sigma(I_i + A_id_i^2) = 321.29$ in^4

$$\text{Distance to centroid} = \bar{Y} = \frac{\Sigma(A_iy_i)}{A_T} = 5.50 \text{ in (by inspection)}$$

Step 4: For each rectangular plate,

$$I_2 = I_3 = bh^3/12 = (6.0)(0.5)^3/12 = 0.063 \text{ in}^4$$

Step 5: Distance from the overall centroid to the centroid of each part:

$$d_1 = 0.0 \text{ in because the centroids are coincident}$$
$$d_2 = 5.50 - 0.25 = 5.25 \text{ in}$$
$$d_3 = 10.75 - 5.50 = 5.25 \text{ in}$$

Step 6: Transfer term for each part:

$$A_1d_1^2 = 0.0 \text{ because } d_1 = 0.0$$
$$A_2d_2^2 = A_3d_3^2 = (3.00)(5.25)^2 = 82.69 \text{ in}^4$$

Step 7: Total moment of inertia:

$$I_T = I_1 + I_2 + A_2d_2^2 + I_3 + A_3d_3^2$$
$$I_T = 155.79 + 0.063 + (3.0)(5.25)^2 + 0.063 + (3.0)(5.25)^2$$
$$I_T = 321.29 \text{ in}^4$$

Comment: Notice that the two added plates more than double the total value of the moment of inertia as compared with the original I-beam shape. Also, virtually all of the added value is due to the transfer terms and not to the basic moment of inertia of the plates themselves.

Example Problem 6–7

Compute the moment of inertia of the fabricated I-beam shape shown in Figure 6–15 with respect to its centroidal axis. The shape is formed by welding four standard L4×4×$\frac{1}{2}$ steel angles to a $\frac{1}{2}$ × 16 vertical plate.

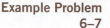

FIGURE 6–15   Data used in Example Problem 6–7.

Solution   Objective: Compute the moment of inertia.

Given: Shape shown in Figure 6–15. For each angle:
I = 5.52 in^4; A = 3.75 in^2 (from Appendix A–5)

Analysis: Use the **General Procedure** listed earlier in this section. As **Step 1,** we can consider the vertical plate to be part 1. Since the angles are all the same and placed an equal distance from the centroid of the composite shape, we can compute key values for one angle and multiply the results by 4. The location of the angles places the flat faces even with the top and bottom of the vertical plate. The location of the centroid of each angle is then 1.18 in from the top or bottom, based on the location of the centroid of the angles themselves, as listed in Appendix A–5. For **Steps 2 and 3,** the centroid is coincident with the centroid of the I-beam because the composite shape is symmetrical. Thus, $\bar{Y}$ = 8.00 in, or one-half of the total height of the composite shape. No separate calculation of $\bar{Y}$ is needed.

Results: The following table summarizes the complete set of data used in steps 4–7 to compute the total moment of inertia with respect to the centroid of the beam shape. Some of the data are shown also in Figure 6–15. Comments are given here to show the manner of arriving at certain data. The second line of the table, shown in italics, gives data for one angle for reference only. Line 3 gives the data for all four angles. Then the final results are found by summing lines 1 and 3.

Part	A_i	y_i	$A_i y_i$	I_i	$d_i = \bar{Y} - y_i$	$A_i d_i^2$	$I_i + A_i d_i^2$
1	8.00	8.00	–	170.67	0	0	170.67
(2) Ref.	*3.75*	*1.18*	*–*	*5.56*	*6.82*	*174.42*	*179.98*
4 × (2)	15.00	–	–	22.24	–	697.68	719.93

$A_T = \Sigma A_i = 23.00$ in^2 $\Sigma(A_i y_i) = -$ $I_T = \Sigma(I_i + A_i d_i^2) = 890.60$ in^4

Distance to centroid = $\bar{Y} = \dfrac{\Sigma(A_i y_i)}{A_T}$ = 8.00 in (by inspection)

Step 4: For the vertical rectangular plate,

$$I_1 = bh^3/12 = (0.5)(16)^3/12 = 170.67 \text{ in}^4$$

Step 5: Distance from the overall centroid to the centroid of each part:

$$d_1 = 0.0 \text{ in because the centroids are coincident}$$
$$d_2 = 8.00 - 1.18 = 6.82 \text{ in}$$

Step 6: Transfer term for each part:

$$A_1 d_1^2 = 0.0 \text{ because } d_1 = 0.0$$
$$A_2 d_2^2 = A_3 d_3^2 = (3.75)(6.82)^2 = 174.42 \text{ in}^4$$

Step 7: Total moment of inertia:

$$I_T = 170.67 + 4[5.56 + 3.75(6.82)^2]$$
$$I_T = 170.67 + 719.93 = 890.60 \text{ in}^4$$

Comment: Approximately 80% of the total value of moment of inertia is contributed by the four angles.

6–9 MOMENT OF INERTIA FOR SHAPES WITH ALL RECTANGULAR PARTS

A method is shown here for computing the moment of inertia of special shapes that can be divided into parts, all of which are rectangles with their sides perpendicular and parallel to the axis of interest. An example would be the tee section analyzed in Example Problem 6–5 and shown in Figure 6–11. The method is somewhat simpler than the method described in Section 6–7 that used the parallel axis theorem, although both methods are based on the same fundamental principles.

The method involves the following steps:

1. Divide the composite section into a convenient number of parts so that each part is a rectangle with its sides perpendicular and parallel to the horizontal axis.

2. For each part, identify the following dimensions:

   b = width

   y_1 = distance from the base of the composite section to the bottom of that part

   y_2 = distance from the base of the composite section to the top of that part

3. Compute the area of each part from the equation,

$$A = b(y_2 - y_1)$$

4. Compute A_T, the sum of the areas for all parts.

5. Compute the moment of the area of each part from the equation,

$$M = b(y_2^2 - y_1^2)/2$$

6. Compute M_T, the sum of the moments for all parts.

7. Compute the location of the centroid relative to the base of the composite section from

$$\overline{Y} = M_T/A_T$$

8. Compute the moment of inertia with respect to the base of the composite section of each part from

$$I_b = b(y_2^3 - y_1^3)/3$$

9. Compute I_{bT}, the sum of the moments of inertia for all parts.

10. Compute the total moment of inertia with respect to the centroid of the composite section from

$$I_c = I_{bT} - A_T \overline{Y}^2$$

This process lends itself very well to automated computation using a programmable calculator, a computer program, or a spreadsheet. As an illustration, Figure 6–16 shows the spreadsheet calculation of the centroidal moment of inertia for the tee section shown in Figure 6–11 and for which the calculation of the moment of inertia was done in Example Problem 6–5 using the parallel axis theorem. The results are, of course, identical. See also Figure 6–17 for data.

Note that there are some blank lines in the spreadsheet because allowance was made for up to six parts for the composite section whereas this one has only two. The spreadsheet could be expanded to include any number of parts.

6–10 RADIUS OF GYRATION

Another property of the area of a cross section is called the *radius of gyration*. It is used most frequently in the study of strength of materials when analyzing or designing columns, relatively long, slender members carrying axial compressive loads, as described in Chapter 16.

Columns are mentioned briefly in the activity in Chapters 1 and 7 when direct compressive stresses are defined. There it is emphasized that the simple direct stress formula, $\sigma = F/A$, applies to compression members only if they are "short." Short compression members fail when the direct compressive stress exceeds the yield strength of the material in compression. Long compression members, called columns, fail by elastic instability. That is, they are unable to maintain their initially straight form when a critical compressive load is reached. Instead they begin to bend elastically and, if the load is increased moderately above the critical load, the column will buckle suddenly and collapse, usually with catastrophic consequences.

MOMENT OF INERTIA FOR A SHAPE WITH ALL RECTANGULAR PARTS

Problem ID: Example Problem 6–5
Enter data for all parts of the composite area in the shaded cells
For each part: b = width; y_1 = distance to bottom of part; y_2 = distance to top of part

| | Dimensions | | | Area | Moment | I with respect to base |
| | b | y_1 | y_2 | A | M | I_b |
	(in)	(in)	(in)	(in²)	(in³)	(in⁴)
Part:						
1	1.000	0.000	4.000	4.000	8.000	21.333
2	4.000	4.000	4.500	2.000	8.500	36.167
3				0.000	0.000	0.000
4				0.000	0.000	0.000
5				0.000	0.000	0.000
6				0.000	0.000	0.000

Totals: A_T = 6.000 in² M_T = 16.500 in³ I_{bT} = 57.500 in⁴

Note: $A = b(y_2 - y_1)$ $M = b(y_2^2 - y_1^2)/2$ $I_b = b(y_2^3 - y_1^3)/3$

Results:

Distance from base to centroid = $M_T/A_T = \overline{Y}$ = 2.75 in

Moment of inertia with respect to the centroid = $I_{bT} - A_T\overline{Y}^2 = I_c$ = 12.125 in⁴

FIGURE 6–16 Spreadsheet for computing moment of inertia for a shape with all rectangular parts.

You may have observed column buckling behavior. Have you ever taken a long, thin rod or a thin strip of wood, plastic, or metal and subjected it to a compressive load? For example, the typical meter stick you probably used in an early science class is a long, slender column. If you supported it on a table or the floor and pushed down on it with a moderate force, it would bend as shown in Figure 6–18. This is an example of elastic instability or buckling. If you release the load just after the onset of buckling, you will notice that the column did not deform permanently. This is evidence that the failure was not in the material, but rather it was because the shape of the column was not able to maintain the straightness of the column. The property of the shape that determines when a column will buckle is its radius of gyration.

However, the method of determining when the compression member is long or short is not given in Chapter 1 because it is dependent on the moment of inertia that is the focus of much of this chapter. In fact, it is the *ratio of the length of the member to its radius of gyration* that determines when a member is long or short. That ratio is called the slenderness ratio. For now, we only define the radius of gyration. Later, in Chapter 16, it will be used in the analysis of columns.

(a) Composite tee shape (b) Part 1 (c) Part 2

FIGURE 6–17 Tee section for illustration of the method of computing moment of inertia described in Section 6–10.

For axis X-X: $r_x = 0.2887w$
$I_x = t\,w^3/12$
(b)

Column cross section

Critical buckling axis
For axis Y-Y: $r_y = 0.2887t$
$I_y = w\,t^3/12$
(c)

FIGURE 6–18   Illustration of a metal stick exhibiting buckling as a column.

Definition of Radius of Gyration, *r*

We define the radius of gyration, *r*, as

$$r = \sqrt{\frac{I}{A}} \qquad (6\text{–}6)$$

where I = moment of inertia of the cross section of the column with respect to one of the principal axes

A = area of the cross section

Because both I and A are geometrical properties of the cross section, the radius of gyration, r, is also. Formulas for computing r for several common shapes are given in Appendix A–1. Also, r is listed with other properties for some of the standard structural shapes in the Appendix. For those for which r is not listed, the values of I and A are available and Equation (6–6) can be used to compute r very simply.

Example Problem 6–8   Compute the radius of gyration for the cross section of a meter stick that has a thickness of 2.5 mm and a width of 30 mm. Consider both principal axes, *X* and *Y*, as illustrated in Figure 6–18. Use Equation (6–6).

Solution   Objective: Compute the radius of gyration with respect to axes *X* and *Y*.

Given: A rectangular cross section. Thickness $= t = 2.5$ mm. Width $= w = 30$ mm.

Analysis: Use Equation (6–6). $r = \sqrt{I/A}$

Results: *r_x with Respect to the X-Axis.*   First, let's compute the moment of inertia with respect to the *X*-axis. From Appendix A–1, we find the general form of the equation for the moment of inertia of a rectangle to be $I = bh^3/12$. Note that b is the dimension parallel to the axis of interest and h is the dimension perpendicular to the axis of interest. In this case, $b = t = 2.5$ mm and $h = w = 30.0$ mm. Then

$$I_X = tw^3/12 = (2.5\,\text{mm})(30.0\,\text{mm})^3/12 = 5625\,\text{mm}^4$$

Also, $A = tw = (2.5\text{ mm})(30.0\text{ mm}) = 75.0\text{ mm}^2$.

Then the radius of gyration is

$$r_X = \sqrt{I_X/A} = \sqrt{(5625\,\text{mm}^4)/(75.0\,\text{mm}^2)} = 8.66\,\text{mm}$$

r_y with Respect to the Y-Axis.   Similarly, let's first compute the moment of inertia with respect to the Y-axis. In this case, $b = w = 30.0$ mm and $h = t = 2.5$ mm. Then

$$I_Y = wt^3/12 = (30.0\,\text{mm})(2.5\,\text{mm})^3/12 = 39.06\,\text{mm}^4$$

Also, $A = tw = (2.5\text{ mm})(30.0\text{ mm}) = 75.0\text{ mm}^2$.

Then the radius of gyration is

$$r_Y = \sqrt{I_Y/A} = \sqrt{(39.06\,\text{mm}^4)/(75.0\,\text{mm}^2)} = 0.722\,\text{mm}$$

Comment: It is obvious that the choice of the axis about which to compute the radius of gyration is critical to the result. Here I_X is much greater than I_Y and therefore, r_X is much greater than r_Y. You should gain an appreciation of the importance of the selection of the axis from this example. You will see the effect of this observation in Chapter 16 on column analysis and design. The result is that the meter stick would buckle about its Y-axis, not the X-axis. It will always buckle about the axis with the smallest radius of gyration.

Alternate Solution   Where a formula for computing the radius of gyration is available, it makes, the computation much simpler. For the rectangular section of this problem, Appendix A–1 gives

$$r_X = h/\sqrt{12} = 0.2887\,h$$

and

$$r_Y = b/\sqrt{12} = 0.2887\,b$$

formulas for r_X and r_Y are easily derived from the basic definition:

$$r_X = \sqrt{I/A} = \sqrt{(bh^3/12)/(bh)} = \sqrt{h^2/12} = h/\sqrt{12}$$
$$r_Y = \sqrt{I/A} = \sqrt{(hb^3/12)/(bh)} = \sqrt{b^2/12} = b/\sqrt{12}$$

In this case, $h = w = 30.0$ mm and $b = t = 2.5$ mm. Then

$$r_X = 0.2887\,w = (0.2887)(30.0\,\text{mm}) = 8.66\,\text{mm}$$

and

$$r_Y = 0.2887\,t = (0.2887)(2.5\,\text{mm}) = 0.722\,\text{mm}$$

These values match those found using Equation (6–6).

Radius of Gyration for Structural Shapes

Review the data in Appendices A–4 through A–13 that give the section properties for a large variety of structural shapes often used in the structures of machines, platforms, or buildings. Some of these tables list the radius of gyration directly. Others list both the moment of inertia, I, and the cross-sectional area, A. Then you can use Equation (6–6) to compute the radius of gyration, r.

It is also necessary to determine the axis for which the radius of gyration is the smallest. For the wide-flange beams (Appendix A–7) and American Standard beams (Appendix A–8), the minimum value of r is that computed with respect to the Y-Y axis; that is,

$$r_{\min} = \sqrt{\frac{I_Y}{A}}$$

Similarly, for rectangular structural tubing (Appendix A–9), the minimum radius of gyration is that with respect to the Y-Y axis. Values for r are listed in the table.

For structural steel angles, called L-shapes, neither the X-X nor the Y-Y axis provides the minimum radius of gyration. As illustrated in Appendix A–5, $r_{\min}$ is computed with respect to the Z-Z axis, with the values listed in the table.

For symmetrical sections, the value of r is the same with respect to any principal axis. Such shapes are the solid or hollow circular section and the solid or hollow square section.

6–11 SECTION MODULUS

In the analysis of bending of beams in Chapter 12, it is shown that the bending stress is directly proportional to the bending moment acting on the beam at a given location and inversely proportional to a property of the cross-sectional area called the *section modulus*. Expressed mathematically,

$$\text{Stress due to bending} = \sigma = M/S$$

where M is the bending moment found using the principles discussed in Chapter 11 and S is the section modulus, defined in Chapter 12 to be

$$S = I/c$$

The symbol, c, is taken to be the distance from the centroidal axis of the cross section to the outermost fiber of the beam. Because both I and c are properties of the cross-sectional area of the beam, the section modulus is also a property of the area, and it would typically be used for analysis and design.

Figure 6–19 shows an example illustrating these concepts. The standard steel W-shape, W8×21, has an area moment of inertia of 75.3 in⁴ and an overall depth of 8.28 in as given in Appendix A–7. Because of symmetry, we see that the horizontal centroidal axis about which bending occurs is at the middle of the depth. Therefore, the distance from the centroid to the outermost fiber of the beam, c, is

$$c = \text{Depth}/2 = (8.28 \text{ in})/2 = 4.14 \text{ in}$$

FIGURE 6–19 Illustration of c distance on W-shape beam cross section.

Finally, we can calculate the value of the section modulus from

$$S = I/c = (75.3 \text{ in}^4)/4.14 \text{ in} = 18.2 \text{ in}^3$$

Now we can use this value to compute the stress due to bending. Assume that the beam is subjected to a bending moment of 18 750 lb · ft. Then

$$\sigma = M/S = \frac{18\,750 \text{ lb} \cdot \text{ft}}{18.2 \text{ in}^3}\frac{12 \text{ in}}{\text{ft}} = 12\,363 \text{ lb/in}^2 = 12.4 \text{ ksi}$$

Bending produces tensile stress on one outside surface and compressive stress on the other, as will be demonstrated in Chapter 12. For this example, the tensile stress and the compressive stress would have equal magnitude because the c distance is equal to either the top or bottom surface.

Not every section is symmetrical, as demonstrated in the following example problem, resulting in unequal stresses at the top and bottom surfaces of the beam.

Example Problem 6–9   Determine the section modulus for the tee shape used in Example Problem 6–5. Then compute the maximum stress due to bending at the top and bottom surfaces if the bending moment is 8650 lb · in.

Solution   The results of Example Problem 6–5, shown in Figure 6–12, include the following:

- Overall height of the section: 4.50 in
- Location of the horizontal centroidal axis: 2.75 in above the base
- Area moment of inertia: 12.125 in⁴

We can now observe that the distances from the centroidal axis to the top and bottom surfaces are unequal.

$$c_b = 2.75 \text{ in from the centroid to the bottom surface}$$
$$c_t = 4.50 \text{ in} - 2.75 \text{ in} = 1.75 \text{ in from the centroid to the top surface}$$

Two values of the section modulus can be computed.

Top surface: $S_t = I/c_t = (12.125 \text{ in}^4)/1.75 \text{ in} = 6.93 \text{ in}^3$

Bottom Surface: $S_b = I/c_b = (12.125 \text{ in}^4)/2.75 \text{ in} = 4.41 \text{ in}^3$

We can now compute the stresses at the top and bottom surfaces.

Top surface: $\sigma = M/S = (8650\,\text{lb}\cdot\text{in})/6.93\,\text{in}^3 = 1248\,\text{psi}$

Bottom surface: $\sigma = M/S = (8650\,\text{lb}\cdot\text{in})/4.41\,\text{in}^3 = 1961\,\text{psi}$

Note that the lower value for section modulus gives the higher value of stress and that would typically be the objective of the stress calculation. However, this will be explored in greater detail in Chapter 12 for special cases where the material of a beam has different strengths in tension and in compression.

REFERENCES

1. American Institute of Steel Construction, *Steel Construction Manual,* 13th ed., Chicago, IL, 2005.

2. Oberg, Erik, et al., *Machinery's Handbook,* 28th ed., New York, Industrial Press, 2008.

3. Young, Warren C., and Richard G. Budynas, *Roark's Formulas for Stress and Strain,* 7th ed., New York, McGraw-Hill, 2002.

INTERNET SITES

1. Profiles, Inc. www.profiles-inc.com Manufacturer of custom rolled or drawn shapes for precision components using carbon and alloy steel, copper, brass, and other nonferrous materials.

2. iLevel by Weyerhaeuser www.ilevel.com/floors Manufacturer of floor joists for building construction made as efficient I-shaped fabrications from simple wood components. Select the TJI® Joists product. The Joist Specifier Guide gives detailed dimensions and application data for the Silent Floor—Trus Joist products.

3. Nucor Corporation—Vulcraft Group www.vulcraft.com Manufacturer of open-web steel joists, joist girders, floor and roof deck shapes, and composite floor joist systems for use in commercial shopping centers, schools, and office buildings.

4. Aluminum Extruders Council www.aec.org/exapps/alumshow .html An industry organization promoting the use of aluminum extrusions in many types of product, building construction, and transportation system. The *Showcase* illustrates in-depth descriptions of successful applications. The *Extrusion Basics* and *Technical Information* tabs describe the extrusion process, alloys used, environmental implications, and a variety of useful resources for instructors and students.

5. American Extruded Plastics, Inc. www.aeplastics.com Manufacturer of extruded plastics for mechanical, architectural, building construction, aerospace, electronics, transportation, and many other industries.

6. Green Bull, Inc. www.greenbullladder.com Manufacturer of ladders made from aluminum, steel, fiberglass, and wood. Also produces extendable planks, platforms, mobile scaffolds, and carts.

7. Altec, Inc. www.altec.com Designer and manufacturer of truck-mounted articulated and extendable aerial platforms for positioning personnel and equipment at elevated sites. Applications to the electrical power distribution, building construction, tree trimming, and other industries.

8. Unistrut Corporation www.unistrut.com Manufacturer of systems of components made from steel or fiberglass that can be assembled in a variety of configurations to produce frames, structures, brackets, storage units, support systems, mobile units, and other devices. Online catalogs provide section property data, application information, assembly instructions, and illustrations of uses.

9. 80/20, Inc. www.8020.net Manufacturer of systems of components made from extruded aluminum that can be assembled in a variety of configurations for use as production equipment, carts, automation devices, workstations, storage units, and a variety of other systems.

10. Paramount Extrusions Company http://paramountextrusions .com/shapes Manufacturer of a variety of standard extrusions such as channels, H-sections, T-sections, hollow tubes (round, square, rectangular), square and rectangular bars, handle sections, and many others for use in office furniture, consumer products, cabinets, and so forth. The catalog gives data that can be useful in problems in this book.

11. Jackson Tube Service, Inc. http://www.jackson-tube .comshapes.htm Manufacturer of a wide variety of tubing shapes and sizes that can be useful in problems in this book.

PROBLEMS

Centroid and Moment of Inertia

For each of the shapes in Figures P6–1 through P6–48, determine the location of the horizontal centroidal axis and the magnitude of the moment of inertia of the shape with respect to that axis using the parallel axis theorem described in Section 6–7.

The method described in Section 6–10 can be applied to compute the moment of inertia for those shapes consisting of two or more rectangular parts whose sides are perpendicular and parallel to the horizontal axis. Included are the shapes in Figures P6–1 through P6–15 and the composite shapes made from standard wood beam shapes in P6–21 through P6–24.

Figures P6–1 through P6–20 are special shapes that might be produced by plastic or aluminum extrusion, by machining from solid bar stock, or by welding separate components.

FIGURE P6–1

FIGURE P6–2

FIGURE P6–3

FIGURE P6–4

FIGURE P6–5

FIGURE P6–6

FIGURE P6–7

FIGURE P6–8

FIGURE P6–9

FIGURE P6–10

FIGURE P6–11

4 mm typical

20 mm

50 mm

FIGURE P6–12

190 mm

250 mm

60 mm

FIGURE P6–17

10 mm

5 mm typical

55 mm

30 mm

10 mm   20 mm   10 mm

FIGURE P6–13

20 mm

20 mm

40 mm

10 mm

20 mm

30 mm

FIGURE P6–18

1.40 in

0.10 typical

0.60

1.70 in

2.60 in

0.20 in

FIGURE P6–14

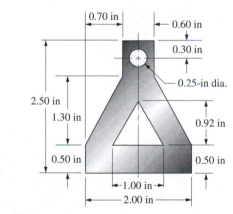

0.70 in

0.60 in

0.30 in

0.25-in dia.

2.50 in

1.30 in

0.92 in

0.50 in

0.50 in

1.00 in

2.00 in

FIGURE P6–19

0.10 typical

2.40

1.50

1.20

Dimensions in inches

FIGURE P6–15

0.375 in

1.5 in

0.75 in

FIGURE P6–16

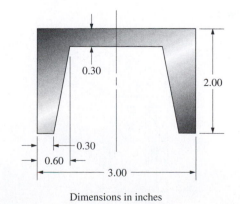

0.30

2.00

0.30

0.60

3.00

Dimensions in inches

FIGURE P6–20

Figures P6–21 through P6–24 are composite beam sections that can be made by fastening standard wood shapes together by nails, screws, or adhesive.

Figures P6–25 through P6–46 are composite beam sections fabricated from standard steel or aluminum structural shapes.

FIGURE P6–21

FIGURE P6–25

FIGURE P6–22

FIGURE P6–26

FIGURE P6–23

FIGURE P6–27

FIGURE P6–24

FIGURE P6–28

$\frac{1}{2} \times 4\frac{1}{2}$ in plates (2)

L2×2×$\frac{3}{8}$ steel angles (4)

$\frac{1}{2} \times 6$ in plate

7 in

$4\frac{1}{2}$ in

FIGURE P6–29

12 in

$\frac{1}{2} \times 12$ plate

L4×3×$\frac{1}{4}$ steel

$4\frac{3}{4}$ in

$\frac{1}{4} \times 10$ plate

10 in

FIGURE P6–33

C6×13 steel

FIGURE P6–34

$3\frac{1}{2}$ in

6 in

$\frac{1}{4}$-in plates (2)

C3×6 steel

FIGURE P6–30

$\frac{3}{8}$-in plate

$\frac{3}{8}$-in plate

C5×9 steel

6 in

L2×2×$\frac{1}{4}$ Steel

FIGURE P6–35

$1\frac{1}{2}$-in schedule 40 steel pipe

6 in

$\frac{1}{2}$ in

FIGURE P6–31

W6×15 steel

L2×2×$\frac{1}{4}$ steel

FIGURE P6–36

3-in schedule 40 steel pipe

Lacing bars (disregard in analysis for I)

24 in

18 in

FIGURE P6–32

$4 \times 2 \times \frac{1}{4}$ steel HSS

FIGURE P6–37

$3 \times 2 \times \frac{1}{4}$ steel HSS

$6 \times 6 \times \frac{1}{2}$ steel HSS

$4 \times 2 \times \frac{1}{4}$ steel HSS

FIGURE P6–38

3.00 in

C6×13 steel

W12×30 steel

FIGURE P6–41

C4×5.4 steel channel

$6 \times 2 \times \frac{1}{4}$ steel HSS

FIGURE P6–39

$4 \times 2 \times \frac{1}{4}$ steel HSS

W4×13 steel

$4 \times 2 \times \frac{1}{4}$ steel HSS

FIGURE P6–42

L4×3×$\frac{1}{4}$ steel

W12×30 steel

FIGURE P6–43

I3×2.030 aluminum I-beam

C5×2.212 aluminum channel

FIGURE P6–40

$\frac{1}{2} \times 2.00$

$6 \times 2 \times \frac{1}{4}$ steel HSS

$\frac{1}{2} \times 2.00$

FIGURE P6–44

C8×4.147
aluminum

FIGURE P6–45

1.25 in

0.50 in   0.50 in

1.50 in

FIGURE P6–48

$\frac{3}{4}$-in plate

10 in

L8×4×$\frac{1}{2}$
steel

L8×4×$\frac{1}{2}$
steel

$\frac{1}{2}$-in plate

18 in

FIGURE P6–46

6–47. Use Figure P6–47.   **6–48.** Use Figure P6–48.

For Problems 6–67 through 6–81, compute the radius of gyration for the shape in the indicated figure. Use the vertical centroidal axis for the computation.

6–67. Use Figure P6–2.	**6–75.** Use Figure P6–23.
6–68. Use Figure P6–3.	**6–76.** Use Figure P6–24.
6–69. Use Figure P6–4.	**6–77.** Use Figure P6–25.
6–70. Use Figure P6–5.	**6–78.** Use Figure P6–26.
6–71. Use Figure P6–16.	**6–79.** Use Figure P6–27.
6–72. Use Figure P6–17.	**6–80.** Use Figure P6–42.
6–73. Use Figure P6–21.	**6–81.** Use Figure P6–44.
6–74. Use Figure P6–22.	

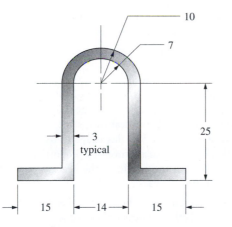

10

7

3
typical

25

15   14   15

Dimensions in millimeters

FIGURE P6–47

Radius of Gyration

For Problems 6–49 through 6–66, compute the radius of gyration for the shape in the indicated figure. Use the horizontal centroidal axis for the computation.

6–49. Use Figure P6–2.	**6–58.** Use Figure P6–14.
6–50. Use Figure P6–3.	**6–59.** Use Figure P6–15.
6–51. Use Figure P6–4.	**6–60.** Use Figure P6–16.
6–52. Use Figure P6–5.	**6–61.** Use Figure P6–17.
6–53. Use Figure P6–6.	**6–62.** Use Figure P6–21.
6–54. Use Figure P6–8.	**6–63.** Use Figure P6–22.
6–55. Use Figure P6–9.	**6–64.** Use Figure P6–23.
6–56. Use Figure P6–11.	**6–65.** Use Figure P6–24.
6–57. Use Figure P6–12.	**6–66.** Use Figure P6–25.

Problems for Centroids and Moments of Inertia Using SI Metric Data

The following problems are given with metric data:

P6–4 through P6–13, P6–17 through P6–18, P6–47, P6–51 through P6–57, P6–61, P6–69 through P6–70, P6–72

Additional problems can be expressed in SI metric data by making the following substitutions to earlier given figures. SI metric data for these sections are listed in the SI Units versions of the Appendix tables for properties of the given shapes. Note that the dimensions for rectangular plates have been modified to use preferred metric dimensions for thickness and width according to the data in Appendix A–2.

Figures P6–21 through P6–24: Use metric data from Appendix A–4 for standard dimension lumber.

Figure P6–25: Specify W360×64 steel beam with added 12 mm × 200 mm plates.

Figure P6–26: Specify S300×74 steel beam with added C300×37 steel channel.

Figure P6–27: Specify I305×23.80 aluminum I-beam with added 12 mm × 180 mm plates.

Figure P6–28: Specify C305×12.31 aluminum channel with added 12 mm × 250 mm plates.

Figure P6–29: Specify L51×51×9.5 steel angles with added 12 mm × 110 mm plates top and bottom and 12 mm × 150 mm web.

Figure P6–30: Specify C80×8.9 steel channels with added 6 mm × 150 mm side plates.

Figure P6–31: Specify PIPE 38 STD steel pipes with 12 mm thick web, 150 mm on center.

Figure P6–32: Specify PIPE 75 STD steel pipes spaced apart 600 mm vertically and 450 mm horizontally.

Figure P6–33: Specify L102×76×6.4 steel angles with added 12 mm × 300 mm top plate and 6.0 mm × 250 mm bottom plate. Overall height = 120 mm.

Figure P6–34: Specify C150×19.3 steel channels.

Figure P6–35: Specify C130×13 steel channels with 10 mm × 150 mm side plates and l51×51×6.4 angles.

Figure P6–36: Specify W150×22.5 steel beam with added l51×51×6.4 steel angles.

Figure P6–37: Specify HSS102×51×6.4 steel rectangular tube.

Figure P6–38: Specify HSS102×51×6.4 in place of HSS 4×2×¼; HSS76×51×6.4 in place of HSS 3×2×¼; and HSS152×152×12.7 in place of HSS 6×6×½ steel tubes.

Figure P6–39: Specify HSS152×51×6.4 and C100×8.

Figure P6–40: Specify C127×3.291 aluminum channel and 176×3.021 aluminum I-beam.

Figure P6–41: Specify W310×44.5 steel beam with added C150×19.3 steel channel.

Figure P6–42: Specify W100×19.3 steel beam with added HSS102×51×6.4 steel tube.

Figure P6–43: Specify W310×44.5 steel beam with added two L102×76×6.4 steel angles.

Figure P6–44: Specify HSS152×51×6.4 and two 12 mm × 50 mm plates.

Figure P6–45: Specify three C203×6.17 aluminum channels.

Figure P6–46: Specify two L203×102×12.7 steel angles with added 12 mm × 450 mm bottom plate and 20 mm × 250 mm web.

COMPUTER ASSIGNMENTS

1. For a generalized I-shape having equal top and bottom flanges similar to that shown in Figure P6–2, write a computer program or spreadsheet to compute the location of the horizontal centroidal axis, the total area, and the moment of inertia with respect to the horizontal centroidal axis for any set of actual dimensions to be input by the user.

2. For the generalized T-shape similar to that shown in Figure P6–4, write a computer program or spreadsheet to compute the location of the horizontal centroidal axis, the total area, and the moment of inertia with respect to the horizontal centroidal axis for any set of actual dimensions to be input by the user.

3. For the generalized I-shape similar to that shown in Figure P6–5, write a computer program or spreadsheet to compute the location of the horizontal centroidal axis, the total area, and the moment of inertia with respect to the horizontal centroidal axis for any set of actual dimensions to be input by the user.

4. For any generalized shape that can be subdivided into some number of rectangular components with horizontal axes, write a computer program or spreadsheet to compute the location of the horizontal centroidal axis, the total area, and the moment of inertia with respect to the horizontal centroidal axis for any set of actual dimensions to be input by the user. Use the parallel axis theorem.

5. For the generalized hat-section shape similar to that shown in Figure P6–11, write a computer program or spreadsheet to compute the location of the horizontal centroidal axis, the total area, and the moment of inertia with respect to the horizontal centroidal axis for any set of actual dimensions to be input by the user.

6. Given a set of standard dimension lumber, compute the area and moment of inertia with respect to the horizontal centroidal axis for the generalized box shape similar to that shown in Figure P6–22. The data for the lumber should be input by the user.

7. Enhance Assignment 6 by building a data file containing the dimensions of a set of standard dimension lumber. Then permit the user to select sizes for the top and bottom plates and the two vertical members for the box shape shown in Figure P6–22.

8. Write a computer program or spreadsheet to compute the area and moment of inertia with respect to the horizontal centroidal axis for a standard W- or S-beam shape with identical plates attached to both the top and bottom flanges similar to that shown in Figure 6–14. The data for the beam shape and the plates are to be input by the user.

9. Enhance Assignment 8 by building a database of standard W- or S-beam shapes. The user will select a beam shape. The data for the plates are to be input by the user.

10. Given a standard W- or S-beam shape and its properties, write a computer program or spreadsheet to compute the required thickness for plates to be attached to the top and bottom flanges to produce a specified moment of inertia of the composite section, as shown in Figure 6–14. Make the width of the plates equal to the width of the flange. For the resulting section, compute the total area.

11. Using the computer program or spreadsheet written for Assignment 1 for the generalized I-shape, perform an analysis of the area (A), the moment of inertia (I), and the ratio of I to A, as the thickness of the web is varied over a specified range. Keep all other dimensions for the shape the same. Note that the ratio of I to A is basically the same as the ratio of the stiffness of a beam having this shape to its weight, because the deflection of a beam is inversely proportional to the moment of inertia and the weight of the beam is proportional to its cross-sectional area.

12. Repeat Assignment 11 but vary the height of the section while keeping all other dimensions the same.

13. Repeat Assignment 11 but vary the thickness of the flange while keeping all other dimensions the same.

14. Repeat Assignment 11 but vary the width of the flange while keeping all other dimensions the same.

15. Write a computer program or spreadsheet to compute the moment of inertia with respect to the horizontal centroidal axis for any composite shape that can be divided into parts, all of which are rectangular having sides perpendicular and parallel to the horizontal axis, using the method described in Section 6–10. Produce output from the program for any of the shapes in Figures P6–1 through P6–15 and Figures P6–21 through P6–24.

CHAPTER SEVEN

BASIC CONCEPTS IN STRENGTH OF MATERIALS

MAJOR TOPICS

7–1 The Big Picture—Basic Concepts in Strength of Materials
7–2 Objective of This Book—To Ensure Safety
7–3 The Concept of Stress
7–4 Direct Normal Stress
7–5 Stress Elements for Direct Normal Stresses
7–6 The Concept of Strain
7–7 Direct Shear Stress
7–8 Stress Elements for Shear Stresses
7–9 Experimental and Computational Stress Analysis

7–1 THE BIG PICTURE—BASIC CONCEPTS IN STRENGTH OF MATERIALS

Discussion Map

- Products, machines, and structures must be designed to be safe and to provide satisfactory performance during the intended use.
- Safety is paramount. The load-carrying components must not fracture during use.
- Excessive deformation is another form of failure.
- Buckling, occurring when the shape of a load-carrying member becomes unstable, must be avoided.
- You will learn about the basic nature of stresses and strains in this course.
- You will be able to recognize several types of stresses created by different loading and support situations.
- You will analyze situations where more than one kind of stress is experienced by a load-carrying member at the same time.
- Design requires that you determine the shape and size of a load-carrying member and specify the material from which it is to be made.
- You will learn how to design safe load-carrying components of machines and structures.

Discover   Think about products, machines, and structures that you are familiar with that contain components that must carry loads safely. For each device that you think of, write down the following information:

The basic function or purpose of the device.

The description and sketches of its primary components, particularly those that are subjected to significant forces.

The material from which each component is made. Is it a metal or plastic? What kind of metal? What kind of plastic? Is it some other material?

How is each component supported within the product, machine, or structure?

How are forces applied to the component?

What would be the consequence if the component broke?

How much deformation would cause the component to be incapable of performing its desired function?

Consider products around your home; parts of your bicycle, car, or motorcycle; buildings; toys; amusement park rides; aircraft and space vehicles; ships; manufacturing machinery; construction equipment; agricultural machinery; and others.

Discuss these products and systems with your colleagues and with your course instructor or facilitator.

Review now the introductory Section 1–1 on page 1 of this book. Figure 1–1 showed five representative applications of the study of statics and strength of materials:

- Building construction
- Consumer product design
- Design of automotive ground transportation vehicles
- Construction equipment design and use
- Aerospace transportation vehicles

Revisit the discussion that accompanied that set of illustrations and consider how the principles of statics that you learned in the first part of this book are used in the analysis of stresses in components for a variety of applications.

Figure 1–2 showed six types of stresses created in load-carrying members and related them to the kinds of applied forces and support conditions they experience. Recall the discussion of the basic principles shown in the following parts of that figure:

a. Tensile forces and the resulting tensile stresses

b. Direct compressive forces and the resulting direct compressive stresses

c. and d. Compressive forces applied to long, slender members called *columns*

e. and f. Punching shearing forces and the resulting shearing stresses

g. Direct shearing stresses

h. Beams and the bending and shearing stresses created in them

In addition, torsional loading is shown in Figures 1–4 to 1–7 and discussed in the related section.

The fundamental concept of analyzing the *load path* through structures and machinery was discussed in Section 1–3. This is an essential part of stress analysis—determining the forces that exist on load-carrying members and visualizing the kinds of stresses that result. Examples were discussed in many applications: building structures in general, wood frame houses, mechanical devices, and power transmission systems. These are but a few of the many examples of situations in which you may use the skills from statics and strength of materials in your career.

The remaining chapters in this book build from the analysis of forces, moments, and torques you learned before, presenting the details of stress analysis. Here we guide your learning about

1. Direct tensile and compressive stress

2. Direct shearing stress

3. Bearing stress

4. Torsional shear stress

5. Shearing forces and bending moments in beams

6. Stress due to bending

7. Shearing stresses in beams

8. Deflection of beams

9. Combinations of the variety of stresses listed above

10. Columns—long, slender compression members

11. Pressure vessels

12. Connections—joints between members using bolts, rivets, and welding

The *Materials* part of the title of this book leads to the discussion in Chapter 8 about the design properties of many kinds of metallic and nonmetallic materials. You will learn about their ability to withstand different stresses, how they deform, and their physical properties. This background is integral to the emphasis on *design* in this book where you will ensure that stress levels are safe and that deformations are within tolerable limits.

Numerous example problems are given for each topic to demonstrate the approach to solving real analysis and design problems. You should develop a high level of skill in setting up such problems and documenting your solution procedure so that others may understand and evaluate your work.

7–2 OBJECTIVE OF THIS BOOK— TO ENSURE SAFETY

It is essential that any product, machine, or structure be safe and stable under the loads exerted on it during any foreseeable use. The analysis and design of such devices or structures to ensure safety is the primary objective of this book.

Failure of a component of a structure can occur in several ways:

1. The material of the component could fracture completely.

2. The material may deform excessively under load so that the component is no longer suitable for its purpose.

3. The structure could become unstable and buckle and, thus, be unable to carry the intended loads.

Examples of these failure modes should help you to understand the importance of your learning the principles of applied strength of materials as presented in this book.

Preventing Failure by Fracture

Figure 7–1 shows two rods carrying a heavy casting. Imagine that you are the person responsible for designing the rods. Certainly you would want to ensure that the rods were

FIGURE 7–1 Two rods supporting a heavy casting.

sufficiently strong so they will not break and allow the casting to fall, possibly causing great damage or injury to people. As the designer of the rods, what information would you need? What design decisions must you make? Following is a partial list of questions you should ask.

1. What is the weight and physical size of the casting?

2. Where is its center of gravity? This is important so you can decide where to place the points of attachment of the rods to the casting.

3. How will the rods be attached to the casting and to the support system at the top?

4. Of what material will the rods be made? What is its strength?

5. What will be the shape and size of the cross section of the rods?

6. How will the load of the casting be initially applied to the rods: slowly, with shock or impact, or with a jerking motion?

7. Will the rods be used for many cycles of loading during their expected life?

Knowing these factors will permit you to design the rods to be safe; that is, so they will not break under the anticipated service conditions. This will be discussed in more detail in this chapter and in Chapter 9.

Preventing Excessive Deformation

Gears are used in mechanical drives to transmit power in such familiar systems as the transmission of a truck, the drive for a conveyor belt, or the spindle of a machine tool. For proper operation of the gears, it is essential that they are properly aligned so the teeth of the driving gear will mesh smoothly with those of the mating gear. Figure 7–2 shows two shafts carrying gears in mesh. The shafts are supported in bearings that are in turn mounted rigidly in the housing of the transmission. When the gears are transmitting power, forces are developed that tend to push them apart. These forces are resisted by the shafts so that they are loaded as shown in Figure 7–3. The action of the forces perpendicular to the shafts tends to cause them to bend, which would cause the gear teeth to become misaligned. Therefore, the shafts must be designed to keep deflections at the gears to a small,

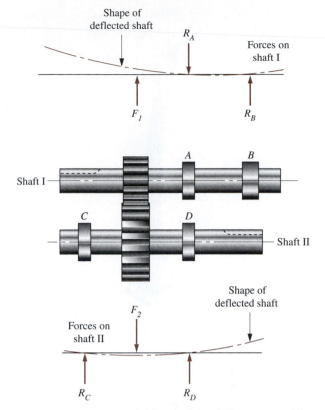

FIGURE 7–3 Forces on shafts I and II of Figure 7–2 with resulting deflection of shafts.

acceptable level. Of course, the shafts must also be designed to be safe under the applied loads. In this type of loading, the shafts are considered to be *beams*. Chapters 12–14 discuss the principles of the design of beams for strength and deflection.

Stability and Buckling

A structure may collapse because one of its critical support members is unable to hold its shape under applied loads, even though the material does not fail by fracturing. An example is a long, slender post or column subjected to a downward, compressive load. At a certain critical load, the column will *buckle*. That is, it will suddenly bend or bow out, losing its preferred straight shape. When this happens, if the load remains applied, the column will totally collapse. Figure 7–4 shows a sketch of such a column that is relatively long with a thin rectangular cross section. You can demonstrate the buckling of this type of column using a simple ruler or meter stick. To prevent buckling, you must be able to specify an appropriate material, shape, and size for the cross section of a compression member of a given length so it will remain straight under the expected loads. Chapter 16 presents the analysis and design of columns.

In summary, design and analysis, using the principles of strength of materials, are required to ensure that a component is safe with regard to *strength, rigidity,* and *stability.* It is the objective of this book to help you gain the ability to design and analyze load-carrying components of structures and machines that will be safe and suitable for their intended functions.

FIGURE 7–2 Two shafts carrying gears in mesh.

Force

Y   *X*
X   *Y*

Base

FIGURE 7–4 A slender column in compression, illustrating elastic instability or buckling.

ACTIVITY   Chapter 7—Basic Concepts of Strength of Materials

Let's break something!

In the Big Picture for Chapter 1 you explored many situations in which the principles of strength of materials are applied to create safe structures and products that perform well in their intended applications. When applied properly, well-designed products will not break.

However, useful learning can happen when you purposely cause something to break. This destruction may give you many insights, such as

1. How loads are applied
2. How load-carrying members are supported
3. How the shape and size of an object affect its ability to withstand forces
4. How different materials behave under load
5. Where the failure of an object initiates
6. How things typically deform significantly before they actually fracture, unless they are made from a brittle material
7. How, sometimes, you would judge a product to have failed because of excessive deformation even if it does not break

Here are some examples of what you can do. You are encouraged to be creative and find others. But be careful—

Ensure that proper personal protection is provided before attempting to break anything. Wear eye protection. Be sure broken pieces cannot fly off and cause injury to yourself or others nearby. Use caution when handling knives, screwdrivers, or other sharp objects. Protect furniture, windows, walls, or other objects that could be damaged.

a. Direct tension failure—Occurs when an axial pulling force overcomes the tensile strength of the material on which the force is exerted.
- Gather a piece of string such as what you might use for flying a kite.
- Attach one end to a strong fixed location such as by tying it securely to a post.
- Arrange a way to take hold of the other end of the string that will permit you to pull it slowly with steadily increasing force until it breaks. Perhaps you could tie the free end around a fairly large wooden dowel, say 20 mm (1.0 in) in diameter and about 250 mm (10 in) long, so you can pull with both hands [Figure 1–2(a)].
- Ask a colleague to observe the string as you increase the force. (Again, be careful!)
- Have the colleague describe any event that indicates when the string is about to fail; for example, when one small strand breaks.
- Try to observe the initiation of the failure yourself by relaxing the force.
- If you have a camera available, take a photograph of the failure area.
- Then resume applying the force, slowly increasing it until the string breaks completely. Observe the progression of the failure.
- After the string breaks, lay it on a table and carefully observe the nature of the broken ends.
- Document your process and the appearance of the failure.

b. Direct compression failure—Occurs when a direct axial pushing force causes the material to crush.
- The piece selected for compression failure should be quite short with a broad cross-sectional area in order for direct compression failure to occur with little or no bending or buckling prior to failure [Figure 1–2(b)]. Note that longer, more slender shapes in compression are called *columns* and fail in a radically different manner.
- Try a short length, say 25 mm (1.0 in) or less, of a cylindrical wooden dowel.
  - A good way to apply the compression load is with a bench vise like those found in a home shop, an automotive garage, or an industrial workplace.
  - Align the dowel carefully so the force is applied along the axis of the cylinder.
  - Observe the failure, typically a significant amount of crushing, probably near one of the ends. The wood fibers tend to keep the dowel from splitting completely.
- Now get a short piece of chalk, say about 25 mm long and 10 mm in diameter.
  - Crush the chalk in a vise as described for the wooden dowel.
  - Note how the chalk does fracture completely because it is quite brittle.

- A short thin-walled plastic tube provides an interesting failure mode.
  - Use the vise to apply the axial compressive force slowly and observe how the tube deforms.
  - You should notice the walls swelling outward, a sign of local instability of the material in this shape.

c. **Column buckling**—Occurs when a long, straight, and slender member bends and buckles by a significant amount before any of the material actually fails.
  - The terms *long* and *slender* are not precise and depend greatly on the shape, cross sectional dimensions, and material from which the column is made. Here are some examples of long slender columns that you should be able to obtain:
    - A wood dowel about 4.5 mm ($\frac{3}{16}$ in) in diameter and longer than about 300 mm (12.0 in).
    - A thin wood meter stick (cross section about 6 mm by 25 mm) or yardstick (cross section about 0.25 in by 1.0 in).
    - A thin strip of plastic, say 1 mm thick by 20 mm wide and longer than 50 mm; even a plastic knife like those for picnics could work.
  - Try to apply slowly an axial compressive force to the column, being careful that it does not skid away from you. Observe what happens [Figure 1–2(c)].
    - You should notice that the column bends noticeably without breaking the material.
    - When the force is relieved, the column returns to its straight form.
  - This kind of failure is called *elastic instability* or *buckling,* and could lead to catastrophic failure if the load is increased beyond that which initially causes the buckling.

d. **Direct shear**—Occurs when a cutting action is applied as you would see with scissors, tin snips, or punches.
  - A simple example is a hole punch for writing paper like those you may see in a classroom or an office [Figure 1–2(e)].
    - As you press a lever, the punch shears the paper as it passes into a die below.
    - The paper is sheared along the circumference of the hole, and the thickness of the paper must be cut.
  - It may be more difficult to demonstrate this phenomenon for material that is thicker or stronger than simple writing paper.
  - Try to find components that have holes or other punched shapes to visualize where direct shear is used in familiar products. Examples are [Figure 1–2(f)]:
    - Some types of brackets, clamps, and straps made from sheet metal
    - Metal cabinets with round holes punched for fasteners or other shapes to permit mounting of instruments or other devices
    - Electrical outlet boxes with "knockouts" that can be removed to insert wires into the box

- Metal striker plates for door latches
- Perforated sheet metal often used for decorative purposes
- Another example of direct shear occurs in a type of hinge called a *clevis* [Figure 1–2(g)]. A cylindrical pin passes through end pieces connected to one component and through a hole in another component.
  - As a force is applied to one of the moving components, the pin is subjected to a direct shearing stress that tends to cut it across its cross section.

e. **Torsional shear stress**—Occurs when a load-carrying member is twisted about its long axis due to an applied torque at one or more points along the member, resisted by a reaction torque at one or more other points.
  - One of the most frequent applications in which torsional shear stress occurs is in a power transmission shaft. For example, the drive shaft for a vehicle or an industrial machine receives power from a prime mover such as an engine or electric motor. The power is transmitted along the shaft, typically having the shape of a relatively long solid or hollow cylinder, and delivers it to an axle to propel the vehicle, or to a piece of rotating machinery such as a conveyor, machine tool, pump, compressor, food mixer, coffee grinder, or any of numerous other types of mechanical devices. See Figure 7–2.
  - The shafts of the mechanical devices listed above are designed to transmit safely the power required to perform the desired function without breaking or twisting a noticeable amount. Therefore, you would not normally be able to apply sufficient torque to break the shaft or to observe significant torsional deformation (twisting). So let's try to find some smaller cylindrical rods or tubes that can be twisted and/or broken when you apply a torque with your hands.
  - When discussing columns earlier, we suggested using a small-diameter wood dowel, about 4.5 mm ($\frac{3}{16}$ in). With a length of dowel approximately 500 mm (20 in) or longer, you should be able to hold one end steady, apply a torque to the other end, and observe a significant angle of twist between the fixed and free ends. This is an example of producing torsional shear stress. Try to find other similar examples made from plastic rod or tubing, or small-diameter steel wire such as from a coat hanger [Figure 1–2(d)]. Long foam cylinders sometimes seen at swimming pools are also good.
  - You may use the vise again to hold one end of a longer, larger diameter steel rod, say about 6 mm (0.25 in) diameter by about 550 mm (22 in) long. Then use a strong tool such as locking pliers to grab onto the opposite end and twist the rod. Notice how the applied torque must increase with the larger diameter and stiffer material.
  - Note that shapes other than cylindrical rods or tubes could also be placed in torsion. However, their behavior and the resulting stresses are radically different,

as will be discussed in Chapter 10. For example, find any flat, thin, rectangular piece such as a ruler, paint stirring stick, or meter stick and twist it with your hands. If the piece is fairly long and thin, you should be able to produce a large deformation without breaking it.

f. **Bending stress**—Occurs when a member carries a load perpendicular to its long axis while being supported in a stable manner, thereby "bending" the member, now called a "beam."

- A simple example is a flat wood stick from an ice cream bar. Set the stick on a pair of simple supports near the ends and press down near the middle. You should be able to deflect the stick (bend the beam) easily. Note the curved, bent shape taken by the beam as it is loaded. With moderate loading the beam should return to its original shape after you remove the load, indicating that the deflection is elastic.

- Other shapes you could use that are easily obtained are a meter stick or yardstick, paint stirring stick, heavy card stock, thin plastic strips, or thin sheet metal such as might be used in a metal ruler. Apply the loads slowly to ensure that the beam does not break or deform permanently if you want to keep it [Figure 1–2(h)].

- However, if you have a beam example that is expendable, go ahead now and increase the load and observe what happens. That depends mainly on the kind of material.

- A wood stick will bend significantly and then begin to shatter or split. Make note of where the failure starts and the appearance of the wood after it breaks.

- A thin metal beam will also bend a lot while still returning to its initial shape. Then you should notice that it takes a permanent deformed shape without breaking into two separate pieces. It is said to have "yielded" where some of the material has been stressed beyond its yield strength, as described in Chapters 8 and 9.

Objectives of This Chapter

In this chapter we present the basic concepts in strength of materials that will be expanded on in later chapters. After completing this chapter, you should be able to:

1. Use correct units for quantities encountered in the study of strength of materials in both the International System of Units (SI) and the U.S. Customary System.

2. Use the terms *mass* and *weight* correctly and be able to compute the value of one when the value of the other is given.

3. Define *stress*.

4. Define *direct normal stress,* both tensile and compressive.

5. Represent normal stresses on stress elements.

6. Define *normal strain.*

7. Define *direct shear stress* and the terms *single shear* and *double shear.*

8. Represent shear stresses on stress elements.

9. Recognize standard structural shapes and standard screw threads and use data for their geometric properties.

10. Describe several approaches to experimental and computational stress analysis.

Importance of Knowledge of Statics An accurate application of the principles of strength of materials requires that the loads to be carried by the member being analyzed or designed be known with a high degree of confidence. The loads may be any combination of gravity forces, direct axial forces, shearing forces, directly applied moments that tend to cause rotation of a member or structure, or torsional moments (torques) that are applied to a shaft to transmit power. It is expected that you have a good background in the study of physics mechanics and its application through the study of statics. You should be able to prepare complete free body diagrams of structures or individual elements, apply equations of equilibrium, and solve for reactions at supports. If needed, refer to Chapters 1–5 to review the fundamental principles of statics.

7-3 THE CONCEPT OF STRESS

The study of strength of materials depends on an understanding of the principles of *stress* and *strain* produced by applied loads on a structure or a machine and the members that make up such systems. These principles are introduced here and applied to relatively simple types of loading with an emphasis on *analysis.* That is, the loads and the geometry of the members are given in the problems, and your task is to analyze the resulting stress within the members and the deformation that is caused by the stress.

We use *direct loading* here to develop the concept of stress and, later, the concept of strain. By the term *direct stress* we refer to cases where the total applied force is shared equally by all parts of the cross section of the load-carrying member. The types of loading considered in this chapter are

- Direct axial loads
- Direct shearing forces
- Bearing loads

Later chapters introduce other types of loading such as bending loads on beams and torsional loads on shafts where the stresses are not uniform across the cross section.

Applications are shown for such real members as

- A rod supporting a heavy load that tends to pull the rod with a tensile force [Figure 7–1]

- Short blocks supporting heavy loads that tend to crush them with compressive forces [Figure 7–5(b)]

- A pin that carries a load acting perpendicular to its axis that tends to cut the pin across one or more of its cross sections (called a *shearing force*) [Figure 1–2(g)]

Punch press

Floor

Support block

(a)

Compressive stress on support block

(b)

Bearing stress on floor

(c)

FIGURE 7–5 Illustration of compressive stress and bearing stress.

- A floor on which a leg of a heavy machine is set, tending to cause an indentation in the floor (called a *bearing load*) [Figure 7–5(c)]

In later chapters, we introduce the concept of *design* in which the objective is to specify the material from which a member is to be made or the detailed dimensions of the member to ensure that it is safe and will perform its intended function. This requires an understanding of the ability of the material to withstand the applied forces without failure (breaking) or excessive deformation. This is really where the term *strength of materials* comes into play. Chapter 8 discusses the design properties of materials and material selection.

Understanding the meaning of *stress* in a load-carrying member, as given below, is of utmost importance in studying strength of materials.

> *Stress is the internal resistance offered by a unit area of the material from which a member is made to an externally applied load.*

We are concerned with what happens *inside* a load-carrying member. We must determine the magnitude of

force exerted on each *unit area* of the material. The concept of stress can be expressed mathematically as

Definition of Stress ▼

$$\text{Stress} = \frac{\text{Force}}{\text{Area}} = \frac{F}{A} \qquad (7\text{–}1)$$

In some cases, as described in the following section on *direct normal stress*, the applied force is shared uniformly by the entire cross section of the member. In such cases stress can be computed by simply dividing the total force by the area of the part that resists the force. Then the level of stress will be the same at any point in any cross section.

In other cases, such as for *stress due to bending* presented in Chapter 12, the stress will vary at different positions within the cross section. Then it is essential that you consider the level of stress *at a point*. Typically, your objective is to determine at which point the maximum stress occurs and what is its magnitude.

In the U.S. Customary System, the typical unit for force is the pound and the most convenient unit of area is the square inch. Thus, stress is indicated in lb/in^2, abbreviated psi. Typically encountered stress levels in machine design and structural analysis are several thousand psi. For this reason, the unit of kip/in^2 is often used, abbreviated ksi. For example, if a stress is computed to be 26 500 psi, it might be reported as

$$\text{Stress} = \frac{26\,500\,\text{lb}}{\text{in}^2} \times \frac{1\,\text{kip}}{1000\,\text{lb}} = \frac{26.5\,\text{kip}}{\text{in}^2} = 26.5\,\text{ksi}$$

In the SI unit system, the standard unit for force is the newton and area is in square meters. Thus the standard unit for stress is the N/m^2, given the name *pascal* and abbreviated Pa. Typical levels of stress are several million pascals, so the most convenient unit for stress is the megapascal or MPa. This is convenient for another reason. In calculating the cross-sectional area of load-carrying members, measurements of dimensions in mm are usually used. Then the stress would be in N/mm^2 and it can be shown that this is numerically equal to the unit of MPa. For example, assume that a force of 15 000 N is exerted over a square area 50 mm on a side. The resisting area would be 2500 mm^2 and the resulting stress would be

$$\text{Stress} = \frac{\text{Force}}{\text{Area}} = \frac{15\,000\,\text{N}}{2500\,\text{mm}^2} = \frac{6.0\,\text{N}}{\text{mm}^2}$$

Converting this to pascals would produce

$$\text{Stress} = \frac{6.0\,\text{N}}{\text{mm}^2} \times \frac{(1000)^2\,\text{mm}^2}{\text{m}^2} = 6.0 \times 10^6\,\text{N/m}^2$$
$$= 6.0\,\text{MPa}$$

In summary, ***the unit N/mm^2 is identical to MPa.***

7–4 DIRECT NORMAL STRESS

One of the most fundamental types of stress that exists is *normal stress*, indicated by the lowercase Greek letter σ (sigma), in which the stress acts perpendicular, or normal, to the cross section of the load-carrying member. If the stress is also uniform across the resisting area, the stress is called a *direct normal stress*.

Normal stresses can be either compressive or tensile.

A compressive stress is one that tends to crush the material of the load-carrying member and to shorten the member itself.

A tensile stress is one that tends to stretch the member and pull the material apart.

The equation for direct normal stress follows from the basic definition of stress because the applied force is shared equally across the entire cross section of the member carrying the force. That is,

Direct Normal Stress ▼

$$\text{Direct normal stress} = \sigma = \frac{\text{Applied force}}{\text{Area of cross section}} = \frac{F}{A}$$

$$(7\text{--}2)$$

The area of the cross section of the load-carrying member is taken perpendicular to the line of action of the force.

An example of a member subjected to compressive stress is shown in Figure 7–6. The support stand is designed to be placed under heavy equipment during assembly, and the weight of the equipment tends to crush the square shaft of the support, placing it in compression.

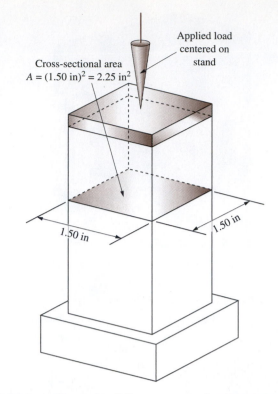

FIGURE 7–6   Example of direct compressive stress.

Example Problem 7–1   Figure 7–6 shows a support stand designed to carry downward loads. Compute the stress in the square shaft at the upper part of the stand for a load of 27 500 lb. The line of action of the applied load is centered on the axis on the shaft, and the load is applied through a thick plate that distributes the force to the entire cross section of the stand.

Solution   Objective: Compute the stress in the upper part of the stand.

Given: Load = F = 27 500 lb; load is centered on the stand.
Cross section is square; dimension of each side is 1.50 in.

Analysis: On any cross section of the stand, there must be an internal resisting force that acts upward to balance the downward applied load. The internal force is distributed over the cross-sectional area, as shown in Figure 7–7. Each small unit area of the cross section would support the same part of the total load. The stress produced in the square shaft tends to crush the material and it is, therefore, a compressive stress. Equation (7–2) can be used to compute the magnitude of the stress.

FIGURE 7–7
Compressive stress on an arbitrary cross section of support stand shaft.

Results: Stress $= \sigma =$ Force/Area $= F/A$ (compressive)

$$A = (1.50 \,\text{in})^2 = 2.25 \,\text{in}^2$$
$$\sigma = F/A = 27\,500 \,\text{lb}/2.25 \,\text{in}^2 = 12\,222 \,\text{lb/in}^2 = 12\,222 \,\text{psi}$$

Comment: This level of stress would be present on any cross section of the square shaft between its ends.

An example of a member subjected to a tensile load is shown in Figure 7–1. The two rods suspend a heavy casting from a crane and the weight of the casting tends to stretch the rods and pull them apart.

Example Problem 7–2   Figure 7–1 shows two circular rods carrying a casting weighing 11.2 kN. If each rod is 12.0 mm in diameter and the two rods share the load equally, compute the stress in the rods.

Solution   Objective: Compute the stress in the support rods.

Given: Casting weighs 11.2 kN. Each rod carries half the load.
Rod diameter $= D = 12.0$ mm.

Analysis: Direct tensile stress is produced in each rod. Use Equation (1–2).

Results:
$$F = 11.2 \,\text{kN}/2 = 5.60 \,\text{kN} \quad \text{or} \quad 5600 \,\text{N on each rod}$$
$$\text{Area} = A = \pi D^2/4 = \pi(12.0 \,\text{mm})^2/4 = 113 \,\text{mm}^2$$
$$\sigma = F/A = 5600 \,\text{N}/113 \,\text{mm}^2 = 49.5 \,\text{N/mm}^2 = 49.5 \,\text{MP}$$

LESSON 10

Comment: Figure 7–8 shows an arbitrarily selected part of the rod with the applied load on the bottom and the internal tensile stress distributed uniformly over the cut section.

FIGURE 7–8 Tensile stress on an arbitrary cross section of a circular rod.

Tensile stress uniformly distributed over an arbitrarily selected cross section

Applied load

7–5 STRESS ELEMENTS FOR DIRECT NORMAL STRESSES

The illustrations of stresses in Figures 7–7 and 7–8 are useful for visualizing the nature of the internal resistance to the externally applied force, particularly for these cases in which the stresses are uniform across the entire cross section. In other cases it is more convenient to visualize the stress condition on a small (infinitesimal) element. Consider a small cube of material anywhere inside the square shaft of the support stand shown in Figure 7–6. There must be a net compressive force acting on the top and bottom faces of the cube, as shown in Figure 7–9(a). If the faces are considered to be unit areas, these forces can be considered to be the stresses acting on the faces of the cube. Such a cube is called a *stress element.*

(a) Three-dimensional
stress element

(b) Two-dimensional
stress element

FIGURE 7–9   Stress element for compressive stresses.

Because the element is taken from a body in equilibrium, the element itself is also in equilibrium and the stresses on the top and bottom faces are the same. A simple stress element like this one is often shown as a two-dimensional square element rather than the three-dimensional cube, as shown in Figure 7–9(b).

Similarly, the tensile stress on any element of the rod in Figure 7–1 can be shown as in Figure 7–10 with the stress vector acting outward from the element. Note that either compressive or tensile stresses are shown acting perpendicular (normal) to the surface of the element.

7–6 THE CONCEPT OF STRAIN

Any load-carrying member deforms under the influence of the load applied. The square shaft of the support stand in Figure 7–6 gets shorter as the heavy equipment is placed on the stand. The rods supporting the casting in Figure 7–1 get longer as the casting is hung onto them.

The total deformation of a load-carrying member can, of course, be measured. It will also be shown later how the deformation can be calculated.

Figure 7–11 shows an axial tensile force of 10 000 lb applied to an aluminum bar that has a diameter of 0.75 in. Before the load was applied, the length of the bar was 10.000 in. After the load is applied, the length is 10.023 in. Thus, the total deformation is 0.023 in.

Strain, also called *unit deformation*, is found by dividing the total deformation by the original length of the bar. The lowercase Greek letter epsilon (ϵ) is used to denote strain:

Definition of Strain ▼

$$\text{Strain} = \epsilon = \frac{\text{Total deformation}}{\text{Original length}} \qquad (7\text{–}3)$$

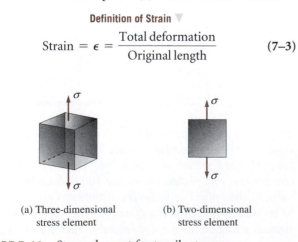

(a) Three-dimensional
stress element

(b) Two-dimensional
stress element

FIGURE 7–10   Stress element for tensile stresses.

0.75-in.-dia. rod

10 000 lb ◄———   ———► 10 000 lb

|← 10 in →| |← 0.023 in
Original   Elongation
length

FIGURE 7–11   Elongation of a bar in tension.

For the case shown in Figure 7–11,

$$\epsilon = \frac{0.023 \text{ in}}{10.000 \text{ in}} = 0.0023 \text{ in/in}$$

Strain could be said to be dimensionless because the units in the numerator and denominator could be canceled. However, it is better to report the units as in/in or mm/mm to maintain the definition of deformation per unit length of the member. Later chapters contain more about strain and deformation.

7–7 DIRECT SHEAR STRESS

Shear refers to a cutting-like action. When you use common household scissors, often called *shears,* you cause one blade of the pair to slide over the other to cut (shear) paper, cloth, or other material. A sheet metal fabricator uses a similar shearing action when cutting metal for ductwork. In these examples, the shearing action progresses along the length of the line to be cut so only a small part of the total cut is being made at any given time. And, of course, the objective of the action is to actually cut the material. That is, you *want* the material to fail.

The examples described in this section along with their accompanying figures illustrate several cases where *direct shear* is produced. That is, the applied shearing force is resisted uniformly by the area of the part in shear, producing a uniform level of shearing force across the area. The symbol used for shear stress is τ, the lowercase Greek letter *tau.* Then the direct shear stress can be computed from

Direct Shear Stress ▼

$$\text{Direct shear stress} = \tau = \frac{\text{Applied force}}{\text{Shear area}} = \frac{F}{A_s} \qquad (7\text{–}4)$$

Figure 7–12 shows a punching operation where the objective is to actually cut one part of the material from the other. The punching action produces a slot in the flat sheet metal. The part removed in the operation is sometimes called a slug. Many different shapes can be produced by punching with either the slug or the sheet having the hole being the desired part. Normally, the punching operation is designed so the entire shape is punched out at the same time. Therefore, the shearing action occurs along the *sides* of the slug. The area in shear for this case is computed by multiplying the length of the perimeter of the cut shape by the thickness of the sheet. That is, for a punching operation,

Shear Area for Punching ▼

$$A_s = \text{Perimeter} \times \text{Thickness} = p \times t \qquad (7\text{–}5)$$

(a) Punching operation

(b) Geometry of slug

(c) Slug

FIGURE 7–12 Illustration of direct shear stress in a punching operation.

Example Problem For the punching operation shown in Figure 7–12, compute the shear stress in the material if a
7–3 force of 1250 lb is applied through the punch. The thickness of the material is 0.040 in.

Solution Objective: Compute the shear stress in the material.

Given: $F = 1250$ lb; shape to be punched shown in Fig. 7–12; $t = 0.040$ in.

Analysis: The sides of slug are placed in direct shear resisting the applied force.
Use Equations (7–4) and (7–5).

Results: The perimeter, p, is

$$p = 2(0.75 \text{ in}) + \pi(0.50 \text{ in}) = 3.07 \text{ in}$$

The shear area,

$$A_s = p \times t = (3.07 \text{ in})(0.040 \text{ in}) = 0.1228 \text{ in}^2$$

Then the shear stress is

$$\tau = F/A_s = 1250 \text{ lb}/0.1228 \text{ in}^2 = 10\,176 \text{ psi}$$

Comment: At this time, we do not know whether or not this level of stress will cause the slug to be punched
out; it depends on the shear strength of the material, which is discussed in Chapters 8 and 9.

**LESSON
12**

Single Shear

A pin or a rivet is often inserted into a cylindrical hole
through mating parts to connect them, as shown in Figure
7–13. When forces are applied perpendicular to the axis of
the pin, there is the tendency to cut the pin across its cross
section, producing a shear stress. This action is often called
single shear because a single cross section of the pin resists
the applied shearing force. In this case, the pin is usually

designed so the shear stress is below the level that would
cause the pin to fail. Chapter 9 contains more about allowable stress levels.

Double Shear

When a pin connection is designed as shown in Figure 7–14,
there are two cross sections to resist the applied force. In this
arrangement the pin is in *double shear*.

Top view

w = width

Shear area, A_s
$A_s = \pi D^2/4$

D

Pin

t = thickness

Retaining ring

Shear plane

Link

Force on link

Reaction force

Bush

Side view

Photograph of a model of a pin in single shear

FIGURE 7–13 Pin connection illustrating single shear.

Source of model: Gary Drigel.

Shear area is *two* cross sections of pin
$A_s = 2(\pi D^2/4)$

Two shear planes

Pin (Diameter = D)

Retaining ring

Reaction force

Force on link

Photograph of a model of a pin in double shear

FIGURE 7–14 Pin connection illustrating double shear.

Source of model: Gary Drigel.

Example Problem 7–4 The force on the link in the simple pin joint shown in Figure 7–13 is 3550 N. If the pin has a diameter of 10.0 mm, compute the shear stress in the pin.

Solution Objective: Compute the shear stress in the pin.

Given: $F = 3550$ N; $D = 10.0$ mm

Analysis: Pin is in direct shear with one cross section of the pin resisting all of the applied force (single shear). Use Equation (7–4).

FIGURE 7–15 Direct shear stress in a pin.

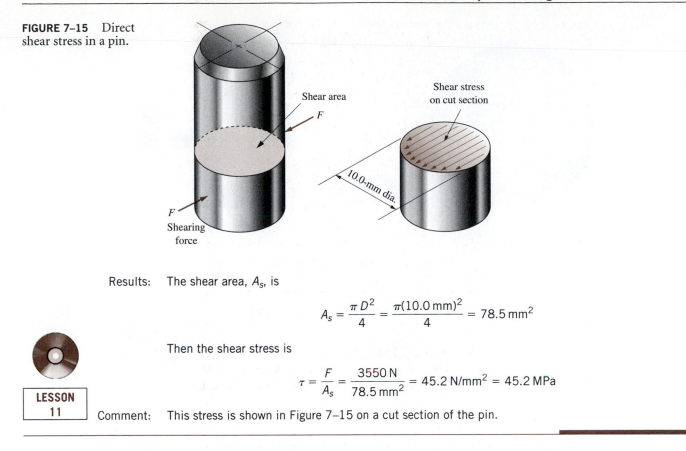

Shear area

F

10.0-mm dia.

Shear stress on cut section

F
Shearing force

Results: The shear area, A_s, is

$$A_s = \frac{\pi D^2}{4} = \frac{\pi(10.0 \text{ mm})^2}{4} = 78.5 \text{ mm}^2$$

Then the shear stress is

$$\tau = \frac{F}{A_s} = \frac{3550 \text{ N}}{78.5 \text{ mm}^2} = 45.2 \text{ N/mm}^2 = 45.2 \text{ MPa}$$

LESSON 11

Comment: This stress is shown in Figure 7–15 on a cut section of the pin.

Example Problem 7–5

If the pin joint just analyzed was designed as shown in Figure 7–14, compute the shear stress in the pin.

Solution   Objective: Compute the shear stress in the pin.

Given: $F = 3550$ N; $D = 10.0$ mm (same as in Example Problem 7–4).

Analysis: Pin is in direct shear with two cross sections of the pin resisting the applied force (double shear). Use Equation (7–4).

Results: The shear area, A_s, is

$$A_s = 2\left(\frac{\pi D^2}{4}\right) = 2\left[\frac{\pi(10.0 \text{ mm})^2}{4}\right] = 157 \text{ mm}^2$$

The shear stress in the pin is

$$\tau = \frac{F}{A_s} = \frac{3550 \text{ N}}{157 \text{ mm}^2} = 22.6 \text{ N/mm}^2 = 22.6 \text{ MPa}$$

Comment: The resulting shear stress is $\frac{1}{2}$ of the value found for single shear.

Keys

Figure 7–16 shows an important application of shear in mechanical drives. When a power transmitting element, such as a gear, chain sprocket, or belt pulley, is placed on a shaft, a key is often used to connect the two and permit the transmission of torque from one to the other.

The torque produces a tangential force at the interface between the shaft and the inside of the hub of the mating element. The torque is reacted by the moment of the force on the key times the radius of the shaft. That is, $T = F(D/2)$. Then the force is $F = 2T/D$.

In Figure 7–16, we have shown the force F_1 exerted by the shaft on the left side of the key. On the right side, an equal force F_2 is the reaction exerted by the hub on the key. This pair of forces tends to cut the key, producing a shear stress. Note that the shear area, A_s, is a rectangle with dimensions $b \times L$. The following example problem illustrates the computation of direct shear stress in a key.

FIGURE 7–16   Direct shearing action on a key between a shaft and the hub of a gear, pulley, or sprocket in a mechanical drive system.

Source of model: Gary Drigel.

Example Problem 7–6   Figure 7–16 shows a key inserted between a shaft and the mating hub of a gear. If a torque of 1500 lb · in is transmitted from the shaft to the hub, compute the shear stress in the key. For the dimensions of the key, use $L = 0.75$ in; $h = b = 0.25$ in. The diameter of the shaft is 1.25 in.

Solution   Objective: Compute the shear stress in the key.

Given: $T = 1500$ lb · in; $D = 1.25$ in; $L = 0.75$ in; $h = b = 0.25$ in

Analysis: The key is in direct shear. Use Equation (7–4).

Results: Shear area: $A_s = b \times L = (0.25\,\text{in})(0.75\,\text{in}) = 0.1875\,\text{in}^2$. The force on the key is produced by the action of the applied torque. The torque is reacted by the moment of the force on the key times the radius of the shaft. That is, $T = F(D/2)$. Then the force is

$$F = 2T/D = (2)(1500\,\text{lb} \cdot \text{in})/(1.25\,\text{in}) = 2400\,\text{lb}$$

Then the shear stress is

$$\tau = F/A_s = 2400\,\text{lb}/0.1875\,\text{in}^2 = 12\,800\,\text{psi}$$

7–8 STRESS ELEMENTS FOR SHEAR STRESSES

An infinitesimally small cubic element of the material from the shear plane of any of the examples shown in Section 7–7 would appear as shown in Figure 7–17 with the shear stresses acting parallel to the surfaces of the cube. For example, an element taken from the shear plane of the key in Figure 7–16 would have a shear stress acting toward the left on its top surface. For equilibrium of the element with regard to horizontal forces, there must be an equal stress acting toward the right on the bottom surface. This is the cutting action characteristic of shear.

But the two stress vectors on the top and bottom surfaces cannot exist alone because the element would tend to rotate under the influence of the couple formed by the two shearing forces acting in opposite directions. To balance that couple, a pair of equal shear stresses is developed on the vertical sides of the stress element, as shown in Figure 7–17(a).

The stress element is often drawn in the two-dimensional form shown in Figure 7–17(b). Note how the stress vectors on adjacent faces tend to meet at the corners. Such stress elements are useful in the visualization of stresses acting at a point within a material subjected to shear.

7–9 EXPERIMENTAL AND COMPUTATIONAL STRESS ANALYSIS

The primary goal of this book is to help you develop skills to determine the magnitude of stresses analytically. But there are cases in which that is difficult or impossible. Examples include the following:

- Components that have a complex shape for which the calculation of areas and other geometrical factors needed for stress analysis is difficult.

- The loading patterns and the manner of support may be complex, leading to an inability to compute the appropriate loads at a point of interest.

- Combinations of complex shape and loading may make it difficult to determine where the maximum stress occurs.

In such situations, experimental or computational stress analysis techniques can be used to determine the magnitude and location of critical stresses. A variety of techniques are available as discussed next.

Photoelastic Stress Analysis

When a component is basically a two-dimensional shape with a constant thickness, a model of the shape can be produced using a special material that permits the visualization of the stress distribution within the component. The material is typically a transparent plastic that is illuminated while being loaded.

Figure 7–18 shows an example in which a flat bar with a centrally located hole is subjected to an axial load. Variations in the stresses within the bar show as variations in black and white lines and areas in the loaded component. The stress in the main part of the bar away from the hole is nearly uniform and that part shows as a dark field. But in the vicinity of the hole, bands of dark and light areas, called *fringes,* appear, indicating that a gradient in stress levels occurs around the hole.

Indeed, the presence of the hole causes higher stress to occur not only because material has been removed from the bar, but also because the change in cross-sectional shape

FIGURE 7–17 Stress element showing shear stress. (a) Three-dimensional stress element. (b) Two-dimensional stress element.

FIGURE 7–18 Photoelastic model of a flat bar with an axial load and a central hole.

Source: Measurements Group, Inc., Raleigh, NC.

causes *stress concentrations* to occur. Any case in which a change in the cross-sectional shape or dimensions occur will produce stress concentrations. Knowing the optical characteristics of the photoelastic material allows the determination of the stress at any point.

Data and analytical techniques for determining the maximum stress in the vicinity of stress concentrations are presented in future chapters. Additional photoelastic models will be shown there.

Photoelastic Coatings

For more complex three-dimensional shapes it is impossible to model the geometry in the rigid photoelastic material. Also, it is often desired to measure the stress distribution on real solid components such as an engine block, a valve body, or the structure of a large machine. For such cases, a special photoelastic coating material can be applied to the surface of the part. It is molded to follow contours and adhered to the surface. See Figure 7–19(a).

When the component is loaded, the strains at the surface are transferred to the coating material. A polarized light is then directed at the material, and fringes are evident that indicate the gradients in stresses in the actual part. See Figure 7–19(b). From these observations, you can determine the areas where the maximum stresses occur and compute their approximate values. Then, knowing the general stress distribution and the locations of the highest stress levels, more precise testing can be done if greater accuracy is desired. Strain gaging, described next, is typically used at this time.

Strain Gaging

One of the most frequently used experimental stress analysis devices is the electrical resistance strain gage. It is a very thin metal foil grid made from a strain-sensitive material, such as constantan, with an insulating backing. One style is shown in Figure 7–20(a). The gage is carefully applied with a special adhesive to the surface of the component where critical stresses are likely to occur. Figure 7–20(b) shows a typical installation.

When the component is loaded, the gage experiences the same strains as the surface. The resistance of the gage changes in proportion to the applied strain. The gage is then connected into an electrical measurement circuit called a *Wheatstone bridge* as illustrated in Figure 7–21. The bridge has four arms, and the strain gage is connected as one arm. The other arms contain dummy resistors of the same nominal resistance. When the strain gage is subjected to stain, its resistance changes causing the voltage measured across the diagonal of the bridge to change. The voltage reading can be converted to strain when a calibration factor called the *gage factor* is applied. For one-directional strain, as due to direct axial tension, you can use Hooke's law to determine stress from

$$\sigma = E\epsilon$$

(a)

FIGURE 7–19 (a) Photoelastic coating being applied to a complex casting.(b) Evaluating a stress pattern in a photoelastic material using polarized light.

Source: Measurements Group, Inc., Raleigh, NC.

(b)

FIGURE 7–19 (*Continued*)

where

    ϵ is the strain

    E is the tensile modulus of elasticity of the material for the part being tested

    σ is the normal stress, either tension or compression

Other styles of strain gages and other patterns of installation are used for cases that involve more complex stress conditions. Some of these will be discussed in later chapters.

Finite Element Analysis

Finite element analysis (FEA) is a computational technique that allows the determination of stresses and deflections in complex computer models of load-carrying components. The model is created on a computer-aided design system using a three-dimensional solid modeling approach. Then the model is subdivided into many small, but finite, elements. Simulated loads can be applied, and support conditions can be established on the computer model. Also, material property data

FIGURE 7–20 Strain gage—typical geometry and installation.

Source: Measurements Group, Inc., Raleigh, NC.

Backing

Encapsulation

Copper-coated tabs

(a) Typical geometry

(b) Strain gage mounted on a flat bar

FIGURE 7–20   (*Continued*)

such as the modulus of elasticity and Poisson's ratio are entered. Figure 7–22 is an example of a finite element model.

The FEA software uses a matrix analysis approach that computes the effect of forces acting at the boundaries of each element. The result is the integrated effect of the externally applied loads and the supports, yielding computed values for stresses and deflections throughout the component. The compiled results are typically displayed in color graphics that clearly show points of maximum stress and deflection.

The designer can then evaluate the relative safety of the stress levels and proceed to optimize the design to take full advantage of the strength and stiffness of the material from which the component is made. Much more efficient, lighter, and cost-effective designs can be made using this technique. And the process can be done fairly quickly without making physical models. Often several design iterations are completed in a single day, leading to rapid product development.

Two-wire circuit for single active gage (quarter bridge)

FIGURE 7–21   Wheatstone bridge with strain gage installed in one arm.

Source: Measurements Group, Inc., Raleigh, NC.

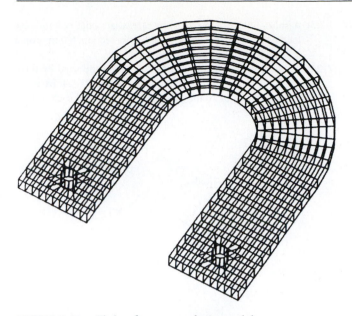

FIGURE 7–22 Finite element analysis model.
Source: Measurements Group, Inc., Raleigh, NC.

REFERENCES

1. Aluminum Association, *Aluminum Design Manual,* Washington, DC, 2005.

2. American Institute of Steel Construction, *Steel Construction Manual,* 13th ed., Chicago, 2005.

3. American Society for Testing and Materials, ASTM Standard A6/A6M-05a, *Standard Specification for General Requirements for Rolled Structural Steel Bars, Plates, Shapes, and Sheet Piling,* West Conshohocken, PA, 2005.

4. American Society for Testing and Materials, ASTM Standard E621-94(1999)e1, *Standard Practice for the Use of Metric (SI) Units in Building Design and Construction,* West Conshohocken, PA, 1999.

5. American Society for Testing and Materials, ASTM Standard SI-10, *Standard for Use of the International System of Units (SI): The Modern Metric System,* West Conshohocken, PA, 2002. (Note: This standard replaces ASTM E380.)

6. Avallone, Eugene A., Theodore Baumeister and Ali Sadegh, eds., *Marks' Standard Handbook for Mechanical Engineers,* 11th ed., New York, McGraw-Hill, 2007.

7. Budynas, R. G. and J. K. Nisbett, *Shigley's Mechanical Engineering Design,* 8th ed., New York, McGraw-Hill, 2007.

8. Mott, R. L., *Applied Fluid Mechanics,* 6th ed., Upper Saddle River, NJ, Prentice-Hall, 2006.

9. Oberg, E., F. D. Jones, H. L. Horton, and H. H. Ryffell, *Machinery's Handbook,* 28th ed., New York, Industrial Press, 2008.

10. Young, W. C. and R. G. Budynas, *Roark's Formulas for Stress and Strain,* 7th ed., New York, McGraw-Hill, 2002.

INTERNET SITES

1. TubeNet www.tubenet.org/tubes.shtml A listing of dimensions, properties, and suppliers of pipe and tubing in steel, copper, titanium, tantalum, and other materials.

2. Stainless Tubular Products www.stainlesstubular.com A supplier of stainless steel pipe and tubing.

3. Copper Development Association www.copper.org A professional association of the copper industry; the site offers a large amount of data on sizes and physical characteristics of copper tubing. A *Copper Tube Handbook* or parts of it can be downloaded from the site.

4. Plastics Pipe Institute www.plasticpipe.org An association representing all segments of the plastics piping industry; the site includes a list of manufacturers of plastic pipe from which much data for sizes, materials, and application information can be found.

5. Charter Plastics www.charterplastics.com A supplier of polyethylene plastic pipe and tubing.

6. Expert Piping Supply www.expertpiping.com A supplier of copper, steel, PVC, CPVC, polyethylene, and polypropylene pipe in a wide range of diameters and wall thicknesses.

7. The Piping Tool Box www.piping-toolbox.com A site containing data for pipe dimensions, pipe standards, and related topics.

8. American Institute of Steel Construction (AISC) www.aisc.org A trade association that serves the structural steel design community and the construction industry. Publisher of the *AISC Steel Construction Manual.*

9. Webco Mechanical Tubing Products www.webcoindustries.com/tubing/mechanical/ Manufacturer of mechanical tubing in round, square, and rectangular forms made from a variety of carbon and alloy steels.

10. Central Steel & Wire Co. www.centralsteel.com Supplier of a variety of metals (carbon and alloy steel, stainless, aluminum, copper, brass, bronze) in many shapes (structural shapes, strip, sheets, plates, bars, tubing [round, square, rectangular], pipe, wire).

11. Ryerson, Inc. www.ryerson.com Supplier of a wide variety of metals (carbon and alloy steel, stainless, nickel alloys, aluminum, copper, brass, bronze) in many shapes (structural shapes, strip, sheets, plates, bars, tubing [round, square, rectangular], pipe, and wire). Also supplies plastics in sheet, rod, and tubing.

12. Vishay Micro-Measurements www.vishay.com Producer of precision strain gages and associated instruments and accessories.

PROBLEMS

Definitions

7–1. Describe three modes of failure for a component of a structure.
7–2. Give examples of components that are likely to fail by buckling.
7–3. Define *stress* and state its units in both systems.
7–4. Define *direct normal stress.*
7–5. Explain the difference between compressive stress and tensile stress.
7–6. Define *direct shear stress.*
7–7. Explain the difference between single shear and double shear.
7–8. Draw a stress element subjected to direct tensile stress.
7–9. Draw a stress element subjected to direct compressive stress.
7–10. Draw a stress element subjected to direct shear stress.
7–11. Define *normal strain* and state its units in both systems.
7–12. Describe the concept of *punching shear* and describe how to compute the shear stress.
7–13. Describe how a key is used to transmit torque to a shaft from a power-transmitting element such as a gear, belt sheave, or chain sprocket. Describe how that action produces shearing stress in the key.

7–14. A stress computation result is 48 625 951 N/m². Express the result in preferred SI notation to four significant numbers.

7–15. A stress computation result is 23.7389 N/mm². Express the result in preferred SI notation to three significant figures.

Direct Tensile and Compressive Stresses

7–16. Compute the stress in a round bar subjected to a direct tensile force of 3200 N if the diameter of the bar is 10 mm.

7–17. Compute the stress in a rectangular bar having cross-sectional dimensions of 10 mm by 30 mm if a direct tensile force of 20 kN is applied.

7–18. A link in a mechanism for an automated packaging machine is subjected to a tensile force of 860 lb. If the link is square, 0.40 in on a side, compute the stress in the link.

7–19. A circular rod, with a diameter of $\frac{3}{8}$ in, supports a heater assembly weighing 1850 lb. Compute the stress in the rod.

7–20. A shelf is being designed to hold crates having a total mass of 1840 kg. Two support rods like those shown in Figure P7–20 hold the shelf. Each rod has a diameter of 12.0 mm. Assume that the center of gravity of the crates is at the middle of the shelf. Compute the stress in the middle portion of the rods.

FIGURE P7–20 Shelf support for Problem 7–20.

7–21. A concrete column base is circular, with a diameter of 8 in, and carries a direct compressive load of 70 000 lb. Compute the compressive stress in the concrete.

7–22. Three short, square, wood blocks, $3\frac{1}{2}$ in on a side, support a machine weighing 29 500 lb. Compute the compressive stress in the blocks.

7–23. A short link in a mechanism carries an axial compressive load of 3500 N. If it has a square cross section, 8.0 mm on a side, compute the stress in the link.

7–24. A machine having a mass of 4200 kg is supported by three solid steel rods arranged as shown in Figure P7–24. Each rod has a diameter of 20 mm. Compute the stress in each rod.

FIGURE P7–24 Support rods for Problem 7–24.

7–25. A centrifuge is used to separate liquids according to their densities using centrifugal force. Figure P7–25 illustrates one arm of a centrifuge having a bucket at its end to hold the liquid. In operation, the bucket and the liquid have a mass of 0.40 kg. The centrifugal force has the magnitude in newtons of

$$F = 0.010\ 97 \cdot m \cdot R \cdot n^2$$

where m = rotating mass of bucket and liquid (kilograms)
R = radius to center of mass (meters)
n = rotational speed (revolutions per minute)
   = 3000 rpm

Compute the stress in the round bar. Consider only the force due to the container.

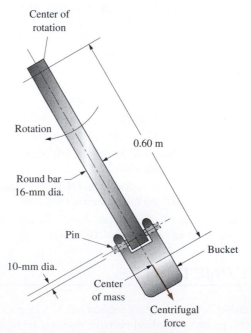

FIGURE P7–25 Centrifuge for Problem 7–25.

FIGURE P7–26   Square bar for Problem 7–26.

7–26. A square bar carries a series of loads as shown in Figure P7–26. Compute the stress in each segment of the bar. All loads act along the central axis of the bar.

7–27. Repeat Problem 7–26 for the circular bar in Figure P7–27.

FIGURE P7–27   Circular bar for Problem 7–27.

7–28. Repeat Problem 7–26 for the pipe in Figure P7–28. The pipe is $1\frac{1}{2}$ in schedule 40 steel pipe.

7–29. Compute the stress in member *BD* shown in Figure P7–29 if the applied force *F* is 2800 lb.

For Problems 7–30 and 7–31 using the trusses shown in Figures P7–30 and P7–31, compute the forces in all members and the stresses in the midsection, away from any joint. Refer to the Appendix for the cross-sectional area of the members indicated in the figures. Consider all joints to be pinned.

7–30. Use Figure P7–30.

7–31. Use Figure P7–31.

FIGURE P7–28   Pipe for Problem 7–28.

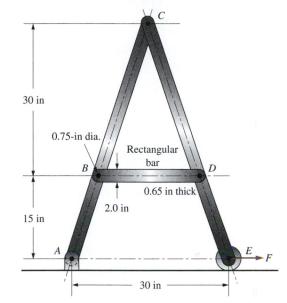

FIGURE P7–29   Frame for Problem 7–29.

(a)

(b) Cross section of members *AB, BC*

(c) Cross section of member *BD*

(d) Cross section of members *AD, CD*

FIGURE P7–30   Truss for Problem 7–30.

6000 lb

12 000 lb

8 ft

A
B
C
D
E
F

6 ft

6 ft

6 ft

Member specifications:

AD, DE, EF L2 × 2 × ⅛ – doubled ⎯⎯ ⌐⌐

BD, CE, BE L2 × 2 × ⅛ – single ⎯⎯ ⌐

AB, BC, CF C3 × 4.1 – doubled ⎯⎯ ⌐⌐

FIGURE P7–31 Truss for Problem 7–31.

7–32. Find the tensile stress in member *AB* shown in Figure P7–32.

Square
20 mm

A

2.5 m

.5 m

C

B

D

12.5 kN

FIGURE P7–32 Support for Problem 7–32.

7–33. Figure P7–33 shows the shape of a test specimen used to measure the tensile properties of metals (as described in Chapter 8). An axial tensile force is applied through the threaded ends, and the test section is the reduced-diameter part near the middle. Compute the stress in the middle portion when the load is 12 600 lb.

$\frac{3}{8}$ in *R*

0.505-in dia.

2 in

FIGURE P7–33 Tensile test specimen for Problem 7–33.

7–34. A short compression member has the cross section shown in Figure P7–34. Compute the stress in the member if a compressive force of 52 000 lb is applied in line with its centroidal axis.

0.5 in

0.5 in

2.65 in

1.40 in

FIGURE P7–34 Short compression member for Problem 7–34.

7–35. A short compression member has the cross section shown in Figure P7–35. Compute the stress in the member if a compressive force of 640 kN is applied in line with its centroidal axis.

15

40

60

80

Dimensions in
millimeters

FIGURE P7–35 Short compression member for Problem 7–35.

Direct Shearing Stresses

7–36. A pin connection like that shown in Figure 7–13 is subjected to a force of 16.5 kN. Determine the shear stress in the 12.0-mm-diameter pin.

7–37. In a pair of pliers, the hinge pin is subjected to direct shear, as indicated in Figure P7–37. If the pin has a diameter of 3.0 mm and the force exerted at the handle, F_h, is 55 N, compute the stress in the pin.

FIGURE P7–40 Shape of a slug for Problem 7–40.

7–41. Figure P7–41 shows the shape of a slug to be punched from a sheet of steel 0.194 in thick. Compute the shear stress in the steel if a punching force of 45 000 lb is applied.

FIGURE P7–41 Shape of a slug for Problem 7–41.

FIGURE P7–37 Pliers for Problem 7–37.

7–38. For the centrifuge shown in Figure P7–25 and the data from Problem 7–25, compute the shear stress in the pin between the bar and the bucket.

7–39. A notch is made in a piece of wood, as shown in Figure P7–39, in order to support an external load F of 1800 lb. Compute the shear stress in the wood.

FIGURE P7–39 Notched wood block loaded in shear for Problem 7–39.

7–40. Figure P7–40 shows the shape of a slug to be punched from a sheet of aluminum 5.0 mm thick. Compute the shear stress in the aluminum if a punching force of 38.6 kN is applied.

7–42. The key in Figure 7–16 has the dimensions $b = 10$ mm, $h = 8$ mm, and $L = 22$ mm. Determine the shear stress in the key when 95 N · m of torque is transferred from the 35-mm-diameter shaft to the hub.

7–43. A key is used to connect a hub of a gear to a shaft, as shown in Figure 7–16. It has a rectangular cross section with $b = \frac{1}{2}$ in and $h = \frac{3}{8}$ in. The length is 2.25 in. Compute the shear stress in the key when it transmits 8000 lb · in of torque from the 2.0-in-diameter shaft to the hub.

7–44. A set of two tubes is connected in the manner shown in Figure P7–44. Under a compressive load of 20 000 lb, the load is transferred from the upper tube through the pin to the connector, then through the collar to the lower tube. Compute the shear stress in the pin and in the collar.

7–45. A small hydraulic crane, like that shown in Figure P7–45, carries an 800-lb load. Determine the shear stress that occurs in the pin at B which is in double shear. The pin diameter is $\frac{3}{8}$ in.

7–46. A ratchet device on a jack stand for a truck has a tooth configuration as shown in Figure P7–46. For a load of 88 kN, compute the shear stress at the base of the tooth.

7–47. Figure P7–47 shows an assembly in which the upper block is brazed to the lower block. Compute the shear stress in the brazing material if the force is 88.2 kN.

7–48. Figure P7–48 shows a bolt subjected to a tensile load. One failure mode would be if the circular shank of the bolt pulled out from the head, a shearing action. Compute the shear stress in the head for this mode of failure if a force of 22.3 kN is applied.

(b) Photograph of model fully assembled

(c) Photograph of model with separate upper and lower components

(a) Cutaway drawing

(d) Photograph of model—exploded view showing all components

FIGURE P7–44 Connector assembly for Problem 7–44.

Source of model: Gary Drigel.

FIGURE P7–45 Hydraulic crane for Problem 7–45.

FIGURE P7–46 Ratchet for a jack stand for Problem 7–46.

7–49. Figure P7–49 shows a riveted lap joint connecting two steel plates. Compute the shear stress in the rivets due to a force of 10.2 kN applied to the plates.

7–50. Figure P7–50 shows a bolted butt joint with cover plates connecting two steel plates. Compute the shear stress in the bolts due to a force of 10.2 kN applied to the plates.

FIGURE P7–47 Brazed components for Problem 7–47.

19.0 mm

Cylindrical
shear area

8.00 mm

12.0 mm

M12×1.75 thread

Shear
area

Applied
force

Reaction force
on underside
of head

FIGURE P7–48 Bolt for Problem 7–48.

Rivets

Top view

Side view

F F

50 mm

12-mm dia.

15 mm

15 mm

F

F

FIGURE P7–49   Riveted lap joint for Problem 7–49.

F F

50 mm

12-mm dia. typical

(a) Top view

10 mm

F F

15 mm

10 mm

(b) Side view

(c) Photograph of a model bolted connection of two plates

FIGURE P7–50   Bolted butt joint for Problem 7–50.

Source of model: Gary Drigel.

DESIGN PROPERTIES OF MATERIALS

MAJOR TOPICS

8–1 The Big Picture—Design Properties of Materials
8–2 Design Properties of Materials
8–3 Steel
8–4 Cast Iron
8–5 Aluminum
8–6 Copper, Brass, and Bronze
8–7 Zinc, Magnesium, Titanium, and Nickel-Based Alloys
8–8 Nonmetals in Engineering Design
8–9 Wood
8–10 Concrete
8–11 Plastics
8–12 Composites
8–13 Materials Selection

8–1 THE BIG PICTURE— DESIGN PROPERTIES OF MATERIALS

Discussion Map

- As you design machines, structures, or consumer products, you must understand how materials react to externally applied loads.

- You will see how the relationships between stress and strain for a given material are critical to this understanding.

- The strength of a material is taken as the value for which it fails by fracture or by excessive deformation called *yielding*.

- You will become familiar with the ways material properties are listed in the Appendix of this book and in numerous other industrial sources of data.

- You will get an overview of the principal design properties of many metals and nonmetals.

Activity

Have you ever observed carefully a product that failed because a critical component had broken? Describe the product, how it failed, and the appearance of the material in the vicinity of the break. What was the material? Was it a metal or nonmetal?

Look for examples made from metals, plastics, concrete, or other materials. Compare the nature of the failures

for each material in terms of the appearance of the fractured material.

Does it look as if it broke suddenly, somewhat how chalk breaks when it is dropped onto a floor? Or did it deform noticeably near the break before actually coming apart?

Why did it break? How was it being used when it broke? Was it a momentary overload, impact, vibration, or repeated flexure?

Study the material properties listed in Appendices A–14 to A–20 of this book. Note the different kinds of steels, cast iron, aluminum, wood, and plastics. Look for the strength ratings called ultimate strength, yield strength, and shear strength. Other properties include modulus of elasticity, percent elongation (a measure of ductility), and density.

Safety, the ultimate goal of design and stress analysis, requires that materials do not fail, thus rendering a product unfit for its intended purpose. Failure can take several forms. The primary types of failure considered in this book are the following:

1. The material of some component could fracture completely.

2. The material may deform excessively under load so the component no longer performs satisfactorily.

3. A structure or one of its components could become unstable and buckle and, thus, be unable to carry the intended loads.

The study of strength of materials requires a knowledge of how external forces and moments affect the stresses and deformations developed in the material of a load-carrying member. In order to put this knowledge to practical use, however, a designer needs to know how such stresses and deformations can be withstood safely by the material. Thus, material properties as they relate to design must be understood along with the analysis required to determine the magnitude of stresses and deformations.

In this chapter we present information concerning the materials most frequently used to make components for structures and mechanical devices, emphasizing the design properties of the materials rather than their metallurgical structure or chemical composition. Although it is true that a thorough knowledge of the structure of materials is an aid to a designer, it is most important to know how the materials behave when carrying loads. This is the behavior on which we concentrate in this chapter. See References 4, 9, 12, and 29.

First, we discuss the design properties of materials that affect the performance of a product. Then a wide variety of metals are described including carbon and alloy steel, stainless steel, structural steel, cast iron, aluminum, copper, brass, bronze, zinc, magnesium, titanium, and some of the nickel alloys.

Nonmetals presented include wood, concrete, plastics, and composites. The manner in which the behavior of these materials differs from that of metals is discussed, along with some of their special properties.

Objectives of This Chapter

After completing this chapter, you should be able to:

1. List typical uses for engineering materials.
2. Define the strength properties of metals, *ultimate tensile strength, yield point, yield strength, elastic limit,* and *proportional limit.*
3. Define *Hooke's law* and the stiffness properties of metals, *modulus of elasticity in tension* and *modulus of elasticity in shear.*
4. Define *Poisson's ratio* and give its value for typical materials.
5. Describe ductile and brittle behavior of materials.
6. Define *percent elongation* and describe its relationship to the ductility of materials.
7. Describe the Unified Numbering System (UNS) for metals and alloys.
8. Describe the four-digit designation system for steels.
9. Describe the important properties of carbon steels, alloy steels, stainless steels, and structural steels.
10. Describe the four-digit designation system for wrought and cast aluminum alloys.
11. Describe the aluminum temper designations.
12. Describe the design properties of copper, brass, bronze, zinc, magnesium, and titanium.
13. Describe the design properties of cast irons, including gray iron, ductile iron, austempered ductile iron, white iron, and malleable iron.
14. Describe the design properties of wood, concrete, plastics, and composites.

8–2 DESIGN PROPERTIES OF MATERIALS

Material selection requires consideration of many factors, such as

strength	stiffness	ductility
weight	fracture toughness	machinability
workability	weldability	appearance
stability	cost	availability

compatibility with the environment in which the product must operate

More is said about the general nature of material selection process later in this chapter. Relative to the study of strength of materials, the primary emphasis is on *strength, stiffness,* and *ductility.* The types of strength considered most frequently are *tensile strength, compressive strength,* and *yield strength.* Tensile and yield strengths are considered first.

Tensile and Yield Strengths

Reference data listing the mechanical properties of metals will almost always include the *tensile strength* and *yield strength* of the metal. Comparison of the actual stresses in a part with the tensile or yield strength of the material from which the part is made is the usual method of evaluating the suitability of the material to carry the applied loads safely. Chapter 9 and subsequent chapters discuss more about the details of stress analysis.

The tensile strength and yield strength are determined by testing a sample of the material in a tensile-testing machine such as the one shown in Figure 8–1. (See Reference 8 for the methods used.) A round bar or flat strip is placed in

FIGURE 8–1   Universal testing machine for obtaining stress–strain data for materials.

Source: Instron Corporation, Norwood, MA.

FIGURE 8–2 Tensile test specimen mounted in a holder.

Source: Tinius Olsen Testing Machine Co., Inc., Willow Grove, PA.

the upper and lower jaws. Figure 8–2 shows a photograph of a typical tensile test specimen. A pulling force is applied slowly and steadily to the sample, stretching it until it breaks. During the test, a graph is made that shows the relationship between the stress in the sample and the strain or unit deformation.

A typical stress–strain diagram for a low-carbon steel is shown in Figure 8–3. It can be seen that during the first phase of loading, the plot of stress versus strain is a straight line, indicating that stress is directly proportional to strain. After point *A* on the diagram, the curve is no longer a straight line. This point is called the *proportional limit.* As the load on the sample is continually increased, a point called the *elastic limit* is reached, marked *B* in Figure 8–3. At stresses below this point, the material will return to its

original size and shape if the load is removed. At higher stresses, the material is permanently deformed. The *yield point* is the stress at which a noticeable elongation of the sample occurs with no apparent increase in load. The yield point is at *C* in Figure 8–3, about 36 000 psi (248 MPa). Applying still higher loads after the yield point has been reached causes the curve to rise again. After reaching a peak, the curve drops somewhat until finally the sample breaks, terminating the plot. The highest apparent stress taken from the stress–strain diagram is called the *tensile strength.* In Figure 8–3 the tensile strength would be about 53 000 psi (365 MPa).

The fact that the stress–strain curve in Figure 8–3 drops off after reaching a peak tends to indicate that the stress level decreases. Actually, it does not; the *true stress* continues to rise until ultimate failure of the material. The reason for the apparent decrease in stress is that the plot taken from a typical tensile test machine is actually *load versus elongation* rather than *stress versus strain.* The vertical axis is converted to stress by dividing the load (force) on the specimen by the *original* cross-sectional area of the specimen. When the specimen nears its breaking load, there is a reduction in diameter and consequently a reduction in the cross-sectional area. The reduced area requires a lower force to continue stretching the specimen, even though the actual stress in the material is increasing. This results in the dropping curve shown in Figure 8–3. Because it is very difficult to monitor the decreasing diameter, and because experiments have shown that there is little difference between the true maximum stress and that found from the peak of the *apparent stress* versus strain curve, the peak is accepted as the tensile strength of the material.

A summary of the definitions of key strength properties of steels follows:

The proportional limit is the value of stress on the stress–strain curve at which the curve first deviates from a straight line.

The elastic limit is the value of stress on the stress–strain curve at which the material has deformed plastically; that is, it will no longer return to its original size and shape after removing the load.

The yield point is the value of stress on the stress–strain curve at which there is a significant increase in strain with little or no increase in stress.

The tensile strength is the highest value of apparent stress on the stress–strain curve.

Many metals do not exhibit a well-defined yield point like that in Figure 8–3. Some examples are high-strength alloy steels, aluminum, and titanium. However, these materials do in fact yield, in the sense of deforming a sizable amount before fracture actually occurs. For these materials, a typical stress–strain diagram would look like the one shown in Figure 8–4. The curve is smooth with no pronounced yield point. For such materials, the yield strength is defined by a line like *M-N* drawn parallel to the straight-line portion of the test curve.

FIGURE 8–3 Typical stress–strain curve for steel.

FIGURE 8–4 Typical stress–strain curve for aluminum.

Point *M* is usually determined by finding that point on the strain axis representing a strain of 0.002 in/in. This point is also called the point of 0.2% offset. The point *N,* where the offset line intersects the curve, defines the yield strength of the material, about 55 000 psi in Figure 8–4. The ultimate strength is at the peak of the curve, as was described before. *Yield strength* is used in place of yield point for these materials.

The units for the offset on the strain axis can be in any system. Recall that strain itself is sometimes called *unit deformation,* a ratio of the elongation of the material at a specific load to its original length in the unloaded condition. Thus, strain is actually dimensionless. If measurements were being taken directly in the SI metric system, the strain would be m/m, meters of elongation per meter of original length. Or mm/mm could be used.

The value of the offset, 0.2%, is typical for most commonly used metals. Other values may be used if 0.2% does not give reliable or convenient results. However, the value of 0.2% is assumed unless otherwise stated.

In summary, for many materials that do not exhibit a pronounced yield point, the definition of *yield strength* is as follows.

The yield strength is the value of stress on the stress–strain curve at which a straight line drawn from a strain value of 0.002 in/in (or m/m) and parallel to the straight portion of the stress–strain curve intersects the curve.

Compressive Strength

The stress–strain behavior of most wrought metals is nearly the same in compression as it is in tension because the material has a nearly uniform, homogeneous structure throughout. When a material behaves in a similar manner regardless of the direction of the loads, it is referred to as an *isotropic* material. For isotropic materials, then, separate compression tests are not usually performed and separate data are not reported for compressive strength.

But many materials exhibit different behavior and strength in compression than in tension. This is called *anisotropic* behavior. Examples are many cast metals, some plastics, concrete, wood, and composites. You should seek data for both compressive strength and tensile strength for such materials.

Stiffness

It is frequently necessary to determine how much a part will deform under load to ensure that excessive deformation does not destroy the usefulness of the part. This can occur at stresses well below the yield strength of the material, especially in very long members or in high-precision devices. Stiffness of a material is a function of its *modulus of elasticity,* sometimes called *Young's modulus.*

The modulus of elasticity, E, *is a measure of the stiffness of a material determined by the slope of the straight-line portion of the stress–strain curve. It is the ratio of the change of stress to the corresponding change in strain.*

This can be stated mathematically as

Modulus of Elasticity ▼

$$E = \frac{\text{Stress}}{\text{Strain}} = \frac{\sigma}{\epsilon} \qquad (8\text{--}1)$$

Therefore, a material having a steeper slope on its stress–strain curve will be stiffer and will deform less under load than a material having a less steep slope. Figure 8–5 illustrates this concept by showing the straight-line portions of the stress–strain curves for steel, titanium, aluminum, and magnesium. It can be seen that if two otherwise identical parts were made of steel and aluminum, respectively, the aluminum part would deform about three times as much when subjected to the same load.

The design of typical load-carrying members in machines and structures is such that the stress is below the proportional limit, that is, in the straight-line portion of the stress–strain curve. Here we define *Hooke's law:*

When the level of stress in a material under load is below the proportional limit and there is a straight-line relationship between stress and strain, it is said that Hooke's law applies.

Many of the formulas used for stress analysis are based on the assumption that Hooke's law applies. This concept is also useful for experimental stress analysis techniques in which strain is measured at a point. The corresponding stress at the point can be computed from a variation of Equation (8–1),

$$\sigma = E\epsilon \qquad (8\text{--}2)$$

Equation (8–2) is valid only where strain occurs in only one direction. This is called *uniaxial strain,* and it applies to

FIGURE 8–5 Modulus of elasticity for different metals.

members subjected to axial tension or compression and to beams in pure bending. When stresses occur in two directions (*biaxial stress*), an additional effect of the second stress must be considered. Biaxial stress is discussed in Chapter 15.

Ductility

When metals break, their fracture can be classified as either ductile or brittle. A ductile material will stretch and yield prior to actual fracture, causing a noticeable decrease in the cross-sectional area at the fractured section. Conversely, a brittle material will fracture suddenly with little or no change in the area at the fractured section. Ductile materials are preferred for parts that carry repeated loads or are subjected to impact loading because they are usually more resistant to fatigue failure and because they are better at absorbing the impact energy.

Ductility in metals is usually measured during the tensile test by noting how much the material has elongated permanently after fracture. At the start of the test, a set of gage marks is scribed on the test sample as shown in Figure 8–6. Most tests use 2.000 in or 50.0 mm for the gage length as shown in the figure. Very ductile structural steels sometimes use 8.000 in or 200.0 mm for the gage length. After the sample has been pulled to failure, the broken parts are fitted back together and the distance between the marks is again measured. See Figure 8–7. From these data, the *percent elongation* is computed from

Percent Elongation ▼

$$\text{Percent elongation} = \frac{\text{Final length} - \text{Gage length}}{\text{Gage length}} \times 100\%$$

(8–3)

Gage length

Typically 2.00 in or 50.0 mm

(a)

(b)

FIGURE 8–6 Gage length for tensile test specimen. (a) Gage length marked on a test specimen. (b) Test specimen in a fixture used to mark gage length.

Source: Tinius Olsen Testing Machine Co., Inc., Willow Grove, PA.

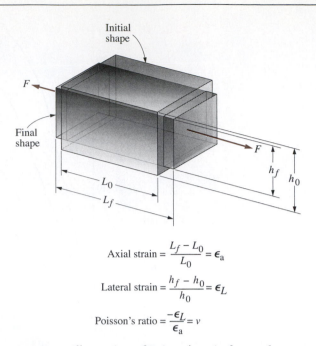

$$\text{Axial strain} = \frac{L_f - L_0}{L_0} = \epsilon_a$$

$$\text{Lateral strain} = \frac{h_f - h_0}{h_0} = \epsilon_L$$

$$\text{Poisson's ratio} = \frac{-\epsilon_L}{\epsilon_a} = \nu$$

FIGURE 8–7 Measurement of percent elongation.

A metal is considered to be *ductile* if its percent elongation is greater than about 5.0%. A material with a percent elongation under 5.0% is considered to be *brittle* and does not exhibit the phenomenon of yielding. Failure of such materials is sudden, without noticeable deformation prior to ultimate fracture. In most structural and mechanical design applications, ductile behavior is desirable, and the percent elongation of the material should be significantly greater than 5.0%. A high-percentage elongation indicates a highly ductile material.

In summary, the following definitions are used to describe ductility for metals:

The percent elongation is the ratio of the plastic elongation of a tensile specimen after ultimate failure within a set of gage marks to the original length between the gage marks. It is one measure of ductility.

A ductile material is one that can be stretched, formed, or drawn to a significant degree before fracture. A metal that exhibits a percent elongation greater than 5.0% is considered to be ductile.

A brittle material is one that fails suddenly under load with little or no plastic deformation. A metal that exhibits a percent elongation less than 5.0% is considered to be brittle.

Virtually all wrought forms of steel and aluminum alloys are ductile. But the higher-strength forms tend to have lower ductility and the designer is often forced to compromise strength and ductility in the specification of a material. Gray cast iron, many forms of cast aluminum, and some high-strength forms of wrought or cast steel are brittle.

Poisson's Ratio

Figure 8–8 shows a more complete understanding of the deformation of a member subjected to normal stress. The element shown is taken from the bar in tension. The tensile force on the bar causes an elongation in the direction of the applied force, as would be expected. But at the same time, the width of the bar is being shortened. Thus a simultaneous elongation and contraction occurs in the stress element. From the elongation, the axial strain can be determined, and from the contraction, the lateral strain can be determined.

FIGURE 8–8 Illustration of Poisson's ratio for an element in tension.

We define the term *Poisson's ratio* as the ratio of the amount of lateral strain to axial strain. That is,

Poisson's Ratio ▼

$$\text{Poisson's ratio} = \nu = \frac{\text{Lateral strain}}{\text{Axial strain}} = \frac{-\epsilon_L}{\epsilon_a} \quad (8–4)$$

The negative sign on the lateral strain is introduced to ensure that Poisson's ratio is a positive number when strains are computed as indicated in Figure 8–8. Poisson's ratio is also said to be the absolute value of the strain ratio.

Most commonly used metallic materials have a Poisson's ratio value between 0.25 and 0.35. For concrete, ν varies widely depending on the grade and on the applied stress, but it usually falls between 0.1 and 0.25. Elastomers and rubber may have Poisson's ratios approaching 0.50. Approximate values for Poisson's ratio are listed in Table 8–1.

Poisson's ratio is used in the calculation of contact stress, sometimes called *Hertz stress,* when forces are transferred across very small areas. Examples shown later in Figure 9–14 include cylinders or spheres acting on flat plates, spherical ball bearings acting on their races, and the contact area between two gear teeth. The complete analysis of such stresses requires a detailed description of the deformation that occurs around the small area of contact. Principles of elasticity must be employed, and the deformations are significantly influenced by the Poisson's ratio of the materials of the two members in contact. See References 7 and 10 in Chapter 7.

Some types of experimental stress analysis also use Poisson's ratio. For example, electrical resistance strain gages sense the local strains on the surface of load-carrying parts. Precise analysis requires that the principal strains in two perpendicular directions be

LESSON
13

TABLE 8–1 Approximate Values of Poisson's Ratio, ν

Concrete	0.10–0.25	Aluminum (most alloys)	0.33
Glass	0.24	Brass	0.33
Ductile iron	0.27	Copper	0.33
Gray cast iron	0.21	Zinc	0.33
Plastics	0.20–0.40	Phosphor bronze	0.35
Carbon and alloy steel	0.29	Magnesium	0.35
Stainless steel (18-8)	0.30	Lead	0.43
Titanium	0.30	Rubber, elastomers	~0.50

Notes: Values are approximate and vary somewhat with specific composition. Rubber and elastomers approach a limiting value of 0.50.

measured with the results used to compute the stress at the location of the gages.

Shearing Strain

Earlier discussions of strain in Chapter 7 described normal strain because it is caused by the normal tensile or compressive stress developed in a load-carrying member. Under the influence of a shear stress, shearing strain would be produced.

Figure 8–9 shows a stress element subjected to shear. The shearing action on parallel faces of the element tends to deform it angularly, as shown to an exaggerated degree. The angle γ (gamma), measured in radians, is the *shearing strain*. Only very small values of shearing strain are encountered in practical problems and, thus, the dimensions of the element are only slightly changed.

Modulus of Elasticity in Shear

The ratio of the shearing stress to the shearing strain is called the modulus of elasticity in shear *or the* modulus of rigidity, *and is denoted by* G.

That is,

Shear Modulus of Elasticity ▼

$$G = \frac{\text{Shearing stress}}{\text{Shearing strain}} = \frac{\tau}{\gamma} \qquad (8\text{–}5)$$

Shearing strain

FIGURE 8–9 Shearing strain shown on a stress element.

G is a property of the material and is related to the tensile modulus and Poisson's ratio by

Relation Between *G*, *E*, and Poisson's Ratio ▼

$$G = \frac{E}{2(1 + \nu)} \qquad (8\text{–}6)$$

The calculation of torsional deflection of shafts in Chapter 10 requires the use of G.

Flexural Strength and Flexural Modulus

Other stiffness and strength measures often reported, particularly for plastics, are called the *flexural strength and flexural modulus*. As the name implies, a specimen of the material is loaded as a beam in flexure (bending) with data taken and plotted for load versus deflection. From these data and from knowledge of the geometry of the specimen, stress and strain can be computed. The ratio of stress to strain is a measure of the flexural modulus. ASTM standard D 790[1] defines the complete method. Note that the values are significantly different from the tensile modulus because the stress pattern in the specimen is a combination of tension and compression. The data are useful for comparing the strength and stiffness of different materials when a load-carrying part is subjected to bending in service. ISO standard 178 describes a similar method for determining flexural properties.

Hardness

The resistance of a material to indentation by a penetrator is an indication of its *hardness*. Several types of devices, procedures, and penetrators measure hardness; the Brinell hardness tester and the Rockwell hardness tester are most frequently used for machine elements. For steels, the Brinell hardness tester employs a hardened steel ball 10 mm in diameter as the penetrator under a load of 3000-kg force. The load causes a permanent indentation in the test material, and the diameter of the indentation is related to the Brinell hardness number, which is abbreviated HB. The actual quantity being measured is the load divided by the contact

[1]ASTM International. *Standard Test Method for Flexural Properties of Unreinforced and Reinforced Plastics and Electrical Insulating Materials, Standard D 790*. West Conshohocken, PA: ASTM International, 2003.

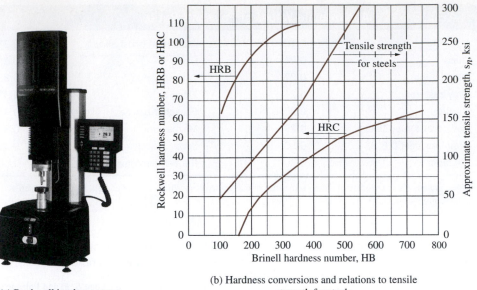

(a) Rockwell hardness tester

(b) Hardness conversions and relations to tensile strength for steels

FIGURE 8–10 Hardness tester and conversions. (a) Rockwell hardness tester. (b) Hardness conversions and relations to tensile strength for steels.

Source: Instron Corporation, Norwood, MA.

area of the indentation. For steels, the value of HB ranges from approximately 100 for an annealed, low-carbon steel to more than 700 for high-strength, high-alloy steels in the asquenched condition. In the high ranges, above HB 500, the penetrator is sometimes made of tungsten carbide rather than steel. For softer metals, a 500-kg load is used.

The Rockwell hardness tester uses a hardened steel ball with a $\frac{1}{16}$-in diameter under a load of 100-kg force for softer metals, and the resulting hardness is listed as Rockwell B, R_B, or HRB. For harder metals, such as heat-treated alloy steels, the Rockwell C scale is used. A load of 150-kg force is placed on a diamond penetrator (a *brale* penetrator) made in a sphero-conical shape. Rockwell C hardness is sometimes referred to as R_C or HRC. Many other Rockwell scales are used.

The Brinell and Rockwell methods are based on different parameters and lead to quite different numbers. However, since they both measure hardness, there is a correlation between them, as noted in Figure 8–10. It is also important to note that, especially for highly hardenable alloy steels, there is a nearly linear relationship between the Brinell hardness number and the tensile strength of the steel, according to the equation

Approximate Relationship Between Hardness and Strength for Steel ▼

$$0.50(\text{HB}) = \text{approximate tensile strength (ksi)} \qquad (8\text{–}7)$$

This relationship is shown in Figure 8–10. Hardness in a steel indicates wear resistance, as well as strength.

To compare the hardness scales with the tensile strength, consider Table 8–2. Note that there is some overlap between the HRB and HRC scales. Normally, HRB is used for the softer metals and ranges from approximately 60 to 100, whereas HRC is used for harder metals and ranges from 20 to 65. Using HRB numbers above 100 or HRC numbers below 20 is not recommended. Those shown in Table 8–2 are for comparison purposes only.

Whereas Rockwell B and C scales are used for substantial components made from steels, other Rockwell hardness scales can be used for different materials as listed in Table 8–3. See Internet site 27 for a downloadable report on Rockwell hardness measurement of metallic materials.

For plastics, rubbers, and elastomers, it is typical to use the Shore or IRHD (International rubber hardness degree)

TABLE 8–2	Comparison of Hardness Scales with Tensile Strength for Steels				
Material and condition	**Hardness**			**Tensile strength**	
	HB	**HRB**	**HRC**	**ksi**	**MPa**
1020 annealed	121	70		60	414
1040 hot-rolled	144	79		72	496
4140 annealed	197	93	13	95	655
4140 OQT 1000	341	109	37	168	1160
4140 OQT 700	461		49	231	1590

TABLE 8–3 Rockwell Hardness Scales and Their Uses

Scale	Symbol	Typical uses
A	HRA	Thin steel, hardened steel with shallow case
B	HRB	Softer steels such as low carbon, annealed or hot rolled; softer aluminum, copper, cast irons
C	HRC	Harder steels such as heat-treated alloy steels, tool steels, titanium
D	HRD	Medium-depth case-hardened steels, harder cast irons
E	HRE	Harder bearing metals, aluminum, magnesium, cast irons
F	HRF	Soft thin sheet metals, annealed copper and zinc alloys
G	HRG	Beryllium copper, phosphor bronze, softer cast irons
H	HRH	Aluminum, zinc, lead
K	HRK	Softer bearing metals, plastics, rubbers, and other soft materials. Scales L, M, P, R, S, V, and α also available
15 N	HR15N	Similar to A scale but for thinner materials or thin case hardening
30 N	HR30N	Similar to C scale but for thinner materials or thin case hardening
45 N	HR45N	Similar to D scale but for thinner materials or thin case hardening
15 T	HR15T	Similar to B scale but for thinner materials
30 T	HR30T	Similar to F scale but for thinner materials
45 T	HR45T	Similar to G scale but for thinner materials

Note: Measurement devices use different indenter sizes and shapes and applied forces. For details, consult standards ASTM E18, ISO 6508.

methods. Table 8–4 lists some of the more popular scales. Also used for some plastics are the Rockwell R, L, M, E, K, and α scales. These vary by the size and geometry of the indenter and the applied force.

Other hardness measurement methods include Vickers, Knoop, Universal, and rebound hardness. See Internet sites 24 and 26 for discussions of these methods and their uses.

Toughness, Impact Energy

Toughness is the ability of a material to absorb applied energy without failure. Parts subjected to suddenly applied loads, shock, or impact need a high level of toughness. Several methods are used to measure the amount of energy required to break a particular specimen made from a material of

TABLE 8–4 Hardness Measurement for Plastics, Rubbers, and Elastomers

Shore method		
Scale of durometer	**Symbol**	**Typical uses**
A	Shore A	Soft vulcanized natural rubbers, elastomers (e.g., neoprene), leather, wax, felt
B	Shore B	Moderately hard rubber as used for printer rolls and platens
C	Shore C	Medium hard rubbers and plastics
D	Shore D	Hard rubbers and plastics such as vinyl sheets, plexiglas, laminate countertops
DO	Shore DO	Very dense textile windings
O	Shore O	Soft rubber such as artgum; textile windings
OO	Shore OO	Low density textile windings; sponge rubber
OOO	Shore OOO	Soft plastic foams
T	Shore T	Medium density textiles on spools
M	Shore M	Rubber O-rings and thin sheet rubber
IRHD method (International rubber hardness degree)		
Name	**Typical uses**	
IRHD Micro	Small: O-rings, small components, thin materials	
IRHD Macro L	Soft: Readings up to 35 IRHD L	
IRHD Macro N	Normal: Readings from 30 IRHD N to 100 IRHD N	

Note: Measurement devices use different indenter sizes and shapes and applied forces. For details, consult standards ASTM D2240, ISO 868, or DIN 53505.

interest. The energy absorption value from such tests is often called *impact energy* or *impact resistance*. However, it is important to note that the actual value is highly dependent on the nature of the test sample, particularly its geometry. It is not possible to use the test results in a quantitative way when making design calculations. Rather, the impact energy for several candidate materials for a particular application can be compared with each other as a qualitative indication of their toughness. The final design should be tested under real service conditions to verify its ability to survive safely during expected use.

For metals and plastics, two methods of determining impact energy, *Izod* and *Charpy*, are popular, with data often reported in the literature from vendors of the material. Figure 8–11(a) and (b) show sketches of the dimensions of standard specimens and the manner of loading. In each method, a pendulum with a heavy mass carrying a specially designed striker is allowed to fall from a known height. The striker contacts the specimen with a high velocity at the bottom of the pendulum's arc; therefore, the pendulum possesses a known amount of kinetic energy. Part (c) shows a tester. The specimen is typically broken during the test, taking some of the energy from the pendulum but allowing it to pass through the test area. The testing machine is configured to measure the final height to which the pendulum swings and to indicate the amount of energy removed. That value is reported in energy units of joules (J) or N · m. Some highly ductile metals and many plastics do not break during the test, and the result is then reported as *No Break*.

The standard *Izod* test employs a square specimen with a V-shaped notch carefully machined 2.0 mm (0.079 in) deep according to specifications in ASTM standard D 256.[2] The specimen is clamped in a special vise with the notch aligned with the top edge of the vise. The striker contacts the specimen at a height of 22 mm above the notch, loading it as a cantilever in bending. When used for plastics, the width dimension can be different from that shown in Figure 8–11. This obviously changes the total amount of energy that the specimen will absorb during fracture. Therefore, the data for impact energy are divided by the actual width of the specimen, and the results are reported in units of N · m/m or ft · lb/in. Also, some vendors and customers may agree to test the material with the notch facing away from the striker rather than toward it as shown in Figure 8–11. This gives a measure of the material's impact energy with less influence from the notch.

The *Charpy* test also uses a square specimen with a 2.0 mm (0.079 in) deep notch, but it is centered along the length. The specimen is placed against a rigid anvil without being clamped. See ASTM standard A 370[3] for the specific geometry and testing procedure. The notch faces away

from the place where the striker contacts the specimen. The loading can be described as the bending of a simply supported beam. The Charpy test is most often used for testing metals.

Another impact testing method used for some plastics, composites, and completed products is the *drop-weight* tester. See Figure 8–11(d). Here a known mass is elevated vertically above the test specimen to a specified height. Thus, it has a known amount of potential energy. Allowing the mass to fall freely imparts a predictable amount of kinetic energy to the specimen clamped to a rigid base. The initial energy, the manner of support, the specimen geometry, and the shape of the striker (called a *tup*) are critical to the results found. One standard method, described in ASTM D 3763,[4] employs a spherical tup with a diameter of 12.7 mm (0.50 in). The tup usually pierces the specimen. The apparatus is typically equipped with sensors that measure and plot the load versus deflection characteristics dynamically, giving the designer much information about how the material behaves during an impact event. Summary data reported typically include maximum load, deflection of the specimen at the point of maximum load, and the energy dissipated up to the maximum load point. The energy is calculated by determining the area under the load-deflection diagram. The appearance of the test specimen is also described, indicating whether fracture occurred and whether it was a ductile or brittle fracture.

Fatigue Strength or Endurance Strength

Parts subjected to repeated applications of loads or to stress conditions that vary with time over several thousands or millions of cycles fail because of the phenomenon of *fatigue*. Materials are tested under controlled cyclic loading to determine their ability to resist such repeated loads. The resulting data are reported as the *fatigue strength,* also called the *endurance strength* of the material. See Reference 25 and Internet site 23.

Creep

When materials are subjected to high loads continuously, they may experience progressive elongation over time. This phenomenon, called *creep*, should be considered for most plastics and metals operating at high temperatures. You should check for creep when the operating temperature of a loaded metal member exceeds approximately 0.3 (T_m), where T_m is the melting temperature expressed as an absolute temperature. Creep can be important for critical members in internal combustion engines, furnaces, steam turbines, gas turbines, nuclear reactors, or rocket engines. The stress can be tension, compression, flexure, or shear.

[2]ASTM International. *Standard Test Methods for Determining the Izod Pendulum Impact Resistance of Plastics, Standard D 256.* West Conshohocken; PA: ASTM International, 2006.

[3]ASTM International. *Standard Test Methods and Definitions for Mechanical Testing of Steel Products, Standard A 370.* West Conshohocken, PA: ASTM International, 2005.

[4]ASTM International. *Standard Test Methods for High Speed Puncture of Plastics Using Load and Displacement Sensors, Standard D 3763.* West Conshohocken, PA: ASTM International, 2002.

(a) Izod test specimen and striker
(side view)

(b) Charpy test specimen and striker
(top view)

(c) Pendulum-type impact tester

(d) Drop-type impact tester

FIGURE 8–11   Impact testing devices and equipment.

Source: Instron Corporation, Norwood, MA.

Figure 8–12 shows the typical behavior of metals that creep. The vertical axis is the creep strain, in units such as in/in or mm/mm, over that which occurs initially as the load is applied. The horizontal axis is time, typically measured in hours because creep develops slowly over a long term. During the primary portion of the creep strain versus time curve, the rate of increase in strain initially rises with a rather steep slope that then decreases. The slope is constant (straight line) during the secondary portion of the curve. Then the slope increases in the tertiary portion that precedes the ultimate fracture of the material.

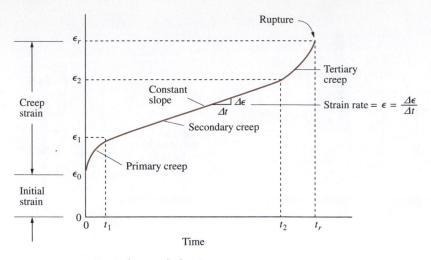

FIGURE 8–12   Typical creep behavior.

Creep is measured by subjecting a specimen to a known steady load, possibly through application of a deadweight, while the specimen is heated and maintained at a uniform temperature. Data for strain versus time are taken at least into the secondary creep stage and possibly all the way to fracture to determine the creep rupture strain. Testing over a range of temperatures gives a family of curves that is useful for design.

Creep can occur for many plastics even at or near room temperature. Figure 8–13 shows one way that creep data are displayed for plastic materials. It is a graph of applied stress versus strain in the member with data shown for a specific temperature of the specimen. The curves show the amount of strain that would be developed within the specified times at increasing stress levels. For example, if this material were subjected to a constant stress of 5.0 MPa for 5000 hours, the total strain would be 1.0%. That is, the specimen would elongate by an amount 0.01 times the original length. If the

stress were 10.0 MPa for 5000 hours, the total strain would be approximately 2.25%. The designer must take this creep strain into account to ensure that the product performs satisfactorily over time.

Density

Density is defined as the mass per unit volume of a material. Its usual units are kg/m^3 in the SI and lb/in^3 in the U.S. Customary System, where the pound unit is taken to be pounds-mass. The Greek letter rho (ρ) is the symbol for density.

In some applications, the term *specific weight* or *weight density* is used to indicate the weight per unit volume of a material. Typical units are N/m^3 in the SI and lb/in^3 in the U.S. Customary System, where the pound is taken to be pounds-force. The Greek letter gamma (γ) is the symbol for specific weight.

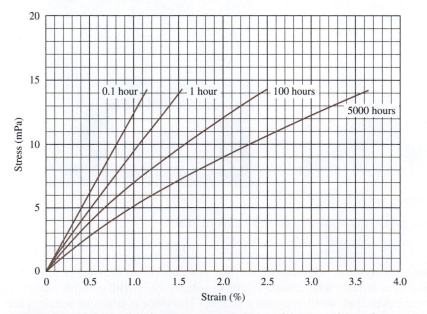

FIGURE 8–13   Example of stress versus strain as a function of time for nylon 66 plastic at 23°C (73°F).

Source: DuPont Polymers, Wilmington, DE.

TABLE 8–5   Unified Numbering System (UNS)

Number series	Types of metals and alloys	Responsible organization
Nonferrous metals and alloys		
A00001–A99999	Aluminum and aluminum alloys	AA
C00001–C99999	Copper and copper alloys	CDA
E00001–E99999	Rare earth metals and alloys	ASTM
L00001–L99999	Low melting metals and alloys	ASTM
M00001–M99999	Miscellaneous nonferrous metals and alloys	ASTM
N00001–N99999	Nickel and nickel alloys	SAE
P00001–P99999	Precious metals and alloys	ASTM
R00001–R99999	Reactive and refractory metals and alloys	SAE
Z00001–Z99999	Zinc and zinc alloys	ASTM
Ferrous metals and alloys		
D00001–D99999	Steels, mechanical properties specified	SAE
F00001–F99999	Cast irons and cast steels	ASTM
G00001–G99999	Carbon and alloy steels (includes former SAE carbon and alloy steels)	AISI
H00001–H99999	H-steels; specified hardenability	AISI
J00001–J99999	Cast steels (except tool steels)	ASTM
K00001–K99999	Miscellaneous steels and ferrous alloys	ASTM
S00001–S99999	Heat and corrosion resistant (stainless) steels	ASTM
T00001–T99999	Tool steels	AISI

Coefficient of Thermal Expansion

The *coefficient of thermal expansion* is a measure of the change in length of a material subjected to a change in temperature. It is defined by the relation

Coefficient of Thermal Expansion ▽

$$\alpha = \frac{\text{Change in length}}{L_o(\Delta T)} = \frac{\text{Strain}}{(\Delta T)} = \frac{\epsilon}{(\Delta T)} \quad \text{(8–8)}$$

where L_o = original length

   ΔT = change in temperature

Virtually all metals and plastics expand with increasing temperature, but different materials expand at different rates. For machines and structures containing parts of more than one material, the different rates can have a significant effect on the performance of the assembly and on the stresses produced. See Chapter 9 for data and applications.

Classification of Metals and Alloys

Various industry associations take responsibility for setting standards for the classification of metals and alloys. Each has its own numbering system, convenient to the particular metal covered by the standard. But this leads to confusion at times when there is overlap between two or more standards and when widely different schemes are used to denote the metals. Order has been brought to the classification of metals by the use of the Unified Numbering System (UNS) as defined

in the Standard E 527-83 (2003), **Standard Practice for Numbering Metals and Alloys (UNS),** by the American Society for Testing and Materials (ASTM). (See Reference 3.) Besides listing materials under the control of ASTM itself, the UNS coordinates designations of

   The Aluminum Association (AA)

   American Iron and Steel Institute (AISI)

   Copper Development Association (CDA)

   Society of Automotive Engineers (SAE)

The primary series of numbers within UNS are listed in Table 8–5 along with the organization having responsibility for assigning numbers within each series.

Many alloys within UNS retain the familiar numbers from the systems used for many years by the individual association. For example, the following section describes the four-digit designation system of the AISI for carbon and alloy steels. Figure 8–14 shows two examples: AISI 1020, a plain carbon steel, and AISI 4140, an alloy steel. These steels would carry the UNS designations G10200 and G41400, respectively.

8–3 STEEL

The term *steel* refers to alloys of iron and carbon and, in many cases, other elements. Because of the large number of steels available, they will be classified in this section as carbon steels, alloy steels, stainless steels, and structural steels. See Reference 5.

FIGURE 8–14 Steel designation system.

For *carbon steels* and *alloy steels,* a four-digit designation code is used to define each alloy. Figure 8–14 shows the significance of each digit. The four digits would be the same for steels classified by the American Iron and Steel Institute (AISI) and the Society of Automotive Engineers (SAE). Classification by the American Society for Testing and Materials (ASTM) will be discussed later.

Usually, the first two digits in a four-digit designation for steel will denote the major alloying elements, other than carbon, in the steel. The last two digits denote the average percent (or points) of carbon in the steel. For example, if the last two digits are 40, the steel would have about 0.4% carbon content. Carbon is given such a prominent place in the alloy designation because, in general, as carbon content increases, the strength and hardness of the steel also increases. Carbon content usually ranges from a low of 0.1% to about 1.0%. It should be noted that while strength increases with increasing carbon content, the steel also becomes less ductile.

Table 8–6 shows the major alloying elements, which correspond to the first two digits of the steel designation. Table 8–7 lists some common alloys along with the principal uses for each.

Conditions for Steels

The mechanical properties of carbon and alloy steels are very sensitive to the manner in which they are formed and to heat-treating processes. Appendix A–14 lists the ultimate strength, yield strength, and percent elongation for a variety of steels in a variety of conditions. Note that these are typical or example properties and may not be relied on for design. Material properties are dependent on many factors, including section size, temperature, actual composition, variables in processing, and fabrication techniques. It is the responsibility of the designer to investigate the possible range of properties for a material and to design load-carrying members to be safe regardless of the combination of factors present in a given situation.

Generally, the more severely a steel is worked, the stronger it will be. Some forms of steel, such as sheet, bar, and structural shapes, are produced by *hot rolling* while still at an elevated temperature. This produces a relatively soft, low-strength steel, which has a very high ductility and is easy to form. Rolling the steel to final form while at or near room temperature is called *cold rolling* and produces a higher strength and somewhat lower ductility. Still higher strength can be achieved by *cold drawing,* drawing the material through dies while it is at or near room temperature. Thus, for these three popular methods of producing steel shapes, the cold-drawn (CD) form results in the highest strength, followed by the cold-rolled (CR) and hot-rolled (HR) forms. This can be seen in Appendix A–14 by comparing the strength of the same steel, say AISI 1040, in the hot-rolled and cold-drawn conditions.

Alloy steels are usually heat-treated to develop specified properties. (See Reference 7.) Heat treatment involves raising the temperature of the steel to above about 1450 to 1650°F (depending on the alloy) and then cooling it rapidly by quenching in either water or oil. After quenching, the steel has a high strength and hardness, but it may also be brittle. For this reason a subsequent treatment called *tempering* (or *drawing*) is usually performed. The steel is reheated to a temperature in the range 400 to 1300°F and then cooled. The effect of tempering an alloy steel can be seen by referring to Figure 8–15. Thus, the properties of a

TABLE 8–6 Major Alloying Elements in Steel Alloys

Steel AISI No.	Alloying elements	Steel AISI No.	Alloying elements
10xx	Plain carbon	46xx	Molybdenum-nickel
11xx	Sulfur (free-cutting)	47xx	Molybdenum-nickel-chromium
13xx	Manganese	48xx	Molybdenum-nickel
14xx	Boron	5xxx	Chromium
2xxx	Nickel	6xxx	Chromium-vanadium
3xxx	Nickel-chromium	8xxx	Nickel-chromium-molybdenum
4xxx	Molybdenum	9xxx	Nickel-chromium-molybdenum (except 92xx)
41xx	Molybdenum-chromium	92xx	Silicon-manganese
43xx	Molybdenum-chromium-nickel		

TABLE 8–7	Common Steel Alloys and Typical Uses
Steel AISI No.	**Typical uses**
1020	Structural steel, bars, plate
1040	Machinery parts, shafts
1050	Machinery parts
1095	Tools, springs
1137	Shafts, screw machine parts (free-cutting alloy)
1141	Shafts, machined parts
4130	General-purpose, high-strength steel; shafts, gears, pins
4140	Same as 4130
4150	Same as 4130
5160	High-strength gears, bolts
8760	Tools, springs, chisels

The properties of heat-treated steels at tempering temperatures of 400 and 1300°F show the total range of properties that the heat-treated steel can have. However, typical practice would specify tempering temperatures no lower than 700°F because steels tend to be too brittle at the lower tempering temperatures. The properties of several alloys are listed in Appendix A–14 at tempering temperatures of 700°F, 900°F, 1100°F, and 1300°F to give you a feeling for the range of strengths available. These alloys are good choices for selecting materials in problems in later chapters. Strengths at intermediate temperatures can be found by interpolation.

Annealing and *normalizing* are thermal treatments designed to soften steel, give it more uniform properties, make it easier to form, or relieve stresses developed in the steel during such processes as welding, machining, or forming. Two of the types of annealing processes used are full anneal and stress-relief anneal. Figure 8–16 illustrates these thermal treatment processes, along with quenching and tempering.

Normalizing of steel starts by heating it to approximately the same temperature (called the *upper critical temperature*) as would be required for through-hardening by quenching, as described before. But rather than quenching, the steel is cooled in still air to room temperature. This results in a uniform, fine-grained structure, improved ductility, better impact resistance, and improved machinability.

heat-treated steel can be controlled by specifying a tempering temperature. In Appendix A–14, the condition of heat-treated alloys is described in a manner like OQT 400. This means that the steel was heat-treated by quenching in oil and then tempered at 400°F. Similarly, WQT 1300 means water quenched and tempered at 1300°F.

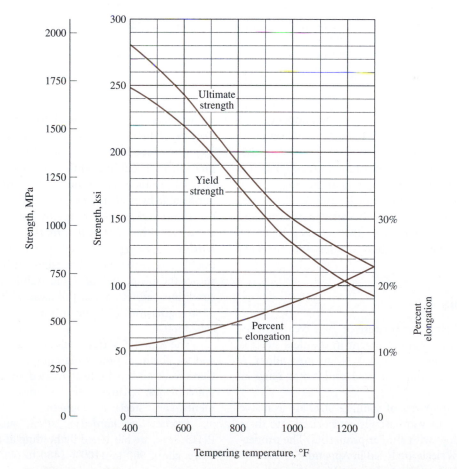

FIGURE 8–15 Effect of tempering temperature on the strength and ductility of an alloy steel.

Note:
RT = room temperature
LC = lower critical temperature
UC = upper critical temperature

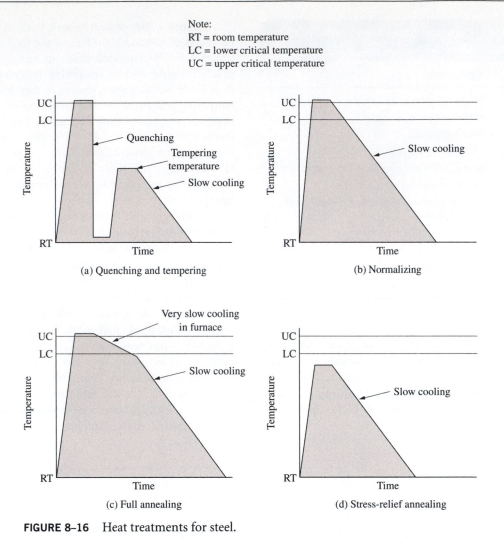

(a) Quenching and tempering

(b) Normalizing

(c) Full annealing

(d) Stress-relief annealing

FIGURE 8–16   Heat treatments for steel.

Full annealing involves heating to above the upper critical temperature followed by very slow cooling to the lower critical temperature and then cooling in still air to room temperature. This is one of the softest forms of the steel and makes it more workable for shearing, forming, and machining.

Stress-relief annealing consists of heating to below the lower critical temperature, holding to achieve uniform temperature throughout the part, and then cooling to room temperature. This relieves residual stresses and prevents subsequent distortion.

Stainless Steels

Stainless steels get their name because of their corrosion resistance. The primary alloying element in stainless steels is chromium, being present at about 17% in most alloys. A minimum of 10.5% chromium is used, and it may range as high as 27%.

Although over 40 grades of stainless steel are available from steel producers, they are usually categorized into three series containing alloys with similar properties. The properties of some stainless steels are listed in Appendix A–15.

The 200 and 300 series steels have high strength and good corrosion resistance. They can be used at temperatures up to about 1200°F with good retention of properties. Because of their structure, these steels are essentially nonmagnetic. Good ductility and toughness and good weldability make them useful in chemical processing equipment, architectural products, and food-related products. They are not hardenable by heat treatment, but they can be strengthened by cold working. The range of cold working is typically given as *quarter hard, half hard, three-quarters hard,* and *full hard,* with strength increasing with higher hardness. But ductility decreases with increasing hardness. Appendix A–15 shows the properties of some stainless steel alloys at two conditions, annealed and full hard, indicating the extremes of strengths available. The annealed condition is sometimes called *soft.*

The AISI 400 series steels are used for automotive trim and for chemical processing equipment such as acid tanks. Certain alloys can be heat-treated so they can be used as knife blades, springs, ball bearings, and surgical instruments. These steels are magnetic.

Precipitation hardening steels, such as 17-4PH and PH13-8Mo, are hardened by holding at an elevated temperature, about 900 to 1100°F (480 to 600°C). Such steels are generally classed as *high-strength stainless steels* having yield strengths about 180 000 psi (1240 MPa) or higher and are used in aerospace, engines, and other applications where the high strength and corrosion resistance are required.

Structural Steels

Structural steels are produced in the forms of sheet, plate, bars, tubing, and structural shapes such as I-beams, wide-flange beams, channels, and angles. The American Society for Testing and Materials (ASTM) assigns a number designation to these steels, which is the number of the standard that defines the required minimum properties. Appendix A–16 lists frequently used grades of structural steels and their properties. See also Reference 3 and Internet site 6.

A very popular steel for structural applications is ASTM A36, a carbon steel used for many commercially available shapes, plates, and bars. It has a minimum yield strength of 36 ksi (248 MPa), is weldable, and is used in bridges, buildings, and for general structural purposes.

The W-shapes used widely in building construction and other industrial structures are now commonly made from ASTM A992, one of several grades of high-strength, low-alloy (HSLA) steels. With a minimum yield strength of 50 ksi (345 MPa), it allows the use of lighter beams, as compared with the formerly used ASTM A36 steel, in many applications with consequent significant cost savings. It should be noted that virtually all steels have the same modulus of elasticity, which is an indication of the stiffness of the material. Therefore, it is critical to assess the deflection of a beam in addition to its strength.

Another HSLA grade of structural steel that is growing in use is ASTM A913, Grade 65, having a minimum yield strength of 65 ksi (448 MPa). Its use in heavy column sections and some critical beam or truss applications has produced weight and cost savings for major structures such as the professional football stadium in Houston, TX and a high-rise office tower in New York City. This steel is also available in Grades 50, 60, and 70 with corresponding minimum yield strengths in ksi.

ASTM A242, in Grades 42, 46, and 50, is another HSLA steel that is produced as shapes, plates, and bars for general structural uses. An additional advantage for this alloy is its corrosion resistance, about four times that of plain carbon steel, leading many to refer to it as a *weathering steel*. All three grades are available in W-shapes; Grade 50 is most commonly available for other rolled shapes. As indicated in Appendix A–16, the available grades for plates and bars depend on thickness.

ASTM A514 is a high-strength alloy steel, heat treated by quenching and tempering, and produced as plates and bars. Thicknesses up to 2.5 in have a yield strength of 100 ksi (690 MPa). Larger thicknesses are rated at a minimum yield strength of 90 ksi (620 MPa).

Another general-purpose HSLA structural steel is ASTM A572, available as all types of shapes, plates, and bars. Grades 42, 50, 55, 60, and 65 are all used for shapes. All plates and bars up to 8 in thick are available in Grade 42; up to 4 in for Grade 50; up to 2 in for Grade 55; and up to 1.25 in for Grades 60 and 65.

Hollow structural sections (HSS), sometimes called *structural tubing,* are either round, square, or rectangular and are typically made from ASTM A501 (hot formed) or ASTM A500 (cold formed) in a variety of strength grades. When produced as *pipe,* ASTM A53 Grade B with a yield strength of 35 ksi (240 MPa) is specified. See Appendices A–9 and A–12 for a sampling of sizes of HSS shapes.

In summary, structural steel products come in many forms and a wide range of properties. Careful selection of a suitable type of steel considers the required strength, cost, and availability. Table 8–8 lists the preferred material specifications along with others for which the availability should be confirmed prior to specification. See Reference 7 in Chapter 1 for a more complete list of preferred and available structural steels for a wider variety of types and sizes of members.

8–4 CAST IRON

The attractive properties of cast iron include low cost, good wear resistance, good machinability, and the ability to be cast into complex shapes. Five varieties will be discussed here: gray iron, ductile iron, austempered ductile iron, white iron, and malleable iron.

Gray Iron

Gray iron is used in automotive engine blocks, machinery bases, brake drums, and large gears. It is usually specified by giving a grade number corresponding to the minimum ultimate tensile strength. For example, grade 20 gray cast iron has a minimum ultimate strength of 20 000 psi (138 MPa); grade 60 has $s_u = 60\,000$ psi (414 MPa), and so on. The usual grades available are from 20 to 60. Gray iron is somewhat brittle, so that yield strength is not usually reported as a property. An outstanding feature of gray iron is that its compressive strength is very high, about three to five times as high as the tensile strength. This should be taken into account in design, especially when a part is subjected to bending stresses, as discussed in Chapter 12.

Because of variations in the rate of cooling after the molten cast iron is poured into a mold, the actual strength of a particular section of a casting is dependent on the thickness of the section. Figure 8–17 illustrates this for grade 40 gray iron. The in-place strength may range from as high as 52 000 psi (359 MPa) to as low as 27 000 psi (186 MPa).

Ductile Iron

Ductile iron differs from gray iron in that it does exhibit yielding, has a greater percent elongation, and has generally higher tensile strength. Grades of ductile iron are designated by a three-number system such as 80-55-6. The first number indicates the minimum ultimate tensile strength in ksi; the second is the yield strength in ksi; and the third is the percent elongation. Thus, grade 80-55-6 has an ultimate strength of 80 000 psi, a yield strength of 55 000 psi, and a percent elongation of 6%. Uses for ductile iron include crankshafts and heavily loaded gears.

TABLE 8–8 Structural Steel Grades Available for Typical Shapes, Plates, and Bars

ASTM designation	Grade	Yield strength ksi	Yield strength MPa	W shapes	S, C, L shapes	Pipe	Rectangular or square HSS	Round HSS	Plates and bars
A36		36	248	A	P	—	—	—	P: up to 8 in
A53		35	241	—	—	P	—	—	—
A242		42	290	A	—	—	—	—	A: 1.5 in to 4 in
		46	317	A	—	—	A	—	A: 0.75 to 1.5 in
		50	345	A	A	—	A	—	A: up to 0.75 in
A500	B	42	290	—	—	—	—	P	—
		46	317	—	—	—	P	—	—
	C	46	317	—	—	—	—	A	—
		50	345	—	—	—	A	—	—
A501		36	248	—	—	—	A	A	—
A514		90	621	—	—	—	—	—	A: 2.5 in to 6 in
		100	690	—	—	—	—	—	A: up to 2.5 in
A572	42	42	290	A	A	—	—	—	A: up to 6 in
	50	50	345	A	A	—	—	—	A: up to 4 in
	55	55	379	A	A	—	—	—	A: up to 2 in
	60	60	414	A	A	—	—	—	A: up to 1.25 in
	65	65	448	A	A	—	—	—	A: up to 1.25 in
A913	65	65	448	A	A	—	—	—	—
A992		50	345	P	—	—	—	—	—

Notes: Adapted from Reference 2, Chapter 7. Other materials, grades, and shapes available.

  P = preferred material specification

  A = Available—Check with supplier

  — = Not available

Austempered Ductile Iron

Austempered ductile iron (ADI) has higher strength and better ductility than standard ductile irons, as can be seen in Appendix A–17. This allows parts to be smaller and lighter and makes ADI desirable for uses such as automotive gears, crankshafts, and structural members for construction and transportation equipment, replacing wrought or cast steels.

Austempering can increase the strength of ductile iron by nearly a factor of 2. First the castings are heated to between 1500°F and 1700°F (816°C to 927°C) and held to achieve a uniform structure. Then they are quenched rapidly to a lower temperature, 450°F to 750°F (232°C to 400°C), and held again. After several hours of isothermal soaking, the castings are allowed to cool to room temperature.

White Iron

White iron is produced by rapidly chilling a casting of either gray iron or ductile iron during the solidification process. The chilling is typically applied to selected areas, which become very hard and have a high wear resistance. The chilling does not allow the carbon in the iron to precipitate out during solidification, resulting in the white appearance. Areas away from the chilling medium solidify more slowly and acquire the normal properties of the base iron. One disadvantage of the chilling process is that the white iron is very brittle.

Malleable Iron

Malleable iron is used in automotive and truck parts, construction machinery, and electrical equipment. It does exhibit yielding, has tensile strengths comparable to ductile iron, and has ultimate compressive strengths that are somewhat higher than ductile iron. Generally, a five-digit number is

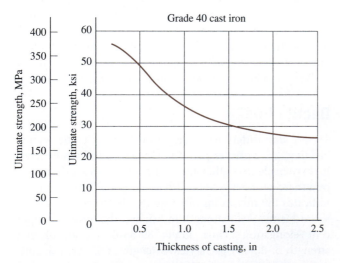

FIGURE 8–17 Strength versus thickness for grade 40 gray cast iron.

used to designate malleable iron grades. For example, grade 40 010 has a yield strength of 40 000 psi (276 MPa) and a percent elongation of 10%.

Appendix A–17 lists the mechanical properties of several grades of gray iron, ductile iron, ADI, and malleable iron.

8–5 ALUMINUM

Alloys of aluminum are designed to achieve optimum properties for specific uses. Some are produced primarily as sheet, plate, bars, or wire. Standard structural shapes and special sections are often extruded. Several alloys are used for forging, while others are special casting alloys. Appendix A–18 lists the properties of selected aluminum alloys. See also References 1 and 6.

Aluminum in wrought form uses a four-digit designation to define the several alloys available. The first digit indicates the alloy group according to the principal alloying element. The second digit denotes a modification of the basic alloy. The last two digits identify a specific alloy within the group. A brief description of the seven major series of aluminum alloys follows.

- *1000 Series*, 99.0% aluminum or greater. Excellent corrosion resistance, workability, and thermal and electrical conductivity. Low mechanical properties. Used in chemical and electrical fields, for automotive trim (alloy 1100), and for extruded condenser tubes and heat exchanger fins (alloy 1200).

- *2000 series*, copper alloying element. Heat-treatable with high mechanical properties. Lower corrosion resistance than most other alloys. Used in aircraft skins and structures, automotive body panels, screw machine parts, fasteners, and seat shells.

- *3000 series*, manganese alloying element. Non-heat-treatable, but moderate strength can be obtained by cold working. Good corrosion resistance and workability. Used in chemical equipment, cooking utensils, residential siding, storage tanks, automotive radiators and heater cores, trim, and condenser tubes.

- *4000 series*, silicon alloying element. Non-heat-treatable with a low melting point. Used as welding wire and brazing alloy. Alloy 4032 used as pistons.

- *5000 series*, magnesium alloying element. Non-heat-treatable, but moderate strength can be obtained by cold working. Good corrosion resistance and weldability. Used in marine service, pressure vessels, auto trim, builder's hardware, welded structures, TV towers, drilling rigs, truck bumpers, heat shields, wheels, and various brackets on engines.

- *6000 series*, silicon and magnesium alloying elements. Heat-treatable to moderate strength. Good corrosion resistance, formability, and weld-ability. Used as heavy-duty structures, truck and railroad equipment, pipe, furniture, architectural extrusions, machined parts, and forgings. Automotive applications include suspension parts, bumper assembly parts, driveshafts, yokes, brake cylinders and pistons, screw machine parts, fasteners, and luggage racks. Alloy 6061 is one of the most versatile available.

- *7000 series*, zinc alloying element. Heat-treatable to very high strength. Relatively poor corrosion resistance and weldability. Used mainly for aircraft structural members. Alloy 7075 is among the highest strength alloys available. It is produced in most rolled, drawn, and extruded forms and is also used in forgings. Automotive applications include seat tracks, bumper reinforcements, headrest support bars, and condenser fins.

Aluminum Temper Designations

Since the mechanical properties of virtually all aluminum alloys are very sensitive to cold working or heat treatment, suffixes are applied to the four-digit alloy designations to describe the temper. The most frequently used temper designations are described as follows:

- *O temper.* Fully annealed to obtain high ductility. Annealing makes most alloys easier to form by bending or drawing. Parts formed in the annealed condition are frequently heat-treated later to improve properties.

- *H temper,* strain-hardened. Used to improve the properties of non-heat-treatable alloys such as those in the 1000, 3000, and 5000 series. The H is always followed by a two- or three-digit number to designate a specific degree of strain hardening or special processing. The second digit following the H ranges from 0 to 8 and indicates a successively greater degree of strain hardening, resulting in higher strength. Appendix A–18 lists the properties of several aluminum alloys. Referring to alloy 3003 in that table shows that the yield strength is increased from 18 000 psi (124 MPa) to 27 000 psi (186 MPa) as the temper is changed from H12 to H18.

- *T temper,* heat-treated. Used to improve strength and achieve a stable condition. The T is always followed by one or more digits indicating a particular heat treatment. For wrought products such as sheet, plate, extrusions, bars, and drawn tubes, the most frequently used designations are T4 and T6. The T6 treatment produces higher strength but generally reduces workability. In Appendix A–18 several heat-treatable alloys are listed in the O, T4, and T6 tempers to illustrate the change in properties.

Cast aluminum alloys are designated by a modified four-digit system of the form XXX.X, in which the first digit indicates the main alloy group according to the major alloying elements. Table 8–9 shows the groups. The second two digits indicate the specific alloy within the group or indicate the aluminum purity. The last digit, after the decimal point, indicates the product form: 0 for castings, and 1 or 2 for ingots.

Aluminum is also sensitive to the manner in which it is produced, the size of the section, and temperature. Appendix A–18 lists typical properties and cannot be relied on for design. References 1, 3, and 6 give extensive data on minimum strengths.

TABLE 8–9 Cast Aluminum Alloy Groups

Group	Major alloying elements
1XX.X	99% or greater aluminum
2XX.X	Copper
3XX.X	Silicon, copper, magnesium
4XX.X	Silicon
5XX.X	Magnesium
6XX.X	(Unused series)
7XX.X	Zinc
8XX.X	Tin
9XX.X	Other elements

Cast aluminum is used for large structural members and housings, engine blocks, cylinder heads, manifolds, internal engine parts, steering gears, transmission housings and internal parts, wheels, and fuel metering devices.

8–6 COPPER, BRASS, AND BRONZE

Copper and copper alloys have good corrosion resistance, are readily fabricated, and have an attractive appearance. They are typically classified as *copper, brass,* or *bronze,* depending on the amounts and types of alloying elements employed. Cold working is used to modify the strength and hardness properties of these metals, ranging from *soft, quarter hard, half hard, three-quarters hard, hard, extra hard, spring,* and *extra spring* tempers, depending on the percent of cold working during processing. The properties of four types of copper-based materials are listed in Appendix A–15 and discussed next. Internet site 11 allows searching through a large database of copper materials.

Copper

The name *copper* is most often used to denote virtually the pure metal having 99% or more copper. The C14500 material has nominally 99.5% copper and 0.5% tellurium, and its typical uses include electrical connectors, motor parts that carry electric current, screw machine products, plumbing fittings, and sprinkler heads.

Beryllium copper (C17200) has nominally 98.1% copper and 1.9% beryllium, thus departing from the strict definition of pure copper. The beryllium adds significant strength, leading to uses such as fuse clips, relay parts, fasteners, Bourdon tubing for pressure gages, pump parts, valves, a variety of machined components, springs, and forgings.

Brass

Brasses are alloys of copper and zinc with lead sometimes added to improve machinability. The C36000 material is a free-cutting brass having 61.5% copper, 35.4% zinc, and 3.1% lead. Automotive uses include thermostat parts, fluid connectors, sensor bodies, and threaded inserts for plastic parts. Other industrial uses include builders' hardware, fasteners, faucet components valve seats, nozzles, gears, and screw machine parts. Other brasses are used in marine applications, ammunition cases, home furnishings, and heat exchanger tubing.

Bronze

Copper, alloyed with tin and a wide variety of elements such as lead, zinc, and phosphorous, fall into the *bronze* family of materials. Phosphor bronze (C54400) has 88% copper, 4% tin, 4% lead, 4% zinc, and 0.25% phosphorous. Its good strength, ductility, and wear properties lead to uses such as gears, shafts, bearings, thrust washers, screw machine parts, electrical connectors, and parts for valves.

8–7 ZINC, MAGNESIUM, TITANIUM, AND NICKEL-BASED ALLOYS

Zinc

Zinc has moderate strength and toughness and excellent corrosion resistance. It is used in wrought forms such as rolled sheet and foil and drawn rod or wire. Dry-cell battery cans, builder's hardware, and plates for photoengraving are some of the major applications.

Many zinc parts are made by die casting because the melting point is less than 800°F (427°C), much lower than other die-casting metals. The as-cast finish is suitable for many applications, such as business machine parts, pump bodies, motor housings, and frames for light-duty machines. Where a decorative appearance is required, electroplating with nickel and chromium is easily done. Such familiar parts as lamp housings and auto body moldings are made in this manner. See Internet site 13.

Zinc alloys typically contain aluminum and a small amount of magnesium. Some alloys include copper or nickel. The performance of the final products can be very sensitive to small amounts of other elements, and maximum limits are placed on the content of iron, lead, cadmium, and tin in some alloys.

The most widely used zinc casting alloy is called *alloy No. 3,* sometimes referred to as *Zamak 3.* It has 4% aluminum and 0.035% magnesium. Another is called *Zamak 5* and it also contains 4% aluminum with 0.055% magnesium and 1.0% copper. A group of alloys having higher aluminum content are the ZA-alloys, with ZA-8, ZA-12, and ZA-27 being very popular. The strength properties of cast ZA-12 zinc alloy are listed in Appendix A–15.

Magnesium

Magnesium is the lightest metal commonly used in load-carrying parts. Its density of only 0.066 lb/in^3 (1830 kg/m^3) is only about one-fourth that of steel and zinc, one-fifth that of copper, and two-thirds that of aluminum. It has moderate strength and lends itself well to applications where the final fabricated weight of the part or structure should be light. Ladders, hand trucks, conveyor parts, portable power tools, and lawn-mower housings use magnesium. In the automotive industry, body parts, blower wheels, pump bodies, and brackets are often made of magnesium. In aircraft, its lightness makes this metal attractive for floors, frames, fuselage skins, and wheels. The stiffness (modulus of elasticity) of magnesium is low, which is an advantage in parts where impact energy must be absorbed. Also, its lightness results in low-weight designs when compared with other metals on an equivalent rigidity basis. See Appendix A–15 for properties of one cast magnesium alloy.

Titanium

Titanium has very high strength, and its density is only about half that of steel. Although aluminum has a lower density, titanium is superior to both aluminum and most steels on a strength-to-weight basis. It retains a high percentage of its strength at elevated temperatures and can be used up to about 1000°F (538°C). Most applications of titanium are in the aerospace industry in engine parts, fuselage parts and skins, ducts, spacecraft structures, and pressure vessels. Because of its corrosion resistance and high-temperature strength, the chemical industries use titanium in heat exchangers and as a lining for processing equipment. High cost is a major factor to be considered.

Appendix A–15 shows the properties of one example of each of the four major types of titanium; pure alpha titanium, alpha alloy, beta alloy, and alpha-beta alloy. In general, the alpha-beta and beta alloys are the stronger forms, and they are heat-treatable to permit close tailoring of properties to specific uses, considering formability, machinability, forgeability, weldability, corrosion resistance, high-temperature strength, and creep resistance in addition to the typical room temperature strength, stiffness, and ductility data. The term *aged* indicates a final heat treatment, and more complete references for aging should be consulted to assess the range of properties that can be achieved.

Nickel-Based Alloys

Nickel-based alloys are often specified for applications where high corrosion resistance and/or high strength at elevated temperatures are required. Examples are turbine engine components, rocket motors, furnace parts, chemical processing systems, marine-based systems, aerospace systems, valves, pumps, heat exchangers, food processing equipment, pharmaceutical processing systems, nuclear power systems, and air pollution equipment.

Appendix A–15 gives typical properties of three nickel-based alloys listed according to their UNS designations in the NXXXXX series. Most nickel-based alloys are not heat treated; however, cold working is employed to increase strength as is evident for the N04400 alloy in Appendix A–15. For alloys N06600 and N06110, strength data are provided for room temperature and several elevated temperatures up to 1200°F (649°C). Some alloys are usable up to 2000°F (1093°C). Alloys are often grouped under the categories of *corrosion resistant* or *heat resistant*. While nickel makes up 50% or more of these alloys, the major alloying elements include chromium, molybdenum, cobalt, iron, copper, and niobium. See Internet sites 28–30 for some producers of these alloys. Note the brand names offered by these companies and the extensive set of property data available on their Web sites.

Creep performance is an important property for the nickel-based alloys because they often operate under significant loads and stresses at high temperatures where creep rupture is a possible failure mode. Also, the modulus of elasticity decreases with increasing temperature, which can result in unacceptable deformations that must be considered in the design phase. Producers publish creep and elastic modulus versus temperature data for their materials in addition to the basic strength data.

8–8 NONMETALS IN ENGINEERING DESIGN

Wood and concrete are widely used in construction. Plastics and composites are found in virtually all fields of design, including consumer products, industrial equipment, automobiles, aircraft, and architectural products. To the designer, the properties of strength and stiffness are of primary importance with nonmetals, as they are with metals. Because of the structural differences in the nonmetals, their behavior is quite different from the metals.

Wood, concrete, composites, and many plastics have structures that are *anisotropic*. This means that the mechanical properties of the material are different, depending on the direction of the loading. Also, because of natural chemical changes, the properties vary with time and often with climatic conditions. The designer must be aware of these factors.

8–9 WOOD

Since wood is a natural material, its structure is dependent on the way it grows and not on manipulation by human beings, as is the case in metals. The long, slender, cylindrical shape of trees results in an internal structure composed of longitudinal cells. As the tree grows, successive rings are added outside the older wood. Thus, the inner core, called heartwood, has different properties than the sapwood, near the outer surface.

The species of the wood also affects its properties, as different kinds of trees produce harder or softer, stronger or weaker wood. Even in the same species variability occurs because of different growing conditions, such as differences in soil and amount of sun and rain.

The cellular structure of the wood gives it the grain that is so evident when sawn into boards and timber. The strength of the wood is dependent on whether it is loaded perpendicular to or parallel to the grain. Also, going across the grain, the strength is different in a radial direction than

TABLE 8–10   Machine-Graded Lumber

Grade	Allowable stresses (parallel to grain)						Modulus of elasticity	
	Bending		Tension		Compressive			
	psi	MPa	psi	MPa	psi	MPa	psi	GPa
Machine-stress-rated lumber (MSR)								
1350f-1.3E	1350	9.3	750	5.2	1600	11.0	1.3×10^6	9.0
1800f-1.6E	1800	12.4	1175	8.1	1750	12.1	1.6×10^6	11.0
2400f-2.0E	2400	16.5	1925	13.3	1975	13.6	2.0×10^6	13.8
2850f-2.3E	2850	19.7	2300	15.9	2150	14.8	2.3×10^6	15.9
Machine-evaluated lumber (MEL)								
M-10	1400	9.7	800	5.5	1600	11.0	1.2×10^6	8.3
M-14	1800	12.4	1000	6.9	1750	12.1	1.7×10^6	11.7
M-21	2300	15.9	1400	9.7	1950	13.4	1.9×10^6	13.1
M-24	2700	18.6	1800	12.4	2100	14.5	1.9×10^6	13.1

Source: Forest Products Society, 1997.

in a tangential direction with respect to the original cylindrical tree stem from which it was cut.

Another important variable affecting the strength of wood is moisture content. Changes in relative humidity can vary the amount of water absorbed by the cells of the wood.

Appendix A–19 gives data for typical properties of three species of wood often used in construction: Douglas fir, hemlock, and southern pine. Notice the three grades listed for each. Grade No. 1 is the highest grade, representing clear wood with few if any defects such as knots. Grade No. 2 is common construction lumber that may have occasional knots or other defects. Grade 3 is a poorer grade with generally noticeable knots and large variations in grain. The allowable strength values decrease dramatically from Grade 1 to Grade 3. The grades and, thus, the allowable strengths are determined by using standard rules adopted by the U.S. Forest Products Laboratory. See References 19 and 31.

The tension and compression properties are used in Chapter 9. Bending and horizontal shear strengths are considered in Chapter 12 on stresses in beams. Notice the very low strength in shear, only 70 to 95 psi (0.48 to 0.66 MPa). This represents the prevalent mode of failure for wood beams as discussed further in Chapter 13.

Machine-Graded Structural Lumber

Alternate grading systems for lumber use machine-grading processes to sort lumber into grades related to strength and stiffness properties of the wood. Two such systems are *machine-stress-rated (MSR)* and *machine-evaluated lumber (MEL)*. Table 8–10 shows representative grades from each system. More grades are available and not all grades are used for all products. Users should discuss details of rating systems with vendors before specifying particular products.

8–10 CONCRETE

The components of concrete are cement and an aggregate. The addition of water and the thorough mixing of the components tend to produce a uniform structure with cement coating all the aggregate particles. After curing, the mass is securely bonded together. Some of the variables involved in determining the final strength of the concrete are the type of cement used, the type and size of aggregate, and the amount of water added.

A higher quantity of cement in concrete yields a higher strength. Decreasing the quantity of water relative to the amount of cement increases the strength of the concrete. Of course, sufficient water must be added to cause the cement to coat the aggregate and to allow the concrete to be poured and worked before excessive curing takes place. The density of the concrete, affected by the aggregate, is also a factor. A mixture of sand, gravel, and broken stone is usually used for construction grade concrete.

Concrete is graded according to its compressive strength, which varies from about 2000 psi (14 MPa) to 7000 psi (48 MPa). The tensile strength of concrete is extremely low, and it is common practice to assume that it is zero. Of course, reinforcing concrete with steel bars allows its use in beams and wide slabs since the steel resists the tensile loads.

Concrete must be cured to develop its rated strength. It should be kept moist for at least 7 days, at which time it has about 60% to 70% of its rated compressive strength. Although its strength continues to increase for years, the strength at 28 days is often used to determine its rated strength.

Testing of concrete can be done using the universal testing machine similar to that shown in Figure 8–1, arranged to apply a compressive load. The test specimen is typically a 6-in diameter cylinder having a length of 12 in

FIGURE 8–18 Typical stress–strain curve for concrete.

FIGURE 8–19 Typical stress–strain curves for a variety of concrete grades.

(150 mm × 300 mm). The load is slowly increased until compressive fracture occurs, tracking the stress and strain continuously during the test. Figure 8–18 shows a typical stress–strain curve for concrete. Note the following:

1. The stress–strain curve has a continuous curvature, reaching a peak at a strain of approximately 0.002.

2. The peak of the curve is taken as the ultimate compressive strength, or rated strength, of the concrete, called f'_c in the construction industry.

3. The lack of a straight-line portion of the stress–strain curve requires a special definition of the modulus of elasticity for concrete. The American Concrete Institute (ACI) defines the *secant modulus*, as shown in Figure 8–18, to be the slope of the line from the origin through the intersection on the curve with the stress level of $f'_c/2$ (point A).

4. Failure occurs at a strain of approximately 0.003 to 0.004, depending on the ultimate strength of the concrete. Failure may occur as shearing along angular lines within the specimen or as splitting along nearly vertical planes.

Figure 8–19 shows typical stress–strain curves for several grades of concrete. An important observation from these curves is that the secant modulus varies significantly, being higher for the higher strength grades.

The specific weight of concrete, sometimes called the *unit weight*, ranges from about 90 lb/ft³ to 160 lb/ft³ (14.1 kN/m³ to 25.1 kN/m³) with the kind of aggregate having a major influence. When using SI units, mass density is often called for, which ranges from 1.44 kg/m³ to 2.56 kg/m³. For common construction-grade concrete the nominal values are

Specific weight: 150 lb/ft³ or 23.6 kN/m³

Density: 150 lb$_m$/ft³ or 2.40 kg/m³

The modulus of elasticity is somewhat dependent on the specific weight and the rated strength. According to the American Concrete Institute, an estimate of the modulus can be computed from

$$E_c = 33\gamma^{3/2}\sqrt{f'_c} \qquad (8\text{–}9)$$

where E_c = modulus of elasticity in compression, psi

 γ = specific weight, lb/ft³

 f'_c = rated compressive strength of the concrete, psi

Using $\gamma = 150$ lb/ft³, the range of expected values for modulus of elasticity, computed from Equation (8–9), is shown in Table 8–11.

The allowable working stresses in concrete are typically 25% of the rated 28-day strength. For example, concrete rated at 2000 psi (14 MPa) would have an allowable stress of 500 psi (3.4 MPa).

TABLE 8–11	Modulus of Elasticity for Concrete		
Rated strength, f'_c		Modulus of elasticity, E_c	
psi	MPa	psi	GPa
2000	13.8	2.7×10^6	18.6
3000	20.7	3.3×10^6	22.7
4000	27.6	3.8×10^6	26.2
5000	34.5	4.3×10^6	29.6
6000	41.4	4.7×10^6	32.4
7000	48.3	5.1×10^6	35.2

8–11 PLASTICS

Plastics are composed of long chainlike molecules called *polymers*. They are synthetic organic materials that can be formulated and processed in literally thousands of ways.

One possible classification is between *thermoplastic* materials and *thermosetting* materials. Thermoplastics can be softened repeatedly by heating with no change in properties or chemical composition. Conversely, after initial curing of thermosetting plastics, they cannot be resoftened. A chemical change occurs during curing with heat and pressure.

Some examples of thermoplastics include ABS, acetals, acrylics, cellulose acetate, TFE fluorocarbons, nylon, polyethylene, polypropylene, polystyrene, and vinyls. Thermosetting plastics include phenolics, epoxies, polyesters, silicones, urethanes, alkyds, allyls, and aminos. Some of these are described next.

Thermoplastics

- **Nylon (Polyamide PA):** Good strength, wear resistance, and toughness; wide range of possible properties depending on fillers and formulations. Used for structural parts, mechanical devices such as gears and bearings, and parts needing wear resistance.

- **Acrylonitrile-Butadiene-Styrene (ABS):** Good impact resistance, rigidity, moderate strength. Used for housings, helmets, cases, appliance parts, pipe, fittings.

- **Polycarbonate:** Excellent toughness, impact resistance, and dimensional stability. Used for cams, gears, housings, electrical connectors, food processing products, helmets, and pump and meter parts.

- **Acrylic:** Good weather resistance and impact resistance; can be made with excellent transparency or translucent or opaque with color. Used for glazing, lenses, signs, and housings.

- **Polyvinyl Chloride (PVC):** Good strength, weather resistance, and rigidity. Used for pipe, electrical conduit, small housings, ductwork, and moldings.

- **Polyimide:** Good strength and wear resistance; very good retention of properties at elevated temperatures up to 500°F. Used for bearings, seals, rotating vanes, and electrical parts.

- **Acetal:** High strength, stiffness, hardness, and wear resistance; low friction; good weather resistance and chemical resistance. Used for gears, bushings, sprockets, conveyor parts, and plumbing products. [Generic name: poly-oxymethylene (POM).]

- **Polyurethane Elastomer:** A rubberlike material with exceptional toughness and abrasion resistance; good heat resistance and resistance to oils. Used for wheels, rollers, gears, sprockets, conveyor parts, and tubing.

- **Thermoplastic Polyester Resin (PET):** Polyethylene terephthalate (PET) resin with fibers of glass and/or mineral. Very high strength and stiffness, excellent resistance to chemicals and heat, excellent dimensional stability, and good electrical properties. Used for pump parts, housings, electrical parts, motor parts, auto parts, oven handles, gears, sprockets, and sporting goods.

- **Polyether-Ester Elastomer:** Flexible plastic with excellent toughness and resilience, high resistance to creep, impact, and fatigue under flexure, good chemical resistance. Remains flexible at low temperatures and retains good properties at moderately elevated temperatures. Used for seals, belts, pump diaphragms, protective boots, tubing, springs, and impact absorbing devices. High modulus grades can be used for gears and sprockets.

Thermosets

- **Phenolic:** High rigidity, good moldability and dimensional stability, very good electrical properties. Used for load-carrying parts in electrical equipment, switchgear, terminal strips, small housings, handles for appliances and cooking utensils, gears, and structural and mechanical parts. Alkyd, allyl, and amino thermosets have properties and uses similar to those of the phenolics.

- **Polyester:** Known as *fiberglass* when reinforced with glass fibers; high strength and stiffness, good weather resistance. Used for housings, structural shapes, and panels.

Special Considerations for Selecting Plastics

A particular plastic is often selected for a combination of properties, such as light weight, flexibility, color, strength, stiffness, chemical resistance, low friction characteristics, or transparency. Table 8–12 lists the primary plastic materials used for six different types of applications. References 4, 11, 14, 16, 29, and 30 provide an extensive comparative study of the design properties of plastics.

While most of the same definitions of design properties described in Section 8–2 of this chapter can be used for plastics, as well as metals, a significant amount of additional information is typically needed to specify a suitable plastic material. Some of the special characteristics of plastics follow. Use data from Appendix A–20 for problem solutions in this book requiring the properties of selected plastics. There is a wide range of properties among the many formulations of plastics even within a given class. Consult the extensive amount of design guidance available from vendors of the plastic materials. See Internet sites 14–19.

1. Most properties of plastics are highly sensitive to temperature. In general, tensile strength, compressive strength, elastic modulus, and impact failure energy decrease significantly as the temperature increases.

2. Many plastics absorb a considerable amount of moisture from the environment and exhibit dimensional changes and degradation of strength and stiffness properties as a result.

3. Components that carry loads continuously must be designed to accommodate creep or relaxation as described in Section 8–2.

TABLE 8–12   Applications of Plastic Materials

Applications	Desired properties	Suitable plastics
Housings, containers, ducts	High impact strength, stiffness, low cost, formability, environmental resistance, dimensional stability	ABS, polystyrene, polypropylene, polyethylene, cellulose acetate, acrylics
Low friction—bearings, slides	Low coefficient of friction; resistance to abrasion, heat, corrosion	TFE fluorocarbons, nylon, acetals
High-strength components, gears, cams, rollers	High tensile and impact strength, stability at high temperatures, machinable	Nylon, phenolics, TFE-filled acetals
Chemical and thermal equipment	Chemical and thermal resistance, good strength, low moisture absorption	Fluorocarbons, polypropylene, polyethylene, epoxies, polyesters, phenolics
Electrostructural parts	Electrical resistance, heat resistance, high impact strength, dimensional stability, stiffness	Allyls, alkyds, aminos, epoxies, phenolics, polyesters, silicones
Light-transmission components	Good light transmission in transparent and translucent colors, formability, shatter resistance	Acrylics, polystyrene, cellulose acetate, vinyls

4. The behavior of plastics under repeated loading (fatigue), shock, and impact is highly variable, and many grades are specially formulated for good performance in such environments. Data for toughness as indicated by Izod, Charpy, or drop testing should be acquired from the vendor along with specific fatigue strengths of the materials.

5. Processing methods can have large effects on the final dimensions and properties of parts made from plastics. Molded plastics shrink significantly during solidification and curing. Parting lines produced where mold halves meet may affect strength. The rate of solidification may be widely different in a given part depending on the section thicknesses, the complexity of the shape and the location of sprues that deliver molten plastic into the mold. The same material can produce different properties depending on whether it is processed by injection molding, extrusion, blow molding, or machining from a solid block or bar. See Reference 18.

6. Resistance to chemicals, weather, direct sunlight, and other environmental conditions must be checked.

7. Plastics may exhibit a change in properties as they age, particularly when subjected to elevated temperatures.

8. Flammability and electrical characteristics must be considered. Some plastics are specially formulated for high flammability ratings as called for by Underwriters Laboratory and other agencies.

9. Plastics used for food storage or processing must meet U.S. Food and Drug Administration standards.

8–12 COMPOSITES

Composites are materials having two or more constituents blended in a way that results in mechanical or adhesive bonding between the materials. To form a composite, a filler material is distributed in a matrix so that the filler reinforces the matrix. Typically, the filler is a strong, stiff material while the matrix has a relatively low density. When the two materials bond together, much of the load-carrying ability of the composite is produced by the filler material. The matrix serves to hold the filler in a favorable orientation relative to the manner of loading and to distribute the loads to the filler. The result is a somewhat optimized composite that has high strength and high stiffness with low weight. See References 10, 15, 17, 20, 21, 23, 26, and 27.

A virtually unlimited variety of composite materials can be produced by combining different matrix materials with different fillers in different forms and in different orientations. Even wood and concrete, discussed earlier in this chapter, are technically examples of composites. However, the common use of the term refers to those materials using polymers, metals, or ceramics as matrix materials with a wide variety of filler materials to form composites.

Examples of Finished Products Made from Composite Materials

The number and variety of applications for composite materials is large and growing. The following items are but a sampling of these applications.

Consumer Products and Recreation: Sporting goods such as tennis rackets, snow skis, snowboards, water skis, surfboards, baseball bats, hockey sticks, vaulting poles, and golf clubs; numerous products having the familiar fiberglass housings and panels; boat hulls and other onboard equipment; medical systems and prosthetic devices. Figure 8–20 shows a knee brace with several components made from composites.

Ground Transportation Equipment: Bicycle frames, wheels, and seats; automotive and truck body panels and support structures, air ducts, air bags, driveshafts,

FIGURE 8–20 Knee brace with several components made from composites.

Source: A&P Technology, Cincinnati, OH.

springs for high-performance sports cars and trucks, floor pans, pickup truck beds, and bumpers. Figure 8–21 shows a high-performance bicycle employing a composite frame and other structural parts.

Aircraft and Aerospace Systems: Fuselage panels and internal structural elements, wings, control surfaces (ailerons, spoilers, tails, rudders), floor systems, engine cowls and nacelles, landing gear doors, cargo compartment structure and fittings, interior sidewalls, trim, partitions, ceiling panels, dividers, environmental control system ducting, stowage bins, lavatory structure systems and fixtures, airfoils (blades) in the compressor

FIGURE 8–21 High-performance bicycle employing a composite frame and other structural parts.

Source: A&P Technology, Cincinnati, OH.

FIGURE 8–22 Stator vanes for a turbine engine made from composite materials.

Source: A&P Technology, Cincinnati, OH.

section of turbine engines, rocket nozzles, helicopter rotors, propellers, and onboard tanks for storing water and wastewater. Figures 8–22 and 8–23 show turbine engine compressor blades and some of the tooling used to manufacture them.

Industrial Facilities: Storage tanks and pressure vessels for chemical, agricultural, and petroleum processing, piping for chemicals and through corrosive environments, septic systems, wastewater treatment facilities, chemical cleaning and plating systems, pulp and paper making equipment, portable tanks for trucks and railroad applications, environmental treatment equipment, protective clothing and helmets, food processing and storage systems, mining, and material handling systems.

Electrical and Electronic Systems: Printed circuit boards, printed wiring boards, surface mount cards,

FIGURE 8–23 Forming press for manufacturing stator vanes for a turbine engine, shown in Figure 8–22.

Source: A&P Technology, Cincinnati, OH.

packaging of electronic components, and switching system components.

Building Construction: Structural shapes, exterior panels, roofing and decking systems, doors, window frames, equipment housings, gutters and downspouts, cooling towers, bridges and walkways, piping systems, and ductwork.

Classifications of Composite Materials by Matrix

One method of classifying composite materials is by the type of matrix material. Three general classifications are used as described next, along with typical matrix materials, uses, and matrix-filler combinations.

Polymer Matrix Composites (PMC)

Thermoplastics: Polyethylenes, Polyamides (nylons), polystyrenes, polypropylenes, polycarbonates, poly-etheretherketones (PEEK), polyphenylene sulfides (PPS), polyvinyl chloride (PVC)

Thermosets: Polyesters, epoxies, phenolics, polyimides (PI), vinyl esters, silicones

PMCs are used for their high strength and stiffness, low density, and relatively low cost in aerospace, automotive, marine, chemical, electrical, and sporting applications. Common PMC composites include polyester-glass (conventional fiberglass), epoxy-glass, polyimide-glass, epoxy-aramid, epoxy-carbon, PEEK-carbon, and PPS-carbon.

Metal Matrix Composites (MMC)

Aluminum (Al), titanium (Ti), magnesium (Mg), iron (Fe), copper (Cu), nickel (Ni), and alloys of these metals with themselves and with molybdenum (Mo), cesium (Ce), boron (B).

MMCs are preferred for high strength, high stiffness, abrasion resistance, dimensional stability, electrical and thermal conductivity, ability to operate in high temperatures, and toughness and are applied typically in aerospace and engine applications. Examples of MMC composites include Al-SiC (silicon carbide), Ti-SiC, Al-B, Al-C (carbon), Al-graphite, Mg-SiC, and Al-Al$_2$O$_3$ (aluminum oxide).

Ceramic Matrix Composites (CMC)

Silicon carbide, silicon nitride, alumina, zirconia, glass-ceramic, glass, carbon, graphite.

CMCs are preferred for high strength, high stiffness, high fracture toughness relative to ceramics alone, ability to operate at high temperatures, and low thermal expansion and are attractive for furnaces, engines, and aerospace applications. Common CMC composites include carbon-carbon (C-C), silicon carbide-carbon (SiC-C), silicon carbide-silicon carbide (SiC-SiC), glass ceramic-silicon carbide, silicon carbide-lithium aluminosilicate (SiC-LAS), and silicon carbide-calcium aluminosilicate (SiC-CAS). Where the same basic material is listed as both the matrix and the filler, the filler is of a different form such as whiskers, chopped fibers, or strands to achieve the preferred properties.

Forms of Filler Materials

Many forms of filler materials are used as listed here.

- Continuous fiber strand consisting of many individual filaments bound together
- Chopped strands in short lengths (0.75 to 50 mm or 0.03 to 2.00 in)
- Chopped longer strands randomly spread in the form of a mat
- Chopped longer strands aligned with the principal directions of the load path
- Roving: A group of parallel strands
- Woven fabric made from roving or strands
- Metal filaments or wires
- Solid or hollow microspheres
- Metal, glass, or mica flakes
- Single crystal whiskers of materials such as graphite, silicon carbide, and copper

Figure 8–24 shows several styles of woven and braided fabric formed into tubular shapes and sleeving adaptable to making structural tubing, driveshafts, turbine blades, bats, housings, components for prosthetic devices, surgical devices, and many other composite-based products. Both glass and carbon fibers can be employed. The products shown in Figures 8–20 to 8–22 are made with these braided sleeves.

Types of Filler Materials

Fillers, also called fibers, come in many types based on both organic and inorganic materials. Some of the more popular fillers appear in the following list.

- Glass fibers in five different types:

  A-glass: good chemical resistance because it contains alkalis such as sodium oxide

FIGURE 8–24 Fabrics and braided sleeves made from glass and carbon fibers used for composite materials. The products shown in Figures 8–20 to 8–22 are made with these filler materials.

Source: A&P Technology, Cincinnati, OH.

C-glass: special formulations for even higher chemical resistance than A-glass

E-glass: widely used glass with good electrical insulating ability and good strength

S-glass: high-strength, high-temperature glass

D-glass: better electrical properties than E-glass

■ Quartz fibers and high-silica glass: good properties at high temperatures up to 2000°F (1095°C)

■ Carbon fibers made from PAN-base carbon (PAN is polyacrylonitrile): approximately 95% carbon with very high modulus of elasticity

■ Graphite fibers: greater than 99% carbon and even higher modulus of elasticity than carbon; the stiffest fibers typically used in composites

■ Boron coated onto tungsten fibers: good strength and higher modulus of elasticity than glass

■ Silicon carbide coated onto tungsten fibers: strength and stiffness similar to boron/tungsten but with higher temperature capability

■ Aramid fibers: a member of the polyamide family of polymers; higher strength and stiffness with lower density as compared with glass; very flexible (aramid fibers produced by the DuPont company carry the name *Kevlar*)

Advantages of Composites

Designers typically seek to produce products that are safe, strong, stiff, lightweight, and highly tolerant of the environment in which the product will operate. Composites often excel in meeting these objectives when compared to alternative materials such as metals, wood, and unfilled plastics. Two parameters that are used to compare materials are *specific strength* and *specific modulus*, defined as

Specific strength is the ratio of the tensile strength of a material to its specific weight.

Specific modulus is the ratio of the modulus of elasticity of a material to its specific weight.

Because the modulus of elasticity is a measure of the stiffness of a material, the specific modulus is sometimes called *specific stiffness*.

Although obviously not a length, both of these quantities have the *unit* of length, derived from the ratio of the units for strength or modulus of elasticity and the units for specific weight. In the U.S. Customary System, the units for tensile strength and modulus of elasticity are lb/in^2 while specific weight (weight per unit volume) is in lb/in^3. Thus, the unit for specific strength or specific modulus is inches. In the SI metric system, strength and modulus are expressed in N/m^2 (Pascals) while specific weight is in N/m^3. Then the unit for specific strength or specific modulus is meters.

Table 8–13 gives comparisons of the specific strength and specific stiffness of selected composite materials with certain steel, aluminum, and titanium alloys. Figure 8–25 shows a comparison of these materials using bar charts. Figure 8–26 is a plot of these data with specific strength on the vertical axis and specific modulus on the horizontal axis. When weight is critical, the ideal material would lie in the upper right part of this chart. Note that data in these charts and figures are for composites having the filler materials aligned in the most favorable direction to withstand the applied loads.

TABLE 8–13 Comparison of Specific Strength and Specific Modulus for Selected Material

Material	Tensile strength, s_u (ksi)	Specific weight, γ (lb/in³)	Specific strength (in)	Specific modulus (in)
Steel ($E = 30 \times 10^6$ psi)				
AISI 1020 HR	55	0.283	0.194×10^6	1.06×10^8
AISI 5160 OQT 700	263	0.283	0.929×10^6	1.06×10^8
Aluminum ($E = 10.0 \times 10^6$ psi)				
6061-T6	45	0.098	0.459×10^6	1.02×10^8
7075-T6	83	0.101	0.822×10^6	0.99×10^8
Titanium ($E = 16.5 \times 10^6$ psi) Ti-6Al-4V quenched and aged at 1000°F	160	0.160	1.00×10^6	1.03×10^8
Glass/epoxy composite ($E = 4.0 \times 10^6$ psi) 34% fiber content	114	0.061	1.87×10^6	0.66×10^8
Aramid/epoxy composite ($E = 11.0 \times 10^6$ psi) 60% fiber content	200	0.050	4.0×10^6	2.20×10^8
Boron/epoxy composite ($E = 30.0 \times 10^6$ psi) 60% fiber content	270	0.075	3.60×10^6	4.00×10^8
Graphite/epoxy composite ($E = 19.7 \times 10^6$ psi) 62% fiber content	278	0.057	4.86×10^6	3.45×10^8
Graphite/epoxy composite ($E = 48 \times 10^6$ psi) Ultrahigh modulus	160	0.058	2.76×10^6	8.28×10^8

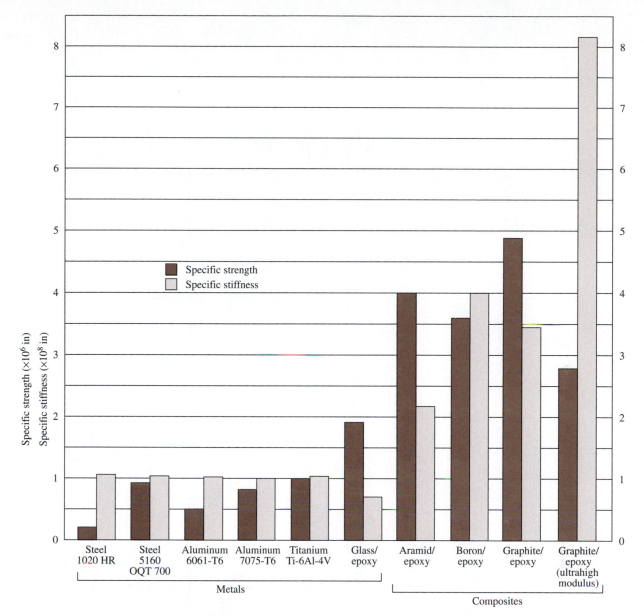

FIGURE 8–25 Comparison of specific strength and specific stiffness for selected materials.

Advantages of composites can be summarized as follows:

1. Specific strengths for composite materials can range as high as five times those of high strength steel alloys.

2. Specific modulus values for composite materials can be as high as eight times those for either steel, aluminum, or titanium alloys.

3. Composite materials typically perform better than steel or aluminum in applications where cyclic loads are encountered leading to the potential for fatigue failure.

4. Where impact loads and vibrations are expected, composites can be specially formulated with materials that provide high toughness and a high level of damping.

5. Some composites have much higher wear resistance than metals.

6. Careful selection of the matrix and filler materials can provide superior corrosion resistance.

7. Dimensional changes due to changes in temperature are typically much less for composites than for metals. More on this topic is include in Chapter 9 in which the property *coefficient of thermal expansion* is defined.

8. Because composite materials have properties that are highly directional, designers can tailor the placement of reinforcing fibers in directions that provide the required strength and stiffness under the specific loading conditions to be encountered.

9. Composite structures can often be made in complex shapes in one piece, thus reducing the number of parts in a product and the number of fastening operations required. The elimination of joints typically improves the reliability of such structures as well.

10. Composite structures are typically made in their final form directly or in a near-net shape, thus reducing the number of secondary operations required.

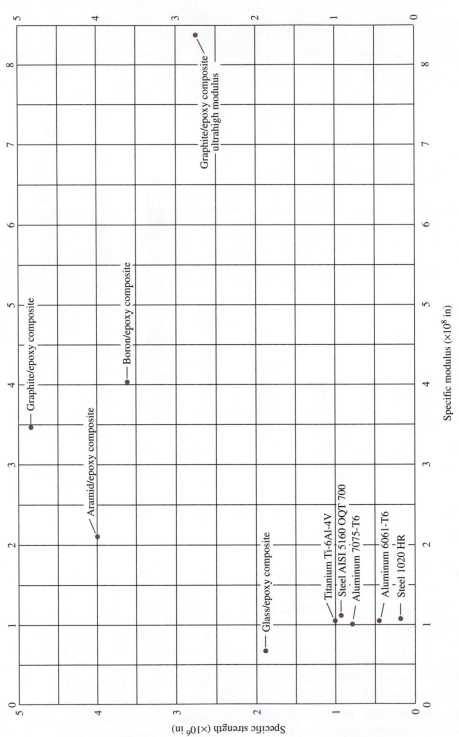

FIGURE 8–26 Specific strength versus specific modulus for selected metals and composites.

Limitations of Composites

Designers must balance many properties of materials in their designs while simultaneously considering manufacturing operations, costs, safety, life, and service of the product. The following list gives some of the major concerns when using composites.

1. Material costs for composites are typically higher than for many alternative materials.

2. Fabrication techniques are quite different from those used to shape metals. New manufacturing equipment may be required along with additional training for production operators.

3. The performance of products made from some composite production techniques is subject to a wider range of variability than for most metal fabrication techniques.

4. The operating temperature limit for composites having a polymeric matrix is typically 500°F (260°C). [But ceramic or metal matrix composites can be used at higher temperatures, up to 1500°C (2700°F), as found in engines.]

5. The properties of composite materials are not isotropic. This means that properties vary dramatically with the direction of the applied loads. Designers must account for these variations to ensure safety and satisfactory operation under all expected types of loading.

6. At this time, many designers lack an understanding of the behavior of composite materials and the details of predicting failure modes. While major advancements have been made in certain industries such as the aerospace and recreational equipment fields, there is a need for more general understanding about designing with composite materials.

7. The analysis of composite structures requires the detailed knowledge of more properties of the materials than would be required for metals.

8. Inspection and testing of composite structures is typically more complicated and less precise than for metal structures. Special nondestructive techniques may be required to ensure that there are no major voids in the final product that could seriously weaken the structure. Testing of the complete structure may be required rather than testing a sample of the material because of the interaction of different parts on each other and because of the directionality of the material properties.

9. Repair and maintenance of composite structures is a serious concern. Some of the initial production techniques require special environments of temperature and pressure that may be difficult to reproduce in the field when damage repair is required. Bonding of a repaired area to the parent structure may also be difficult.

Directional Nature of the Properties of Composites

In general, a given material can have three kinds of behaviors with regard to the relationships among its strength and stiffness properties and the directions of load applications. These behaviors are called *isotropic, anisotropic, and orthotropic.*

Isotropic behavior means that the elastic response of the material is the same regardless of the direction of the applied load. Homogeneous materials such as most wrought metals (steel, aluminum, copper, titanium, etc.) are typically analyzed as isotropic materials. Many of the stress and deformation analysis techniques used in this book are based on the assumption of isotropic behavior. Material properties data for isotropic materials typically include a single value for tensile strength, modulus of elasticity, Poisson's ratio, and other properties.

Anisotropic behavior means that the elastic response of the material is different in all directions. A material that is a random, highly nonuniform aggregate of constituents would be expected to exhibit anisotropic behavior. Thus, it would not be correct to use the basic analysis techniques developed in this book because they are based on the assumption of isotropic behavior. For components and structures made from such materials, testing under actual loading conditions is advised to verify suitability for a particular application.

Orthotropic behavior means that the material properties are different in three mutually perpendicular directions. This is the behavior that is most often assumed for composite materials, particularly those made from laminated construction as described next. To fully describe the behavior of an orthotropic material requires the determination of strength, modulus of elasticity (stiffness), and Poisson's ratio in each of the mutually perpendicular directions. Typically, these directions are called 1, 2, and 3. See Reference 27.

Figure 8–27 shows a segment of a composite material made with the reinforcing fibers aligned in a particular direction. In such cases, the 1-direction is aligned in that direction. Then the 2-direction is typically taken perpendicular to the direction of the fibers in the plane of the surface of the component. The 3-direction is then perpendicular to that plane. You should be able to see that the properties of the material in the three directions would be quite different from one another, based on the way in which the fibers contribute to the resistance of the applied loads.

Laminated Composite Construction

Many structures made from composite materials are made from several layers of the basic material containing both the matrix and the reinforcing fibers. The manner in which

FIGURE 8–27   Unidirectional composite showing direction of principal axes.

TABLE 8–14 Examples of the Effect of Laminate Construction on Strength and Stiffness

| | Tensile strength | | | | Modulus of elasticity | | | |
| | Longitudinal | | Transverse | | Longitudinal | | Transverse | |
Laminate type	ksi	MPa	ksi	MPa	10^6 psi	GPa	10^6 psi	GPa
Unidirectional	200	1380	5	34	21	145	1.6	11
Quasi-isotropic	80	552	80	552	8	55	8	55

the layers are oriented relative to one another affects the final properties of the completed structure.

As an illustration, consider that each layer is made from a set of parallel strands of the reinforcing filler material, such as E-glass fibers, embedded in the resin matrix, such as polyester. In this form, the material is sometimes called a *prepreg,* indicating that the filler has been preimpregnated with the matrix prior to forming the structure and curing the assembly.

To produce the maximum strength and stiffness in a particular direction, several layers or plies of the prepreg could be laid on top of one another with all of the fibers aligned in the direction of the expected tensile load. This is called a *unidirectional laminate.* After curing, the laminate would have a very high strength and stiffness when loaded in the direction of the strands, called the *longitudinal* direction. However, the resulting product would have a very low strength and stiffness in the direction perpendicular to the fiber direction, called the *transverse* direction. If any off-axis loads are encountered, the part may fail or deform significantly. Table 8–14 gives sample data for a unidirectional laminated carbon/epoxy composite.

To overcome the lack of off-axis strength and stiffness, laminated structures should be made with a variety of orientations of the layers. One popular arrangement is shown in Figure 8–28. Naming the longitudinal direction of the surface layer the *0° ply,* this structure is referred to as

$$0°, 90°, +45°, -45°, -45°, +45°, 90°, 0°$$

The symmetry and balance of this type of layering technique results in more nearly uniform properties in two directions. The term *quasi-isotropic* is sometimes used to describe such a structure. Note that the properties perpendicular to the faces of the layered structure (through the thickness) are still quite low because fibers do not extend in that direction. Also, the strength and stiffness in the primary directions are somewhat lower than if the plies were aligned in the same direction. Table 8–14 shows sample data for a quasi-isotropic laminate compared with one having unidirectional fibers in the same matrix.

Processing of Composites

One method that is frequently used to produce composite products is first to place layers of sheet-formed fabrics on a form having the desired shape and then to impregnate the fabric with wet resin. Each layer of fabric can be adjusted in its orientation to produce special properties of the finished article. After the lay-up and resin impregnation are completed, the entire system is subjected to heat and pressure while a curing agent reacts with the base resin to produce cross-linking that binds all of the elements into a three-dimensional, unified structure. The polymer binds to the

FIGURE 8–28 Multilayer laminated composite construction designed to produce quasi-isotropic properties.

fibers and holds them in their preferred position and orientation during use.

An alternative method of fabricating composite products starts with a process of preimpregnating the fibers with the resin material to produce strands, tape, braids, or sheets. The resulting form, called a *prepreg,* can then be stacked into layers or wound onto a form to produce the desired shape and thickness. The final step is the curing cycle as described for the wet process.

Polyester-based composites are often produced as *sheet-molding compounds (SMC)* in which preimpregnated fabric sheets are placed into a mold and shaped and cured simultaneously under heat and pressure. Large body panels for automotive applications can be produced in this manner.

Pultrusion is a process in which the fiber reinforcement is coated with resin as it is pulled through a heated die to produce a continuous form in the desired shape. This process is used to produce rod, tubing, structural shapes (I-beams, channels, angles, and so on), tees, and hat sections used as stiffeners in aircraft structures.

Filament winding is used to make pipe, pressure vessels, rocket motor cases, instrument enclosures, and odd-shaped containers. The continuous filament can be placed in a variety of patterns, including helical, axial, and circumferential, to produce desired strength and stiffness characteristics.

Predicting Composite Properties

The following discussion summarizes some of the important variables needed to define the properties of a composite. The subscript c refers to the composite, m refers to the matrix, and f refers to the fibers. The strength and stiffness of a composite material depend on the elastic properties of the fiber and matrix components. But another parameter is the relative volume of the composite composed of fibers, V_f, and that composed of the matrix material, V_m. That is,

V_f = Volume fraction of fiber in the composite

V_m = Volume fraction of matrix in the composite

Note that for a unit volume, $V_f + V_m = 1$. Then $V_m = 1 - V_f$.

We will use an ideal case to illustrate the way in which the strength and stiffness of a composite can be predicted. Consider a composite with unidirectional continuous fibers aligned in the direction of the applied load. The fibers are typically much stronger and stiffer than the matrix material. Furthermore, the matrix will be able to undergo a larger strain before fracture than the fibers can. Figure 8–29 shows these phenomena on a plot of stress versus strain for the fibers and the matrix. We will use the following notation for key parameters from Figure 8–29.

s_{uf} = Ultimate strength of fiber

ϵ_{uf} = Strain in the fiber corresponding to its ultimate strength

σ'_m = Stress in the matrix at the same strain as ϵ_{uf}

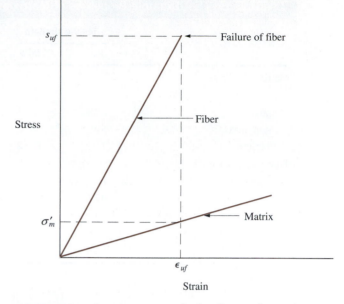

FIGURE 8–29 Stress versus strain for fiber and matrix materials.

The ultimate strength of the composite, s_{uc}, is at some intermediate value between s_{uf} and σ'_m, depending on the volume fraction of fiber and matrix in the composite. That is,

$$s_{uc} = s_{uf}V_f + \sigma'_m V_m \qquad (8\text{--}10)$$

At any lower level of stress, the relationship among the overall stress in the composite, the stress in the fibers, and the stress in the matrix follows a similar pattern.

$$\sigma_c = \sigma_f V_f + \sigma_m V_m \qquad (8\text{--}11)$$

Figure 8–30 illustrates this relationship on a stress–strain diagram.

FIGURE 8–30 Relationship among stresses and strains for a composite and its fiber and matrix materials.

TABLE 8–15   Example Properties of Matrix and Filler Materials

	Tensile strength		Tensile modulus		Specific weight	
	ksi	MPa	10^6 psi	GPa	lb/in^3	kN/m^3
Matrix materials:						
Polyester	10	69	0.40	2.76	0.047	12.7
Epoxy	18	124	0.56	3.86	0.047	12.7
Aluminum	45	310	10.0	69	0.100	27.1
Titanium	170	1170	16.5	114	0.160	43.4
Filler materials:						
S-glass	600	4140	12.5	86.2	0.090	24.4
Carbon–PAN	470	3240	33.5	231	0.064	17.4
Carbon–PAN (high-strength)	820	5650	40	276	0.065	17.7
Carbon (high-modulus)	325	2200	100	690	0.078	21.2
Aramid	500	3450	19.0	131	0.052	14.1

Both sides of Equation (8–6) can be divided by the strain at which these stresses occur. And, since for each material $\sigma/\epsilon = E$, the modulus of elasticity for the composite can be shown as

$$E_c = E_f V_f + E_m V_m \qquad (8\text{–}12)$$

The density of a composite can be computed in a similar fashion.

$$\rho_c = \rho_f V_f + \rho_m V_m \qquad (8\text{–}13)$$

Density is defined as *mass per unit volume*. A related property, specific weight, is defined as *weight per unit volume* and is denoted by the symbol γ (Greek letter gamma). The relationship between density and specific weight is simply $\gamma = \rho g$, where g is the acceleration due to gravity. Multiplying each term in Equation (8–8) by g gives the formula for the specific weight of a composite,

$$\gamma_c = \gamma_f V_f + \gamma_m V_m \qquad (8\text{–}14)$$

The form of Equations (8–12) to (8–14) is often called the *rule of mixtures*.

Table 8–15 lists example values for the properties of some matrix and filler materials. Remember that wide variations can occur in such properties, depending on the exact formulation and the condition of the materials.

Example Problem 8–1   Compute the expected properties of ultimate tensile strength, modulus of elasticity, and specific weight of a composite made from unidirectional strands of carbon-PAN fibers in an epoxy matrix. The volume fraction of fibers is 30%. Use data from Table 8–15.

Solution   Objective: Compute the expected values of s_{uc}, E_c, and γ_c for the composite.

Given: Matrix–epoxy: $s_{um} = 18$ ksi; $E_m = 0.56 \times 10^6$ psi; $\gamma_m = 0.047$ lb/in^3.

Fiber–carbon-PAN: $s_{uf} = 470$ ksi; $E_f = 33.5 \times 10^6$ psi; $\gamma_f = 0.064$ lb/in^3.

Volume fraction of fiber, $V_f = 0.30$. And $V_m = 1.0 - 0.30 = 0.70$.

Analysis and Results: Ultimate tensile strength, s_{uc}, computed from Equation (8–5),

$$s_{uc} = s_{uf} V_f + \sigma'_m V_m$$

To find σ'_m we first find the strain at which the fibers would fail at s_{uf}. Assume that the fibers are linearly elastic to failure. Then

$$\epsilon_f = s_{uf}/E_f = (470 \times 10^3 \text{ psi})/(33.5 \times 10^6 \text{ psi}) = 0.014$$

At this same strain, the stress in the matrix is

$$\sigma'_m = E_m \epsilon = (0.56 \times 10^6 \text{ psi})(0.014) = 7840 \text{ psi}$$

Then, in Equation (8–5),

$$s_{uc} = (470\,000 \text{ psi})(0.30) + (7840 \text{ psi})(0.70) = 146\,500 \text{ psi}$$

Modulus of elasticity, computed from Equation (8–7),

$$E_c = E_f V_f + E_m V_m = (33.5 \times 10^6)(0.30) + (0.56 \times 10^6)(0.70)$$
$$E_c = 10.4 \times 10^6 \, \text{psi}$$

Specific weight, computed from Equation (8–9),

$$\gamma_c = \gamma_f V_f + \gamma_m V_m = (0.064)(0.30) + (0.047)(0.70) = 0.052 \, \text{lb/in}^3$$

Summary of Results:

$$s_{uc} = 146\,500 \, \text{psi}$$
$$E_c = 10.4 \times 10^6 \, \text{psi}$$
$$\gamma_c = 0.052 \, \text{lb/in}^3$$

Comment: Note that the resulting properties for the composite are intermediate between those for the fibers and the matrix.

8–13 MATERIALS SELECTION

One of the most important tasks for a designer is the specification of the material from which any individual component of a product is to be made. The decision must consider a huge number of factors, many of which have been discussed in this chapter.

The process of material selection must commence with a clear understanding of the functions and design requirements for the product and the individual component. Then the designer should consider the interrelationships among the following:

- The functions of the component
- The component's shape
- The material from which the component is to be made
- The manufacturing process used to produce the component

Overall requirements for the performance of the component must be detailed. This includes, for example,

- The nature of the forces applied to the component
- The types and magnitudes of stresses created by the applied forces
- The allowable deformation of the component at critical points
- Interfaces with other components of the product
- The environment in which the component is to operate
- Physical size and weight of the component
- Aesthetics expected for the component and the overall product
- Cost targets for the product as a whole and this component in particular
- Anticipated manufacturing processes available

A much more detailed list may be made with more knowledge of specific conditions.

From the results of the exercises described previously, you should develop a list of key material properties that are important. Examples often include

1. Strength as indicated by ultimate tensile strength, yield strength, compressive strength, fatigue strength, shear strength, and others

2. Stiffness as indicated by the tensile modulus of elasticity, shear modulus of elasticity, or flexural modulus

3. Weight and mass as indicated by specific weight or density

4. Ductility as indicated by the percent elongation

5. Toughness as indicated by the impact energy (Izod, Charpy, etc.)

6. Creep performance data

7. Corrosion resistance and compatibility with the environment

8. Cost for the material

9. Cost to process the material

A list of candidate materials should then be created using your knowledge of the behavior of several material types, successful similar applications, and emerging materials technologies. A rational decision analysis should be applied to determine the most suitable types of materials from the list of candidates. This could take the form of a matrix in which data for the properties just listed for each candidate material are entered and ranked. An analysis of the complete set of data will aid in making the final decision.

More comprehensive materials selection processes are described in References 4–6, 9, 16, and 29.

Figure 8–31 is a broad overview of the types of materials from which a designer may choose for a particular application. Basically it shows five primary classes: metals, polymers, ceramics, glasses, and elastomers, along with hybrids that combine two or more materials to achieve specialized properties. Composites, discussed in Section 8–12, are an obvious and increasingly popular example of hybrid materials. Also included as hybrids are fabricated structures such as sandwich

FIGURE 8–31   Classifications of materials.

Source: Granta Design, Cambridge, United Kingdom.

panels employing very light but relatively stiff inner cores between strong outer skins that carry much of the load. See Figure 8–32. The cores may be foams, honeycombs, corrugated sheets, or other such materials. The resulting structure is an example of selecting materials to optimize the performance of a given component.

However, the sheer number of different materials from which to choose makes material selection a daunting task. Specialized approaches, as described in Reference 4, offer

significant guidance in the selection process. Furthermore, computer software is available to permit rapid searching on many parameters to produce lists of candidate materials with quantitative data about their performance, cost, producibility, or other important criteria. Two relatively simple examples of selection processes help to illustrate the method. See Internet site 31.

Some components of a structure or a machine may be limited by strength where avoiding failure is primary. In addition, it may be desirable to minimize the weight of the component as in aerospace applications or where a building designer wants a low total weight of a structure for cost savings and to decrease loads on a foundation. In this case, arranging material property data in the form shown in Figure 8–33 would help to visualize possible choices. Here the vertical axis is the yield strength of the materials and the horizontal axis is the density. Desiring a high strength and low weight, the designer would look toward the upper left of the chart. The shaded areas show the range of strength and density properties for the different classes of materials.

An alternate approach may be applied where stiffness of a component is primary, as in designing a floor for a passenger aircraft or an apartment building. Occupants want a stiff, rigid floor that does not flex noticeably. Again, the designer may want to minimize weight. Figure 8–34 shows a chart of modulus of elasticity, *E*, versus density for a variety of materials. Here, again, the designer would look to the upper left for desirable materials. More detailed charts show more specific examples within these areas, such as specific steel alloys within the larger "Metals" region.

Numerous other parametric searches can be constructed using the selection software.

(a) Curved panel with foam core and composite skins

(b) Flat panel with honeycomb core and composite skins

FIGURE 8–32   Laminated panels with lightweight cores.

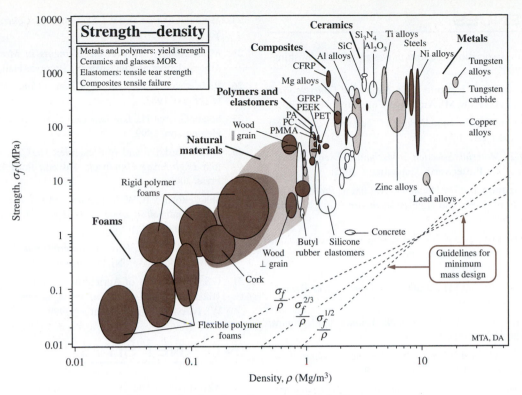

FIGURE 8–33 Strength versus density chart for materials selection.

Source: Granta Design, Cambridge, United Kingdom.

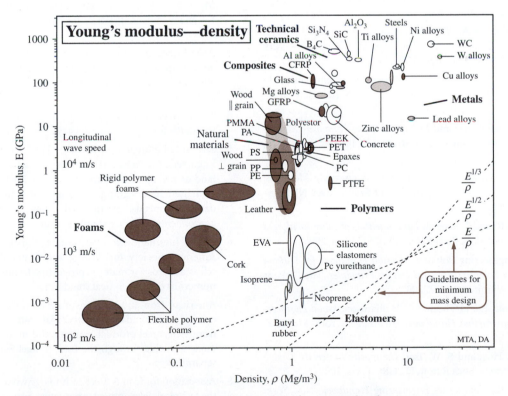

FIGURE 8–34 Young's modulus versus density chart for materials selection.

Source: Granta Design, Cambridge, United Kingdom.

REFERENCES

1. Aluminum Association, *Aluminum Design Manual*, Washington, DC, 2005.

2. American Concrete Institute, *2006 Manual of Concrete Practice*, Farmington Hills, MI, American Concrete Institute, 2006.

3. ASTM International, *Annual Book of Standards*, 2006, West Conshohocken, PA, 2006.

4. Ashby, M. F., *Materials Selection in Mechanical Design*, 3rd ed., Burlington, MA, Butterworth-Heinemann, 2005.

5. ASM International, *ASM Handbook, Volume 1: Properties and Selection: Irons, Steels, and High-Performance Alloys*, Materials Park, OH, 1990.

6. ASM International, *ASM Handbook, Volume 2: Properties and Selection: Nonferrous Alloys and Special-Purpose Materials*, Materials Park, OH, 1990.

7. ASM International, *ASM Handbook, Volume 4: Heat Treating*, Materials Park, OH, 1991.

8. ASM International, *ASM Handbook, Volume 8: Mechanical Testing*, Materials Park, OH, 2000.

9. ASM International, *ASM Handbook, Volume 20: Materials Selection and Design*, Materials Park, OH, 1997.

10. ASM International, *ASM Handbook, Volume 21: Composites*, Materials Park, OH, 2001.

11. ASM International, *Engineering Plastics, Engineered Materials Handbook, Volume 2*, Materials Park, OH, 1988.

12. ASM International, *Metals Handbook Desk Edition*, 2nd ed., Materials Park, OH, 1998.

13. Avallone, Eugene A., Theodore Baumeister, and Ali Sadegh eds., *Marks' Standard Handbook for Mechanical Engineers*, 11th ed., New York, McGraw-Hill, 2007.

14. Berins, Michael L., ed., *Plastics Engineering Handbook of the Society of the Plastics Industry, Inc.*, 5th ed., New York, Chapman and Hall, 1995.

15. Bertholet, Jean-Marie and Michael J. Cole, *Composite Materials: Mechanical Behavior and Structural Analysis*, New York, Springer Verlag, 1999.

16. Budinski, K. G. and M. K. Budinski, *Engineering Materials: Properties and Selection*, 8th ed., Upper Saddle River, NJ, Prentice Hall, 2005.

17. Bunsell, A. R. and J. Renard, *Fundamentals of Fibre Reinforced Composite Materials*, Boca Raton, FL, CRC Press, 2005.

18. DuPont Engineering Polymers, *Design Handbook for DuPont Engineering Polymers: General Design Principles*, The DuPont Company, Wilmington, DE, 2002.

19. Faherty, Keith F. and Thomas G. Williamson, *Wood Engineering and Construction Handbook*, 3rd ed., New York, McGraw-Hill, 1999.

20. Gay, D., D. Hoa, and S. W. Tsai, *Composite Materials: Design and Applications*, Boca Raton, FL, CRC Press, 2002.

21. Mallick, P. K., *Composites Engineering Handbook*, New York, Marcel Dekker, 1997.

22. The Masonry *Society, Masonry Designer's Guide*, 4th ed., Boulder, CO, The Masonry Society, 2003.

23. Mazumdar, S., *Composites Manufacturing: Materials, Product, and Process Engineering*, Boca Raton, FL, CRC Press, 2001.

24. Mehta, P. Kumar and P. J. M. Monteiro, *Concrete*, 3rd ed., New York, McGraw-Hill, 2006.

25. Mott, Robert L., *Machine Elements in Mechanical Design*, 4th ed., Upper Saddle River, NJ, Prentice-Hall, 2004.

26. Richardson, T., *Composites: A Design Guide*, New York, Industrial Press, 1987.

27. Staab, George H., *Laminar Composites*, Boston, Butterworth-Heinemann, 1999.

28. Shackelford, J. F. and W. Alexander, *The CRC Materials Science and Engineering Handbook*, 3rd ed., Boca Raton, FL, CRC Press, 2000.

29. Shackelford, J. F., W. Alexander, and J. S. Park, *The CRC Practical Handbook of Materials Selection*, Boca Raton, FL, CRC Press, 1995.

30. Strong, A. Brent., *Plastics: Materials and Processing*, 3rd ed., Upper Saddle River, NJ, Prentice-Hall, 2006.

31. U.S. Department of Agriculture Forest Products Laboratory, *Wood Handbook—Woods as an Engineering Material*, Madison, WI, Forest Products Laboratory, 1999.

INTERNET SITES

1. AZoM.com (The A to Z of Materials) www.azom.com Materials information resource for the design community. No cost, searchable databases for metals, ceramics, polymers, and composites. Can also search by keyword, application, or industry type.

2. Matweb www.matweb.com Database of material properties for many metals, plastics, ceramics, and other engineering materials.

3. ASM International www.asm-intl.org The society for materials engineers and scientists; a worldwide network dedicated to advancing industry, technology, and applications of metals and other materials.

4. TECHstreet www.techstreet.com A store for purchasing standards for the metals industry.

5. SAE International www.sae.org The Society of Automotive Engineers, the engineering society for advancing mobility on land or sea, in air or space. A resource for technical information used in designing self-propelled vehicles. Offers standards on metals, plastic, and other materials along with components and subsystems of vehicles.

6. ASTM International www.astm.org Formerly known as the American Society for Testing and Materials. Develops and sells standards for material properties, testing procedures, and numerous other technical standards.

7. American Iron and Steel Institute www.steel.org AISI develops industry standards for steel materials and products made from steel. Steel product manuals and industry standards are made available through the Iron & Steel Society (ISS), listed separately.

8. Association for Iron & Steel Technology www.aist.org Advances the technical development, production, processing and application of iron and steel. Provides publications, conferences, and technical meetings for the iron and steel community.

9. Aluminum Association www.aluminum.org The association of the aluminum industry. Provides numerous publications that can be purchased.

10. Alcoa, Inc. www.alcoa.com A producer of aluminum and fabricated products. Web site can be searched for properties of specific alloys.

11. Copper Development Association www.copper.org Provides a large searchable database of properties of wrought and cast copper, copper alloys, brasses, and bronzes. Allows searching for appropriate alloys for typical industrial uses based on several performance characteristics.

12. Metal Powder Industries Federation www.mpif.org The international trade association representing the powder metal producers. Standards and publications related to the design and production of products using powder metals.

13. INTERZINC www.interzinc.com A market development and technology transfer group dedicated to increasing awareness of zinc casting alloys. Provides design assistance, alloy selection guide, alloy properties, and descriptions of casting alloys.

14. RAPRA Technology Limited www.rapra.net Comprehensive information source for the plastics and rubber industries. Formerly Rubber and Plastics Research Association. This site also hosts the Cambridge Engineering Selector, a computerized resource using the materials selection methodology of M. F. Ashby. See Reference 4.

15. DuPont Plastics www.plastics.dupont.com Information and data on DuPont plastics and their properties. Searchable database by type of plastic or application.

16. PolymerPlace.com www.polymerplace.com Information resource for the polymer industry.

17. PlasticsUSA.com www.plasticsusa.com An Internet portal providing technical information for the polymer industry and end-user industries including automotive, construction, electrical, medical, packaging, and polyurethane sectors. Includes a plastics materials database, information on trade names, and links to suppliers of plastics materials and processing equipment.

18. Plastics Technology Online www.ptonline.com An online source of information about the plastics industry and processes such as injection molding, extrusion, blow molding, thermoforming, foam, materials, tooling, and auxiliary equipment.

19. Society of Plastics Engineers www.4spe.org SPE promotes scientific and engineering knowledge and education about plastics and polymers worldwide.

20. U.S. Department of Agriculture, Forest Products Laboratory www.fpl.fs.fed.us Research organization devoted to producing information useful to the forest products industry and users of wood products.

21. American Concrete Institute www.aci-int.org Publisher of technical documents related to advancing concrete knowledge, including the 2006 Manual of Concrete Practice.

22. The Masonry Society www.masonrysociety.org A professional association dedicated to the advancement of knowledge on masonry and publisher of Masonry Designer's Guide.

23. MTS Systems Corporation www.mts.com A supplier of materials testing products including universal testing machines, dynamic and fatigue testing machines for metals, plastics, biomaterials, wood, fibers, and others.

24. Instron Corporation www.instron.com A supplier of materials testing products including universal testing machines, dynamic and fatigue testing machines, thermal shock testers, hardness (Wilson and Shore brands), impact testers, crack propagation and fracture toughness testers.

25. Tinius Olsen, Inc. www.tiniusolsen.com A supplier of materials testing products including static universal testing machines, impact testers, torsion testers, and special equipment for the plastics and food industries. The site has a useful Resource Center tab providing quick reference information.

26. Zwick/Roell Group www.zwickroell.com A supplier of materials testing products including universal testing machines, dynamic and fatigue testing machines, impact testers, and specialized automated systems. The Indentec Hardness Testing Machines, Ltd., unit provides a wide range of hardness testers.

27. National Institute of Standards and Technology (NIST)—Special Publication 960–5, *Rockwell Hardness Measurement of Metallic Materials,* by Samuel R. Low. May be downloaded from: http://www.metallurgy.nist.gov/reports/recommendedpractice/SP9605.pdf

28. Special Metals Corporation www.specialmetals.com Producer of nickel-based alloys with brand names INCONEL, INCOLOY, NIMONIC, UDIMET, MONEL, and NILO.

29. Allegheny Ludlum Corporation www.alleghenyludlum.com Producer of nickel-based alloys with brand names AL, ALL-CORR, and ALTEMP.

30. Haynes International, Inc. www.haynesintl.com Producer of nickel- and cobalt-based alloys with brand names HASTELLOY, HAYNES, and ULTIMET.

31. Granta Material Intelligence www.grantadesign.com Producer and marketer of the EduPack materials selection software using concepts developed by Professor Michael F. Ashby.

PROBLEMS

8–1. Name four kinds of metals commonly used for load-carrying members.

8–2. Name 11 factors that should be considered when selecting a material for a product.

8–3. Define *ultimate tensile strength*.

8–4. Define *yield point*.

8–5. Define *yield strength*.

8–6. When is yield strength used in place of yield point?

8–7. Define *stiffness*.

8–8. What material property is a measure of its stiffness?

8–9. State Hooke's law.

8–10. What material property is a measure of its ductility?

8–11. How is a material classified as to whether it is ductile or brittle?

8–12. Name four types of steels.

8–13. What does the designation AISI 4130 for a steel mean?

8–14. What are the ultimate strength, yield strength, and percent elongation of AISI 1040 hot-rolled steel? Is it a ductile or a brittle material?

8–15. Which has a greater ductility: AISI 1040 hot-rolled steel or AISI 1020 hot-rolled steel?

8–16. What does the designation AISI 1141 OQT 700 mean?

8–17. If the required yield strength of a steel is 150 ksi, could AISI 1141 be used? Why?

8–18. What is the modulus of elasticity for AISI 1141 steel? For AISI 5160 steel?

8–19. A rectangular bar of steel is 1.0 in by 4.0 in by 14.5 in. How much does it weigh in pounds?

8–20. A circular bar is 50 mm in diameter and 250 mm long. How much does it weigh in newtons?

8–21. If a force of 400 N is applied to a bar of titanium and an identical bar of magnesium, which would stretch more?

8–22. Name four types of structural steels and list the yield point for each.

8–23. What does the aluminum alloy designation 6061-T6 mean?

8–24. List the ultimate strength, yield strength, modulus of elasticity, and density for 6061-O, 6061-T4, and 6061-T6 aluminum.

8–25. List five uses for bronze.

8–26. List three desirable characteristics of titanium as compared with aluminum or steel.

8–27. Name five varieties of cast iron.

8–28. Which type of cast iron is usually considered to be brittle?

8–29. What are the ultimate strengths in tension and in compression for ASTM A48 grade 40 cast iron?

8–30. How does a ductile iron differ from gray iron?

8–31. List the allowable stresses in bending, tension, compression, and shear for No. 2 grade Douglas fir.

8–32. What is the normal range of compressive strengths for concrete?

8–33. Describe the difference between thermoplastic and thermosetting materials.

8–34. Name three suitable plastics for use as gears or cams in mechanical devices.

8–35. Describe the term *composite*.

8–36. Name five basic types of materials that are used as a matrix for composites.

8–37. Name five different thermoplastics used as a matrix for composites.

8–38. Name three different thermosetting plastics used as a matrix for composites.

8–39. Name three metals used as a matrix for composites.

8–40. Describe nine forms that filler materials take when used in composites.

8–41. Discuss the differences among *strands, roving,* and *fabric* as different forms of fillers for composites.

8–42. Name seven types of filler materials used for composites.

8–43. Name five different types of glass fillers used for composites and describe the primary features of each.

8–44. Which of the commonly used filler materials has the highest stiffness?

8–45. Which filler materials should be considered for high temperature applications?

8–46. What is a common brand name for aramid fibers?

8–47. Define *specific strength* of a composite.

8–48. Define *specific modulus* of a composite.

8–49. List ten advantages of composites when compared to metals.

8–50. List nine limitations of composites.

8–51. From the data for selected materials in Table 8–13, list the ten materials in order of specific strength from highest to lowest. For each, compute the ratio of its specific strength to that for AISI 1020 HR steel.

8–52. From the data for selected materials in Table 8–13, list the ten materials in order of specific modulus from highest to lowest. For each, compute the ratio of its specific modulus to that for AISI 1020 HR steel.

8–53. Describe a unidirectional laminate and its general strength and stiffness characteristics.

8–54. Describe a quasi-isotropic laminate and its general strength and stiffness characteristics.

8–55. Compare the generally expected specific strength and stiffness characteristics of a quasi-isotropic laminate with a unidirectional laminate.

8–56. Describe a laminated composite that carries the designation 0°, +45°, −45°, −45°, +45°, 0°.

8–57. Describe a laminated composite that carries the designation 0°, +30°, +45°, +45°, +30°, 0°.

8–58. Define the term *volume fraction of fiber* for a composite.

8–59. Define the term *volume fraction of matrix* for a composite.

8–60. If a composite has a volume fraction of fiber of 0.60, what is the volume fraction of matrix?

8–61. Write the equation for the expected ultimate strength of a composite in terms of the properties of its matrix and filler materials.

8–62. Write the equations for the rule of mixtures as applied to a unidirectional composite for the stress in the composite, its modulus of elasticity, its density, and its specific weight.

8–63. Compute the expected properties of ultimate strength, modulus of elasticity, and specific weight of a composite made from unidirectional strands of high-strength carbon-PAN fibers in an epoxy matrix. The volume fraction of fibers is 50%. Compute the specific strength and specific stiffness. Use data from Table 8–15.

8–64. Repeat Problem 8–63 with high modulus carbon fibers.

8–65. Repeat Problem 8–63 with aramid fibers.

Problems 8–66 to 8–77

Use Figure P8–66 for all problems. For the given data in the problem statement and data you read from the indicated stress–strain curve, determine the following properties of the material:

a. Yield strength. State whether the yield point is used to determine this value or if the 0.2% offset method is used.

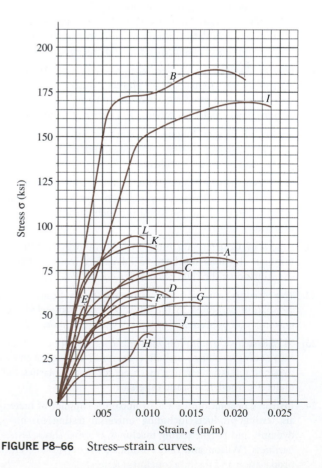

FIGURE P8–66   Stress–strain curves.

b. Ultimate tensile strength.
c. Proportional limit.
d. Elastic limit.
e. Modulus of elasticity for the range of stress in which Hooke's law applies.
f. Percent elongation. (The gage length for each test is 2.00 in.)
g. State whether the material is ductile or brittle.
h. Examine the results and judge the kind of metal used to determine the test data.
i. Compare your results and find a particular alloy from Appendix tables A–14 to A–18 that has similar properties.

8–66. Use curve A from Figure P8–66. Final length between gage marks = 2.22 in.

8–67. Use curve B from Figure P8–66. Final length between gage marks = 2.30 in.

8–68. Use curve C from Figure P8–66. Final length between gage marks = 2.30 in.

8–69. Use curve D from Figure P8–66. Final length between gage marks = 2.72 in.

8–70. Use curve E from Figure P8–66. Final length between gage marks = 2.01 in.

8–71. Use curve F from Figure P8–66. Final length between gage marks = 2.42 in.

8–72. Use curve G from Figure P8–66. Final length between gage marks = 2.10 in.

8–73. Use curve H from Figure P8–66. Final length between gage marks = 2.10 in.

8–74. Use curve I from Figure P8–66. Final length between gage marks = 2.16 in.

8–75. Use curve J from Figure P8–66. Final length between gage marks = 2.34 in.

8–76. Use curve K from Figure P8–66. Final length between gage marks = 2.30 in.

8–77. Use curve L from Figure P8–66. Final length between gage marks = 2.04 in.

CHAPTER NINE

DIRECT STRESS, DEFORMATION, AND DESIGN

MAJOR TOPICS

9–1 The Big Picture—Direct Stresses, Deformation, and Design

9–2 Design of Members Under Direct Tension or Compression

9–3 Design Normal Stresses

9–4 Design Factor

9–5 Design Approaches and Guidelines for Design Factors

9–6 Methods of Computing Design Stress

9–7 Elastic Deformation in Tension and Compression Members

9–8 Deformation Due to Temperature Changes

9–9 Thermal Stress

9–10 Members Made of More Than One Material

9–11 Stress Concentration Factors for Direct Axial Stresses

9–12 Bearing Stress

9–13 Design Bearing Stress

9–14 Design Shear Stress

9–1 THE BIG PICTURE—DIRECT STRESSES, DEFORMATION, AND DESIGN

Discussion Map

- You will now build on your ability to compute stresses that you learned in Chapter 7 and develop skills in the *design of load-carrying members.*

- In design, you either specify or determine by calculation a suitable material from which to make the member, its shape, and the dimensions required to carry a given load safely.

- You will consider the design of members under direct stresses: axial tensile stress, axial compressive stress, bearing stress, and direct shear stress.

- You will learn how to select a reasonable *design factor, N,* and to apply it to the appropriate material properties to ensure that the member experiences a safe level of stress during its use.

- Different kinds of loading will be discussed: static loads, repeated loads, impact, and shock.

- You will learn to consider stress concentrations for axially loaded members in which abrupt changes in cross section occur.

- In addition, you will learn how to compute the deformation of axially loaded members due to both stress and thermal expansion.

Discover

Reflect back on the Big Picture discussion you had for Chapter 7. There you identified components of consumer products, structures, and machines that you are familiar with.

You thought about the structure of a home, furniture, appliances, bicycles, automobiles, construction equipment, commercial buildings, aircraft, and space vehicles. What else did you think of?

Now focus on the kinds of loads some of those examples are subjected to.

Which are subjected to loads that do not vary significantly with time, called static loads? An example might be a beam in the basement of your home that holds the house structure above. The deadweight of the structure does not vary over time. What other examples can you find that are subjected to static loads?

Which are subjected to repeated loads? This is the case when a load is applied and removed many times during the expected life of the component. Some components may also experience reversing loads that are alternately stressed in tension and then in compression. Consider parts of your car. For example, the functional parts of the door latch experience high loads and stresses each time the latch is engaged. The loads are removed when the latch is disengaged. That cycle of loading and unloading is repeated many thousands of times during the life of the car. What other examples can you identify?

Which items on your list are subjected to shock and impact? Here the load is applied suddenly and sharply. An ideal example is a nail being struck by a hammer or a portable music player being dropped on the floor. Consider how a baseball bat or tennis racket responds when you hit a long line drive or a strong baseline shot. Can you think of more examples?

Can you see how the three types of loads just described require different criteria for designing safe structures and components? In this chapter you will learn how to specify suitable materials to carry such loads. You will also learn how to compute a safe design stress and how to determine the required shape and dimensions of load-carrying parts so they will not experience stresses above those levels.

In Chapter 7 the concept of direct stress was presented together with examples of the calculation of direct tensile stress, direct compressive stress, and direct shear stress. The emphasis was on the understanding of the basic phenomena, units, terminology, and the magnitude of stresses encountered in typical structural and mechanical applications. Nothing was said about the acceptability of the stress levels that were computed or about the design of members to carry a given load.

In this chapter the primary emphasis is on *design* in which you, as the designer, must make decisions about whether or not a proposed design is satisfactory, what the shape and size of the cross section of a load-carrying member should be, and what material the member should be made from.

ACTIVITY Chapter 3—Direct Stress, Deformation, and Design

Set up a system for holding a fine metal wire firmly by one end while providing a means of applying a direct axial tension force to the other. Figure 9–1 shows a commercial test device that accomplishes these functions. A universal tensile testing machine as shown in Figure 8–1 could be used if available. Perform the following:

1. Measure the diameter of the wire and the initial, unloaded length between the point of support and the place where the load is applied.
2. Apply loads in small increments, measuring the amount that the wire elongates for each load.
3. Compute the stress in the wire at each total load applied.
4. Divide each elongation measurement by the initial length of the wire to determine the strain in the wire.
5. Plot a graph of the stress in the wire on the vertical axis versus the elongation, showing also the strain on the horizontal axis.
6. After several points have been plotted (and before the wire breaks!), determine the slope of the line on the

FIGURE 9–1   Testing device for tensile loading of wire.

Source: P. A. Hilton Ltd/Hi-Tech, Hampshire, England.

graph by dividing the change in stress by the change in strain over some convenient part of the graph where the plot is best approximated as a straight line.

7. The slope computed is a measure of the tensile modulus of elasticity, E, for the metal.
8. Compare the computed value of E with that reported in tables if you know what the wire material is. If not, try to identify the type of metal by comparing it with values listed for some of the metals in Appendixes 14–18.
9. Observe that the straight-line relationship on the graph can be stated mathematically as
   a. $E = \Delta\sigma/\Delta\epsilon$ = Change in stress/Change in strain
   b. $\sigma = E\epsilon$ or Stress = Modulus of elasticity $\times$ Strain
10. Now continue to add more load until the wire breaks.
11. Using the maximum load that the wire held before breaking, compute the maximum stress in the wire.
12. Of course, we do not normally want to break the wire. Therefore, assuming that we want to limit the load to a safe value no more than one-half of the stress at which it broke, compute that stress.
13. Summarize the results by reporting the following:
    a. Modulus of elasticity of the metal from which the wire is made.
    b. Maximum stress at which the wire broke.

c. Stress at one-half of the maximum, calling this the *allowable stress* or the *design stress*.

d. Total elongation at the load for which the design stress is produced in the wire.

This simple test illustrates many of the main concepts you use in this chapter. In addition to stress and deformation caused by direct axial loads, you will study related topics dealing with bearing stress, contact stress, and direct shearing stresses.

Preferred Sizes and Screw Threads

One responsibility of a designer is to specify the final dimensions for load-carrying members. After completing the analyses for stress and deformation (strain), minimum acceptable values for dimensions are known that will ensure that the member will meet performance requirements. The designer then typically specifies the final dimensions to be standard or convenient values that will facilitate the purchase of materials and the manufacture of the parts. This section presents some guides to aid in these decisions.

Preferred Basic Sizes

When the component being designed will be made to the designer's specifications, it is recommended that final dimensions be specified from a set of *preferred basic sizes*. Appendix A–2 lists such data for fractional inch dimensions, decimal inch dimensions, and metric dimensions.

American Standard Screw Threads

Threaded fasteners and machine elements having threaded connections are manufactured according to standard dimensions to ensure interchangeability of parts and to permit convenient manufacture with standard machines and tooling. Appendix A–3 gives the dimensions of American Standard Unified threads. Sizes smaller than $\frac{1}{4}$ in are given numbers from 0 to 12, while fractional-inch sizes are specified for $\frac{1}{4}$ in and larger sizes. Two series are listed: UNC is the designation for coarse threads and UNF designates fine threads. Standard designations are illustrated as follows with a few examples.

6–32 UNC	(number size 6, 32 threads per inch, coarse thread)
12–28 UNF	(number size 12, 28 threads per inch, fine thread)
$\frac{1}{2}$–13 UNC	(fractional size $\frac{1}{2}$ in, 13 threads per inch, coarse thread)
$1\frac{1}{2}$–12 UNF	(fractional size $1\frac{1}{2}$ in, 12 threads per inch, fine thread)

Given in the tables are the basic major diameter (D), number of threads per inch (n), and the tensile stress area found from

$$A_t = 0.7854\left(D - \frac{0.9743}{n}\right)^2$$

When a threaded member is subjected to direct tension, the tensile stress area is used to compute the average tensile stress. It corresponds to the smallest area that would be produced by a transverse cut across the threaded rod. For convenience, some standards use the root area or gross area and adjust the value of the allowable stress.

Metric Screw Threads

Appendix A–3 gives similar dimensions for metric threads. Standard metric thread designations are of the form

$$M10 \times 1.5$$

where   M stands for metric
10 is the basic major diameter in mm
1.5 is the pitch between adjacent threads in mm

Thus, the designation shown would denote a metric thread with a basic major diameter of 10.0 mm and a pitch of 1.5 mm. Note that pitch = $1/n$.

Objectives of This Chapter

After completing this chapter, you should be able to:

1. Describe the conditions that must be met for satisfactory application of the direct stress formulas.

2. Define *design stress* and tell how to determine an acceptable value for it.

3. Define *design factor* and select appropriate values for it depending on the conditions present in a particular design.

4. Discuss the relationship among the terms *design stress*, *allowable stress*, and *working stress*.

5. Discuss the relationship among the terms *design factor*, *factor of safety*, and *margin of safety*.

6. Describe 11 factors that affect the specification of the design factor.

7. Describe various types of loads experienced by structures or machine members, including static load, repeated load, impact, and shock.

8. Design members subjected to direct tensile stress, direct compressive stress, direct shear stress, and bearing stress.

9. Evaluate deformations caused by axial stress and thermal expansion and include them in design and analysis.

10. Determine when stress concentrations exist and specify suitable values for stress concentration factors.

11. Use stress concentration factors in design.

9–2 DESIGN OF MEMBERS UNDER DIRECT TENSION OR COMPRESSION

In Chapter 7 the direct stress formula was developed and stated as follows:

$$\sigma = \frac{F}{A} \tag{9–1}$$

where σ = direct normal stress: tension or compression
         F = direct axial load
         A = cross-sectional area of member subjected to F

For Equation (9–1) to be valid, the following conditions must be met:

1. The loaded member must be straight.
2. The loaded member must have a uniform cross section over the length under consideration.
3. The material from which the member is made must be homogeneous.
4. The load must be applied along the centroidal axis of the member so there is no tendency to bend it.
5. Compression members must be short so that there is no tendency to buckle. (See Chapter 16 for the special analysis required for long, slender members under compressive stress and for the method to decide when a member is to be considered long or short.)

It is important to recognize that the concept of stress refers to the internal resistance provided by a *unit area*, that is, an infinitely small area. Stress is considered to act at a point and may, in general, vary from point to point in a particular body. Equation (9–1) indicates that for a member subjected to direct axial tension or compression, the stress is uniform across the entire area if the five conditions are met. In many practical applications the minor variations that could occur in the local stress levels are accounted for by carefully selecting the allowable stress, as discussed later.

9–3 DESIGN NORMAL STRESSES

Failure occurs in a load-carrying member when it breaks or deforms excessively, rendering it unacceptable for the intended purpose. Therefore, it is essential that the level of applied stress never exceed the ultimate tensile strength or the yield strength of the material. Consideration of excessive deformation without yielding is discussed later in this chapter.

> *Design stress is that level of stress which may be developed in a material while ensuring that the loaded member is safe.*

To compute design stress, two factors must be specified: the *design factor N* and the *property of the material on which the design will be based.* Usually, for metals, the design stress is based on either the yield strength s_y or the ultimate strength s_u of the material.

> *The design factor N is a number by which the reported strength of a material is divided to obtain the design stress σ_d.*

A variety of symbols are used in different fields for the strength properties of materials. In this book we use the following:

s_y = yield strength of a material

s_u = ultimate tensile strength of a material, or simply tensile strength

The term *yield strength* will be used regardless of whether the value was obtained from observing the yield point or using the offset technique as described in Chapter 8.

Other references may use the symbols σ_y and σ_u for these values. In the building construction field, particularly for steel building frames and members governed by the American Institute of Steel Construction (AISC), the symbols F_y and F_u are used.

The following equations can be used to compute the design stress for a certain value of *N:*

Design Stress ▼

$$\sigma_d = \frac{s_y}{N} \quad \text{based on yield strength} \qquad \text{(9–2)}$$

or

$$\sigma_d = \frac{s_u}{N} \quad \text{based on ultimate strength} \qquad \text{(9–3)}$$

The value of the design factor is normally determined by the designer, using judgment and experience. In some cases, codes, standards, or company policy may specify design factors or design stresses to be used. When the designer must determine the design factor, his or her judgment must be based on an understanding of how parts may fail and the factors that affect the design factor. Sections 9–4, 9–5, and 9–6 give additional information about the design factor and about the choice of methods for computing design stresses.

Other references may use the term *factor of safety* in place of *design factor.* Also, *allowable stress* or *working stress* may be used in place of *design stress.* The choice of terms used in this book emphasizes the role of the designer in specifying the design stress.

Theoretically, a material could be subjected to a stress up to s_y before yield would occur. This condition corresponds to a value of the design factor of $N = 1$ in Equation (9–2). Similarly, with a design factor of $N = 1$ in Equation (9–3), the material would be on the brink of ultimate fracture. Thus, $N = 1$ is the lowest value we can consider.

In this book we use the concept of design stresses and design factors as opposed to the margin of safety.

9–4 DESIGN FACTOR

Many different aspects of the design problem are involved in the specification of the design factor. In some cases the precise conditions of service are not known. The designer must then make conservative estimates of the conditions, that is, estimates that would cause the resulting design to be on the safe side when all possible variations are considered. The final choice of a design factor depends on the following 12 conditions.

Codes and Standards

If the member being designed falls under the jurisdiction of an existing code or standard, obviously the design factor or design stress must be chosen to satisfy the code or standard. Examples of standard-setting bodies are:

American Institute of Steel Construction (AISC): buildings, bridges, and similar structures using steel

Aluminum Association (AA): buildings, bridges, and similar structures using aluminum

American Society of Mechanical Engineers (ASME): boilers, pressure vessels, shafting, and numerous other mechanical components.

State building codes: buildings, bridges, and similar structures affecting the public safety

Department of Defense—Military Standards: aerospace vehicle structures and other military products

American National Standards Institute (ANSI): a wide variety of products

American Gear Manufacturers Association (AGMA): gears and gear systems

It is the designer's responsibility to determine which, if any, standards or codes apply to the member being designed and to ensure that the design meets those standards.

Material Strength Basis

Most designs using metals are based on either yield strength or ultimate strength or both, as stated previously. This is because most theories of metal failure show a strong relationship between the stress at failure and these material properties. Also, these properties will almost always be reported for materials used in engineering design. The value of the design factor will be different, depending on which material strength is used as the basis for design, as will be shown later.

Type of Material

A primary consideration with regard to the type of material is its ductility. The failure modes for brittle materials are quite different from those for ductile materials. Since brittle materials such as gray cast iron do not exhibit yielding, designs are always based on ultimate strength. Generally a metal is considered to be brittle if its percent elongation in a 2-in gage length is less than 5%. Except for highly hardened alloys, virtually all steels are ductile. Except for castings, aluminum is ductile. Other material factors that can affect the strength of a part are its uniformity and the confidence in the stated properties.

Manner of Loading

Three main types of loading can be identified. A *static load* is one that is applied to a part slowly and gradually and that remains applied, or at least is applied and removed only infrequently during the design life of the part. *Repeated loads* are those that are applied and removed several thousand times during the design life of the part. Significant fluctuations in load without completely unloading the part are also considered to be repeated loads. Under repeated loading a part fails by the mechanism of fatigue at a stress level much lower than that which would cause failure under a static load. This calls for the use of a higher design factor for repeated loads than for static loads. Parts subject to *impact* or *shock* require the use of a large design factor for two reasons. First, a suddenly applied load causes stresses in the part that are several times higher than those that would be computed by standard formulas. Second, under impact loading the material in the part is usually required to absorb energy from the impacting body. The certainty with which the designer knows the magnitude of the expected loads also must be considered when specifying the design factor.

Possible Misuse of the Part

In most cases the designer has no control over actual conditions of use of the product he or she designs. Legally, it is the responsibility of the designer to consider any reasonably foreseeable use or *misuse* of the product and to ensure the safety of the product. The possibility of an accidental overload on any part of a product must be considered.

Complexity of Stress Analysis

As the manner of loading or the geometry of a structure or a part becomes more complex, the designer is less able to perform a precise analysis of the stress condition. Thus the confidence in the results of stress analysis computations has an effect on the choice of a design factor.

Environment

Materials behave differently in different environmental conditions. Consideration should be given to the effects of temperature, humidity, radiation, weather, sunlight, and corrosive atmospheres on the material during the design life of the part.

Size Effect, Sometimes Called *Mass Effect*

Metals exhibit different strengths as the cross-sectional area of a part varies. Most material property data were obtained using standard specimens about 0.50 in (12.7 mm) in diameter. Parts with larger sections usually have lower strengths. Parts of smaller size, for example drawn wire, have significantly higher strengths. An example of the size effect is shown in Table 9–1.

Quality Control

The more careful and comprehensive a quality control program is, the better a designer knows how the product will actually appear in service. With poor quality control, a larger design factor should be used.

Hazard Presented by a Failure

The designer must consider the consequences of a failure to a particular part. Would a catastrophic collapse occur? Would people be placed in danger? Would other equipment be damaged? Such considerations may justify the use of a higher than normal design factor.

TABLE 9–1   Size Effect for AISI 4140 OQT 1100 Steel

Specimen size		Tensile strength		Yield strength		Percent elongation
in	mm	ksi	MPa	ksi	MPa	% in 2 in
0.50	12.7	158	1089	149	1027	18
1.00	25.4	140	965	135	931	20
2.00	50.8	128	883	103	710	22
4.00	101.6	117	807	87	600	22

Cost

Compromises must usually be made in design in the interest of limiting cost to a reasonable value under market conditions. Of course, where danger to life or property exists, compromises should not be made that would seriously affect the ultimate safety of the product or structure.

Market Segment in Which the Part Is to Be Used

Normally you will be aware of the use for the part you are designing, and this can affect your decision on the appropriate design factor. To use a low design factor requires that the loads, material properties, and manufacturing considerations are well known. Lack of confidence in any of these parameters should lead you to specify a larger design factor. Gaining that confidence may require a significant amount of additional research, stress analysis, quality control, and testing, all of which are expensive. The aerospace industry typically invests in the necessary research and analysis to justify a low design factor so the resulting part is as small and light as practical. Conversely, designers of special manufacturing equipment and some heavy-duty construction or agricultural equipment sometimes use larger design factors because of the inability of finding accurate data on the conditions of use.

9–5 DESIGN APPROACHES AND GUIDELINES FOR DESIGN FACTORS

Experience in design and knowledge about the conditions discussed in the preceding section must be applied to determine a design factor for a particular situation. Ultimately, it is the designer's responsibility to set the design factor to ensure safety of the component being designed while also achieving a cost-effective design.

In this chapter you will find several guidelines for specifying a design factor for direct stresses that will be used in this book. The guidelines are not precise and are based on average conditions. Additional guidelines are presented in future chapters for other kinds of stresses such as torsional shear stresses and stresses due to bending.

It is wasteful to purposely overdesign a component. However, there are times when uncertainty about the actual conditions of service warrant using a more conservative choice for a design factor than those given in the guidelines.

Also, there are numerous codes and standards that should be consulted in certain industries. Among these codes are those pertaining to building and construction, piping and pressure vessels, military, and aerospace. It is your responsibility to investigate whether the product or system you are designing is controlled by such codes and standards. A small sample of the codes for the application of steel or aluminum in buildings is presented in the next section.

The guidelines presented here are somewhat simplified in order to concentrate on the basic kinds of stresses dealt with in this book. They are generally applicable to homogeneous, isotropic metals. Additional study is advised to extend your understanding of more complex components and structures, nonisotropic materials, and more complex kinds of loading. Particular additional study should be given to repeated loads (called fatigue loading), shock, and impact. All of the references listed at the end of this chapter provide information for such additional study.

Design Factor Guidelines for Direct Normal Stresses

Table 9–2 includes guidelines for selecting design factors for problems in this book for which the component being designed or analyzed is subjected to direct normal stresses, tension, or compression.

The use of design factors and outlines of typical design approaches are summarized here. The specific approach used depends on the goal of the problem. Is the goal to evaluate the relative safety of a given design? Is it to specify a suitable material from which to make a component? Is it to determine the required shape and dimensions of the component when the loading is known and the material has been specified?

TABLE 9–2   Design Stress Guidelines—Direct Normal Stresses

Manage of loading	Ductile material	Brittle material
Static	$\sigma_d = s_y/2$	$\sigma_d = s_u/6$
Repeated	$\sigma_d = s_u/8$	$\sigma_d = s_u/10$
Impact or shock	$\sigma_d = s_u/12$	$\sigma_d = s_u/15$

Case A   To Evaluate the Safety of a Given Design.

Given: a. The magnitude and type of loading on the component of interest.
b. The material, including its condition, from which the component is made.
c. The shape and dimensions of critical geometry of the component.

Find: Whether or not the component is reasonably safe.

Method: 1. Identify the kind of stress produced by the given loading.
2. Determine the applicable stress analysis technique.
3. Complete the stress analysis to determine the maximum expected stress, σ_{max}, in the component.
4. Determine the yield strength, ultimate tensile strength, and percent elongation for the material. Decide if the material is ductile (percent elongation >5%) or brittle (percent elongation <5%).
5. Determine the appropriate design stress relationship. For direct normal stresses, use σ_d from either Equation (9–2) or (9–3).
6. Set $\sigma_{max} = \sigma_d$ and solve for the resulting design factor, N.
7a. When the design is based on yield strength:

$$\sigma_{max} = \sigma_d = s_y/N$$
$$N = s_y/\sigma_{max}$$

7b. When the design is based on ultimate tensile strength:

$$\sigma_{max} = \sigma_d = s_u/N$$
$$N = s_u/\sigma_{max}$$

8. Compare the resulting value of the design factor with recommended guidelines, considering Table 9–2 and all factors discussed in the preceding section.
9. If the actual design factor is lower than the recommended value, redesign should be done to increase the resulting design factor.
10. If the actual design factor is significantly higher than the recommended value, you should redesign the component to achieve a more cost-effective design that uses less material.

Case B   To Specify a Suitable Material from Which to Make a Component.

Given: a. The magnitude and type of loading on the component of interest.
b. The shape and dimensions of critical geometry of the component.

Find: The material, including its condition, from which to make the component.

Method: 1. Identify the kind of stress produced by the given loading.
2. Determine the applicable stress analysis technique.
3. Complete the stress analysis to determine the maximum expected stress, σ_{max}, in the component.
4. Specify a reasonable design factor from recommended guidelines, considering all factors discussed in the preceding section.
5. Determine the appropriate design stress relationship. For direct normal stresses, use σ_d from either Equation (9–2) or (9–3) using the guidelines in Table 9–2.
6. Set $\sigma_{max} = \sigma_d$ and solve for the required strength of the material.
7a. When the design is based on yield strength:

$$\sigma_{max} = \sigma_d = s_y/N$$
$$\text{Required } s_y = N(\sigma_{max})$$

7b. When the design is based on ultimate tensile strength:

$$\sigma_{max} = \sigma_d = s_u/N$$
$$\text{Required } s_u = N(\sigma_{max})$$

8. Specify a suitable material that has the required strength. Consider also the ductility of the material. If the loading is repeated, shock, or impact, a highly ductile material is recommended.

Case C   To Determine the Shape and Dimensions of the Component.

Given: a. The magnitude and type of loading on the component of interest.
        b. The material, including its condition, from which the component is to be made.

Find: The shape and dimensions of critical geometry of the component.

Method: 1. Determine the yield strength, ultimate strength, and percent elongation for the selected material. Decide if the material is ductile (percent elongation $>5\%$) or brittle (percent elongation $<5\%$).
     2. Specify an appropriate design factor considering the type of loading, the type of material, the conditions listed in the preceding section, and recommended guidelines. For direct normal stresses, use Table 9–2.
     3. Compute the design stress from Equation (9–2) or (9–3).

$$\sigma_d = s_y/N \text{ based on yield strength}$$
$$\sigma_d = s_u/N \text{ based on ultimate tensile strength}$$

     4. Write the equation for the expected maximum stress in the component. For direct normal stresses,

$$\sigma_{max} = F/A$$

     5. Set $\sigma_{max} = \sigma_d$ and solve for the required cross-sectional area.

$$\sigma_{max} = \sigma_d = F/A$$
$$\text{Required } A = F/\sigma_d$$

     6. Determine the minimum required dimensions of the cross-sectional area to achieve the necessary total area. This is dependent on the shape you choose to make the component. It may be solid circular, square, or rectangular, a hollow tube, a standard structural shape such as an angle, or some special shape of your own design.
     7. Specify convenient dimensions from the list of preferred basic sizes listed in Appendix A–2.

Case D   To Determine the Allowable Load on a Component.

Given: a. The type of loading on the component of interest.
        b. The material, including its condition, from which the component is to be made.
        c. The shape and dimensions of the component.

Find: The allowable load on the component.

Method: 1. Determine the yield strength, ultimate strength, and percent elongation for the selected material. Decide if the material is ductile (percent elongation $>5\%$) or brittle (percent elongation $<5\%$).
     2. Specify an appropriate design factor considering the type of loading, the type of material, the conditions listed in the preceding section, and recommended guidelines. For direct normal stresses, use Table 9–2.
     3. Compute the design stress from Equation (9–2) or (9–3).

$$\sigma_d = s_y/N \text{ based on yield strength}$$
$$\sigma_d = s_u/N \text{ based on ultimate tensile strength}$$

     4. Write the equation for the expected maximum stress in the component. For direct normal stresses,

$$\sigma_{max} = F/A$$

     5. Set $\sigma_{max} = \sigma_d$ and solve for the maximum allowable load.

$$\sigma_{max} = \sigma_d = F/A$$
$$\text{Maximum allowable } F = \sigma_d(A)$$

9–6 METHODS OF COMPUTING DESIGN STRESS

In this section we discuss failure modes relevant to parts subjected to tensile and compressive loads. Other kinds of loading are discussed later.

Ductile and brittle materials behave differently. Static loads, repeated and shock loads produce different modes of failure. It is strongly advised that documented data from suppliers be sought for the strength of materials to be specified for critical applications, particularly where fatigue and/or shock and impact loads are expected. Data for yield strength, ultimate tensile strength, endurance strength, and toughness should be acquired for the condition of the material to be used, such as hot rolled, cold drawn, or a specific heat treatment. The metallurgical structure of a material can affect its endurance strength significantly so attention must be paid to grain structures, inclusions, casting voids, residual stresses, types of alloying elements, and impurities.

Ductile Materials under Static Loads

Ductile materials will undergo large plastic deformations when the stress reaches the yield strength of the material. Under most conditions of use, this would render the part unfit for its intended use. Therefore, for ductile materials subjected to static loads, the design stress is usually based on yield strength. That is,

$$\sigma_d = \frac{s_y}{N}$$

As indicated in Table 9–2, a design factor of $N = 2$ would be a reasonable choice under average conditions.

Ductile Materials under Repeated Loads

Under repeated loads, ductile materials fail by a mechanism called *fatigue*. The level of stress at which fatigue occurs is lower than the yield strength. By testing materials under repeated loads, the stress at which failure will occur can be measured. The terms *fatigue strength* and *endurance strength* are used to denote this stress level. However, fatigue-strength values are often not available. Also, factors such as surface finish, the exact pattern of loading, and the size of a part have a marked effect on the actual fatigue strength. To overcome these difficulties, it is often convenient to use a high value for the design factor when computing the design stress for a part subjected to repeated loads. It is also recommended that the ultimate strength be used as the basis for the design stress because tests show that there is a good correlation between fatigue strength and the ultimate strength. Therefore, for ductile materials subjected to repeated loads, the design stress can be computed from

$$\sigma_d = \frac{s_u}{N}$$

A design factor of $N = 8$ would be reasonable under average conditions. Also, stress concentrations, which are discussed

in Section 9–11, must be accounted for since fatigue failures often originate at points of stress concentrations.

Where data are available for the endurance strength of the material, the design stress can be computed from

$$\sigma_d = \frac{s_n}{N}$$

where s_n is the symbol for endurance strength. A design factor N of 3 to 4 is recommended. See References 6 and 7.

Ductile Materials Under Impact or Shock Loading

The failure modes for parts subjected to impact or shock loading are quite complex. They depend on the ability of the material to absorb energy and the flexibility of the part. Large design factors are recommended, because of the general uncertainty of stresses under shock loading. In this book we will use

$$\sigma_d = \frac{s_u}{N}$$

with $N = 12$ for ductile materials subjected to impact or shock loads.

Brittle Materials

Since brittle materials do not exhibit yielding, the design stress must be based on ultimate strength. That is,

$$\sigma_d = \frac{s_u}{N}$$

with $N = 6$ for static loads, $N = 10$ for repeated loads, and $N = 15$ for impact or shock loads. These design factors are higher than for ductile materials because of the sudden manner of failure exhibited by brittle materials.

Design Stresses from Selected Codes

Table 9–3 gives a summary of specifications for design stresses for structural steel as defined by the American Institute of Steel Construction (AISC) and by the Aluminum Association for aluminum alloys. These data pertain to members loaded in tension under static loads such as those found in building-type structures. See References 1 and 2 for a more detailed discussion of these specifications.

TABLE 9–3	Design Stress from Selected Codes—Direct Normal Stresses—Static Loads on Building-Like Structures

Structural Steel (AISC): Allowable Stress Design (ASD)
  $\sigma_d = s_y/1.67 = 0.60\,s_y$ or $\sigma_d = s_u/2.00 = 0.50\,s_u$ whichever is lower

Aluminum (Aluminum Association):
  $\sigma_d = s_y/1.65 = 0.61\,s_y$ or $\sigma_d = s_u/1.95 = 0.51\,s_u$ whichever is lower

The design stresses for steel listed in Table 9–3 relate to the *allowable stress design* method (ASD) that had been the standard for many years. In recent years, the AISC has completed the implementation of a revised approach for the design of structural members for bridges, buildings, and other structures called *load and resistance factor design* (LRFD). Complete information about this approach is included in Reference 2, and the details are included in courses that follow strength of materials in programs in civil engineering, civil engineering technology, construction technology, construction management, architecture, and related programs. This book continues the use of ASD because it introduces the fundamental concepts of design without the very large amount of detail required to fully apply the LRFD method. The basic philosophy of LRFD includes the application of factors, γ, to the various types of loads that can occur singly or in combination in buildings, bridges, and other such structures. The factors were developed after extensive research on how dead loads (D), live loads (L), wind loads (W), earthquake loads (E), and roof, snow, or rainwater loads (L_r, S, R) interact on a statistical basis over a 50-year lifetime. Also included is a set of *resistance factors*, ϕ, for different types of loading such as tensile yielding, tensile fracture, compression, flexure, and shear yielding. These are accumulated and considered in the mathematical inequality,

$$\sum \gamma_i Q_i \le \phi R_n$$

where R_n is the nominal strength or resistance provided by the load-carrying member. The left side of this equation represents the summation of all of the load effects (forces, moments, etc.) Q_i experienced by the member multiplied by the corresponding factor that depends on the specific combination of loads. Therefore, the sum of the factored loads must be less than or equal to the resistance modified by specified factors that are defined in the LRFD manual.

Similarly, for programs such as mechanical engineering and engineering technology, manufacturing engineering and engineering technology, and industrial technology, additional detail on design methodology is typically included in courses that follow strength of materials. For example, see Reference 7.

Example Problem 9–1

A structural support for a machine will be subjected to a static tensile load of 16.0 kN. It is planned to fabricate the support from a square rod made from AISI 1020 hot-rolled steel. Specify suitable dimensions for the cross section of the rod.

Solution   Objective: Specify the dimensions of the cross section of the rod.

Given: $F = 16.0$ kN $= 16\,000$ N static load.
Material: AISI 1020 HR; $s_y = 331$ MPa; 36% elongation (ductile).
(Data from Appendix A–14)

Analysis: Use Case C from Section 9–5.
Let $\sigma = \sigma_d = s_y/2$ (Table 9–2; ductile material, static load).
Stress analysis: $\sigma = F/A$; then required area $= A = F/\sigma_d$.
But $A = a^2$ ($a =$ dimension of side of square).
Minimum allowable dimension $a = \sqrt{A}$.

Results: $\sigma_d = s_y/2 = 331$ MPa$/2 = 165.5$ MPa $= 165.5$ N/mm^2.
Required area: $A = F/\sigma_d = (16\,000$ N$)/(165.5$ N/mm$^2) = 96.7$ mm^2.
Minimum dimension a: $a = \sqrt{A} = \sqrt{96.7 \text{ mm}^2} = 9.83$ mm.
Specify: $a = 10$ mm (Appendix A–2; preferred size).

Example Problem 9–2

A tensile member for a roof truss for a building is to carry a static axial tensile load of 19 800 lb. It has been proposed that a standard, equal-leg structural steel angle be used for this application using ASTM A36 structural steel. Use the AISC code. Specify a suitable angle from Appendix A–5.

Solution   Objective: Specify a standard equal-leg steel angle.

Given: $F = 19\,800$ lb static load.
Material: ASTM A36; $s_y = 36\,000$ psi; $s_u = 58\,000$ psi.
(Data from Appendix A–16)

Analysis: Use Case C from Section 9–5.
Let $\sigma = \sigma_d = 0.60 s_y$ or $\sigma_d = 0.50 s_u$ (Table 9–3).
Stress analysis: $\sigma = F/A$; then required area $= A = F/\sigma_d$.

Results: $\sigma_d = 0.60 s_y = 0.60 (36\,000$ psi$) = 21\,600$ psi
or $\sigma_d = 0.50 s_u = 0.50 (58\,000$ psi$) = 29\,000$ psi
Use lower value; $\sigma_d = 21\,600$ psi.
Required area: $A = F/s_d = (19\,800$ lb$)/(21\,600$ lb/in$^2) = 0.917$ in^2.

This is the minimum allowable area.
Specify: L2 × 2 × 1/4 steel angle (Appendix A–5; lightest section).
$A = 0.944\ in^2$; weight = 3.21 lb/ft.

Example Problem 9–3

A machine element in a packaging machine is subjected to a tensile load of 36.6 kN, which will be repeated several thousand times over the life of the machine. The cross section of the element is 12 mm thick and 20 mm wide. Specify a suitable material from which to make the element.

Solution

Objective: Specify a material for a machine element.

Given: $F = 36.6\ kN = 36\ 600\ N$ repeated load.
Cross section of machine element: rectangle; 12 mm × 20 mm.

Analysis: Use Case B from Section 9–5.
Ductile material desirable for repeated loading.
Let $\sigma = \sigma_d = s_u/8$ (Table 9–2). Then required $s_u = 8\sigma$.
Stress analysis: $\sigma = F/A$.

Results: Area $= A = (12\ mm)(20\ mm) = 240\ mm^2$
$\sigma = F/A = (36\ 600\ N)/(240\ mm^2) = 152.5\ N/mm^2 = 152.5\ MPa$
Required ultimate strength: $s_u = 8\sigma = 8\,(152.5\ MPa) = 1220\ MPa$.
Specify: AISI 4140 OQT 900 steel (Appendix A–14).
$s_u = 1289\ MPa$; 15% elongation; adequate strength, good ductility.

Comment: Other materials could be selected. Required strength indicates that a heat-treated alloy steel is required. The one selected has the highest percent elongation of any listed in Appendix A–14. If the size of the element could be made somewhat larger, the required strength would be lower and a less costly steel may be found.

Example Problem 9–4

Figure 9–2 shows a design for the support for a heavy machine that will be loaded in axial compression. Gray cast iron, grade 20, has been selected for the support. Specify the allowable static load on the support.

FIGURE 9–2 Machine support for Example Problem 9–4.

Top view

0.75
3.00
1.50

0.50 radius typical

1.38
1.25
4.00

Side view

1.25

Dimensions in inches

Solution Objective: Specify the allowable static axial compression load on the support.

Given: Material: Gray cast iron, grade 20; $s_u = 80$ ksi in compression (Table A–17); material is brittle. Assume load will be static. Shape of support in Figure 9–2. Compression member is *short* so no buckling occurs.

Analysis: Use Case D from Section 9–5.
Stress analysis: $\sigma = F/A$; area computed from Figure 9–2.
Let $\sigma = \sigma_d = s_u/N$; use $N = 6$ (Table 9–2).
Then, allowable, $F = \sigma_d(A)$

Results: $\sigma_d = s_u/6 = 80\,000$ psi$/6 = 13\,300$ psi
The cross section of the support is the same as the top view. The net area can be calculated by taking the area of a 3.00 in by 4.00 in rectangle and subtracting the area of the slot and the four corner fillets.

$$\text{Rectangle: } A_R = (3.00\,\text{in})(4.00\,\text{in}) = 12.00\,\text{in}^2$$

$$\text{Slot: } A_S = (0.75)(1.25) + \frac{\pi(0.75)^2}{4} = 1.38\,\text{in}^2$$

The area of each fillet can be computed by the difference between the area of a square with sides equal to the radius of the corner (0.50 in) and a quarter circle of the same radius. Then

$$\text{Fillet: } A_F = r^2 - \frac{1}{4}(\pi r^2)$$

$$A_F = (0.50)^2 - \frac{1}{4}[\pi(0.50)^2] = 0.0537\,\text{in}^2$$

Then the total area is

$$A = A_R - A_S - 4A_F = 12.00 - 1.38 - 4(0.0537) = 10.41\,\text{in}^2$$

We now have the data needed to compute the allowable load.

$$F = A\sigma_d = (10.41\,\text{in}^2)(13\,300\,\text{lb/in}^2) = 138\,500\,\text{lb}$$

This completes the example problem.

9–7 ELASTIC DEFORMATION IN TENSION AND COMPRESSION MEMBERS

Deformation refers to some change in the dimensions of a load-carrying member. Being able to compute the magnitude of deformation is important in the design of precision mechanisms, machine tools, building structures, and machine structures.

An example of where deformation is important is shown in Figure 9–3, in which circular steel tie rods are attached to a C-frame punch press. The tie rods are subjected to tension when in operation. Since they contribute to the rigidity of the press, the amount that they deform under load is something the designer needs to be able to determine.

To develop the relationship from which deformation can be computed for members subjected to axial tension or compression, some concepts from Chapter 7 must be reviewed. *Strain* is defined as the ratio of the total deformation to the original length of a member. (See Figure 9–4.) Using the symbols ϵ for strain, δ for total deformation, and

L for length, the formula for strain becomes

$$\epsilon = \frac{\delta}{L} \tag{9–4}$$

The stiffness of a material is a function of its modulus of elasticity, E, defined as

$$E = \frac{\text{Stress}}{\text{Strain}} = \frac{\sigma}{\epsilon} \tag{9–5}$$

Solving for strain gives

$$\epsilon = \frac{\sigma}{E} \tag{9–6}$$

Now Equations (9–4) and (9–6) can be equated:

$$\frac{\delta}{L} = \frac{\sigma}{E} \tag{9–7}$$

Solving for deformation gives

$$\delta = \frac{\sigma L}{E} \tag{9–8}$$

FIGURE 9–3   C-frame press for Example Problem 9–5.

$$\text{Strain} = \epsilon = \frac{\text{Total deformation}}{\text{Original length}} = \frac{\delta}{L}$$

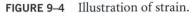

FIGURE 9–4   Illustration of strain.

Arial Deformation ▼

$$\delta = \frac{\sigma L}{E} = \frac{FL}{AE} \tag{9–9}$$

Equation (9–9) can be used to compute the total deformation of any load-carrying member, provided that it meets the conditions defined for direct tensile and compressive stress. That is,

- The member must be straight and have a constant cross section.
- The material must be homogeneous.
- The load must be directly axial.
- The stress must be below the proportional limit of the material. Recall that the value of the proportional limit is close to the yield strength, s_y.

Since this formula applies to members that are subjected to either direct tensile or compressive forces, the direct-stress formula can be used to compute the stress σ. That is, $\sigma = F/A$, where F is the applied load and A is the cross-sectional area of the member. Substituting this into Equation (9–8) gives

---

Example Problem 9–5

The tie rods in the press in Figure 9–3 are made of the steel alloy AISI 5160 OQT 900. Each rod has a diameter of 2.00 in and an initial length of 68.5 in. An axial tensile load of 40 000 lb is exerted on each rod during operation of the press. Compute the deformation of the rods. Check also if the strength of the material is adequate.

Solution   Objective: Compute the deformation of the tie rods.

Given: Rods are steel, AISI 5160 OQT 900; s_y = 179 ksi, s_u = 196 ksi, 12% elongation
Diameter = D = 2.00 in. Length = L = 68.5 in. Axial force = F = 40 000 lb.

Analysis: Equation (9–9) will be used to compute deformation. The stress in the rods must be checked to ensure that it is below the proportional limit and safe under repeated shock loading.

Results: **Axial Tensile Stress:** $\sigma = F/A$.

$$\text{Area} = A = \frac{\pi D^2}{4} = \frac{\pi (2.0 \text{ in})^2}{4} = 3.14 \text{ in}^2$$

$$\text{Then, } \sigma = \frac{40\ 000 \text{ lb}}{3.14 \text{ in}^2} = 12\ 700 \text{ psi}$$

Therefore, the stress is well below the proportional limit.

For shock loading in a ductile material, Table 9–2 recommends the following design stress. The percent elongation of 12% is in the ductile range.

$$\sigma_d = s_u/12 = 196 \text{ ksi}/12 = 16.33 \text{ ksi} = 16\ 330 \text{ psi}$$

Because the actual expected stress is below the design stress, the bar should be safe.

Axial Deformation: Use Equation (9–9). All data are known except the modulus of elasticity, *E*. From the footnotes of Appendix A–14 we find $E = 30 \times 10^6$ psi. Then

$$\delta = \frac{FL}{AE} = \frac{(40\,000\text{ lb})(68.5\text{ in})}{(3.14\text{ in}^2)(30 \times 10^6\text{ lb/in}^2)} = 0.029\text{ in}$$

Comment: For a precision mechanical press, the deformation of 0.029 in (0.74 mm) may be high. An analysis of the complete press system would be advisable. If found to be excessive, the diameter of the bar could be increased, noting that the deformation is inversely proportional to the area of the bar.

Example Problem 9–6

A large pendulum is composed of a 10.0-kg ball suspended by an aluminum wire having a diameter of 1.00 mm and a length of 6.30 m. The aluminum is the alloy 7075-T6. Compute the elongation of the wire due to the weight of the 10-kg ball.

Solution Objective: Compute the elongation of the wire.

Given: Wire is aluminum alloy 7075-T6; diameter = D = 1.00 mm.

Length = L = 6.30 m; ball has a mass of 10.0 kg.

Analysis: The force on the wire is equal to the weight of the ball, which must be computed from $w = mg$. Then the stress in the wire must be checked to ensure that it is below the proportional limit. Finally, because the stress will then be known, Equation (9–8) will be used to compute the elongation of the wire.

Result: **Force on the Wire:** $F = w = m \cdot g = (10.0\text{ kg})(9.81\text{ m/s}^2) = 98.1\text{ N}$

Axial Tensile Stress: $\sigma = F/A$

$$A = \frac{\pi D^2}{4} = \frac{\pi (1.00\text{ mm})^2}{4} = 0.785\text{ mm}^2$$

$$\sigma = \frac{F}{A} = \frac{98.1\text{ N}}{0.785\text{ mm}^2} = 125\text{ N/mm}^2 = 125\text{ MPa}$$

Appendix A–18 lists the yield strength of 7075-T6 aluminum alloy to be 503 MPa. The stress is well below the proportional limit.

The stress can be considered to be steady for a slow-moving pendulum, and the aluminum is ductile, having 11% elongation. From Table 9–2, the design stress can be computed from

$$\sigma_d = s_y/2 = 503\text{ MPa}/2 = 251\text{ MPa}$$

Therefore, the wire is safe.

Elongation: For use in Equation (9–8), all data are known except the modulus of elasticity, *E*. The footnote of Appendix A–18 lists the value of $E = 72$ GPa $= 72 \times 10^9$ Pa. Then

$$\delta = \frac{\sigma L}{E} = \frac{(125\text{ MPa})(6.30\text{ m})}{72\text{ GPa}} = \frac{(125 \times 10^6\text{ Pa})(6.30\text{ m})}{72 \times 10^9\text{ Pa}}$$

$$\delta = 10.9 \times 10^{-3}\text{ m} = 10.9\text{ mm}$$

Comment: What do you see around you right now that has a dimension similar to 10.9 mm (0.429 in)? Measure the thickness of one of your fingers. Certainly the design of the system containing the pendulum in this example problem would have to take this deflection into account.

Example Problem 9–7

A tension link in a machine must have a length of 610 mm and will be subjected to a repeated axial load of 3000 N. It has been proposed that the link be made of steel and that it have a square cross section. Determine the required dimensions of the link if the elongation under load must not exceed 0.05 mm.

Solution   Objective: Determine the required dimensions of the square cross section of the link to limit the elongation, δ, to 0.05 mm or less.

Given: Axial loading on the link = F = 3000 N; length = L = 610 mm.

Link will be steel; then E = 207 GPa = 207×10^9 N/m². (Appendix A–14)

Analysis: In Equation (9–9) for the axial deformation, let δ = 0.05 mm. Then all other data are known except for the cross-sectional area, A. We can solve for A, which is the minimum acceptable cross-sectional area of the link. Let each side of the square cross section be d. Then $A = d^2$ and we can compute the minimum acceptable value for d from $d = \sqrt{A}$. After specifying a convenient size for d, we must check to ensure that the stress is safe and below the proportional limit.

Results: **Required Area:** Solving for A from Equation (9–9) and substituting values gives

$$A = \frac{FL}{E\delta} = \frac{(3000\ \text{N})(610\ \text{mm})}{(207 \times 10^9\ \text{N/m}^2)(0.05\ \text{mm})} = 176.8 \times 10^{-6}\ \text{m}^2$$

Converting to mm²

$$A = 176.8 \times 10^{-6}\ \text{m}^2 \times \frac{(10^3\text{mm})^2}{\text{m}^2} = 176.8\ \text{mm}^2$$

and

$$d = \sqrt{A} = \sqrt{176.8\ \text{mm}^2} = 13.3\ \text{mm}$$

Appendix A–2 lists the next larger preferred size to be 14.0 mm.
The actual cross-sectional area is $A = d^2 = (14.0\ \text{mm})^2 = 196\ \text{mm}^2$.

Stress: $\sigma = F/A = 3000\ \text{N}/196\ \text{mm}^2 = 15.3\ \text{N/mm}^2 = 15.3\ \text{MPa}$.
For a repeated load, Table 9–2 recommends that the design stress be $\sigma_d = s_u/8$. Letting $\sigma_d = \sigma$, the required value for the ultimate strength is

$$s_u = 8(\sigma) = 8(15.3\ \text{MPa}) = 123\ \text{MPa}$$

Comment: Referring to Appendix A–14, we can see that virtually any steel has an ultimate strength far greater than 123 MPa. Unless there were additional design requirements, we should specify the least expensive steel. Then we will specify

d = 14.0 mm
AISI 1020 hot-rolled steel should be low cost. s_u = 448 MPa

Note that this design was limited by the allowable elongation and that the resulting stress is relatively low.

Deformation for Members Carrying Multiple Loads or Having Differing Properties

In Example Problems 9–5 through 9–7, the entire member of interest was uniform in material and cross section, and it was subjected to the same magnitude of axial load throughout. Under such conditions, Equation (9–9) can be used directly to compute the total deformation. In fact, we can say that this equation can *only* be used when all factors, F, L, E, and A, are constant over the section of interest.

When any factor in Equation (9–9) is different over the length of a given member, the member must be divided into segments for which all factors are the same. Then you can use *superposition* to determine the total deformation of the member. The principle of superposition states that the overall effect of multiple actions on a member is the algebraic sum of the effects of individual components of those actions.

Example Problem 9–8 that follows illustrates the approach to such problems.

Example Problem 9–8   Figure 9–5 shows a steel pipe being used as a bracket to support equipment through cables attached as shown. Select the smallest standard schedule 40 steel pipe that will limit the stress to no more than 18 000 psi. Then for the pipe selected, determine the total downward deflection of point C at the bottom of the pipe as the loads are applied.

FIGURE 9–5   Pipe for Example
Problem 9–8.

Solution   Objective: Specify a suitable standard schedule 40 steel pipe size and determine the elongation of the pipe.

Given: Loading in Figure 9–5; $F_1 = F_2 = 8000$ lb (two forces); $F_3 = 2500$ lb.
Pipe length from A to B: $L_{A-B} = 4.00$ ft(12 in/ft) = 48.0 in.
Pipe length from B to C: $L_{B-C} = 3.00$ ft(12 in/ft) = 36.0 in.
Maximum allowable stress = 18 000 psi; $E = 30 \times 10^6$ psi (steel).

Analysis: The maximum axial tensile load on the pipe is the sum of F_3 plus the vertical components of each of the 8000-lb forces. This occurs throughout the part of the pipe from A to B. The size of pipe, and the resulting cross-sectional area, must result in a stress in that section of 18 000 psi or less. From B to C, the axial tensile load is $F_{B-C} = 2500$ lb. Because the load is different in the two sections, the computation of the elongation of the pipe must be done in two separate calculations. That is,

$$\delta_C = \delta_{total} = \delta_{A-B} + \delta_{B-C}$$

Results: **Axial Loads:**

$$F_{A-B} = F_3 + F_1 \cos 30° + F_2 \cos 30°$$

But

$$F_1 = F_2 = 8000 \text{ lb}$$

Then

$$F_{A-B} = 2500 \text{ lb} + 2(8000 \text{ lb}) \cos 30°$$
$$F_{A-B} = 16\,400 \text{ lb}$$
$$F_{B-C} = F_3 = 2500 \text{ lb}$$

Stress Analysis and Calculation of Required Cross-Sectional Area: Letting $\sigma = 18\,000$ psi, the required cross-sectional area of the metal in the pipe is

$$A = \frac{F_{A-B}}{\sigma} = \frac{16\,400 \text{ lb}}{18\,000 \text{ lb/in}^2} = 0.911 \text{ in}^2$$

From Appendix A–12, listing the properties of steel pipe, the standard size with the next larger cross-sectional area is the 2-in schedule 40 pipe with $A = 1.075$ in². As discussed in Chapter 1, this same pipe would be called PIPE2STD in the construction industry, indicating a 2-in nominal pipe size and standard wall thickness.

Deflection:

$$\delta_C = \delta_{total} = \delta_{A-B} + \delta_{B-C}$$

$$\delta_{A-B} = \left(\frac{F_{A-B}\, L_{A-B}}{AE}\right) = \frac{(16\,400 \text{ lb})(48 \text{ in})}{(1.075 \text{ in}^2)(30 \times 10^6 \text{ lb/in}^2)} = 0.024 \text{ in}$$

$$\delta_{B-C} = \left(\frac{F_{B-C}\, L_{B-C}}{AE}\right) = \frac{(2500 \text{ lb})(36 \text{ in})}{(1.075 \text{ in}^2)(30 \times 10^6 \text{ lb/in}^2)} = 0.003 \text{ in}$$

Then

$$\delta_C = \delta_{A-B} + \delta_{B-C} = 0.027 \text{ in}$$

Comment: In summary, when the bracket shown in Figure 9–5 is made from a standard 2-in schedule 40 steel pipe, point C moves downward 0.027 in under the influence of the applied loads.

9–8 DEFORMATION DUE TO TEMPERATURE CHANGES

A machine or a structure could undergo deformation or be subjected to stress by changes in temperature in addition to the application of loads. Bridge members and other structural components see temperatures as low as $-30°F(-34°C)$ to as high as $110°F(43°C)$ in some areas. Vehicles and machinery operating outside experience similar temperature variations. Frequently, a machine part will start at room temperature and then become quite hot as the machine operates. Examples are parts of engines, furnaces, metal-cutting machines, rolling mills, plastics molding and extrusion equipment, food-processing equipment, air compressors, hydraulic and pneumatic devices, and high-speed automation equipment.

As a metal part is heated, it tends to expand. If the expansion is unrestrained, the dimensions of the part will grow but no stress will be developed in the metal. However, in some cases the part is restrained, preventing the change in dimensions. Under such circumstances, stresses will occur.

Different materials change dimensions at different rates when subjected to temperature changes. Most materials expand with increasing temperature, but a few contract and some virtually stay the same size. The *coefficient of thermal expansion* governs the thermal deformation and thermal stress experienced by a material.

The coefficient of thermal expansion, α, is the property of a material that indicates the amount of unit change in dimension with a unit change in temperature.

The lowercase Greek letter alpha, α, is used as the symbol for the coefficient of thermal expansion.

The units for α are derived from its definition. Stated slightly differently, α is the measure of the change in length of a material per unit length for a 1.0-degree change in temperature. The units, then, for α in the U.S. Customary System would be

$$\text{in/(in} \cdot °F) \quad \text{or} \quad 1/°F \quad \text{or} \quad °F^{-1}$$

In SI units, α would be

$$\text{m/(m} \cdot °C) \quad \text{or} \quad \text{mm/(mm} \cdot °C) \quad \text{or} \quad 1/°C \quad \text{or} \quad °C^{-1}$$

For use in computations, the last form of each unit type is most convenient. However, the first form will help you remember the physical meaning of the term.

It follows from the definition of the coefficient of thermal expansion that the change in length δ of a member can be computed from the equation

Thermal Expansion ▼

$$\delta = \alpha \cdot L \cdot \Delta t \qquad (9–10)$$

where L = original length of the member
      Δt = change in temperature

Table 9–4 gives representative values for the coefficient of thermal expansion for several metals, plate glass, pine wood, and concrete. The actual value for any material varies somewhat with temperature. Those reported in Table 9–4 are approximately average values over the range of temperatures from 32°F (0°C) to 212°F (100°C).

Table 9–5 gives values for α for selected plastic materials. Note that the actual values depend strongly on temperature

TABLE 9–4	Coefficients of Thermal Expansion, α, for Some Metals, Plate Glass, Wood, and Concrete	
	α	
Material	**°F^{-1}**	**°C^{-1}**
Steel, AISI		
1020	6.5×10^{-6}	11.7×10^{-6}
1040	6.3×10^{-6}	11.3×10^{-6}
4140	6.2×10^{-6}	11.2×10^{-6}
Structural steel	6.5×10^{-6}	11.7×10^{-6}
Gray cast iron	6.0×10^{-6}	10.8×10^{-6}
Stainless steel		
AISI 301	9.4×10^{-6}	16.9×10^{-6}
AISI 430	5.8×10^{-6}	10.4×10^{-6}
AISI 501	6.2×10^{-6}	11.2×10^{-6}
Aluminum alloys		
2014	12.8×10^{-6}	23.0×10^{-6}
6061	13.0×10^{-6}	23.4×10^{-6}
7075	12.9×10^{-6}	23.2×10^{-6}
Brass, C36000	11.4×10^{-6}	20.5×10^{-6}
Bronze, C22000	10.2×10^{-6}	18.4×10^{-6}
Copper, C14500	9.9×10^{-6}	17.8×10^{-6}
Magnesium, ASTM AZ63A-T6	14.0×10^{-6}	25.2×10^{-6}
Titanium, Ti-6 A1-4V	5.3×10^{-6}	9.5×10^{-6}
Plate glass	5.0×10^{-6}	9.0×10^{-6}
Wood (pine)	3.0×10^{-6}	5.4×10^{-6}
Concrete	6.0×10^{-6}	10.8×10^{-6}

TABLE 9-5   Coefficients of Thermal Expansion, α, for Selected Plastics

Material	α °F^{-1}	α °C^{-1}
ABS–resin	53×10^{-6}	95.4×10^{-6}
ABS/glass	16×10^{-6}	28.8×10^{-6}
Acetal–resin	45×10^{-6}	81.0×10^{-6}
Acetal/glass	22×10^{-6}	39.6×10^{-6}
Nylon 66–resin	45×10^{-6}	81.0×10^{-6}
Nylon 66/glass	13×10^{-6}	23.4×10^{-6}
Polycarbonate–resin	37×10^{-6}	66.6×10^{-6}
Polycarbonate/glass	13×10^{-6}	23.4×10^{-6}
Polyester–resin	53×10^{-6}	95.4×10^{-6}
Polyester/glass	12×10^{-6}	21.6×10^{-6}
Polystyrene–resin	36×10^{-6}	64.8×10^{-6}
Polystyrene/glass	19×10^{-6}	34.2×10^{-6}

and on the inclusion of any filler material in the plastic resin. For each plastic listed, the approximate values for α are given for the unfilled resin and for one filled with 30% glass.

Composites were described in Chapter 8 as materials that combine a matrix with reinforcing fibers made from a variety of materials such as glass, aramid polymer, carbon, or graphite. The matrix materials can be polymers such as polyester or epoxy, ceramics, or some metals such as aluminum. The value for α for the fibers is typically much smaller than for the matrix. In addition, there are numerous ways in which the fibers can be placed in the matrix. The coefficient of thermal expansion for composites, therefore, is very difficult to generalize.

Table 9–6 gives representative values for a few forms of composites. Refer to Chapter 8 for the description of the terms *unidirectional* and *quasi-isotropic*. Particularly with unidirectional placement of the fibers in the matrix, there is a dramatic difference in the value of the coefficient of thermal expansion as a function of the orientation of the material. In the longitudinal direction, aligned with the fibers, the low value of α for the fibers tends to produce a low overall value. However, in the transverse direction, the fibers are not very effective, and the overall value of α is much higher. Notice, too, that for the particular unidirectional aramid/epoxy composite listed, the value of α is actually negative, meaning that this composite gets *smaller* with increasing temperature.

Example Problem 9–9   A rod made from AISI 1040 steel is used as a link in a steering mechanism of a large truck. If its nominal length is 56 in, compute its change in length as the temperature changes from -30°F to 110°F.

Solution   Objective: Compute the change in length for the link.

Given: Link is made from AISI 1040 steel; length $= L = 56$ in.
Original temperature $= t_1 = -30$°F.
Final temperature $= t_2 = 110$°F.

Analysis: Use Equation (9–10). From Table 9–4, $\alpha = 6.3 \times 10^{-6}$°F^{-1}.

$$\Delta t = t_2 - t_1 = 110°F - (-30°F) = 140°F$$

Results: $\delta = \alpha \cdot L \cdot \Delta t = (6.3 \times 10^{-6}°F^{-1})(56\text{ in})(140°F) = 0.049\text{ in}$

Comment: The significance of this amount of deformation would have to be evaluated within the overall design of the steering mechanism for the truck. Approximately 12 sheets of standard bond paper stacked together have a thickness of 0.049 in.

TABLE 9-6   Coefficients of Thermal Expansion, α, for Selected Composites

Material	α Longitudinal °F^{-1}	α Longitudinal °C^{-1}	α Transverse °F^{-1}	α Transverse °C^{-1}
E-glass/epoxy–unidirectional	3.5×10^{-6}	6.30×10^{-6}	11.0×10^{-6}	19.8×10^{-6}
Aramid/epoxy–unidirectional	-1.1×10^{-6}	-1.98×10^{-6}	38.0×10^{-6}	68.4×10^{-6}
Carbon/epoxy–unidirectional	0.05×10^{-6}	0.09×10^{-6}	9.0×10^{-6}	16.2×10^{-6}
Carbon/epoxy–quasi-isotropic	1.6×10^{-6}	2.88×10^{-6}	1.6×10^{-6}	2.88×10^{-6}

Example Problem 9–10

A pushrod in the valve mechanism of an automotive engine has a nominal length of 203 mm. If the rod is made of AISI 4140 steel, compute the elongation due to a temperature change from −20°C to 140°C.

Solution Objective: Compute the change in length for the pushrod.

Given: Link is made from AISI 4140 steel; length = L = 203 mm.
Original temperature = t_1 = −20°C.
Final temperature = t_2 = 140°C.

Analysis: Use Equation (9–10). From Table 9–4, $\alpha = 11.2 \times 10^{-6}°C^{-1}$.

$$\Delta t = t_2 - t_1 = 140°C - (-20°C) = 160°C$$

Results: $\delta = \alpha \cdot L \cdot \Delta T = (11.2 \times 10^{-6}°C^{-1})(203 \text{ mm})(160°C) = 0.364 \text{ mm}$

Comment: It would be important to account for this expansion in the design of the valve mechanism. It could cause noise due to loose-fitting parts or high stresses if the expansion is restrained.

Example Problem 9–11

An aluminum frame of 6061 alloy for a window is 4.350 m long and holds a piece of plate glass 4.347 m long when the temperature is 35°C. At what temperature would the aluminum and glass be the same length?

Solution Objective: Compute the temperature at which the aluminum frame and the glass would be the same length.

Given: Aluminum is 6061 alloy; from Table 9–4 we find that $\alpha_a = 23.4 \times 10^{-6}°C^{-1}$.
For plate glass, $\alpha_g = 9.0 \times 10^{-6}°C^{-1}$.
At t_1 = 35°C; L_{a1} = 4.350 m; L_{g1} = 4.347 m.

Analysis: The temperature would have to decrease in order for the aluminum and glass to reach the same length, since aluminum contracts at a greater rate than glass. As the temperature decreases, the change in temperature Δt would be the same for both the aluminum and the glass. After the temperature change, the length of the aluminum would be

$$L_{a2} = L_{a1} - \alpha_a \cdot L_{a1} \cdot \Delta t$$

where the subscript a refers to the aluminum, 1 refers to the initial condition, and 2 refers to the final condition. The length of the glass would be

$$L_{g2} = L_{g1} - \alpha_g \cdot L_{g1} \cdot \Delta t$$

But when the glass and the aluminum have the same length,

$$L_{a2} = L_{g2}$$

Then

$$L_{a1} - \alpha_a \cdot L_{a1} \cdot \Delta t = L_{g1} - \alpha_g \cdot L_{g1} \cdot \Delta t$$

Solving for Δt gives

$$\Delta t = \frac{L_{a1} - L_{g1}}{\alpha_a \cdot L_{a1} - \alpha_g \cdot L_{g1}}$$

Results: $\Delta t = \dfrac{4.350 \text{ m} - 4.347 \text{ m}}{(23.4 \times 10^{-6}°C^{-1})(4.350 \text{ m}) - (9.0 \times 10^{-6}°C^{-1})(4.347 \text{ m})}$

$$\Delta t = \frac{0.003}{(0.000102) - (0.000039)}°C = 48°C$$

Then; $t_2 = t_1 - \Delta t = 35°C - 48°C = -13°C$

Comment: Since this is well within the possible ambient temperature for a building, a dangerous condition could be created by this window. The window frame and glass would contract without stress until a temperature of $-13°C$ was reached. If the temperature continued to decrease, the frame would contract faster than the glass and would generate stress in the glass. Of course, if the stress was great enough, the glass would fracture, possibly causing injury. The window should be reworked so there is a larger difference in size between the glass and the aluminum frame.

9–9 THERMAL STRESS

In the preceding section, parts that were subjected to changes in temperature were unrestrained, so that they could grow or contract freely. If the parts were held in such a way that deformation was resisted, stresses would be developed.

Consider a steel structural member in a furnace that is heated while the members to which it is attached are kept at a lower temperature. Assuming the ideal case, the supports would be considered rigid and immovable. Thus, all expansion of the steel member would be prevented.

If the steel part were allowed to expand, it would elongate by an amount $\delta = \alpha \cdot L \cdot \Delta t$. But since it is restrained, this represents the apparent total deformation in the steel. Then the unit strain would be

$$\epsilon = \frac{\delta}{L} = \frac{\alpha \cdot L \cdot \Delta t}{L} = \alpha(\Delta t) \qquad (9\text{–}11)$$

The resulting stress in the part can be found from

$$\sigma = E\epsilon$$

or

Thermal Stress ▼
$$\sigma = E\alpha(\Delta t) \qquad (9\text{–}12)$$

This stress would occur without the addition of external forces. You must check to ensure that the resulting stress level will not cause yielding or fracture of the material. Also, for relatively long, slender members in compression, the restraint of expansion can cause the members to buckle due to column action (see Chapter 16).

Example Problem 9–12   A steel structural member in a furnace is made from AISI 1020 steel and undergoes an increase in temperature of 95°F while being held rigid at its ends. Compute the resulting stress in the steel.

Solution   Objective: Compute the thermal stress in the steel.

Given: Steel is AISI 1020; from Table 9–4, $\alpha = 6.5 \times 10^{-6}°F^{-1}$.

$E = 30 \times 10^6$ psi; $\Delta t = 95°F$.

Analysis: Use Equation (9–12); $\sigma = E\alpha(\Delta t)$.

Results: $\sigma = (30 \times 10^6 \text{ psi})(6.5 \times 10^{-6}°F^{-1})(95°F) = 18\,500$ psi

Comment: Appendix A–14 shows the yield strength of annealed AISI 1020 steel, its weakest form, to be 43 000 psi. Therefore, the structural member would be safe from yielding. But column buckling should also be checked because the stress is compressive.

Example Problem 9–13   An aluminum rod of alloy 2014–T6 in a machine is held at its ends while being cooled from 95°C. At what temperature would the tensile stress in the rod equal half of the yield strength of the aluminum if it is originally at zero stress?

Solution   Objective: Compute the temperature when $\sigma = s_y/2$.

Given: Aluminum is alloy 2014-T6; from Table 9–4, $\alpha = 23.0 \times 10^{-6}°C^{-1}$.
From Appendix A–18, $s_y = 414$ MPa; $E = 73$ GPa; $t_1 = 95°C$.

Analysis: Use Equation (9–12) and solve for Δt.

$$\sigma = E\alpha(\Delta t)$$

$$\Delta t = \frac{\sigma}{E\alpha}$$

Let the stress be

$$\sigma = \frac{s_y}{2} = \frac{414 \text{ MPa}}{2} = 207 \text{ MPa}$$

Results:

$$\Delta t = \frac{\sigma}{E\alpha} = \frac{207 \text{ MPa}}{(73 \text{ GPa})(23.0 \times 10^{-6}\text{°C}^{-1})}$$

$$\Delta t = \frac{207 \times 10^6 \text{ Pa}}{(73 \times 10^9 \text{ Pa})(23.0 \times 10^{-6}\text{°C}^{-1})} = 123\text{°C}$$

Since the rod had zero stress when its temperature was 95°C, the temperature at which the stress would be 207 MPa would be

$$t = 95\text{°C} - 123\text{°C} = -28\text{°C}$$

Comment: The stress level of one-half of the yield strength is a reasonable design stress for a statically loaded member. Thus, if the temperature were to be lower than −28°C, the safety of the rod would be jeopardized.

Thermal Deformation and Stress Under Partial Restraint

There are many practical cases in which a member is originally free to expand due to temperature changes, but after some increase in temperature, external restraints begin to be applied.

Consider the situation shown in Figure 9–6. A brass bar made from C36000 brass, hard, is initially installed in a rigid frame with a small clearance. When the temperature is increased, a series of two actions takes place. First, the bar will expand as the temperature increases until the bar just touches the inside of the frame. For this part of the process, there is no stress produced in the bar. As the bar becomes restrained by the frame, no further expansion can occur and thermal stresses develop during the final part of the temperature rise.

Example Problem 9–14 illustrates how such a problem can be approached.

FIGURE 9–6 Brass bar for Example Problem 9–14.

Example Problem 9–14: The brass bar shown in Figure 9–6 is part of a conveyor that transports components into an oven. Initially, when the temperature is 15°C, there is a total clearance of 0.25 mm between the end of the bar and the inside of the frames on both sides. Describe what happens when the temperature increases from 15°C to 90°C. Consider the frames to be rigid and that they do not change dimension as the temperature rises.

Solution

Objective: Describe the behavior of the bar as the temperature rises.

Given: Initial length of bar: $L = 250$ mm
Initial gap: $\delta = 0.25$ mm
Material of bar: Brass C36000, hard. $\alpha = 20.5 \times 10^{-6}\text{°C}^{-1}$ (Table 9–4).

$$s_y = 310 \text{ MPa.} \quad E = 110 \text{ GPa (Appendix A–15)}$$

Analysis: **Step 1:** First determine what temperature rise will cause the bar to expand by 0.25 mm, bringing the end of the bar just into contact with the frame. Equation (9–10) can be used.

Step 2: Then determine how much additional temperature rise occurs from that point up to when the temperature is 90°C.

Step 3: Equation (9–12) can then be used to compute the stress developed in the bar during the final temperature rise. The safety of this stress should be evaluated.

Results: **Step 1:** The amount of temperature rise to elongate the bar 0.25 mm:

$$\delta = \alpha L(\Delta t_1)$$

Solving for Δt_1 gives

$$\Delta t_1 = \frac{\delta}{\alpha L} = \frac{0.25\,\text{mm}}{(20.5 \times 10^{-6}\text{°C}^{-1})(250\,\text{mm})} = 48.8°\text{C}$$

Step 2: The temperature at which this occurs is

$$t_2 = t_1 + \Delta t_1 = 15°\text{C} + 48.8°\text{C} = 63.8°\text{C}$$

Then the additional temperature rise to the maximum temperature is

$$\Delta t_2 = 90°\text{C} - 63.8°\text{C} = 26.2°\text{C}$$

Step 3: For this final temperature rise, the bar is considered to be fully restrained. Therefore a compressive stress is induced into the bar. Using Equation (9–12),

$$\sigma = E\alpha(\Delta t) = (110 \times 10^9\,\text{Pa})(20.5 \times 10^{-6}\text{°C}^{-1})(26.2°\text{C}) = 59.08\,\text{MPa}$$

Let's check this stress against yielding. Let $\sigma = \sigma_d = s_y/N$. Then

$$N = s_y/\sigma = 310\,\text{MPa}/59.08\,\text{MPa} = 5.25$$

LESSON 16

This is a sufficiently high design factor to ensure that yielding will not occur. However, column buckling should be checked (see Chapter 16).

9–10 MEMBERS MADE OF MORE THAN ONE MATERIAL

When two or more materials in a load-carrying member share the load, a special analysis is required to determine what portion of the load each material takes. Consideration of the elastic properties of the materials is required.

Figure 9–7 shows a steel pipe filled with concrete and used to support part of a large structure. The load is distributed evenly across the top of the support. We want to determine the stress in both the steel and the concrete.

Two concepts must be understood in deriving the solution to this problem.

1. The total load F is shared by the steel and the concrete such that $F = F_s + F_c$.

2. Under the compressive load F, the composite support deforms and the two materials deform in equal amounts. That is, $\delta_s = \delta_c$.

Now since the steel and the concrete were originally the same length,

$$\frac{\delta_s}{L} = \frac{\delta_c}{L}$$

But

$$\frac{\delta_s}{L} = \epsilon_s \quad \text{and} \quad \frac{\delta_c}{L} = \epsilon_c$$

Then the strains in the two materials are equal.

$$\epsilon_s = \epsilon_c$$

Also, from the definition of modulus of elasticity,

$$\epsilon_s = \frac{\sigma_s}{E_s} \quad \text{and} \quad \epsilon_c = \frac{\sigma_c}{E_c}$$

Then

$$\frac{\sigma_s}{E_s} = \frac{\sigma_c}{E_c}$$

FIGURE 9–7   Steel and concrete post.

Solving for σ_s yields

$$\sigma_s = \frac{\sigma_c E_s}{E_c} \qquad \text{(9–13)}$$

This equation gives the relationship between the two stresses.

Now consider the loads,

$$F_s + F_c = F$$

Since both materials are subject to axial stress,

$$\sigma_s = F_s/A_s \quad \text{and} \quad \sigma_c = F_c/A_c$$

or,

$$F_s = \sigma_s A_s \quad \text{and} \quad F_c = \sigma_c A_c$$

where A_s and A_c are the areas of the steel and concrete, respectively. Then

$$\sigma_s A_s + \sigma_c A_c = F \qquad \text{(9–14)}$$

Substituting Equation (9–13) into Equation (9–14) gives

$$\frac{A_s \sigma_c E_s}{E_c} + \sigma_c A_c = F$$

Now, solving for σ_c gives

$$\sigma_c = \frac{FE_c}{A_s E_s + A_c E_c} \qquad \text{(9–15)}$$

Equations (9–13) and (9–15) can now be used to compute the stresses in the steel and the concrete.

Example Problem 9–15   For the support shown in Figure 9–7, the pipe is a standard 6-in schedule 40 steel pipe completely filled with concrete. If the load F is 155 000 lb, compute the stress in the concrete and the steel. For steel use $E = 30 \times 10^6$ psi. For concrete use $E = 3.3 \times 10^6$ psi for a rated strength of $s_c = 3000$ psi (see Section 8–10).

Solution   **Objective:**   Compute the stress in the concrete and the steel.

Given: Load $= F = 155\,000$ lb; $E_s = 30 \times 10^6$ psi; $E_c = 3.3 \times 10^6$ psi.
From Appendix A–12, for a 6-in schedule 40 pipe:
$A_s = 5.581$ in^2; inside diameter $= d = 6.065$ in.

Analysis: Use Equation (9–15) to compute the stress in the concrete, σ_c. Then use Equation (9–13) to compute σ_s. All data are known except A_c. But

$$A_c = \frac{\pi d^2}{4} = \frac{\pi (6.065 \text{ in})^2}{4} = 28.89 \text{ in}^2$$

Results: Then in Equation (9–15)

$$\sigma_c = \frac{(155\,000 \text{ lb})(3.3 \times 10^6 \text{ psi})}{(5.581 \text{ in}^2)(30 \times 10^6 \text{ psi}) + (28.89 \text{ in}^2)(3.3 \times 10^6 \text{ psi})} = 1946 \text{ psi}$$

Using Equation (9–13) gives

$$\sigma_s = \frac{\sigma_c E_s}{E_c} = \frac{(1946 \text{ psi})(30 \times 10^6 \text{ psi})}{3.3 \times 10^6 \text{ psi}} = 17\,696 \text{ psi}$$

Comment: These stresses are fairly high. If it were desired to have at least a design factor of 2.0 based on the yield strength of the steel and 4.0 on the rated strength of the concrete, the required strengths would be

$$\text{Steel: } s_y = 2(17\,696 \text{ psi}) = 35\,392 \text{ psi}$$

$$\text{Concrete: Rated } \sigma_c = 4(1946 \text{ psi}) = 7784 \text{ psi}$$

LESSON 17

The steel could be similar to AISI 1020 annealed or any stronger condition. A rated strength of 3000 psi for the concrete would not be satisfactory.

Summary and Generalization

The analysis presented for Example Problem 9–14 can be generalized for any situation in which loads are shared by two or more members of two different materials provided all members undergo equal strains. Replacing the subscripts s and c in the preceding analysis by the more general subscripts 1 and 2,

we can state Equations (9–15) and (9–13) in the forms

$$\sigma_2 = \frac{FE_2}{A_1 E_1 + A_2 E_2} \qquad \text{(9–16)}$$

$$\sigma_1 = \frac{\sigma_2 E_1}{E_2} \qquad \text{(9–17)}$$

Composite Compression Members in Construction

The use of concrete-filled HSS members or concrete-encased column sections is common in building construction. AISC reference manuals list the design strength in axial compression (in kips) of such members as a function of their effective length in extensive sets of tables. Both direct compressive stress and column behavior (discussed later in this book) are considered. Covered are

- Square and rectangular steel HSS shapes from $2\frac{1}{2}$ in $\times$ $2\frac{1}{2}$ in to 20 in $\times$ 12 in (63.5 mm $\times$ 63.5 mm to 508 mm $\times$ 304.8 mm) and all common wall thicknesses. Included are four combinations of the strength of steel (ASTM A500, Grade B) and the rated strength of concrete:
  - Yield strength of steel = 42 ksi (290 MPa) and f'_c = 4 ksi (27.6 MPa)
  - Yield strength of steel = 42 ksi (290 MPa) and f'_c = 5 ksi (34.5 MPa)
  - Yield strength of steel = 46 ksi (317 MPa) and f'_c = 4 ksi (27.6 MPa)
  - Yield strength of steel = 46 ksi (317 MPa) and f'_c = 5 ksi (34.5 MPa)
- Round steel HSS (ASTM A500, Grade B) from 4.00 in to 20.00 in outside diameter (101.6 mm to 508 mm) and all common wall thicknesses. Two combinations of steel and concrete strength are included.
  - Yield strength of steel = 42 ksi (290 MPa) and f'_c = 4 ksi (27.6 MPa)
  - Yield strength of steel = 42 ksi (290 MPa) and f'_c = 5 ksi (34.5 MPa)
- Steel pipe (ASTM A53 Grade B) from 3 in to 12 in nominal size (75 mm to 310 mm) and all common wall thicknesses; Standard, XS, and XXS. These are combined with two rated strengths of concrete:
  - Yield strength of steel = 35 ksi (241 MPa) and f'_c = 4 ksi (27.6 MPa)
  - Yield strength of steel = 35 ksi (241 MPa) and f'_c = 5 ksi (34.5 MPa)

9–11 STRESS CONCENTRATION FACTORS FOR DIRECT AXIAL STRESSES

In defining the method for computing stress due to a direct tensile or compressive load on a member, it was emphasized that the member must have a uniform cross section in order for the equation $\sigma = F/A$ to be valid. The reason for this restriction is that where a change in the geometry of a loaded member occurs, the actual stress developed is higher than would be predicted by the standard equation. This phenomenon is called *stress concentration* because detailed studies reveal that localized high stresses

FIGURE 9–8 Stress distribution near a change in geometry.

appear to concentrate around sections where geometry changes occur.

Figure 9–8 illustrates the case of stress concentration for the example of an axially loaded round bar in tension that has two diameters with a step between them. Note that there is a small fillet at the base of the step. Its importance will be discussed later. Below the sketch of the stepped bar is a graph of stress versus position on the bar. At section 1, where the bar diameter is D and well away from the step, the stress can be computed from

$$\sigma_1 = F/A_1 = F/(\pi D^2/4)$$

At section 2, where the bar diameter has the smaller value of d, the stress is

$$\sigma_2 = F/A_2 = F/(\pi d^2/4)$$

Then the plot of stress versus position might be expected to appear as the straight lines with an abrupt step at the location of the change in diameter. But tests would show that the actual stress distribution would look more like the curved line: blending with the two straight lines well away from the step, but rising much higher near the step itself.

To account for the higher-than-predicted stress at the step, we modify the direct stress formula to include a *stress concentration factor*, K_t, to produce the form shown in Equation (9–18).

$$\sigma_{max} = K_t \sigma_{nom} \qquad (9\text{–}18)$$

where, in this case, the nominal stress is based on the smaller section 2. That is,

$$\sigma_{nom} = \sigma_2 = F/A_2 = F/(\pi d^2/4)$$

Then the value of K_t represents the factor by which the actual stress is higher than the nominal stress computed by the standard formula.

FIGURE 9–9 Photoelastic model showing stress concentration at a hole in an axially loaded flat bar.

Source: Measurements Group, Inc., Raleigh, NC.

Figure 9–9 shows a photoelastic model of another case that demonstrates the phenomenon of stress concentration. The transparent plastic model is a straight rectangular bar of constant thickness carrying an axial load. Near the middle of the bar there is a hole on the axis of the member. The nominal stress in the vicinity of the hole naturally increases because of the removal of material. But the actual maximum stress near the hole is higher than the nominal stress as indicated by the concentration of dark fringes. Appendix A–22–4, Curve A, provides data for the magnitude of K_t for this case. The value is dependent on the ratio of the hole diameter to the full width of the bar.

Stress concentrations are most damaging for dynamic loading such as repeated loads, impact, or shock. Indeed, fatigue failures most often occur near locations of stress concentrations with small local cracks that grow with time until the remaining section is unable to withstand the load. Under static loading, the high stress near the discontinuity may cause local yielding that would result in a redistribution of the stress to an average value less than the yield strength, and thus the part would still be safe.

Values of Stress Concentration Factors

The magnitude of the stress concentration factor, K_t, depends on the geometry of the member in the vicinity of the discontinuity. Most data have been obtained experimentally by carefully measuring the maximum stress, σ_{max}, using experimental stress analysis techniques such as strain gaging or photoelasticity. (See Section 7–9 in Chapter 7.) Computerized approaches using finite element analysis could also be used. Then, the value of K_t is computed from

Stress Concentration Factor ▼

$$K_t = \sigma_{max}/\sigma_{nom} \qquad (9-19)$$

where σ_{nom} is the stress that would be computed at the section of interest without considering the stress concentration. In the case now being discussed, direct tensile stress, $\sigma_{nom} = F/A$.

Appendix A–22 contains several charts that can be used to determine the value of K_t for a variety of geometries. See References 8 and 10 for many more cases.

A–22–1: Axially loaded round bar in tension with a circular groove

A–22–2: Axially loaded round bar in tension with a step and a fillet

A–22–3: Axially loaded flat plate in tension with a step and a fillet

A–22–4: Flat plate with a central hole

A–22–5: Round bar with a transverse hole

The chart in Appendix A–22–1 shows the typical pattern for presenting values for stress concentration factors. The vertical axis gives the value of K_t itself. The pertinent geometry factors are the diameter of the full round section, D, the diameter at the base of the groove, d_g, and the radius of the circular groove, r. From these data, two parameters are computed. The horizontal axis is the ratio of r/d_g. The family of curves in the chart is for different values of the ratio of D/d_g. The normal use of such a chart when the complete geometry is known is to enter the chart at the value of r/d_g, then project vertically to the curve of D/d_g, then horizontally to the vertical axis to read K_t. Interpolation between curves in the chart is often necessary. Note that the nominal stress for the grooved round bar is based on the stress at the *bottom of the groove*, the smallest area in the vicinity. While this is typical, it is important for you to know the basis of the nominal stress for any given stress concentration chart. The grooves with a circular bottom are often used to distribute oil or other lubricants to a shaft.

The chart in Appendix A–22–2 for the stepped round bar has three geometry factors: the larger diameter, D, the smaller diameter, d, and the radius of the fillet, r, at the step where the change of diameter takes place. Note that the value of K_t rises rapidly for small values of the fillet radius. As a designer, you should provide the largest practical radius for such a fillet to maintain a relatively low maximum stress at the step.

An important use for the chart in Appendix A–22–2 is the analysis of stress concentration factors for round bars having retaining ring grooves, as shown in Figure 9–10. The typical geometry of the groove, specified by the ring manufacturer, is shown in Figure 9–11. The bottom of the groove is flat and the fillet at each end is quite small to provide a sizeable vertical surface against which to seat the ring. The result is that the ring groove acts like two steps close together. Then Appendix A–22–2 can be used to determine the value of K_t. Sometimes the geometry of the ring groove will result in K_t values that are off the top of the chart. In such cases, an estimate of $K_t = 3.0$ is reasonable but additional data should be sought.

FIGURE 9–10 Stepped shaft with a ring groove.

FIGURE 9–11 Sample geometry for a retaining ring groove in a round bar.

The chart in Appendix A–22–4 contains three curves, all related to a flat plate with a central hole. Curve A is for the case where the plate is subjected to direct tensile stress across its entire cross section in the vicinity of the hole (see Figure 9–9). Curve B is for the case where a close-fitting pin is inserted in the hole and the tensile load is applied through the pin. The resulting stress concentration factors are somewhat larger because of the more concentrated load. Curve C is for the case of the plate in bending, and this will be dis-

cussed in Chapter 12. In each case, however, note that the nominal stress is based on the *net section* through the plate at the location of the hole. For tensile loading, the *net area* is used for σ_{nom}. That is,

$$\sigma_{nom} = F/A_{net} = F/(w - d)t$$

where w = width of the plate
t = thickness
d = diameter of the hole

The chart in Appendix A–22–5 for the round bar with a transverse hole contains a variety of data for different loading patterns: tension, bending, and torsion. For now we are concerned with only Curve A for the case of axial tension. Torsion is discussed in Chapter 10 and bending in Chapter 12. Also note that the stress concentration factor is based on the *gross section*, not on the net section at the hole. This means that K_t includes the effects of both the removal of material and the discontinuity, resulting in values that are quite high. However, it makes the use of the chart much easier for you, the designer, because you do not have to compute the area of net section.

Example Problem 9–16   The stepped bar shown in Figure 9–8 is subjected to an axial tensile force of 12 500 lb. Compute the maximum tensile stress in the bar for the following dimensions:

$$D = 1.50 \text{ in}; \qquad d = 0.75 \text{ in}; \qquad r = 0.060 \text{ in}$$

Solution   Objective: Compute the maximum tensile stress.

Given: F = 12 500 lb; D = 1.50 in; d = 0.75 in; r = 0.060 in

Analysis: Because of the change in diameter, use Equation (9–18).
Use the chart in Appendix A–22–2 to find the value of K_t using r/d and D/d as parameters.

Results: $\sigma_{max} = K_t \sigma_{nom}$
$\sigma_{nom} = \sigma_2 = F/A_2 = F/(\pi d^2/4) = (12\,500 \text{ lb})/[\pi(0.75 \text{ in})^2/4]$
$\sigma_{nom} = 28\,294 \text{ lb/in}^2$
$r/d = 0.06/0.75 = 0.080$ and $D/d = 1.50/0.75 = 2.00$
Read $K_t = 2.12$ from Appendix A–22–2.
Then $\sigma_{max} = K_t \sigma_{nom} = 2.12\,(28\,294 \text{ psi}) = 59\,983 \text{ psi}$.

LESSON 14

Comment: The actual maximum stress of approximately 60 000 psi is more than double the value that would be predicted by the standard formula.

9–12 BEARING STRESS

When one solid body rests on another and transfers a load normal to it, the form of stress called *bearing stress* is developed at the surfaces in contact. Similar to direct compressive stress, the bearing stress, called σ_b, is a measure of the tendency for the applied force to crush the supporting member.

Bearing stress is computed in a manner similar to direct normal stresses:

Bearing Stress ▼

$$\sigma_b = \frac{\text{Applied load}}{\text{Bearing area}} = \frac{F}{A_b} \qquad (9\text{--}20)$$

For flat surfaces in contact, the bearing area is simply the area over which the load is transferred from one member to the other. If the two parts have different areas, the smaller area is used. Another requirement is that the materials transmitting the loads must remain nearly rigid and flat in order to maintain their ability to carry the loads. Excessive deflection will reduce the effective bearing area.

Figure 9–12 shows an example from building construction in which bearing stress is important. A hollow steel column 4.00 in square rests on a thick steel plate 6.00 in square. The plate rests on a concrete pier, which in turn rests on a base of gravel. These successively larger areas are necessary to limit bearing stresses to reasonable levels for the materials involved.

FIGURE 9–12 Bearing stress example.

Example Problem 9–17

Refer to Figure 9–12. The square steel tube carries 30 000 lb of axial compressive force. Compute the compressive stress in the tube and the bearing stress between each mating surface. Consider the weight of the concrete pier to be 338 lb.

Solution

Objectives: Compute the compressive stress in the tube. Compute the bearing stress at each surface.

Given: Load $F = 30\,000$ lb compression. Weight of pier = 338 lb. Geometry of members shown in Figure 9–12.

Analysis: Tube: Use direct compressive stress formula. Bearing stresses: Use Equation (9–20) for each pair of mating surfaces.

Results: *Compressive Stress in the Tube:* (area $= A = 3.37$ in^2)

$$\sigma = F/A = 30\,000\ \text{lb}/3.37\ \text{in}^2 = 8902\ \text{psi}$$

Bearing Stress Between Tube and Square Plate: This will be equal in magnitude to the compressive stress in the tube because the cross-sectional area of the tube is the smallest area in contact with the plate. Then

$$\sigma_b = 8902\ \text{psi}$$

Bearing Stress Between Plate and Top of Concrete Pier: The bearing area is that of the square plate because it is the smallest area at the surface.

$$\sigma_b = F/A_b = 30\,000\ \text{lb}/(6.00\ \text{in})^2 = 833\ \text{psi}$$

Bearing Stress Between the Pier and the Gravel: The bearing area is that of a square, 24 in on a side. Add 338 lb for the weight of the pier.

$$\sigma_b = F/A_b = 30\,338\ \text{lb}/(24.00\ \text{in})^2 = 52.7\ \text{psi}$$

Comment: Allowable bearing stresses are discussed later.

Bearing Stresses in Pin Joints

Frequently in mechanical and structural design, cylindrical pins are used to connect components together. One design for such a connection is shown in Figure 9–13. See also the photograph of this connection in Figure 7–13. When transferring a load across the pin, the bearing stress between the pin and each mating member should be computed.

The effective bearing area for the cylindrical pin in a close-fitting hole requires that the *projected area* be used, computed as the product of the diameter (D) of the pin and the length (L) of the surface in contact. That is,

Bearing Area for a Pin Joint ▽

$$A_b = D \times L \qquad\qquad (9\text{–}21)$$

Top view

w = width

D

Pin

t = thickness

Retaining ring

Link

Force on link

Reaction force

Base

Side view

Pictorial view

FIGURE 9–13 Pin connection.

Example Problem 9–18

Refer to Figure 9–13. Compute the bearing stress between the 10.0-mm-diameter pin and the mating hole in the link. The force applied to the link is 3550 N. The thickness of the link is 15.0 mm and its width is 25.0 mm.

Solution

Objective: Compute the bearing stress between the mating surfaces of the pin and the inside of the hole in the link.

Given: Load = F = 3550 N. t = 15.0 mm; w = 25.0 mm; D = 10.0 mm. Geometry of members shown in Figure 9–13.

Analysis: Bearing stresses: Use Equation (9–20) for each pair of mating surfaces. Use the projected area of the hole for the bearing area.

Results: **Between the Pin and the Link:** $L = t$ = thickness of the link. From Equation (9–21),

$$A_b = D \times t = (10.0 \, \text{mm})(15.0 \, \text{mm}) = 150 \, \text{mm}^2$$

Then the bearing stress is

$$\sigma_b = \frac{3550 \, \text{N}}{150 \, \text{mm}^2} = 23.7 \, \text{N/mm}^2 = 23.7 \, \text{MPa}$$

Contact Stress

Cases of bearing stress considered earlier in this section are those in which *surfaces* are in contact and the applied force is distributed over a relatively large area. When the load is applied over a very small area, the concept of *contact stress* must be used.

Examples of contact stress situations are shown in Figure 9–14 and include the following examples:

- Cylindrical roller on a flat plate such as a steel railroad wheel on a rail
- Spherical ball on a flat plate such as a ball transfer device to move heavy loads
- Spherical ball on curved plate such as a ball bearing rolling in its outer race

- Two convex curved surfaces such as gear teeth in contact

The detailed analyses of contact stresses, sometimes called *Hertz stresses*, are not developed in this book. But it is important to recognize that the magnitude of contact stresses can be very high.

Consider the case of a spherical ball on a flat plate carrying a downward load. A perfectly spherical surface will contact a flat plane at only a single, infinitesimally small point. Then, when applying the bearing stress relationship, $\sigma_b = F/A_b$, the magnitude of the area approaches *zero*. Then the stress approaches *infinity*. Actually, because of the elasticity of the materials in contact, there will be some deformation and the contact area becomes a finite, but small, circular area. But the

Line of contact

Cylinder on a flat surface

Point of contact

Sphere on a flat surface

Side view

Contact points

Part of a ball bearing

Line of contact — Length of line equals face width of gear teeth perpendicular to the page.

Two gear teeth in contact

FIGURE 9–14   Examples of load-carrying members subjected to contact stress.

local stress will still be very large. For this reason, load-carrying members subjected to contact stresses are typically made from very hard, high-strength materials.

Similarly, when a cylindrical roller contacts a flat plate, the contact is theoretically a line having a zero width. Therefore, the bearing area is theoretically zero. The elasticity of the materials will produce an actual bearing area that is a narrow rectangle, again resulting in a finite, but large, contact stress. The special case of steel rollers on steel plates is discussed next. See Reference 10 for more detailed analysis.

9–13 DESIGN BEARING STRESS

Bearing stress, defined in Section 9–12, is a localized phenomenon created when two load-carrying parts are placed in contact. This section describes several special cases in-

volving different materials and different geometries of the mating surfaces. Steel, aluminum, concrete, masonry, and soils are discussed.

For some applications a *design bearing stress*, σ_{bd}, is used, defined as

$$\sigma_{bd} = s_y/N \quad \text{or} \quad \sigma_{bd} = Cs_y$$

where N is a design factor and C is a specified coefficient. Note that $C = 1/N$. If the mating parts have different strengths, the lower yield strength should be used. The actual bearing stress must be less than the design bearing stress.

For some applications it is more convenient to specify an *allowable bearing load*, R_a. Then the actual contact load must be lower than the allowable bearing load.

Steel

According to the AISC (see Reference 2), the allowable bearing stress in steel for flat surfaces or on the projected area of pins in reamed, drilled, or bored holes is

Design Bearing Stress for Steel ▼

$$\sigma_{bd} = 0.90 s_y \qquad (9\text{--}22)$$

When rollers or rockers are used to support a beam or other load-carrying member to allow for expansion of the member, the bearing stress is dependent on the diameter of the roller or rocker, d, and its length, L. The stress is inherently very high because the load is carried on only a small rectangular area. Theoretically, the contact between the flat surface and the roller is simply a line, but because of the elasticity of the materials the actual area is rectangular.

Instead of specifying an allowable bearing stress, the AISC standard (see Reference 2) calls for the computation of the allowable bearing load, R_a, from

Allowable Bearing Load for Steel Roller ▼

$$R_a = (s_y - 13)(0.03)(d)(L) \qquad (9\text{--}23)$$

where s_y is in ksi
d and L are in inches and $d \leq 25$ in
R_a is in kips

When using SI metric units, Equation (9–23) becomes

$$R_a = (s_y - 90)(3.0 \times 10^{-5})(d)(L) \qquad (9\text{--}24)$$

where s_y is in MPa
d and L are in mm and $d \leq 635$ mm
R_a is in kN

Example Problem 9–19   A short beam, shown in Figure 9–15, is made from a rectangular steel bar, 1.25 in thick and 4.50 in high. At each end, the length resting on a steel plate is 2.00 in. If both the bar and the plate are made from ASTM A36 structural steel, compute the maximum allowable load, W, which could be carried by the beam, based on only the bearing stress at the supports. The load is centered between the supports.

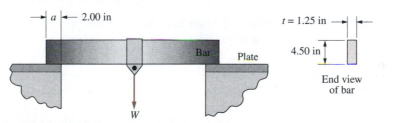

FIGURE 9–15   Beam for Example Problem 9–19.

Solution   Objective: Compute the allowable load, W, based on only bearing stress.

Given: Loading in Figure 9–15.
Bearing area at each end: 2.00 in by 1.25 in (that is, $A_b = t \cdot a$).
Material: ASTM A36 structural steel ($s_y = 36\,000$ psi).

Analysis: Design bearing stress: $\sigma_{bd} = R_a / A_b$
Where R_a is the allowable reaction at the support and $R_a = W/2$.
Design stress Equation (9–22):

$$\sigma_{bd} = 0.90\, s_y = 0.90(36\,000 \text{ psi}) = 32\,400 \text{ psi}$$

Then $R_a = A_b \sigma_{bd}$ and $W = 2R_a$.

Results: Bearing area: $A_b = t \cdot a = (1.25 \text{ in})(2.00 \text{ in}) = 2.50 \text{ in}^2$

$$R_a = A_b \sigma_{bd} = (2.50 \text{ in}^2)(32\,400 \text{ lb/in}^2) = 81\,000 \text{ lb}$$
$$W = 2R_a = 2(81\,000 \text{ lb}) = 162\,000 \text{ lb} = 162 \text{ kip}$$

Comments: This is a very large load, so it is unlikely that bearing is the mode of failure for this beam.

Example Problem 9–20

An alternative proposal for supporting the bar described in Example Problem 9–19 is shown in Figure 9–16. To allow for expansion of the bar, the left end is supported on a 2.00-in-diameter roller made from AISI 1040 CD steel. Compute the allowable load, W, for this arrangement.

FIGURE 9–16 Beam for Example Problem 9–20.

Roller
2.00-in dia.

Bar

Plate

W

Solution Objective: Compute the allowable load on the beam.

Given: Loading in Figure 9–16.
Width of the beam ($t = 1.25$ in) rests on the roller ($d = 2.00$ in).
Beam material: ASTM A36 structural steel ($s_y = 36$ ksi).
Roller material: AISI 1040 CD ($s_y = 71$ ksi).

Analysis: At the roller support, Equation (9–23) applies to determine the allowable load. Note that $W = 2R_a$. Use $s_y = 36$ ksi, the weaker of the two materials in contact.

Results: The allowable bearing load is

$$R_a = (36 - 13)(0.03)(2.00)(1.25) = 1.73 \text{ kip}$$

This would be the allowable reaction at each support. The total load is

$$W = 2R_a = 2(1.73 \text{ kip}) = 3.45 \text{ kip}$$

Comment: Note that this is significantly lower than the allowable load for the surfaces found in Example Problem 9–19. Bearing at the roller support may, indeed, limit the load that could be safely supported.

Aluminum

The Aluminum Association (see Reference 1) bases the allowable bearing stresses on aluminum alloys for flat surfaces and pins on the *bearing yield strength*.

$$\sigma_{bd} = \frac{\sigma_{by}}{2.48} \qquad (9\text{–}25)$$

The minimum values for bearing yield strength are listed in Reference 1. But many references, including the appendix tables in this book, do not include these data. An analysis of the data shows that for most aluminum alloys, the bearing yield strength is approximately 1.60 times larger than the tensile yield strength. Then Equation (9–25) can be restated as

Design Bearing Stress for Aluminum ▼

$$\sigma_{bd} = \frac{1.60s_y}{2.48} = 0.65s_y \qquad (9\text{–}26)$$

We will use this form for design bearing stress for aluminum in this book.

Example Problem 9–21

A rectangular bar is used as a hanger, as shown in Figure 9–17. Compute the allowable load on the basis of bearing stress at the pin connection if the bar and the clevis members are made from 6061-T4 aluminum. The pin is to be made from a stronger material.

Solution Objective: Compute the allowable load on the hanger.

Given: Loading in Figure 9–17. Pin diameter $= d = 18$ mm.
Thickness of the hanger $= t_1 = 25$ mm; width $= w = 50$ mm.
Thickness of each part of clevis $= t_2 = 12$ mm.
Hanger and clevis material: aluminum 6061-T4 ($s_y = 145$ MPa).
Pin is stronger than hanger or clevis.

FIGURE 9–17 Hanger for Example Problem 9–21.

Analysis: For cylindrical pins in close-fitting holes, the bearing stress is based on the *projected* area in bearing, found from the diameter of the pin times the length over which the load is distributed.

$$\sigma_b = \frac{F}{A_b} = \frac{F}{dL}$$

Let $\sigma_b = \sigma_{bd} = 0.65\, s_y$ for aluminum 6061-T4.
Bearing area for hanger: $A_{b1} = t_1 d = (25\,\text{mm})(18\,\text{mm}) = 450\,\text{mm}^2$.
This area carries the full applied load, W.
For each side of clevis: $A_{b2} = t_2 d = (12\,\text{mm})(18\,\text{mm}) = 216\,\text{mm}^2$.
This area carries 1/2 of the applied load, W/2.
Because A_{b2} is less than 1/2 of A_{b1}, bearing on the clevis governs.

Results: $\sigma_{bd} = 0.65\, s_y = 0.65\,(145\,\text{MPa}) = 94.3\,\text{MPa} = 94.3\,\text{N/mm}^2$
$\sigma_b = \sigma_{bd} = (W/2)/A_{b2}$
Then $W = 2(A_{b2})(\sigma_{bd}) = 2(216\,\text{mm}^2)(94.3\,\text{N/mm}^2) = 40\,740\,\text{N}$.

Comment: This is a very large force, and other failure modes for the hanger would have to be analyzed. Failure could occur by shear of the pin or tensile failure of the hanger bar or the clevis.

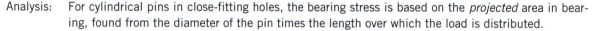

Masonry

The lower part of a support system for a building is often made from concrete, brick, or stone. Loads transferred to such supports usually require consideration of bearing stresses because the strengths of these materials are relatively low compared with metals. It should be noted that the actual strengths of the materials should be used whenever possible because of the wide variation of properties. Also, some building codes list allowable bearing stresses for certain kinds of masonry.

In the absence of specific data, this book uses the allowable bearing stresses shown in Table 9–7.

Note that the allowable bearing stress on concrete is dependent on the rated strength of the concrete and on how the load is applied. The bearing area is taken to be just the area of the steel plate, A_1, that transfers the load to the concrete. The area of the concrete is called A_2 and it can be equal to or greater than A_1. The ratio of A_2/A_1 is a factor in calculating the allowable bearing stress as shown in Table 9–7.

When bearing is on the full area of the concrete (A_2/A_1 is 1.0), the allowable bearing stress is

$$\sigma_{bd} = 0.34\, f_c'$$

where f_c' = rated strength of the concrete typically 2000 psi to 7000 psi (14 MPa to 48 MPa).

A higher allowable bearing stress can be used if the area of the concrete is larger than the area of the steel bearing plate. This is because the concrete for a moderate distance

TABLE 9–7 Allowable Bearing Stresses on Masonry and Soils for Use in This Book

| Material | Allowable bearing stress, σ_{bd} | |
	psi	MPa
Sandstone and limestone	400	2.76
Brick in cement mortar	250	1.72
Solid hard rock	350	2.41
Shale or medium rock	140	0.96
Soft rock	70	0.48
Hard clay or compact gravel	55	0.38
Soft clay or loose sand	15	0.10

Concrete: $\sigma_{bd} = Kf'_c = (0.34\sqrt{A_2/A_1})f'_c$ (But maximum $\sigma_{bd} = 0.68f'_c$)

where f'_c = rated strength of concrete
A_1 = bearing area
A_2 = full area of the support

$A_2/A_1 = 1.0$

$A_2/A_1 > 1.0$
A_1 = Bearing area
A_2 = Support area

beyond the plate is stressed and resists some of the load. When $A_2/A_1 > 1.0$,

$$\sigma_{bd} = 0.34\sqrt{A_2/A_1 f'_c} = Kf'_c$$

Table 9–7 shows a graph of K versus the ratio A_2/A_1. Note that the maximum allowable bearing stress is $0.68\, f'_c$. This corresponds to the area ratio $A_2/A_1 = 4.0$. The support area A_2 must be symmetrical around the plate.

Soils

The masonry or concrete supports are often placed on soils to transfer the loads directly to the earth. *Marks' Standard Handbook for Mechanical Engineers* (see Reference 3) lists the values for the safe bearing capacity of soils, as shown in Table 9–7. Variations should be expected, and test data should be obtained where possible.

Example Problem 9–22   Figure 9–18 shows a column resting on a foundation and carrying a load of 26 000 lb. Determine if the bearing stresses are acceptable for the concrete and the soil. The concrete has a specified rated strength of 2000 psi, and the soil is compact gravel.

Solution **Objective:**   Determine if bearing stresses on the concrete and the soil are safe.

Given: Foundation shown in Figure 9–18: Load = F = 26 000 lb.
For concrete: f'_c = 2000 psi.
For soil (compact gravel): σ_{bd} = 55 psi (Table 9–7).

FIGURE 9–18   Column foundation for Example Problem 9–22.

Analysis and Results: For concrete: The load is transferred from the column to the concrete through the 12-in-square steel plate ($A_1 = 144\,in^2$). But the concrete pier is 18 in square ($A_2 = 324\,in^2$). Therefore, the bearing load is on less than the full area of the concrete. Then, from Table 9–7,

$$\sigma_{bd} = Kf'_c = 0.34\sqrt{\frac{A_2}{A_1}}f'_c = 0.34\sqrt{\frac{324}{144}}f'_c = 0.510f'_c$$

Note that $K < 0.68$. Then

$$\sigma_{bd} = 0.510(2000\,psi) = 1020\,psi$$

The bearing stress exerted on the concrete by the steel plate at the base of the column is

$$\sigma_b = \frac{F}{A_b} = \frac{26\,000\,lb}{(12\,in)^2} = 180\,psi \; < \sigma_{bd}$$

Thus, the bearing stress on the concrete is acceptable.
For the soil (gravel) at the base of the foundation:

$$\sigma_b = \frac{F}{A_b} = \frac{26\,000\,lb}{(36\,in)^2} = 20.1\,psi$$

This is acceptable because the allowable bearing stress for compact gravel is 55 psi as shown in Table 9–7.

9–14 DESIGN SHEAR STRESS

When members are subjected to shear stresses, as discussed in Chapter 7, Section 7–7, design must be based on the design shear stress, τ_d.

Design Shear Stress ▼

$$\tau_d = \frac{s_{ys}}{N} \quad \text{based on the yield strength in shear} \quad \text{(9–27)}$$

Yield strength in shear.

The yield strength in shear, s_{ys}, is the level of shear stress at which the material would exhibit the phenomenon of yield. That is, it would undergo a significant amount of shear deformation with little or no increase in applied shear-type loading.

Virtually all designs for members in shear would require that the actual shear stress be well below the value of s_{ys}, as indicated by Equation (9–27). The selection of the design factor is taken from Table 9–8. Also consult Section 9–4 for other considerations in the selection of a design factor. Conditions that are more severe than normally encountered or where there is a significant amount of uncertainty about the magnitude of loads or material properties would justify higher design factors.

Of course, if the values of the yield strength in shear are available, they can be used in the design stress equations. But unfortunately, such values are frequently not reported, and it is necessary to rely on estimates. For the yield strength in shear, a frequently used estimate is

Estimate for Yield Strength in Shear ▼

$$s_{ys} = \frac{s_y}{2} = 0.5\, s_y \quad \text{(9–28)}$$

This value is taken from the observation of a typical tensile test in which the shear stress is one-half of the direct tensile stress. This phenomenon, related to the *maximum shear stress theory of failure*, is somewhat conservative and will be discussed further in Chapter 15.

Ultimate strength in shear.

The ultimate strength in shear, s_{us}, is the level of shear stress at which the material would actually fracture.

There are some practical applications of shear stress where fracture of the shear-loaded member is *intended* and, therefore, an estimate of s_{us} is needed. Examples include the *shear pin* often used as an element in the drive train of machines

(a) Cutaway drawing of a propeller shaft employing a shear pin

(b) Photograph of a model of a pinned hub

FIGURE 9–19   Drive pin thought a propeller and hub for Example Problem 9–23.

Source of model: Gary Drigel.

having expensive components. Figure 9–19 shows a propeller drive shaft for a boat in which the torque from the drive shaft is transmitted through the pin to the hub of the propeller. The pin must be designed to transmit a level of torque that is typically encountered in moving the boat through the water. However, if the propeller should encounter an obstruction such as a submerged log, it would be desirable to have the inexpensive pin fail rather than the costly propeller. See Example Problem 9–23.

Another example where an estimate of the ultimate strength in shear is needed is the case of the punching operation as described in Chapter 7 and shown in Section 7–7. Here, the punch is expected to completely cut (shear) the desired part from the larger sheet of material. Therefore, the sheared sides of the part must be stressed up to the ultimate strength in shear.

When data for the ultimate strength in shear are known, they should be used. For example, Appendix A–17 gives some values for cast irons and Appendix A–18 gives data for aluminum alloys. But, for occasions when published data are not available, estimates can be computed from the relationships in Table 9–9, taken from Reference 5.

TABLE 9–8	Design Stress Guidelines for Shear	
	Design stress–Ductile materials	
Manner of loading	$t_d = s_{ys}/N = 0.5s_y/N = s_y/2N$	
Static	Use $N = 2$	$\tau_d = s_y/4$
Repeated	Use $N = 4$	$\tau_d = s_y/8$
Impact	Use $N = 6$	$\tau_d = s_y/12$

Formula	Material
$s_{us} = 0.65\,s_u$	Aluminum alloys
$s_{us} = 0.82\,s_u$	Steel
$s_{us} = 0.90\,s_u$	Malleable iron and copper alloys
$s_{us} = 1.30\,s_u$	Gray cast iron

TABLE 9–9 Estimates for the Ultimate Strength in Shear

Brittle Materials

Design shear stresses for brittle materials are based on the ultimate strength in shear because they do not exhibit yielding. A higher design factor should be used than for ductile materials because the materials are often less consistent in structure. However, published data on acceptable design factors are lacking. It is recommended that testing be performed on actual prototypes of shear-loaded members made from brittle materials.

Design Approaches

The general design approaches for components subjected to shearing stresses are very similar to those outlined in Cases A–D in Section 9–5. But the design stress in shear, τ_d, should be used in place of the design normal stress, σ_d. You should adapt the design approaches from Section 9–5 to outline the solution procedure depending on whether your goal is to

a. Evaluate the safety of a given design
b. Specify a material for a component in shear
c. Specify the shape and size of the component to resist applied shearing forces
d. Determine the allowable shearing force on a component

See Table 9–8 for guidelines for design factors when ductile materials are used.

When doing an analysis to determine the shearing force that will actually cause failure in shear, the design factor is taken to be 1.0. This applies to the case of shearing a plug of material from a larger sheet (see Section 7–7) or the case of a shear pin as analyzed in Example Problem 9–24.

Example Problem 9–23

Figure 9–19 shows a boat propeller mounted on a shaft with a cylindrical drive pin inserted through the hub and the shaft. The torque required to drive the propeller is 1575 lb·in and the shaft diameter inside the hub is 1.75 in. Usually, the torque is steady, and it is desired to design the pin to be safe for this condition. Specify a suitable material and the diameter for the pin.

Solution **Objective:** Specify a material and the diameter for the pin.

Given: Torque = T = 1575 lb·in (steady).
Shaft diameter = D = 1.75 in.

Analysis:
1. The pin would be subjected to direct shear at the interface between the shaft and the inside of the hub, as shown in Figure 9–20. The applied torque from the shaft results in two equal forces acting perpendicular to the axis of the pin on opposite sides of the shaft, forming a couple. That is,

$$T = FD$$

Then $F = T/D$.

2. A material with a moderate-to-high strength is desirable so that the pin will not be overly large. Also, it should have good ductility because of the likelihood of mild shock loading from time to time. Several materials could be chosen.

FIGURE 9–20 Cross section through propeller hub and shaft.

3. Stress analysis: $\tau = F/A$, where $A = \pi d^2/4$. This is one cross-sectional area for the pin.
4. Design shear stress: $\tau_d = s_y/4$ (Table 9–8).
5. Let $\tau = \tau_d$. Then the required $A = F/\tau_d$ and the required d is

$$d = \sqrt{4A/\pi}$$

6. Specify a convenient standard size for the pin.

Results: 1. $F = T/D = (1575\,\text{lb}\cdot\text{in})/(1.75\,\text{in}) = 900\,\text{lb}$
2. Specify AISI 1020 cold drawn as a trial ($s_y = 64\,000\,\text{psi}$; 15% elongation). Note that the pin would have to be protected from corrosion.
3, 4, 5. $\tau_d = s_y/4 = 64\,000\,\text{psi}/4 = 16\,000\,\text{psi}$

$$A = F/\tau_d = (900\,\text{lb})/(16\,000\,\text{lb/in}^2) = 0.05625\,\text{in}^2$$

$$d = \sqrt{4A/\pi} = \sqrt{4(0.05626\,\text{in}^2)/\pi} = 0.260\,\text{in}$$

6. From Appendix A–2, specify $d = 0.3125\,\text{in}$ ($\frac{5}{16}$ in).
Material: AISI 1020 cold drawn steel.

Comment: The size seems reasonable compared to the diameter of the shaft.

When the drive pin designed in Example Problem 9–23 is subjected to an overload such as that seen when striking a log, it is desirable for the pin to shear rather than to damage the propeller. The following example problem considers this situation.

Example Problem 9–24   Compute the torque required to shear the pin designed in Example Problem 9–23 and shown in Figures 9–19 and 9–20.

Solution Objective: Compute the torque required to shear the pin.

Given: Design as shown in Figures 9–19 and 9–20.

$D = 1.75\,\text{in}$; $d = 0.3125\,\text{in}$.

Material: AISI 1020 cold drawn steel. $s_u = 75\,000\,\text{psi}$.

Analysis: The analysis would be similar to the reverse of that for Example Problem 9–23, as summarized below.
1. The pin material would fail when $\tau = s_{us} = 0.82\,s_u$ (Table 9–9).
2. $\tau = F/A$. Then $F = \tau A = (0.82\,s_u)A$.
Also, $A = \pi d^2/4$.
3. $T = FD$

Results: 1. $s_{us} = 0.82\,s_u = 0.82\,(75\,000\,\text{psi}) = 61\,500\,\text{psi}$

2. $A = \pi d^2/4 = A = \pi\,(0.3125\,\text{in})^2/4 = 0.0767\,\text{in}^2$

$F = \tau A = (61\,500\,\text{lb/in}^2)(0.0767\,\text{in}^2) = 4717\,\text{lb}$

3. $T = FD = (4717\,\text{lb})(1.75\,\text{in}) = 8255\,\text{lb}\cdot\text{in}$

Comment: Compared with the normally applied torque, this value is quite high. The ratio of normal torque to that required to shear the pin is

$$\text{Ratio} = 8255/1575 = 5.24$$

This indicates that the pin would not likely shear under anticipated conditions. However, it may be too high to protect the propeller. Testing of the propeller should be done.

REFERENCES

1. Aluminum Association, *Aluminum Design Manual*, Washington, DC, 2005.

2. American Institute of Steel Construction, *Steel Construction Manual*, 13th ed., Chicago, IL, 2005.

3. Avallone, E. A., T. Baumeister, and Ali Sadegh eds., *Marks' Standard Handbook for Mechanical Engineers*, 11th ed., 2007.

4. Budynas, R. G. and J. K. Nisbett, *Shigley's Mechanical Engineering Design*, 8th ed., New York, McGraw-Hill, 2007.

5. Deutschman, A. D., W. J. Michels, and C. E. Wilson, *Machine Design Theory and Practice*, New York, Macmillan, 1975.

6. Juvinall, R. C. and K. M. Marshek, *Fundamentals of Machine Component Design*, 4th ed., New York, John Wiley & Sons, Inc., 2005.

7. Mott, R. L., *Machine Elements in Mechanical Design*, 4th ed., Upper Saddle River, NJ, Prentice Hall, 2004.

8. Pilkey, W., *Peterson's Stress Concentration Factors*, New York, John Wiley & Sons, Inc., 1997.

9. Spotts, M. F., T. E. Shoup and L. E. Hornberger, *Design of Machine Elements*, 8th ed., Upper Saddle River, NJ, Prentice Hall, 2004.

10. Young, W. C. and R. G. Budynas, *Roark's Formulas for Stress and Strain*, 7th ed., New York, McGraw-Hill, 2002.

PROBLEMS

Direct Tension or Compression

9–1. Specify a suitable aluminum alloy for a round bar having a diameter of 10 mm subjected to a static direct tensile force of 8.50 kN.

9–2. A rectangular bar having cross-sectional dimensions of 10 mm by 30 mm is subjected to a direct tensile force of 20.0 kN. If the force is to be repeated many times, specify a suitable steel material.

9–3. A link in a mechanism for an automated packaging machine is subjected to a direct tensile force of 1720 lb, repeated many times. The link is square, 0.40 in on a side. Specify a suitable steel for the link.

9–4. A circular steel rod having a diameter of $\frac{3}{8}$-in supports a heater assembly and carries a static tensile load of 1850 lb. Specify a suitable structural steel for the rod.

9–5. A tension member in a wood roof truss is to carry a static tensile force of 5200 lb. It has been proposed to use a standard 2 × 4 made from southern pine, No. 2 grade. Would this be acceptable?

9–6. For the data in Problem 9–5, suggest an alternative design that would be safe for the given loading. A different size member or a different material may be specified.

9–7. A guy wire for an antenna tower is to be aluminum, having an allowable stress of 12 000 psi. If the expected maximum load on the wire is 6400 lb, determine the required diameter of the wire.

9–8. A hopper having a mass of 1150 kg is designed to hold a load of bulk salt having a mass of 6350 kg. The hopper is to be suspended by four rectangular straps, each carrying one-fourth of the load. Steel plate with a thickness of 8.0 mm is to be used to make the straps. What should be the width in order to limit the stress to 70 MPa?

9–9. A shelf is being designed to hold crates having a total mass of 1840 kg. Two support rods like that shown in Figure P9–9 will hold the shelf. Assume that the center of gravity of the crates is at the middle of the shelf. Specify the required diameter of the circular rods to limit the stress to 110 MPa.

FIGURE P9–9   Shelf support rods for Problem 9–9.

9–10. A concrete column base is circular, with a diameter of 8.0 in, and carries a static direct compressive load of 70 000 lb. Specify the required rated strength of the concrete according to the recommendations in Section 8–10.

9–11. Three short wood blocks made from standard 4 × 4 posts support a machine weighing 29 500 lb and share the load equally. Specify a suitable type of wood for the blocks.

9–12. A circular pier to support a column is to be made of concrete having a rated strength of 3000 psi (20.7 MPa). Specify a suitable diameter for the pier if it is to carry a direct compressive load of 1.50 MN.

9–13. An aluminum ring has an outside diameter of 12.0 mm and an inside diameter of 10 mm. If the ring is short and is made of 2014-T6, compute the force required to produce ultimate compressive failure in the ring. Assume that s_u is the same in both tension and compression.

9–14. A wooden cube 40 mm on a side is made from No. 2 hemlock. Compute the allowable compressive force that could be applied to the cube either parallel or perpendicular to the grain.

9–15. A round bar of ASTM A242 structural steel is to be used as a tension member to stiffen a frame. If a maximum static load of 4000 lb is expected, specify a suitable diameter of the rod.

9–16. A portion of a casting made of ASTM A48, grade 20 gray cast iron has the shape shown in Figure P9–16 and is subjected to compressive force in line with the centroidal

FIGURE P9–16   Short compression member for Problem 9–16.

Dimensions in millimeters

axis of the section. If the member is short and carries a load of 52 000 lb, compute the stress in the section and the design factor.

9–17. A part for a truck suspension system is to carry a compressive load of 135 kN with the possibility of shock loading. Malleable iron ASTM A220 grade 45008 is to be used. The cross section is to be rectangular with the long dimension twice the short dimension. Specify suitable dimensions for the part.

9–18. A rectangular plastic link in an office printer is to be made from glass-filled acetal copolymer (see Appendix A–20). It is to carry a tensile force of 110 lb. Space limitations permit a maximum thickness for the link of 0.20 in. Specify a suitable width of the link if a design factor of 8 based on the tensile strength of the plastic is expected.

9–19. Figure P9–19 shows the cross section of a short compression member that is to carry a static load of 640 kN. Specify a suitable material for the member.

9–20. Figure P9–20 shows a bar carrying several static loads. If the bar is made from ASTM A36 structural steel, is it safe?

9–21. In Figure P9–21, specify a suitable aluminum alloy for the member AB if the load is to be repeated many times. Consider only the square part near the middle of the member.

9–22. A standard steel angle, L2 × 2 × $\frac{1}{4}$, serves as a tension member in a truss carrying a static load. If the angle is made from ASTM A36 structural steel, compute the allowable tensile load based on the AISC specifications.

Elastic Deformation

9–23. A post of No. 2 grade hemlock is 6.0 ft long and has a square cross section with side dimensions of 3.50 in. How much would it be shortened when it is loaded in compression up to its allowable load parallel to the grain?

9–24. Determine the elongation of a strip of plastic 0.75 mm thick by 12 mm wide by 375 mm long if it is subjected to a load of 90 N and is made of (a) medium-impact ABS or (b) phenolic (see Appendix A–20).

FIGURE P9–19   Short compression member for Problem 9–19.

FIGURE P9–20   Bar carrying axial loads for Problem 9–20.

FIGURE P9–21   Support for Problem 9–21.

9–25. A hollow aluminum cylinder made of 2014-T4 has an out-side diameter of 2.50 in and a wall thickness of 0.085 in. Its length is 14.5 in. What axial compressive force would cause the cylinder to shorten by 0.005 in? What is the resulting stress in the aluminum?

9–26. A metal bar is found in a stock bin and it appears to be made from either aluminum or magnesium. It has a square cross section with side dimensions of 0.25 in. Discuss two ways that you could determine what the material is.

9–27. A tensile member is being designed for a car. It must with-stand a repeated load of 3500 N and not elongate more than 0.12 mm in its 630-mm length. Use a design factor of 8 based on ultimate strength, and compute the required diameter of a round rod to satisfy these requirements using (a) AISI 1020 hot-rolled steel, (b) AISI 4140 OQT 700 steel, and (c) alu-minum alloy 6061-T6. Compare the mass of the three options.

9–28. A steel bolt has a diameter of 12.0 mm in the unthreaded portion. Determine the elongation in a length of 220 mm if a force of 17.0 kN is applied.

9–29. In an aircraft structure, a rod is designed to be 1.25 m long and have a square cross section 8.0 mm on a side. Deter-mine the amount of elongation that would occur if it is made of (a) titanium Ti-6A1-4V and (b) AISI 501 OQT 1000 stainless steel. The load is 5000 N.

9–30. A tension member in a welded steel truss is 13.0 ft long and subjected to a force of 35 000 1b. Choose an equal-leg angle made of ASTM A36 steel that will limit the stress to 21 600 psi. Then compute the elongation in the angle due to the force. Use $E = 29.0 \times 10^6$ psi for structural steel.

9–31. A link in a mechanism is a rectangular steel bar that is sub-jected alternately to a tensile load of 450 lb and a compressive load of 50 lb. Its dimensions are: length $= 8.40$ in, width $= 0.25$ in, thickness $= 0.125$ in. Compute the elon-gation and compression of the link.

9–32. A cylindrical steel bar is attached at the top and is subjected to three axial forces, as shown in Figure P9–32. Its cross sec-tion has an area of 0.50 in². Determine the deflection of the free end.

9–33. A link in an automated packaging machine is a hollow tube made from 6061-T6 aluminum. Its dimensions are: outside diameter $= 1.250$ in, inside diameter $= 1.126$ in, length $= 36.0$ in. Compute the force required to produce a deflection of the bar of 0.050 in. Would the stress produced by the force just found be safe if the load is applied repeatedly?

9–34. A tension member in a truss is subjected to a static load of 2500 lb. Its dimensions are: length $= 8.75$ ft, outside diam-eter $= 0.750$ in, inside diameter $= 0.563$ in. First specify a suitable aluminum alloy that would be safe. Then compute the elongation of the member.

9–35. A hollow 6061-T4 aluminum tube, 40 mm long, is used as a spacer in a machine and is subjected to an axial compressive force of 18.2 kN. The tube has an outside diameter of 56.0 mm and an inside diameter of 48.0 mm. Compute the de-flection of the tube and the resulting compressive stress.

9–36. A guy wire is made from AISI 1020 CD steel and has a length of 135 ft. Its diameter is 0.375 in. Compute the stress in the wire and its deflection when subjected to a tensile force of 1600 lb.

9–37. Compute the total elongation of the bar shown in Figure P9–37 if it is made from titanium Ti-6AI-4V.

FIGURE P9–37 Bar carrying axial loads for Problem 9–37.

9–38. During a test of a metal bar it was found that an axial tensile force of 10 000 lb resulted in an elongation of 0.023 in. The bar had these original dimensions: length $= 10.000$ in, di-ameter $= 0.750$ in. Compute the modulus of elasticity of the metal. What kind of metal was it probably made from?

9–39. The bar shown in Figure P9–39 carries three loads. Com-pute the deflection of point D relative to point A. The bar is made from standard acrylic plastic.

FIGURE P9–39 Bar carrying axial loads for Problem 9–39.

9–40. A column is made from a cylindrical concrete base supporting a standard HSS 4 × 4 × $\frac{1}{2}$ steel tube, with a length of 8.60 ft. The base is 3.0 ft long and has a diameter of 8.00 in. First, specify the concrete from Section 8–10 with a rated strength suitable to carry a compressive load of 64 000 lb. Then, assum-ing that no buckling occurs, compute the total amount that the column would be shortened.

FIGURE P9–32 Bar in axial tension for Problem 9–32.

9–41. A 10.5 ft length of 14-gage copper electrical wire (C14500, hard) is rigidly attached to a beam at its top. The wire diameter is 0.064 in. How much would it stretch if a person weighing 120 lb hung from the bottom? How much would it stretch if the person weighs 200 lb?

9–42. A measuring tape used by carpenters is 25.00 ft long and is made from flat strip steel with these dimensions: width = 0.750 in, thickness = 0.006 in. Compute the elongation of the tape and the stress in the steel if a tensile force of 25.0 lb is applied.

9–43. A wooden post is made from a standard 4 × 4 (Appendix A–4). If it is made from No. 2 grade southern pine, compute the axial compressive load it could carry before reaching its allowable stress in compression parallel to the grain. Then, if the post is 10.75 ft long, compute the amount that it would be shortened under that load.

9–44. Ductile iron, ASTM A536, grade 60-40-18, is formed into a hollow square shape, 200 mm outside dimension and 150 mm inside dimension. Compute the load that would produce an axial compressive stress in the iron of 200 MPa. Then, for that load, compute the amount that the member would be shortened from its original length of 1.80 m.

9–45. A brass wire (C36000, hard) has a diameter of 3.00 mm and is initially 3.600 m long. At this condition, the lower end, with a plate for applying a load, is 6.0 mm from the floor. How many kilograms of lead would have to be added to the plate to cause it to just touch the floor? What would be the stress in the wire at that time?

9–46. Compute the elongation of the square bar *AB* in Figure P9–46 if it is 1.25 m long and made from aluminum 6061-T6.

Thermal Deformation

9–47. A concrete slab in a highway is 80 ft long. Determine the change in length of the slab if the temperature changes from −30°F to +110°F.

9–48. A steel rail for a railroad siding is 12.0 m long and made from AISI 1040 HR steel. Determine the change in length of the rail if the temperature changes from −34°C to +43°C.

9–49. Determine the stress that would result in the rail described in Problem 9–48 if it were completely restrained from expanding.

9–50. The pushrods that actuate the valves on a six-cylinder engine are AISI 1040 steel and are 625 mm long and 8.0 mm in diameter. Calculate the change in length of the rods if their temperature varies from −40°C to +116°C and the expansion is unrestrained.

9–51. If the pushrods described in Problem 9–50 were installed with zero clearance with other parts of the valve mechanism at 25°C, compute the following:
a. The clearance between parts at −40°C.
b. The stress in the rod due to a temperature rise to 116°C. Assume that mating parts are rigid.

9–52. A bridge deck is made as one continuous concrete slab to 140 ft long at 30°F. Determine the required width of expansion joints at the ends of the bridge if no stress is to be developed when the temperature varies from +30°F to +110°F.

9–53. When the bridge deck of Problem 9–52 was installed, the width of the expansion joint at each end was only 0.25 in. What stress would be developed if the supports are rigid? For the concrete use f'_c = 4000 psi and find *E* from Section 8–10.

9–54. For the bridge deck in Problem 9–52, assume that the deck is to be just in contact with its support at the temperature of 110°F. If the deck is to be installed when the temperature is 60°F, what should the gap be between the deck and its supports?

9–55. A ring of AISI 301 stainless steel is to be placed on a shaft having a temperature of 20°C and a diameter of 55.200 mm. The inside diameter of the ring is 55.100 mm. To what temperature must the ring be heated to make it 55.300 mm in diameter and thus allow it to be slipped onto the shaft?

9–56. When the ring of Problem 9–55 is placed on the shaft and then cooled back to 20°C, what tensile stress will be developed in the ring?

9–57. A heat exchanger is made by arranging several brass (C36000) tubes inside a stainless steel (AISI 430) shell. Initially, when the temperature is 10°C, the tubes are 4.20 m long and the shell is 4.50 m long. Determine how much each will elongate when heated to 85°C.

9–58. In Alaska, a 40-ft section of AISI 1020 steel pipe may see a variation in temperature from −50°F when it is at ambient temperature to +140°F when it is carrying heated oil. Compute the change in the length of the pipe under these conditions.

9–59. A square bar of magnesium is 30 mm on a side and 250.0 mm long at 20°C. It is placed between two rigid supports set 250.1 mm apart. The bar is then heated to 70°C while the supports do not move. Compute the resulting stress in the bar.

FIGURE P9–46 Support for Problem 9–46.

FIGURE P9–64   Problem 9–64.

9–60. A square rod, 8.0 mm on a side, is made from AISI 1040 cold-drawn steel and has a length of 175 mm. It is placed snugly between two unmoving supports with no stress in the rod. Then the temperature is increased by 90°C. What is the final stress in the rod?

9–61. A square bar is made from 6061-T4 aluminum alloy. At 75°F its length is 10.500 in. It is placed between rigid supports with a distance between them of 10.505 in. If the supports do not move, describe what would happen to the bar when its temperature is raised to 400°F.

9–62. A straight carpenter's level is supported on two bars; one made of a polyester resin and the other made of titanium Ti-6A1-4V. The distance between the bars is 24.00 in. At a temperature of 65°F, the level is perfectly level and each bar has a length of 30.00 in. What would be the angle of tilt of the level when the temperature is increased to 212°F?

9–63. When manufactured, a steel (AISI 1040) measuring tape was exactly 25.000 ft long at a temperature of 68°F. Compute the error that would result if the tape is used at −15°F.

9–64. Figure P9–64 shows two bars of different materials separated by 0.50 mm when the temperature is 20°C. At what temperature would they touch?

9–65. A stainless steel (AISI 301) wire is stretched between rigid supports so that a stress of 40 MPa is induced in the wire at a temperature of 20°C. What would be the stress at a temperature of −15°C?

9–66. For the conditions described in Problem 9–65, at what temperature would the stress in the wire be zero?

Members Made from Two Materials

9–67. A short post is made by welding steel plates into a square, as shown in Figure P9–67, and then filling the area inside with concrete. Compute the stress in the steel and in the concrete

if $b = 150$ mm, $t = 10$ mm, and the post carries an axial load of 900 kN. See Section 8–10 for concrete properties. Use $f'_c = 6000$ psi.

9–68. A short post is made by filling a standard $6 \times 6 \times \frac{1}{2}$ steel tube with concrete, as shown in Figure P9–67. The steel has an allowable stress of 21 600 psi. The concrete has a rated strength of 6000 psi but, in this application, the stress is to be limited to 1500 psi. See Section 8–10 for the modulus of elasticity for the concrete. Compute the allowable load on the post.

9–69. A short post is being designed to support an axial compressive load of 500 000 lb. It is to be made by welding $\frac{1}{2}$-in-thick plates of A36 steel into a square and filling the area inside with concrete, as shown in Figure P9–67. It is required to determine the dimension of the side of the post b in order to limit the stress in the steel to no more than 21 600 psi and in the concrete to no more than 1500 psi. See Section 8–10 for concrete properties. Use $f'_c = 6000$ psi.

9–70. Two disks are connected by four rods, as shown in Figure P9–70. All rods are 6.0 mm in diameter and have the same length. Two rods are steel ($E = 207$ GPa), and two are aluminum ($E = 69$ GPa). Compute the stress in each rod when an axial force of 11.3 kN is applied to the disks.

FIGURE P9–70   Problem 9–70.

9–71. An array of three wires is used to suspend a casting having a mass of 2265 kg in such a way that the wires are symmetrically loaded (see Figure P9–71). The outer two wires are AISI 430 stainless steel, full hard. The middle wire is hard beryllium copper, C17200. All three wires have the same diameter and length. Determine the required diameter of the wires if none is to be stressed beyond one-half of its yield strength.

9–72. Figure P9–72 shows a load being applied to an inner cylindrical member that is initially 0.12 mm longer than a second concentric hollow pipe. What would be the stress in both members if a total load of 350 kN is applied?

9–73. Figure P9–73 shows an aluminum cylinder being capped by two end plates that are held in position with four steel tie

FIGURE P9–67   Post for Problems 9–67, 9–68, and 9–69.

FIGURE P9–71 Problem 9–71.

F = 350 kN

0.12 mm

140-mm-dia. solid
aluminum cylinder

Standard 6-in
schedule 40
steel pipe
(Pipe 152 STD)

1.65 m

FIGURE P9–72 Aluminum bar in a steel pipe under an axial compression load for Problem 9–72.

rods. A clamping force is created by tightening the nuts on the ends of the tie rods. Compute the stress in the cylinder and the tie rods if the nuts are turned one full turn from the hand-tight condition.

9–74. A column for a building is made by encasing a W6 × 15 wide-flange shape in concrete, as shown in Figure P9–74. The concrete helps to protect the steel from the heat of a fire and also shares in carrying the load. What stress would be produced in the steel and the concrete by a total load of

M10 × 1.25 thread

Rigid plate

10-mm-dia. bolt

450 mm

Aluminum tube
150 mm O.D.
138 mm I.D.

FIGURE P9–73 Tie rods on a cylinder for Problem 9–73.

50 kip? See Section 8–10 for concrete properties. Use $f'_c = 2000$ psi.

Additional Deformation and Design Problems

9–75. A 4 × 4 wood post having a length of 4.25 m is made from No. 2 southern pine wood. Specify the maximum load in newtons the column would support and not exceed the allowable stress in compression parallel to the grain. Assume no buckling occurs. Then compute the total deformation of the post when subjected to this magnitude of load.

9–76. An aluminum bar (6061-T4) is initially 225.00 mm long at a temperature of 20°C. It is placed between two immovable end plates that are 225.50 mm apart. The bar is then heated to 205°C. Describe the sequence of events that occurs as the heating takes place in terms of deformation and stress. Evaluate the acceptability of the final condition.

9–77. A rectangular bar of aluminum (2014-T4) has a length of 2.400 m and cross-sectional dimensions of 25 mm by 50 mm when the temperature is 20°C.
   a. At what temperature would the length be 2.405 m if the bar is unrestrained?
   b. If the temperature is increased an additional 30°C while the bar is restrained from growing lengthwise, compute the resulting stress in the bar.
   c. Is the stress level safe?

FIGURE P9–74 Steel column encased in concrete for Problem 9–74.

9–78. A tensile rod in a mechanism is made from AISI 4140 OQT 1300 steel. It has a rectangular cross section, 30 mm × 20 mm and a length of 700 mm. It is to be used in the following manner: (a) it must not elongate more than 0.50 mm under load and (b) it must be safe when subjected to a repeated axial tensile load. Specify the maximum permissible load on the rod.

9–79. Refer to Figure P9–79. Compute the total elongation of the bar if it is made from the aluminum alloy 2014-T6.

9–80. Figure P9–80 shows two rods, AC and BC, supporting a load at C. Each rod is steel and has a diameter of 8.00 mm. Compute the elongation of rods AC and BC when a load having a mass of 680 kg is suspended from C.

9–81. Figure P9–81 shows two rods, AB and BC, supporting a load at B. Each rod is steel and has a diameter of 10.00 mm.

Compute the elongation of rods AB and BC when a load having a mass of 4200 kg is suspended from B.

9–82 to 9–99. A standard tensile test specimen has a straight cylindrical center section with a diameter of 0.505 in. Gage marks are inscribed 2.000 in apart when the specimen is in the unloaded condition. Compute the distance between the gage marks when the given material is subjected to an axial stress equal to 90% of the listed yield strength from the Appendix tables. Assume that these stresses are below the proportional limit for the materials. Also compute force required to create this stress level and the strain in the area between the gage marks.

9–82. AISI 1040 cold-drawn steel
9–83. AISI 5160 OQT 700 steel
9–84. AISI 501 OQT 1000 stainless steel
9–85. C17200 hard beryllium copper
9–86. ASTM AZ 63A-T6 cast magnesium
9–87. ZA 12 cast zinc
9–88. ASTM A572 Grade 65 structural steel
9–89. Grade 4 austempered ductile iron
9–90. 5154-H38 aluminum alloy
9–91 to 9–99. A tensile test specimen has a straight rectangular center section with a thickness of 12.5 mm and a width of 16.0 mm. Gage marks are inscribed 50.0 mm apart when the specimen is in the unloaded condition. Compute the distance between the gage marks when the given material is subjected to an axial tensile stress equal to 50% of the listed tensile strength from the Appendix tables or tables in Chapter 8. Assume that the materials obey Hooke's law and that these stresses are below the proportional limit for the materials. Also compute the force required to create this stress level and the strain in the area between the gage marks.

9–91. Nylon 66 plastic, dry
9–92. ABS, high-impact plastic
9–93. Acetal copolymer plastic
9–94. Polyurethane elastomer
9–95. Phenolic plastic
9–96. Glass/epoxy composite (Table 8–13)
9–97. Aramid/epoxy composite (Table 8–13)
9–98. Graphite/epoxy, 62% fiber (Table 8–13)
9–99. Graphite/epoxy composite ultrahigh modulus (Table 8–13)

Stress Concentrations for Direct Axial Stresses

9–100. A circular rod with a diameter of 40.0 mm has a groove cut to a diameter of 35 mm. A full radius of 3.0 mm is produced at the bottom of the groove. Compute the maximum stress in the rod when an axial tensile force of 46 kN is applied.

FIGURE P9–79 Bar for Problem 9–79.

FIGURE P9–80 Support system for Problem 9–80.

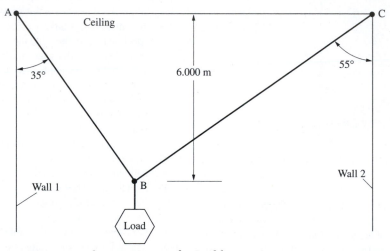

FIGURE P9–81 Support system for Problem 9–81.

9–101. A circular rod with a diameter of 1.50 in has a groove cut to a diameter of 1.25 in. A full radius of 0.12 in is produced at the bottom of the groove. Compute the maximum stress in the rod when an axial tensile force of 10 300 lb is applied.

9–102. A circular rod with a diameter of 0.40 in has a groove cut to a diameter of 0.35 in. A full radius of 0.040 in is produced at the bottom of the groove. Compute the maximum stress in the rod when an axial tensile force of 1250 lb is applied.

9–103. A circular rod with a diameter of 10.0 mm has a groove cut to a diameter of 8.0 mm. A full radius of 1.20 mm is produced at the bottom of the groove. Compute the maximum stress in the rod when an axial tensile force of 5500 N is applied.

9–104. A flat plate has a width of 2.50 in and a thickness of 0.400 in. Part of the plate is machined to a width of 2.20 in using a milling cutter having a diameter of 0.250 in. Compute the maximum stress in the plate when an axial tensile force of 17 500 lb is applied.

9–105. A flat plate has a width of 60.0 mm and a thickness of 10.0 mm. Part of the plate is machined to a width of 55 mm using a milling cutter having a diameter of 6.00 mm. Com-

pute the maximum stress in the plate when an axial tensile force of 75 kN is applied.

9–106. A flat plate has a width of 25.0 mm and a thickness of 3.0 mm. Part of the plate is machined to a width of 22 mm using a milling cutter having a diameter of 5.00 mm. Compute the maximum stress in the plate when an axial tensile force of 6800 N is applied.

9–107. A flat plate has a width of 0.80 in and a thickness of 0.120 in. Part of the plate is machined to a width of 0.50 in using a milling cutter having a diameter of 0.200 in. Compute the maximum stress in the plate when an axial tensile force of 1800 lb is applied.

9–108. A circular rod with a diameter of 50.0 mm is cut down to a diameter of 40 mm with a cutter having a nose radius of 6.0 mm. Compute the maximum stress in the rod when an axial tensile force of 230 kN is applied.

9–109. A circular rod with a diameter of 2.50 in is cut down to a diameter of 1.75 in with a cutter having a nose radius of 0.25 in. Compute the maximum stress in the rod when an axial tensile force of 48.0 kips is applied.

9–110. A circular rod with a diameter of 0.38 in is cut down to a diameter of 0.32 in with a cutter having a nose radius of

0.02 in. Compute the maximum stress in the rod when an axial tensile force of 375 lb is applied.

9–111. A circular rod with a diameter of 10.0 mm is cut down to a diameter of 8.00 mm with a cutter having a nose radius of 0.50 mm. Compute the maximum stress in the rod when an axial tensile force of 1600 N is applied.

9–112. A flat plate has a width of 2.50 in and a thickness of 0.400 in. A hole with a diameter of 1.75 in is bored through the plate. Compute the maximum stress in the plate when an axial tensile force of 14 200 lb is applied.

9–113. A flat plate has a width of 60 mm and a thickness of 8.00 mm. A hole with a diameter of 40.0 mm is bored through the plate. Compute the maximum stress in the plate when an axial tensile force of 65.0 kN is applied.

9–114. A flat plate has a width of 18.0 mm and a thickness of 2.50 mm. A hole with a diameter of 8.00 mm is bored through the plate. Compute the maximum stress in the plate when an axial tensile force of 2250 N is applied.

9–115. A flat plate has a width of 0.60 in and a thickness of 0.088 in. A hole with a diameter of 0.25 in is bored through the plate. Compute the maximum stress in the plate when an axial tensile force of 475 lb is applied.

9–116. A circular rod with a diameter of 50.0 mm has a 20 mm diameter hole cut transversely through it. Compute the maximum stress in the rod when an axial tensile force of 120 kN is applied.

9–117. A circular rod with a diameter of 2.0 in has a 0.75 in diameter hole cut transversely through it. Compute the maximum stress in the rod when a 22 500-lb axial tensile force is applied.

9–118. A circular rod with a diameter of 0.63 in has a 0.35 in diameter hole cut transversely through it. Compute the maximum stress in the rod when a 2800-lb axial tensile force is applied.

9–119. A circular rod with a diameter of 12.0 mm has a 7.25 mm diameter hole cut transversely through it. Compute the maximum stress in the rod when an axial tensile force of 7500 N is applied.

9–120. Figure P9–120 shows a circular shaft subjected to a repeated axial tensile load of 25 kN. The shaft is made from AISI 4140 OQT 1100 steel. Determine the design factor at the hole and at the fillet.

FIGURE P9–120 Shaft for Problem 9–120.

9–121. A valve stem in an automotive engine is subjected to an axial tensile load of 900 N due to the valve spring, as shown in Figure P9–121. Compute the maximum stress in the stem at the place where the spring force acts against the shoulder.

FIGURE P9–121 Valve stem for Problem 9–121.

9–122. A part of a tie rod in a steering linkage has the shape shown in Figure P9–122. As the machine cycles, it exerts a repeated direct tensile load of 8.25 kN on the tie rod. Compute the expected maximum tensile stress on the rod and specify a suitable material from which to make it.

9–123. A round shaft has two grooves in which rings are placed to retain a gear in position, as shown in Figure P9–123. If the shaft is subjected to an axial tensile force of 36 kN, compute the maximum tensile stress in the shaft.

9–124. Figure P9–124 shows the proposed design for a tensile rod. The larger diameter is known, $D = 1.00$ in, along with the hole diameter, $a = 0.50$ in. It has also been decided that the stress concentration factor at the fillet is to be 1.7. The smaller diameter, d, and the fillet radius, r, are to be specified such that the stress at the fillet is the same as that at the hole.

9–125. A machine member shown in Figure P9–125 is made from AISI 1141 OQT 1100 steel. Determine the allowable repeated axial tensile force that can be applied. The force is applied through a pin in the end holes.

9–126. Refer to Figure P9–126. Specify a suitable material for the bar shown if the applied force F is 12.6 kN. Mild shock is expected.

Bearing Stress

9–127. Compute the bearing stresses at the mating surfaces A, B, C, and D, in Figure P9–127.

9–128. A 2-in schedule 40 steel pipe is used as a leg for a machine. The load carried by the leg is 2350 lb.
    a. Compute the bearing stress on the floor if the pipe is left open at its end.

FIGURE P9–122 Part of a tie-rod for Problem 9–122.

FIGURE P9–123   Shaft for Problem 9–123.

Rod subjected to direct tensile load

FIGURE P9–124   Tensile rod for Problem 9–124.

FIGURE P9–125   Machine member for Problem 9–125.

FIGURE 9–126   Bar for Problem 9–126.

b. Compute the bearing stress on the floor if a flat plate is welded to the bottom of the pipe having a diameter equal to the outside diameter of the pipe.

9–129. A bolt and washer are used to fasten a wooden board to a concrete foundation as shown in Figure P9–129. A tensile force of 385 lb is created in the bolt as it is tightened. Compute the bearing stress (a) between the bolt head and the steel washer, and (b) between the washer and the board.

9–130. For the data of Problem 7–43, compute the bearing stress on the side of the key.

9–131. For the data of Problem 7–44, compute the bearing stress on the tube at the interfaces with the pin and the collar.

9–132. For the data of Problem 7–49, compute the bearing stress on the rivets.

9–133. For the data of Problem 7–50, compute the bearing stress on the rivets.

9–134. The heel of a woman's shoe has the shape shown in Figure P9–134. If the force on the heel is 535 N, compute the bearing stress on the floor.

9–135. A machine weighs 90 kN and rests on four legs. Two concrete supports serve as the foundation. The load is sym-

metrical, as shown in Figure P9–135. Evaluate the proposed design with regard to the safety of the steel plate, the concrete foundation, and the soil for bearing.

9–136. A column for a building is to carry 160 kN. If the column is to be supported on a square concrete foundation that sits on soft rock, determine the required dimensions of the foundation block.

9–137. A base for moving heavy machinery is designed as shown in Figure P9–137. How much load could the base support based on the bearing capacity of the steel plate if it is 1.25 in thick and made from ASTM A36 structural steel?

9–138. Repeat Problem 9–137 but use ASTM A242 high-strength low-alloy steel plate.

9–139. Figure P9–139 shows an alternative design for the machinery moving base described in Problem 9–137. Compute the allowable load for this design if it rests on (a) ASTM A36 steel or (b) ASTM A242 steel.

9–140. A heavy table for industrial use along with its load weighs 10 000 lb. It has four legs that share the weight equally. Compute the bearing stress on the floor for each of the

26 kip

12 × 12 steel plate

W6×15 steel wide-flange beam

A

B

12 × 12 steel plate

18-in square concrete

C

3 ft 0 in square concrete

D

FIGURE P9–127 Column foundation for Problem 9–127.

10 mm

6 mm

FIGURE P9–134 Shoe heel for Problem 9–134.

Machine weighs 90 kN

Machine

0.10 m square leg

0.20 m square steel plate ASTM A36

0.3 m

1.0 m

Compact gravel

Concrete: 2000 psi rated strength

0.8 m

2.5 m

FIGURE P9–135 Machine supports for Problem 9–135.

Bolt head

Washer

Wall

0.750

Board

Concrete

Earth

I.D. 0.562

1.375 O.D.

Washer dimensions (in)

FIGURE P9–129 Bolt and washer for Problem 1–129.

3.00-in dia.

Base

Steel plate

8.00 in

1.25 in thick

FIGURE P9–137 Roller base for moving machinery for Problems 9–137 and 9–138.

1.25 in thick

FIGURE P9–139 Roller base for moving machinery for Problem 9–139.

(a) Photograph of a bearing stress model—hollow square tube

(b) Photograph of a bearing stress model—channel

FIGURE P9–140 Bearing stress models.

Source of models: Gary Drigel.

following proposed designs for the size and shape of the legs. See Figure P9–140.

a. Each leg is made from a hollow square steel tube 1.50 in on a side and a wall thickness of 1/8 in as shown in Figure P9–140(a). The cross section of the tube contacts the floor.

b. Each leg is made from an aluminum channel made to the same dimensions as a standard steel channel, C3 × 4.1,

as listed in Appendix A-6. Refer to Figure P9–140(b). The cross section of the channel contacts the floor.

c. The square steel tubes from (a) above are welded to a square flat plate 3.50 in on a side, as shown in Figure P9–140(a), with the plate resting on the floor.

d. Specify the minimum dimension of the sides of the square plate from (c) if the bearing stress could be as high as 400 psi.

9–141. One end of a beam is supported on a rocker having a radius of 200 mm and a width of 150 mm. If the rocker and the plate on which it rests are made of ASTM A36 structural steel, specify the maximum allowable reaction at this end of the beam.

9–142. The base of a special machine has four legs made from standard L3 × 3 × $\frac{1}{4}$ steel angles. Compute the bearing stress exerted by the legs on the floor if the total weight of the machine is 28 500 lb and the load is equally divided among the four legs.

9–143. Redesign the bottom of the machine legs of Problem 9–142 to permit them to sit on a concrete floor and to satisfy the bearing stress requirements shown in Table 9–7. The concrete has a rated strength of 3000 psi.

9–144. A wood block having a cross section of 3.50 × 7.50 in is to be placed on a bed of compact gravel. Compute the maximum allowable compressive load that can be applied to the block so as not to exceed the bearing strength of the gravel.

9–145. A column is made from a standard 4-inch schedule 40 steel pipe. Compute the maximum axial compressive load it can take and not exceed the allowable bearing stress on a concrete floor. The concrete has a rated strength of 4000 psi.

9–146. Add a steel bearing plate to the bottom of the column described in Problem 9–145 to allow an axial compressive load at least 10.0 times as much as that determined for the column without a plate.

Shearing Stress

9–147. A support for a beam is made as shown in Figure P9–147. Determine the required thickness of the projecting ledge a if the maximum shear stress is to be 6000 psi. The load on the support is 21 000 lb.

9–148. The lower control arm on an automotive suspension system is connected to the frame by a round steel pin 16 mm in diameter. Two sides of the arm transfer loads from the

FIGURE P9–147 Beam support for Problem 9–147.

FIGURE P9–148 Pin for automotive suspension system for Problem 9–148.

frame to the arm, as sketched in Figure P9–148. How much shear force could the pin withstand if it is made of AISI 1040 cold-drawn steel and a design factor of 6 based on the yield strength in shear is desired?

9–149. A centrifuge is used to separate liquids according to their densities using centrifugal force. Figure P9–149 illustrates one arm of a centrifuge having a bucket at its end to hold the liquid. In operation, the bucket and the liquid have a mass of 0.40 kg. The centrifugal force has the magnitude in newtons of

$$F = 0.01097 \cdot m \cdot R \cdot n^2$$

where m = rotating mass of bucket and liquid (kg)
   R = radius to center of mass (meters)
   n = rotational speed (rpm)

The centrifugal force places the pin holding the bucket in direct shear. Compute the stress in the pin due to a rotational speed of 3000 rpm. Then specify a suitable steel for the pin, considering the load to be repeated.

FIGURE P9–149 Centrifuge for Problem 9–149.

9–150. A circular punch is used to punch a 20.0-mm-diameter hole in a sheet of AISI 1020 hot-rolled steel having a thickness of 8.0 mm. Compute the force required to punch out the slug.

9–151. Repeat Problem 9–150, using aluminum 6061-T4.

9–152. Repeat Problem 9–150, using C14500 copper, hard.

9–153. Repeat Problem 9–150, using AISI 430 full hard stainless steel.

9–154. Determine the force required to punch a slug the shape shown in Figure P9–154 from a sheet of AISI 1020 hot-rolled steel having a thickness of 5.0 mm.

FIGURE P9–154 Shape of a slug for Problem 9–154.

9–155. Determine the force required to punch a slug the shape shown in Figure P9–155 from a sheet of aluminum 3003-H18 having a thickness of 0.194 in.

FIGURE P9–155 Shape of a slug for Problem 9–155.

9–156. A notch is made in a piece of wood, as shown in Figure P9–156, to support an external load of 1800 lb. Compute the shear stress in the wood. Is the notch safe? (See Appendix A–19.)

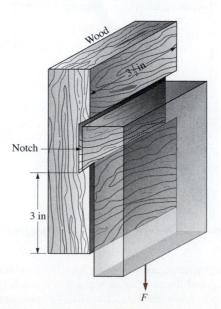

FIGURE P9–156 Notched wood block loaded in shear for Problem 9–156.

9–157. Compute the force required to shear a straight edge of a sheet of AISI 1040 cold-drawn steel having a thickness of 0.105 in. The length of the edge is 7.50 in.

9–158. Repeat Problem 9–157 using AISI 5160 OQT 700 steel.

9–159. Repeat Problem 9–157 using AISI 301 full hard stainless steel.

9–160. Repeat Problem 9–157 using C36000 brass, hard.

9–161. Repeat Problem 9–157, using aluminum 5154-H32.

9–162. For the lever shown in Figure P9–162, called a bellcrank, compute the required diameter of the pin A if the load is repeated and the pin is made from C17200 beryllium copper, hard. The loads are repeated many times.

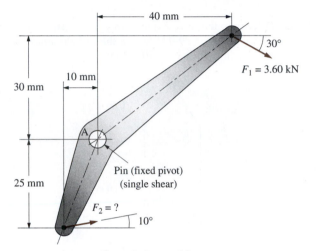

FIGURE P9–162 Bellcrank for Problem 9–162.

9–163. For the structure shown in Figure P9–163, determine the required diameter of each pin if it is made from AISI 1020 cold-drawn steel. Each pin is in double shear and the load is static.

FIGURE P9–163 Pin-connected structure for Problem 9–163.

9–164. For the structure shown in Figure P9–164, determine the required diameter of each pin if it is made from ASTM A572 high-strength, low-alloy columbium-vanadium structural steel, grade 50. Each pin is in double shear and the load is static.

FIGURE P9–164 Truss for Problem 9–164.

9–165. A pry bar as shown in Figure P9–165 is used to provide a large mechanical advantage for lifting heavy machines. An operator can exert a force of 280 lb on the handle. Compute the lifting force and the shear stress on the wheel axle.

FIGURE P9–165 Pry bar for Problem 9–165.

9–166. Figure P9–166 shows a steel strap with holes and slots punched from it. The entire strap is also punched from a larger sheet of 3003-H12 aluminum alloy, 1.40 mm thick. Compute the total force required to produce the part if all shearing is done in one stroke.

FIGURE P9–166 Steel Strap for Problem 9–166.

9–167. Figure P9–167 shows an anvil for an impact hammer held in a fixture by a circular pin. If the force is to be 500 lb, specify a suitable diameter for the steel pin if it is to be made from AISI 1040 WQT 900.

9–168. Figure P9–168 shows a steel flange forged integrally with the shaft, which is to be loaded in torsion. Eight bolts serve to couple the flange to a mating flange. Assume that each bolt carries an equal load. Compute the maximum permissible

FIGURE P9–167 Impact hammer for Problem 9–167.

FIGURE P9–168 Flange coupling for Problem 9–168.

torque on the coupling if the shear stress in the bolts is not to exceed 6000 psi.

9–169. Compute the force required to completely punch out the shape shown in Figure P9–169 from a large sheet of steel having a thickness of 0.085 in. The material is AISI 1020 cold-drawn steel.

FIGURE P9–169 Punched shape for Problem 9–169.

9–170. Compute the force required to punch the shape shown in Figure P9–170 from a larger sheet of 6061-T4 aluminum sheet having a thickness of 0.10 in.

FIGURE P9–170 Punched shape for Problem 9–170.

9–171. Compute the force required to punch out a slug having the shape shown in Figure P9–171 from a sheet of the aluminum alloy 3003-H18. The thickness of the sheet is 3.0 mm.

Dimensions in millimeters

FIGURE P9–171 Punched shape for Problem 9–171.

9–172. Compute the force required to punch out a slug having the shape shown in Figure P9–172 from a sheet of AISI 1040 cold-drawn steel having a thickness of 1.60 mm.

FIGURE 9–172 Punched shape for Problem 9–172.

Dimensions in millimeters

FIGURE P9-173 Utility knife blade for Problem 9-173.

9-173. Figure P9-173 shows the shape of the blade for a utility knife to be sheared from a sheet of AISI 1080 OQT 900 steel. The blade is 0.80 mm thick. Compute the force required to punch the blade shape from a larger sheet.

9-174. Compute the force required to punch out the shaded shape shown in Figure P9-174 from the larger sheet of the aluminum alloy 5154-H38 having a thickness of 2.00 mm.

9-175. Use the shape shown in Figure P9-174 and other data from Problem 9-174. However, the design of the punch is different. It first punches the central hole and the two rectangular slots. Then it punches the outside edges of the shape. Determine the press force required to produce the part with this punch design.

Problems with More Than One Kind of Direct Stress and Design Problems

9-176. A gear transmits 5000 lb·in of torque to a circular shaft having a diameter of 2.25 in. A square key 0.50 in on a side connects the shaft to the hub of the gear, as shown in Figure P9-176. The key is made from AISI 1020 cold-drawn steel. Determine the required length of the key, L, to be safe for shear and bearing. Use a design factor of 2.0 based on yield for shear and the AISC allowable bearing stress.

9-177. A section of pipe is supported by a saddle-like structure that, in turn, is supported on two steel pins, as illustrated in Figure P9-177. If the load on the saddle is 42 000 lb, determine the required diameter and length of the pins. Use AISI 1040 cold-drawn steel. Consider both shear and bearing.

9-178. Figure P9-178 shows the design of the lower end of the member *AB* in Figure P9-21. Use the load shown in Figure P9-21 and assume that it is static. Member 1 is made from aluminum alloy 6061-T4; member 2 is made from aluminum alloy 2014-T4; pin 3 is made from aluminum alloy 2014-T6. Perform the following analyses:
   a. Evaluate member 1 for safety in tension in the area of the pin holes.
   b. Evaluate member 1 for safety in bearing at the pin.
   c. Evaluate member 2 for safety in bearing at the pin.
   d. Evaluate the pin 3 for safety in bearing.
   e. Evaluate the pin 3 for safety in shear.

9-179. Figure P9-179 illustrates a type of engineering chain used for conveying applications. All components are made from AISI 1040 cold-drawn steel. Evaluate the allowable tensile force (repeated) on the chain with respect to:
   a. Shear of the pin
   b. Bearing of the pin on the side plates
   c. Tension in the side plates.

9-180. Figure P9-180 shows two plates connected by two rivets. The plates are made from aluminum 6061-T6 and the rivets are made from aluminum 2014-T4. Evaluate the maximum allowable force on the connection to meet the following design criteria:
   a. The shear stress in a rivet cannot exceed $\frac{1}{4}$ of the ultimate strength in shear.
   b. The tensile stress in the plates cannot exceed $\frac{1}{3}$ of the yield strength.
   c. The bearing stress on either the plates or the rivets cannot exceed the design bearing stress from Equation 9-26.

FIGURE P9-174 Punched shape for Problems 3-174 and 3-175.

Shear plane

h

b

L

Shear area = $A_s = b \times L$

Enlarged view of key

FIGURE P9–176   Key for Problem 9–176.

Hub

Key

Shaft

Pictorial sketch of key,
shaft, and hub

Pipe

Saddle

Pin

Pin

L

L

FIGURE P9–177   Pipe saddle for Problem 9–177.

①

20

20

②

③

12-mm dia.

14

20

14

50

Dimensions in millimeters

FIGURE P9–178   Clevis for beam support for Problem 9–178.

Bushing

0.38 in

0.63-in dia.

Side plate

Pin

1.50 in

FIGURE P9–179   Conveyor chain for Problem 9–179.

FIGURE P9–180   Riveted plates for Problem 9–180.

FIGURE P9–182   Riveted plates for Problem 9–182.

9–181. Repeat Problem 9–180 for the connection shown in Figure P9–181. The materials and the design criteria are the same.

FIGURE P9–181   Riveted plates for Problem 9–181.

9–182. Repeat Problem 9–180 for the connection shown in Figure P9–182. The materials and the design criteria are the same.

9–183. Link A in Figure P9–183 is connected to a mating link through a pin joint. Link A is made from AISI 4140 OQT 1100. The pin has a diameter of 0.75 in and is made from AISI 1141 OQT 1100. Compute the allowable repeated tensile force *F* that can be applied to link A for safe operation in regard to tensile stress in the link, shearing stress in the pin, and bearing stress between the pin and the hole in link A.

9–184. For the truss sketched in Figure P9–184, compute the forces in all members. Then, for those members subjected to tensile forces, propose a design for their material and cross-sectional shape and dimensions. Consider how each member will be connected to adjacent members at the pin joints. Also consider how the loads can be applied at joints *C* and *E* and how the supports can be provided at joints *A* and *B*.

9–185. For the truss sketched in Figure P9–185, compute the forces in all members. Then, for those members subjected to tensile forces, propose a design for their material and cross-sectional shape and dimensions. Consider how each member will be connected to adjacent members at the pin joints. Also consider how the loads can be applied at joints *F* and *G* and how the supports can be provided at joints *A* and *D*.

9–186. See Figure P9–186. The load is 34.0 kN and it will be applied with moderate shock.
   a. Compute the forces in the support rods AB and AE.
   b. Design the support rods, specifying the material and a suitable diameter.
   c. Compute the shearing forces in all six pins.

FIGURE P9–183   Connection for Problem 9–183.

FIGURE P9–184   Truss for Problem 9–184.

FIGURE P9–185   Truss for Problem 9–185.

FIGURE P9–186   Structure for Problem 9–186.

d. Design the pins to be safe in shear, specifying the material and a suitable diameter.

e. Check the pins you have designed for acceptable bearing stress. Assume that the mating material

through which the holes are produced is stronger than the pin material. Specify the minimum required thickness of the mating part to provide adequate bearing area.

TORSIONAL SHEAR STRESS AND TORSIONAL DEFORMATION

MAJOR TOPICS

10–1   The Big Picture—Torsional Shear Stress and Torsional Deformation
10–2   Torque, Power, and Rotational Speed
10–3   Torsional Shear Stress in Members with Circular Cross Sections
10–4   Development of the Torsional Shear Stress Formula
10–5   Polar Moment of Inertia for Solid Circular Bars
10–6   Torsional Shear Stress and Polar Moment of Inertia for Hollow Circular Bars
10–7   Design of Circular Members Under Torsion
10–8   Comparison of Solid and Hollow Circular Members
10–9   Stress Concentrations in Torsionally Loaded Members
10–10 Twisting—Elastic Torsional Deformation
10–11 Torsion in Noncircular Sections

10–1 THE BIG PICTURE— TORSIONAL SHEAR STRESS AND TORSIONAL DEFORMATION

Discussion Map

- In this chapter, you will learn how to compute the level of shearing stress experienced by members carrying a torque.

- You will learn how to use the torsional shear stress equation for this calculation.

- You will also learn how to compute the amount of torsional deformation in a torsionally loaded member.

- You will use the angle of twist equation.

Discover

Think about products or machines that you are familiar with that have rotating shafts that transmit power. Make a list, describing the source of the power and the path that the power takes as it is delivered to the point of ultimate use. See Figure 10–1, the gear-type speed reducer example, on the next page.

Sometimes torque is applied to a member that does not rotate. Can you think of an example? Consider the socket wrench shown in Figure 10–4 that is being used to tighten a bolt. The bolt rotates at first when it is being driven down to clamp two members together. Then, when the bolt is finally tight, you are applying a torque through the extension shaft to the head of the bolt. In fact, proper assembly of bolted connections requires that a specified torque be applied to the head of each bolt.

When any member is subjected to a torque, it demonstrates some amount of twisting where one end is rotated with respect to the other. The amount of this elastic deformation must often be determined by analytical methods to ensure that the member has sufficient rigidity.

Can you think of other examples where torsional rigidity is an important design and performance parameter?

Torsion refers to the loading of a member that tends to cause it to rotate or twist. Such a load is called a *torque, rotational moment, twisting moment,* or *couple.* When a torque is applied to a member, *shearing stress* is developed within the member and *torsional deformation* is created, resulting in an angle of twist of one end of the member relative to the other.

A familiar example of a torsional member is a circular shaft being driven by an electric motor and delivering power through a gear to a mating gear. Figure 10–1 shows a drawing of a possible layout. Its operation is described next.

The motor uses electrical energy to produce a driving torque in its output shaft. The motor shaft is connected to the input shaft of the gear-type speed reducer through a flexible coupling. The coupling allows the transmission of torque while eliminating any tendency to bend the shaft or to cause axial loads to develop. When the motor rotates while developing torque, it transmits power to the shafts of the system shown.

The input shaft is supported on two antifriction bearings and it carries the small gear A, called a *pinion,* between the two bearings. The torque applied to the shaft from the coupling is transmitted down the shaft to the gear, where a

FIGURE 10–1   General layout of mechanical drive with a single-stage gear reducer.

reaction torque exists in the form of a resistance offered by the mating gear B.

The power is then transferred to gear B, causing it to rotate. When the gears are operating at a constant speed and transmitting power, there is a uniform level of torque in the input shaft from the coupling to the gear. That part of the input shaft from gear A to the outboard bearing experiences virtually no torque because the bearing provides very little resistance.

The mating gear B receives the power from the pinion A and delivers it into the output shaft. The power then travels down the output shaft, through another flexible coupling, to the driven machine. This may be a fan, a conveyor, a machine tool, or some other device that uses the power to perform some useful work. Thus, the output shaft experiences a torque from gear B to the coupling. The driveshaft for the driven machine also is subjected to that same level of torque.

You should try to visualize the flow of power and torque described here. Each part of each shaft that experiences an applied torque will develop internal shearing stresses. You must learn how to compute the level of such

stresses as a function of the applied torque and the geometry of the shaft using the torsional shear stress equation.

You will also learn how to compute the amount of torsional deformation in a torsionally loaded member. You will use the *angle of twist equation* that relates the applied torque, the length of the member, its geometry, and the modulus of elasticity of its material to the angle of twist.

The goal of these analyses is to ensure that the loaded member is safe for the applied torsional load and sufficiently rigid to perform properly in service.

ACTIVITY   Chapter 10—Torsional Shear Stress and Torsional Deformation

A. Nonquantitative Activity

Gather some materials that will help you to get a good "feel" for how a torque is applied to an object and how the object behaves when resisting the torque. Most of the examples discussed in this chapter are either solid or hollow cylinders. However, some attention will be given to noncircular shapes

FIGURE 10–2   Twisting of a cylindrical rod.

such as solid or hollow squares, rectangles, hexagons, or special rolled or extruded shapes. Here are some examples:

1. A relatively thin cardboard tube such as that on which paper towels or gift-wrapping paper are rolled

2. A flexible foam cylinder such as a swim toy or a cushion

3. A long, small-diameter plastic rod or tube, or a wood dowel

4. A long, rolled, or extruded shape such as might be used for a closet door rail, a curtain rod, or a part of the frame of a toy, a cabinet, or an appliance

Recognizing that the terms *long, small-diameter, flexible,* and *thin* are imprecise, look for items that deform somewhat easily when manipulated with your own hands.

Grasp one of your cylindrical items by the ends as illustrated in Figure 10–2. Then twist it in opposing directions with each hand while observing how it deforms. The cylinder in part (a) of Figure 10–2 is not subjected to twisting. A rectangular grid, made from circumferential and longitudinal lines, is illustrated on its surface. Part (b) of the figure shows what happens when twisting is applied. Note how the longitudinal lines warp as the cylinder twists, while the circumferential lines remain mostly unchanged. If feasible, make a similar grid on your item. These observations are useful when considering the formulas we use to calculate stresses and deformations in members subjected to torsion.

Using a thin cardboard tube and applying the twisting action, notice how it is quite stiff in torsion relative to the weight or general strength of the cardboard. If you do not have a commercially made tube, find a piece of stiff paper card stock about 5 in × 11 in (125 mm × 275 mm) and roll it into a tube approximately 1.0 in to 1.5 in (25 mm to 35 mm) in diameter. You might roll it around a cylindrical broom handle to form the tube. Apply tape along the entire length. The tube will be stiff in torsion even though you may be able to crush it easily. Now, carefully cut the tube along its length and notice how it becomes extremely flexible in torsion. This illustrates the importance of a continuous circular cross section for good torsional rigidity.

B. Quantitative Activity

For quantitative measurements, a rigid structure is required to hold fixed one end of a member such as a cylindrical rod or a tube while allowing the other end to be held in place but free to rotate as a torque is applied to that end. A means of measuring the applied torque and the torsional deformation (angle of twist) is required.

Figure 10–3 shows a commercially available apparatus that facilitates such a test and demonstration. Torque is applied by hanging known weights at the end of an arm attached to the end of the rod that is allowed to rotate. The applied torque is calculated simply from the basic definition of torque:

$$T = Fr$$

where T = applied torque
   F = applied force at the end of the arm
   r = perpendicular distance from the line of action of the force to the center of the rod

The angular deformation is measured on a fixed protractor scale with an indicator affixed to the rod.

Load on hanger
applies torque to rod

FIGURE 10–3   Torsion testing device.
Source: P. A. Hilton Ltd./Hi-Tech, Hampshire, England.

Design an experiment from which you can record the following data:

1. General description of the rod or tube being subjected to the twisting action
2. Material from which the rod is made
3. Diameter of the rod along with the inside diameter if a hollow tube is used
4. Length between where the rod is held fixed and where the torque is applied
5. A table of data listing the following for each data point:
   a. Incremental applied force on the arm
   b. Length of the torque arm
   c. Computed torque ($T = Fr$)
   d. Angle of rotation

A variety of deliverables from the testing can be produced:

1. Graph of angle of twist versus the applied torque
2. Record and description of any failure of the rod or tube that may have occurred, such as excessive bending, fracture, tearing, crushing, or wrinkling, and the load at which the failure occurred
3. Maintenance of a record of the data for use after the principles of torsional shear stress and torsional deformation analysis are developed later in this chapter. Then appropriate calculations of geometrical and material properties of the rod or tube can be made, and computed stress and deformation can be compared with the measured data.

Objectives of This Chapter

After completing this chapter, you should be able to:

1. Define *torque* and compute the magnitude of torque exerted on a member subjected to torsional loading.
2. Define the relationship among the three critical variables involved in power transmission: power, torque, and rotational speed.
3. Manipulate the units for power, torque, and rotational speed in both the SI metric system and the U.S. Customary System.
4. Compute the maximum shear stress in a member subjected to torsional loading.
5. Define the *polar moment of inertia* and compute its value for solid and hollow round shafts.
6. Compute the shear stress at any point within a member loaded in torsion.
7. Specify a suitable design shear stress for a member loaded in torsion.
8. Define the *polar section modulus* and compute its value for solid and hollow round shafts.
9. Determine the required diameter of a shaft to carry a given torque safely.
10. Compare the design of solid and hollow shafts on the basis of the mass of the shafts required to carry a certain torque while limiting the torsional shear stress to a certain design value.
11. Apply stress concentration factors to members in torsion.
12. Compute the angle of a twist of a member loaded in torsion.
13. Define the *shear modulus of elasticity*.
14. Discuss the method of computing torsional shear stress and torsional deflection for members with noncircular cross sections.
15. Describe the general shapes of members having relatively high torsional stiffness.

10–2 TORQUE, POWER, AND ROTATIONAL SPEED

A necessary task in approaching the calculation of torsional shear stress and deflection is the understanding of the concept of *torque* and the relationship among the three critical variables involved in power transmission: *torque, power,* and *rotational speed.*

Torque

Figure 10–4 shows a socket wrench with an extension shaft being used to tighten a bolt. The *torque,* applied to both the bolt and the extension shaft, is the product of the applied force and the distance from the line of action of the force to the axis of the bolt. That is,

Torque ▼
$$\text{Torque} = T = F \times d \qquad (10\text{–}1)$$

Thus, torque is expressed in the units of *force times distance,* which is N · m in the SI metric system and lb · in or lb · ft in the U.S. Customary System.

FIGURE 10–4   Wrench applying a torque to a bolt.

Example Problem
10–1

For the wrench in Figure 10–4, compute the magnitude of the torque applied to the bolt if a force of 50 N is exerted at a point 250 mm out from the axis of the socket.

Solution Objective: Compute the torque on the wrench.

Given: Setup in Figure 10–4. $F = 50$ N. $d = 250$ mm = torque arm length

Analysis: Use Equation (10–1), $T = F \times d$

Results: Torque $= T = (50 \text{ N})(250 \text{ mm}) \times \dfrac{1 \text{ m}}{1000 \text{ mm}} = 12.5 \text{ N} \cdot \text{m}$

Comment: The wrench is applying a torque of 12.5 N·m to the bolt.

Power

Power is defined as the rate of transferring energy.

Figure 10–5 shows a drive system for a boat. Power developed by the engine flows through the transmission and the driveshaft to the propeller, where it drives the boat forward. The crankshaft inside the engine, the various shafts in the transmission, and the driveshaft are all subjected to torsion.

The magnitude of the torque in a power transmission shaft is dependent on the amount of power it carries and on the speed of rotation, according to the following relation:

Power ▽

$$\text{Power} = \text{Torque} \times \text{Rotational speed}$$
$$P = T \times n \qquad (10\text{–}2)$$

This is a very useful relationship because if any two values, P, n, or T, are known, the third can be computed.

Careful attention must be paid to units when working with torque, power, and rotational speed. Appropriate units in the SI metric system and the U.S. Customary System are reviewed next.

Power in SI Metric System Units

In the SI metric system, the *joule* (J) is the standard unit for energy and it is equivalent to the N · m, the standard unit

for torque. That is,

$$1.0 \text{ J} = 1.0 \text{ N} \cdot \text{m}$$

Then power is defined as

SI Units for Power ▽

$$\text{Power} = \frac{\text{Energy}}{\text{Time}} = \frac{\text{Joule}}{\text{Second}} = \frac{\text{J}}{\text{s}} = \frac{\text{N} \cdot \text{m}}{\text{s}} = \text{Watt} = \text{W}$$

$$(10\text{–}3)$$

Note that 1.0 J/s is defined to be 1.0 watt (1.0 W). The watt is a rather small unit of power, so the kilowatt (1.0 kW = 1000 W) is often used.

The standard unit for rotational speed in the SI metric system is *radians per second* (rad/s). Frequently, however, rotational speed is expressed in revolutions per minute (rpm). The conversion required is illustrated below, converting 1750 rpm to rad/s.

$$n = \frac{1750 \text{ rev}}{\text{min}} \times \frac{2\pi \text{ rad}}{\text{rev}} \times \frac{1 \text{ min}}{60 \text{ s}} = 183 \text{ rad/s}$$

When using n in rad/s in Equation (10–2), the radian is considered to be *no unit at all*, as illustrated in the following example problem.

FIGURE 10–5 Drive system for a boat.

Example Problem
10–2

The driveshaft for the boat shown in Figure 10–5 transmits 95 kW of power while rotating at 525 rpm. Compute the torque in the shaft.

Solution   Objective: Compute the torque in the shaft.

Given: $P = 95$ kW $= 95\,000$ W $= 95\,000$ N$\cdot$m/s; $n = 525$ rpm

Analysis: Equation (10–2) will be solved for T and used to compute torque.

$$P = Tn; \quad \text{then } T = P/n$$

But n must be in rad/s, found as follows:

$$n = \frac{525\text{ rev}}{\text{min}} \times \frac{2\pi\text{ rad}}{\text{rev}} \times \frac{1\text{ min}}{60\text{ s}} = 55.0\text{ rad/s}$$

Results: The torque is

$$T = \frac{P}{n} = \frac{95\,000\text{ N}\cdot\text{m}}{\text{s}} \times \frac{1}{55.0\text{ rad/s}} = 1727\text{ N}\cdot\text{m}$$

Comment: Note that the radian unit is ignored in such calculations.

Power in U.S. Customary System Units

Typical units for torque, power, and rotational speed in the U.S. Customary System are

$T = $ torque (lb$\cdot$in)

$n = $ rotational speed (rpm)

$P = $ power (horsepower, hp)

Note that 1.0 hp $= 6600$ lb$\cdot$in/s. Then the unit conversions required to ensure consistent units are

$$\text{Power} = T(\text{lb}\cdot\text{in}) \times n\left(\frac{\text{rev}}{\text{min}}\right) \times \frac{1\text{ min}}{60\text{ s}} \times \frac{2\pi\text{ rad}}{\text{rev}}$$
$$\times \frac{1\text{ hp}}{6600\text{ lb}\cdot\text{in/s}}$$

or

U.S. Customary System Units for Power ▼

$$\text{Power} = \frac{Tn}{63\,000} \tag{10–4}$$

Example Problem
10–3

Compute the power, in the unit of horsepower, being transmitted by a shaft if it is developing a torque of 15 000 lb$\cdot$in and rotating at 525 rpm.

Solution   Objective: Compute the power transmitted by the shaft.

Given: $T = 15\,000$ lb$\cdot$in; $n = 525$ rpm

Analysis: Equation (10–4) will be used directly because T and n are in the proper units of lb$\cdot$in and rpm. Power will be in horsepower.

Results: The power is

$$P = \frac{Tn}{63\,000} = \frac{(15\,000)(525)}{63\,000} = 125\text{ hp}$$

Torque Sensors

There are times when mechanical designers and engineers need to measure torque in product development laboratories or to build torque sensing into operating equipment. Internet site 4 lists some commercial manufacturers of torque sensors that cover a wide range of torque values from 5 mN$\cdot$m to over 225 000 N$\cdot$m (0.044 lb$\cdot$in to 2×10^6 lb$\cdot$in). The sensors incorporate transducers to produce an electronic digital readout or analog signal to indicate torque values. Internet site 5 is an article discussing the history, current status, and future projections for torque-sensing technology.

10–3 TORSIONAL SHEAR STRESS IN MEMBERS WITH CIRCULAR CROSS SECTIONS

When a member is subjected to an externally applied torque, an internal resisting torque must be developed in the material from which the member is made. The internal resisting torque is the result of stresses developed in the material.

Figure 10–6 shows a circular bar subjected to a torque, T. Section N would be rotated relative to section

FIGURE 10–6   Torsional shear stress in a circular bar.

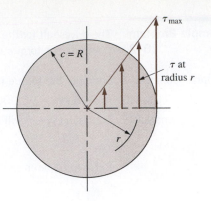

FIGURE 10–7   Distribution of shear stress on a cross section of the bar.

M as shown. If an element on the surface of the bar were isolated, it would be subjected to shearing forces on the sides parallel to cross sections M and N, as shown. These shearing forces result in shear stresses on the element. For the stress element to be in equilibrium, equal shearing stresses must exist on the top and bottom faces of the element.

The shear stress element shown in Figure 10–6 is fundamentally the same as that shown in Figure 7–17 in the discussion of direct shear stress. While the manner in which the stresses are created differs, the nature of torsional shear stress is the same as direct shear stress when an infinitesimal element is considered.

When the circular bar is subjected to the externally applied torque, the material in each cross section is deformed in a manner such that the fibers on the outside surface experience the maximum strain. At the central axis of the bar, no strain at all is produced. Between the center and the outside, there is a linear variation of strain with radial position r. Because stress is directly proportional to strain, we can say that the maximum stress occurs at the outside surface, that there is a linear variation of stress with radial position r, and that a zero stress level occurs at the center. Figure 10–7 illustrates these observations.

The derivation will be shown in the next section for the formula for the maximum shear stress on the outer surface of the bar. For now, we state the *torsional shear stress formula* as

Torsional Shear Stress Formula ▽

$$\tau_{max} = \frac{Tc}{J} \qquad (10\text{–}5)$$

where T = applied torque at the section of interest
c = radius of the cross section
J = polar moment of inertia of the circular cross section

The formula for J for a solid circular cross section is developed in Section 10–6. At this time, we show the results of the derivation as,

Polar Moment of Inertia for Circular Bar ▽

$$J = \frac{\pi D^4}{32} \qquad (10\text{–}6)$$

where D is the diameter of the shaft; that is, $D = 2R$.

Because of the linear variation of stress and strain with position in the bar as shown in Figure 10–7, the stress, τ, at any radial position, r, can be computed from

Shear Stress at Any Radius ▽

$$\tau = \tau_{max}\frac{r}{c} \qquad (10\text{–}7)$$

Equations (10–5), (10–6), and (10–7) can be used to compute the shear stress at any point in a circular bar subjected to an externally applied torque. The following example problems illustrate the use of these equations.

Example Problem 10–4   For the socket wrench extension shown in Figure 10–4, compute the maximum torsional shear stress in the middle portion where the diameter is 9.5 mm. The applied torque is 10.0 N · m.

Solution   Objective: Compute the maximum torsional shear stress in the extension.

Given: Torque = T = 10.0 N · m; diameter = D = 9.5 mm

Analysis: Use Equation (10–6) to compute J and Equation (10–5) to compute the maximum shear stress. Also, $c = D/2 = 9.5$ mm/2 = 4.75 mm.

Results: $J = \dfrac{\pi D^4}{32} = \dfrac{\pi (9.5\,\text{mm})^4}{32} = 800\,\text{mm}^4$

$\tau_{max} = \dfrac{Tc}{J} = \dfrac{(100\,\text{N}\cdot\text{m})(4.75\,\text{mm})}{800\,\text{mm}^4} \times \dfrac{10^3 \text{mm}}{\text{m}} = 59.4\,\text{N/mm}^2 = 59.4\,\text{MPa}$

Comment: This level of stress would occur at all points on the surface of the circular part of the extension.

Example Problem 10–5   Calculate the maximum torsional shear stress that would develop in a solid circular shaft, having a diameter of 1.25 in, if it is transmitting 125 hp while rotating at 525 rpm.

Solution   Objective: Compute the maximum torsional shear stress in the shaft.

Given: Power $= P = 125$ hp; rotational speed $= n = 525$ rpm
Shaft diameter $= D = 1.25$ in

Analysis: Solve Equation (10–4) for the torque, T. Use Equation (10–6) to compute J and Equation (10–5) to compute the maximum shear stress. Also, $c = D/2 = 1.25\,\text{in}/2 = 0.625$ in.

Results: Equation (10–4),

$$\text{Power} = P = \frac{Tn}{63\,000}$$

Solving for the torque T gives

$$T = \frac{63\,000\,P}{n}$$

Recall that this equation will give the value of the torque directly in lb·in when P is in horsepower and n is in rpm. Then

$$T = \frac{63\,000(125)}{525} = 15\,000\,\text{lb}\cdot\text{in}$$

$$J = \frac{\pi D^4}{32} = \frac{\pi(1.25\,\text{in})^4}{32} = 0.240\,\text{in}^4$$

Then

$$\tau_{max} = \frac{Tc}{J} = \frac{(15\,000\,\text{lb}\cdot\text{in})(0.625\,\text{in})}{0.240\,\text{in}^4} = 39\,100\,\text{psi}$$

Comment: This level of stress would occur at all points on the surface of the shaft.

10–4 DEVELOPMENT OF THE TORSIONAL SHEAR STRESS FORMULA

The standard form of the torsional shear stress formula for a circular bar subjected to an externally applied torque was shown as Equation (10–5), and its use was illustrated in Example Problems 10–4 and 10–5. This section will show the development of that formula. Reference should be made to Figures 10–6 and 10–7 for the general nature of the torsional loading and a visualization of the effect the torque has on the behavior of the circular bar.

In this development, it is assumed that the material for the bar behaves in accordance with Hooke's law; that is, stress is directly proportional to strain. Also, the properties of the bar are homogeneous and isotropic; that is, the material reacts the same regardless of the direction of the applied loads. Also, it is assumed that the bar is of constant cross section in the vicinity of the section of interest.

Considering two cross sections M and N at different places on the bar, section N would be rotated through an angle θ relative to section M. The fibers of the material would undergo a strain that would be maximum at the outside surface of the bar and vary linearly with radial position to zero at the center of the bar. Because, for elastic materials obeying Hooke's law, stress is proportional to strain, the maximum stress would also occur at the outside of the bar, as shown in Figure 10–7. The linear variation of stress, τ, with radial position in the cross section, r, is also shown. Then, by proportion using similar triangles,

$$\frac{\tau}{r} = \frac{\tau_{max}}{c} \qquad (10\text{–}8)$$

Then the shear stress at any radius can be expressed as a function of the maximum shear stress at the outside of the shaft,

$$\tau = \tau_{max} \times \frac{r}{c} \qquad (10\text{–}9)$$

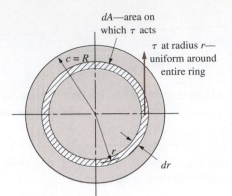

FIGURE 10–8   Shear stress τ at radius r acting on the area dA.

It should be noted that the shear stress τ acts uniformly on a small ring-shaped area, dA, of the shaft, as illustrated in Figure 10–8. Now since force equals stress times area, the force on the area dA is

$$dF = \tau dA = \underbrace{\tau_{max}\frac{r}{c}}_{\text{stress}} \times \underbrace{dA}_{\text{area}}$$

The next step is to consider that the torque dT developed by this force is the product of dF and the radial distance to dA. Then

$$dT = dF \times r = \underbrace{\tau_{max}\frac{r}{c}dA}_{\text{force}} \times \underbrace{r}_{\text{radius}} = \tau_{max}\frac{r^2}{c}dA$$

This equation is the internal resisting torque developed on the small area dA. The total torque on the entire area would be the sum of all the individual torques on all areas of the cross section. The process of summing is accomplished by the mathematical technique of integration, illustrated as follows:

$$T = \int_A dT = \int_A \tau_{max}\frac{r^2}{c}dA$$

In the process of integration, constant terms such as τ_{max} and c can be brought outside the integral sign, leaving

$$T = \frac{\tau_{max}}{c}\int_A r^2 dA \qquad \textbf{(10–10)}$$

In mechanics, the term $\int r^2 dA$ is given the name *polar moment of inertia* and is identified by the symbol J. Equation (10–10) can then be written

$$T = \tau_{max}\frac{J}{c}$$

or

$$\tau_{max} = \frac{Tc}{J} \qquad \textbf{(10–11)}$$

The method of evaluating J is developed in the next section.

Equation (10–11), which is identical to Equation (10–5), can be used to compute the maximum shear stress on a circular bar subjected to torsion. The maximum shear stress occurs anywhere on the outer surface of the bar.

10–5 POLAR MOMENT OF INERTIA FOR SOLID CIRCULAR BARS

Refer to Figure 10–8 showing a solid circular cross section. To evaluate J from

$$J = \int r^2 dA$$

it must be seen that dA is the area of a small ring located at a distance r from the center of the section and having a thickness dr.

For a small magnitude of dr, the area is that of a strip having a length equal to the circumference of the ring times the thickness.

$$dA = 2\pi r \times dr$$

where the bracket above indicates *thickness of the ring* and the bracket below indicates *circumference of a ring at the radius r*.

Then the polar moment of inertia for the entire cross section can be found by integrating from $r = 0$ at the center of the bar to $r = R$ at the outer surface.

$$J = \int_0^R r^2 dA = \int_0^R r^2(2\pi r)dr = \int_0^R 2\pi r^3 dr$$

$$= \frac{2\pi R^4}{4} = \frac{\pi R^4}{2}$$

It is usually more convenient to use diameter rather than radius. Then since $R = D/2$,

$$J = \frac{\pi(D/2)^4}{2} = \frac{\pi D^4}{32} \qquad \textbf{(10–12)}$$

10–6 TORSIONAL SHEAR STRESS AND POLAR MOMENT OF INERTIA FOR HOLLOW CIRCULAR BARS

It will be shown later that there are many advantages to using a hollow circular bar, as compared with a solid bar, to carry a torque. This section covers the method of computing the maximum shear stress and the polar moment of inertia for a hollow bar.

Figure 10–9 shows the basic geometry for a hollow bar. The variables are

R_i = inside radius

D_i = inside diameter

R_o = outside radius = c

D_o = outside diameter

The logic and details of the development of the torsional shear stress formula as shown in Section 10–4 apply as well to a hollow bar as to the solid bar. The difference between them is in the evaluation of the polar moment of inertia, as will be shown later. Therefore, Equation (10–5) or (10–11)

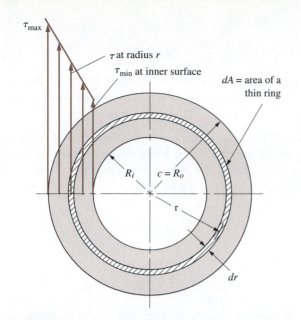

FIGURE 10–9 Notation for variables used to derive J for a hollow round bar.

can be used to compute the maximum torsional shear stress in either the solid or the hollow bar.

Also, as illustrated in Figure 10–9, the maximum shear stress occurs at the outer surface of the bar, and there is a linear variation of stress with radial position inside the bar. The minimum shear stress occurs at the inside surface. The shear stress at any radial position can be computed from Equation (10–7) or (10–9).

Polar Moment of Inertia for a Hollow Bar

The process of developing the formula for the polar moment of inertia for a hollow bar is similar to that used for the solid bar. Refer again to Figure 10–9 for the geometry.

Starting with the basic definition of the polar moment of inertia,

$$J = \int r^2 dA$$

as before, $dA = 2\pi r dr$. But for the hollow bar, r varies only from R_i to R_o. Then

$$J = \int_{R_i}^{R_o} r^2 (2\pi r) dr = 2\pi \int_{R_i}^{R_o} r^3 dr = \frac{2\pi (R_o^4 - R_i^4)}{4}$$

$$J = \frac{\pi}{2}(R_o^4 - R_i^4)$$

Substituting $R_o = D_o/2$ and $R_i = D_i/2$ gives

Polar Moment of Inertia for a Hollow Bar ▼

$$J = \frac{\pi}{32}(D_o^4 - D_i^4) \qquad (10\text{–}13)$$

This is the equation for the polar moment of inertia for a hollow circular bar.

Summary of Relationships for Torsional Shear Stresses in Hollow Circular Bars

Maximum Shear Stress ▼

$$\tau_{max} = \frac{Tc}{J} \qquad (10\text{–}11)$$

τ_{max} occurs at the outer surface of the bar, where c is the radius of the bar.

Shear Stress at Any Radial Position r ▼

$$\tau = \tau_{max}\frac{r}{c} = \frac{Tr}{J} \qquad (10\text{–}9)$$

Polar Moment of Inertia for Hollow Bars ▼

$$J = \frac{\pi}{32}(D_o^4 - D_i^4) \qquad (10\text{–}13)$$

Example Problem 10–6

For the propeller driveshaft of Figure 10–5, compute the torsional shear stress when it is transmitting a torque of 1.76 kN·m. The shaft is a hollow tube having an outside diameter of 60 mm and an inside diameter of 40 mm. Find the stress at both the outer and inner surfaces.

Solution

Objective: Compute the torsional shear stress at the outer and inner surfaces of the hollow propeller driveshaft.

Given: Shaft shown in Figure 10–5. Torque $= T = 1.76$ kN·m $= 1.76 \times 10^3$ N·m.
Outside diameter $= D_o = 60$ mm; inside diameter $= D_i = 40$ mm.

Analysis: The final calculation for the torsional shear stress at the outer surface will be made using Equation (10–11). Equation (10–9) will be used to compute the stress at the inner surface. The polar moment of inertia will be computed using Equation (10–13). And $c = D_o/2 = 30$ mm.

Results: At the outer surface,

$$\tau_{max} = \frac{Tc}{J}$$

$$J = \frac{\pi}{32}(D_o^4 - D_i^4) = \frac{\pi}{32}(60^4 - 40^4)\,\text{mm}^4 = 1.02 \times 10^6\,\text{mm}^4$$

$$\tau_{max} = \frac{Tc}{J} = \frac{(1.76 \times 10^3\,\text{N·m})(30\,\text{mm})}{1.02 \times 10^6\,\text{mm}^4} \times \frac{10^3\,\text{mm}}{\text{m}}$$

$$\tau_{max} = 51.8\,\text{N/mm}^2 = 51.8\,\text{MPa}$$

At the inner surface, $r = D_i/2 = 40 \text{ mm}/2 = 20 \text{ mm}$.

$$\tau = \tau_{max}\frac{r}{c} = 51.8 \text{ MPa} \times \frac{20 \text{ mm}}{30 \text{ mm}} = 34.5 \text{ MPa}$$

Comment: You should visualize these stress values plotted on the cross section shown in Figure 10–9.

10–7 DESIGN OF CIRCULAR MEMBERS UNDER TORSION

In a design problem, the loading on a member is known, and what is required is to determine the geometry of the member to ensure that it will carry the loads safely. Material selection and the determination of design stresses are integral parts of the design process. *The techniques developed in this section are for circular members only, subjected only to torsion.* Of course, both solid and hollow circular members are covered. Torsion in noncircular members is covered in a later section of this chapter. The combination of torsion with bending and axial loads is presented in later chapters.

The basic torsional shear stress equation, Equation (10–11), was expressed as

$$\tau_{max} = \frac{Tc}{J} \tag{10–11}$$

In design, we can substitute a certain design stress τ_d for τ_{max}. As in the case of members subjected to direct shear stress and made of ductile materials, the design stress is related to the yield strength of the material in shear. That is,

$$\tau_d = \frac{s_{ys}}{N}$$

where N is the design factor chosen by the designer based on the manner of loading. Table 10–1 can be used as a guide to determine the value of N.

Where the data for s_{ys} are not available, the value can be estimated as $s_y/2$. This will give a reasonable, and usually conservative, estimate for ductile metals, especially steel. Then

Design Shear Stress ▼

$$\tau_d = \frac{s_{ys}}{N} = \frac{s_y}{2N} \tag{10–14}$$

The torque, T, would be known in a design problem. Then, in Equation (10–11), only c and J are left to be deter-

mined. Notice that both c and J are properties of the geometry of the member that is being designed. For solid circular members (shafts), the geometry is completely defined by the diameter. It has been shown that

$$c = \frac{D}{2}$$

and

$$J = \frac{\pi D^4}{32}$$

It is now convenient to note that if the quotient J/c is formed, a simple expression involving D is obtained.

In the study of strength of materials, the term J/c is given the name *polar section modulus,* and the symbol Z_p is used to denote it.

Polar Section Modulus—Solid Shafts ▼

$$Z_p = \frac{J}{c} = \frac{\pi D^4}{32} \times \frac{1}{D/2} = \frac{\pi D^3}{16} \tag{10–15}$$

Substituting Z_p for J/c in Equation (10–11) gives

Maximum Shear Stress ▼

$$\tau_{max} = \frac{T}{Z_p} \tag{10–16}$$

To use this equation in design, we can let $\tau_{max} = \tau_d$ and then solve for Z_p.

Required Polar Section Modulus ▼

$$Z_p = \frac{T}{\tau_d} \tag{10–17}$$

Equation (10–17) will give the required value of the polar section modulus of a circular shaft to limit the torsional shear stress to τ_d when subjected to a torque T. Then Equation (10–15) can be used to find the required diameter of a solid circular shaft. Solving for D gives us

Required Diameter ▼

$$D = \sqrt[3]{\frac{16Z_p}{\pi}} \tag{10–18}$$

If a hollow shaft is to be designed,

Polar Section Modulus—Hollow Shafts ▼

$$Z_p = \frac{J}{c} = \frac{\pi}{32}(D_o^4 - D_i^4) \times \frac{1}{D_o/2}$$

$$Z_p = \frac{\pi}{16}\frac{D_o^4 - D_i^4}{D_o} \tag{10–19}$$

In this case, one of the diameters *or* the relationship between the two diameters would have to be specified in order to solve for the complete geometry of the hollow shaft.

TABLE 10–1	Design Factors and Design Shear Stresses for Ductile Metals	
Manner of loading	**Design factor**	**Design shear stress $\tau_d = s_y/2N$**
Static or steady torsion	2	$\tau_d = s_y/4$
Repeated torsion	4	$\tau_d = s_y/8$
Torsional impact or shock	6	$\tau_d = s_y/12$

PROGRAMMED
EXAMPLE PROBLEM

**PROGRAMMED
EXAMPLE PROBLEM**

The completion of this problem is presented in a programmed format. You should answer each question as it is posed before looking at the next panel beyond the line across the page. This process is intended to involve you in the decision-making activities required of a designer.

**Example Problem
10–7**

The final drive to a conveyor that feeds coal to a railroad car is a shaft loaded in pure torsion and carrying 800 N·m of torque. A proposed design calls for the shaft to have a solid circular cross section. Complete the design by first specifying a suitable steel for the shaft and then specifying the diameter.

Solution **Objective:**

1. Specify a suitable steel for the shaft.
2. Specify the shaft diameter.

Given:

Applied torque = T = 800 N·m
Shaft drives a coal conveyor.

Analysis:

First, as an aid in selecting a suitable material, what manner of loading will the shaft experience in service?

The drive for a coal conveyor is likely to experience very rough service as coal is dumped onto the conveyor. Therefore, the design should be able to accommodate impact and shock loading. What properties should the steel for the shaft possess?

A highly ductile material should be used because such materials withstand shock loading much better than more brittle materials. The steel should have a moderately high strength so that the required diameter of the shaft is reasonable. It may be important to choose a steel with good machinability because the shaft is likely to require machining during its manufacture. What is a typically used measure of ductility for steels?

Chapter 8 mentions that the *percent elongation* for a steel is an indication of its ductility. To withstand impact and shock, a steel having a value somewhat higher than 10% elongation should be specified. Now specify a suitable steel.

There are many steels that can be used satisfactorily. Let's specify AISI 1141 OQT 1300. List pertinent data from Appendix A–14.

You should have found that s_y = 469 MPa and that the 28% elongation indicates a high ductility. Also note, as stated in Chapter 8, the 1100 series steels have good machinability because of a relatively high sulfur content in the alloy.

We will be using Equations (10–16), (10–17), and (10–18) to continue the design process with the ultimate goal of specifying a suitable diameter for the shaft. We know the applied torque is 800 N·m. The next task should be to determine an acceptable design shear stress. How will you do that?

Table 10–1 calls for $\tau_d = s_y/2N$ with N = 6; that is, $\tau_d = s_y/12$. Then

$$\tau_d = s_y/12 = 469\ \text{MPa}/12 = 39.1\ \text{MPa} = 39.1\ \text{N/mm}^2$$

What is the next step?

We can use Equation (10–17) to compute the required value of the polar section modulus for the cross section of the shaft. Do that now.

You should have the required $Z_p = 20.5 \times 10^3\ \text{mm}^3$, found from

$$Z_p = \frac{T}{\tau_d} = \frac{800\ \text{N·m}}{39.1\ \text{N/mm}^2} \times \frac{10^3\ \text{mm}}{\text{m}} = 20.5 \times 10^3\ \text{mm}^3$$

What is the next step?

We can compute the minimum acceptable diameter for the shaft using Equation (10–18). Do that now.

You should have $D_{min} = 47.1$ mm, found from

$$D = \sqrt[3]{\frac{16Z_p}{\pi}} = \sqrt[3]{\frac{16(20.5 \times 10^3)\text{mm}^3}{\pi}} = 47.1 \text{ mm}$$

It would be appropriate to specify a convenient size for the shaft diameter that is slightly larger than this value. Use Appendix A–2 as a guide and specify a diameter.

Specifying $D = 50$ mm is preferred.

Summary of Results: The shaft will be made from AISI 1141 OQT 1300 steel with a diameter of 50 mm.

Comment: The maximum shear stress at the outer surface of the 50-mm diameter shaft is actually less than the design stress because we specified a preferred diameter slightly greater than the minimum required diameter of 47.1 mm. Let's now compute the actual maximum stress in the shaft. First, we will compute the polar section modulus for the 50-mm diameter shaft.

$$Z_p = \frac{\pi D^3}{16} = \frac{\pi (50)^3 \text{ mm}^3}{16} = 24.5 \times 10^3 \text{ mm}^3$$

Then the maximum shear stress is

$$\tau_{max} = \frac{T}{Z_p} = \frac{800 \text{ N} \cdot \text{m}}{24.5 \times 10^3 \text{ mm}^3} \times \frac{10^3 \text{ mm}}{1 \text{ m}} = 32.6 \text{ N/mm}^2 = 32.6 \text{ MPa}$$

10–8 COMPARISON OF SOLID AND HOLLOW CIRCULAR MEMBERS

We will now demonstrate, by example, that hollow shafts are more efficient than solid shafts. Here the term *efficiency* is used as a measure of the mass of material in a shaft required to carry a given torque with a given shear stress level. The following example problem shows the design of a hollow shaft with a slightly larger outside diameter that has the same maximum shear stress as the 50-mm diameter solid shaft just designed. Then the mass of the hollow shaft is compared with that of the solid shaft.

PROGRAMMED EXAMPLE PROBLEM

The completion of this problem is presented in a programmed format. You should answer each question as it is posed before looking at the next panel beyond the line across the page. This process is intended to involve you in the decision-making activities required of a designer.

Example Problem 10–8

An alternative design for the shaft described in Example Problem 10–7 would be to use a hollow tube for the shaft. Assume that a tube having an outside diameter of 60 mm is available in the same material as specified for the solid shaft (AISI 1141 OQT 1300). Compute what maximum inside diameter the tube can have that would result in the same stress in the steel as the 50-mm solid shaft.

Solution   Objective: Compute the maximum allowable inside diameter for the hollow shaft.

Given: From Example Problem 10–7, maximum shear stress $= \tau_{max} = 32.6$ MPa.
$D_o = 60$ mm. Applied torque $= T = 800$ N $\cdot$ m.

Analysis: Because torsional shear stress is inversely proportional to the polar section modulus, it is necessary that the hollow tube have the same value for Z_p as does the 50-mm diameter solid shaft. That is, $Z_p = 24.5 \times 10^3$ mm^3. What is the formula for Z_p for a hollow shaft?

$$Z_p = \frac{\pi}{16} \frac{D_o^4 - D_i^4}{D_o}$$

The outside diameter, D_o, is known to be 60 mm. We can then solve for the required inside diameter, D_i. Do that now.

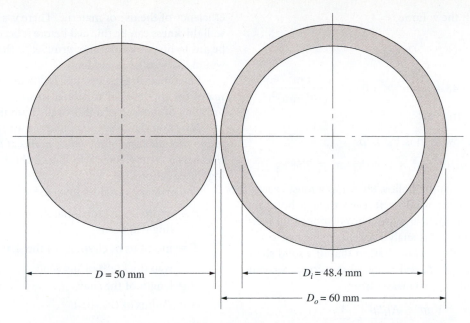

FIGURE 10–10 Comparison of solid and hollow shafts for Example Problem 10–8.

You should have

$$D_i = \left(D_o^4 - \frac{16 Z_p D_o}{\pi} \right)^{1/4}$$

Now compute the maximum allowable inside diameter.

$$D_i = \left[(60)^4 - \frac{(16)(24.5 \times 10^3)(60)}{\pi} \right]^{1/4} \text{mm} = 48.4 \text{ mm}$$

Summary of Results: Final design of the hollow shaft:
$D_o = 60$ mm; $D_i = 48.4$ mm.
Material: AISI 1141 OQT 1300 steel.
Maximum shear stress at outer surface = τ_{max} = 32.6 MPa.

Comment: Figure 10–10 shows a comparison of the hollow shaft with the solid shaft having the same maximum shear stress. The designs are drawn full size. It appears to the eye that the hollow shaft uses less material. The following section will demonstrate that this is true.

Masses of Solid and Hollow Shafts

In many design situations, economy of material usage is a major criterion of performance for a product. In aerospace applications, every reduction in the mass of the aircraft or space vehicle allows increased payload. Automobiles achieve higher fuel economy when they are lighter. Also, since raw materials are purchased on a price per unit mass basis, a lighter part generally costs less.

Providing economy of material usage for load-carrying members requires that all the material in the member be stressed to a level approaching the safe design stress. Then every portion is carrying its share of the load.

Example Problems 10–7 and 10–8 can be used to illustrate this point. Recall that both designs shown in Figure 10–10 result in the same maximum torsional shear stress in the steel shaft. The hollow shaft is slightly larger in outside

diameter, but it is the *volume* of metal that determines the mass of the shaft. Consider a length of shaft 1.0 m long. For the solid shaft, the volume is the cross-sectional area times the length.

$$V_S = A_S L = \frac{\pi D^2}{4} L$$

$$V_S = \frac{\pi (50 \text{ mm})^2}{4} \times 1.0 \text{ m} \times \frac{1 \text{ m}^2}{(10^3 \text{ mm})^2}$$

$$= 1.96 \times 10^{-3} \text{ m}^3$$

The mass is the volume times the density, ρ. Appendix A–14 gives the density of steel to be 7680 kg/m^3. Then the mass of the solid shaft is $V_S \times \rho$.

$$M_S = 1.96 \times 10^{-3} \text{ m}^3 \times 7680 \text{ kg/m}^3 = 15.1 \text{ kg}$$

For the hollow shaft the volume is

$$V_H = A_H L = \frac{\pi}{4}(D_o^2 - D_i^2)(L)$$

$$V_H = \frac{\pi}{4}(60^2 - 48.4^2)\,\text{mm}^2 \times 1.0\,\text{m} \times \frac{1\,\text{m}^2}{(10^3\,\text{mm})^2}$$

$$V_H = 0.988 \times 10^{-3}\,\text{m}^3$$

The mass of the hollow shaft is $V_H \times \rho$.

$$M_H = 0.988 \times 10^{-3}\,\text{m}^3 \times 7680\,\text{kg/m}^3 = 7.58\,\text{kg}$$

Thus it can be seen that the hollow shaft has almost exactly *one-half the mass* of the solid shaft, even though both are subjected to the same stress level for a given applied torque.

A generalization of this analysis can be made. For comparison of the mass of a hollow shaft to that of a solid shaft having the same length, L, and material density, ρ, we can compute the ratio of the two masses from,

$$\frac{M_H}{M_S} = \frac{V_H \rho}{V_S \rho} = \frac{A_H L \rho}{A_S L \rho} = \frac{A_H}{A_S}$$

The ratio of the masses is equal to the ratio of the cross-sectional areas of the shafts.

The reason for the hollow shaft being lighter is that a greater portion of its material is being stressed to a higher level than in the solid shaft. Figure 10–7 shows a sketch of the stress distribution in the solid shaft. The maximum stress, 32.6 MPa, occurs at the outer surface. The stress then varies linearly with the radius for other points within the shaft to *zero* at the center. Therefore the material near the middle of the shaft is not being used efficiently.

Contrast this with the sketch of the hollow shaft in Figure 10–9. Again the stress at the outer surface is the maximum, 32.6 MPa. The stress at the inner surface of the hollow shaft can be found from Equation (10–6).

$$\tau = \tau_{max}\frac{r}{c}$$

At the inner surface, $r = R_i = D_i/2 = 48.4\,\text{mm}/2 = 24.2\,\text{mm}$. Also, $c = R_o = D_o/2 = 60\,\text{mm}/2 = 30\,\text{mm}$. Then

$$\tau = 32.6\,\text{MPa}\frac{24.2}{30} = 26.3\,\text{MPa}$$

The stress at points between the inner and outer surfaces varies linearly with the radius to each point. Thus it can be seen that all of the material in the hollow shaft shown in Figure 10–9 is being stressed to a fairly high but safe level. This illustrates why the hollow section requires less material.

Of course, the specific data used in the previous illustration cannot be generalized to all problems. However, for torsional loading of circular members, a hollow section can be designed that is lighter than a solid section while subjecting the material to the same maximum torsional shear stress.

Caution for Thin Hollow Shafts

You should approach with care the tendency for making the wall thickness of hollow shafts thinner and thinner to gain

efficiency of the use of material. There is a limit to which the wall thickness can be thinned before it becomes unstable and begins to buckle locally and wrinkle. As that occurs, the shaft would likely collapse suddenly.

Reference 5 presents the following equations as estimates for the level of torsional shear stress that will cause buckling of the wall of a thin-walled tubular shaft.

Note that the assumption that the wall is thin typically requires that the ratio of the mean radius of the wall to the wall thickness is 10 or greater. You should check this before applying the equations.

The primary variables involved are

τ' = approximate torsional shear stress at which buckling will occur

E = modulus of elasticity of the shaft material

t = wall thickness of the shaft

l = length of the shaft

r = Radius of the shaft

The choice of which equation to use is based on the value of the ratio of l/r.

a. If $l/r > 6.6 \sqrt[4]{1 - v^2}\,\sqrt{r/t}$:

$$\tau' = 0.272\frac{E}{(1 - v^2)^{3/4}}(t/r)^{3/2}$$

See Table 8–1 in Chapter 8 for estimates of the values for v of common metals. If v is approximately 0.3 as it is for steel, stainless steel, titanium, aluminum, and brass, this equation becomes

$$\tau' = 0.292E(t/r)^{3/2}$$

b. If $5 < l/r < 6.6 \sqrt[4]{1 - v^2}\,\sqrt{r/t}$:

$$\tau' = E(t/l)^2[1.8 + \sqrt{1.2 + 0.201(l/\sqrt{tr})^3}]$$

c. If $l/r < 5$:

$$\tau' = 0.10E(t/r) + 5E(t/l)^2$$

10–9 STRESS CONCENTRATIONS IN TORSIONALLY LOADED MEMBERS

Torsionally loaded members, particularly power transmission shafts, are often made with changes in the geometry at various positions. Figure 10–11 shows an example. This is a part of a shaft where a power transmission element, such as a gear, would be mounted. The bore in the hub of the gear would have a diameter that would allow it to slide over the right part of the shaft where the shaft diameter is $d = 25$ mm. A square or rectangular key would be placed in the keyseat and there would be a corresponding keyway in the hub of the gear so it could pass over the key. The gear would then be moved onto the shaft from the right until it stopped against the shoulder at Section 2, created by the increase in the shaft diameter to $D = 40$ mm. To keep the gear in position, a retaining ring is inserted into the ring groove at Section 4.

FIGURE 10–11   Shaft with stress concentrations.

Changes in the cross section of a member loaded in torsion cause the local stress near the changes to be higher than would be predicted by using the torsional shear stress formula. The actual level of stress in such cases is determined experimentally. Then a *stress concentration factor* is determined, which allows the maximum stress in similar designs to be computed from the relationship

$$\tau_{max} = K_t \tau_{nom} = K_t(T/Z_p) \qquad (10\text{–}20)$$

The term τ_{nom} is the nominal stress due to torsion, which would be developed in the parts if the stress concentration were not present. Thus the standard torsional shear stress formulas [Equations (10–5) and (10–16)] can be used to compute the nominal stress. The value of K_t is a factor by which the actual maximum stress is greater than the nominal stress.

Referring again to Figure 10–11, note that there would be several levels of stress at different places along the length of the bar, even if the applied torque is the same throughout. The differing diameters and the presence of stress concentrations cause the varying stress levels. The stress at Section 1, where $D = 40$ mm, would be relatively low because there is a large diameter and a correspondingly large polar section modulus. At Section 2, the diameter of the shaft reduces to $d = 25$ mm, and the step produces a stress concentration that tends to raise the local stress level. Then the keyseat at Section 3 sets up a different stress concentration. At Section 4, two major factors occur that both tend to increase the local stress. Cutting the ring groove reduces the diameter to $d_g = 16$ mm and also produces two closely spaced steps with relatively small fillet radii at the bottom of the groove. At Section 5, well away from the ring groove, the stress would be equal to the nominal stress in the 25-mm-diameter shaft. Example Problem 10–10 illustrates all of these situations by performing actual calculations of the stresses at all five sections of the shaft.

Before the example problem, let's look at the nature of stress concentration factors. The following list of Appendix charts gives data for several typical cases.

Appendix A–22–5: Round bar with transverse hole in torsion

Appendix A–22–6: Grooved round bar in torsion

Appendix A–22–7: Stepped round bar in torsion

Appendix A–22–11: Shafts with keyseats

Round Bar with Transverse Hole

One purpose for drilling a hole in a shaft is to insert a pin through the hole and through the corresponding hole in the hub of a machine element such as a gear, pulley, or chain sprocket. See Figure 10–12. The pin serves to locate the machine element axially on the shaft while also transmitting torque from the shaft to the element or from the element to the shaft. The hole in the shaft is an abrupt change in geometry and it causes a stress concentration. Appendix A–22–5 is a chart for this case from which K_t can be determined. Curve C is for the case of torsionally loaded shafts. Note that the formula for the nominal stress in the shaft is based on the full, gross, circular cross section of the shaft.

Grooved Round Bar

Round-bottomed grooves are cut into round bars for the purpose of installing seals or for distributing lubricating oil around a shaft. See Figure 10–13. The stress concentration factor is dependent on the ratio of the shaft diameter to the diameter of the groove, and on the ratio of the groove radius

Cylindrical pin

FIGURE 10–12   Round shaft with a transverse hole used to allow connection of a gear to the shaft with a cylindrical pin.

FIGURE 10–13   Shaft with a circular groove.

to the base diameter of the groove. The groove is cut with a tool having a rounded nose to produce the round-bottomed groove. The radius should be as large as possible to minimize the stress concentration factor. Note that the nominal stress is based on the diameter *at the base of the groove*. See Appendix A–22–6.

Stepped Round Bar

Shafts are often made with two or more diameters, resulting in a stepped shaft like that shown in Appendix A–22–7 and Figure 10–11. The face of the step provides a convenient means of locating one side of an element mounted on the shaft, such as a bearing, gear, pulley, or chain sprocket. Care should be exercised in defining the radius at the bottom of the step, called the *fillet radius*. Sharp corners are to be

avoided, as they cause extremely high stress concentration factors. The radius should be as large as possible while being compatible with the elements mounted on the shaft.

Retaining rings seated in grooves cut into the shaft are often used to locate machine elements, as shown in Figure 10–11. The grooves are typically flat bottomed with small radii at the sides. Some designers treat such grooves as two steps on the shaft close together and use the stepped shaft chart (Appendix A–22–7) to determine the stress concentration factor. Because of the small radius at the base of the groove, the relative radius is often quite small, resulting in values of K_t so high they are off the chart. In such cases a value of $K_t = 3.0$ is used in this book.

Shafts with Keyseats

Power transmission elements typically transmit torque to and from shafts through keys fitted into keyseats cut into the shaft, as shown in Figure 10–14. The V-belt pulley mounted on the end of the motor shaft shown is an example. Two types of keyseats are in frequent use: the *sled-runner* and the *profile* keyseats.

A circular milling cutter having a thickness equal to the width of the keyseat is used to cut the sled-runner keyseat, typically on the end of a shaft, as shown in Figure 10–14(b). As the cutter ends its cut, it leaves a gentle radius, as shown in the side view, resulting in $K_t = 1.6$ as a design value.

A profile keyseat is cut with an end mill having a diameter equal to the width of the keyseat. Usually used at a location

$$\text{Stress in shaft} = \tau_{max} = K_t\,\tau_{nom}$$

$$\tau_{nom} = T/(\pi D^3/16)$$

(a) Shaft with keyseat

(b) Sled-runner keyseat
$K_t = 1.6$

(c) Profile keyseat
$K_t = 2.0$

FIGURE 10–14   Stress concentration factors for keyseats. (a) Typical application. (b) Sled-runner-type keyseat made with a circular milling cutter. (c) Profile-type keyseat made with an end mill.

away from the ends of the shaft, it leaves a square corner at the ends of the keyseat when viewed from the side, as shown in Figure 10–14(c). This is more severe than the sled-runner and a value of $K_t = 2.0$ is used. Note that the stress concentration factors account for both the removal of material from the shaft and the change in geometry.

Commercially Available Keys, Pins, and Retaining Rings

Internet sites 1–3 show commercially available keys, pins, retaining rings, and shafts with keyseats that have been discussed in this section as causes of geometric discontinuities that lead to stress concentrations in shafting. For the retaining rings site (2), pages can be accessed from an online catalog giving dimensions for numerous styles of rings for shaft sizes ranging from approximately $\frac{1}{8}$ in to 10 in (3.2 mm to 254 mm). Style Basic 5100 is commonly used on shafts to secure active elements such as bearings, gears, sprockets, pulleys, sheaves, and other devices. The overall geometry and detailed dimensions are given for the grooves in the shaft in which the rings seat. Note the small fillet radius called out for the base of the groove that gives rise to relatively high stress concentration factors.

The use of stress concentration factors is illustrated in the following example problems.

Example Problem 10–9

Figure 10–13 shows a portion of a shaft in which a circular groove has been machined. For an applied torque of 4500 lb·in, compute the torsional shear stress at Section 1 in the full-diameter part of the shaft and at Section 2 where the groove is located. Use $r = 0.10$ in, $D = 1.50$ in, $d_g = 1.25$ in.

Solution Objective: Compute the stress at Sections 1 and 2.

Given: Applied torque = $T = 4500$ lb·in. Shaft geometry in Figure 10–13.
$r = 0.10$ in, $D = 1.50$ in, $d_g = 1.25$ in.

Analysis: Assuming that Section 1 is well away from the groove, there is no significant stress concentration. Then the standard torsional shear stress formula [Equation (10–16)] can be used. At Section 2 where the groove is located, Equation (10–20) must be used. The value of the stress concentration factor can be determined from Appendix A–22–6.

Results: At Section 1: $\tau_{max} = T/Z_p$.

$$Z_p = \pi D^3/16 = \pi(1.50\text{ in})^3/16 = 0.663\text{ in}^3$$

$$\tau_{max} = \frac{4500\text{ lb·in}}{0.663\text{ in}^3} = 6790\text{ psi}$$

At Section 2: $\tau_{max} = K_t T/Z_p$

$$Z_p = \pi d_g^3/16 = \pi(1.25\text{ in})^3/16 = 0.383\text{ in}^3$$

To evaluate K_t, two ratios must be computed, as called for in Appendix A–22–6.

$$D/d_g = (1.50\text{ in})/(1.25\text{ in}) = 1.20$$
$$r/d_g = (0.10\text{ in})/(1.25\text{ in}) = 0.08$$

Then, reading from the chart in Appendix A–22–6, $K_t = 1.55$.
We can now compute the maximum shear stress.

$$\tau_{max} = \frac{K_t T}{Z_p} = \frac{(1.55)(4500\text{ lb·in})}{0.383\text{ in}^3} = 18\,200\text{ psi}$$

Comment: Note that the stress at the groove is 2.6 times higher than that in the full-diameter part of the shaft. One reason is the reduction in the diameter at the groove. Also, the use of the stress concentration factor is essential to predict the actual maximum stress level at the groove.

PROGRAMMED EXAMPLE PROBLEM

The stress in each section will be analyzed separately in a panel set off by a horizontal line across the page. You should perform the indicated calculations before looking at the results shown.

Example Problem 10–10

Figure 10–11 shows a portion of a shaft where a gear is to be mounted. The gear will be centered over the keyseat at Section 3. It will rest against the shoulder at Section 2 and be held in position with a retaining ring placed in the groove at Section 4. A repeated torque of 20 N·m is applied throughout the shaft. Compute the maximum shear stress in the shaft at Sections 1, 2, 3, 4, and 5. Then specify a suitable steel material for the shaft.

Solution   Objective: 1. Compute the stresses at Sections 1, 2, 3, 4, and 5.
2. Specify a suitable steel for the shaft.

Given: Shaft geometry shown in Figure 10–11. $T = 20$ N·m repeated.

Analysis: In each case, the analysis requires the application of Equation (10–20).

$$\tau_{max} = K_t T / Z_p$$

The torque will always be taken to be 20 N·m. You must evaluate the stress concentration factor and the appropriate polar section modulus for each section. Note that $K_t = 1.0$ where there is no change in the geometry. Compute the stress at Section 1.

Section 1.   There is no change in geometry, so $K_t = 1.0$. The shaft diameter is $D = 40$ mm. Then

$$Z_p = \frac{\pi D^3}{16} = \frac{\pi (40 \text{ mm})^3}{16} = 12\,570 \text{ mm}^3$$

$$\tau_1 = \frac{20 \text{ N} \cdot \text{m}}{12\,570 \text{ mm}^3} \times \frac{10^3 \text{ mm}}{\text{m}} = 1.59 \frac{\text{N}}{\text{mm}^2} = 1.59 \text{ MPa}$$

Compute the stress at Section 2.

Section 2.   The stepped shaft and the shoulder fillet produce a stress concentration here that must be evaluated using Appendix A–22–7. The polar section modulus must be based on the smaller diameter; $d = 25$ mm. The results are

$$\tau_{nom} = \frac{T}{\pi d^3 / 16} = \frac{20 \text{ N} \cdot \text{m}}{[\pi (25)^3 / 16] \text{ mm}^3} \times \frac{10^3 \text{ mm}}{\text{m}}$$

$$= 6.52 \text{ N/mm}^2 = 6.52 \text{ MPa}$$

The value of K_t depends on the ratios D/d and r/d.

$$\frac{D}{d} = \frac{40 \text{ mm}}{25 \text{ mm}} = 1.60$$

$$\frac{r}{d} = \frac{2 \text{ mm}}{25 \text{ mm}} = 0.08$$

Then from Appendix A–22–7, $K_t = 1.45$. Then

$$\tau_2 = (1.45)(6.52 \text{ MPa}) = 9.45 \text{ MPa}$$

Now compute the stress at Section 3.

Section 3.   The profile-type keyseat presents a stress concentration factor of 2.0. The nominal stress is the same as that computed at the shoulder fillet. Then

$$\tau_3 = K_t \tau_{nom} = (2.0)(6.52 \text{ MPa}) = 13.04 \text{ MPa}$$

Compute the stress at Section 4.

Section 4.   Section 4 is the location of the ring groove. Here the nominal stress is computed on the basis of the root diameter of the groove.

$$\tau_{nom} = \frac{T}{\pi d_g^3 / 16} = \frac{20 \text{ N} \cdot \text{m}}{[\pi (16)^3 / 16] \text{ mm}^3} \times \frac{10^3 \text{ mm}}{\text{m}} = 24.9 \frac{\text{N}}{\text{mm}^2}$$

$$\tau_{nom} = 24.9 \text{ MPa}$$

The value of K_t depends on d/d_g and r/d_g.

$$\frac{d}{d_g} = \frac{25 \text{ mm}}{16 \text{ mm}} = 1.56$$

$$\frac{r}{d_g} = \frac{0.2 \text{ mm}}{16 \text{ mm}} = 0.013$$

Referring to Appendix A–22–7, the stress concentration factor is off the chart. This is the type of case for which $K_t = 3.0$ is reasonable.

$$\tau_4 = K_t \tau_{nom} = (3.0)(24.9 \text{ MPa}) = 74.7 \text{ MPa}$$

Compute the stress at Section 5.

Section 5.   Section 5 is in the smaller portion of the shaft, where no stress concentration occurs. Then $K_t = 1.0$ and

$$\tau = \frac{T}{Z_p} = \frac{T}{\pi d^3/16}$$

Notice that this is identical to the nominal stress computed for Sections 2 and 3. Then at Section 5,

$$\tau_5 = 6.52 \, \text{MPa}$$

Summary of Results: A wide range of stress levels exists in the vicinity of the place on the shaft where the gear is to be mounted.

$\tau_1 = 1.59 \, \text{MPa}$	$D = 40 \, \text{mm}.$	$K_t = 1.0.$
$\tau_2 = 9.45 \, \text{MPa}$	$d = 25 \, \text{mm}.$	$K_t = 1.45.$ Step.
$\tau_3 = 13.04 \, \text{MPa}$	$d = 25 \, \text{mm}.$	$K_t = 2.0.$ Keyseat.
$\tau_4 = 74.7 \, \text{MPa}$	$d_g = 16 \, \text{mm}.$	$K_t = 3.0.$ Ring groove.
$\tau_5 = 6.52 \, \text{MPa}$	$d = 25 \, \text{mm}.$	$K_t = 1.0.$

The specification of a suitable material must be based on the stress at Section 4 at the ring groove. Let the design stress, τ_d, be equal to that stress level and determine the required yield strength of the material.

You should have a required yield strength of $s_y = 598$ MPa. For the repeated torque, $N = 4$ is recommended in Table 10–1, resulting in

$$\tau_d = s_y/2N = s_y/8$$

Then solving for s_y gives

$$s_y = 8(\tau_d) = 8(74.7 \, \text{MPa}) = 598 \, \text{MPa}$$

Specify a suitable material.

From Appendix A–14, two suitable steels for this requirement are AISI 1040 WQT 900 and AISI 4140 OQT 1300. Both have adequate strength and a high ductility as measured by the percent elongation. Certainly, other alloys and heat treatments could be used.

Comment: Review the results of this example problem. It illustrates the importance of considering the details of the design of a shaft at any local area where stress concentrations may occur.

10–10 TWISTING—ELASTIC TORSIONAL DEFORMATION

Stiffness in addition to strength is an important design consideration for torsionally loaded members. The measure of torsional stiffness is the angle of twist of one part of a shaft relative to another part when a certain torque is applied.

In mechanical power transmission applications, excessive twisting of a shaft may cause vibration problems, which would result in noise and improper synchronization of moving parts. One guideline for torsional stiffness is related to the desired degree of precision, as listed in Table 10–2 (see References 1 and 3).

In structural design, load-carrying members are sometimes loaded in torsion, as well as tension or bending. The rigidity of the structure then depends on the torsional stiffness of the components. Any load applied off from the axis of a member and transverse to the axis will produce torsion. This section will discuss twisting of circular members, both solid

TABLE 10–2   Recommended Torsional Stiffness: Angle of Twist Per Unit Length

	Torsional deflection	
Application	deg/in	rad/m
General machine part	1×10^{-3} to 1×10^{-2}	6.9×10^{-4} to 6.9×10^{-3}
Moderate precision	2×10^{-5} to 4×10^{-4}	1.4×10^{-5} to 2.7×10^{-4}
High precision	1×10^{-6} to 2×10^{-5}	6.9×10^{-7} to 1.4×10^{-5}

FIGURE 10–15   Torsional deformation in a circular bar.

and hollow. Noncircular sections will be covered in a later section. It is very important to note that the behavior of an open-section shape such as a channel or angle is much different from that of a closed section such as a pipe or rectangular tube. In general, the open sections have very low torsional stiffness.

Angle of Twist of a Circular Member

Consider the shaft shown in Figure 10–15. One end of the shaft, Section M, is held fixed while a torque T is applied to the other end. Under these conditions the shaft will twist between the two ends through an angle θ.

The derivation of the angle-of-twist formula depends on some basic assumptions about the behavior of a circular member when subjected to torsion. As the torque is applied, an element of length L along the outer surface of the member, which was initially straight, rotates through a small angle γ (gamma). Likewise, a radius of the member in a cross section rotates through a small angle θ. In Figure 10–15, the rotations γ and θ are both related to the arc length AB on the surface of the bar. From geometry, for small angles, the arc length is the product of the angle in radians and the distance from the center of the rotation. Therefore, the arc length AB can be expressed as either

$$AB = \gamma L$$

or

$$AB = \theta c$$

where c is the outside radius of the bar. These two expressions for the arc length AB can be equated to each other,

$$\gamma L = \theta c$$

Solving for γ gives

$$\gamma = \frac{\theta c}{L} \qquad (10\text{–}21)$$

The angle γ is a measure of the maximum shearing strain in an element on the outer surface of the bar. It was discussed in Chapter 8 that the shearing strain, γ, is related to the shearing stress, τ, by the modulus of elasticity in shear, G. That was expressed as Equation (8–5),

$$G = \frac{\tau}{\gamma} \qquad (8\text{–}5)$$

At the outer surface, then,

$$\tau = G\gamma$$

But the torsional shear stress formula [Equation (10–11)] states

$$\tau = \frac{Tc}{J}$$

Equating these two expressions for τ gives

$$G\gamma = \frac{Tc}{J}$$

Now, substituting from Equation (10–21) for γ, we obtain

$$\frac{G\theta c}{L} = \frac{Tc}{J}$$

We can now cancel c and solve for θ:

Angle of Twist ▼

$$\theta = \frac{TL}{JG} \qquad (10\text{–}22)$$

The resulting angle of twist, θ, is in radians. When consistent units are used for all terms in the calculation, all units will cancel, leaving a dimensionless number. This should be interpreted as the angle, θ, in radians.

Use of Angle of Twist Formula

- Equation (10–22) can be used to compute the angle of twist of one section of a circular bar, either solid or hollow, with respect to another section where L is the distance between them, provided that the torque, T, the polar moment of inertia, J, and the shear modulus of elasticity, G, are the same over the entire length, L.

- If any of these factors vary in a given problem, the bar can be subdivided into segments over which they are constant to compute angles of rotation for those segments. Then the computed angles can be combined algebraically to get the total angle of twist. This principle, called *superposition*, will be illustrated in Example Problem 10–13.

- The shear modulus of elasticity, G, is a measure of the torsional stiffness of the material of the bar. Table 10–3 gives values for G for selected materials.

TABLE 10–3 Shear Modulus of Elasticity, *G*		
	Shear modulus, *G*	
Material	**GPa**	**psi**
Plain carbon and alloy steels	80	11.5×10^6
Stainless steel type 304	69	10.0×10^6
Aluminum 6061-T6	26	3.75×10^6
Beryllium copper	48	7.0×10^6
Magnesium	17	2.4×10^6
Titanium alloy	43	6.2×10^6

**Example Problem
10–11**

Determine the angle of twist in degrees between two sections 250 mm apart in a steel rod having a diameter of 10 mm when a torque of 15 N·m is applied. Figure 10–15 shows a sketch of the arrangement.

Solution Objective: Compute the angle of twist in degrees.

Given: Applied torque = $T = 15$ N·m. Circular bar: diameter = $D = 10$ mm.
Length = $L = 250$ mm. (Figure 10–15)

Analysis: Equation (10–22) can be used. Compute $J = \pi D^4/32$.
$G = 80$ GPa $= 80 \times 10^9$ N/m^2 (Table 10–3).

Result: $\theta = \dfrac{TL}{JG}$

The value of J is

$$J = \frac{\pi D^4}{32} = \frac{\pi (10 \text{ mm})^4}{32} = 982 \text{ mm}^4$$

Then

$$\theta = \frac{TL}{JG} = \frac{(15 \text{ N} \cdot \text{m})(250 \text{ mm})}{(982 \text{ mm}^4)(80 \times 10^9 \text{ N/m}^2)} \times \frac{(10^3 \text{ mm})^3}{1 \text{ m}^3} = 0.048 \text{ rad}$$

Note that all units cancel. Expressing the angle in degrees,

$$\theta = 0.048 \text{ rad} \times \frac{180 \text{ deg}}{\pi \text{ rad}} = 2.73 \text{ deg}$$

**Example Problem
10–12**

Determine the required diameter of a round shaft made of aluminum alloy 6061-T6 if it is to twist not more than 0.08 deg in 1.0 ft of length when a torque of 75 lb·in is applied.

Solution Objective: Compute the required diameter, D, of the round shaft.

Given: Applied torque = $T = 75$ lb·in. Length = $L = 1.0$ ft = 12 in.
Maximum angle of twist = 0.08 deg. Aluminum 6061-T6.

Analysis: Equation (10–22) can be solved for J because J is the only term involving the unknown diameter, D. Then solve for D from $J = \pi D^4/32$.
$G = 3.75 \times 10^6$ psi (Table 10–3).

Results: $\theta = \dfrac{TL}{JG}$

$J = \dfrac{TL}{\theta G}$

The angle of twist must be expressed in radians.

$$\theta = 0.08 \text{ deg} \times \frac{\pi \text{ rad}}{180 \text{ deg}} = 0.0014 \text{ rad}$$

Then

$$J = \frac{TL}{\theta G} = \frac{(75 \text{ lb} \cdot \text{in})(12 \text{ in})}{(0.0014)(3.75 \times 10^6 \text{lb/in}^2)} = 0.171 \text{ in}^4$$

Now since $J = \pi D^4/32$,

$$D = \left(\frac{32J}{\pi}\right)^{1/4} = \left[\frac{(32)(0.171 \text{ in}^4)}{\pi}\right]^{1/4} = 1.15 \text{ in}$$

Comment: This is the minimum acceptable diameter. You should specify a convenient, preferred size, say 1.25 in. The resulting angle of twist will then be less than 0.08 deg over a 1.0-ft length.

Deformation in Shafts with Multiple Levels of Torque

When a single shaft carries two or more elements that apply torque to it, it is necessary to determine the level of torque in all segments of the shaft. The use of free body diagrams of parts of the shaft is helpful for this process.

We can use the principle of equilibrium to determine the level of torque that must exist internal to the shaft at any section by imagining that the shaft is cut at a section of interest. Then the externally applied torque values on the remaining part of the shaft to either side of the cut must be balanced by the internal torque in the shaft at the cut. This technique is illustrated in Example Problem 10–13.

Let's define a notation convention for such problems. First, to report the torque applied to a shaft by an active element such as a gear or pulley, refer to that torque with a single subscript that designates the section where the torque is applied to the shaft. For example, the torque applied to a section called A will be called T_A.

To report the internal torque in the shaft between two sections of interest, refer to that torque with a double subscript that identifies the end points of the segment for which that value applies. For example, to report the torque in the segment between sections B and C on a shaft, call it T_{BC}.

The problem also illustrates the calculation of the angle of twist for each segment of the shaft for which the values of torque and polar moment of inertia are constant. Then the net resulting angle of twist over the total length of the shaft is computed from the algebraic sum of the angles of twist in the various segments.

Stresses in Shafts with Multiple Levels of Torque

The torsional shear stress also varies with position in the shaft when either the torque in the shaft or the size of the shaft changes. Example Problem 10–14 demonstrates the method of analyzing such shafts.

PROGRAMMED EXAMPLE PROBLEM

The completion of this example problem will be shown in programmed format. You should perform each indicated operation before moving to the next panel.

Example Problem 10–13

Figure 10–16 shows a steel shaft to which three disks are attached. The shaft is fixed against rotation at its left end but free to rotate in a bearing at its right end. Each disk is 300 mm in diameter. Downward forces act at the outer surfaces of the disks so that torques are applied to the rod. Determine the angle of twist of section A relative to the fixed section E.

Solution Objective: Compute the angle of twist of the shaft at A relative to E.

Given: Shaft is steel. $G = 80$ GPa. (Table 10–3)
Geometry of shaft and loading is shown in Figure 10–16.
For each disk, diameter $= D = 300$ mm. Radius $= R = 150$ mm.

Analysis: The shaft shown in Figure 10–16 has a total length of 1000 mm or 1.0 m. But there are four segments of the shaft having different lengths, diameters, or applied levels of torque. Therefore, Equation (10–22) must be applied to each segment separately to compute the angle of twist for each. Then, the total angle of twist in section A relative to E will be the algebraic sum of the four angles.

The first operation is to compute the magnitude and direction of the applied torques at each disk, B, C, and D. Do that now. Recall the definition of torque from Equation (10–1).

For directions, assume a view point along the rod from the right end. The magnitude of the torque on each disk is the product of the force acting on the periphery of the disk times the radius of the disk. Then, considering clockwise to be positive, torque on disk B, clockwise:

$$T_B = (100 \text{ N})(150 \text{ mm}) = 15\,000 \text{ N} \cdot \text{mm} = 15 \text{ N} \cdot \text{m (CW)}$$

FIGURE 10–16 Shaft for Example Problem 10–13.

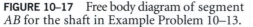

FIGURE 10–17 Free body diagram of segment *AB* for the shaft in Example Problem 10–13.

FIGURE 10–18 Free body diagram of segment to right of section *C* for shaft in Example Problem 10–13.

torque on disk *C*, counterclockwise:

$$T_C = -(250\text{ N})(150\text{ mm}) = -37\,500\text{ N} \cdot \text{mm} = -37.5\text{ N} \cdot \text{m (CCW)}$$

torque on disk *D*, counterclockwise:

$$T_D = -(50\text{ N})(150\text{ mm}) = -7500\text{ N} \cdot \text{mm} = -7.5\text{ N} \cdot \text{m (CCW)}$$

Now determine the level of torque in each segment of the shaft. You should visualize a free body diagram of any part of the shaft between the ends of each segment by "cutting" the shaft and computing the magnitude of the torque applied to the shaft to the right of the cut section. The internal torque in the shaft must be equal in magnitude and opposite in direction to the externally applied torque to maintain equilibrium.

It is suggested that you start at the right end at *A*. The bearing allows free rotation of the shaft at that end. Then move to the left and compute the torque for segments *AB*, *BC*, *CD*, and *DE*. What is the level of torque in segment *AB*?

For the segment *AB*, up to, but not including *B*, the torque in the shaft is zero because the bearing allows free rotation. This is illustrated in Figure 10–17 that shows the part of the shaft from section *B* to the end of the shaft at *A*. Visualize a cut in the shaft anywhere between *B* and *A* and isolate the cut section to the right as a free body. Because there is no externally applied torque, the internal torque in the shaft everywhere between *B* and *A* must be zero. That is, $T_{AB} = 0$.

Now, consider the torque applied by the disk at *B* and determine the torque in the segment *BC*.

Cutting the shaft anywhere to the right of *C* in the segment *BC* would result in an externally applied torque of 15 N · m clockwise, due to the torque on disk *B*. Figure 10–18 shows the free body diagram of that part of the shaft to the right of section *C* after it is assumed to be cut. Only disk *B* applies an external torque to the shaft. The internal torque in the shaft at the cut must balance the torque on disk *B*.

Therefore, the torque throughout the segment *BC* is

$$T_{BC} = 15\text{ N} \cdot \text{m (CW)}$$

We will consider this torque to be clockwise (CW) and positive because it tends to cause a clockwise rotation of the rod.

Now determine the torque in segment *CD*, called T_{CD}.

Cutting the shaft anywhere between *C* and *D* would result in both T_C and T_B acting on the shaft to the right of the cut section. See Figure 10–19. But they act in opposite sense, one clockwise and one counterclockwise. Thus, the net torque applied to the shaft is the difference between them. That is,

$$T_{CD} = -T_C + T_B = -37.5\text{ N} \cdot \text{m} + 15\text{ N} \cdot \text{m} = -22.5\text{ N} \cdot \text{m (CCW)}$$

FIGURE 10–19 Free body diagram of segment to right of section *D* for shaft in Example Problem 10–13.

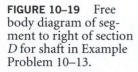

FIGURE 10–20 Free body diagram for segment to right of section E for shaft in Example Problem 10–13.

Continue this process for the final segment, *DE*.

Between D and E in the shaft, the torque is the resultant of all the applied torques at D, C, and B. See Figure 10–20. Then

$$T_{DE} = -T_D - T_C + T_B = -7.5\,\text{N}\cdot\text{m} - 37.5\,\text{N}\cdot\text{m} + 15\,\text{N}\cdot\text{m} = -30\,\text{N}\cdot\text{m}\,(\text{CCW})$$

The fixed support at E must be capable of providing a reaction torque of 30 N · m to maintain the shaft in equilibrium.

In summary, the distribution of torque in the shaft can be shown in graphical form as in Figure 10–21. Notice that the applied torques T_B, T_C, and T_D are the *changes* in torque that occur at B, C, and D but that they are not necessarily the magnitudes of the torque *in the shaft* at those points.

Now compute the angle of twist in each segment by applying Equation (10–22) successively. Start with segment *AB*.

Segment AB.

$$\theta_{AB} = T_{AB}\left(\frac{L}{JG}\right)_{AB}$$

Since $T_{AB} = 0$, $\theta_{AB} = 0$. There is no twisting of the shaft between A and B.

Now continue with segment *BC*.

Segment BC.

$$\theta_{BC} = T_{BC}\left(\frac{L}{JG}\right)_{BC}$$

We know $T_{BC} = 15\,\text{N}\cdot\text{m}$, $L = 200\,\text{mm}$, and $G = 80\,\text{GPa}$ for steel. For the 10–mm-diameter shaft,

$$J = \frac{\pi D^4}{32} = \frac{\pi(10\,\text{mm})^4}{32} = 982\,\text{mm}^4$$

Then

$$\theta_{BC} = \frac{(15\,\text{N}\cdot\text{m})(200\,\text{mm})}{(982\,\text{mm}^4)(80\times10^9\,\text{N/m}^2)} \times \frac{(10^3)^3\,\text{mm}^3}{\text{m}^3} = 0.038\,\text{rad}$$

FIGURE 10–21 Torque distribution in shaft for Example Problem 10–13.

This means that section B is rotated 0.038 rad clockwise relative to section C, since θ_{BC} is the total angle of twist in the segment BC.
Continue with segment CD.

Segment CD.

$$\theta_{CD} = T_{CD}\left(\frac{L}{JG}\right)_{CD}$$

Here $T_{CD} = -22.5$ N·m, $L = 300$ mm, and the shaft diameter is 15 mm. Then

$$J = \frac{\pi D^4}{32} = \frac{\pi (15\,\text{mm})^4}{32} = 4970\,\text{mm}^4$$

$$\theta_{CD} = \frac{-(22.5\,\text{N}\cdot\text{m})(300\,\text{mm})}{(4970\,\text{mm}^4)(80 \times 10^9\,\text{N/m}^2)} \times \frac{(10^3)^3\,\text{mm}^3}{\text{m}^3} = -0.017\,\text{rad}$$

Section C is rotated 0.017 rad counterclockwise relative to section D.
Finally, complete the analysis for segment DE.

Segment DE.

$$\theta_{DE} = T_{DE}\left(\frac{L}{JG}\right)_{DE}$$

Here $T_{DE} = -30$ N·m, $L = 300$ mm, and $D = 20$ mm. Then

$$J = \frac{\pi D^4}{32} = \frac{\pi (20\,\text{mm})^4}{32} = 15\,700\,\text{mm}^4$$

$$\theta_{DE} = \frac{-(30\,\text{N}\cdot\text{m})(300\,\text{mm})}{(15\,700\,\text{mm}^4)(80 \times 10^9\,\text{N/m}^2)} \times \frac{(10^3)^3\,\text{mm}^3}{\text{m}^3} = -0.007\,\text{rad}$$

Section D is rotated 0.007 rad counterclockwise relative to E. The final operation is to compute the total angle of twist from E to A by summing the angles of twist for all segments algebraically. Do that now.

Total angle of twist from E to A

$$\theta_{AE} = \theta_{AB} + \theta_{BC} - \theta_{CD} - \theta_{DE}$$
$$= 0 + 0.038 - 0.017 - 0.007 = 0.014\,\text{rad}$$

Summary and Comment: It should help your visualization of what is happening in the shaft throughout its length by plotting a graph of the angle of twist as a function of position. This is done in Figure 10–22 by setting the zero reference point as E. The straight line from E to D shows the linear change in angle with position to the value of -0.007 rad (counterclockwise). From there, the angle grows by an additional -0.017 rad between D and C. In segment BC, the relative rotation is clockwise with a magnitude of 0.038 rad, ending at the final value of 0.014 rad at B. And, because there is no torque in the segment AB, the angle of rotation remains at that value.

FIGURE 10–22 Angle of twist versus position on the shaft for Example Problem 10–13.

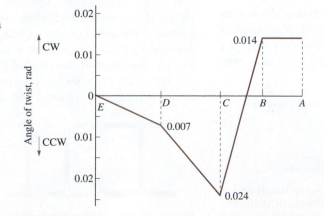

Example Problem 10–14   Compute the torsional shear stress that occurs in each segment of the shaft shown in Figure 10–16 and described in Example Problem 10–13.

Solution   Objective: Compute the torsional shear stress.

Given: Shaft in Figure 10–16 having multiple levels of torque and different diameters.

Analysis: The torsional shear stress equation is used.

$$\tau = Tc/J$$

We will apply this equation over each segment where all factors are the same. The shaft will be broken into segments *AB*, *BC*, *CD*, and *DE*. Some data are taken from the results of Example Problem 10–13.

Results: **Segment AB.**   No torque is transmitted in this segment. Therefore,

$$\tau_{AB} = 0$$

Segment BC. $T_{BC} = 15 \text{ N} \cdot \text{m}$. $D = 10$ mm. $c = 5$ mm. $J_{BC} = 982$ mm^4. Then

$$\tau_{BC} = \frac{T_{BC}\, c}{J_{BC}} = \frac{(15\,\text{N}\cdot\text{m})(5\,\text{mm})}{982\,\text{mm}^4}\frac{1000\,\text{mm}}{\text{m}} = 76.4\,\text{N/mm}^2 = 76.4\,\text{MPa}$$

Segment CD. $T_{CD} = 22.5 \text{ N} \cdot \text{m}$. $D = 15$ mm. $c = 7.5$ mm. $J_{CD} = 4970$ mm^4. Then

$$\tau_{CD} = \frac{T_{CD}\, c}{J_{CD}} = \frac{(22.5\,\text{N}\cdot\text{m})(7.5\,\text{mm})}{4970\,\text{mm}^4}\frac{1000\,\text{mm}}{\text{m}} = 34.0\,\text{N/mm}^2 = 34.0\,\text{MPa}$$

Segment DE. $T_{DE} = 30 \text{ N} \cdot \text{m}$. $D = 20$ mm. $c = 10$ mm. $J_{DE} = 15\ 700$ mm^4. Then

$$\tau_{DE} = \frac{T_{DE}\, c}{J_{DE}} = \frac{(30\,\text{N}\cdot\text{m})(10\,\text{mm})}{15700\,\text{mm}^4}\frac{1000\,\text{mm}}{\text{m}} = 19.1\,\text{N/mm}^2 = 19.1\,\text{MPa}$$

LESSON 18

Comment: A summary of the stress distribution in the shaft follows.

$$\tau_{AB} = 0 \qquad \tau_{BC} = 76.4\,\text{MPa} \qquad \tau_{CD} = 34.0\,\text{MPa} \qquad \tau_{DE} = 19.1\,\text{MPa}$$

10–11 TORSION IN NONCIRCULAR SECTIONS

The behavior of noncircular sections when subjected to torsion is vastly different from that of circular sections, for which the discussions earlier in this chapter applied. There is a large variety of shapes that can be imagined, and the analysis of stiffness and strength is different for each. The development of the relationships involved will not be done here. Compilations of the pertinent formulas occur in References 1 to 5, and a few are given in this section.

Some generalizations can be made. Solid sections having the same cross-sectional area are stiffer when their shape more closely approaches a circle (see Figure 10–23). Conversely, a member made up of long, thin sections that do not form a closed, tube-like shape are very weak and flexible in torsion.

Examples of flexible sections are common structural shapes such as wide-flange beams, standard I-beams, channels, angles, and tees, as illustrated in Figure 10–24. Pipes, solid bars, and structural rectangular tubes have high rigidity, or stiffness (see Figure 10–25).

An interesting illustration of the lack of stiffness of open, thin sections is shown in Figure 10–26. The thin plate (a), the angle (b), and the channel (c) have the same thickness and cross-sectional area, and all have nearly the same torsional stiffness. Likewise, if the thin plate were formed into a circular shape (d), but with a slit remaining, its stiffness would remain low. However, closing the tube completely as in Figure 10–25(a) by welding or by drawing a seamless tube would produce a relatively stiff member. Understanding these comparisons is an aid to selecting a reasonable shape for members loaded in torsion.

Figure 10–27 shows torsional properties for seven cases of noncircular cross sections that are commonly encoun-

FIGURE 10–23   Comparison of stiffness for rectangle and square sections in torsion. Square is two times stiffer than rectangle even though both have the same area: $ab = h^2$.

FIGURE 10–24   Torsionally flexible sections.

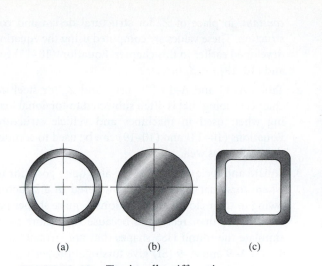

(a)   (b)   (c)

FIGURE 10–25 Torsionally stiff sections.

(a)   (d)

(b)   (c)

FIGURE 10–26 Sections having nearly equal (and low) torsional stiffness.

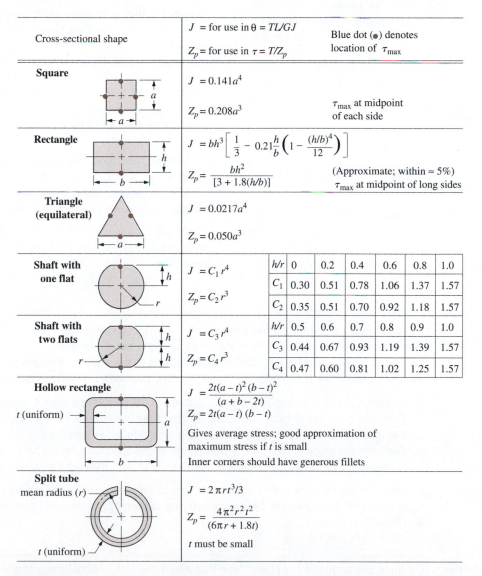

Cross-sectional shape	J = for use in $\theta = TL/GJ$ Z_p = for use in $\tau = T/Z_p$	Blue dot (•) denotes location of τ_{max}
Square	$J = 0.141a^4$ $Z_p = 0.208a^3$	τ_{max} at midpoint of each side
Rectangle	$J = bh^3\left[\dfrac{1}{3} - 0.21\dfrac{h}{b}\left(1 - \dfrac{(h/b)^4}{12}\right)\right]$ $Z_p = \dfrac{bh^2}{[3 + 1.8(h/b)]}$	(Approximate; within $\approx 5\%$) τ_{max} at midpoint of long sides
Triangle (equilateral)	$J = 0.0217a^4$ $Z_p = 0.050a^3$	

Shaft with one flat							
$J = C_1 r^4$ $Z_p = C_2 r^3$	h/r	0	0.2	0.4	0.6	0.8	1.0
	C_1	0.30	0.51	0.78	1.06	1.37	1.57
	C_2	0.35	0.51	0.70	0.92	1.18	1.57

Shaft with two flats							
$J = C_3 r^4$ $Z_p = C_4 r^3$	h/r	0.5	0.6	0.7	0.8	0.9	1.0
	C_3	0.44	0.67	0.93	1.19	1.39	1.57
	C_4	0.47	0.60	0.81	1.02	1.25	1.57

Hollow rectangle t (uniform)	$J = \dfrac{2t(a-t)^2(b-t)^2}{(a+b-2t)}$ $Z_p = 2t(a-t)(b-t)$ Gives average stress; good approximation of maximum stress if t is small Inner corners should have generous fillets
Split tube mean radius (r) t (uniform)	$J = 2\pi rt^3/3$ $Z_p = \dfrac{4\pi^2 r^2 t^2}{(6\pi r + 1.8t)}$ t must be small

FIGURE 10–27 Methods for determining values for J and Z_p for several types of cross sections.

Source: Machine Elements in Mechanical Design, 4th ed., Robert L. Mott, Copyright © Prentice Hall, Upper Saddle River, N.J., 2004. Reprinted by permission of the publisher.

tered in machine design and structural analysis. We will use the same equations for the maximum torsional shear stress, τ_{max} [Equation (10–16)], and torsional deformation, δ [Equation (10–22)].

$$\tau_{max} = \frac{T}{Z_p} \qquad (10\text{–}16)$$

$$\theta = \frac{TL}{GJ} \qquad (10\text{–}22)$$

Therefore, although the strict definition of J and Z_p may not be met, we will use those symbols in calculations. In summary, the polar moment of inertia J is used to represent the stiffness of a shape subjected to torsion, and the polar section modulus Z_p is used to represent the factor related to strength. Note that the construction industry uses the symbol C, called the *torsional constant*, in place of Z_p for structural shapes as discussed later in this chapter.

Note in Figure 10–27 that the points of maximum shear stress are indicated by a large dot and that they typically fall at the midpoints of flat sides. This is radically different from circular cross sections where the entire outer surface experiences the maximum shear stress. See References 2 and 4 for more detail about this phenomenon. Reference 5 gives equations for torsional properties for many more noncircular sections.

Torsional Properties of Structural Shapes

Data for the torsional properties of some structural shapes are given in Appendix tables, as listed here:

1. Round hollow structural shapes (HSS or pipes) have excellent torsional resistance. Values for J and Z_p are included in Appendix Tables A–12 and A–12 (SI). AISC data for such shapes use the symbol C, called a *torsional constant*, in place of Z_p for structural design and construction. These values are computed using the equations developed earlier in this chapter, Equation (10–13) for J and (10–19) for Z_p or C.

2. Tables A–13 and A–13 (SI) list J and Z_p for steel mechanical tubing that is often subjected to torsional loading when used in machines and vehicle structures. Equations (10–13) and (10–19) can be used to compute these values as well.

3. Square and rectangular HSS are also quite good for use where torsional loading is encountered in a structure, owing to their closed cross sections. Square shapes perform better than rectangles because they more nearly simulate the round HSS shapes that are virtually ideal. Tables A–9 and A–9 (SI) give torsional property values using the AISC symbols of J and C. Note that the square and rectangular HSS are of the type shown in Figure 10–27. The equations for J and Z_p (or C) are approximate, yet they give good estimates, particularly when the wall thickness is thin. Comparing the tabular values in Tables A–9 and A–9 (SI) with values calculated using the equations in Figure 10–27 shows that values for J are within less than 3.0% and the values of Z_p are within 4.0%. For most of the shapes with $\frac{1}{4}$ in wall thickness, the values are within less than 1.0%. Differences may be caused by variations in the shape of the rounded corners.

4. Values are not given for torsional properties of open shapes such as L, C, W, and S shapes because they are very flexible in torsion, and it is advised that they not be used in applications where significant torques are encountered.

Example Problem 10–15 illustrates the high degree of flexibility of an open shape, a split tube, in comparison with a closed tube.

Example Problem 10–15   A tube is made by forming a flat steel sheet, 4.0 mm thick, into the circular shape having an outside diameter of 90 mm. The final step is to weld the seam along the length of the tube. Figures 10–26(a), 10–26(d), and 10–25(a) show the stages of the process. Perform the following calculations to compare the behavior of the closed, welded tube with that of the open tube.

    a. Compute the torque that would create a stress of 10 MPa in the closed, welded tube.
    b. Compute the angle of twist of a 1.0-m length of the closed tube for the torque found in part (a).
    c. Compute the stress in the open tube for the torque found in part (a).
    d. Compute the angle of twist of a 1.0-m length of the open tube for the torque found in part (a).
    e. Compare the stress and deflection of the open tube with those of the closed tube.

Solution   *The solution will be shown in a programmed format with each part of the solution, (a)–(e), given as a separate panel. Each part can be approached as a different problem with* Objective, Given, Analysis, *and* Results *sections.*
    Complete part (a) now.

Objective: Compute the torque on the closed tube that would create a torsional shear stress of 10 MPa.

Given: Tube is steel. $D_o = 90$ mm. Wall thickness $= t = 4.0$ mm.

$$D_i = D_o - 2t = 90 \text{ mm} - 2(4.0 \text{ mm}) = 82 \text{ mm}.$$

Analysis: Use the maximum shear stress Equation (10–11) and solve for T.

Results:

$$\tau_{max} = \frac{Tc}{J} \qquad (10\text{--}11)$$

Then

$$T = \frac{\tau_{max}J}{c}$$

We can compute J from Equation (10–13),

$$J = \frac{\pi}{32}(D_o^4 - D_i^4) \qquad (10\text{--}13)$$

Using $D_o = 90$ mm $= 0.09$ m and $D_i = 82$ mm $= 0.082$ m,

$$J = \frac{\pi}{32}(0.09^4 - 0.082^4)\,m^4 = 2.00 \times 10^{-6}\,m^4$$

Now, letting $\tau_{max} = 10$ MPa $= 10 \times 10^6$ N/m^2, we have

$$T = \frac{\tau_{max}J}{c} = \frac{(10 \times 10^6\,N/m^2)(2.00 \times 10^{-6}\,m^4)}{0.045\,m} = 444\,N \cdot m$$

That is, a torque of 444 N·m applied to the closed welded tube would produce a maximum torsional shear stress of 10 MPa in the tube. Note that this is a very low stress level for steel.
    Complete part (b) of the problem.

Objective: For the closed tube used in (a), compute the angle of twist.

Given: $J = 2.00 \times 10^{-6}$ m^4. Length $= L = 1.0$ m. Torque $= T = 444$ N·m. $G = 80$ GPa.

Analysis: Use Equation (10–22) to compute the angle of twist, θ.

Results: $$\theta = \frac{TL}{GJ} = \frac{(444\,N \cdot m)(1.0\,m)}{(80 \times 10^9\,N/m^2)(2.00 \times 10^{-6}\,m^4)} = 0.00278\,rad$$

Converting θ to degrees gives

$$\theta = 0.00278\,rad\frac{180\,deg}{\pi\,rad} = 0.159\,deg$$

Again, note that this is a very small angle of twist.
    Consider a tube shaped as in parts (a) and (b) except that it is not closed, as shown in Figure 10–26(d). Compute the torsional shear stress in the open tube due to the torque of 444 N·m.

Objective: Compute the shear stress in the open tube.

Given: Torque $= T = 444$ N·m. $D_o = 90$ mm. Wall thickness $= t = 4.0$ mm.

Analysis: Equation (10–16) can be used to compute the maximum shear stress for the open tube before it is welded, treating it as a noncircular cross section. The formula for Z_p is given in Figure 10–27.

$$Z_p = \frac{4\pi^2 r^2 t^2}{6\pi r + 1.8t}$$

Results: The mean radius is

$$r = \frac{D_o}{2} - \frac{t}{2} = \frac{90\,mm}{2} - \frac{4\,mm}{2} = 43\,mm$$

Then

$$Z_p = \frac{4\pi^2(43)^2(4)^2}{6\pi(43) + 1.8(4)}\,mm^3 = 1428\,mm^3$$

The stress in the open tube is, then,

$$\tau_{max} = \frac{T}{Z_p} = \frac{444\,N \cdot m}{1428\,mm^3}\frac{1000\,mm}{m} = 311\,MPa$$

Now, complete part (d) by computing the angle of twist of the open tube.

Objective: Compute the angle of twist of the open tube for $T = 444$ N·m.

Given: Length $= L = 1.0$ m. Torque $= T = 444$ N·m. $G = 80$ GPa.
$D_o = 90$ mm. Wall thickness $= t = 4.0$ mm.

Analysis: The angle of twist for the open tube can be computed using Equation (10–22). Figure 10–27 gives us the formula for the torsional rigidity constant, J.

Results: Using $r = 43$ mm $= 0.043$ m and $t = 4$ mm $= 0.004$ m yields

$$J = \frac{2\pi r t^3}{3} = \frac{2\pi (0.043)(0.004)^3}{3} = 5.764 \times 10^{-9} \text{ m}^4$$

Then the angle of twist is

$$\theta = \frac{TL}{GJ} = \frac{(444 \text{ N·m})(1.0 \text{ m})}{(80 \times 10^9 \text{ N/m}^2)(5.764 \times 10^{-9} \text{ m}^4)} = 0.963 \text{ rad}$$

Converting θ to degrees, we obtain

$$\theta = 0.963 \text{ rad}\frac{180 \text{ deg}}{\pi \text{ rad}} = 55.2 \text{ deg}$$

Now complete the final part (e), comparing the open and closed tubes.

Objective: Compare the stress and deflection of the open tube with those of the closed tube.

Given: From parts (a), (b), (c), and (d):

Stress in the closed tube $= \tau_c = 10$ MPa.

Angle of twist of the closed tube $= \theta_c = 0.159$ deg.

Stress in the open tube $= \tau_o = 311$ MPa.

Angle of twist of the open tube $= \theta_o = 55.2$ deg.

Analysis: Compute the ratio of stresses τ_o/τ_c and the ratio of the angles of twist θ_o/θ_c.

Results: Ratio of stresses:

$$\frac{\tau_o}{\tau_c} = \frac{311 \text{ MPa}}{10 \text{ MPa}} = 31.1 \text{ times greater}$$

Ratio of angle of twist:

$$\frac{\theta_o}{\theta_c} = \frac{55.2 \text{ deg}}{0.159 \text{ deg}} = 347 \text{ times greater}$$

Comment: These comparisons dramatically show the advantage of using closed sections to carry torsion.
In addition, the actual stress level computed for the open tube is probably greater than the allowable stress for many steels. Recall that the design shear stress is

$$\tau_d = \frac{0.5 s_y}{N}$$

Then the required yield strength of the material to be safe at a stress level of 311 MPa and a design factor of 2.0 is

$$s_y = \frac{N\tau_d}{0.5} = \frac{2(311 \text{ MPa})}{0.5} = 1244 \text{ MPa}$$

Only a few highly heat-treated steels listed in Appendix A–14 have this level of yield strength.

REFERENCES

1. Blodgett, O. W., *Design of Weldments*, Cleveland, OH, James F. Lincoln Arc Welding Foundation, 1963.

2. Boresi, A. P. and R. J. Schmidt, *Advanced Mechanics of Materials*, 6th ed., New York, John Wiley & Sons, Inc., 2002.

3. Mott, R. L., *Machine Elements in Mechanical Design*, 4th ed., Upper Saddle River, NJ, Prentice Hall, 2004.

4. Young, W. C. and R. D. Cook, *Advanced Mechanics of Materials*, 2nd ed., Upper Saddle River, NJ, Prentice Hall, 1999.

5. Young, W. C. and R. G. Budynas, *Roark's Formulas for Stress and Strain*, 7th ed., New York, McGraw-Hill, 2002.

INTERNET SITES

1. Driv-Lok, Inc. www.driv-lok.com Manufacturer of a wide variety of parallel keys and press-fit fastening devices such as grooved pins, dowel pins, spring pins, and studs. Includes data for key size and materials available for keys.

2. Truarc Company, LLC. www.truarc.com Manufacturer of a wide variety of retaining rings for industrial, commercial, military, and consumer products. Online catalog data give groove dimensions and details for a large range of sizes. Model 5100 Basic rings are suggested for shaft applications.

3. Keystone Manufacturing www.trukey.com Manufacturer of keys and shafting with premilled keyways. Includes tables of key sizes for different sizes of shafts and material properties for shafting.

4. Sensors Portal www.sensorsportal.com/HTML/SENSORS/ TorqueSens_Manuf.htm The Sensors Portal site identifies numerous manufacturers of sensors for many kinds of measurement, including torque measurement.

5. S&T e-Digest http://www.sensorsportal.com/HTML/DIGEST/ march_06/P_59.pdf An electronic version of *Sensors & Transducers Magazine*. The listed page is an article called "Evolution and Future of Torque Measurement Technology," by Dr. W. Krimmel, published March 27, 2006.

FIGURE P10–9 Shaft for Problem 10–9.

FIGURE P10–10 Shaft for Problem 10–10.

PROBLEMS

10–1. Compute the torsional shear stress that would be produced in a solid circular shaft having a diameter of 20 mm when subjected to a torque of 280 N · m.

10–2. A hollow shaft has an outside diameter of 35 mm and an inside diameter of 25 mm. Compute the torsional shear stress in the shaft when it is subjected to a torque of 560 N · m.

10–3. Compute the torsional shear stress in a shaft having a diameter of 1.25 in when carrying a torque of 1550 lb · in.

10–4. A steel tube is used as a shaft carrying 5500 lb · in of torque. The outside diameter is 1.75 in, and the wall thickness is $\frac{1}{8}$ in. Compute the torsional shear stress at the outside and the inside surfaces of the tube.

10–5. A movie projector drive mechanism is driven by a 0.08-kW motor whose shaft rotates at 180 rad/s. Compute the torsional shear stress in its 3.0-mm-diameter shaft.

10–6. The impeller of a fluid agitator rotates at 42 rad/s and requires 35 kW of power. Compute the torsional shear stress in the shaft that drives the impeller if it is hollow and has an outside diameter of 40 mm and an inside diameter of 25 mm.

10–7. A driveshaft for a milling machine transmits 15.0 hp at a speed of 240 rpm. Compute the torsional shear stress in the shaft if it is solid with a diameter of 1.44 in. Would the shaft be safe if the torque is applied with shock and it is made from AISI 4140 OQT 1300 steel?

10–8. Repeat Problem 10–7 if the shaft contains a profile keyseat.

10–9. Figure P10–9 shows the end of the vertical shaft for a rotary lawn mower. Compute the maximum torsional shear stress in the shaft if it is transmitting 7.5 hp to the blade when rotating 2200 rpm. Specify a suitable steel for the shaft.

10–10. Figure P10–10 shows a stepped shaft in torsion. The larger section also has a hole drilled through.

   a. Compute the maximum shear stress at the step for an applied torque of 7500 lb · in.

   b. Determine the largest hole that could be drilled in the shaft and still maintain the stress near the hole at or below that at the step.

10–11. Compute the torsional shear stress and the angle of twist in degrees in an aluminum tube, 600 mm long, having an inside diameter of 60 mm and an outside diameter of 80 mm when subjected to a steady torque of 4500 N · m. Then specify a suitable aluminum alloy for the tube.

10–12. Two designs for a shaft are being considered. Both have an outside diameter of 50 mm and are 600 mm long. One is solid but the other is hollow with an inside diameter of 40 mm. Both are made from steel. Compare the torsional shear stress, the angle of twist, and the mass of the two designs if they are subjected to a torque of 850 N · m.

10–13. Determine the required inside and outside diameters for a hollow shaft to carry a torque of 1200 N · m with a maximum torsional shear stress of 45 MPa. Make the ratio of the outside diameter to the inside diameter approximately 1.25.

10–14. A gear driveshaft for a milling machine transmits 7.5 hp at a speed of 240 rpm. Compute the torsional shear stress in the 0.860-in-diameter solid shaft.

10–15. The input shaft for the gear drive described in Problem 10–14 also transmits 7.5 hp but rotates at 1140 rpm. Determine the required diameter of the input shaft to give it the same stress as the output shaft.

10–16. Determine the stress that would result in a $1\frac{1}{2}$-in schedule 40 steel pipe if a plumber applies a force of 80 lb at the end of a wrench handle 18 in long.

10–17. A rotating sign makes 1 rev every 5 s. In a high wind, a torque of 30 lb · ft is required to maintain rotation. Compute the power required to drive the sign. Also compute the stress in the final driveshaft if it has a diameter of 0.60 in. Specify a suitable steel for the shaft to provide a design factor of 4 based on yield strength in shear.

10–18. A short, cylindrical bar is welded to a rigid plate at one end, and then a torque is applied at the other. If the bar has a diameter of 15 mm and is made of AISI 1020 cold-drawn steel, compute the torque that must be applied to it to subject it to a stress equal to its yield strength in shear. Use $s_{ys} = s_y/2$.

10–19. A propeller driveshaft on a ship is to transmit 2500 hp at 75 rpm. It is to be made of AISI 1040 WQT 1300 steel. Use a design factor of 6 based on the yield strength in shear. The shaft is to be hollow with the inside diameter equal to 0.80 times the outside diameter. Determine the required diameter of the shaft.

10–20. If the propeller shaft of Problem 10–19 was to be solid instead of hollow, determine the required diameter. Then compute the ratio of the weight of the solid shaft to that of the hollow shaft.

10–21. A power screwdriver uses a shaft with a diameter of 5.0 mm. What torque can be applied to the screwdriver if the limiting stress due to torsion is 80 MPa?

10–22. An extension for a socket wrench similar to that shown in Figure 10–4 has a diameter of 6.0 mm and a length of 250 mm. Compute the stress and angle of twist in the extension when a torque of 5.5 N · m is applied. The extension is steel.

10–23. Compute the angle of twist in a steel shaft 15 mm in diameter and 250 mm long when a torque of 240 N · m is applied.

10–24. Compute the angle of twist in an aluminum tube that has an outside diameter of 80 mm and an inside diameter of 60 mm when subjected to a torque of 2250 N · m. The tube is 1200 mm long.

10–25. A steel rod with a length of 8.0 ft and a diameter of 0.625 in is used as a long wrench to unscrew a plug at the bottom of a pool of water. If it requires 40 lb · ft of torque to loosen the plug, compute the angle of twist of the rod.

10–26. For the rod described in Problem 10–25, what must the diameter be if only 2.0 deg of twist is desired when 40 lb · ft of torque is applied?

10–27. Compute the angle of twist of the free end relative to the fixed end of the steel bar shown in Figure P10–27.

FIGURE P10–27   Bar for Problem 10–27.

10–28. A meter for measuring torque uses the angle of twist of a shaft to indicate torque. The shaft is to be 150 mm long and made of 6061-T6 aluminum alloy. Determine the required diameter of the shaft if it is desired to have a twist of 10.0 deg when a torque of 5.0 N · m is applied to the meter. For the shaft thus designed, compute the torsional shear stress and then compute the resulting

design factor for the shaft. Is it satisfactory? If not, what would you do?

10–29. A beryllium copper wire having a diameter of 1.50 mm and a length of 40 mm is used as a small torsion bar in an instrument. Determine what angle of twist would result in the wire when it is stressed to 250 MPa.

10–30. A fuel line in an aircraft is made of a titanium alloy. The tubular line has an outside diameter of 18 mm and an inside diameter of 16 mm. Compute the stress in the tube if a length of 1.65 m must be twisted through an angle of 40 deg during installation. Determine the design factor based on the yield strength in shear if the tube is Ti-6Al-4V, aged.

10–31. For the shaft shown in Figure P10–31, compute the angle of twist of pulleys B and C relative to A. The steel shaft has a diameter of 35 mm throughout its length. The torques are $T_1 = 1500$ N · m, $T_2 = 1000$ N · m, $T_3 = 500$ N · m. The lengths are $L_1 = 500$ mm, $L_2 = 800$ mm.

FIGURE P10–31   Shaft for Problem 10–31.

10–32. A torsion bar in a light truck suspension is to be 820 mm long and made of steel. It is subjected to a torque of 1360 N · m and must be limited to 2.2 deg of twist. Determine the required diameter of the solid round bar. Then compute the stress in the bar.

10–33. A steel driveshaft for an automobile is a hollow tube 1525 mm long. Its outside diameter is 75 mm, and its inside diameter is 55 mm. If the shaft transmits 120 kW of power at a speed of 225 rad/s, compute the torsional shear stress in the shaft and the angle of twist of one end relative to the other.

10–34. A part of a rear axle of an automobile is a solid steel shaft having the configuration shown in Figure P10–34. Considering the stress concentration due to the shoulder, compute the torsional shear stress in the axle when it rotates at 70.0 rad/s, transmitting 60 kW of power.

FIGURE P10–34   Axle for Problem 10–34.

10–35. Part of an output shaft from an automotive transmission has the configuration shown in Figure P10–35. If the shaft is transmitting 105 kW at 220 rad/s, compute the maximum torsional shear stress in the shaft. Account for the stress concentration at the place where the speedometer gear is located.

FIGURE P10–35   Shaft for Problem 10–35.

The indicated figures for Problems 10–36 through 10–39 show portions of shafts from power transmission equipment. Compute the maximum repeated torque that could be safely applied to each shaft if it is to be made from AISI 1141 OQT 1100 steel.

10–36. Use Figure P10–36.

FIGURE P10–36   Shaft for Problem 10–36.

10–37. Use Figure P10–37.

FIGURE P10–37   Shaft for Problem 10–37.

10–38. Use Figure P10–38.

FIGURE P10–38   Shaft for Problem 10–38.

10–39. Use Figure P10–39.

FIGURE P10–39   Shaft for Problem 10–39.

Noncircular Sections

10–40. Compute the torque that would produce a torsional shear stress of 50 MPa in a square steel rod 20 mm on a side.

10–41. For the rod described in Problem 10–40, compute the angle of twist that would be produced by the torque found in the problem over a length of 1.80 m.

10–42. Compute the torque that would produce a torsional shear stress of 7500 psi in a square aluminum rod 1.25 in on a side.

10–43. For the rod described in Problem 10–42, compute the angle of twist that would be produced by the torque found in the problem over a length of 48 in.

10–44. Compute the torque that would produce a torsional shear stress of 7500 psi in a rectangular aluminum bar 1.25 in thick by 3.00 in wide.

10–45. For the bar described in Problem 10–44, compute the angle of twist that would be produced by the torque found in the problem over a length of 48 in.

10–46. An extruded aluminum bar is in the form of an equilateral triangle 30 mm on a side. What torque is required to cause an angle of twist in the bar of 0.80 deg over a length of 2.60 m?

10–47. For the triangular bar described in Problem 10–46, what stress would be developed in the bar when carrying the torque found in the problem?

10–48. As shown in Figure P10–48, a portion of a steel shaft has a flat machined on one side. Compute the torsional shear stress in both the circular section and the one with the flat when a torque of 850 lb · in is applied.

10–49. For the steel shaft shown in Figure P10–48, compute the angle of twist of one end relative to the other if a torque of 850 lb · in is applied uniformly along the length.

Section A–A   Section B–B

FIGURE P10–48   Problems 10–48 to 10–51.

10–50. Repeat Problem 10–48 with all the data the same except that two flats are machined on the shaft, resulting in a total measurement across the flats of 1.25 in.

10–51. Repeat Problem 10–49 for a shaft having two flats, resulting in a total measurement across the flats of 1.25 in.

10–52. A square stud 200 mm long and 8 mm on a side is made from titanium, Ti-6Al-4V, aged. What angle of twist will result when a wrench is applying a pure torque that causes the stress to equal the yield strength of the material in shear?

10–53. A standard square structural steel tube has cross-sectional dimensions of $4 \times 4 \times \frac{1}{4}$ in and is 8.00 ft long. Compute the torque required to twist the tube 3.00 deg.

10–54. For the tube in Problem 10–53, compute the maximum stress in the tube when it is twisted 3.00 deg. Would this be safe if the tube is made from ASTM A501 structural steel and the load was static?

10–55. Repeat Problem 10–53 for a rectangular tube $6 \times 4 \times \frac{1}{4}$.

10–56. Repeat Problem 10–54 for a rectangular tube, $6 \times 4 \times \frac{1}{4}$.

10–57. A standard 6-in schedule 40 steel pipe has approximately the same cross-sectional area as a square tube, $6 \times 6 \times \frac{1}{4}$, and thus both would weigh about the same for a given length. If the same torque were applied to both, compare the resulting torsional shear stress and angle of twist for the two shapes.

ADDITIONAL PRACTICE, REVIEW, AND DESIGN PROBLEMS

10–58. Compute the torsional shear stress in a solid round shaft carrying 125 kW of power while rotating at 1150 rpm. The shaft has a diameter of 35 mm.

10–59. For the shaft in Problem 10–58, specify a suitable steel if the power is smooth and steady.

10–60. For the shaft in Problem 10–58, specify a suitable steel if the power is repeated.

10–61. For the shaft in Problem 10–58, specify a suitable steel if the power is applied with shock.

10–62. A power transmission shaft is to carry a steady 12.0 horsepower while rotating at 1150 rpm. Specify a suitable diameter for the solid circular shaft if it is to be made from AISI 1040 WQT 1100 steel.

10–63. A power transmission shaft is to carry a steady 20.0 horsepower while rotating at 3450 rpm. The shaft drives a special machine that forms sheet steel and some shock loading is expected. Specify a suitable diameter for the solid circular shaft if it is to be made from AISI 4140 OQT 1300 steel.

10–64. An alloy steel shaft has an outside diameter of 100 mm. A central hole of 60 mm diameter is bored in part of its length as shown in Figure P10–64. Compute the shearing stress in the hollow section if the stress in the solid section is 200 MPa.

10–65. For the shaft described in Problem 10–64, compute the angle of twist of the right end with respect to the left end.

10–66. Figure P10–66 shows a shaft carrying three pulleys that rotates at 1750 rpm. Pulley A delivers 15 kW to a mating pulley that drives a fan. Pulley C delivers 20 kW to a different mating pulley that drives a conveyor. All power comes into the shaft through pulley B. Considering only torsion, compute the shearing stress in each part of the shaft. Consider stress concentrations.

10–67. Figure P10–67 shows a shaft carrying three gears that rotates at 1150 rpm. Gear A delivers 20 kW to a mating gear that drives a mixer. Gear C delivers 12 kW to a different mating gear that drives a circular saw. All power comes into the shaft through gear B. Considering only torsion, compute the shearing stress in each part of the shaft. Consider stress concentrations.

10–68. Design a hollow steel shaft to transmit 225 kW of power at 80 rpm without exceeding a shearing stress of 60 MPa. Consider only torsion and assume that there are no stress concentrations. Make the ratio of the outside diameter to the inside diameter to be approximately 1.25.

10–69. A solid aluminum rod having a diameter of 4.0 mm is to be twisted through $\frac{1}{2}$ revolution (180 degrees) while acting as a torsion bar. The torsional shear stress must not exceed 150 MPa. Determine the required length of the rod.

10–70. A torsion bar is to be made from a hollow titanium bar that has a length of 200 mm. The ratio of the outside diameter to the inside diameter is to be approximately 1.50. The desired torsional stiffness is to be 0.015 degree of rotation per 1.0 N · m of applied torque. Specify the outside diameter and the inside diameter of the bar.

10–71. For the bar designed in Problem 10–70, compute the torsional shear stress that would occur in the bar when it is rotated 10.0 degrees.

FIGURE P10–64   Shaft for Problems 10–64 and 10–65.

Fillet radius = 0.50 mm at each step
Dimensions in mm
Sled-runner keyseat at A, C
Profile keyseat at B

FIGURE P10–66   Shaft for Problem 10–66.

Profile keyseats at A, B, C
Fillet radius = 1.00 mm at each step
Dimensions in millimeters

FIGURE P10–67   Shaft for Problem 10–67.

10–72. For the bar designed in Problem 10–70, compute the angle of twist when the torsional shear stress in the bar has a design factor of 3 based on yield strength in shear. The material for the bar is AISI 4140 OQT 1100 steel.

COMPUTER ASSIGNMENTS

Note: Any of these assignments can be completed as a computer program in a technical language such as QBASIC or C. Or they could be designed as a spreadsheet or solved with a computer algebra system or some other mathematics analysis software. Be sure to

consider the units of all factors entered into the problems and indicate the units of all results.

1. Given the need to design a solid circular shaft for a given torque, a given material yield strength, and a given design factor, compute the required diameter for the shaft.

Enhancements to Assignment 1

a. For a given power transmitted and speed of rotation, compute the applied torque.

b. Include a table of materials from which the designer can select. Then automatically look up the yield strength.

c. Include the design factor table, Table 10–1. Then prompt the designer to specify the manner of loading only and determine the appropriate design factor from the table within the program.

2. Repeat assignment 1, except design a hollow circular shaft. Three possible solution procedures exist:
   a. For a given outside diameter, compute the required inside diameter.
   b. For a given inside diameter, compute the required outside diameter.
   c. For a given ratio of D_i/D_o, find both D_i and D_o.

Enhancements to Assignment 2

a. Compute the mass of the resulting design for a given length and material density.

b. If the computer has graphics capability, draw the resulting cross section and dimension it.

3. Enter the stress concentration factor curves into a program, allowing the automatic computation of K_t for given factors such as fillet radius, diameter ratio, hole diameter, and so on. Any of the cases shown in Appendix A–22–5, A–22–6, or A–22–7 could be used. This program could be run by itself or made an enhancement of other stress analysis programs.

4. Compute the angle of twist from Equation (10–22) for given T, L, G, and J.

Enhancements to Assignment 4

a. Compute J for given dimensions for the shaft, either solid or hollow.

b. Include a table of values for G, from Table 10–3, in the program.

5. Compute the required diameter of a solid circular shaft to limit the angle of twist to a specified amount.

6. Compute the angle of twist for one end of a multisection shaft relative to the other, similar to Example Problem 10–13. Allow the lengths, diameters, materials, and torques to be different in each section.

7. Write a program to compute the values for the effective section modulus, Z_p, and the torsional stiffness constant, J, from Figure 10–27 for any or all cases.

Enhancement to Assignment 7

Determine the equations for C_1, C_2, C_3, and C_4 in terms of the ratio, h/r, for shafts with flats. Use a curve-fitting routine.

SHEARING FORCES AND BENDING MOMENTS IN BEAMS

MAJOR TOPICS

11–1 The Big Picture—Shearing Forces and Bending Moments in Beams
11–2 Beam Loading, Supports, and Types of Beams
11–3 Reactions at Supports
11–4 Shearing Forces and Bending Moments for Concentrated Loads
11–5 Guidelines for Drawing Beam Diagrams for Concentrated Loads
11–6 Shearing Forces and Bending Moments for Distributed Loads
11–7 General Shapes Found in Bending Moment Diagrams
11–8 Shearing Forces and Bending Moments for Cantilever Beams
11–9 Beams with Linearly Varying Distributed Loads
11–10 Free Body Diagrams of Parts of Structures
11–11 Mathematical Analysis of Beam Diagrams
11–12 Continuous Beams—Theorem of Three Moments

11–1 THE BIG PICTURE— SHEARING FORCES AND BENDING MOMENTS IN BEAMS

Discussion Map

■ Much of the discussion in the next six chapters deals with beams. You should be familiar with the following definition of a beam:

> *A beam is a member that carries loads transversely, that is, perpendicular to its long axis.*

■ In this chapter you will learn how to characterize the types of beams and the loadings placed on them.

■ In later chapters, you will learn how to compute the magnitude of stresses in beams and to design them to ensure that the beams are safe.

■ You will also learn in later chapters how to compute the deflection of a beam under a given loading pattern.

Discover

Imagine that you are trying to cross a small stream of flowing water. You find a wide plank of wood that is long enough to span from your side of the stream to the other side. You lay it across and get ready to step onto the plank. Then you might think to yourself, "Is this plank strong enough to carry my weight as I cross the creek?" You recognize that when you walk onto the plank it is going to bend downward. You likely know that it will bend more as

you move toward the middle of the span. Will it bend so much that your feet will dip into the water? Will it actually break under your weight and plunge you in all the way?

Unless you know how to analyze the stresses in beams, and how to compute their deflections, you are left to your own judgment about the safety of the plank. This simple example should give you some understanding of the need for learning how to analyze the behavior of a beam.

What other examples of beams can you think of?

Consider the construction of your home, commercial buildings, or bridges.

Consider recreation equipment and exercise devices.

Look carefully at children's toys and see if some of their parts carry loads perpendicular to their long axes.

Check out a car or a truck and see if you can tell which parts will bend under the load of the vehicle itself or from loads applied to them during operation.

Consider chairs and other seating systems. Look at the flat, plank-like seating in the bleachers at a football, basketball, or baseball game.

Look at equipment used in highway construction or in building any kind of structure. Most equipment must lift and carry heavy loads or push or pull with large forces.

How does your list of beam examples compare with the following items?

■ In the basement of your home you might have a long steel beam extending across the room or along the middle of the length of the building to hold the joists of the floor above.

- The joists themselves are beams carrying the load from the floor to the supporting steel beam or to the walls of the foundation.
- Even the flooring itself acts as a beam carrying your weight and the weight of furniture and appliances that apply loads between the joists.
- The tops of doors and windows have strong beams, called lintels, to hold up the structure above these openings in walls.
- The sides of stairs, called stringers, act as beams as they carry the weight of a person climbing the steps.
- A baseball bat or a tennis racket acts like a special kind of beam, called a cantilever, that is held fixed at one end by your hands and that applies a force to the ball out near the end of the racket or bat.
- The balance beam in gymnastics is an obvious example, and you could also include the parallel bars.
- Did you consider the basketball hoop as a beam? When a player stuffs a dunk and hangs on to the rim, it certainly bends down. It has to be strong and stiff enough to withstand that load without breaking or permanently bending.
- Some people exercise by rowing, pulling on oars to move the boat through the water. You apply a pulling force on the handles against the resistance of the oar blade in the water, and the oar bends with respect to the oarlock attached to the boat.
- Did you look for examples of beams at a swimming pool? The diving board is a beam that must flex with just the right feel while remaining strong for thousands of dives. The steps of the ladder used to climb to the high board are beams.
- Certain kinds of car, truck, and trailer axles are beams that transfer the loads from the vehicle to the wheels and then to the ground.
- At construction sites for new buildings, you should be able to see numerous horizontal beams being erected that will carry the loads of each floor of the building to the vertical columns and on to the foundation.

In this chapter you will learn how to compute how the loads on a beam result in internal forces and moments that must be resisted. The stress and deflection analyses follow in later chapters.

ACTIVITY Chapter 11—Shearing Forces and Bending Moments in Beams

To comprehend clearly the behavior of beams in bending, you must visualize what is happening *within the fibers of the beam.* This is difficult because you cannot observe directly how each small particle of the material of the beam is affected by the action of loads on the beam. However, a simple experiment will provide some insight and allow you to deduce the nonvisible phenomena of *shearing forces and bending moments* to which

this chapter is devoted. It will lead to further understanding of related phenomena in the following several chapters.

Setup. Acquire or fabricate a simple beam and two supports for it. The beam should be sufficiently flexible that you can bend it with small forces using only your hands, but it should be stiff enough to demonstrate the natural curvature that a beam exhibits under load. Possibilities are beams with lengths of approximately 200 mm to 1000 mm long (8 in to 36 in), 25 mm wide (1.0 in), and made from

1. A thin flat piece of wood about 2 mm to 3 mm thick (0.08 in to 0.12 in)
2. Dense cardboard of similar dimensions as described in item 1. You may glue several thinner strips together to achieve the proper thickness and stiffness. Spread the glue evenly across the entire surfaces of contact to ensure that the composite piece acts as a single unit.
3. A strip of flat sheet aluminum about 1.6 mm thick (0.06 in)
4. A strip of flat sheet steel about 0.5 mm thick (0.02 in)
5. A strip of flat plastic sheet about 2.0 mm thick (0.08 in)

Several of these shapes and materials are approximated by inexpensive rulers, straightedges, and meter sticks available in office supply or home improvement stores.

The two supports should be stable when laid on a flat surface and should support the beam at least 15 mm (0.6 in) off the surface. An ideal example is a small angle with leg lengths from 20 mm to 30 mm (0.75 in to 1.25 in) made from wood, plastic, or metal. Place the legs down to create a "knife-edge" on which to support the beam.

Also acquire a good straightedge about 75 mm to 150 mm long (3.0 in to 6.0 in).

Activity A

Start by simply picking up the beam and applying forces that tend to bend it while observing the shape that the beam assumes and the direction of the forces required to cause the beam to bend. Bring the beam to a steady equilibrium after each action. If possible, sketch the beam in its bent shape and indicate where the supporting forces or grasping actions were applied. How many ways can you do that? Possibilities are

1. Grasp the beam with the thumb and two or more fingers of each hand somewhat apart from each other and rotate the beam to a bent shape.
2. Place the thumb of each hand on the bottom surface of the beam and one finger of each hand outboard from the thumbs. Then push down with the fingers.
3. Place the thumb of each hand on the bottom surface of the beam and one finger of each hand inboard from the thumbs. Then push down with the fingers.

Steps 4–7 may be easier if you have two rigid blocks about 75 mm to 100 mm long (3.0 in to 4.0 in) on each side of the beam and grasp the beam like a "sandwich."

4. Grasp one end of the beam firmly with one hand. Then push up or down with a finger of the other hand.

5. Grasp both ends of the beam firmly and rotate the beam by rotating your wrists. Try different combinations of the direction of rotation of each hand.

6. Grasp both ends of the beam firmly and rotate the beam by moving one hand up or down relative to the other, keeping your wrists steady.

7. After performing step 6, release your grasp of one hand and observe what happens to the shape of the beam. As you release your grip, try to maintain the position of the beam where it was.

Observations and Conclusions.   You should observe the following:

1. Some actions produced a continuous curvature of the beam in one direction.

2. Other actions produced curvatures that are partly concaved upward and partly concaved downward.

3. When parts of the beam were outside the points of support or grasp, that part of the beam was straight.

4. The firm grasping controlled not only the position of the beam but also the curvature so that the slope of the beam was inline with the grasp at the point where it was clamped.

5. When a part of a bent beam was released from a firm grasp, its shape assumed a continuous, one-direction curvature regardless of its shape while being grasped.

6. If a supporting force or a firm grasp was removed completely, the beam would continue to move back to its originally straight form.

Activity B

1. Set two simple supports on a flat surface and apart so that approximately $\frac{1}{4}$ of the length of the beam overhangs each support.

2. Place the beam on the supports.

3. Apply a downward force with one finger to the midpoint of the beam between the supports.

4. Use the straightedge to observe the curvature of the beam between the supports and in each overhang.

Observations and Conclusions.   What do you observe from this simple experiment? Here are some things to consider:

1. The downward force applied to the beam at its middle causes the beam to bend in a curved shape between the supports that is concaved upward. This illustrates *positive bending.*

2. The parts of the beam overhanging the supports are straight, not curved.

3. The maximum deflection occurs at the middle of the beam where the load is applied.

4. The deflection is below the level of the supports between the two supports but upward in the overhangs outside of the supports.

Activity C

1. Set the supports on a flat surface and apart so that approximately $\frac{1}{4}$ of the length of the beam overhangs each support.

2. Apply a downward force at the end of both overhangs and note the shape of the beam.

3. Now apply only one force at one end and observe that the reaction force at the opposite support must be redirected to be downward rather than upward as before.

4. Apply one force between the supports and one at the end of one of the overhangs. Observe the shape of the deflected beam and sketch it.

5. Now apply three forces, one downward on each overhang and one downward between the supports. Observe the shape of the deflected beam and sketch it.

Activity D

This activity requires that you now think about how the material within the beam must act to cause the beam to behave in the observed manner, depending on the way it was supported and loaded.

1. Repeat Activity B and hold the downward force, F, in position. Figure 11–1(a) shows the beam in its original straight, undeflected form with the supports at points A and C and the force, F, applied at B.

2. Sketch the shape of the deflected beam as shown in Figure 11–1(b). With a small straightedge, verify that the deflected beam shape is curved between A and C while the overhang to the left of A remains straight.

3. What direction must the forces act at each support? (***Answer:*** upward)

4. What is the magnitude of the force at each support? (***Answer:*** $F/2$)

5. If the beam were to be cut at any place, D, between the left support and the middle of the beam where the load is applied, what would have to happen at the cut end of each part of the beam if the beam was to be kept in the same position and shape as before it was cut? Refer to Figure 11–1(c) and (d) showing the deflected beam with a cut made at point D, located an arbitrary distance x from the left support.
   a. ***Answer*** for what happens to the *left cut section* using that part only as a free body diagram [Figure 11–1(c)]:
      i. The upward reaction force of $F/2$ remains applied at the support. Then to maintain equilibrium of forces in the vertical direction, a downward force of $F/2$ must be exerted on the cut section. This force is shown and called V_{DL}, the shearing force at D for the left cut section. In the complete beam, this force had been applied by the part of the beam to the right that was cut away.

FIGURE 11–1   Simply supported beam with concentrated load at midspan and an overhang—illustration of internal shearing force and bending moment.

ii. To maintain the part of the beam to the left of the cut in its formerly curved shape, a counterclockwise moment must be applied at the cut, called M_{DL}, the moment at D for the left cut section.

iii. To maintain rotational equilibrium of the part of the beam to the left of the cut, the moment must be equal in magnitude to the couple created by the reaction $(F/2)$ and the internal shearing force at the cut $(F/2)$ acting at a distance of x apart. That moment is

$$M_{DL} = (F/2)(x) = Fx/2 \text{ counterclockwise}$$

iv. Now, the cut section is in equilibrium for forces in the vertical direction and for moments. Furthermore, the moment at the cut section maintains the curved shape of the deflected beam between the support and the cut section, while allowing the part of the beam beyond the overhang to remain straight.

b. *Answer* for what happens to the *right cut section* using that part only as a free body diagram [Figure 11–1(d)]:

i. The same logic applies as for the left cut section. To maintain equilibrium of vertical forces, an upward shearing force must exist in the beam,

called V_{DR}, the shearing force at D for the right cut section.

ii. The magnitude of the shearing force can be found by summing forces in the vertical direction for the right cut section.

$$\Sigma F_v = 0 = F/2 + V_{DR} - F$$
$$V_{DR} = F - F/2 = F/2$$

iii. To maintain the curved shape of the beam, a clockwise moment must be applied at the cut, called M_{DR}, the moment at D for the right section.

iv. To maintain rotational equilibrium of the part of the beam to the right of the cut, the moment must be equal to the sum of the moments due to the other two forces: F acting at $(L/2 - x)$ from the cut, and $F/2$ acting at $(L - x)$ from the cut. Then

$$M_{DR} = F(L/2 - x) - (F/2)(L - x)$$
$$= Fx/2 \text{ clockwise}$$

c. In summary, parts (c) and (d) of Figure 11–1 show the free body diagrams of the left and right cut sections. Note that the shearing forces and bending moments at the cut on the two parts form action–reaction pairs, equal in magnitude but opposite in direction.

In this chapter, you will apply processes such as this to more complex beams with a variety of loading and support conditions. You will learn procedures to determine the magnitude of shearing forces and bending moments in a more direct fashion and to draw complete shearing force and bending moment diagrams that display their variations across the entire length of the beam. The ability to perform these analyses is a critical *first step* to computing the stresses and deformations in a beam, as will be discussed in the following chapters.

Objectives of This Chapter

After completing this chapter, you should be able to:

1. Define the term *beam* and recognize when a load-carrying member is a beam.

2. Describe several kinds of beam loading patterns: *concentrated loads, uniformly distributed loads, linearly varying distributed loads,* and *concentrated moments.*

3. Describe several kinds of beams according to the manner of support: *simple beam, overhanging beam, cantilever,* and *composite beam* having more than one part.

4. Draw free body diagrams for beams and parts of beams showing all external forces and reactions.

5. Compute the magnitude of reaction forces and moments and determine their directions.

6. Define *shearing force* and determine the magnitude of shearing force anywhere within a beam.

7. Draw free body diagrams of *parts* of beams and show the internal shearing forces.

8. Draw complete shearing force diagrams for beams carrying a variety of loading patterns and with a variety of support conditions.

9. Define *bending moment* and determine the magnitude of bending moment anywhere within a beam.

10. Draw free body diagrams of *parts* of beams and show the internal bending moments.

11. Draw complete bending moment diagrams for beams carrying a variety of loading patterns and with a variety of support conditions.

12. Use the *laws of beam diagrams* to relate the load, shearing force, and bending moment diagrams to each other and to draw the diagrams.

13. Draw free body diagrams of parts of compound beams and structures and draw the shearing force and bending moment diagrams for each part.

14. Properly consider *concentrated moments* in the analysis of beams.

15. Use the *theorem of three moments* to analyze continuous beams having three or more supports and carrying any combination of concentrated and distributed loads.

11–2 BEAM LOADING, SUPPORTS, AND TYPES OF BEAMS

Recall the definition of a beam.

A beam is a member that carries loads transversely, that is, perpendicular to its long axis.

When analyzing a beam to determine reactions, internal shearing forces, and internal bending moments, it is helpful to classify the manner of loading, the type of supports, and the type of beam.

Beams are subjected to a variety of loading patterns, including

Normal concentrated loads

Inclined concentrated loads

Uniformly distributed loads

Varying distributed loads

Concentrated moments

Support types include

Simple, roller-type support

Pinned support

Fixed support

Beam types include

Simply supported beams, or simple beams

Overhanging beams

Cantilever beams, or cantilevers

Compound beams

Continuous beams

Understanding all of these terms will help you communicate features of beam designs and perform the required analyses. Following are brief descriptions of each with illustrations to help you visualize them.

Loading Patterns

In later sections we will show that the nature of the loading pattern determines the variation of the shearing force and bending moment along the length of the beam. Here we define the five most frequently encountered loading patterns and give examples of each. More complex loading patterns can often be analyzed by considering them to be combinations of two or more of these basic types.

Normal concentrated loads

A normal concentrated load is one that acts perpendicular (normal) to the major axis of the beam at only a point or over a very small length of the beam.

Figure 11–2(a) shows the typical manner of representing a beam carrying normal concentrated loads. Each load is shown as a vector acting on the beam, perpendicular to its long axis. Part (b) is a sketch of a situation that would produce concentrated loads. The weight of the pipes and their contents determines the magnitudes of the loads.

(a) Schematic representation of
beam, loads, and reactions

(b) Pictorial representation
of beam and loads

FIGURE 11–2 Simple beam with normal concentrated loads.

While we often visualize loads acting downward due to gravity, actual loads can act in any direction. Particularly in mechanical machinery, forces produced by linkages, actuators, springs, clamps, and other devices can act in any direction. Figure 11–3 shows a simple example.

Normal concentrated loads tend to cause pure bending of the beam. Most of the problems in this chapter will feature this type of loading. The analysis of the bending stresses produced is presented in Chapter 12.

Inclined concentrated loads

An inclined concentrated load is one that acts effectively at a point but whose line of action is at some angle to the main axis of the beam.

Figure 11–4 shows an example of an inclined concentrated load. The inclined load exerted by the spring causes a combination of bending and axial stresses to be developed in the beam. Chapter 15 presents the analysis techniques for such a loading.

Uniformly distributed loads

Loads of constant magnitude acting perpendicular to the axis of a beam over a significant part of the length of the beam are called uniformly distributed loads.

An example of this type of load would be the weight of wet snow of uniform depth on a roof carried by a horizontal roof beam. Also, the materials that make up the roof structure itself are often uniformly distributed. Figure 11–5 illustrates such a loading pattern and shows the manner we use in this book to represent uniformly distributed loads in problems. The blue-shaded rectangular area defines the extent of the load along the length of the beam. The magnitude of the load is called w, in units of force per unit length. Typical units would be lb/in, kN/m, or K/ft. Recall that 1 K = 1 kip = 1000 lb.

For example, if the loading on the beam shown in Figure 11–5 was $w = 150$ lb/ft, you should visualize that each 1.0 ft of length of the beam carries 150 lb of load. If the span is 10.0 ft, the total load is 1500 lb.

Varying distributed loads

Loads of varying magnitude acting perpendicular to the axis of a beam over a significant part of the length of the beam are called varying distributed loads.

Figures 11–6 and 11–7 show examples of structures carrying varying distributed loads. When the load varies linearly, we quantify such loads by giving the value of w at each end of the sloped line representing the load. For more complex, nonlinear variations, other schemes of giving the magnitude of the load would have to be devised.

FIGURE 11–3 Machine lever as a simple beam carrying normal concentrated loads.

FIGURE 11–4   Machine lever as a simple beam carrying normal concentrated loads and an inclined load.

(a) Schematic representation of beam, load, and reactions

(b) Pictorial example

FIGURE 11–5   Simple beam with uniformly distributed load.

(a) Schematic representation of beam, load, reaction force, and moment with sample data

(b) Pictorial example—snow load on a projecting roof

FIGURE 11–6   Example of linearly varying distributed load on a cantilever.

(a) Schematic representation of beam, load, and reactions with sample data.

(b) Pictorial example—gravel on a platform

FIGURE 11–7 Example of linearly varying distributed load on a simple beam.

Concentrated Moments A moment is an action that tends to cause rotation of an object. Moments can be produced by a pair of parallel forces acting in opposite directions; this is called a *couple*. The action on a crank or a lever also produces a moment.

> *When a moment acts on a beam at a point in a manner that tends to cause it to undergo pure rotation, it is called a concentrated moment.*

Figure 11–8 shows an example of a concentrated moment. The forces acting at the ends of the vertical arms form a couple and tend to twist the beam into the shape shown. The fact that the two forces making up the couple are equal and opposite results in no net horizontal force being applied to the beam.

Concentrated moments can also be applied to a beam by any force acting parallel to its axis with a line of action some distance from the axis. This is illustrated in Figure 11–9, which represents a shaft carrying a helical gear. The force F acts parallel to the axis of its shaft and tends to bend

it in the plane of the page. The difference here is that there is an unbalanced horizontal force also applied to the beam that must be reacted to at one of the supports.

Support Types

All beams must be supported in a stable manner to hold them in equilibrium. All externally applied loads and moments must be reacted to by one or more supports. Different types of supports offer different types of reaction capability.

Simple support or roller support

A simple support is one that can resist only forces acting perpendicular to the beam.

One of the best illustrations of simple supports is the pair of theoretically frictionless rollers shown at the ends of the beam in Figure 11–10(a). They provide upward support against the downward action of the load on the beam. As the beam tends to bend under the influence of the applied loading and the reactions, the bending deformation would not be resisted by the rollers. But if there were any horizon-

(a) Schematic representation of horizontal part of compound beam carrying a concentrated moment

(b) Compound beam—pictorial example

FIGURE 11–8 Concentrated moment on a compound beam.

(a) Schematic representation of horizontal part of compound beam showing concentrated moment and horizontal reaction

(b) Compound beam—pictorial example

FIGURE 11–9 Concentrated moment on a compound beam.

(a) Beam on two rollers (b) Beam with pinned support (c) Free body diagram
and one roller for (a) or (b)

FIGURE 11–10 Examples of simple supports.

tal components of the load, the rollers would roll and the beam would be unrestrained. Therefore, using two rollers alone is not adequate.

Pinned Support

An example of a pinned support is a hinge that can resist forces in two directions but which allows rotation about the axis of the pin in the hinge.

Figure 11–10(b) shows the same beam as in Figure 11–10(a) with the roller at the left end replaced by a pinned support. This system would provide adequate support while allowing the beam to bend freely. Any horizontal force would be resisted in the pinned joint.

Fixed support

A fixed support is one that is held solidly such that it resists forces in any direction and also prohibits rotation of the beam at the support.

One manner of creating a fixed support is to produce a close-fitting, socket-like hole in a rigid structure into which the end of the beam is inserted. The fixed support resists moments as well as forces because it prevents rotation. Figure 11–11 shows two examples of the use of fixed supports.

Beam Types

The type of beam is indicated by the types of supports and their placement.

Simple Beam A simple beam is one that carries only loads acting perpendicular to its axis and that is supported only at its ends by simple supports acting perpendicular to the axis. Figure 11–1 is an example of a simple beam. When all loads act downward, the beam would deflect in the classic bent, concave upward shape. This is referred to as positive bending.

Overhanging Beam An overhanging beam is one in which the loaded beam extends outside the supports. Figure 11–12 is an example of an overhanging beam. The loads on the overhangs tend to bend them downward, producing negative bending.

Cantilever A cantilever has only one end supported, as shown in Figure 11–13 depicting a crane boom attached rigidly to a stiff vertical column. It is essential that the support be fixed because it must provide vertical support for the externally applied loads along with a moment reaction to resist the moment produced by the loads. Figures 11–6 and 11–11(a) are other examples of cantilevers.

(a) Schematic representation

(a) Schematic representation of fixed support for a cantilever

(c) Schematic representation of beam with two fixed supports

(b) Pictorial representation

(d) Pictorial representation

(b) Pictorial example

FIGURE 11–11 Beams with fixed supports.

FIGURE 11–12 Overhanging beam.

FIGURE 11–13 Cantilever beam.

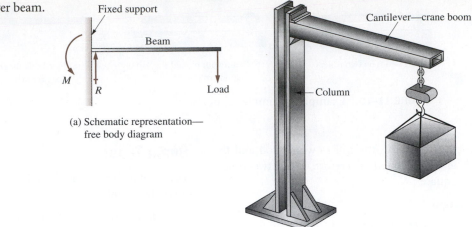

(a) Schematic representation—
free body diagram

(b) Pictorial example

Compound Beam   While the beams depicted so far were single, straight members, we will use the term *compound beam* to refer to one having two or more parts extending in different directions. Figures 11–8 and 11–9 are examples of compound beams. Such beams are typically analyzed in parts to determine the internal shearing forces and bending moments throughout. Often the place where one part joins another is a critical point of interest.

Continuous Beam   Except for the beam shown in Figure 11–11, parts (c) and (d), beams discussed before had one or two supports and only two unknown reactions. The principles of statics allow us to compute all reaction forces and moments from the classical equations of equilibrium because there are two unknowns and two independent equations available from which to solve for those unknowns. Such beams are called *statically determinate*. In contrast, *continuous beams* have additional supports, requiring different approaches to analyze the reaction forces and moments. These are called *statically indeterminate* beams. Figure 11–14 shows an example of a continuous beam on three supports. Figure 11–15 shows a beam with one fixed end and one simply supported end. Notice that there are two unknown forces and one unknown moment. Figures 11–11(c) and (d) show a beam with two fixed ends that is also statically indeterminate because there are two reaction forces and two moments that must be determined. Sections 11–12 and 14–6 present the analysis techniques for statically indeterminate beams.

11–3 REACTIONS AT SUPPORTS

The first step in analyzing a beam to determine its safety under a given loading arrangement is to show completely the loads and support reactions on a free body diagram. It is very important to be able to construct free body diagrams from the physical picture or description of the loaded beam. This was done in each case for Figures 11–1 through 11–15.

After constructing the free body diagram, it is necessary to compute the magnitude of all support reactions. It is assumed that the methods used to find reactions were studied previously. Therefore, only a few examples are shown here as a review and as an illustration of the techniques used throughout this book.

The following general procedure is recommended for solving for reactions on simple or overhanging beams.

GUIDELINES FOR SOLVING FOR REACTIONS ▼

1. Draw the free body diagram.
2. Use the equilibrium equation $\Sigma M = 0$ by summing moments about the point of application of one support reaction. The resulting equation can then be solved for the other reaction.
3. Use $\Sigma M = 0$ by summing moments about the point of application of the second reaction to find the first reaction.
4. Use $\Sigma F = 0$ to check the accuracy of your calculations.

FIGURE 11–14   Continuous beam on three supports.

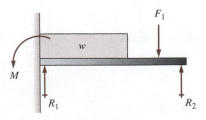

FIGURE 11–15   Supported cantilever.

Example Problem 11–1   Figure 11–16 shows the free body diagram for the beam carrying pipes, which was originally shown in Figure 11–2. Compute the reaction forces in the support rods.

FIGURE 11–16 Beam loading.

Solution   Objective: Compute the reaction forces at the ends of the beam.

Given: Pictorial of the beam shown in Figure 11–2. Free body diagram showing the loading is Figure 11–16. Loads are applied at points labeled B, C, D, and E. Reactions act at points A and F and are called R_A and R_F.

Analysis: The *guidelines for solving for reactions* will be used to compute the reactions. Figure 11–16 is the free body diagram, so we will start with Step 2.

Results: To find the reaction R_F, sum moments about points A.

$$\sum M_A = 0 = 3.5(400) + 4.3(800) + 1.2(1200) + 2.8(1500) - R_F(1800)$$

Note that all forces are in kN and distances in mm. Each moment term has the units of kN · mm. Now solve for R_F.

$$R_F = \frac{[3.5(400) + 4.3(800) + 1.2(1200) + 2.8(1500)]\text{kN} \cdot \text{mm}}{1800 \text{ mm}} = 5.82 \text{ kN}$$

Now, to find R_A, sum moments about point F.

$$\sum M_F = 0 = 2.8(300) + 1.2(600) + 4.3(1000) + 3.5(1400) - R_A(1800)$$

$$R_A = \frac{[2.8(300) + 1.2(600) + 4.3(1000) + 3.5(1400)]\text{kN} \cdot \text{mm}}{1800 \text{ mm}} = 5.98 \text{ kN}$$

Now use $\sum F = 0$ for the vertical direction as a check.

Downward forces: $(3.5 + 4.3 + 1.2 + 2.8) \text{ kN} = 11.8 \text{ kN}$

Upward reactions: $(5.82 + 5.98) \text{ kN} = 11.8 \text{ kN}$ (check)

Comment: Remember to show the reaction forces R_A and R_F at their proper points on the beam and in the proper directions.

Example Problem 11–2   Compute the reactions on the beam shown in Figure 11–17(a).

FIGURE 11–17 Beam loading.

Solution Objective: Compute the reaction forces at the ends of the beam.

Given: Pictorial of the beam shown in Figure 11–17(a). Distributed load of 2200 lb/ft is applied over 10 ft at the left end of the beam. Reactions act at points *A* and *C* and are called R_A and R_C.

Analysis: The *guidelines for solving for reactions* will be used to compute the reactions. Figure 11–17(b) is an equivalent free body diagram with the resultant of the distributed load shown acting at the centroid of the load.

Results: $\sum M_A = 0 = 22\,000\,\text{lb}(5\,\text{ft}) - R_C(12\,\text{ft})$

$$R_C = \frac{22\,000\,\text{lb}(5\,\text{ft})}{12\,\text{ft}} = 9167\,\text{lb}$$

$\sum M_C = 0 = 22\,000\,\text{lb}(7\,\text{ft}) - R_A(12\,\text{ft})$

$$R_A = \frac{22\,000\,\text{lb}(7\,\text{ft})}{12\,\text{ft}} = 12\,833\,\text{lb}$$

Finally, as a check, in the vertical direction,

$$\sum F = 0$$

Downward forces: 22 000 lb

Upward forces: $R_A + R_C = 12\,833 + 9167 = 22\,000\,\text{lb}$ (check)

Comment: *Note that the resultant is used only to find reactions.* Later, when we find shearing forces and bending moments, the distributed load itself must be used.

Example Problem 11–3 Compute the reactions for the overhanging beam in Figure 11–18.

FIGURE 11–18 Beam loading.

Solution Objective: Compute the reaction forces at points *B* and *D*.

Given: Loading on the beam shown in Figure 11–18. Reactions are R_B and R_D.

Analysis: The *guidelines for solving for reactions* will be used to compute the reactions.

Results: First, summing moments about point *B*,

$$\sum M_B = 0 = 1000(200) - R_D(250) + 1200(400) - 800(100)$$

Notice that forces that tend to produce clockwise moments about *B* are considered positive in this calculation. Now solving for R_D gives

$$R_D = \frac{[1000(200) + 1200(400) - 800(100)]\text{N}\cdot\text{mm}}{250\,\text{mm}} = 2400\,\text{N}$$

Summing moments about *D* will allow computation of R_B.

$$\sum M_D = 0 = 1000(50) - R_B(250) + 800(350) - 1200(150)$$

$$R_B = \frac{[1000(50) + 800(350) - 1200(150)]\text{N}\cdot\text{mm}}{250\,\text{mm}} = 600\,\text{N}$$

Check with $\Sigma F = 0$ in the vertical direction:

Downward forces: $(800 + 1000 + 1200)\text{N} = 3000\,\text{N}$

Upward forces: $R_B + R_D = (600 + 2400)\text{N} = 3000\,\text{N}$ (check)

Comment: In summary, the left reaction, R_B, is 600 N and the right reaction, R_D, is 2400 N.

Example Problem   Compute the reactions for the cantilever beam in Figure 11–19.
11–4

FIGURE 11–19   Beam
loading.

(a)

(b) Free body diagram

Solution   Objective: Compute the reactions at point *A* at the wall.

Given: Loading on the beam shown in Figure 11–19(a).

Analysis: The *guidelines for solving for reactions* will be used.
   In the case of cantilever beams, the reactions at the wall are composed of an upward force R_A that must balance all downward forces on the beam and a reaction moment M_A that must balance the tendency for the applied loads to rotate the beam. These are shown in Figure 11–19(b). Also shown is the resultant, 60 kN, of the distributed load.

Results: Then, by summing forces in the vertical direction, we obtain

$$R_A = 60\,\text{kN} + 4\,\text{kN} = 64\,\text{kN}$$

Summing moments about point *A* yields

$$M_A = 60\,\text{kN}(1.0\,\text{m}) + 4\,\text{kN}(2.5\,\text{m}) = 70\,\text{kN} \cdot \text{m}$$

Comment: The reaction at the support for the cantilever shown in Figure 11–19 includes a vertical force, $R_A =$ 64 kN and a counterclockwise moment, $M_A = 70\,\text{kN} \cdot \text{m}$.
   Again, note that the resultant of the distributed load is used only for finding reactions. The original distributed load acting over its entire length must be considered when analyzing shearing forces and bending moments in later sections.

11–4 SHEARING FORCES AND BENDING MOMENTS FOR CONCENTRATED LOADS

Chapters 12 and 13 show that two kinds of stresses are developed in a beam, *shearing stresses* and *bending stresses*. Shearing stress is developed within the beam because of internal shearing forces that tend to shear, or cut, the beam. Bending stress is developed by internal bending moments that tend to bend the beam into a curved shape. The internal shearing forces and bending moments are produced in reaction to external forces and moments applied to the beam. Recall what you did in the activity.

Prior to studying the actual shearing stresses and bending stresses, this chapter helps you to visualize and to compute the values for the internal shearing forces and bending moments themselves. Each is dependent on the nature of the applied loads on the beam and the way it is supported.
   We define shearing forces as follows:

Shearing forces are internal forces developed in the material of a beam to balance externally applied forces in order to secure equilibrium of all parts of the beam.

While internal shearing forces can act in any direction, we need to consider first those that act perpendicular to the long axis of the beam. Because we typically visualize beams to be oriented horizontally and to carry loads acting vertically downward, these internal shearing forces will usually act vertically. Thus, we refer to them as *vertical shearing forces*, indicated by the symbol, *V*. Of course, beams can actually be oriented in any direction as illustrated in Figures 11–3 and 11–4.

Bending moments are defined as follows:

> *Bending moments are internal moments developed in the material of a beam to balance the tendency for external forces to cause rotation of any part of the beam.*

Bending moments cause the beam to assume its characteristic curved, or "bent," shape. As you did in the activity at the start of this chapter, it is desirable to use a flexible flat beam to visualize the gross behavior of a beam in response to different loading patterns and support conditions.

Figure 11–20 shows a simply supported beam with a single concentrated load acting at the midpoint of its length, similar to that shown in Figure 11–1 and used in the activity. Our goal now is to develop data and observations from which we can prepare complete plots of the variations in shearing forces and bending moments across the entire length of the beam. We use the free body diagram approach as we did in the activity, considering segments equally spread along the length of the beam at points *A*, *B*, *C*, *D*, and *E*. Parts (b), (c), (d), and (e) of Figure 11–20 show the

(a) Beam loading

(b)

(c)

(d)

(e)

FIGURE 11–20   Free body diagrams used to find shearing forces and bending moments.

resulting free body diagrams, using only segments to the left of the cuts at the given positions.

The diagrams in parts (b) and (c) result in vertical shearing forces acting downward at the cut section. We now establish the following sign convention for shearing forces:

> Internal shearing forces acting *downward* are considered to be *positive*.
> Internal shearing forces acting *upward* are considered to be *negative*.

We can also generalize the method used to find the magnitude of the shearing force at any section of a beam as follows:

> The magnitude of the shearing force in any part of a beam is equal to the algebraic sum of all external forces acting to the left of the section of interest.

As you study parts (b) and (c) of Figure 11–20, it should be clear that if you consider a cut section anywhere between position *A* and position *C*, the internal shearing force would be 500 N. The general rule demonstrated here is

> On any segment of a beam where no loads are applied, the value of the shearing force remains constant.

Note that diagrams (c) and (d) are both taken at position *C*. However, one is just before the point of application of the 1000-N force, and the other is just beyond where the force is applied. The result is that the internal vertical shearing force abruptly changes direction (sign) from (downward) positive to (upward) negative at *C* where the concentrated force is applied. The following rule can now be stated:

> A concentrated load (or reaction) on a beam causes an abrupt change in the shearing force in the beam by an amount equal to the magnitude of the load and in the direction of the load.

Now let's consider how bending moments vary at different points along the beam's length. If we cut the beam at a position quite close to *A* at the left support, the moment inside the beam would be zero because, even though the reaction force acts there, the moment arm for this force is zero. This leads to the following rule for bending moments:

> The bending moments at the ends of a simply supported beam are zero.

We now establish the following sign convention for bending moments:

> Counterclockwise internal bending moments are considered to be positive. Clockwise internal bending moments are considered to be negative.

Study diagrams (b), (c), (d), and (e) in Figure 11–20 and demonstrate to yourself how the values of the bending moment were determined by summing moments about the position of the cut section as was done in the activity earlier. The general rule is then

> The magnitude of the internal bending moment at any section of a beam is equal to the algebraic sum of the moments, taken with respect to the cut section, of all forces acting to the left of the section of interest.

We can summarize the results from the analyses at the various cut sections of the beam.

Section A: $V_A = 500\ \text{N}$ $M_A = 0$

Section B: $V_B = 500\ \text{N}$ $M_B = 250\ \text{N} \cdot \text{m}$

Section C to left of load: $V_C = 500\ \text{N}$ $M_C = 500\ \text{N} \cdot \text{m}$

Section C to right of load: $V_C = -500\ \text{N}$ $M_C = 500\ \text{N} \cdot \text{m}$

Section D: $V_D = -500\ \text{N}$ $M_D = 250\ \text{N} \cdot \text{m}$

Section E: $V_E = -500\ \text{N}$ $M_E = 0\ \text{N} \cdot \text{m}$

It is useful to plot these values on diagrams as shown in Figure 11–21. The top diagram is called the *load diagram*. The middle one is the *shearing force diagram* and the lower one is the *bending moment diagram*. The three diagrams should be aligned as shown to emphasize the relationships of the shearing forces and the bending moments to positions along the length of the beam. *You should also label the vertical axes as shown to communicate to any user of the diagrams exactly what values are being plotted and in what units they are expressed.*

Review, now, the various rules given in shaded boxes during this analysis. Note how the effects of those rules are illustrated in the diagrams of Figure 11–21. Also notice that the bending moment plots as a straight line with a positive slope between A and C. It then plots as a straight line with a negative slope between C and E. These observations lead to other general rules:

> The plot of the bending moment curve will be a straight line over those segments of the beam where the shearing force has a constant value. If the shearing force in the segment is positive, the bending moment curve will have a constant positive slope. Conversely, if the shearing force in the segment is negative, the bending moment curve will have a constant negative slope.

One more rule can be developed that allows the preparation of the bending moment diagram directly from the data in the shearing force diagram. We state the rule first and then we apply it to the beam loading shown in Figure 11–21. Then we show the mathematical basis for the rule.

FIGURE 11–21 Shearing force and bending moment diagrams.

The change in bending moment between two points on a beam is equal to the area under the shearing force curve between the same two points.

In most cases, the areas required to implement this rule are simply computed. This rule, called the *area rule,* can be applied over a segment of any length in a beam to determine the change in bending moment.

Consider the segment *A–C* in Figure 11–21. The shearing force is a constant value of 500 N over the 1.0-m length of this segment. Then the area under the shearing force curve is

$$\text{Area}_{A-C} = (500\,\text{N})(1.0\,\text{m}) = 500\,\text{N} \cdot \text{m}$$

We first note that the bending moment at *A* at the left end of the beam is zero based on a rule developed earlier. Then the *change in bending moment* from *A* to *C* is 500 N · m. Then

$$M_C = M_A + \text{Area}_{A-C} = 0 + 500\,\text{N} \cdot \text{m} = 500\,\text{N} \cdot \text{m}$$

We can plot the values of M_A and M_C on the bending moment diagram. Then an earlier rule stated that the bending moment curve is a straight line with a positive slope when the shearing force is a constant positive value, as it is in the segment *A–C*. You can then simply draw the straight line from 0 at *A* to 500 N · m at *C*.

Continuing to the right half of the beam, segment *C–E*, we compute the area to be

$$\text{Area}_{C-E} = (-500\,\text{N})(1.0\,\text{m}) = -500\,\text{N} \cdot \text{m}$$

This value is the *change in bending moment* from *C* to *E*. However, the bending moment at C starts as 500 N · m. Then

$$M_E = M_C + \text{Area}_{C-E} = 500\,\text{N} \cdot \text{m} - 500\,\text{N} \cdot \text{m} = 0\,\text{N} \cdot \text{m}$$

Also demonstrated here is the rule that the bending moment at the ends of a simply supported beam is zero.

Plotting M_C and M_E and joining them with a straight line with a negative slope completes the bending moment diagram.

After some practice, you should be able to prepare the three beam diagrams by directly applying the rules developed in this section without drawing the individual free body diagrams.

Development of the Area Rule for Bending Moment Diagrams

In general, the area rule can be stated as

$$dM = V\,dx \qquad (11\text{–}1)$$

where dM is the change in moment due to a shearing force V acting over a small length segment dx. Figure 11–22 illustrates Equation (11–1).

Over a larger length segment, the process of integration can be used to determine the total change in moment over the segment. Between two points, *A* and *C*,

$$\int_{M_A}^{M_C} dM = \int_{x_A}^{x_C} V\,dx \qquad (11\text{–}2)$$

If the shearing force V is constant over the segment, as it is in segment *A–C* in Figure 11–22, Equation (11–2) becomes

$$\int_{M_A}^{M_C} dM = V\int_{x_A}^{x_C} dx \qquad (11\text{–}3)$$

FIGURE 11–22 Shearing force and bending moment diagrams— area rule.

Completing the integration gives

$$M_C - M_A = V(x_C - x_A) \qquad (11\text{--}4)$$

This result is consistent with the rule stated previously. Note that $M_C - M_A$ is the *change* in moment between points A and C. The right side of Equation (11–4) is the area under the shearing force curve between A and C.

11–5 GUIDELINES FOR DRAWING BEAM DIAGRAMS FOR CONCENTRATED LOADS

The principles developed and illustrated in the preceding section can be generalized into a set of guidelines that can be used to draw the load, shearing force, and bending moment diagrams for virtually any simply supported beam carrying concentrated loads.

The guidelines are listed in the following shaded box. You should review them now and relate them to the rules developed in Section 11–4. Then we will apply the guidelines to new and more complex problems.

Guidelines for drawing the load, shearing force, and bending moment diagrams for simply supported beams carrying only concentrated loads.

A. Load diagram.
   1. Draw the beam and show the loads and reaction forces acting on it.
   2. Label points on the beam where loads or reactions act with letters.
   3. Solve for the magnitudes and directions of the reaction forces and show them on the load diagram.
B. Shearing force diagram.
   1. Draw vertical and horizontal axes for the diagram in relation to the drawing of the beam-loading diagram in a manner similar to that shown in Figure 11–21.
   2. Label the vertical axis as *Shearing force, V,* and give the units for force.
   3. Project lines from each applied load or reaction force on the beam loading diagram down to the shearing force diagram. Label points of interest for reference. Label points where loads or reaction forces act as done in the load diagram.

4. Construct the shearing force graph by starting at the left end of the beam and proceeding to the right, applying the following rules.
   a. *Shearing force diagrams start and end at zero at the ends of the beam.*
   b. *A concentrated load or reaction on a beam causes an abrupt change in the shearing force in the beam by an amount equal to the magnitude of the load or reaction and in the same direction.*
   c. *On any segment of the beam where no loads are applied, the value of the shearing force remains constant, resulting in a straight horizontal line on the shearing force diagram.*
   d. Show the value of the shearing force at key points on the diagram, generally at points where loads or reactions act.
C. Bending moment diagram.
   1. Draw vertical and horizontal axes for the diagram in relation to the shearing force diagram in a manner similar to that shown in Figure 11–21.
   2. Label the vertical axis as *Bending Moment, M,* and give the units for moment.
   3. Project vertical lines at points of interest from the shearing force diagram, including all points used for the shearing force diagram, down to the bending moment axis.
   4. Construct the bending moment graph by starting at the left end of the beam and proceeding to the right, applying the following rules.
      a. *At the ends of a simply supported beam, the bending moment is zero.*
      b. *The change in bending moment between two points on a beam is equal to the area under the shearing force curve between those two points. Thus, when the area under the shearing force curve is positive (above the axis), the bending moment is increasing, and vice versa.*
      c. *The maximum bending moment occurs at a point where the shearing force curve crosses its zero axis.*
      d. *On a section of the beam where no loads are applied, the bending moment diagram will be a straight line.*
      e. *The slope of the bending moment curve at any point is equal to the magnitude of the shearing force at that point.*
      f. Label the values for the bending moment on the graph at each point of interest.

Example Problem 11–5   Draw the complete shearing force and bending moment diagrams for the beam shown in Figure 11–23.

FIGURE 11–23 Beam loading.

Solution Objective: Draw the complete shearing force and bending moment diagrams.

Given: The beam loading shown in Figure 11–23. It is a simply supported beam carrying all concentrated loads. This is the same beam for which the reaction forces were computed in Example Problem 11–1. The results of Part A of the guidelines are shown in Figure 11–23.

Analysis: The *guidelines for drawing shearing force and bending moment diagrams for simply supported beams carrying only concentrated loads* will be used.

FIGURE 11–24 Shearing force diagram.

Results Part B: The completed shearing force diagram is shown in Figure 11–24. The process used to determine each part of the diagram is described.

Point A: The reaction R_A is encountered immediately. Then

$$V_A = R_A = 5.98 \, \text{kN}$$

Between A and B: Because there are no loads applied, the shearing force remains constant. That is,

$$V_{A-B} = 5.98 \, \text{kN}$$

Point B: The applied load of 3.5 kN causes an abrupt decrease in V.

$$V_B = 5.98 \, \text{kN} - 3.5 \, \text{kN} = 2.48 \, \text{kN}$$

Between B and C: The shearing force remains constant.

$$V_{B-C} = 2.48 \, \text{kN}$$

Point C: The applied load of 4.3 kN causes an abrupt decrease in V.

$$V_C = 2.48 \, \text{kN} - 4.3 \, \text{kN} = -1.82 \, \text{kN}$$

Between C and D: The shearing force remains constant.

$$V_{C-D} = -1.82 \, \text{kN}$$

Point D: The applied load of 1.2 kN causes an abrupt decrease in V.

$$V_D = -1.82 \, \text{kN} - 1.2 \, \text{kN} = -3.02 \, \text{kN}$$

Between D and E: The shearing force remains constant.

$$V_{D-E} = -3.02 \text{ kN}$$

Point E: The applied load of 2.8 kN causes an abrupt decrease in V.

$$V_E = -3.02 \text{ kN} - 2.8 \text{ kN} = -5.82 \text{ kN}$$

Between E and F: The shearing force remains constant.

$$V_{E-F} = -5.82 \text{ kN}$$

Point F: The reaction force of 5.82 kN causes an abrupt increase in V.

$$V_F = -5.82 \text{ kN} + 5.82 \text{ kN} = 0$$

Comment: Note that values for the shearing forces at key points are shown right on the diagram at those points.

Results: **Part C:**   The bending moment diagram is drawn directly below the shearing force diagram so the relationship between points on the beam loading diagram can be related to both. See Figure 11–25. It is helpful to extend vertical lines from the points of interest on the beam (points A to F in this example) down to the horizontal axis of the bending moment diagram.

It is most convenient to start drawing the bending moment diagram at the left end of the beam and work toward the right, considering each segment separately. For beams like this one having all concentrated loads, the segments should be chosen as those for which the shearing force is constant. In this case, the segments are AB, BC, CD, DE, and EF.

At Point A: We use the rule that the bending moment is zero at the ends of a simply supported beam. That is, $M_A = 0$.

Point B: For this and each subsequent point, we apply the area rule. The general pattern is

$$M_B = M_A + [\text{Area}]_{AB}$$

where $[\text{Area}]_{AB}$ = area under the shearing force curve between A and B. Using data from the shearing force curve,

$$[\text{Area}]_{AB} = V_{AB} \times \text{Width of the segment } AB$$

FIGURE 11–25   Shearing force and bending moment diagrams.

But, $V_{AB} = 5.98$ kN over the segment AB that is 0.40 m long. Then, as shown in Figure 11–25,

$$[\text{Area}]_{AB} = 5.98 \text{ kN}(0.40 \text{ m}) = 2.39 \text{ kN} \cdot \text{m}$$

Finally, then,

$$M_B = M_A + [\text{Area}]_{AB} = 0 + 2.39 \text{ kN} \cdot \text{m} = 2.39 \text{ kN} \cdot \text{m}$$

This value is plotted at point B on the bending moment diagram. Then a straight line is drawn from M_A to M_B because the shearing force is constant over that segment. Bending moment values at C, D, E, and F are found in a similar manner. Refer to Figure 11–25.

Point C:
$$M_C = M_B + [\text{Area}]_{BC}$$
$$[\text{Area}]_{BC} = 2.48 \text{ kN}(0.40 \text{ m}) = 0.99 \text{ kN} \cdot \text{m}$$
$$M_C = 2.39 \text{ kN} \cdot \text{m} + .99 \text{ kN} \cdot \text{m} = 3.38 \text{ kN} \cdot \text{m}$$

Point D:
$$M_D = M_C + [\text{Area}]_{CD}$$
$$[\text{Area}]_{CD} = -1.82 \text{ kN}(0.40 \text{ m}) = -0.73 \text{ kN} \cdot \text{m}$$
$$M_D = 3.38 \text{ kN} \cdot \text{m} - 0.73 \text{ kN} \cdot \text{m} = 2.65 \text{ kN} \cdot \text{m}$$

Note that $[\text{Area}]_{CD}$ is negative because it is below the axis.

Point E:
$$M_E = M_D + [\text{Area}]_{DE}$$
$$[\text{Area}]_{DE} = -3.02 \text{ kN}(0.30 \text{ m}) = -0.91 \text{ kN} \cdot \text{m}$$
$$M_E = 2.65 \text{ kN} \cdot \text{m} - 0.91 \text{ kN} \cdot \text{m} = 1.74 \text{ kN} \cdot \text{m}$$

Point F:
$$M_F = M_E + [\text{Area}]_{EF}$$
$$[\text{Area}]_{EF} = -5.82 \text{ kN}(0.30 \text{ m}) = -1.74 \text{ kN} \cdot \text{m}$$
$$M_F = 1.74 \text{ kN} \cdot \text{m} - 1.74 \text{ kN} \cdot \text{m} = 0 \text{ kN} \cdot \text{m}$$

Summary and Comment: The bending moment values are shown on the diagram at their respective points so users of the diagram can see the relative values. The fact that $M_F = 0$ is a check on the calculations because the rule for simply supported beams states that the bending moment at F must be zero. The objective of drawing the bending moment diagram is often to locate the point where the maximum bending moment exists. Here we see that $M_C = 3.38$ kN · m is the maximum value.

Example Problem 11–6

Draw the complete shearing force and bending moment diagrams for the beam shown in Figure 11–26.

Solution **Objective:** Draw the complete shearing force and bending moment diagrams.

Given: The beam loading shown in Figure 11–26. It is an overhanging beam carrying all concentrated loads. This is the same beam for which the reaction forces were computed in Example Problem 11–3. The results of Part A of the guidelines are shown in Figure 11–26. $R_B = 600$ N. $R_D = 2400$ N.

Analysis: The *guidelines for drawing shearing force and bending moment diagrams for beams carrying only concentrated loads* will be used.

FIGURE 11–26 Beam loading.

FIGURE 11–27 Shearing force and bending moment diagrams for Example Problem 11–6.

Results: The steps of Part B of the guidelines were used to prepare the completed shearing force diagram shown in Figure 11–27. The process is described here.

Point A: The applied downward force of 800 N causes the shearing force diagram to drop immediately to −800 N. That is, $V_A = -800$ N.

Between A and B: No loads are applied in this segment. Then $V_{A-B} = -800$ N.

Point B: The upward reaction force of 600 N acts at B. Then

$$V_B = -800\,\text{N} + 600\,\text{N} = -200\,\text{N}$$

Between B and C: No loads are applied in this segment. Then $V_{B-C} = -200$ N.

Point C: The applied downward force of 1000 N causes the shearing force diagram to drop immediately by that amount. Then

$$V_C = -200\,\text{N} - 1000\,\text{N} = -1200\,\text{N}$$

Between C and D: No loads are applied in this segment. Then $V_{C-D} = -1200$ N.

Point D: The upward reaction force of 2400 N acts at D. Then

$$V_D = -1200\,\text{N} + 2400\,\text{N} = +1200\,\text{N}$$

Between D and E: No loads are applied in this segment. Then $V_{D-E} = +1200$ N.

Point E: The applied downward force of 1200 N causes the shearing force diagram to drop immediately by that amount. Then

$$V_E = 1200\,\text{N} - 1200\,\text{N} = 0\,\text{N}$$

The steps of Part C of the guidelines were used to prepare the completed bending moment diagram shown in Figure 11–27. The process is described here.

At Point A: This is a free end of the beam. Therefore, $M_A = 0$ N·m.
We apply the area rule for each subsequent point on the beam.

Point B: $M_B = M_A + [\text{Area}]_{AB} = 0 + (-800\,\text{N})(0.10\,\text{m}) = -80\,\text{N}\cdot\text{m}$

Point C: $M_C = M_B + [\text{Area}]_{BC} = -80\,\text{N}\cdot\text{m} + (-200\,\text{N})(0.20\,\text{m}) = -120\,\text{N}\cdot\text{m}$

Point D: $M_D = M_C + [\text{Area}]_{CD} = -120\,\text{N}\cdot\text{m} + (-1200\,\text{N})(0.05\,\text{m}) = -180\,\text{N}\cdot\text{m}$

Point E: $M_E = M_D + [\text{Area}]_{DE} = -180\,\text{N}\cdot\text{m} + (1200\,\text{N})(0.15\,\text{m}) = 0\,\text{N}\cdot\text{m}$

Summary and Comment:
When seeking the maximum bending moment, we really look for the maximum absolute value. The maximum bending moment is $-180\,\text{N}\cdot\text{m}$ at point *D*. Note that the resulting bending moment diagram for this problem lies below the axis. This is indicative of *negative bending* for which the beam would bend in a concave downward fashion. The forces at A and E at the ends of the overhangs are the cause. Contrast this with the result of Example Problem 11–5 for which the entire bending moment diagram is above the axis, indicating a positive bending moment. That beam would bend in a concave upward fashion.

11–6 SHEARING FORCES AND BENDING MOMENTS FOR DISTRIBUTED LOADS

We now consider beams that carry uniformly distributed loads and analyze them to determine the shearing force and bending moment that exist in any part of the beam. Let's use the free body diagram approach similar to that used in Section 11–5 for beams with only concentrated loads.

Figure 11–28 shows a simply supported beam carrying a uniformly distributed load of 1500 N/m over a 6-m length from *A* to *B*. The 3-m length from *B* to *C* carries no load. A method similar to that in Example Problem 11–2 was used to determine the magnitude of the reaction forces at *A* and *C* as shown. You should check your ability to verify these values.

FIGURE 11–28 Beam loading.

Now refer to Figure 11–29 for a set of four free body diagrams of portions of the beam in length increments of 2.0 m. Any increment could have been chosen. The 2.0-m increments provide a sufficient number of points to eventually plot graphs for the shearing force and bending moment diagrams using a method similar to that used in the preceding sections. But the shape of the curves will be

FIGURE 11–29 Free body diagrams used to find shearing forces and bending moments.

different because of the distributed load. Follow along with the development of the values for shearing force and bending moment. Then we generalize the method of producing the diagrams directly without the use of free body diagrams.

Part (a) of Figure 11–29: This is the free body diagram of the first 2.0-m segment. The reaction force, $R_A = 6000$ N, acts at the left end of the beam, point A. Then we determine that the 2.0-m length of the 1500 N/m distributed load exerts a resultant force of 3000 N and it can be visualized to act midway on the segment. Then

At A:

$$\text{Shearing force} = V_A = 6000 \text{ N}$$
$$\text{Bending moment} = M_A = 0$$

At the Cut Section:

$$V_2 = 6000 \text{ N} - 3000 \text{ N} = 3000 \text{ N (downward)}$$
$$M_2 = (6000 \text{ N})(2.0 \text{ m}) - (3000 \text{ N})(1.0 \text{ m})$$
$$= 9000 \text{ N} \cdot \text{m (CCW)}$$

The subscript 2 indicates that the cut section is at 2.0 m from point A.

Part (b) of Figure 11–29: This is the segment having a length of 4.0 m. The resultant of the distributed load is 6000 N shown acting midway on the segment. Then

At the Cut Section:

$$V_4 = 6000 \text{ N} - 6000 \text{ N} = 0$$
$$M_4 = (6000 \text{ N})(4.0 \text{ m}) - (6000 \text{ N})(2.0 \text{ m})$$
$$= 12\,000 \text{ N} \cdot \text{m (CCW)}$$

Part (c) of Figure 11–29: This 6.0-m length has a resultant load of 9000 N acting at 3.0 m from A. This is the entire distributed load.

At the Cut Section:

$$V_6 = 6000 \text{ N} - 9000 \text{ N} = -3000 \text{ N (upward)}$$
$$M_6 = (6000 \text{ N})(6.0 \text{ m}) - (9000 \text{ m})(3.0 \text{ m})$$
$$= 9000 \text{ N} \cdot \text{m (CCW)}$$

Part (d) of Figure 11–29: This 8.0-m length of the beam carries the entire distributed load having a resultant of 9000 N acting 3.0 m from A as before. There are no loads acting on the final 2.0 m of the segment.

At the Cut Section:

$$V_8 = 6000 \text{ N} - 9000 \text{ N} = -3000 \text{ N (upward)}$$
$$M_8 = (6000 \text{ N})(8.0 \text{ m}) - (9000 \text{ N})(5.0 \text{ m})$$
$$= 3000 \text{ N} \cdot \text{m (CCW)}$$

Remember that the bending moment at C at the right end of the beam is zero.

Now we can plot these points as shown in Figure 11–30. We can make some observations about the shape of the shearing force and bending moment curves on the part of the beam that carries the distributed load.

FIGURE 11–30 Complete beam loading, shearing force, and bending moment diagrams.

- The shearing force varies linearly with position. The slope of the straight line is equal to the load per unit length of the distributed load. That is, the slope is -1500 N/m. You should observe that each 1.0-m length of the beam carries a load of 1500 N.

- The bending moment increases along a curve over the first 4.0-m length of the beam while the shearing force is positive. Then it decreases along a curve for the next 2.0 m where the shearing force is negative.

These observations are consistent with the guidelines for drawing beam diagrams in Section 11–6. We could have used the *area rule* to produce the values for bending moment at any point. See Rule 4b in Section C of the guidelines. For example, consider the first 4.0-m length of the beam. The shearing force starts at 6000 N at A and decreases to zero at the 4.0-m point. The area under that part of the curve is all positive, above the axis of the shearing force diagram. Therefore, the bending moment will increase from zero at the free end of the beam at A by an amount equal to the area of the triangle with a height of 6000 N and a length of 4.0 m. That is,

$$M_4 = M_A + [\text{Area}]_{0-4} = 0 + (1/2)(4.0 \text{ m})(6000 \text{ N})$$
$$= 12\,000 \text{ N} \cdot \text{m}$$

We could have used the area rule to find the value of the bending moment at B where the distributed load ends. We called this value M_6. Note that the shearing force decreases from zero to -3000 N over this length. Then the bending moment at B is

$$M_B = M_6 = M_4 + [\text{Area}]_{4-6}$$
$$= 12\,000 \text{ N} \cdot \text{m} + (1/2)(2.0 \text{ m})(-3000 \text{ N})$$
$$= 9000 \text{ N} \cdot \text{m}$$

The final segment of the beam diagrams can be drawn using the guidelines given before because there are no loads applied over the last 3.0 m of the beam. The shearing force is a constant value of -3000 N. Therefore, the bending moment plots as a straight line with a constant negative slope. The bending moment at point 9 is

$$M_9 = M_6 + [\text{Area}]_{6-9} = 9000\ \text{N} \cdot \text{m}$$
$$+ (-3000\ \text{N})(3.0\ \text{m}) = 0$$

We now state new rules that can be appended to the guidelines for drawing shearing force and bending moment diagrams.

Whenever uniformly distributed loads act on a beam:

1. Over the length of a beam carrying a uniformly distributed load, the shearing force curve is a straight line having a negative slope equal to the amount of the load per unit length.

2. The change in shearing force between any two points on a beam where a uniformly distributed load acts is equal to the resultant of the distributed load over that length. That is, the change in shearing force is

   $\Delta V = (\text{Load per unit length})(\text{Length of the segment})$

3. Over the length of a beam carrying a uniformly distributed load, the bending moment plots as a curved line where the slope of the line at any point is equal to the magnitude of the shearing force at that point.

---

4. The change in bending moment between any two points on a beam is equal to the area under the shearing force curve between those two points.

Let's discuss Rule 3, defining the slope of the bending moment curve. Refer to Figure 11–30 again and observe the shape of the bending moment curve.

Starting at the left end of the beam, point A, we observe that the shearing force at that point is 6000 N, the largest value on the beam. Rule 3 states that the bending moment curve should have very high positive slope at that point, and that is how the curve is drawn.

Now consider point 2, 2.0 m from A. The magnitude of the shearing force has decreased to 3000 N. Therefore, the slope of the bending moment curve should be less steep but still positive. In fact, if we progress from A to point 4, the slope of the bending moment curve should start with a high positive slope and gradually decrease in slope to zero at point 4 where the shearing force crosses the axis. Recall that a line drawn tangent to a point on a curve having a zero slope is horizontal. This identifies the peak of the curve and the value of the bending moment there is a maximum for that particular segment. This logic illustrates Rule 4c in Section C of the guidelines from before. *The maximum bending moment occurs at a point where the shearing force crosses its zero axis.* It is necessary for you to find the point where the shearing force curve crosses the axis. The example problem that follows illustrates the process.

Example Problem 11–7   Draw the complete shearing force and bending moment diagrams for the beam shown in Figure 11–31.

Solution   **Objective:**   Draw the complete shearing force and bending moment diagrams.

Given: Beam loading in Figure 11–31. The beam has a concentrated load at the end of an overhang to the left of the left support and a uniformly distributed load over the span between the two supports.

Analysis: We will use the guidelines for drawing beam diagrams first presented in Section 11–5 and supplemented with additional guidelines for distributed loads developed in this section.

Results: The completed beam diagrams are shown in Figure 11–32. The details are explained next.

Reactions at Supports: The resultant of the distributed load is

$$(2500\ \text{N/m})(3.0\ \text{m}) = 7500\ \text{N}$$

FIGURE 11–31   Beam loading for Example Problem 11–7.

FIGURE 11–32   Results for Example Problem 11–7.

The resultant acts at the midpoint of the distributed load, 1.5 m from B.

$$\Sigma M_B = 0 = (7500\,\text{N})(1.5\,\text{m}) - (1200\,\text{N})(1.2\,\text{m}) - R_C(3.0\,\text{m})$$
$$R_C = 3270\,\text{N upward}$$
$$\Sigma M_C = 0 = (7500\,\text{N})(1.5\,\text{m}) + (1200\,\text{N})(4.2\,\text{m}) - R_B(3.0\,\text{m})$$
$$R_B = 5430\,\text{N upward}$$

Shearing Force Diagram: Only a concentrated load acts from A to B. The shearing force is a constant negative 1200 N to the point just before the application of the reaction force at B.

At B: The reaction at B causes the shearing force to rise to

$$V_B = -1200\,\text{N} + 5430\,\text{N} = 4230\,\text{N}$$

Between B and C: The shearing force curve decreases 2500 N over each 1.0 m, a total of 7500 N for the 3.0-m span. Then

$$V_C = V_B - 7500\,\text{N} = 4230\,\text{N} - 7500\,\text{N} = -3270\,\text{N}$$

The shearing force curve is a straight line with a constant negative slope of −2500 N/m, equal to the amount of load per unit length applied to the beam.

At C: The reaction force at C, 3270 N, brings the shearing force curve back to zero at C.

At D Where the Shearing Force Curve Crosses the Axis: It is necessary to find the location of point D because that is where the maximum bending moment on the beam occurs. The distance from B to D is called x_1 in the figure. This is the distance required for the shearing force to decrease from $V_B = 4230\,\text{N}$ to zero.

We use the fact that the curve is decreasing 2500 N over each 1.0 m of distance from B in a straight-line fashion. Let's define the magnitude of the distributed load as w. Then

$$w = 2500\,\text{N/m}$$

We can solve for x_1 from

$$x_1 = \frac{V_B}{w} = \frac{4230\,\text{N}}{2500\,\text{N/m}} = 1.692\,\text{m}$$

It is appropriate to determine the distance from D to C now also because it will be needed for the bending moment calculations. This is called x_2 in the figure.

$$x_2 = 3.0\,\text{m} - x_1 = 3.0\,\text{m} - 1.692\,\text{m} = 1.308\,\text{m}$$

This completes the development of the shearing force curve.

Bending Moment Diagram:

Between A and B: The moment at A is zero because it is at the free end of the beam. The change in moment from A to B is equal to the rectangular area under the shearing force curve between A and B.

$$M_B = M_A + [\text{Area}]_{A-B} = 0 + (-1200\,\text{N})(1.2\,\text{m}) = -1440\,\text{N} \cdot \text{m}$$

The curve is a straight line because the shearing force has a constant value in the segment from A to B.

Between B and D: We need to determine the bending moment at D because that will be the maximum value anywhere on the beam. The change in moment from B to D is equal to the triangular area under the shearing force curve between B and D. But the moment has the value of $-1440\,\text{N} \cdot \text{m}$ at B. Therefore, noting that the base of the triangle is $x_1 = 1.692\,\text{m}$,

$$M_D = M_B + [\text{Area}]_{B-D} = -1440\,\text{N} \cdot \text{m} + (1/2)(4230\,\text{N})(1.692\,\text{m}) = 2139\,\text{N} \cdot \text{m}$$

Study carefully the shape of the curve. At B the shearing force has a high positive value. It then decreases to zero at D. Then the curve starts at B with a high positive slope as indicated by the tangent to the curve. The slope decreases steadily until it becomes zero (horizontal) at D.

Between D and C: The triangular area under the shearing force curve between D and C is the change in bending moment between those two points. Then, noting that the height of the triangle is $-3270\,\text{N}$ and the base is $x_2 = 1.308\,\text{m}$,

$$M_C = M_D + [\text{Area}]_{D-C} = 2139\,\text{N} \cdot \text{m} + (1/2)(-3270\,\text{N})(1.308\,\text{m}) = 0\,\text{N} \cdot \text{m}$$

This is actually a check on the calculations because the bending moment at C, a free end of the simply supported beam, *must* be zero. If any other result is found, there has been some error in the calculations.

The shape of the curve from D to C should, again, be studied carefully. At point D the value of the shearing force is zero, and so the bending moment curve at D has a zero slope. Therefore, it continues from the end of the previous curve drawn from B to D. Then the shearing force becomes negative and gets increasingly negative as it progresses to point C. The curve is concave downward, ending with a rather steep negative slope.

This completes the development of the bending moment diagram.

Comment: You should observe some key data from Figure 11–32. You should note the maximum values of the shearing force and the bending moment and where they occur. In summary, then,

Maximum shearing force: $V_B = 4230\,\text{N}$
located at point B at the left support.

Maximum bending moment: $M_D = 2139\,\text{N} \cdot \text{m}$
located 1.692 m from the left support.

For some beam types, it is important to know both the maximum positive and the maximum negative values of the bending moment because the material or the shape of the beam may perform differently under each type of moment. In this beam,

Maximum negative bending moment: $M_B = -1440\,\text{N} \cdot \text{m}$

11-7 GENERAL SHAPES FOUND IN BENDING MOMENT DIAGRAMS

You will likely need practice to become proficient at drawing shearing force and bending moment diagrams and applying the guidelines described in Sections 11–4 to 11–6. One technique that is helpful is to learn the general relationships between the shape of the bending moment curve and the shearing force curve in the same segment of the beam.

Figure 11–33 should help you to visualize those relationships. One key skill is to apply correctly guideline Rule 4e in Section C listed in the shaded box in Section 11–5. That guideline states

> *The slope of the bending moment curve at any point is equal to the magnitude of the shearing force at that point.*

We have seen examples of many of these shapes for segments of the bending moment diagrams in Figures 11–21, 11–25, 11–27, 11–30, and 11–32.

For example, the shapes from parts 1 and 3 of Figure 11–33 appear in any beam having only concentrated loads such as Figures 11–21, 11–25, and 11–27. These same shapes apply for parts of beams that have only concentrated loads such as the last segment of Figure 11–30 and the first segment of Figure 11–32. In each of these cases, the shearing force has a constant value, resulting in a bending moment curve that has a constant slope. That is, the bending moment curve will be a straight line.

Parts 4, 5, 6, and 7 of Figure 11–33 apply to segments of a beam carrying uniformly distributed loads. For example, the first 6 m of the beam in Figure 11–30 looks like parts 5 and 6. Similarly, in Figure 11–32, the 3.0-m length between the supports looks like parts 5 and 6. In each case, the uniformly distributed load causes the shearing force to

decrease by a constant amount per unit length, producing a shearing force curve that starts with a positive value and continues as a straight line with a negative slope. The corresponding bending moment diagram starts with a fairly large positive slope and then curves with a concave downward shape as shown in part 5. If the shearing force curve crosses the zero axis, the bending moment curve reaches a maximum value where the slope is zero. Then, if the shearing force continues into a negative region, as in part 6, the corresponding bending moment will have an increasingly negative slope.

In most applications, and in problems in this book, it is adequate to sketch the general shape of the shearing force and bending moment curves and to indicate the magnitudes at key points of interest. Typically these are the following:

- Points where concentrated loads are applied
- Points where distributed loads begin and end
- At any point where a maximum or minimum value of shearing force or bending moment may occur

11-8 SHEARING FORCES AND BENDING MOMENTS FOR CANTILEVER BEAMS

The manner of support of a cantilever beam causes the analysis of its shearing forces and bending moments to be somewhat different from that for simply supported beams. The most notable difference is that, at the place where the beam is supported, it is fixed and can therefore resist moments. Thus, at the fixed end of the beam, the bending moment is not zero, as it was for simply supported beams. In fact, the bending moment at the fixed end of the beam is usually the *maximum.*

FIGURE 11–33   General shapes for moment curves relating to corresponding shearing force curves.

FIGURE 11–34 Beam loading and reactions.

Consider the cantilever beam shown in Figure 11–34. Earlier, in Example Problem 11–4, it was shown that the support reactions at point A are a vertical force $R_A = 64$ kN and a counterclockwise moment $M_A = 70$ kN $\cdot$ m. These are equal to the values of the shearing force and bending moment at the left end of the beam. According to convention, the upward reaction force R_A is positive, and the counterclockwise moment M_A is negative, giving the starting values for the shearing force diagram and bending moment diagram shown in Figure 11–35. The rules developed earlier about shearing force and bending moment diagrams can then be used to complete the diagrams.

The shearing force decreases in a straight-line manner from 64 kN to 4 kN in the interval A to B. Note that the change in shearing force is equal to the amount of the distributed load, 60 kN. The shearing force remains constant from B to C, where no loads are applied. The 4-kN load at C returns the curve to zero.

The bending moment diagram starts at -70 kN $\cdot$ m because of the reaction moment M_A. Between points A and B, the curve has a positive but decreasing slope (type 5 in Figure 11–33). The change in moment between A and B is equal to the area under the shearing force curve be-

tween A and B. The area is a combination of a rectangle and a triangle.

$$[\text{Area}]_{A-B} = 4 \text{ kN}(2 \text{ m}) + \frac{1}{2}(60 \text{ kN})(2 \text{ m}) = 68 \text{ kN} \cdot \text{m}$$

Then the moment at B is

$$M_B = M_A + [\text{Area}]_{A-B} = -70 \text{ kN} \cdot \text{m} + 68 \text{ kN} \cdot \text{m}$$
$$= -2 \text{ kN} \cdot \text{m}$$

Finally, from B to C,

$$M_C = M_B + [\text{Area}]_{B-C} = -2 \text{ kN} \cdot \text{m} + 4 \text{ kN}(0.5 \text{ m}) = 0$$

Since point C is a *free* end of the beam, the moment there must be zero.

11–9 BEAMS WITH LINEARLY VARYING DISTRIBUTED LOADS

Figures 11–6 and 11–7 in Section 11–2 show two examples of beams carrying linearly varying distributed loads. We demonstrate here the method of drawing the general shape of the shearing force and bending moment diagrams for such beams and how to determine the magnitude of the maximum shearing force and maximum bending moment. For many practical problems these are the primary objectives. Later, in Section 11–11 a mathematical approach is shown that yields a more complete definition of the shape of the shearing force and bending moment diagrams.

Refer now to Figure 11–36, which shows the load diagram for the cantilever from Figure 11–6. The amount of loading per unit length varies linearly from $w = -200$ lb/ft (downward) at the support point A to $w =$ zero at the right end B. This straight-line curve is called a *first-degree curve* because the loading varies directly with position on the beam, x. For such a loading, the reaction at A, called R_A, is the resultant of the total distributed load, found by computing the area under the triangular-shaped load curve. That is,

$$R_A = \tfrac{1}{2}(-200 \text{ lb/ft})(8 \text{ ft}) = -800 \text{ lb}$$

The bending moment at the support, called M_A, must equal the moment of all of the applied load to the right of A. This can be found by considering the resultant to act at the centroid of the distributed load. For the triangular-shaped load

FIGURE 11–35 Complete load, shearing force, and bending moment diagrams.

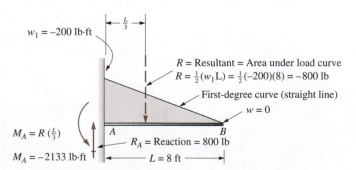

FIGURE 11–36 Load diagram, reaction, and moment for cantilever carrying a linearly varying distributed load.

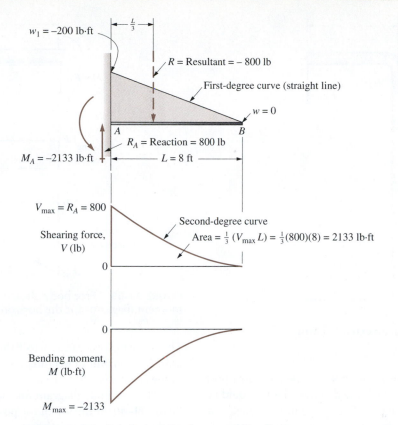

FIGURE 11–37   Load, shearing force, and bending moment diagrams for beam loading in Figure 11–35.

curve, the centroid is 1/3 of the length of the beam from point A. Calling this distance x, we can say

$$x = L/3 = (8\,\text{ft})/3 = 2.667\,\text{ft}$$

Then the moment at A is the product of the resultant times x. That is,

$$M_A = R_A x = (800\,\text{lb})(2.667\,\text{ft}) = 2133\,\text{lb} \cdot \text{ft}$$

These values, $R_A = 800$ lb and $M_A = 2133$ lb·ft, are the maximum values for shearing force and bending moment, respectively. In most cases, that is the objective of the analysis. If so, the analysis can be concluded.

However, if the shapes of the shearing force and bending moment diagrams are desired, they can be sketched using the principles developed earlier in this chapter. Figure 11–37 shows the results. The shearing force diagram starts at A with the value of 800 lb, equal to the reaction, R_A. The value of the shearing force then decreases for points to the right of A as additional loads are applied. Note that the shearing force curve is not a straight line because the amount of loading decreases as we proceed from A toward B. At B the loading is zero, resulting in the zero value of shearing force at B. The slope of the shearing force curve at any point is equal to the amount of loading per unit length at the corresponding point on the load diagram. Thus, the shearing force curve starts with a relatively large negative slope at A and then a progressively smaller negative slope as we approach B. This is generally called a *second-degree curve* because the value varies with the *square* of the distance x.

The bending moment diagram can be sketched by first noting that $M_A = -2133$ lb · ft. The curve has a relatively large positive slope at A because of the large positive value for shearing force at that point. Then the slope progressively decreases, with increasing distance, to zero at point B. The fact that the value of bending moment equals zero at B can be demonstrated, also, by computing the area under the shearing force curve from A to B. Appendix A–1 includes the formulas for computing the area under a second-degree curve of the type shown in the shearing force diagram. That is,

$$\text{Area} = (1/3)(800\,\text{lb})(8\,\text{ft}) = 2133\,\text{lb} \cdot \text{ft}$$

This is the *change* in bending moment from A to B, bringing the bending moment curve to zero at B.

11–10 FREE BODY DIAGRAMS OF PARTS OF STRUCTURES

Examples considered thus far have been for generally straight beams with all transverse loads, that is, loads acting perpendicular to the main axis of the beam. Many machine elements and structures are more complex, having parts that extend away from the main beam-like part.

For example, consider the simple post with an extended arm, shown in Figure 11–38, composed of vertical and horizontal parts. The vertical post is rigidly secured at its base. At the end of the extended horizontal arm, a downward load is applied. An example of such a loading is a support system

FIGURE 11–38 Post with an extended arm.

FIGURE 11–39 Free body, shearing force, and bending moment diagrams for the horizontal arm.

for a sign over a highway. Another would be a support post for a basketball hoop in which the downward force could be a player hanging from the rim after a slam dunk. A mechanical design application is a bracket supporting machine parts during processing.

Under such conditions, it is convenient to analyze the structure or machine element by considering each part separately and creating a free body diagram for each part. At the actual joints between parts, one part exerts forces and moments on the other. Using this approach, you will be able to design each part on the basis of how it is loaded, using the basic principles of beam analysis in this chapter and those that follow.

Post with an Extended Arm

The objective of the analysis is to draw the complete shearing force and bending moment diagrams for the horizontal and vertical parts of the post/arm structure shown in Figure 11–38. The first step is to "break" the arm from the post at the right-angle corner.

Figure 11–39 shows the horizontal arm as a free body with the applied load, F, acting at its right end. The result appears similar to the cantilever analyzed earlier in this chapter. We know that the arm is in equilibrium as a part of the total structure and, therefore, it must be in equilibrium when considered by itself. Then, at the left end where it joins the vertical post, there must be a force equal to F acting vertically upward to maintain the sum of the vertical forces equal to zero. But the two vertical forces form a couple that tends to rotate the arm in a clockwise direction. To maintain rotational equilibrium, there must be a counterclockwise moment internal to the arm at its left end with a magnitude of $M = F \cdot a$, where a is the length of the arm. Having completed the free body diagram, the shearing force and bending moment diagrams can be drawn as shown in Figure 11–39.

The shearing force is equal to F throughout the length of the arm. The maximum bending moment occurs at the left end of the arm where $M = F \cdot a$.

The free body diagram for the vertical post is shown in Figure 11–40. At the top of the post, a downward force and a clockwise moment are shown, exerted on the vertical post by the horizontal arm. *Notice the action–reaction pair that exists at joints between parts. Equal but oppositely directed forces and moments act on the two parts.* Completing the free body diagram for the post requires an upward force and a counterclockwise moment at its lower end, provided by the attachment means at its base. Finally, Figure 11–40 shows the shearing force and bending moment diagrams for the post, drawn vertically to relate the values to positions on the post. No shearing force exists because there are no transverse

FIGURE 11–40 Free body, shearing force, and bending moment diagrams for the vertical post.

FIGURE 11–41   Beam with an L-shaped bracket.

forces acting on the post. Where no shearing force exists, no change in bending moment occurs and there is a uniform bending moment throughout the post.

Beam with an L-Shaped Bracket

Figure 11–41 shows an L-shaped bracket extending below the main beam carrying an inclined force. The main beam is supported by simple supports at A and C. Support C is designed to react to any unbalanced horizontal force. The objective is to draw the complete shearing force and bending moment diagrams for the main beam and the free body diagrams for all parts of the bracket.

Three free body diagrams are convenient to use here: one for the horizontal part of the bracket, one for the vertical part of the bracket, and one for the main beam itself. But first it is helpful to resolve the applied force into its vertical and horizontal components, as indicated by the dashed vectors at the end of the bracket.

Figure 11–42 shows the three free body diagrams. Starting with the part DE shown in (a), the applied forces at E must be balanced by the oppositely directed forces at D for

FIGURE 11–42   Free body diagrams. (a) Free body diagram for part DE. (b) Free body diagram for part BD. (c) Free body diagram for part ABC, the main beam.

equilibrium in the vertical and horizontal directions. But rotational equilibrium must be produced by an internal moment at D. Summing moments with respect to point D shows that

$$M_D = F_{Ey} \cdot d = (16.4 \, \text{kN})(0.6 \, \text{m}) = 9.84 \, \text{kN} \cdot \text{m}$$

In Figure 11–42(b) the forces and moments at D have the same values but opposite directions from those at D in part (a) of the figure. Vertical and horizontal equilibrium conditions show the forces at B to be equal to those at D. The moment at B can be found by summing moments about B as follows:

$$\left(\sum M \right)_B = 0 = M_D - F_{Dx} \cdot c - M_B$$

Then

$$\begin{aligned} M_B &= M_D - F_{Dx} \cdot c = 9.84 \, \text{kN} \cdot \text{m} - (11.5 \, \text{kN})(0.4 \, \text{m}) \\ &= 5.24 \, \text{kN} \cdot \text{m} \end{aligned}$$

Now the main beam ABC can be analyzed. Use Figure 11–42(c). The forces and moment are shown applied at B with the values taken from point B on part BD. We must now solve for the reactions at A and C. First, summing moments about point C yields

$$\left(\sum M \right)_C = 0 = F_{By} \cdot b - F_{Ay} \cdot (a + b) - M_B$$

Notice that the moment M_B applied at B must be included. Solving for F_{Ay} gives

$$\begin{aligned} F_{Ay} &= \frac{(F_{By} \cdot b) - M_B}{a + b} = \frac{(16.4 \, \text{kN})(1.2 \text{m}) - 5.24 \, \text{kN} \cdot \text{m}}{2.0 \, \text{m}} \\ &= 7.22 \, \text{kN} \end{aligned}$$

Similarly, summing moments about point A gives

$$\left(\sum M \right)_A = 0 = F_{By} \cdot a - F_{Cy} \cdot (a + b) + M_B$$

Notice that the moment M_B applied at B is positive because it acts in the same sense as the moment due to F_{By}. Solving for F_{Cy} gives

$$\begin{aligned} F_{Cy} &= \frac{(F_{By} \cdot a) - M_B}{a + b} = \frac{(16.4 \, \text{kN})(0.8 \, \text{m}) + 5.24 \, \text{kN} \cdot \text{m}}{2.0 \, \text{m}} \\ &= 9.18 \, \text{kN} \end{aligned}$$

A check on the calculation for these forces can be made by summing forces in the vertical direction and noting that the sum equals zero.

The completion of the free body diagram for the main beam requires the inclusion of the horizontal reaction at C equal to the horizontal force at B.

Figure 11–43 shows the shearing force and bending moment diagrams for the main beam ABC. The shearing force diagram is drawn in the conventional manner with changes in shearing force occurring at each point of load application. The difference from previous work is in the moment diagram.

The following steps were used:

1. The moment at A equals zero because A is a simple support.

2. The increase in moment from A to B equals the area under the shearing force curve between A and B, $5.78 \, \text{kN} \cdot \text{m}$.

FIGURE 11–43 Shearing force and bending moment diagrams for main beam of Figure 11–41.

3. At point B the moment M_B is considered to be a *concentrated* moment resulting in an abrupt change in the value of the bending moment by the amount of the applied moment, 5.24 kN · m, thus resulting in the peak value of 11.02 kN · m. The convention used here is the following:

---

  a. When a concentrated moment is *clockwise,* the moment diagram *rises.*

  b. When a concentrated moment is *counterclockwise,* the moment diagram *drops.*

---

4. Between B and C, the moment decreases to zero because of the negative shearing force and the corresponding negative area under the shearing force curve.

11–11 MATHEMATICAL ANALYSIS OF BEAM DIAGRAMS

For most practical problems, the preparation of the load, shearing force, and bending moment diagrams using the techniques shown earlier in this chapter are adequate and convenient. A wide variety of beam types and loadings can be analyzed with sufficient detail to permit the logical design of the beams to ensure safety and to limit deflections to acceptable values. The methods of accomplishing these objectives are presented in Chapters 12–16.

However, there are some types of loading and some types of design techniques that can benefit from the representation of the load, shearing force, and bending moment diagrams by mathematical equations. This section presents the methods of creating such equations.

The following guidelines are for writing sets of equations that completely define the load, shearing force, and bending moment as a function of the position on the beam.

GUIDELINES FOR WRITING BEAM DIAGRAM EQUATIONS ▼

1. Draw the load diagram showing all externally applied loads and reactions.

2. Compute the values for all reactions.

3. Label points along the length of the beam where concentrated loads are applied or where distributed loads begin or end.

4. Draw the shearing force and bending moment diagrams using the techniques shown earlier in this chapter, noting values at the critical points defined in Step 3.

5. Establish conventions for denoting positions on the beam and signs of shearing forces and bending moment. In most cases, we will use the following convention (see Figures 11–20 to 11–32).
   a. Position on the beam will be denoted by the variable x measured from the left end of the beam.
   b. Downward loads will be negative.

  c. A positive shearing force is one that acts downward within the beam at a given section. An alternative way of determining this is to analyze the net external vertical force acting on that part of the beam to the left of the section of interest. If the net external force is upward, the internal shearing force in the beam is positive.

  d. A positive bending moment is one that acts counterclockwise within the beam at a given section. A positive bending moment will tend to cause a beam to bend in a concave upward shape, typical of a simply supported beam carrying downward loads between the supports.

6. Consider separately each segment of the beam between the points defined in Step 3. The shearing force curve should be continuous within each segment.

7. If the shearing force diagram consists of all straight lines caused by concentrated or uniformly distributed loads, the fundamental principles of analytic geometry can be used to write equations for the shearing force versus position on the beam for each segment. The resulting equations will be of the form

$$V_{AB} = \text{Constant} \quad \text{(zero-degree equation)}$$
$$V_{BC} = ax + b \quad \text{(first-degree equation)}$$

The subscripts define the beginning and end of the segment of interest.

8. If the shearing force diagram contains segments that are curved caused by varying distributed loads, first write equations for the load versus position on the beam. Then derive the equations for the shearing force versus position on the beam from

$$V_{AB} = \int w_{AB}\,dx + C$$

where w_{AB} is the equation for load in the segment AB as a function of x, and C is a constant of integration. The resulting shearing force equation will be of the second degree or higher, depending on the complexity of the loading pattern. Compute the value of the constants of integration using known values of V at given locations x.

9. Derive equations for the bending moment as a function of position on the beam for each segment, using the method.

$$M_{AB} = \int V_{AB}\,dx + C$$

This is the mathematical equivalent of the *area rule* for beam diagrams used earlier because the process of integration determines the area under the shearing force curve. Compute the value of the constants of integration using known values of M at given locations x.

10. The result at this point is a set of equations for shearing force and bending moment for each segment of the beam. It would be wise to check the equations for accuracy by substituting key values of x for which the shearing force and bending moment are known into the equations to ensure that the correct values for V and M are computed.

11. Determine the maximum values of shearing force and bending moment if they are not already known by substituting values for x in the appropriate equations where the maximum values are expected. Recall the rule that the maximum bending moment will occur at a point where the shearing force curve crosses the x-axis—that is, where $V = 0$.

FIGURE 11–44 Simply supported beam with a concentrated load.

This procedure is demonstrated by the following four examples.

Simply Supported Beam with a Concentrated Load

The objective is to write the equations for the shearing force and bending moment diagrams for the beam and loading shown in Figure 11–44, using the guidelines given in this section.

Steps 1 to 4 have been completed and shown in Figure 11–44. Points of interest are labeled A at the left support, B at the point of application of the load, and C at the right support. Equations will be developed for the two segments, AB and BC, where AB applies from $x = 0$ to $x = 3$ m and BC applies from $x = 3$ m to $x = 5$ m. Steps 5 and 6 will be as defined in the guidelines.

Step 7 can be applied to write the equations for the shearing force curve as follows:

$$V_{AB} = 16$$
$$V_{BC} = -24$$

The units for shearing force are taken to be kN.

Step 8 does not apply to this example.

Step 9 is now applied to derive the equations for the bending moment in the two segments.

$$M_{AB} = \int V_{AB}\,dx + C = \int 16\,dx + C = 16x + C$$

To evaluate the constant of integration C, we can note that at $x = 0$, $M_{AB} = 0$. Substituting these values into the moment equation gives

$$0 = 16(0) + C$$

Then $C = 0$. The final equation can now be written as

$$M_{AB} = 16x$$

As a check, we can see that at $x = 3$ m, the bending moment $M_B = 48$ kN · m, as shown in the bending moment diagram. We can now derive the equation for the bending moment in the segment BC.

$$M_{BC} = \int V_{BC}\,dx + C = \int -24\,dx + C = -24x + C$$

To evaluate C for this segment, we can use the condition at $x = 5$, $M_{BC} = 0$. Then

$$0 = -24(5) + C$$

Then $C = 120$. The final equation is

$$M_{BC} = -24x + 120$$

To check this equation, substitute $x = 3$.

$$M_{BC} = -24(3) + 120 = -72 + 120 = 48 \quad \text{(check)}$$

In summary, the equations for the shearing force and bending moment diagrams are

In the segment AB from $x = 0$ to $x = 3$ m:

$$V_{AB} = 16$$
$$M_{AB} = 16x$$

In the segment BC from $x = 3$ m to $x = 5$ m:

$$V_{BC} = -24$$
$$M_{BC} = -24x + 120$$

The maximum values for the shearing force and bending moment are obvious from the diagrams.

$$V_{max} = -24 \text{ kN throughout the segment } BC$$
$$M_{max} = 48 \text{ kN · m at point } B\,(x = 3 \text{ m})$$

FIGURE 11–45 Simply supported beam with a partial uniformly distributed load.

Simply Supported Beam with a Partial Uniformly Distributed Load

The objective is to write the equations for the shearing force and bending moment diagrams for the beam and loading shown in Figure 11–45, using the guidelines given in this section. Note that this is the same beam and loading that was shown earlier in Figure 11–30.

Steps 1 to 4 have been completed and shown in Figure 11–45. Points of interest are labeled A at the left support, B at the point where the distributed load ends, and C at the right support. Equations will be developed for the two segments, AB and BC, where AB applies from $x = 0$ to $x = 6$ m and BC applies from $x = 6$ m to $x = 9$ m.

Step 5(b) can be used to write an equation for the loading in the segment AB:

$$w_{AB} = -1500 \text{ N/m}$$

Step 7 can be applied to write the equations for the shearing force curve. In the segment AB, the curve is a straight line, so we can write it in the form

$$V_{AB} = ax + b$$

where a is the slope of the line and b is the intercept of the line with the V axis at $x = 0$. A convenient way to determine the slope is to observe that the slope is equal to the amount of loading per unit length for the distributed load. That is, $a = -1500$ N/m. The value of the intercept b can be observed from the shearing force diagram; $b = 6000$ N. Then the final form of the shearing force equation is

$$V_{AB} = -1500x + 6000$$

We can check the equation by substituting $x = 6$ m and computing V_B.

$$V_{AB} = -1500(6) + 6000 = -3000 = V_B$$

This checks with the known value of shearing force at point B.

Note that we could have used Step 8 to determine the equation for V_{AB}. Note that

$$w_{AB} = -1500 \text{ N/m}$$

Then

$$V_{AB} = \int w_{AB}\, dx + C = \int -1500\, dx + C$$
$$= -1500x + C$$

The value of C can be found by substituting $V_{AB} = 6000$ at $x = 0$.

$$6000 = -1500(0) + C$$

Then $C = 6000$. Finally,

$$V_{AB} = -1500x + 6000$$

This is identical to the previous result.

In the segment BC the shearing force is a constant value,

$$V_{BC} = -3000$$

Before proceeding to determine the equations for the bending moment diagram, recall that a critical point occurs where the shearing force curve crosses the zero axis. That will be a point of maximum bending moment. Let's call this point D and find the value of x_D where $V = 0$ by setting the equation for V_{AB} equal to zero and solving for x_D.

$$V_{AB} = 0 = -1500x_D + 6000$$
$$x_D = 6000/1500 = 4.0 \text{ m}$$

We will use this value later to find the bending moment at D.

Step 9 of the guidelines is now used to determine the equations for the bending moment diagram. First, in the segment AB,

$$M_{AB} = \int V_{AB} \, dx + C = \int (-1500\,x + 6000) \, dx + C$$

$$M_{AB} = -750x^2 + 6000x + C$$

To evaluate C, note that at $x = 0$, $M_{AB} = 0$. Then $C = 0$. And

$$M_{AB} = -750x^2 + 6000x$$

We can check the equation by finding M_B at $x = 6$ m.

$$M_B = -750(6)^2 + 6000(6) = 9000 \quad \text{(check)}$$

Also, we need the value of the maximum moment at D where $x = 4.0$ m.

$$M_D = -750(4)^2 + 6000(4) = 12\,000 \quad \text{(check)}$$

For the segment BC:

$$M_{BC} = \int V_{BC} \, dx + C = \int -3000 \, dx + C$$

$$= -3000x + C$$

But, at $x = 9$ m, $M_{BC} = 0$. Then

$$0 = -3000(9) + C$$

and $C = 27\,000$. Finally,

$$M_{BC} = -3000x + 27\,000$$

Check this equation at point B where $x = 6$ m.

$$M_B = -3000(6) + 27\,000 = -18\,000 + 27\,000$$

$$= 9000 \quad \text{(check)}$$

In summary, the equations for the shearing force and bending moment diagrams follow. In the segment AB from $x = 0$ to $x = 6$ m:

$$V_{AB} = -1500x + 6000$$

$$M_{AB} = -750x^2 + 6000x$$

In the segment BC from $x = 6$ m to $x = 9$ m:

$$V_{BC} = -3000$$

$$M_{BC} = -3000x + 27\,000$$

The maximum values for the shearing force and bending moment are obvious from the diagrams.

$$V_{max} = 6000 \text{ N at the left end at point } A$$

$$M_{max} = 12\,000 \text{ N} \cdot \text{m at point } D \, (x = 4 \text{ m})$$

Cantilever with a Varying Distributed Load

The objective is to write the equations for the shearing force and bending moment diagrams for the beam and loading shown in Figure 11–46, using the guidelines given in this section. Note that this is the same beam and loading shown earlier in Figure 11–37.

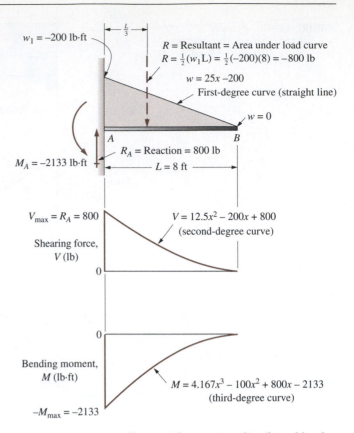

FIGURE 11–46   Cantilever with a varying distributed load.

There will be only one segment for this example, encompassing the entire length of the beam, because the load, shearing force, and bending moment curves are continuous.

We must first write an equation for the loading that varies linearly from a value of -200 lb/ft at the left end at A to zero at point B where $x = 8$ ft. You must note that the loading is shown on top of the beam acting downward in the manner we are accustomed to viewing loads. But the downwardly directed load is actually negative. To aid in writing the equation, you could redraw the loading diagram in the form of a *graph* of load versus position x, as shown in Figure 11–47. Then we can write the equation for the straight line,

$$w_{AB} = ax + b$$

The slope a can be evaluated by the ratio of the change in w over a given distance x. Using the entire length of the beam gives

$$a = \frac{w_1 - w_2}{x_1 - x_2} = \frac{-200 - 0}{0 - 8} = 25$$

FIGURE 11–47   Alternate display of loading on the beam in Figure 11–46.

The value of $b = -200$ can be observed from the diagram in Figure 11–47. Then the final equation for the loading is

$$w_{AB} = 25x - 200$$

We can check this equation by evaluating w at $x = 8$ ft at the end of the beam.

$$w_{AB} = 25(8) - 200 = 0 \quad \text{(check)}$$

Now we can derive the equation for the shearing force diagram.

$$V_{AB} = \int w_{AB}\,dx + C = \int (25x - 200)\,dx + C$$
$$= 12.5x^2 - 200x + C$$

Use the condition that at $x = 0$, $V_{AB} = 800$ to evaluate C.

$$800 = 12.5(0)^2 - 200(0) + C$$

Then $C = 800$. And the final equation for the shearing force is

$$V_{AB} = 12.5x^2 - 200x + 800$$

We can check this equation by evaluating V at $x = 8$ ft at the end of the beam.

$$V_{AB} = 12.5(8)^2 - 200(8) + 800 = 0 \quad \text{(check)}$$

Now we can derive the equation for the bending moment diagram.

$$M_{AB} = \int V_{AB}\,dx + C = \int (12.5x^2 - 200x + 800)\,dx + C$$
$$M_{AB} = 4.167x^3 - 100x^2 + 800x + C$$

Using the condition that at $x = 0$, $M_{AB} = -2133$, we can evaluate C.

$$-2133 = 4.167(0)^3 - 100(0)^2 + 800(0) + C$$

Then $C = -2133$. The final equation for bending moment is

$$M_{AB} = 4.167x^3 - 100x^2 + 800x - 2133$$

We can check this equation by evaluating M at $x = 8$ ft at the end of the beam.

$$M_{AB} = 4.167(8)^3 - 100(8)^2 + 800(8) - 2133$$
$$= 0 \quad \text{(check)}$$

In summary, the equations for the load, shearing force, and bending moment diagrams for the beam shown in Figure 11–46 are

$$w_{AB} = 25x - 200$$
(a first-degree curve; straight line)

$$V_{AB} = 12.5x^2 - 200x + 800$$
(a second-degree curve)

$$M_{AB} = 4.167x^3 - 100x^2 + 800x - 2133$$
(a third-degree curve)

Simply Supported Beam with a Varying Distributed Load

The objective is to write the equations for the shearing force and bending moment diagrams for the beam and loading

shown in Figure 11–48, using the guidelines given in this section. A pictorial of one manner of creating this loading pattern is shown in Figure 11–7.

Because of the symmetry of the loading, the two reactions will be of equal magnitude. Each will be equal to the area under one-half of the load diagram. Breaking that into a rectangle 0.2 kN/m high by 2.30 m wide and a triangle 1.0 kN/m high and 2.30 m wide, we can compute:

$$R_A = R_C = (0.2)(2.30) + 0.5(1.0)(2.30)$$
$$= 0.46 + 1.15 = 1.61 \text{ kN}$$

The general shapes of the shearing force and bending moment diagrams are sketched in Figure 11–48. We can reason that the shearing force curve crosses the zero axis at the middle of the beam at $x = 2.30$ m. Therefore, the maximum bending moment will occur at that point also. In principle, the magnitude of the maximum bending moment is equal to the area under the shearing force curve between points A and B. However, the calculation of that area is difficult because the curve is of the second degree and it does not start at its vertex. Then the formulas in Appendix A–1 cannot be used directly. This is one reason for developing the equations for the shearing force and bending moment diagrams.

Let's first write the equation for the load on the left half of the beam from A to B. The amount of loading per unit length starts at -0.20 kN/m (downward) and increases in magnitude to -1.20 kN/m. Again, as was done in the preceding example, it may help to draw the load diagram as a graph, as shown in Figure 11–49. Then we can write the equation for the straight line in the form

$$w_{AB} = ax + b$$

The slope a can be evaluated by the ratio of the change in w over a given distance x. Using half the length of the beam gives

$$a = \frac{w_1 - w_2}{x_1 - x_2} = \frac{-0.20 - (-1.20)}{0 - 2.30} = -0.4348$$

The value of $b = -0.20$ can be observed from the diagram in Figure 11–48. Then the final equation for the loading is

$$w_{AB} = ax + b = -0.4348x - 0.20$$

We can check this equation at $x = 2.30$ at the middle of the beam.

$$w_{AB} = -0.4348x - 0.20 = -0.4348(2.30) - 0.20$$
$$= -1.20 \text{ kN/m} \quad \text{(check)}$$

Now we can derive the equation for the shearing force diagram for the segment AB.

$$V_{AB} = \int w_{AB}\,dx + C = \int (-0.4348x - 0.20)\,dx + C$$
$$= -0.2174x^2 - 0.20x + C$$

Use the condition that at $x = 0$, $V_{AB} = 1.61$ to evaluate C. Then $C = 1.61$ and the final form of the equation is

$$V_{AB} = -0.2174x^2 - 0.20x + 1.61$$

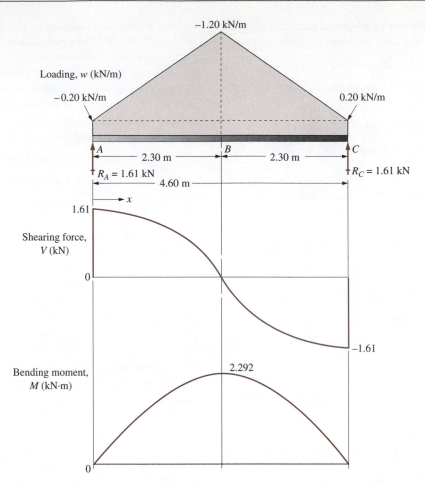

FIGURE 11–48 Simply supported beam with a varying distributed load.

We can check this equation at the middle of the beam by substituting $x = 2.30$ m.

$$V_{AB} = -0.2174(2.30)^2 - 0.20(2.30) + 1.61 = 0 \quad \text{(check)}$$

Now, we can derive the equation for the bending moment diagram.

$$M_{AB} = \int V_{AB}\, dx + C$$

$$= \int (-0.2174x^2 - 0.20x + 1.61)\, dx + C$$

$$M_{AB} = -0.07246x^3 - 0.10x^2 + 1.61x + C$$

Using the condition that at $x = 0$, $M_{AB} = 0$, we can evaluate $C = 0$. And

$$M_{AB} = -0.07246x^3 - 0.10x^2 + 1.61x$$

Checking at $x = 2.30$ m, gives $M_B = 2.292$ kN · m. The equations for the right side of the diagrams could be derived similarly. However, because of the symmetry of the diagrams, the completion of the curves is the mirror image of those already computed. In summary, the equations for the

FIGURE 11–49 Alternate display of loading on the beam in Figure 11–48.

left half of the load, shearing force, and bending moment diagrams are

$$w_{AB} = -0.4348x - 0.20$$
$$V_{AB} = -0.2174x^2 - 0.20x + 1.61$$
$$M_{AB} = -0.07246x^3 - 0.10x^2 + 1.61x$$

The maximum shearing force is 1.61 kN at each support, and the maximum bending moment is 2.292 kN · m at the middle of the beam.

11–12 CONTINUOUS BEAMS— THEOREM OF THREE MOMENTS

A continuous beam on any number of supports can be analyzed by using the *theorem of three moments*. The theorem actually relates the bending moments at three successive supports to each other and to the loads on the beam. For a beam with only three supports, the theorem allows the direct computation of the moment at the middle support. Known end conditions provide data for computing moments at the ends. Then the principles of statics can be used to find reactions.

For beams on more than three supports, the theorem is applied successively to sets of three adjacent supports (two spans), yielding a set of equations that can be solved simultaneously for the unknown moments.

The theorem of three moments can be used for any combination of loads. Special forms of the theorem have been developed for uniformly distributed loads and concentrated loads. These forms will be used in this chapter.

Uniformly Distributed Loads on Adjacent Spans

Figure 11–50 shows the arrangement of loads and the definition of terms applicable to Equation (11–5).

Three-Moment Equation—Distributed Loads ▼

$$M_A L_1 + 2M_B(L_1 + L_2) + M_C L_2 = \frac{-w_1 L_1^3}{4} - \frac{w_2 L_2^3}{4}$$

(11–5)

The values of w_1 and w_2 are expressed in units of force per unit length such as N/m, lb/ft, and so on. The bending moments at the supports A, B, and C are M_A, M_B, and M_C. If M_A and M_C at the ends of the beam are known, M_B can be found from Equation (11–5) directly. Example problems will demonstrate the application of this equation.

FIGURE 11–50 Uniformly distributed loads on a continuous beam of two spans.

The special case in which two equal spans carry equal uniform loads allows the simplification of Equation (11–5). If $L_1 = L_2 = L$ and $w_1 = w_2 = w$, then

$$M_A + 4M_B + M_C = \frac{-wL^2}{2}$$

(11–6)

Concentrated Loads on Adjacent Spans

If adjacent spans carry only one concentrated load each, as shown in Figure 11–51, then Equation (11–7) applies.

Three-Moment Equation—Concentrated Loads ▼

$$M_A L_1 + 2M_B(L_1 + L_2) + M_C L_2$$
$$= \frac{-P_1 a}{L_1}(L_1^2 - a^2) - \frac{P_2 b}{L_2}(L_2^2 - b^2)$$

(11–7)

The special case of two equal spans carrying equal, centrally placed loads allows the simplification of Equation (11–7). If $P_1 = P_2 = P$ and $a = b = L/2$, then

$$M_A + 4M_B + M_C = 3PL/4$$

(11–8)

Combinations of Uniformly Distributed Loads and Several Concentrated Loads

This is a somewhat general case, allowing each span to carry a uniformly distributed load and any number of concentrated loads, as suggested in Figure 11–52. The general equation for such a loading is a combination of Equations (11–5) and (11–7), given as Equation (11–9).

Three-Moment Equation—General Form ▼

$$M_A L_1 + 2M_B(L_1 + L_2) + M_C L_2$$
$$= -\sum \left[\frac{P_i a_i}{L_1}(L_1^2 - a_i^2)\right]_1 - \sum \left[\frac{P_i b_i}{L_2}(L_2^2 - b_i^2)\right]_2$$
$$- \frac{w_1 L_1^3}{4} - \frac{w_2 L_2^3}{4}$$

(11–9)

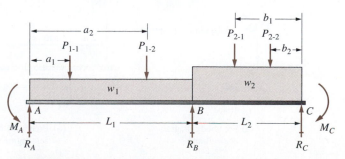

FIGURE 11–52 General notation for terms in Equation (11–9).

The term in the bracket with the subscript *1* is to be evaluated for *each* concentrated load on span 1 and then summed together. Similarly, the term having the subscript *2* is repeatedly applied for all loads on span 2. Notice that the distances a_i are measured from the reaction at A for each load on span 1, and the distances b_i are measured from the reaction at C for each load on span 2. The moments at the ends A and C

can be due to concentrated moments applied there or to loads on overhangs beyond the supports. Any of the terms in Equation (11–9) may be left out of a problem solution if there is no appropriate load or moment existing at a particular section for which the equation is being written. Furthermore, other concentrated loads could be included in addition to those actually shown in Figure 11–52.

Example Problem 11–8

The loading composed of a combination of distributed loads and concentrated loads, shown in Figure 11–53, is to be analyzed to determine the reactions at the three supports and the complete shearing force and bending moment diagrams. The 17-m beam is to be used as a floor beam in an industrial building.

FIGURE 11–53 Beam for Example Problem 11–8.

Solution **Objective:** Determine the support reactions and draw the shearing force and bending moment diagrams.

Given: Beam loading in Figure 11–53.

Analysis: Because the loading pattern contains both concentrated loads and uniformly distributed loads, we must use Equation (11–9). The subscript *1* refers to span 1 between supports A and B, and subscript *2* refers to span 2 between supports B and C. We must evaluate the actual magnitude of M_A and M_C to facilitate the solution of Equation (11–9). Because point C is at the end of a simply supported span, $M_C = 0$. At point A we can consider the overhanging part of the beam to the left of A to be a free body as shown in Figure 11–54(a). Then sum moments about A to find the internal moment in the beam at A. Along with the 12 kN load 2.0 m from A, a resultant of 60 kN of load due to the distributed load acts a distance of 1.0 m from A. Then

$$M_A = -12 \text{ kN}(2\text{m}) - 60 \text{ kN}(1\text{m}) = -84 \text{ kN} \cdot \text{m}$$

Each remaining term in Equation (11–9) will be evaluated.

$$M_A L_1 = -84 \text{ kN} \cdot \text{m} (8 \text{ m}) = -672 \text{ kN} \cdot \text{m}^2$$
$$2 M_B(L_1 + L_2) = 2 M_B(8 + 7) = M_B(30 \text{ m})$$

(a) Free body diagram of overhang

(b) Free body diagram of left segment

(c) Free body diagram of right segment

FIGURE 11–54 Free body diagrams of beam segments used to find reactions R_A, R_B, and R_C.

$$-\sum\left[\frac{P_i a_i}{L_1}(L_1^2 - a_i^2)\right]_1 = -\frac{15(2)}{8}(8^2 - 2^2) - \frac{18(6)}{8}(8^2 - 6^2)$$

$$= -603 \text{ kN} \cdot \text{m}^2$$

$$-\sum\left[\frac{P_i b_i}{L_2}(L_2^2 - b_i^2)\right]_2 = -\frac{20(4)}{7}(7^2 - 4^2) = -377 \text{ kN} \cdot \text{m}^2$$

$$-\frac{w_1 L_1^3}{4} = -\frac{30(8)^3}{4} = -3840 \text{ kN} \cdot \text{m}^2$$

$$-\frac{w_2 L_2^3}{4} = -\frac{50(7)^3}{4} = -4288 \text{ kN} \cdot \text{m}^2$$

Now putting these values into Equation (11–9) gives

$$-672 \text{ kN} \cdot \text{m}^2 + M_B(30 \text{ m}) + 0 = (-603 - 377 - 3840 - 4288)\text{kN} \cdot \text{m}^2$$

Solving for M_B yields

$$M_B = -281 \text{ kN} \cdot \text{m}$$

Support Reactions. Having the three moments at the locations of the support points A, B, and C allows the computation of the reactions at those supports. The solution procedure starts by considering each of the two spans as separate free bodies, as shown in Figure 11–54(b) and (c). In each case, when the beam is divided at B, the moment M_B is shown acting at the cut section to maintain equilibrium. Then, using the left segment, we can sum moments about point B and solve for the left reaction, R_A.

$$\sum M_B = 0 = 12 \text{ kN}(10 \text{ m}) + 15 \text{ kN}(6 \text{ m}) + 300 \text{ kN}(5 \text{ m}) + 18 \text{ kN}(2 \text{ m})$$

$$-281 \text{ kN} \cdot \text{m} - R_A(8 \text{ m})$$

$$R_A = 183 \text{ kN}$$

Similarly, using the right segment and summing moments about point B allows the determination of the right reaction, R_C.

$$\sum M_B = 0 = 20 \text{ kN}(3 \text{ m}) + 350 \text{ kN}(3.5 \text{ m}) - 281 \text{ kN} \cdot \text{m} - R_C(7 \text{ m})$$

$$R_C = 143 \text{ kN}$$

Now we can use the $\sum F_v = 0$ to find the middle reaction, R_B.

$$\sum F_v = 0 = 12 \text{ kN} + 15 \text{ kN} + 18 \text{ kN} + 20 \text{ kN} + 300 \text{ kN} + 350 \text{ kN}$$

$$-183 \text{ kN} - 143 \text{ kN} - R_B$$

$$R_B = 389 \text{ kN}$$

Shearing Force and Bending Moment Diagrams. We now have the necessary data to draw the complete diagrams, as shown in Figure 11–55.

Comment: In summary, the reaction forces are

$$R_A = 183 \text{ kN}$$
$$R_B = 389 \text{ kN}$$
$$R_C = 143 \text{ kN}$$

Figure 11–55 shows that the local maximum positive bending moments occur between the supports, and the local maximum negative bending moments occur at the supports. The overall maximum positive bending moment is 204 kN·m at a point 2.86 m from C where the shearing force curve crosses the zero axis. The actual maximum negative bending moment is −281 kN·m at support B. If a beam of uniform cross section is used, it would have to be designed to withstand a bending moment of 281 kN·m. But notice that this is a rather localized peak on the bending moment diagram. It may prove economical to design the beam to withstand the 204 kN·m bending moment and then add reinforcing plates in the vicinity of support B to increase the section modulus there to a level that would be safe for the 281 kN·m moment. You might notice many highway overpasses designed in this manner.

FIGURE 11–55 Shearing force and bending moment diagrams for Example Problem 11–8.

Formulas for Common Loading and Support Patterns

Appendix A–25 gives illustrations and formulas for shearing force and bending moment diagrams for ten special cases for continuous beams. These cases apply to frequently encountered loading and support patterns in building construction, bridge design, structures for machinery or vehicles, and some consumer products. Included are

1. Supported cantilevers—four loading patterns
2. Fixed end beams—three loading patterns
3. Continuous beams—uniformly distributed loads for two spans, three spans, and four spans.

The tables give

- Sketches of the resulting shearing force and bending moment diagrams
- The general shape of the beam deflection curves
- Equations for reactions at supports, and shearing forces and bending moments at critical points
- Equations for beam deflection for most loadings.

See Chapter 14 for more discussion and analysis techniques for beam deflections.

It is important to understand the general shape of the shearing force and bending moment diagrams so you will be able to design safe and efficient beams as discussed in Chapters 12 and 13.

PROBLEMS

Problems for Figures P11–1 to P11–76

Figures P11–1 to P11–76 show a variety of beam types and loading conditions. For the beam in each figure, any or all of the following problem statements can be applied:

1. Compute the reactions at the supports using the techniques shown in Section 11–3.
2. Draw the complete shearing force and bending moment diagrams using the techniques shown in Sections 11–4 to 11–9.
3. Determine the magnitude and location of the maximum absolute value of the shearing force and bending moment.
4. Use the free body diagram approach shown in Sections 11–4 to 11–6 to determine the internal shearing force and bending moment at any specified point in a beam.
5. Write equations for all segments of the shearing force and bending moment diagrams using the guidelines presented in Section 11–11.

FIGURE P11–1

FIGURE P11–2

FIGURE P11–3

FIGURE P11–4

FIGURE P11–5

FIGURE P11–6

FIGURE P11–7

FIGURE P11–8

FIGURE P11–9

FIGURE P11–10

FIGURE P11–11

FIGURE P11–12

FIGURE P11–13

FIGURE P11–14

8.4 kN   20 kN

2.5 m | 1 | 2.0 | 2.0 | 2.5 m

12.5 kN   15 kN

FIGURE P11–15

85 kN   70 kN   45 kN   85 kN

2 m | 2 m | 4 m | 2 m | 2 m

FIGURE P11–16

120 lb

8 in

FIGURE P11–17

85 lb   75 lb

8 in   4 in

FIGURE P11–18

12 K   12 K

6 ft | 2 ft

FIGURE P11–19

20 K   15 K   10 K

4 ft   4 ft   4 ft

FIGURE P11–20

600 N   1200 N

300 mm   400 mm

FIGURE P11–21

2000 N   400 N   600 N

0.3 m   0.2 m   0.2 m

FIGURE P11–22

40 kN   40 kN

1.2 m   2.4 m

FIGURE P11–23

90 kN   60 kN   30 kN

1.4 m   1.4 m

0.8 m

FIGURE P11–24

20 lb/in

18 in

FIGURE P11–25

4 ft | 2 K/ft | 4 ft

12 ft

FIGURE P11–26

45 lb/in

8 in | 4 in

FIGURE P11–27

2 ft | 4 K/ft

10 ft

FIGURE P11–28

500 N/m

200 N/m

0.25 m | 0.2 m

FIGURE P11–29

FIGURE P11–30

FIGURE P11–31

FIGURE P11–32

FIGURE P11–33

FIGURE P11–34

FIGURE P11–35

FIGURE P11–36

FIGURE P11–37

FIGURE P11–38

FIGURE P11–39

FIGURE P11–40

FIGURE P11–41

FIGURE P11–42

FIGURE P11–43

FIGURE P11–44

FIGURE P11–45

FIGURE P11–46

FIGURE P11–47

FIGURE P11–48

FIGURE P11–49

FIGURE P11–50

FIGURE P11–51

FIGURE P11–52

FIGURE P11–53

FIGURE P11–54

FIGURE P11–55

FIGURE P11–56

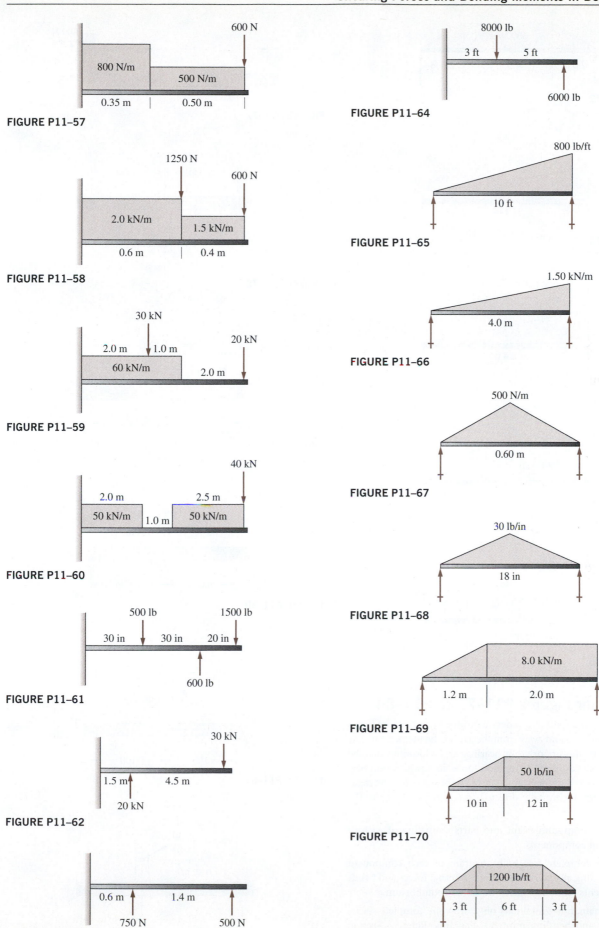

FIGURE P11–57

FIGURE P11–58

FIGURE P11–59

FIGURE P11–60

FIGURE P11–61

FIGURE P11–62

FIGURE P11–63

FIGURE P11–64

FIGURE P11–65

FIGURE P11–66

FIGURE P11–67

FIGURE P11–68

FIGURE P11–69

FIGURE P11–70

FIGURE P11–71

FIGURE P11–72

FIGURE P11–73

FIGURE P11–74

FIGURE P11–75

FIGURE P11–76

Problems for Figures P11–77 to P11–84

Each Figure shows a mechanical device in which one or more forces are applied parallel to and away from the axis of the main, beam-like part. The devices are supported by bearings at the locations marked with an ×, which can provide reaction forces in any direction perpendicular to the axis of the beam. One of the bearings has the capability of resisting horizontally directed forces. For each figure the objectives are

1. Break the compound beam into parts consisting of each of the straight components.

2. Show the complete free body diagram of each component part including all external loads and internal forces and bending moments required to keep the part in equilibrium.

3. For the main, horizontal part only, draw the complete shearing force and bending moment diagrams. Refer to Section 11–10 for examples.

FIGURE P11–77

FIGURE P11–78

FIGURE P11–79

FIGURE P11–80

FIGURE P11–81

FIGURE P11–82

FIGURE P11–83

FIGURE P11–84

ADDITIONAL PRACTICE AND REVIEW PROBLEMS

11–85. For the beam shown in P11–8, determine the shearing force and bending moment at a point 4.0 m from the left support using the free body diagram method.

11–86. For the beam shown in P11–15, determine the shearing force and bending moment at a point 1.0 m to the right of the right support using the free body diagram method.

11–87. For the beam shown in P11–22, determine the shearing force and bending moment at a point 0.45 m from the wall using the free body diagram method.

11–88. For the beam shown in P11–35, determine the shearing force and bending moment at a point 0.9 m from the left end of the beam using the free body diagram method.

11–89. For the beam shown in P11–53, determine the shearing force and bending moment at a point 6.0 m from the left end of the beam using the free body diagram method.

11–90. For the beam shown in P11–58, determine the shearing force and bending moment at a point 0.8 m from the wall using the free body diagram method.

11–91. For the beam shown in P11–69, determine the shearing force and bending moment at a point 2.2 m from the left support using the free body diagram method.

11–92. For the beam shown in P11–76, determine the shearing force and bending moment at a point 4.0 m from the wall using the free body diagram method.

Problems 11–93 to 11–110

Draw the complete shearing force and bending moment diagrams for the beams shown in the accompanying Figures. Report the maximum shearing force, the maximum positive bending moment, and the maximum negative bending moment. Indicate the location of those values on the beam.

Concentrated Loads

11–93. Use Figure P11–93
11–94. Use Figure P11–94
11–95. Use Figure P11–95
11–96. Use Figure P11–96
11–97. Use Figure P11–97
11–98. Use Figure P11–98

FIGURE P11–93

FIGURE P11–94

FIGURE P11–95

FIGURE P11–96

FIGURE P11–97

FIGURE P11–98

Distributed Loads

11–99. Use Figure P11–99
11–100. Use Figure P11–100
11–101. Use Figure P11–101
11–102. Use Figure P11–102
11–103. Use Figure P11–103
11–104. Use Figure P11–104
11–105. Use Figure P11–105
11–106. Use Figure P11–106
11–107. Use Figure P11–107
11–108. Use Figure P11–108
11–109. Use Figure P11–109
11–110. Use Figure P11–110

3200 lb

60 in   60 in

25 lb/in

FIGURE P11–99

1200 lb

500 lb/ft

A   B   C

10.0 ft   3.0 ft

FIGURE P11–100

800 lb   600 lb

50 lb/in   4 in   6 in   4 in

10 in

FIGURE P11–101

2500 N

0.45 m   0.20 m   0.10 m

7500 N/m

FIGURE P11–102

12 kN

4 kN/m

1.5 m

6.0 m

FIGURE P11–103

300 lb

100 lb/in

12 in   4 in

FIGURE P11–104

3000 N

50 kN/m

0.50 m   0.30 m

FIGURE P11–105

20 K   15 K

4 K/ft   4   4   4 ft

8 ft

FIGURE P11–106

800 lb

8 in   24 in

20 lb/in

FIGURE P11–107

60 kN

30 kN/m

2.50 m   0.75 m   0.75 m

FIGURE P11–108

200 N

500 N/m

0.30 m   0.20 m

FIGURE P11–109

6.20 kN

1.0 m   0.8 m   4.50 kN

4.80 kN/m

0.60 m

FIGURE P11–110

Problems 11–111 to 11–116

Figures P11–111 to P11–116 show completed shearing force diagrams for beams. Develop the loading diagram and the bending moment diagram for each beam.

FIGURE P11–111

FIGURE P11–112

FIGURE P11–113

FIGURE P11–114

FIGURE P11–115

FIGURE P11–116

Problems 11–117 to 11–122

Figures P11–117 to P11–122 show completed bending moment diagrams for beams. Develop the shearing force diagram and the loading diagram for each beam.

FIGURE P11–117

FIGURE P11–118

FIGURE P11–119

FIGURE P11–120

FIGURE P11–121

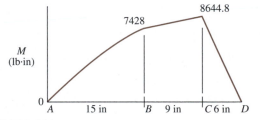

FIGURE P11–122

Continuous Beams—Theorem of Three Moments

For Problems 11–123 through 11–130, use the theorem of three moments to determine the reactions at all supports and draw the complete shearing force and bending moment diagrams. Indicate the maximum shearing force and bending moment for each beam.

11–123. Use Figure P11–123. **11–127.** Use Figure P11–127.
11–124. Use Figure P11–124. **11–128.** Use Figure P11–128.
11–125. Use Figure P11–125. **11–129.** Use Figure P11–129.
11–126. Use Figure P11–126. **11–130.** Use Figure P11–130.

FIGURE P11–123

FIGURE P11–124

FIGURE P11–125

FIGURE P11–126

FIGURE P11–127

FIGURE P11–128

FIGURE P11–129

FIGURE P11–130

STRESS DUE TO BENDING

MAJOR TOPICS

12–1   The Big Picture—Stress Due to Bending
12–2   The Flexure Formula
12–3   Conditions on the Use of the Flexure Formula
12–4   Stress Distribution on a Cross Section of a Beam
12–5   Derivation of the Flexure Formula
12–6   Applications—Beam Analysis
12–7   Applications—Beam Design and Design Stresses
12–8   Section Modulus and Design Procedures
12–9   Stress Concentrations
12–10 Flexural Center or Shear Center
12–11 Preferred Shapes for Beam Cross Sections
12–12 Design of Beams to Be Made from Composite Materials

12–1 THE BIG PICTURE— STRESS DUE TO BENDING

Discussion Map

- Beams were defined in Chapter 11 to be load-carrying members on which the loads act perpendicular to the long axis of the beam.

- You have already learned how to determine the bending moment diagram for beams carrying a variety of load types.

- The stress due to bending in the beam is directly proportional to the bending moment.

- The properties of the cross-sectional area of the beam (overall dimensions, centroid location, area moment of inertia, and section modulus) play important roles in the magnitude of the bending stress. In general, an efficient beam cross section has a high area moment of inertia relative to its area and, therefore, it has a low weight.

- In this chapter you develop your ability to determine the stress due to bending and to design efficient beams.

ACTIVITY   Chapter 12—Stress Due to Bending

Gather some materials that you can use to demonstrate the variety of ways that a beam can fail. Activities from earlier chapters, particularly Chapters 6, 7, and 11, called for collecting a similar set of sample beams, and some of them may be useful here. You will be installing them in a simple fixture that supports the beams at two points and that allows a load to be applied.

Alternatively, you could clamp the beam at one end to solid surface and load it as a cantilever. Figure 12–1 shows a commercially available loading frame that would serve well.

The beams that you gather should demonstrate some of the six failure modes illustrated in Figure 12–2, briefly

(a) Cantilever

(b) Simply supported beam

FIGURE 12–1   Beam loading frame.

Source: P. A. Hilton Ltd./Hi-Tech, Hampshire, England.

(a) Complete fracture

(b) Yielding of part of beam

(c) Local crippling

(d) Web wrinkling

(e) Fastener failure

(f) Interlaminar shear

FIGURE 12–2   Failure modes for beams.

described here. The beams will be destroyed during the tests, so use small, simple, inexpensive samples.

1. *Complete fracture*—Figure 12–2(a): The material for the beam should be somewhat brittle and the cross section should be small enough to break under low to moderate loads. Chalk is a classic example of such a beam. You could also use a wood pencil, a small wood dowel, or small parts made from some alloys of aluminum, zinc, or magnesium, particularly those that are cast. The beam will break completely and usually suddenly.

2. *Yielding of part of the beam*—Figure 12–2(b): The material should be ductile so that it will deform prior to any actual tearing of the material. Small metal rods with a variety of cross-sectional shapes, thin flat metal beams, and small plastic beams are examples. You should stop the loading after significant yielding is evident.

3. *Local crippling*—Figure 12–2(c): This typically occurs when thin extended flanges of a beam are subjected to compressive stress as would be produced by a downward load on the top of a simply supported beam or on the bottom of a cantilever. You should look for a rolled thin sheet metal beam such as a curtain rod as an example.

4. *Web wrinkling*—Figure 12–2(d): A relatively tall I-shape or box beam with thin vertical webs or sides may fail by wrinkling. This may be difficult to find, but rolled thin sheet metal shapes are good examples.

5. *Fastener failure*—Figure 12–2(e): Fabricated beams built up from two or more parts may fail by shearing fasteners such as rivets, screws, bolts, pegs, or nails or by exceeding the shear strength of adhesives, brazed joints, or welds.

6. *Interlaminar shear*—Figure 12–2(f): Components made from composite materials such as fiberglass, carbon/epoxy, or metal-matrix composites are often built up in layers impregnated with the matrix material that is then cured into a cohesive structure. An important failure mode for these materials is the separation of the layers under load such as by bending a panel or a beam. This is called *interlaminar shear*. Try to find such a beam, load it, and watch for the separation of the layers.

You will see during these demonstrations that some of the beam failures are, in fact, *bending failures*, which are discussed in this chapter, whereas some others are *shearing failures*, discussed in the following chapter. Bending and shearing stresses are both inherently present in any beam. Which type of failure actually occurs first depends on the loading pattern, the materials from which the beam is made, the shape of the cross section, and the actual dimensions of that shape.

Objectives of This Chapter

After completing this chapter, you should be able to:

1. Learn the statement of the *flexure formula* and apply it properly to compute the maximum stress due to bending at the outer fibers of the beam.

2. Compute the stress at any point within the cross section of the beam and describe the variation of stress with position in the beam.

3. Understand the conditions on the use of the flexure formula.

4. Recognize that it is necessary to ensure that the beam does not twist under the influence of the bending loads.

5. Define the *neutral axis* and understand that it is coincident with the centroidal axis of the cross section of the beam.

6. Understand the derivation of the flexure formula and the effect of the moment of inertia on bending stress.

7. Determine the appropriate design stress for use in designing beams.

8. Design beams to carry a given loading safely.

9. Define the *section modulus* of the cross section of the beam.

10. Select standard structural shapes for use as beams.

11. Recognize when it is necessary to use stress concentration factors in the analysis of stress due to bending and apply appropriate factors properly.

12. Define the *flexural center* and describe its proper use in the analysis of stress due to bending.

12–2 THE FLEXURE FORMULA

Beams must be designed to be safe. When loads are applied perpendicular to the long axis of a beam, bending moments are developed inside the beam, causing it to bend. Observe a thin beam. The characteristically curved shape shown in Figure 12–3 is evident. The fibers of the beam near its top surface are shortened and placed in compression. Conversely, the fibers near the bottom surface are stretched and placed in tension.

Taking a short segment of the beam from Figure 12–3, we show in Figure 12–4 how the shape would change under the influence of the bending moments inside the beam. In part (a) the segment is in its initially straight form when it is not carrying a load. Part (b) shows the same segment as it is deformed by the application of the bending moments. Lines that were initially horizontal become curved. The ends of the segment, which were initially straight and vertical, remain straight. However, now they are inclined, having rotated about the

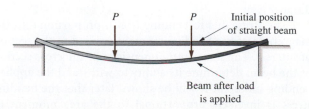

FIGURE 12–3 Example of a beam.

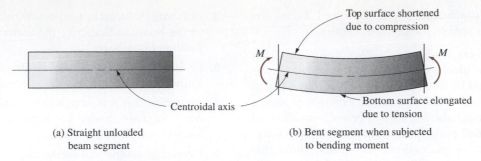

FIGURE 12–4 Influence of bending moment on beam segment.

centroidal axis of the cross section of the beam. The result is that the material along the top surface has been placed under compression and consequently shortened. Also, the material along the bottom surface has been placed under tension and has elongated.

In fact, all of the material above the centroidal axis is in compression. The maximum shortening (compressive strain) occurs at the top. Because stress is proportional to strain, it can be reasoned that the maximum compressive stress occurs at the top surface. Similarly, all of the material below the centroidal axis is in tension. The maximum elongation (tensile strain) occurs at the bottom, producing the maximum tensile stress.

We can also reason that, if the upper part of the beam is in compression and the lower part is in tension, then there must be some place in the beam where there is no strain at all. That place is called the *neutral axis* and it will be shown later that it is coincident with the *centroidal axis* of the beam. In summary, we can conclude that

> *In a beam subjected to a bending moment of the type shown in Figure 12–4, material above the centroidal axis will be in compression with the maximum compressive stress occurring at the top surface.*
>
> *Material below the centroidal axis will be in tension with the maximum tensile stress occurring at the bottom surface.*
>
> *Along the centroidal axis itself, there is zero strain and zero stress due to bending. This is called the neutral axis.*

In designing or analyzing beams, it is usually the objective to determine the maximum tensile and compressive stress. It can be concluded from this discussion that these maximums are dependent on the distance from the neutral axis (centroidal axis) to the top and bottom surfaces. We will call that distance c.

The stress due to bending is also proportional to the magnitude of the bending moment applied to the section of interest. The shape and dimensions of the cross section of the beam determine its ability to withstand the applied bending moment. It will be shown later that the bending stress is inversely proportional to the area moment of inertia of the cross section with respect to its horizontal centroidal axis.

We now state the *flexure formula*, which can be used to compute the maximum stress due to *bending*.

Flexure Formula ▼

$$\sigma_{max} = \frac{Mc}{I} \qquad (12\text{--}1)$$

where σ_{max} = maximum stress at the outermost fiber of the beam

M = bending moment at the section of interest

c = distance from the centroidal axis of the beam to the outermost fiber

I = area moment of inertia of the cross section with respect to its centroidal axis

The following steps are typically taken to apply Equation (12–1).

GUIDELINES FOR APPLYING THE FLEXURE FORMULA ▼

1. Determine the maximum bending moment on the beam by drawing the shearing force and bending moment diagrams.

2. Locate the centroid of the cross section of the beam.

3. Compute the moment of inertia of the area of the cross section with respect to its centroidal axis.

4. Compute the distance c from the centroidal axis to the top or bottom of the beam, whichever is greater.

5. Compute the stress from the flexure formula,

$$\sigma_{max} = \frac{Mc}{I}$$

The flexure formula is discussed in greater detail later. A sample problem will now be shown to illustrate the application of the formula.

Example Problem
12–1

For the beam shown in Figure 12–5, compute the maximum stress due to bending. The cross section of the beam is a rectangle 100 mm high and 25 mm wide. The load at the middle of the beam is 1500 N, and the length of the beam is 3.40 m.

FIGURE 12–5 Beam data for Example Problem 12–1.

(a)

(b) Section A-A

Solution   Objective: Compute the maximum stress due to bending.

Given: Beam and loading shown in Figure 12–5.

Analysis: The guidelines defined in this section will be used.

Results: **Step 1:** The shearing force and bending moment diagrams have been drawn and included in Figure 12–5. The maximum bending moment is 1275 N · m at the middle of the beam.

Step 2: The centroid of the rectangular cross section is at the intersection of the two axes of symmetry, 50 mm from either the top or the bottom surface of the beam.

Step 3: The moment of inertia of the area for the rectangular shape with respect to the centroidal axis is

$$I = \frac{bh^3}{12} = \frac{25(100)^3}{12} = 2.08 \times 10^6 \, mm^4$$

Step 4: The distance $c = 50$ mm from the centroidal axis to either the top or the bottom surface.

Step 5: The maximum stress due to bending occurs at the top or the bottom of the beam at the point of maximum bending moment. Applying Equation (12–1) gives

$$\sigma_{max} = \frac{Mc}{I} = \frac{(1275 \, N \cdot m)(50 \, mm)}{2.08 \times 10^6 \, mm^4} \times \frac{10^3 \, mm}{m}$$

$$\sigma_{max} = 30.6 \, N/mm^2 = 30.6 \, MPa$$

12–3 CONDITIONS ON THE USE OF THE FLEXURE FORMULA

The proper application of the flexure formula requires the understanding of the conditions under which it is valid, listed as follows:

1. The beam must be straight or very nearly so.

2. The cross section of the beam must be uniform.

3. All loads and support reactions must act perpendicular to the axis of the beam.

4. The beam must not twist while the loads are being applied.

5. The beam must be relatively long and narrow in proportion to its depth.

6. The material from which the beam is made must be homogeneous, and it must have an equal modulus of elasticity in tension and compression.

7. The stress resulting from the loading must not exceed the proportional limit of the material.

8. No part of the beam may fail from instability, that is, from the buckling or crippling of thin sections.

9. The section where the stress is to be computed must not be close to the point of application of concentrated loads.

Although the list of conditions appears to be long, the flexure formula still applies to a wide variety of real cases. Beams violating some of the conditions can be analyzed by using a modified formula or by using a combined stress approach. For example, for condition 2, a change in cross section will cause stress concentrations, which can be handled as described in Section 12–9.

The combined bending and axial stress or bending and torsional stress produced by violating condition 3 are discussed in Chapter 15. If the other conditions are violated, special analyses are required, which are not covered in this book.

Condition 4 is important, and attention must be paid to the shape of the cross section to ensure that twisting does not occur. In general, if the beam has a vertical axis of symmetry and if the loads are applied through that axis, no twisting will result. Figure 12–6 shows some typical shapes used for beams that satisfy condition 4. Conversely, Figure 12–7 shows several that do not satisfy condition 4. In each of these cases, the beam would tend to twist as well as bend as the load is applied in the manner shown. Of course, these sections can support some load, but the actual stress condition in them is different from that which would be predicted from the flexure formula. More about these kinds of beams is presented in Section 12–10.

Condition 5 is difficult to quantify in general because of the variety of cross-sectional shapes and support conditions used for beams. Some designers define a relatively long beam as one for which the ratio of length to depth is greater than about 10.

Condition 8 is important because long, thin members and, sometimes, thin sections of members tend to buckle at stress levels well below the yield strength of the material. Such failures are called *instability* and are to be avoided. Frequently,

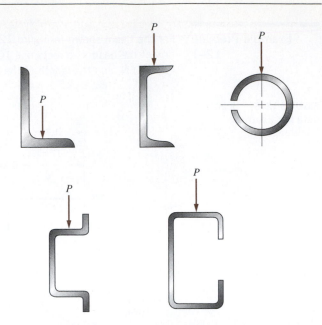

FIGURE 12–7 Example beam shapes with loads not acting through an axis of symmetry, resulting in twisting of the beam.

cross braces or local stiffeners are added to beams to relieve the problem of instability. An example can be seen in the wood joist floor construction of many homes and commercial buildings. The relatively slender wood joists are braced near the midpoints to avoid buckling. See References 1–4 and 16 for additional analysis of long slender beams or thin sections of beams.

The need for condition 9 can be visualized by referring to Figure 12–8, which shows a photoelastic model of a simply supported beam carrying a single concentrated load at the middle of the span. Recall the discussion of photoelasticity from Chapter 7. The beam cross section in this example is a simple rectangle.

First, look at the beam between either support and the point of application of the load. Notice how the black and white fringe lines are smooth and approximately evenly spaced from top to bottom on the beam. Also, note that there are more fringe lines near the middle of the span and fewer as you approach the loads. The greater number of fringe lines indicates a higher stress gradient. This demonstrates that the bending stress is proportional to the bending moment that would be maximum under the load and zero at the supports for this loading pattern.

Now look at any cross section of the beam. The next section shows that the stress distribution on a cross section of such a beam has the following characteristics:

- It is zero at the centroidal axis of the cross section of the beam, halfway from the top or bottom of the beam.
- It is a maximum tensile stress at the bottom surface.
- It is a maximum compressive stress at the top surface.
- It varies linearly from the top to the bottom.

Starting at the midpoint of the cross section and counting the number of fringes from there to the top or bottom

FIGURE 12–6 Example beam shapes with loads acting through an axis of symmetry.

FIGURE 12–8 Photoelastic model of a simply supported beam with a concentrated load.
Source: Measurements Group, Inc., Raleigh, NC.

surface would allow you to determine the stress level at that section if you knew the characteristics of the photoelastic material.

Now look at the parts of the beam that are close to the point of application of the load or near the supports. You can see a more complex fringe pattern that indicates high local stresses that result from the concentrated force. You must be sure that the material of the beam can withstand these local stresses by using relatively large bearing areas at these points. Notice that these local effects near the supports dissipate quickly as you move away from them. This is referred to as the *principle of St. Venant*, in recognition of the work this renowned French scientist reported in 1855.

12–4 STRESS DISTRIBUTION ON A CROSS SECTION OF A BEAM

Refer again to Figure 12–4 showing the manner in which a segment of a beam deforms under the influence of a bending moment. The segment assumes the characteristic "bent" shape as the upper fibers are shortened and the lower fibers are elongated. The neutral axis, coincident with the centroidal axis of the cross section of the beam, bends but it is not strained. Therefore, at the neutral axis the stress due to bending is zero.

Figure 12–4 also shows that the ends of the beam segment, which were initially straight and vertical, remain straight. However, they rotate as the bending moment is applied. The linear distance from a point on the initial vertical end line to the corresponding point on the rotated end line is an indication of the amount of strain produced at that point in the cross section. It can be reasoned, therefore, that there is a linear variation of strain with position in the cross section as a function of the distance away from the neutral axis. Moving from the neutral axis toward the top of the section results in greater compressive strain while moving downward toward the bottom results in greater tensile strain. For materials following Hooke's law, stress is proportional to strain. The resulting stress distribution, then, is as shown in Figure 12–9.

To represent the stress at some point within the cross section, we can express it in terms of the maximum stress by noting the linear variation of stress with distance away from the neutral axis. Calling that distance y, we can write an equation for the stress, σ, at any point as

$$\sigma = \sigma_{max}\frac{y}{c} \qquad (12\text{–}2)$$

The general form of the stress distribution shown in Figure 12–9 would occur in any beam section having a centroidal axis equidistant from the top and bottom surfaces. For such cases, the magnitude of the maximum compressive stress would equal the maximum tensile stress.

If the centroidal axis of the section is not the same distance from both the top and bottom surfaces, the stress distribution shown in Figure 12–10 would occur. Still the stress at the neutral axis would be zero. Still the stress would vary linearly with distance from the neutral axis. Now the maximum stress at the bottom of the section is greater than that at the top because it is farther from the neutral axis. Using the distances c_b and c_t as indicated in Figure 12–10, the stresses would be

$$\sigma_{max} = \frac{Mc_b}{I} \quad \text{(tension at the bottom)}$$

$$\sigma_{max} = \frac{Mc_t}{I} \quad \text{(compression at the top)}$$

FIGURE 12–9 Bending stress distribution on a symmetrical section.

FIGURE 12–10   Bending stress distribution on a nonsymmetrical section.

12–5 DERIVATION OF THE FLEXURE FORMULA

We can better understand the basis for the flexure formula by following the analysis used to derive it. The principles of static equilibrium are used here to show two concepts that were introduced earlier in this chapter but that were stated without proof. One is that the *neutral axis* is coincident with the *centroidal axis* of the cross section. The second is the flexure formula itself and the significance of the moment of inertia of the area of the cross section.

Refer to Figure 12–9, which shows the distribution of stress over the cross section of a beam. The shape of the cross section is not relevant to the analysis and the I-shape is shown merely for example. The figure shows a portion of a beam, cut at some arbitrary section, with an internal bending moment acting on the section. The stresses, some tensile and some compressive, would tend to produce forces on the cut section in the axial direction. Equilibrium requires that the net sum of these forces must be zero. In general, force equals stress times area. Because the stress varies with position on the cross section, it is necessary to look at the force on any small elemental area, dA, and then sum these forces over the entire area using the process of integration. These concepts can be shown analytically as

Equilibrium condition: $\sum F = 0$

Force on any element of area: $dA = \sigma \, dA$

Total force on the cross-sectional area:

$$\sum F = \int_A \sigma \, dA = 0 \qquad \text{(12–3)}$$

Now we can express the stress σ at any point in terms of the maximum stress by using Equation (12–2):

$$\sigma = \sigma_{max}\frac{y}{c}$$

where y is the distance from the neutral axis to the point where the stress is equal to σ. Substituting this into Equation (12–3) gives

$$\sum F = \int_A \sigma \, dA = \int_A \sigma_{max}\frac{y}{c}dA = 0$$

Because σ_{max} and c are constants, they can be taken outside the integral sign.

$$\sum F = \frac{\sigma_{max}}{c}\int_A y \, dA = 0$$

Neither σ_{max} nor c is zero, so the other factor, $\int_A y dA$, must be zero. By definition and as illustrated in Chapter 6,

$$\int_A y dA = \overline{Y}(A)$$

where $\overline{Y}$ is the distance to the centroid of the area from the reference axis and A is the total area. Again, A cannot be zero, so, finally, it must be true that $\overline{Y} = 0$. Because the reference axis is the neutral axis, this shows that the neutral axis is coincident with the centroidal axis of the cross section.

The derivation of the flexure formula is based on the principle of equilibrium, which requires that the sum of the moments about any point must be zero. Figure 12–9 shows that a bending moment M acts at the cut section. This must be balanced by the net moment created by the stress on the cross section. Moment is the product of force times the distance from the reference axis to the line of action of the force. As used above,

$$\sum F = \int_A \sigma \, dA = \int_A \sigma_{max}\frac{y}{c}dA$$

Multiplying this by distance y gives the resultant moment of the force that must be equal to the internal bending moment M. That is,

$$M = \sum F(y) = \int_A \sigma_{max}\frac{y}{c}dA(y)$$

where the terms are labeled: Force, Stress, Area, Moment arm.

Simplifying, we obtain

$$M = \frac{\sigma_{max}}{c}\int_A y^2 dA$$

By definition, and as illustrated in Chapter 6, the last term in this equation is the moment of inertia I of the area of the cross section with respect to its centroidal axis.

$$I = \int_A y^2 dA$$

Then

$$M = \frac{\sigma_{max}}{c}I$$

Solving for σ_{max} yields

$$\sigma_{max} = \frac{Mc}{I}$$

This is the form of the flexure formula shown earlier as Equation (12–1).

Two example problems are shown next illustrating the use of the flexure formula.

12–6 APPLICATIONS—BEAM ANALYSIS

Example Problem 12–2

The tee section shown in Figure 12–11 is from a simply supported beam that carries a bending moment of 100 000 lb·in due to a load on the top surface. It has been determined that $I = 18.16$ in^4. The centroid of the section is 3.25 in up from the bottom of the beam. Compute the stress due to bending in the beam at the six axes a to f indicated in the figure. Then plot a graph of stress versus position in the cross section.

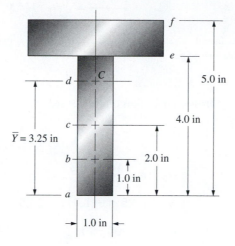

FIGURE 12–11 Tee section for beam in Example Problem 12–2.

Solution **Objective:** Compute the bending stress at six axes a–f. Plot a graph of stress versus position in the cross section.

Given: $M = 100\,000$ lb·in. Tee shape of cross section shown in Figure 12–11. $I = 18.16$ in^4. $\bar{Y} = 3.25$ in from bottom of beam.

Analysis: Equation (12–1) will be used to compute σ_{max}, which occurs at the bottom of the beam (axis a), because that is the location of the outermost fiber of the beam, farthest from the centroidal axis. Then the stress at other axes will be computed using Equation (12–2), giving results to four significant figures to demonstrate the principle. See Figure 12–12 for values of y.

Results: **At Axis a:** In Equation (12–1), use $c = \bar{Y} = 3.25$ in.

$$\sigma_{max} = \sigma_a = \frac{Mc}{I} = \frac{(100\,000 \text{ lb·in})(3.25 \text{ in})}{18.16 \text{ in}^4}$$

$$\sigma_a = 17\,900 \text{ psi} \quad \text{(tension)}$$

FIGURE 12–12 Data for Example Problem 12–2.

FIGURE 12–13   Stress distribution on the tee section for Example Problem 12–2.

At Axis b:

$$\sigma_b = \sigma_{max}\frac{y_b}{c} = \sigma_a\frac{y_b}{c}$$

$$y_b = 3.25\,\text{in} - 1.0\,\text{in} = 2.25\,\text{in}$$

$$\sigma_b = 17\,900\,\text{psi} \times \frac{2.25}{3.25} = 12\,390\,\text{psi} \quad \text{(tension)}$$

At Axis c:

$$y_c = 3.25\,\text{in} - 2.0\,\text{in} = 1.25\,\text{in}$$

$$\sigma_c = 17\,900\,\text{psi} \times \frac{1.25}{3.25} = 6883\,\text{psi} \quad \text{(tension)}$$

At Axis d: At the centroid $y_d = 0$ and $\sigma_d = 0$.

At Axis e:

$$y_e = 4.0\,\text{in} - 3.25\,\text{in} = 0.75\,\text{in}$$

$$\sigma_e = 17\,900\,\text{psi} \times \frac{0.75}{3.25} = 4130\,\text{psi} \quad \text{(compression)}$$

At Axis f:

$$y_f = 5.0\,\text{in} - 3.25\,\text{in} = 1.75\,\text{in}$$

$$\sigma_f = 17\,900\,\text{psi} \times \frac{1.75}{3.25} = 9637\,\text{psi} \quad \text{(compression)}$$

The graph of these data is shown in Figure 12–13.

LESSON 19

Comment: Notice the linear variation of stress with distance from the neutral axis and that the stresses above the neutral axis are compressive while those below are tensile.

Example Problem 12–3   Figure 12–14 shows the bending moment diagram for a 25-ft-long beam in a large machine structure. It has been proposed that the beam be made from a standard W14×43 steel shape. Compute the maximum stress due to bending in the beam.

Solution   Objective: Compute the maximum stress due to bending.

Given: Bending moment diagram shown in Figure 12–14. W14×43 beam shape.

FIGURE 12–14
Bending moment
diagram for beam in
Example Problem 12–3.

Analysis: Use Equation (12–1). In Figure 12–14, identify the maximum bending moment on the beam to be 91 113 lb · ft at point *F*. Find the values of *I* and *c* from the table of properties for W-beam shapes in Appendix A–7.

$$I = 428 \text{ in}^4$$

$$c = \text{depth}/2 = 13.70 \text{ in}/2 = 6.85 \text{ in}$$

Results: $$\sigma_{max} = \frac{Mc}{I} = 91\,113 \text{ lb} \cdot \text{ft} \times \frac{12 \text{ in}}{\text{ft}} \times \frac{6.85 \text{ in}}{428 \text{ in}^4} = 17\,500 \text{ psi}$$

Comment: This maximum stress would occur as a tensile stress on the bottom surface of the beam and as a compressive stress on the top surface at the position *F*. Because this is a relatively long beam, it must be laterally braced as described in Reference 2.

12–7 APPLICATIONS—BEAM DESIGN AND DESIGN STRESSES

To design a beam, its material, length, placement of loads, placement of supports, and the size and shape of its cross section must be specified. Normally, the length, placement of loads, and placement of supports are given by the requirements of the application. The material specification and the size and shape of the cross section are determined by the designer.

The primary duty of the designer is to ensure the safety of the design. This requires a stress analysis of the beam and a decision concerning the allowable or design stress to which the chosen material may be subjected. The examples presented here will concentrate on these items. Also of interest to the designer are cost, appearance, physical size, weight, compatibility of the design with other components of the machine or structure, and the availability of the material or beam shape.

Two basic approaches will be shown for beam design. One involves the specification of the *material* from which the beam will be made and the general *shape* of the beam (circular, rectangular, W-beam, etc.), with the subsequent determination of the required dimensions of the cross sec-

tion of the beam. The second requires specifying the *dimensions* and *shape* of the beam and then computing the required strength of a material from which to make the beam. Then the actual material is specified.

Design Stress for Metals— General Guidelines

When specifying design stresses, it is important to keep in mind that both tensile and compressive stresses are produced in beams. If the material is reasonably homogeneous and isotropic having the same strength in either tension or compression, then design is based on the highest stress developed in the beam. When a material has different strengths in tension and compression, as is the case for cast iron or wood, then both the maximum tensile and the maximum compressive stresses must be checked.

The approach used most often in this book to determine design stresses is similar to that first described in Chapter 9, and it would be wise to review that discussion at this time. Table 12–1 lists the design stress guidelines we will use for beams in machines and special structures under conditions where loads and material properties are well known. Larger design factors may be used where greater

TABLE 12–1 Design Stress Guidelines—Bending Stresses

Manner of loading	Ductile material	Brittle material
Static	$\sigma_d = s_y/2$	$\sigma_d = s_u/6$
Repeated	$\sigma_d = s_u/8$	$\sigma_d = s_u/10$
Impact or shock	$\sigma_d = s_u/12$	$\sigma_d = s_u/15$

uncertainty exists. We will use Table 12–1 for problems in this book involving metals, unless stated otherwise.

Design Stresses from Selected Codes

Table 12–2 gives a summary of specifications for design bending stresses as defined by the American Institute of Steel Construction (AISC) for structural steel and by the Aluminum Association (AA) for aluminum alloys. These data pertain to beams under static loads such as those found in building-type structures and use the allowable stress design method.

Additional analysis is required for the parts of beams under compressive stresses because of the possibility of local buckling, especially in shapes having thin sections or extended flanges. Long beams must also be checked for the possibility of twisting. Lateral supports for the compression flanges of long beams are often required to resist the tendency for the beam to twist. See References 1 and 2 for more detailed discussions of these specifications. Wide-flange beam shapes (W-shapes) should be checked to ensure that they are compact as defined in Reference 2. Noncompact sections require the use of a lower design stress.

Load and Resistance Factor Design (LRFD)

LRFD was first discussed in Chapter 9 in the context of design of axially loaded members. A major objective of LRFD is to account for the probability that the maximum dead load and the maximum live load would occur at the same time over the life of the member. As a result, loading on beams must be separated carefully between that portion that is dead load (primarily the weight of the structure itself) and that which is live load (occupancy, snow, rain,

TABLE 12–2 Design Stresses from Selected Codes—Bending Stresses-Static Loads on Building-Like Structures

Structural steel (AISC):
$\sigma_d = s_y/1.5 = 0.66\,s_y$

Aluminum (Aluminum Association):
$\sigma_d = s_y/1.65 = 0.61\,s_y$ or $\sigma_d = s_u/1.95 = 0.51\,s_u$
whichever is lower

wind, and seismic). Each type is treated with different load factors. Factors are also applied to the basic capacity of the beam, typically bending moment. Then the factored loads are compared with the factored resistance factor to ensure the safety of the member. The LRFD beam design process also includes organized steps to check for the possibility of local buckling of the flanges, crippling of the web, the need for lateral bracing, and details of the beam near support points and where concentrated loads are applied. Details of this process are described in References 2 and 3 and are typically taught in following courses such as structural steel design.

Beam Deflection

Failure of a beam to perform to its expected level may occur because of excessive deflection even though the beam has not failed any strength criterion. This topic will be discussed in Chapter 14. At this point, we will only mention that beam deflection limits are often stated in terms of a proportion of the span (L) of the beam between supports. For example, some applications limit deflection to $L/180$, $L/240$, or $L/360$, depending on the desired rigidity of the structure. You will learn from Chapter 14 that beam deflections are, in general, inversely proportional to the moment of inertia (I) of the area of the cross section of the beam. Equation (12–1) shows that the stress is also dependent on I, as well as the depth of the beam as it relates to the distance c from the centroidal axis to the outside of the beam. Therefore, you should consider beam design based on strength to be tentative until the deflection can be evaluated.

Design Stresses for Nonmetals

When problems involve nonmetals such as wood, plastics, and composites, the concept of yield strength is not typically used. Furthermore, the strengths listed in tables are often based on statistical averages of many tests. Variations in material composition and structure can lead to variations in strength properties. Whenever possible, the actual material to be used in a structure should be tested to determine its strength.

Appendix A–19 lists *allowable* stress values for three species of wood according to the listed grades for applications in building structures and similar static load uses. If load conditions are very well known, a beam can be loaded up to the listed bending stress values. To the extent that there are uncertainties in the conditions of loading, design factors can be applied to the listed values resulting in lower design stresses. No firm guidelines are given here and testing is advised. We will use the listed allowable stresses unless otherwise stated. See also Reference 7.

The properties of plastics listed in Appendix A–20 can be considered typical for the listed types. It should be noted that there are many variables involved in producing plastics, and it is important to obtain more complete data from manufacturers or by testing the actual material to be used.

Also, plastics differ dramatically from one to another in their ability to withstand repeated loading, shock, and impact. In this chapter, we will take the *flexural strength* from Appendix A–20 to be the representative strength of the listed plastics when used in beams. We will assume that failure is imminent at these stress levels. For general, static load cases, we will apply a design factor of $N = 2$ to those values to determine the design stress.

Composites have many advantages when applied to the design of beams because the placement of material can be optimized to provide efficient, lightweight beams. The resulting structure is typically not homogeneous, so the properties are highly anisotropic. Therefore, the flexure formula as stated in Equations (12–1) and (12–2) cannot be relied on to give accurate values of stress. General approaches to using composites in beams will be discussed later in this chapter.

12–8 SECTION MODULUS AND DESIGN PROCEDURES

The stress analysis will require the use of the flexure formula

$$\sigma_{max} = \frac{Mc}{I}$$

A modified form is desirable for cases in which the determination of the dimensions of a section is to be done. Notice that both the moment of inertia I and the distance c are geometrical properties of the cross-sectional area of the beam. Therefore, the quotient I/c is also a geometrical property. For convenience, we can define a new term, *section modulus*, denoted by the letter S.

Section Modulus ▽

$$S = \frac{I}{c} \qquad (12\text{--}4)$$

The flexure formula then becomes

$$\sigma_{max} = \frac{M}{S} \qquad (12\text{--}5)$$

This is the most convenient form for use in design. Example problems will demonstrate the use of the section modulus. Appendix A–1 gives formulas for S for some shapes. Appendices A–4 through A–13 include the value of S for structural shapes.

Design Procedures

Here are two approaches to design problems. The first applies when the loading pattern and material are known and the shape and dimensions of the cross section of the beam are to be determined. The second procedure applies when the loading pattern, the shape of the beam cross section, and its dimensions have been specified and the objective is to specify a suitable material for the beam to ensure safety.

A. Design procedure to determine the required dimensions for a beam.

Given: Loading pattern and material from which the beam is to be made.

1. Determine the maximum bending moment in the beam, typically by drawing the complete shearing force and bending moment diagrams.
2. Determine the applicable approach for specifying the design stress from Section 12–7.
3. Compute the value of the design stress.
4. Using the flexure formula expressed in terms of the section modulus, Equation (12–5), solve for the section modulus, S. Then let the maximum stress equal the design stress and compute the required minimum value of the section modulus to limit the actual stress to no more than the design stress.
5. For a specially designed beam shape, determine the required minimum dimensions of the shape to achieve the required section modulus. Then specify the next larger convenient dimensions using the tables of preferred sizes in Appendix A–2.
6. To select a standard structural shape such as those listed in Appendices A–4 through A–13, consult the appropriate table of data and specify one having at least the value of the section modulus, S, computed in step 4. Typically, it is recommended that the lightest suitable beam shape be specified because the cost of the beam made from a given material is generally directly related to its weight. Reference 3 includes extensive tables for beam shapes with their section modulus values ordered by the weight of the section to facilitate the selection of the lightest beam. Where space limitations exist, the actual dimensions of the shape must be considered.

B. Design procedure for specifying a material for a given beam.

Given: Loading pattern, shape, and dimensions for the beam.

1. Determine the maximum bending moment in the beam, typically by drawing the complete shearing force and bending moment diagrams.
2. Compute the section modulus for the beam cross section.
3. Compute the maximum bending stress from the flexure formula, Equation (12–5).
4. Determine the applicable approach for specifying the design stress from Section 12–7 and specify a suitable design factor.
5. Set the computed maximum stress from step 3 equal to the formula for the design stress.
6. Solve for the required minimum value of the strength of the material, either s_y or s_u.
7. Select the type of material from which the beam is to be made such as steel, aluminum, cast iron, titanium, or copper.
8. Consult the tables of data for material properties such as those in Appendices A–14 through A–20, and identify a set of candidate materials having at least the required strength.
9. Specify the material to be used, considering any factor appropriate to the application, such as ductility, cost, corrosion potential, ease of fabrication, or weight. For metals, it is essential to specify the condition of the material in addition to the alloy.

Example Problems 12–4 through 12–6 illustrate these approaches.

Example Problem 12–4   A beam is to be designed to carry the static loads shown in Figure 12–15. The cross section of the beam will be rectangular and made from ASTM A36 structural steel plate having a thickness of 1.25 in. Specify a suitable height for the cross section.

Cross section of beam—Section A-A

FIGURE 12–15   Loading and cross section for beam in Example Problem 12–4.

Solution   Objective: Specify the height of the rectangular cross section.

Given: Loading pattern shown in Figure 12–15. ASTM A36 structural steel. Width of the beam to be 1.25 in. Static loads.

Analysis: We will use the design procedure A from this section.

Results: **Step 1:** Figure 12–16 shows the completed shearing force and bending moment diagrams. The maximum bending moment is $45\,900\ \text{lb}\cdot\text{in}$ between the loads, in the middle of the beam span from $x = 3.0$ ft to 9 ft.

Step 2: From Table 12–1, for static load on a ductile material,

$$\sigma_d = S_y/2$$

Step 3: From Appendix A–16, $s_y = 36\,000$ psi for ASTM A36 steel. For a static load, a design factor of $N = 2$ based on yield strength is reasonable. Then

$$\sigma_d = \frac{S_y}{N} = \frac{36\,000\ \text{psi}}{2} = 18\,000\ \text{psi}$$

Step 4: The required S is

$$S = \frac{M}{\sigma_d} = \frac{45\,900\ \text{lb}\cdot\text{in}}{18\,000\ \text{lb/in}^2} = 2.55\ \text{in}^3$$

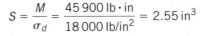

FIGURE 12–16   Load, shearing force, and bending moment diagrams for Example Problem 12–4.

Step 5: The formula for the section modulus for a rectangular section with a height h and a thickness b is

$$S = \frac{I}{c} = \frac{bh^3}{12(h/2)} = \frac{bh^2}{6}$$

For the beam in this design problem, b will be 1.25 in. Then solving for h gives

$$S = \frac{bh^2}{6}$$

$$h = \sqrt{\frac{6S}{b}} = \sqrt{\frac{6(2.55\ \text{in}^3)}{1.25\ \text{in}}}$$

$$h = 3.50\ \text{in}$$

Comment: The computed minimum value for h is a convenient size. Specify $h = 3.50$ in. The beam will then be rectangular in shape, with dimensions of 1.25 in $\times$ 3.50 in. Note that because the beam is rather long, 12 ft, there may be a tendency for it to deform laterally because of elastic instability. Lateral bracing may be required. Also, deflection should be checked using the methods discussed in Chapter 14.

Example Problem 12–5

The roof of an industrial building is to be supported by wide-flange beams spaced 4 ft on centers across a 20-ft span, as sketched in Figure 12–17. The roof will be a poured concrete slab, 4 in thick. The design live load on the roof is 200 lb/ft². Specify a suitable wide-flange beam that will limit the stress in the beam to the design stress for ASTM A36 structural steel using the AISC specification.

Solution   Objective: Specify a suitable wide-flange beam.

Given: Loading pattern in Figure 12–17. Design stress from AISC specifications for ASTM A36 structural steel.

Analysis: Design procedure A from this section will be used.

Results: **Step 1:** We must first determine the load on each beam of the roof structure. Dividing the load evenly among adjacent beams would result in each beam carrying a 4-ft-wide portion of the roof load. In addition to the 200-lb/ft² live load, the weight of the concrete slab offers a sizeable load. In Section 8–10 we find that the concrete weighs 150 lb/ft³. Then each square foot of the roof, 4.0 in thick, would weigh 50 lb. This is called the *dead load*. Then the total loading due to the roof is 250 lb/ft². Now, notice that each foot of length of the beam carries 4 ft² of the roof. Therefore, the load on the beam is a uniformly distributed load of 1000 lb/ft. Figure 12–18 shows the loaded beam and the shearing force and bending moment diagrams. The maximum bending moment is 50 000 lb·ft.

Step 2 and 3: Table 12–2 calls for the design stress to be

$$\sigma_d = 0.66\, s_y$$

FIGURE 12–17   Roof structure for building in Example Problem 12–5.

FIGURE 12–18 Load, shearing force, and bending moment diagrams for Example Problem 12–5.

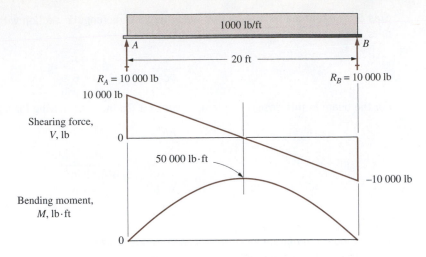

From Appendix A–16, the yield strength of ASTM A36 structural steel is 36 000 psi. Then

$$\sigma_d = 0.66 \, s_y = 0.66(36\,000 \text{ psi}) = 23\,760 \text{ psi}$$

Step 4: In order to select a wide-flange beam, the required section modulus must be calculated.

$$\sigma = \frac{M}{S}$$

$$S = \frac{M}{\sigma_d} = \frac{50\,000 \text{ lb} \cdot \text{ft}}{23\,760 \text{ lb/in}^2} \times \frac{12 \text{ in}}{\text{ft}} = 25.3 \text{ in}^3$$

Step 5: This step does not apply in this problem.

Step 6: A beam must be found from Appendix A–7 that has a value of S greater than 25.3 in³. In considering alternatives, of which there are many, you should search for the lightest beam that will be safe, since the cost of the beam is based on its weight. Some possible beams are

$$W14\times26: \quad S = 35.3 \text{ in}^3, 26 \text{ lb/ft}$$
$$W12\times30: \quad S = 38.6 \text{ in}^3, 30 \text{ lb/ft}$$

Of these, the W14×26 would be preferred because it is the lightest.

Comment: In the final design of the structure, lateral bracing would be required as defined in Reference 2. Also, deflection of the beam should be checked using the methods described in Chapter 14.

Example Problem 12–6

A support beam for a conveyor system carries the loads shown in Figure 12–19. Support points are at points A and C. The 20 kN load at B and the 10 kN load at D are to be applied repeatedly many thousands of times. It has been proposed to use a 50-mm-diameter circular steel bar for the beam. Specify a suitable steel for the beam.

Solution   Objective: Specify a suitable steel.

Given: Loading pattern in Figure 12–19. Loads are repeated.
Beam is to be circular, $D = 50$ mm.

Analysis: Use the type B procedure given in this section.

Results: **Step 1:** Figure 12–20 shows the completed shearing force and bending moment diagrams. The maximum bending moment is 2.00 kN · m at the support at C.

FIGURE 12–19 Beam loading and cross section for beam in Example Problem 12–6.

FIGURE 12–20
Shearing force and
bending moment dia-
grams for Example
Problem 12–6.

Step 2: Appendix A–1 gives the formula for S for a round bar.

$$S = \pi D^3/32 = \pi(50\text{ mm})^3/32 = 12\,272\text{ mm}^3$$

Step 3: Using Equation (12–5),

$$\sigma_{max} = \frac{M}{S} = \frac{2.0\text{ kN}\cdot\text{m}}{12\,272\text{ mm}^3} \times \frac{10^3\text{ mm}}{\text{m}} \times \frac{10^3\text{ N}}{\text{kN}}$$

$$\sigma_{max} = 163\text{ N/mm}^2 = 163\text{ MPa}$$

Step 4: We can use Table 12–1 to determine an appropriate formula for design stress. The steel selected should be highly ductile because of the repeated load. Then we will use $\sigma_d = s_u/8$.

Step 5: Let $\sigma_{max} = 163\text{ MPa} = \sigma_d = s_u/8$

Step 6: Solving for s_u gives

$$s_u = 8(\sigma_{max}) = 8(163\text{ MPa}) = 1304\text{ MPa}$$

Step 7: It was decided to use steel.

Step 8: Appendix A–14 lists several common steel alloys. From that table, we can select candidate
materials that have good ductility and an ultimate strength of at least 1304 MPa. Four are listed as
follows.

AISI 1080 OQT 700; $s_u = 1303$ MPa; 12% elongation

AISI 1141 OQT 700; $s_u = 1331$ MPa; 9% elongation

AISI 4140 OQT 700; $s_u = 1593$ MPa; 12% elongation

AISI 5160 OQT 900; $s_u = 1351$ MPa; 12% elongation

Step 9: For applications to beams carrying repeated loads, it is typical to use a medium carbon
steel. Either the AISI 4140 or the AISI 5160 could be used. With 12% elongation, ductility should
be adequate.

Comment: Note in Appendix A–14 that AISI 4140 OQT 900 has an ultimate strength of 1289 MPa and 15%
elongation. The strength is within 2% of the computed value. It may be suitable to specify this material to gain better ductility. A slight reduction in the design factor would result. Because the values in
Table 12–1 are somewhat conservative this would normally be justified. Alternatively, a larger diameter
bar could be used resulting in a lower design stress. It may then be possible to use a lower cost steel.

12–9 STRESS CONCENTRATIONS

The conditions specified for valid use of the flexure formula in Section 12–3 included the statement that the beam must have a uniform cross section. Changes in cross section result in higher local stresses than would be predicted from the direct application of the flexure formula. Similar observations were made in earlier chapters concerning direct axial stresses and torsional shear stresses. The use of *stress concentration factors* will allow the analysis of beams that do include changes in cross section.

To visualize one example of stress concentration, refer to Figure 12–21, which shows a photoelastic model of a simply supported beam carrying two identical loads. The loading pattern is similar to that used in Example Problem 12–4 and shown in Figures 12–15 and 12–16. The bending moment is the same on any section in the middle of the beam between the two loads. At the very middle of the beam are two semicircular notches that present a local change in geometry. Two effects occur. The removal of material by the notches reduces the section modulus of the beam at that location and causes the increase in stress. In addition, the sudden change in geometry produces higher stresses in the near vicinity of each notch. This is evidenced by the tightly spaced fringe lines around the notches. Note that the effect of the notches is very local, dissipating a short distance to either side of the notches. Published stress concentration factors allow the calculation of such locally increased stresses.

You can also see evidence of the local contact stresses at points of support and at the locations of the applied loads. This was discussed in Section 12–4 as the St. Venant principle in conjunction with the photoelastic model shown in Figure 12–7.

Another example of stress concentrations for a beam is the portion of a circular shaft shown in Figure 12–22. In the design of shafts for carrying power-transmitting elements, the use of steps in the diameter is encountered frequently. Examples were shown in Chapter 10, where torsional shear stresses were discussed. Considering the shaft as a beam subjected to bending moments, there would be stress concentrations at the shoulder (2), the keyseat (3), and the groove (4) as shown in Figure 12–22.

At sections where stress concentrations occur, the stress due to bending would be calculated from a modified form of the flexure formula.

Flexure Formula with Stress Concentration ▼

$$\sigma_{max} = \frac{McK_t}{I} = \frac{MK_t}{S} \tag{12–6}$$

The stress concentration factor K_t is found experimentally, with the values reported in graphs such as those in Appendix A–22, cases 4, 5, 8, 9, 10, and 11. See also Reference 10 in Chapter 9 for additional cases.

FIGURE 12–22 Portion of a shaft with several changes in cross section producing stress concentrations.

Example Problem 12–7

Figure 12–22 shows a portion of a round shaft where a gear is mounted. A bending moment of 30 N·m is applied at this location. Compute the stress due to bending at sections 1, 2, 3, 4, and 5.

Solution

Objective: Compute the stress due to bending at sections 1, 2, 3, 4, and 5.

Given: Beam geometry in Figure 12–22. $M = 30$ N·m.

Analysis: Stress concentrations must be considered because of the several changes in geometry in the area of interest. Equation (12–6) will be used to compute the maximum stress at each section. Appendix A–22 is the source of data for stress concentration factors, K_t.

Results: **Section 1.** This is the part of the shaft where the diameter is 40 mm and no changes in geometry occur. Therefore, the stress concentration factor is 1.0 and the stress is the same as that which would be computed from the flexure formula alone. That is, $\sigma = M/S$.

$$S_1 = \frac{\pi(D)^3}{32} = \frac{\pi(40 \text{ mm})^3}{32} = 6283 \text{ mm}^3 \quad \text{[Appendix A–1]}$$

$$\sigma_1 = \frac{MK_{t1}}{S_1} = \frac{(30 \text{ N·m})(1.0)}{6283 \text{ mm}^3} \times \frac{10^3 \text{ mm}}{m} = 4.77 \text{ N/mm}^2 = 4.77 \text{ MPa}$$

Section 2. The step in the shaft causes a stress concentration to occur. Then the stress is

$$\sigma_2 = \frac{MK_{t2}}{S_2}$$

The smaller of the diameters at section 2 is used to compute S_2. From Appendix A–1,

$$S_2 = \frac{\pi d^3}{32} = \frac{\pi(25 \text{ mm})^3}{32} = 1534 \text{ mm}^3$$

The value of K_{t2} depends on the ratios r/d and D/d. (See Appendix A–22–9.)

$$\frac{r}{d} = \frac{2 \text{ mm}}{25 \text{ mm}} = 0.08$$

$$\frac{D}{d} = \frac{40 \text{ mm}}{25 \text{ mm}} = 1.60$$

From Appendix A–22–9, $K_{t2} = 1.87$. Then

$$\sigma_2 = \frac{MK_t}{S_2} = \frac{(30 \text{ N·m})(1.87)}{1534 \text{ mm}^3} \times \frac{10^3 \text{ mm}}{m} = 36.6 \text{ N/mm}^2$$

$$\sigma_2 = 36.6 \text{ MPa}$$

Section 3. The keyseat causes a stress concentration factor of 2.0 as listed in Appendix A–22–11. S_3 is based on the full 25-mm diameter of the shaft. Therefore, $S_3 = S_2 = 1534 \text{ mm}^3$. Then

$$\sigma_3 = \frac{MK_{t3}}{S_3} = \frac{(30 \text{ N·m})(2.0)}{1534 \text{ mm}^3} \times \frac{10^3 \text{ mm}}{m} = 39.1 \text{ N/mm}^2$$

$$\sigma_3 = 39.1 \text{ MPa}$$

Section 4. The groove requires the use of Appendix A–22–9 again to find K_{t4}. Note that the nominal stress is based on the root diameter of the groove, d_g. For the groove,

$$\frac{r}{d_g} = \frac{1.2 \text{ mm}}{20 \text{ mm}} = 0.06$$

$$\frac{d}{d_g} = \frac{25 \text{ mm}}{20 \text{ mm}} = 1.25$$

Then $K_{t4} = 1.93$. The section modulus at the root of the groove is

$$S_4 = \frac{\pi d_g^3}{32} = \frac{\pi(20 \text{ mm})^3}{32} = 785 \text{ mm}^3$$

FIGURE 12–23 Variation in bending stress in shaft from Example Problem 12–7.

Now the stress at section 4 is

$$\sigma_4 = \frac{MK_{t4}}{S_4} = \frac{(30\ \text{N} \cdot \text{m})(1.93)}{785\ \text{mm}^3} \times \frac{10^3 \text{mm}}{\text{m}} = 73.8\ \text{N/mm}^2$$

$$\sigma_4 = 73.8\ \text{MPa}$$

Section 5. This is the part of the shaft where the diameter is 25 mm and no changes in geometry occur. Therefore, the stress concentration factor is 1.0 and $S_5 = 1534\ \text{mm}^3$ as computed for Section 2.

$$\sigma_5 = \frac{MK_{t5}}{S_5} = \frac{(30\ \text{N} \cdot \text{m})(1.0)}{(1534\ \text{mm}^3)} \times \frac{10^3\ \text{mm}}{\text{m}} = 19.6\ \text{N/mm}^2 = 19.6\ \text{MPa}$$

Comment: Observe the large variation in stress levels that exist over this relatively small portion of the shaft. Figure 12–23 is a graph showing that variation. The stress at Section 4 is by far the largest because of the small diameter at the base of the groove and the rather high stress concentration factor, K_{t4}. Design of the shaft to determine a suitable material must use this level of stress as the actual maximum stress.

Example Problem 12–8   Figure 12–24 shows a cantilever bracket carrying a partial uniformly distributed load over the left 10 in and a concentrated load at its right end. The geometry varies at sections *A*, *B*, and *C* as shown. The bracket is made from aluminum 2014-T6, and it is desired to have a minimum design factor of 8 based on the ultimate strength. Evaluate the acceptability of the given design. If any section is unsafe, propose a redesign that will result in a satisfactory stress level. Consider stress concentrations at sections *B* and *C*. The attachment at *A* is blended smoothly such that it can be assumed that $K_t = 1.0$.

Solution   Objective: Evaluate the beam shown in Figure 12–24 to ensure that the minimum design factor is 8 based on ultimate strength. If not, redesign the beam.

Given: Loading and beam geometry in Figure 12–24; Aluminum 2014-T6.

Analysis: 1. The shearing force and bending moment diagrams will be drawn.
2. The design stress will be computed from $\sigma_d = s_u/8$.
3. The stress will be computed at sections *A*, *B*, and *C*, considering stress concentrations at *B* and *C*. These are the three likely points of failure because of either bending moment or stress concentration. Everywhere else will have a lower bending stress. At each section, the stress will be computed from $\sigma = MK_t/S$, and the values of *M*, K_t, and *S* must be determined at each section.

FIGURE 12–24 Beam and loading for Example Problem 12–8.

4. The computed stresses will be compared with the design stress.
5. For any section having a stress higher than the design stress, a redesign will be proposed and the stress will be recomputed to verify that it is safe as redesigned.

Results: **Step 1:** Figure 12–25 shows the completed shearing force and bending moment diagrams. Note that the values of the bending moment at sections B and C have been computed. You should verify the given values.

Step 2: From Appendix A–18, we find $s_u = 70\,\text{ksi} = 70\,000\,\text{psi}$. Then

$$\sigma_d = s_u/8 = (70\,000)/8 = 8750\,\text{psi}$$

Step 3 and 4: At each section, $\sigma = MK_t/S = K_t\sigma_{nom}$.

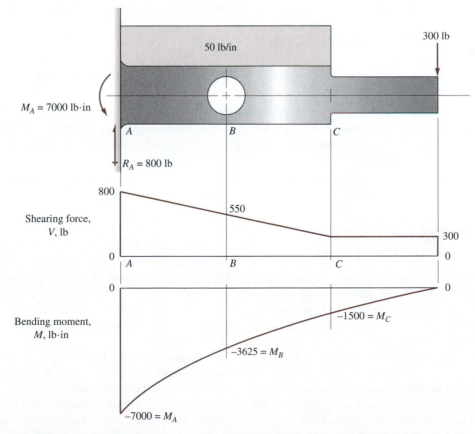

FIGURE 12–25 Shearing force and bending moment diagrams for Example Problem 12–8.

Section A. $K_t = 1.0$ (given). $M_A = 7000$ lb·in.
Dimensions: $b = 0.50$ in; $h = 3.25$ in; rectangle.

$$S = bh^2/6 = (0.50 \text{ in})(3.25 \text{ in})^2/6 = 0.880 \text{ in}^3$$

$$\sigma = \frac{(7000 \text{ lb·in})(1.0)}{0.880 \text{ in}^3} = 7953 \text{ psi} < 8750 \text{ psi} \quad \textbf{OK}$$

Section B. $M_B = 3625$ lb·in. Find K_t from Appendix A–22–4.
Dimensions: $t = b = 0.50$ in; $w = h = 3.25$ in; $d = 2.25$ in
$d/w = 2.25/3.25 = 0.692$; then $K_t = 1.40$ (curve C).

$$\sigma = K_t \sigma_{nom} = \frac{K_t(6Mw)}{(w^3 - d^3)t} = \frac{(1.40)(6)(3625)(3.25)}{[(3.25)^3 - (2.25)^3](0.50)}$$

$$\sigma = 8629 \text{ psi} < 8750 \text{ psi} \quad \textbf{OK}$$

Section C. $M_C = 1500$ lb·in. Find K_t from Appendix A–22–10.
Dimensions: $t = 0.50$ in; $H = 3.25$ in; $h = 2.00$ in; $r = 0.08$ in
$H/h = 3.25/2.0 = 1.625$; $r/h = 0.08/2.00 = 0.04$. Then $K_t = 2.40$.

$$\text{Section modulus} = S = th^2/6 = (0.50)(2.00)^2/6 = 0.333 \text{ in}^3$$

$$\sigma = \frac{MK_t}{S} = \frac{(1500 \text{ lb·in})(2.40)}{0.333 \text{ in}^3} = 10\,800 \text{ psi} \quad \textbf{too high}$$

Step 5 Proposed Redesign at C: Because the stress concentration factor at section C is quite high, increase the fillet radius. The maximum allowable stress concentration factor is found by solving the stress equation for K_t and letting $\sigma = \sigma_d = 8750$ psi. Then

$$K_t = \frac{S\sigma_d}{M} = \frac{(0.333 \text{ in}^3)(8750 \text{ lb/in}^2)}{1500 \text{ lb·in}} = 1.94$$

From Appendix A–22–10, the minimum value of $r/h = 0.08$ to limit K_t to 1.94. Then $r_{min} = 0.08(h) = 0.08(2.00) = 0.16$.
Let $r = 0.20$ in; $r/h = 0.20/2.00 = 0.10$; $K_t = 1.80$. Then

$$\sigma = \frac{MK_t}{S} = \frac{(1500 \text{ lb·in})(1.80)}{0.333 \text{ in}^3} = 8100 \text{ psi} \quad \textbf{OK}$$

Comment: This problem is a good illustration of the necessity of analyzing any point within a beam where high stress may occur because of high bending moment, high stress concentration, small section modulus, or some combination of these. It also demonstrates one method of redesigning a beam to ensure safety.

12–10 FLEXURAL CENTER OR SHEAR CENTER

The flexure formula is valid for computing the stress in a beam provided the applied loads pass through a point called the *flexural center*, or sometimes, the *shear center*. If a section has an axis of symmetry and if the loads pass through that axis, then they also pass through the flexural center. The beam sections shown in Figure 12–6 are of this type.

For sections loaded away from an axis of symmetry, the position of the flexural center, indicated by Q, must be found. Such sections were identified in Figure 12–7.

In order to result in pure bending, the loads must pass through Q, as shown in Figure 12–26. If they don't, then a condition of *unsymmetrical bending* occurs and other analyses would have to be performed, which are not discussed in this book. The sections of the type shown in Figure 12–26 are used frequently in structures. Some lend themselves nicely to production by extrusion and are therefore very economical. Because of the possibility of producing unsymmetrical bending, care must be taken in their application.

An example of a beam application in which unsymmetrical bending would occur is shown in Figure 12–27. An American Standard channel is installed as a cantilever with its web vertical. When unloaded, the channel is straight as shown in part (a) of the figure. However, with a vertical load applied to the top flange as shown in part (b) of the figure, the channel would tend to twist in addition to bending downward. Clearly this is not pure bending and the

FIGURE 12–26   Location of flexural center Q.

flexure formula, Equation (12–1), would not give an accurate prediction of the stress condition in the channel.

Refer to Figure 12–26 to see that the flexural center, point Q, for a channel with its web vertical, is well to the left of the web, outside the channel itself. In order for pure bending to be produced, the line of action of the applied load must pass through this point. You might design a bracket like that shown in Figure 12–27(c) through which to apply the load.

Notice that you may apply the channel as a beam with the legs down or up and pass the load through the axis of symmetry as shown in part (d) of the figure. Then the bending action would not produce twisting and the flexure formula can be used to compute the bending stress.

The following two example problems demonstrate the method of locating the flexural center, Q.

(a) Unloaded channel

(b) Loaded channel showing twisting due to unsymmetrical bending

Load

Q

Flexural center

C — Centroid

(c) Load applied through flexural center to avoid twisting

Load

+ C

(d) Load applied to web through axis of symmetry to avoid twisting

FIGURE 12–27   Illustration of unsymmetrical bending and two ways to avoid it.

Example Problem 12–9

Determine the location of the flexural center for the two sections shown in Figure 12–28.

Solution   Objective: Locate the shear center, Q, for the two shapes.

Given: Shapes in Figure 12–28: channel in 12–28(a); hat section in 12–28(b).

Analysis: The general location of the shear center for each shape is shown in Figure 12–26 along with the means of computing the value of e that locates Q relative to specified features of the shapes.

Results: **Channel Section (a):** From Figure 12–26, the distance e to the flexural center is

$$e = \frac{b^2 h^2 t}{4 I_x}$$

Note that the dimensions b and h are measured to the middle of the flange or web. Then b = 40 mm and h = 50 mm. Because of symmetry about the X axis, I_x can be found by the difference

FIGURE 12–28   Beam sections for which the flexural centers are computed in Example Problem 12–9.

(a) Channel

(b) Hat section

between the value of *I* for the large outside rectangle (54 mm by 42 mm) and the smaller rectangle removed (46 mm by 38 mm).

$$I_x = \frac{(42)(54)^3}{12} - \frac{(38)(46)^3}{12} = 0.243 \times 10^6 \text{ mm}^4$$

Then

$$e = \frac{(40)^2(50)^2(4)}{4(0.243 \times 10^6)} \text{ mm} = 16.5 \text{ mm}$$

This dimension is drawn to scale in Figure 12–28(a).

Hat Section (b): Here the distance *e* is a function of the ratios *c/h* and *b/h*.

$$\frac{c}{h} = \frac{10}{30} = 0.3$$

$$\frac{b}{h} = \frac{30}{30} = 1.0$$

Then from Figure 12–26, *e/h* = 0.45. Solving for *e* yields

$$e = 0.45h = 0.45(30 \text{ mm}) = 13.5 \text{ mm}$$

This dimension is drawn to scale in Figure 12–28(b).

Comment: Now, can you devise a design for using either section as a beam and provide for the application of the load through the flexural center *Q* to produce pure bending?

12–11 PREFERRED SHAPES FOR BEAM CROSS SECTIONS

Recall the discussion earlier in this chapter of the stress distribution in the cross section of a beam characterized by the equations

$\sigma_{max} = Mc/I = M/S$ at the outermost fiber of the beam

$\sigma = \sigma_{max}(y/c)$ at any point at a distance *y* from neutral axis

Figures 12–9 and 12–10 illustrate the stress distribution. It must be understood that these equations apply strictly only to beams made from homogeneous, isotropic materials, that is, those having equal properties in all directions.

Because the larger stresses occur near the top and bottom of the cross section, the material there provides more of the resistance to the externally applied bending moment than material nearer to the neutral axis. It follows that it is desirable to place the larger part of the material away from the neutral axis to obtain efficient use of the material. In this discussion, efficiency refers to maximizing the moment of inertia and section modulus of the shape for a given amount of material, as indicated by the area of the cross section.

Figure 12–29 shows several examples of efficient shapes for beam cross sections. These illustrations are based on the assumption that the most significant stress is a bending stress caused by loads acting on top of the beam perpendicular to the neutral axis. Examples are shown in Figures 12–5, 12–15, and 12–18. In such cases, it is said that the bending is positive

about the horizontal neutral axis. It is also assumed that the material is equally strong in compression and tension.

Starting with the simple rectangular shape in Figure 12–29(a), it is preferred to orient the long dimension vertically as shown because the moment of inertia is proportional to the *cube* of the height of the rectangle, where the height is the dimension perpendicular to the neutral axis. For example, consider the case of a rectangle having dimensions of 40 mm by 125 mm and compare the resulting values of I and S.

FIGURE 12–29   Efficient shapes for beams.

	125 mm dimension vertical	40 mm dimension vertical
$I = bh^3/12$	$I_1 = (40)(125)^3/12 = 6.51 \times 10^6 \, \text{mm}^4$	$I_2 = (125)(40)^3/12 = 0.667 \times 10^6 \, \text{mm}^4$
$S = bh^2/6$	$S_1 = (40)(125)^2/6 = 1.04 \times 10^5 \, \text{mm}^3$	$S_2 = (125)(40)^2/6 = 0.333 \times 10^5 \, \text{mm}^3$

Comparing these results gives

$$\frac{I_1}{I_2} = \frac{6.51 \times 10^6 \, \text{mm}^4}{0.667 \times 10^6 \, \text{mm}^4} = 9.76$$

$$\frac{S_1}{S_2} = \frac{1.04 \times 10^5 \, \text{mm}^3}{0.333 \times 10^5 \, \text{mm}^3} = 3.12$$

The comparison of the values of the section modulus, S, is the most pertinent for comparing stresses in beams because it contains both the moment of inertia, I, and the distance, c, to the outermost fiber of the beam cross section. While the section with the long dimension vertical has a moment of inertia almost ten times that with the long dimension horizontal, it is over three times taller, resulting in the improvement in the section modulus of approximately three times. Still, that is a significant improvement.

A related factor in the comparison of beam shapes is that the *deflection* of a beam is inversely proportional to the moment of inertia, I, as will be shown in Chapter 14. Therefore, the taller rectangular beam in the previous example would be expected to deflect only $1/9.76$ times as much as the shorter one, only about 10% as much.

The shape shown in Figure 12–29(b) is the familiar "I-beam." Placing most of the material in the horizontal flanges at the top and bottom of the section puts them in the regions of the highest stresses for maximum resistance to the bending moment. The relatively thin vertical web serves to hold the flanges in position and provides resistance to shearing forces, as described in Chapter 13. It would be well to study the proportions of standard steel and aluminum I-shaped sections listed in Appendices A–7, A–8, and A–11 to get a feel for reasonable flange and web thicknesses. The thickness of the flange that is in compression is critical with regard to buckling when the beam is relatively long. References 1–3 give data on proper proportions.

The tall, rectangular tube shown in Figure 12–29(c) is very similar to the I-shape in its resistance to bending moments caused by vertical loads. The two vertical sides serve a similar function as the web of the I-shape. In fact, the area moment of inertia with respect to the horizontal centroidal axis for the tube in (c) would be identical to that for the I-shape in (b) if the thickness of the top and bottom horizontal parts were equal and if the vertical sides of the tube were each $\frac{1}{2}$ the thickness of the web of the I-shape. The tube is superior to the I-shape when combinations of loads are encountered that cause bending about the vertical axis in addi-

tion to the horizontal axis, because the placement of the vertical sides away from the Y-Y axis increases the moment of inertia with respect to that axis. The tube is also superior when any torsion is applied, as discussed in Chapter 10. When torsion or bending about the vertical axis is significant, it may be preferred to use the square tube shape shown in Figure 12–29(d). See Appendix A–9 for square and rectangular steel hollow structural sections (HSS), often called simply *tubes*.

Circular pipe or tubes as shown in Figure 12–29(e) make very efficient beams for the same reasons listed earlier for square tubes. They are superior to square tubes when both bending and torsion are present in significant amounts. An obvious example of where a circular tube is preferred is the case of a rotating shaft carrying both bending and torsional loads such as the drive shaft and the axles of a car or truck. See Appendix A–12 for steel pipe and Appendix A–13 for circular mechanical tubing.

Shapes Made from Thin Materials

Economical production of beams having moderate dimensions can be done using rollforming or pressforming of relatively thin flat sheet materials. Aluminum and many plastics are extruded to produce shapes having a uniform cross section, often with thin walls and extended flanges. Examples are shown in Chapter 6, Figures P6–10 through P6–20. Such shapes can be specially adapted to the use of the beam. See if you can identify beam-like members with special shapes around you. In your home you might find such beams used as closet door rails, curtain rods, structures for metal furniture, patio covers or awnings, ladders, parts for plastic toys, tools in the workshop, or parts for appliances or lawn maintenance tools. In your car, look at the windshield wiper arms, suspension members, gearshift levers, linkages or brackets in the engine compartment, and the bumpers. Aircraft structures contain numerous examples of thin-walled shapes designed to take advantage of their very light weight. See also Internet sites for this chapter.

Figure 12–30 shows three examples of extruded or rollformed shapes found around the home. Part (a) shows a closet door rail where the track for the roller that supports the door is produced as an integral part of the aluminum extrusion. The extruded side rail of an aluminum extension ladder is sketched in part (b). Part (c) shows a portion of the roll-formed decking for a patio cover made from aluminum sheet only 0.025 in (0.64 mm) thick. The shape is specially designed to link together to form a continuous panel to

(a) Closet door rail

(b) Side rail for ladder

(c) Panel for patio cover

FIGURE 12–30   Examples of thin-walled beam sections.

cover a wide area. Some design features of these sections should be noted. Extended flanges are reinforced with bulb-like projections to provide local stiffness that resists wrinkling or buckling of the flanges. Broad flat areas are stiffened by ribs or roll-formed corrugations, also to inhibit local buckling. References 1–3 provide guidelines for the design of such features.

Beams Made from Anisotropic Materials

The design of beams to be made from materials having different strengths in tension and compression requires special care. Most types of cast iron, for example, have a much higher compressive strength than tensile strength. Appendix A–17 lists the properties of malleable iron ASTM A220, grade 80002 as

Ultimate tensile strength: $s_u = 655$ MPa (95 ksi)

Ultimate compressive strength:

$s_{uc} = 1650$ MPa (240 ksi)

An efficient beam shape that would account for this difference is the modified I-shape shown in Figure 12–31. Because the typical, positive bending moment places the

bottom flange in tension, providing a larger bottom flange moves the neutral axis down and tends to decrease the resulting tensile stress in the bottom flange relative to the compressive stress in the top flange. Example Problem 12–10 illustrates the result with the design factor based on tensile strength being nearly equal to that based on compressive strength.

FIGURE 12–31   Nonsymmetrical shape for cross section of beam in Example Problem 12–10.

Example Problem 12–10   Figure 12–31 shows the cross section of a beam that is to be made from malleable iron, ASTM A220, grade 80002. The beam is subjected to a maximum bending moment of 1025 N·m, acting in a manner to place the bottom of the beam in tension and the top in compression. Compute the resulting design factor for the beam based on the ultimate strength of the iron. The area moment of inertia for the cross section is 1.80×10^5 mm^4.

Solution   Objective: Compute the design factor based on the ultimate strength.

Given: Beam shape shown in Figure 12–31. $I = 1.80 \times 10^5$ mm^4. $M = 1025$ N·m. Material is malleable iron, ASTM A220, grade 80002.

Analysis: Because the beam cross section is not symmetrical, the value of the maximum tensile stress at the bottom of the beam, σ_{tb}, will be lower than the maximum compressive stress at the top, σ_{ct}. We will compute:

$$\sigma_{tb} = Mc_b/I \quad \text{and} \quad \sigma_{ct} = Mc_t/I$$

where $c_b = \bar{Y} = 14.04$ mm and $c_t = 50 - 14.04 = 35.96$ mm. The tensile stress at the bottom will be compared with the ultimate tensile strength to determine the design factor based on tension, N_t, from

$$\sigma_{tb} = s_u/N_t \quad \text{or} \quad N_t = s_u/\sigma_{tb}$$

where $s_u = 655$ MPa from Appendix A–17. Then the compressive stress at the top will be compared with the ultimate compressive strength to determine the design factor based on compression, N_c, from

$$\sigma_{ct} = s_{uc}/N_c \quad \text{or} \quad N_c = s_{uc}/\sigma_{ct}$$

where $s_{uc} = 1650$ MPa from Appendix A–17. The lower of the two values of N will be the final design factor for the beam.

Results: At the bottom of the beam,

$$\sigma_{tb} = \frac{Mc_b}{I} = \frac{(1025 \text{ N}\cdot\text{m})(14.04 \text{ mm})}{1.80 \times 10^5 \text{ mm}^4} \cdot \frac{(1000 \text{ mm})}{\text{m}} = 79.95 \text{ MPa}$$

$$N_t = s_u/\sigma_{tb} = 655 \text{ MPa}/79.95 \text{ MPa} = 8.19$$

At the top of the beam,

$$\sigma_{ct} = \frac{Mc_t}{I} = \frac{(1025 \text{ N}\cdot\text{m})(35.96 \text{ mm})}{1.80 \times 10^5 \text{ mm}^4} \cdot \frac{(1000 \text{ mm})}{\text{m}} = 204.8 \text{ MPa}$$

$$N_c = s_{uc}/\sigma_{ct} = 1650 \text{ MPa}/204.8 \text{ MPa} = 8.06$$

Comment: The compressive stress at the top of the beam is the limiting value in this problem because the smaller design factor exists there. Note that the two values of the design factor were quite close to being equal, indicating that the shape of the cross section has been reasonably well optimized for the different strengths in tension and compression.

12–12 DESIGN OF BEAMS TO BE MADE FROM COMPOSITE MATERIALS

Composite materials, discussed in Chapter 8, have superior properties when applied to beam design because of the ability to tailor the constituents for the composite and their placement in the beam. Composite processing often allows for unique shapes to be designed that optimize the geometry of the structure with regard to the magnitude and direction of loads to be carried. Combining these features with the inherent advantages of composites in terms of high strength-to-weight and stiffness-to-weight ratios makes them very desirable for use in beams. See References 5, 6, 8, 11, 12, 15, and 17.

The discussion of Section 12–10 applies equally well to the design of composite beams. The designer should select a shape for the beam cross section that is, itself, efficient in resisting bending moments. In addition, the designer can call for the placement of a higher concentration of the stronger, stiffer fibers in the regions where the higher stresses would be expected: namely, at the outermost fibers of the beam farthest from the neutral axis. More plies of a fabric-type filler can be placed in the high-stress regions.

An effective technique for composite beam design is to employ a very light core material for a structure made from a rigid foam or a honeycomb material, covered by relatively thin layers of the strong, stiff fibers in a polymer matrix. If it is known that the bending moments will always act in the same direction, the fibers of the composite can be aligned with the direction of the tensile and compressive stresses in the beam. If the bending moments are expected to act in a variety of directions, a more disperse placement of fibers can be specified or fabric plies can be placed at a variety of angles, as suggested in Figure 8–28.

Care must be exercised in designing and testing composite beam structures because of the multiple possible modes of failure. The structure may fail in the high tensile stress region by failure of the fibers or the matrix or by the disengagement of the fibers from the matrix. Perhaps a more likely failure mode for a laminated composite is interlaminar shear failure in regions of high shear stress near the neutral axis, as sketched in Figure 12–2(f). Failure could also occur in the compressive stress region by local buckling of the shape or by delamination.

When the beam has been designed with the assumption of bending in a certain plane, it is essential that loads are properly applied and that the shape itself promotes pure bending rather than a combination of bending and torsion. The discussion of *flexural center,* Section 12–10, should be reviewed.

The shape and dimensions of the beam cross section can be varied to correspond to the magnitude of the bending moment at various positions in a beam. For example, a cantilever carrying a concentrated load at its end experiences the highest bending moment at the support point, and the magnitude of the bending moment decreases linearly out to the end of the beam. Then the cross section can be deeper at the support and progressively smaller toward the end. A simply supported beam with a load at the center has its highest bending moment at the center, decreasing toward each support. Then the beam can be thicker at the center and thinner toward the ends.

Beams with broad, flat or curved surfaces, such as the wings of an aircraft, must be designed for stiffness of the broad panels, as well as for adequate strength. The skin of the panel may have to be supported by internal ribs to break it into smaller areas.

Penetrations in a composite beam must be carefully designed to ensure smooth transfer of loads from one part of the beam to another. If feasible, the placement of penetrations should be in regions of low stress. Similarly, fasteners must be carefully designed to ensure proper engagement in the fibrous composite material. Thickened bosses may be provided where fasteners are to be located. It may be possible to minimize the number of fasteners by clever shaping of the structure such as by molding in brackets integral with the main structure.

In summary, the designer of composite beams must carefully analyze the stress distribution in the beam and attempt to optimize the placement of material to optimize the shape and dimensions of the beam. The designer must visualize the path of load transfer from its point of application to the ultimate point of support.

REFERENCES

1. Aluminum Association, *Aluminum Design Manual,* Washington, DC, 2005.

2. American Institute of Steel Construction, *Specification for Structural Steel Buildings,* Chicago, IL, AISC, 2005.

3. American Institute of Steel Construction, *Manual of Steel Construction: Load and Resistance Factor Design,* 3rd ed., Chicago, IL, AISC, 2001.

4. American Institute of Steel Construction, *Manual of Steel Construction: Allowable Stress Design,* 9th ed., Chicago, IL, AISC, 1989.

5. ASM International, *ASM Handbook, Volume 21: Composites,* Materials Park, OH, 2001.

6. ASTM International, *The Composite Materials Handbook-MIL 17,* West Conshohocken, PA, ASTM, 2002.

7. Halperin, D. A. and T. G. Bible, *Principles of Timber Design for Architects and Builders,* New York, John Wiley & Sons, Inc., 1994.

8. Jang, Bor Z., *Advanced Polymer Composites: Principles and Applications,* Materials Park, OH, ASM International, 1994.

9. Mallick, P. K., *Composites Engineering Handbook,* New York, Marcel Dekker, 1997.

10. McCormac, J. C. and J. Nelson, *Structural Steel Design—LRFD Method,* 4th ed., Prentice Hall, Upper Saddle River, NJ, 2007.

11. Miravete, A., *Optimisation of Composite Structures Design,* Cambridge, England, Woodhead Publishing, 1996.

12. Quinn, J. A., *Composites Design Manual,* Cambridge, England, Woodhead Publishing, 1999.

13. Segui, W. T., T. Ziolkowski, and B. Stenquist, *LRFD Steel Design,* 3rd ed., Toronto, Ontario, Canada, Thomson-Engineering, 2002.

14. Speigel, L. and G. F. Limbrunner, *Applied Structural Steel Design,* 4th ed., Englewood Cliffs, NJ, Prentice Hall, 2001.

15. Staab, George H., *Laminar Composites,* Boston, Butterworth-Heinemann, 1999.

16. Young, W. C. and R. G. Budynas, *Roark's Formulas for Stress and Strain,* 7th ed., New York, McGraw-Hill, 2002.

17. Zweben, Carl, Composite Materials, Part 1, Section 10, *Mechanical Engineers Handbook, Materials and Mechanical Design,* 3rd ed., Myer Kutz, Ed., New York, John Wiley & Sons, Inc., 2005.

INTERNET SITES

Any open-ended design problem, where the objective is to specify a suitable beam shape and size of a beam's cross section, could make use of the wide variety of shapes and sizes available from the following sites, repeated from Chapter 6, to supplement data given in the Appendix.

1. Profiles, Inc. www.profiles-inc.com Manufacturer of custom rolled or drawn shapes for precision components using carbon and alloy steel, copper, brass, and other nonferrous materials.

2. iLevel by Weyerhaeuser www.ilevel.com/floors Manufacturer of floor joists for building construction made as efficient I-shaped fabrications from simple wood components. Select the TJI® Joists product. The Joist Specifier Guide gives detailed dimensions and application data for the Silent Floor—Trus Joist products.

3. Nucor Corporation—Vulcraft Group www.vulcraft.com Manufacturer of open-web steel joists, joist girders, floor and roof deck shapes, and composite floor joist systems for use in commercial shopping centers, schools, and office buildings.

4. Unistrut Corporation www.unistrut.com Manufacturer of systems of components made from steel or fiberglass that can be assembled in a variety of configurations to produce frames, structures, brackets, storage units, support systems, mobile units, and other devices. Online catalogs provide section property data, application information, assembly instructions, and illustrations of uses.

5. 80/20, Inc. www.8020.net Manufacturer of systems of components made from extruded aluminum that can be assembled in a variety of configurations for use as production equipment, carts, automation devices, workstations, storage units, and a variety of other systems.

6. Paramount Extrusions Company http://paramountextrusions .com/shapes Manufacturer of a variety of standard extrusions such as channels, H-sections, T-sections, hollow tubes (round, square, rectangular), square and rectangular bars, handle sections, and many others for use in office furniture, consumer products, cabinets, and so forth. The catalog gives data that can be useful in problems in this book.

7. Jackson Tube Service, Inc. www.jackson-tube.comshapes.htm. Manufacturer of a wide variety of tubing shapes and sizes that can be useful for problems in this book.

PROBLEMS

Analysis of Bending Stresses

12–1. A square bar 30 mm on a side is used as a simply supported beam subjected to a bending moment of 425 N · m. Compute the maximum stress due to bending in the bar.

12–2. Compute the maximum stress due to bending in a round rod 20 mm in diameter if it is subjected to a bending moment of 120 N · m.

12–3. A bending moment of 5800 lb · in is applied to a beam having a rectangular cross section with dimensions of 0.75 in × 1.50 in. Compute the maximum bending stress in the beam (a) if the vertical side is 1.50 in, and (b) if the vertical side is 0.75 in.

12–4. A wood beam carries a bending moment of 15 500 lb · in. It has a rectangular cross section 1.50 in wide by 7.25 in high. Compute the maximum stress due to bending in the beam.

12–5. The loading shown in Figure P12–5 is to be carried by a W12×16 steel beam. Compute the stress due to bending.

FIGURE P12–5

12–6. An American Standard beam, S12×35, carries the load shown in Figure P12–6. Compute the stress due to bending.

FIGURE P12–6

12–7. The 24-in-long beam shown in Figure P12–7 is an aluminum channel, C4×2.331, positioned with the legs down so that the flat 4-in surface can carry the applied loads. Compute the maximum tensile and maximum compressive stresses in the channel.

FIGURE P12–7

12–8. The 650-lb load at the center of the 28-in-long bar shown in Figure P12–8 is carried by a standard steel pipe, $1\frac{1}{2}$-in schedule 40. Compute the stress in the pipe due to bending.

FIGURE P12–8

12–9. The loading shown in Figure P12–9(a) is to be carried by the fabricated beam shown in Figure P12–9(b). Compute the stress due to bending in the beam.

40 kN   10 kN   10 kN   10 kN

1.2   2.5 m   2.5 m   2.5 m   1.2

(a) Beam loading

$\frac{1}{2} \times 10$ in plates (2)

C12 × 8.274 aluminum

(b) Cross section

FIGURE P12–9

12–10. An aluminum I-beam, I9×8.361, carries the load shown in Figure P12–10. Compute the stress due to bending in the beam.

27 kN   14 kN

2.2 m   3.0 m   1.8 m   2.0 m

30 kN

FIGURE P12–10

12–11. A part of a truck frame is composed of two channel-shaped members, as shown in Figure P12–11. If the moment at the section is 60 000 lb·ft, compute the bending stress in the frame. Assume that the two channels act as a single beam.

3 in

12 in

$\frac{1}{2}$ in

FIGURE P12–11   Truck frame members for Problem 12–11.

Design of Beams

12–12. Compute the minimum required diameter of a round bar used as a beam to carry a bending moment of 240 N·m with a stress no greater than 125 MPa.

12–13. A rectangular bar is to be used as a beam subjected to a bending moment of 145 N·m. If its height is to be three times its width, compute the minimum required dimensions of the bar to limit the stress to 55 MPa.

12–14. The tee section shown in Figure P12–14 is to carry a bending moment of 28.0 kN·m. It is to be made of steel plates welded together. If the load on the beam is a dead load, would AISI 1020 hot-rolled steel be satisfactory for the plates?

175 mm

25 mm

200 mm

25 mm

FIGURE P12–14

12–15. The modified I-section shown in Figure P12–15 is to be made from extruded aluminum. Specify a suitable aluminum alloy if the beam is to carry a repeated load resulting in a bending moment of 275 N·m.

40 mm

5 mm

50 mm

5 mm

5 mm

20 mm

FIGURE P12–15

12–16. A standard steel pipe is to be used as a chinning bar for personal exercise. The bar is to be 42 in long and simply supported at its ends. Specify a suitable size pipe if the bending stress is to be limited to 10 000 psi and a 280-lb man hangs by one hand in the middle.

12–17. A pipeline is to be supported above ground on horizontal beams, 14 ft long. Consider each beam to be simply supported at its ends. Each beam carries the combined weight of 50 ft of 48-in-diameter pipe and the oil flowing through it, about 42 000 lb. Assuming the load acts at the center of the beam, specify the required section modulus of the beam to limit the bending stress to 20 000 psi. Then specify a suitable steel wide flange or American Standard beam.

12–18. A wood platform is to be made of standard plywood and finished lumber using the cross section shown in Figure P12–18(a). Would the platform be safe if four men, weighing 250 lb each, were to stand 2 ft apart, as shown in Figure P12–18(b)? Consider only bending stresses (see Chapter 13 for shear stresses).

(a) Cross section

(b) Beam loading

FIGURE P12-18

12-19. A diving board has a hollow rectangular cross section 30 in wide and 3.0 in thick and is supported as shown in Figure P12-19. Compute the maximum stress due to bending in the board if a 300-lb person stands at the end. Would the board be safe if it were made of extruded 6061-T4 aluminum and the person landed at the end of the board with an impact?

(a) Loads on diving board

(b) Section A-A through board

FIGURE P12-19   Diving board for Problem 12-19.

12-20. The loading shown in Figure P12-20(a) is to be carried by an extruded aluminum hat-section beam having the cross section shown in Figure P12-20(b). Compute the maximum stress due to bending in the beam. If it is made of extruded 6061-T4 aluminum and the loads are dead loads, would the beam be safe?

12-21. The extruded shape shown in Figure P12-21(a) is to be used to carry the loads shown in Figure P12-21(b), which is a part of a business machine frame. The loads are due to a motor mounted on the frame and can be considered dead loads. Specify a suitable aluminum alloy for the beam.

(b)

(a)

FIGURE P12-20

(a)

(b)

FIGURE P12-21

12-22. A beam is being designed to support the loads shown in Figure P12-22. The four shapes proposed are (a) a round bar, (b) a square bar, (c) a rectangular bar with the height made four times the thickness, and (d) the lightest American Standard beam. Determine the required dimensions of each proposed shape to limit the maximum stress due to

FIGURE P12-22

FIGURE P12–23

bending to 80 MPa. Then compare the magnitude of the cross-sectional areas of the four shapes. Since the weight of the beam is proportional to its area, the one with the smallest area will be the lightest.

12–23. A children's play gym includes a cross beam carrying four swings, as shown in Figure P12–23 Assume that each swing carries 300 lb. It is desired to use a standard steel pipe for the beam, keeping the stress due to bending below 10 000 psi. Specify the suitable size pipe for the beam.

12–24. A 60-in-long beam simply supported at its ends is to carry two 4800 lb loads, each placed 14 in from an end. Specify the lightest suitable steel tube for the beam, either square or rectangular, to produce a design factor of 4 based on yield strength. The tube is to be cold formed from ASTM A500, grade B steel.

12–25. Repeat Problem 12–24, but specify the lightest standard aluminum I-beam from Appendix A–11. The beam will be extruded using alloy 6061-T6.

12–26. Repeat Problem 12–24, but specify the lightest wide-flange steel shape from Appendix A–7. The beam will be made from ASTM A992 structural steel.

12–27. Repeat Problem 12–24, but specify the lightest structural steel channel from Appendix A–6. The channel is to be installed with the legs down so that the loads can be applied to the flat back of the web of the channel. The channel will be made from ASTM A36 structural steel.

12–28. Repeat Problem 12–24, but specify the lightest standard schedule 40 steel pipe from Appendix A–12. The pipe is to be made from ASTM A501 hot-formed steel.

12–29. Repeat Problem 12–24, but design the beam using any material and shape of your choosing to achieve a safe beam that is lighter than any of the results from Problems 12–24 through 12–28.

12–30. The shape shown in Figure P12–30 is to be made from extruded plastic and used as a simply supported beam, 12 ft long, to carry two electric cables weighing a total of 6.5 lb/ft of length. Specify a suitable plastic for the extrusion to provide a design factor of 4.0 based on flexural strength.

Dimensions in inches

FIGURE P12–30

12–31. The loading shown in Figure P12–31 represents the load on a floor beam of a commercial building. Determine the maximum bending moment on the beam, and then specify a wide-flange shape that will limit the stress to 150 MPa.

FIGURE P12–31

12–32. Figure P12–32 represents the loading on a motor shaft; the two supports are bearings in the motor housing. The larger load between the supports is due to the rotor plus dynamic forces. The smaller overhung load is due to externally applied loads. Using AISI 1141 OQT 1300 steel for the shaft, specify a suitable diameter based on bending stress only. Use a design factor of 8 based on ultimate strength.

FIGURE P12–32

12–33 to 12–42. Using the indicated loading, specify the lightest standard wide-flange beam shape (W-shape) that will limit the stress due to bending to the allowable design stress from the AISC specification. All loads are static and the beams are made from ASTM A992 structural steel. Each problem number is the same as the corresponding figure number.

12–33. Use Figure P12–33. **12–38.** Use Figure P12–38.
12–34. Use Figure P12–34. **12–39.** Use Figure P12–39.
12–35. Use Figure P12–35. **12–40.** Use Figure P12–40.
12–36. Use Figure P12–36. **12–41.** Use Figure P12–41.
12–37. Use Figure P12–37. **12–42.** Use Figure P12–42.

FIGURE P12–33

FIGURE P12–34

FIGURE P12–35

FIGURE P12–36

FIGURE P12–37

FIGURE P12–38

FIGURE P12–39

FIGURE P12–40

FIGURE P12–41

FIGURE P12–42

FIGURE P12–64   Bench and load for Problems 12–64, 12–65, 12–66, and 12–67.

12–43 to 12–52. Repeat Problems 12–33 through 12–42 but specify the lightest American Standard Beam (S-shape).

12–53 to 12–62. Repeat Problems 12–33 through 12–42 but use ASTM A572 Grade 60 high-strength, low-alloy structural steel.

12–63. A floor joist for a building is to be made from a standard wooden beam selected from Appendix A–4. If the beam is to be simply supported at its ends and carry a uniformly distributed load of 125 lb/ft over the entire 10-ft length, specify a suitable beam size. The beam will be made from No. 2 grade southern pine. Consider only bending stress.

12–64. A bench for football players is to carry the load shown in Figure P12–64 approximating the case when 10 players, each weighing 300 lb, sit close together, each taking 18 in of the length of the bench. If the cross section of the bench is made as shown in Figure P12–64, would it be safe for bending stress? The wood is No. 2 grade hemlock.

12–65. A bench is to be designed for football players. It is to carry the load shown in Figure P12–64 approximating the case when 10 players, each weighing 300 lb, sit close together, each taking 18 in of the length of the bench. The bench is to be T-shaped, made from No. 2 grade hemlock, as shown with a 2×12 top board. Specify the required vertical member of the tee if the bench is to be safe for bending stress.

12–66. Repeat Problem 12–65 but use the cross-sectional shape shown in Figure P12–66.

FIGURE P12–66   Shape for bench cross section for Problem 12–66.

12–67. Repeat Problem 12–65, but use any cross-sectional shape of your choosing made from standard wooden beams

from Appendix A–4. Try to achieve a lighter design than in Problem 12–65 or 12–66. Note that a lighter design would have a smaller cross-sectional area.

12–68. A wood deck is being designed to carry a uniformly distributed load over its entire area of 100 lb/ft². Joists are to be used as shown in Figure P12–68, set 16 in on center. If the deck is to be 8 ft by 12 ft in size, determine the required size for the joists. Use standard wooden beam sections from Appendix A–4 and No. 2 hemlock.

12–69. Repeat Problem 12–68, but run the joists across the 12-ft length rather than the 8-ft width.

12–70. Repeat Problem 12–68, but set the support beam in 18 in from the ends of the joists instead of at the ends.

12–71. Repeat Problem 12–69, but set the support beams in 18 in from the ends of the joists instead of at the ends.

12–72. For the deck design shown in Figure P12–68 specify a suitable size for the cross beams that support the joists.

12–73. Design a bridge to span a small stream. Assume that rigid supports are available on each bank, 10.0 ft apart. The bridge is to be 3.0 ft wide and carry a uniformly distributed load of 60 lb/ft² over its entire area. Design only the deck boards and beams. Use two or more beams of any size from Appendix A–4 or others of your own design.

12–74. Would the bridge you designed in Problem 12–73 be safe if a horse and rider weighing 2200 lb walked slowly across it?

12–75. Millwrights in a factory need to suspend a machine weighing 10 500 lb from a beam having a span of 12.0 ft so that a truck can back under it. Assume that the beam is simply supported at its ends. The load is applied by two cables, each 3.0 ft from a support. Design a suitable beam. Consider standard wooden or steel beams or one of your own design.

12–76. In an amateur theater production, a pirate is to "walk the plank." If the pirate weighs 220 lb, would the design shown in Figure P12–76 be safe? If not, design one to be safe.

12–77. A branch of a tree has the approximate dimensions shown in Figure P12–77. Assuming the bending strength of the wood to be similar to that of No. 3 grade hemlock, would it be safe for a person having a mass of 135 kg to sit in the swing?

FIGURE P12–68 Deck design for Problem 12–68.

12 ft

16 in typical

8 ft

Decking

Joists

Beam

Decking

2.0 ft

Post

Post

2.0 ft

Side view

Decking

Joist

Beam

Beam

End view

Post

Post

2.0 ft

2 × 12 No. 2 southern pine

Plank

8.5 ft

Ship

Water

FIGURE P12–76 Pirate walking the plank in Problem 12–76.

180 mm 140 mm 90 mm

Section A-A Section B-B Section C-C

0.60 m 0.80 m

1.60 m 0.50 m

FIGURE P12–77 Branch and swing for Problem 12–77.

12–78. Would it be safe to use a standard 2×4 made from No. 2 grade southern pine as a lever as shown in Figure P12–78 to lift one side of a machine? If not, what would you suggest be used?

w = ?

0.8 m

Machine – 1600 lb

P

2×4 lever

3.5 in

80 in

8 in

FIGURE P12–78 2×4 used as a lever in Problem 12–78.

10 mm 20 mm 10 mm

20 mm

30 mm

FIGURE P12–79

12–79. Figure P12–79 shows the cross section of an extruded plastic beam made from nylon 66 dry. Specify the largest uniformly distributed load in N/mm the beam could carry if it is simply supported with a span of 0.80 m. The maximum stress due to bending is not to exceed one-half of the flexural strength of the nylon.

12–80. The I-beam shape in Figure P12–80 is to carry two identical concentrated loads of 2.25 kN each, symmetrically placed on a simply supported beam with a span of 0.60

m. Each load is 0.2 m from an end. Which of the plastics from Appendix A–20 would carry these loads with a stress due to bending no more than one-third of their flexural strength?

12–81. A bridge for a toy construction set is to be made from acetal copolymer with the flexural strength listed in Appendix A–20. The cross section of one beam is shown in Figure P12–81. What maximum concentrated load could be applied to the middle of the beam if the span was 1.25 m and the ends are simply supported? Do not exceed one-half of the flexural strength of the plastic.

2.25 kN   2.25 kN

0.60 mm

40 mm

5 mm

30 mm

5 mm

5 mm

FIGURE P12–80

40 mm

5 mm

50 mm

5 mm

5 mm

20 mm

FIGURE P12–81

220 N   280 N

0.20 m   0.20 m   0.40 m

40 N/m

0.8 m

(a)

4 mm typical

20 mm

50 mm

(b)

FIGURE P12–82   Beam in a computer printer for Problem 12–82.

0.30

2.00

0.30

0.60

3.00

Dimensions in inches

FIGURE P12–83

0.10 typical

2.40

1.50

1.20

Dimensions in inches

FIGURE P12–84

12–82. A structural element in a computer printer is to carry the load shown in Figure P12–82(a) with the uniformly distributed load representing electronic components mounted on a printed circuit board and the concentrated loads applied from a power supply. It is proposed to make the beam from polycarbonate with the cross section shape shown in Figure P12–82(b). Compute the stress in the beam and compare it with the flexural strength for polycarbonate from Appendix A–20.

12–83. It is proposed to make the steps for a child's slide from molded polycarbonate with the cross section shown in Figure P12–83. The steps are to be 14.0 in wide and will be simply supported at the ends. What maximum weight can be applied at the center of the step if the stress due to bending must not exceed one-third of the flexural strength of the polycarbonate listed in Appendix A–20?

12–84. The shape shown in Figure P12–84 is used as a beam for a carport. The span of the beam will be 8.0 ft. Compute the maximum uniformly distributed load that can be carried by the beam if it is extruded from 6061-T4 aluminum. Use a design factor of 2 based on yield strength.

12–85. Figure P12–85 shows the cross section of an aluminum beam made by fastening a flat plate to the bottom of a roll-formed hat section. If the beam is used as a cantilever, 24 in long, compute the maximum allowable concentrated load that can be applied at the end if the maximum stress is to be no more than one-eighth of the ultimate strength of 2014-T4 aluminum.

12–86. Repeat Problem 12–85 but use only the hat section without the cover plate.

FIGURE P12–85

FIGURE P12–90

12–87. Figure P12–87 shows the cross section of a beam that is to be extruded from aluminum 6061-T6 alloy. If the beam is to be used as a cantilever, 42 in long, compute the maximum allowable uniformly distributed load it could carry while limiting the stress due to bending to one-sixth of the ultimate strength of the aluminum.

12–88. Repeat Problem 12–87 but use Figure P12–83 for the cross section and aluminum casting alloy 356.0-T6.

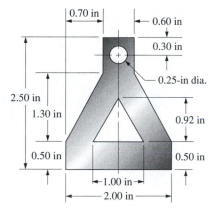

FIGURE P12–87

Beams with Stress Concentrations and Varying Cross Sections

12–89. In Figure P12–89 the 4-in pipe mates smoothly with its support so that no stress concentration exists at D. At C the $3\frac{1}{2}$-in pipe is placed inside the 4-in pipe with a spacer ring to provide a good fit. Then a $\frac{1}{4}$-in smoothly radiused fillet weld is used to secure the section together. Accounting for the stress concentration at the joint, determine how far out point C must be to limit the stress to 20000 psi. Use Appendix A–22–9 for the stress concentration factor. Is the 4-in pipe safe at D?

FIGURE P12–89

12–90. Figure P12–90 shows a round shaft from a gear transmission. Gears are mounted at points A, C, and E. Supporting bearings are at B and D. The forces transmitted from the gears to the shaft are shown, all acting downward. Compute the maximum stress due to bending in the shaft, accounting for stress concentrations.

12–91. The forces shown on the shaft in Figure P12–91 are due to gears mounted at B and C. Compute the maximum stress due to bending in the shaft.

FIGURE P12–91

12–92. Figure P12–92 shows a machine shaft supported by two bearing at its ends. The two forces are exerted on the shaft by gears. Considering only bending stresses, compute the maximum stress in the shaft and tell where it occurs.

12–93. Figure P12–93 shows a lever made from a rectangular bar of steel. Compute the stress due to bending at the fulcrum, 20 in from the pivot and at each of the holes in the bar. The diameter of each hole is 0.75 in.

12–94. Repeat Problem 12–93 but use the diameter of the holes as 1.38 in.

12–95. In Figure P12–93, the holes in the bar are provided to permit the length of the lever to be changed relative to the pivot. Compute the maximum bending stress in the lever as the pivot is moved to each hole. Use the diameter of the holes as 1.25 in.

12–96. The bracket shown in Figure P12–96 carries the opposing forces created by a spring. If the force, F, is 2500 N, compute the bending stress at a section such as A-A, away from the holes.

12–97. If the force, F, in Figure P12–96 is 2500 N, compute the bending stress at a section through the holes, such as B-B. Use $d = 12$ mm for the diameter of the holes.

12–98. Repeat Problem 12–97 but use $d = 15$ mm for the diameter of the holes.

12–99. For the resulting stress computed in Problem 12–98, specify a suitable steel for the bracket if the force is repeated many thousands of times.

$D_1 = 0.68$ in
$D_2 = 1.00$ in
$D_3 = 0.94$ in
$D_4 = 1.38$ in

FIGURE P12–92

Section A-A

200 lb

FIGURE P12–93   Lever for Problems 12–93 through 12–95.

Section B-B   Section A-A

FIGURE P12–96   Bracket for Problems 12–96 through 12–99.

12–100. Figure P12–100 shows a stepped flat bar in bending. If the bar is made from AISI 1040 cold-drawn steel, compute the maximum repeated force, F, that can safely be applied to the bar.

12–101. Repeat Problem 12–100, but use $r = 2.0$ mm for the fillet radius.

12–102. For the stepped flat bar shown in Figure P12–100, change the 75-mm dimension that locates the step to a value that makes the bending stress at the step equal to that at the point of application of the load.

12–103. For the stepped flat bar shown in Figure P12–100, change the size of the fillet radius to make the bending stress at the fillet equal to that at the point of application of the load.

12–104. Repeat Problem 12–100, but change the depth of the bar from 60 mm to 75 mm.

12–105. For the stepped flat bar in Figure P12–100, would it be possible to drill a hole in the middle of the 60-mm depth of the bar between the two forces without increasing the maximum bending stress in the bar? If so, what is the maximum size hole that can be put in?

12–106. Figure P12–106 shows a stepped flat bar carrying three concentrated loads.   Let $P = 200$ N, $L_1 = 180$ mm,

FIGURE P12–100   Stepped flat bar for Problems 12–100 through 12–105.

FIGURE P12–106   Stepped flat bar for Problems 12–106 through 12–110.

$L_2 = 80$ mm, and $L_3 = 40$ mm. Compute the maximum stress due to bending and state where it occurs. The bar is braced against lateral bending and twisting. Note that the length dimensions in the figure are not drawn to scale.

12–107. For the data of Problem 12–106, specify a suitable material for the bar to produce a design factor of 8 based on ultimate strength.

12–108. Repeat Problem 12–107 except use $r = 1.50$ mm for the fillet radius.

12–109. For the stepped flat bar shown in Figure P12–106, let $P = 400$ N. The bar is to be made from titanium, Ti-6Al-4V, and a design factor of 8 based on ultimate strength is desired. Specify the maximum permissible lengths, L_1, L_2, and L_3, that would be safe.

12–110. The stepped flat bar in Figure P12–106 is to be made from AISI 4140 OQT 1100 steel. Use $L_1 = 180$ mm, $L_2 = 80$ mm, and $L_3 = 40$ mm. Compute the maximum allowable force P that could be applied to the bar if a design factor of 8 based on ultimate strength is desired.

12–111. Figure P12–111 shows a flat bar that has a uniform thickness of 20 mm. The depth tapers from $h_1 = 40$ mm to $h_2 = 20$ mm in order to save weight. Compute the stress due to bending in the bar at points spaced 40 mm apart from the support to the load. Then create a graph of stress versus distance from the support. The bar is symmetrical with respect to its middle. Let $P = 5.0$ kN.

12–112. For the bar shown in Figure P12–111, let $h_1 = 60$ mm and $h_2 = 20$ mm. The bar is to be made from polycarbonate plastic. Compute the maximum permissible load P that will produce a design factor of 4 based on the flexural strength of the plastic. The bar is symmetrical with respect to its middle.

12–113. In Figure P12–111, the load $P = 1.20$ kN and the bar is to be made from AISI 5160 OQT 1300 steel. Compute the required dimensions h_1 and h_2 that will produce a design factor of 8 based on ultimate strength. The bar is symmetrical with respect to its middle.

FIGURE P12–111   Tapered flat bar for Problems 12–111 through 12–113.

12–114. A rack is being designed to support large sections of pipe, as shown in Figure P12–114. Each pipe exerts a force of 2500 lb on the support arm. The height of the arm is to be tapered as suggested in the figure, but the thickness will be a constant 1.50 in. Determine the required height of the arm at sections B and C, considering only bending stress. Use AISI 1040 hot-rolled steel for the arms and a design factor of 4 based on yield strength.

FIGURE P12–114 Pipe storage rack for Problem 12–114.

12–115. Specify the lightest wide-flange beam shape (W-shape) that can carry a uniformly distributed static load of 2.5 kip/ft over the entire length of a simply supported span, 12.0 ft long. Use the AISC specification and ASTM A992 structural steel.

12–116. A proposal is being evaluated to save weight for the beam application of Problem 12–115. The result for that problem required a W14×26 beam that would weigh 312 lb for the 12-ft length. A W12×16 beam would only weigh 192 lb but does not have sufficient section modulus S. To increase S, it is proposed to add steel plates, 0.25 in thick and 3.50 in wide, to both the top and the bottom flange over a part of the middle of the beam. Perform the following analyses:
   a. Compute the section modulus of the portion of the W12×16 beam with the cover plates.
   b. If the result of part (a) is satisfactory to limit the stress to an acceptable level, compute the required length over which the plates would have to be applied to the nearest 0.5 ft.
   c. Compute the resulting weight of the composite beam and compare it to the original W14×26 beam.

12–117. Figure P12–117 shows a composite beam made by adding a channel to an American Standard beam shape. Both are made from ASTM A36 structural steel. If the beam is simply supported and carries a uniformly distributed load over a span of 15.0 ft, compute the allowable load for the composite beam and for the S-shape by itself. The load is static and the AISC specification for design stress is to be used.

FIGURE P12–117

Flexural Center

12–118. Compute the location of the flexural center of a channel-shaped member shown in Figure P12–118 measured from the left of the vertical web.

FIGURE P12–118

12–119. A company plans to make a series of three channel-shaped beams by rollforming them from flat-sheet aluminum. Each channel is to have the same outside dimensions as shown in Figure P12–118, but they will have different material thicknesses, 0.50, 1.60, and 3.00 mm. For each design, compute the moment of inertia with respect to the horizontal centroidal axis and the location of the flexural center, measured from the left face of the vertical web.

12–120. Compute the location of the flexural center for the hat section shown in Figure P12–120 measured from the left face of the vertical web.

12–121. A company plans to make a series of three hat sections by rollforming them from flat-sheet aluminum. Each hat section is to have the same outside dimensions as shown in Figure P12–120, but they will have different material thicknesses, 0.020, 0.063, and 0.125 in. For each design, compute the location of the flexural center, measured from the left face of the vertical web.

12–122. Compute the location of the flexural center for the lipped channel shown in Figure P12–122, measured from the left face of the vertical web.

FIGURE P12–120

FIGURE P12–122

12–123. A company plans to make a series of three lipped channels by rollforming them from flat-sheet aluminum. Each channel section is to have the same outside dimensions as shown in Figure P12–122, but they will have different material thicknesses, 0.50, 1.60, and 3.00 mm. For each design, compute the location of the flexural center, measured from the left face of the vertical web.

12–124. Compute the location of the flexural center of a split, thin tube if it has an outside diameter of 50 mm and a wall thickness of 4 mm.

12–125. For an aluminum channel C2×0.577 with its web oriented vertically, compute the location of its flexural center. Neglect the effect of the fillets between the flanges and the web.

12–126. If the hat section shown in Figure P12–126 were turned 90 deg from the position shown, compute the location of its flexural center.

Beams Made from Anisotropic Materials

12–127. The beam section shown in Figure P12–127 is to be extruded from 6061-T6 aluminum. The allowable tensile

FIGURE P12–126

Dimensions in inches

FIGURE P12–127

strength is 19 ksi. Because of the relatively thin extended legs on the top, the allowable compressive strength is only 14 ksi. The beam is to span 6.5 ft and will be simply supported at its ends. Compute the maximum allowable uniformly distributed load on the beam.

12–128. Repeat Problem 12–127 but turn the section upside down. With the legs pointed downward, they are in tension and can withstand 19 ksi. The part of the section in compression at the top is now well supported and can withstand 21 ksi.

12–129. The shape in Figure P12–129 is to carry a single concentrated load at the center of a 1200-mm span. The allowable strength in tension is 100 MPa, while the allowable strength in compression anywhere is 70 MPa. Compute the allowable load.

FIGURE P12–129

12–130. Repeat Problem 12–129 with the section turned up-side down.

12–131. Repeat Problem 12–129 with the shape shown in Figure P12–131.

FIGURE P12–131

12–132. Repeat Problem 12–129 with the shape shown in Figure P12–132.

FIGURE P12–132

12–133. The T-shaped beam cross section shown in Figure P12–133 is to be made from gray cast iron, ASTM A48 Grade 40. It is to be loaded with two equal loads P, 1.0 m from the ends of the 2.80 m long beam. Specify the largest static load P that the beam could carry. Use N = 4.

FIGURE P12–133

12–134. The modified I-beam shape shown in Figure P12–134 is to carry a uniformly distributed static load over its entire 1.20-m length. Specify the maximum allowable load if the beam is made from malleable iron, ASTM A220, class 80002. Use N = 4.

12–135. Repeat Problem 12–134 but turn the beam upside down.

12–136. A wide beam is made as shown in Figure P12–136 from ductile iron, ASTM A536, Grade 120-90-2. Compute the maximum load P that can be carried with a resulting design factor of 10 based on either tensile or compressive ultimate strength.

FIGURE P12–134

Section A-A—beam cross section

FIGURE P12–136

12–137. Repeat Problem 12–136 but increase the depth of the vertical ribs by a factor of 2.0.

12–138. Problems 12–133 through 12–137 illustrate that a beam shape made as a modified I-shape more nearly optimizes the use of the available strength of a material having different strengths in tension and compression. Design an I-shape that has a nearly uniform design factor of 6 based on ultimate strength in either tension or compression when made from gray iron, Grade 20, and that carries a uniformly distributed load of 20 kN/m over its 1.20-m length. (*Note:* You may want to use the computer program written for Assignment 3 at the end of Chapter 6 to facilitate the computations. A trial-and-error solution may be used.)

ADDITIONAL PRACTICE AND REVIEW PROBLEMS

12–139. A solid circular bar has a diameter of 50 mm and a length of 350 mm. It is installed in the frame of a machine as a cantilever and carries a single, static concentrated load of 2.40 kN at its end. Would the bar be safe if it is made from ASTM A48 gray cast iron, Grade 40? Consider only stress due to bending.

FIGURE P12–141

12–140. A beam is to be designed to carry the load shown in Figure 12–15 in Section 12–9. The load is to be applied and removed repeatedly. Specify the lightest standard aluminum I-beam shape for the beam if it is to be made from 6061 T6 aluminum.

12–141. Compute the maximum stress due to bending for the beam shown in Figure P12–141.

12–142. The load shown in Figure P12–142 is carried by a rotating solid circular bar with a diameter of 30.0 mm. The material is AISI 1040 WQT 1300. Is the bar safe in bending?

FIGURE P12–142

12–143. Specify the optimum wide flange beam shape to carry the load shown in Figure P12–143 if the beam is made from ASTM A992 structural steel. Use the AISC specifications.

FIGURE P12–143

12–144. If the beam described in Problem 12–143 is to be made from ASTM A572 HSLA, Grade 65, specify the optimum beam. Is there an advantage to using the higher-strength steel?

12–145. For the beam loading shown in Figure P12–145, specify a suitable square or rectangular hollow structural shape (HSS) if it is made from ASTM A500 Grade C steel. Use the AISC specifications.

FIGURE P12–145

12–146. The loading shown in Figure P12–146 is to be carried by a beam made from a standard schedule 40 steel pipe. It is desired to have a design factor of 4.0 based on yield strength with the pipe made from cold-formed ASTM A500 Grade C steel. Specify a pipe size.

FIGURE P12–146

12–147. The beam loading shown in Figure P12–147 is to be carried by a standard square or rectangular hollow structural steel shape (HSS) made from ASTM A501 hot-formed structural steel. The load will be applied and removed repeatedly. Specify the lightest suitable shape.

FIGURE P12–147

FIGURE P12–148

12–148. A cantilever beam has the cross-sectional shape shown in Figure P12–148. It is 36.0 inches long and it will carry a single concentrated load at its end. A design factor of 4.0 is desired when using ASTM A48 Grade 60 gray cast iron. Compute the allowable load on the beam.

12–149. Redesign the shape of the beam from Problem 12–148 so that it will carry a minimum of 6000 lb using the same material and design factor.

12–150. For the beam loading shown in Figure P12–150, specify the lightest standard wide-flange beam shape if it is to be made from ASTM A992 structural steel. The load is static. Use the AISC specifications. Compute the total weight of the beam.

FIGURE P12–150

12–151. An alternate design for the beam to carry the load shown in Figure P12–150 is shown in Figure P12–151. The material for the hollow tube is ASTM A501 hot-formed structural steel and the plates are made from ASTM A36

FIGURE P12–151

structural steel. Evaluate whether or not this design is safe. Also, compute its total weight and compare it to the W-shape you specified in Problem 12–150.

12–152. Propose a different design for the beam to carry the loading shown in Figure P12–150 that will be safe but lighter than either design described in Problem 12–150 or 12–151.

COMPUTER ASSIGNMENTS

1. Write a program or spreadsheet to compute the maximum bending stress for a simply supported beam carrying a single concentrated load at its center. Allow the operator to input the load, span, and beam section properties. The output should include the maximum bending moment and the maximum bending stress and indicate where the maximum stress occurs.

Enhancements for Assignment 1

a. For the computed stress, compute the required strength of the material for the beam to produce a given design factor.

b. In addition to (a), include a table of properties for a selected material such as the data for steel in Appendix A–14. Then search the table for a suitable steel from which the beam can be made.

2. Repeat Assignment 1 except use a uniformly distributed load.

3. Repeat Assignment 1 except the beam is a cantilever with a single concentrated load at its end.

4. Write a program or spreadsheet to compute the maximum bending moment for a simply supported beam carrying a single concentrated load at its center. Allow the operator to input the load and span. Then compute the required section modulus for the cross section of the beam to limit the maximum bending stress to a given level or to achieve a given design factor for a given material. The output should include the maximum bending moment and the required section modulus.

Enhancements for Assignment 4

a. After computing the required section modulus, have the program complete the design of the beam cross section for a given general shape, such as rectangular with a given ratio of thickness to depth (see Problem 12–13), or circular.

b. Include a table of properties for standard beam sections such as any of those in Appendixes A–4 through A–13 and have the program search for a suitable beam section to provide the required section modulus.

5. Repeat Assignment 4 but use a uniformly distributed load.

6. Repeat Assignment 4 but use the load described in Problem 12–22.

7. Repeat Assignment 4 but use any loading pattern assigned by the instructor.

8. Write a computer program or spreadsheet to facilitate the solution of Problem 12–138, including the computation of section properties for the modified I-shape using the techniques of Chapter 6.

9. Write a computer program or spreadsheet to facilitate the solution of problems of the type given in Problem 12–116. Make the program general, permitting the user to input the loading on the beam, the desired beam section properties, and the dimensions of the plates to be added to the basic beam section.

10. Write a computer program or spreadsheet to perform the computations called for in Problem 12–111 but make the program more general, permitting the user to input values for the load, span, beam cross-section dimensions, and the interval for computing the bending stress. Have the program produce the graph of stress versus position on the beam.

11. Write a computer program or spreadsheet to compute the location for the flexural center for the generalized channel shape in Figure P12–118. Permit the user to input data for all dimensions.

12. Write a computer program or spreadsheet to compute the location for the flexural center for the generalized hat section shown in Figure P12–120. Permit the user to input data for all dimensions. Curve-fitting techniques and interpolation may be used to interpret the graph in Figure 12–26.

13. Write a computer program or spreadsheet to compute the location for the flexural center for the generalized lipped channel shown in Figure P12–122. Permit the user to input data for all dimensions. Curve-fitting techniques and interpolation may be used to interpret the graph in Figure 12–26.

SHEARING STRESSES IN BEAMS

MAJOR TOPICS

13–1   The Big Picture—Shearing Stresses in Beams
13–2   Importance of Shearing Stresses in Beams
13–3   The General Shear Formula
13–4   Distribution of Shearing Stress in Beams
13–5   Development of the General Shear Formula
13–6   Special Shear Formulas
13–7   Design for Shear
13–8   Shear Flow

13–1 THE BIG PICTURE— SHEARING STRESSES IN BEAMS

Discussion Map

- In Chapter 13 your goal is to continue to develop your ability to analyze and design beams. You will build on your knowledge of shearing forces and bending moments from Chapter 11, moment of inertia from Chapter 6, and bending stresses from Chapter 12. Review all of the examples of beams and loading patterns you discussed in the Big Picture from Chapter 11 and the types of beam cross-sectional shapes from Chapter 6.

- For a beam to be safe to carry a given load, it must satisfy design stress limits for *both bending stresses and shearing stresses.* In most cases the bending stress is the most critical. But shearing stresses must always be checked and, as discussed in Section 13–2, there are some important situations where shearing stresses are the predominant mode of failure. Of course, stiffness, or resistance to deflection, is also an important goal when designing a beam.

- Continuing the analysis of beams, this chapter is concerned with the stresses created within a beam due to the presence of shearing forces. As shown in Figure 13–1, shearing forces are visualized to act within the beam on its cross section and to be directed transverse, that is perpendicular, to the axis of the beam. Thus, they would tend to create *transverse shearing stresses,* sometimes called *vertical shearing stresses.*

- But if a small stress element subjected to such shearing stresses is isolated, as shown in Figure 13–2, it can be seen that horizontal shearing stresses must also exist in order to cause the element to be in equilibrium. Thus,

both vertical and horizontal shearing stresses, having the same magnitude at a given point, are created by shearing stresses in beams.

Activity Demonstrate for yourself the existence of horizontal shearing stresses in a beam by completing the following exercise.

1. Make a beam from several thin, flat strips as illustrated in Figure 13–3(a). The width should be narrow relative to the length of the strip. For example, make the length about 125 mm (5.0 inches) and the width about 25 mm (1.0 in).

2. An ideal set of material would be five strips cut from standard index cards that measure about 75 mm by 125 mm (3.0 in $\times$ 5.0 in). Make at least three sets of five strips each. Other suitable materials are thin plastic, playing cards, or thin sheet metal. You will also need a stapler and some white glue.

3. Arrange two simple supports that are about 25 mm (1.0 in) high. Place them approximately 100 mm (4.0 in) apart.

4. Obtain a simple weight that is small and that will cause one of the strips of material to deflect all the way to the table, at least 25 mm (1.0 in). A stack of six quarters works well.

5. It should not be surprising that a stack of six quarters will deflect one thin strip of card stock at least an inch. But what if you make your beam with two, three, four, or five strips, each placed loosely on top of the others? Try that and document what happens.

6. The test in step 5 likely showed that four or five strips would hold the quarters above the table, but there would still be a significant amount of deflection.

7. Now take another stack of five strips and staple them together. Place one staple about 12 mm (0.25 in) from

(a) Load, shearing force, and bending
moment diagrams

$M_a = 7.2$ kN·m

$V_a = 12$ kN = Shearing force

$R_A = 12$ kN

(b) Free body diagram at a-a

$P = 20$ kN

1.6 m

2.0 m

$V_b = 8$ kN = Shearing force

$M_b = 16.0$ kN·m

$R_A = 12$ kN

(c) Free body diagram at b-b

FIGURE 13–1 Shearing forces in beams.

(a) Shearing stress on a cut section of a beam

FIGURE 13–2 Shearing stress in a beam.

9. From this exercise so far, you should be able to visualize that there are horizontal forces that act between the strips of card stock. The loose stack can only resist those forces by friction and that is generally insufficient to keep the strips from sliding. But the staples have adequate strength to resist the horizontal forces.

10. In the process, the material of the staples is placed in shear. The stapled beam still deflects a bit because you used only four staples. Some parts of the beam can still slide a small amount. Adding more staples should produce an even stiffer beam.

11. Now consider what would happen if the tendency for the strips to slide was restrained over the entire surface

FIGURE 13–3 Illustration of the presence of shearing stress in a beam.

each end and one about 45 mm (1.75 in) from each end, for a total of four staples. This leaves an area in the middle to load this beam design with the stack of quarters. What happens?

8. Hey! That works pretty good. The beam deflects just a small amount. How is this beam different from the stack of five loose strips? Figure 13–3(b) demonstrates that the loose strips slide on one another when carrying the load. Each added strip carries a bit more load but the strips are not acting together. The staples in the second stack of five strips hold each strip tightly to the next one and prohibit sliding. Thus, the strips act as a unit.

of each strip in contact with its adjacent strip. Try that using glue. Now test this beam design and compare its performance with the stapled beam.

12. For the glued beam you should notice virtually no deflection when loaded with six quarters. The continuous film of glue prohibits the sliding of the individual strips so the beam acts as a solid, thick bar that is significantly more resistant to deformation than the stacked or stapled designs. In the glued design, the glue must have sufficient shear strength to resist the horizontal shearing stresses created.

13. Now, what if you were able to make the beam from a single strip of card stock that has a thickness equal to the glued stack of five thinner strips? You could demonstrate the equivalence of the two designs. See if you can find a thick piece of card stock and try this.

14. In conclusion, you should have discovered from this exercise that there are shearing stresses developed within the material of a beam in bending.

Your primary goal as you study this chapter is to develop the competence to analyze a beam to determine the shearing stresses created in it. You will also apply this skill to evaluate the safety of the beam with respect to shear and to design a safe beam. When using an adhesive, welding, brazing, or mechanical fasteners, such as nails, rivets, bolts, pins, or even staples to hold components of a beam together, you should be able to analyze them to ensure that they have adequate strength.

Objectives of This Chapter

After completing this chapter, you should be able to:

1. Describe the conditions under which shearing stresses are created in beams.

2. Compute the magnitude of shearing stresses in beams by using the general shear formula.

3. Define and evaluate the *first moment of the area* required in the analysis of shearing stresses.

4. Specify where the maximum shearing stress occurs on the cross section of a beam.

5. Compute the shearing stress at any point within the cross section of a beam.

6. Describe the general distribution of shearing stress as a function of position within the cross section of a beam.

7. Understand the basis for the development of the general shearing stress formula.

8. Describe four design applications where shearing stresses are likely to be critical in beams.

9. Develop and use special shear formulas for computing the maximum shearing stress in beams having rectangular or solid circular cross sections.

10. Understand the development of approximate relationships for estimating the maximum shearing stress in beams having cross sections with tall thin webs or those with thin-walled, hollow, tubular shapes.

11. Specify a suitable design shearing stress and apply it to evaluate the acceptability of a given beam design.

12. Define *shear flow* and compute its value.

13. Use the shear flow to evaluate the design of fabricated beam sections held together by nails, bolts, rivets, adhesives, welding, or other means of fastening.

13–2 IMPORTANCE OF SHEARING STRESSES IN BEAMS

Several situations exist in practical design in which the mode of failure is likely to be shearing of a part of a beam or of a means of fastening a composite beam together. Five such situations are described here.

Wooden Beams

Wood is inherently weak in shear along the planes parallel to the grain of the wood. Consider the beam shown in Figure 13–4, which is similar to the joists used in floor and roof structures for wood-frame construction. The grain runs generally parallel to the long axis in commercially available lumber. When subjected to transverse loads, the initial failure in a wooden beam is likely to be by separation along the grain of the wood, due to excessive horizontal shearing stress. Note in Appendix A–19 that the allowable shearing stress in common species of wood ranges from only 70 to 95 psi (0.48 to 0.66 MPa), very low values.

FIGURE 13–4 Shear failure in a wood beam.

(a) W12×30 steel beam

0.440 in | 6.52 in

Flange

Web

X ——— X

12.3 in

0.260 in

Steel W12×30 $I_x = 238$ in^4

(b) Data in SI units for W12×30 steel beam; designated W310×44.5

11.2 mm | 166 mm

Flange

Web

X ——— X

312 mm

6.60 mm

Steel W310×44.5 $I_x = 9.91×10^7$ mm^4

FIGURE 13–5 Example of thin-webbed beam shape.

Thin-Webbed Beams

An efficient beam cross section would be one with relatively thick horizontal flanges on the top and bottom with a thin vertical web connecting them together. This generally describes the familiar "I-beam," the wide-flange beam, or the American Standard beam. A specific example is sketched in Figure 13–5. Actual dimensions for other beam sections are given in Appendixes A–7, A–8, and A–11.

If the web is excessively thin, it would not have sufficient stiffness and stability to hold its shape, and it would fail due to shearing stress in the thin web. The American Institute of Steel Construction (AISC) defines the allowable shearing stress in the webs of steel beams. See Reference 5. See also the *web shear formula*, defined later in this chapter, Section 13–6.

Short Beams

In very short beams, the bending moment, and therefore the bending stress, is likely to be small. In such beams, the shearing stress may be the limiting stress.

Fastening Means in Fabricated Beams

As shown in Figure 13–3, the fasteners in a composite beam section are subjected to shearing stresses. The concept of *shear flow*, developed later, can be used to evaluate the safety of such beams or to specify the required type, number, and spacing of fasteners to use. Also, beams made of composite materials are examples of fabricated beams. Separation of the layers of the composite, called *interlaminar shear,* is a potential mode of failure.

Stressed Skin Structures

Aircraft and aerospace structures and some ground-based vehicles and industrial equipment are made using a *stressed skin* design. Sometimes called *monocoque* structures, they are designed to carry much of the load in the thin skins of the structure. The method of shear flow is typically used to evaluate such structures, but this application is not developed in this book.

13–3 THE GENERAL SHEAR FORMULA

Presented here is the general shear formula from which you can compute the magnitude of the shearing stress at any point within the cross section of a beam carrying a vertical shearing force. In Section 13–5, the formula itself is developed. You may wish to study the development of the formula along with this section.

The general shear formula is stated as follows:

General Shear Formula ▼

$$\tau = \frac{VQ}{It} \quad (13\text{–}1)$$

where V = *vertical shearing force* at the section of interest. The value of V can be found from the shearing force diagram developed as described in Chapter 11. Generally, the maximum absolute value of V, positive or negative, is used.

    I = *area moment of inertia* of the entire cross section of the beam with respect to its centroidal axis. This is the same value of I used in the flexure formula ($\sigma = Mc/I$) to compute the bending stress.

    t = *thickness* of the cross section taken at the axis where the shearing stress is to be computed.

    Q = *first moment*, with respect to the overall centroidal axis, *of the area* of that part of the cross section that lies away from the axis where the shearing stress is to be computed.

In some books Q is called the *statical moment*. We will use the term *first moment of the area* to refer to Q in this book. But you should remember all of the parts of the definition given. To calculate the value of Q, we define it mathematically as

First Moment of the Area ▼

$$Q = A_p \bar{y} \quad (13\text{–}2)$$

where A_p = area of that *part* of the cross section that lies away from the axis where the shearing stress is to be computed.

    $\bar{y}$ = distance to the centroid of A_p from the centroidal axis of the entire cross section.

A_p

Centroid of A_p

$h = 10$ a——a

$\bar{y}$

C

$\bar{Y} = 5$

——$t = 2$

Dimensions in inches

FIGURE 13–6   Example of A_p and $\bar{y}$ for use in computing Q.

Note that Q is the *moment of an area;* that is, area times distance. Therefore, it will have the units of length cubed, such as in^3, m^3, or mm^3.

Careful evaluation of Q is critical to proper use of the general shear formula. It is helpful to draw a sketch of the beam cross section and to highlight the partial area, A_p. Then show the location of the centroid of the partial area on the sketch. Figure 13–6 shows an example for which this has been done. In this example, the objective is to calculate the shearing stress at the axis labeled *a-a*. The shaded area is A_p, shown as that part away from the axis *a-a*.

The following three example problems illustrate the method of computing Q. In each, this is the procedure used.

METHOD OF COMPUTING Q, THE FIRST MOMENT OF THE AREA ▼

1. Locate the centroidal axis for the entire cross section.
2. Draw in the axis where the shearing stress is to be calculated.
3. Identify the partial area A_p away from the axis of interest and shade it for emphasis.

If the partial area A_p is a simple area for which the centroid is readily found by simple calculations, use steps 4–7 to compute Q. Otherwise, use steps 8–11.

4. Compute the magnitude of A_p.
5. Locate the centroid of the partial area.
6. Compute the distance $\bar{y}$ from the centroidal axis of the full section to the centroid of the partial area.
7. Compute $Q = A_p\bar{y}$.

For cases in which the partial area is itself a composite area made up of several component parts, steps 8–11 are used.

8. Divide A_p into component parts that are simple areas and label them A_1, A_2, A_3, and so on. Compute their values.
9. Locate the centroid of each component area.
10. Determine the distances from the centroidal axis of the full section to the centroid of each component area, calling them y_1, y_2, y_3, and so on.
11. Compute $Q = A_p\bar{y}$ from

$$Q = A_p\bar{y} = A_1y_1 + A_2y_2 + A_3y_3 + \cdots \qquad \textbf{(13–3)}$$

Example Problem 13–1   For the rectangular section in Figure 13–6, compute the first moment of the area Q as it would be used in the general shear formula to compute the vertical shearing stress at the section marked *a-a*.

Solution   **Objective:**   Compute the value of Q.

Given: Shape and dimensions of cross section in Figure 13–6.

Analysis: Use the method defined in this section.

Results: ***Step 1:*** The centroidal axis for this section is at its midheight, $h/2$, from the bottom. For this problem, $h/2 = 5.00$ in.

Step 2: The axis of interest is *a-a*, coincident with the centroidal axis for this example.

Step 3: The partial area, A_p, is shown shaded in the figure to be the upper half of the rectangle.

Because the partial area is itself a simple rectangle, steps 4–7 are used to compute Q.

Step 4: The partial area is

$$A_p = t(h/2) = (2.0\,\text{in})(5.0\,\text{in}) = 10\,\text{in}^2$$

Step 5: The centroid of the partial area is at its midheight, 2.5 in above *a-a*.

Step 6: Because the centroidal axis is coincident with the axis *a-a*, $\bar{y} = 2.5$ in.

Step 7: Now Q can be computed.

$$Q = A_p\bar{y} = (10.0\,\text{in}^2)(2.5\,\text{in}) = 25.0\,\text{in}^3$$

Example Problem
13–2

For the I-shaped section in Figure 13–7, compute the first moment of the area Q as it would be used in the general shear formula to compute the vertical shearing stress at the section marked a-a.

Solution   Objective: Compute the value of Q.

Given: Shape and dimensions of cross section in Figure 13–7.

Analysis: Use the method defined in this section.

Results: *Step 1:* The I-shape is symmetrical and, therefore, the centroidal axis lies at half the height from its base, 5.0 in.

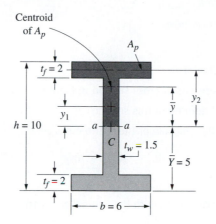

Dimensions in inches

FIGURE 13–7   I-shape for Example Problem 13–2.

Step 2: The axis of interest is a-a, coincident with the centroidal axis for this example.

Step 3: The partial area A_p is shown shaded in the figure to be the upper half of the I-shape. Because the partial area is in the form of a "T," steps 8–11 are used to compute Q.

Step 8: The T-shape is divided into two parts: the upper half of the vertical web is part 1 and the entire top flange is part 2. The magnitudes of these areas are

$$A_1 = \left(\frac{h}{2} - t_f\right)(t_w) = (5.0 \text{ in} - 2.0 \text{ in})(1.5 \text{ in}) = 4.5 \text{ in}^2$$

$$A_2 = bt_f = (6.0 \text{ in})(2.0 \text{ in}) = 12.0 \text{ in}^2$$

Step 9: Each part is a rectangle for which the centroid is at its midheight, as shown in the figure.

Step 10: The required distances are

$$y_1 = \frac{1}{2}\left(\frac{h}{2} - t_f\right) = \frac{1}{2}(5.0 \text{ in} - 2.0 \text{ in}) = 1.5 \text{ in}$$

$$y_2 = \left(\frac{h}{2} - \frac{t_f}{2}\right) = (5.0 \text{ in} - 1.0 \text{ in}) = 4.0 \text{ in}$$

Step 11: Using Equation (13–3) gives us

$$Q = A_1y_1 + A_2y_2$$
$$Q = (4.5 \text{ in}^2)(1.5 \text{ in}) + (12.0 \text{ in}^2)(4.0 \text{ in}) = 54.75 \text{ in}^3$$

Comment: Note that it is not necessary to compute the total area A_p or the location of its centroid of the partial area in the case of composite areas. Only the sum of the products of Ay for all parts of A_p is needed.

Example Problem
13–3

For the T-shaped section in Figure 13–8, compute the first moment of the area Q as it would be used in the general shear formula to compute the vertical shearing stress at the section marked a-a at the very top of the web, just below where it joins the flange.

FIGURE 13–8   T-shape
for Example Problem
13–3.

Dimensions in inches

Solution   Objective: Compute the value of Q.

Given: Shape and dimensions of cross section in Figure 13–8.

Analysis: Use the method defined in this section.

Results: **Step 1:** Locate the centroid of the entire cross section.

$$\bar{Y} = \frac{A_w y_w + A_f y_f}{A_w + A_f}$$

where the subscript w refers to the vertical web and the subscript f refers to the top flange. Then

$$\bar{Y} = \frac{(12)(4) + (16)(9)}{12 + 16} = 6.86 \text{ in}$$

Step 2: The axis of interest, a-a, is at the very top of the web, just below the flange.

Step 3: The partial area above a-a is the entire flange.

Step 4: $A_p = (8 \text{ in})(2 \text{ in}) = 16 \text{ in}^2$

Step 5: The centroid of A_p is 1.0 in down from the top of the flange, which is 9.0 in above the base of the tee.

Step 6: $\bar{y} = 9.0 \text{ in} - \bar{Y} = 9.0 \text{ in} - 6.86 \text{ in} = 2.14 \text{ in}$

Step 7: $Q = A_p \bar{y} = (16 \text{ in}^2)(2.14 \text{ in}) = 34.2 \text{ in}^3$

Comment: It should be noted that the value of Q would be the same if the axis of interest a-a were to be taken at the very bottom of the flange just above the web. But the resulting shearing stresses would be markedly different. The thickness of the section, t, would be equal to the entire width of the flange, whereas for the axis a-a used in this problem, the thickness of the web is used. This will be shown later.

Use of the General Shear Formula

Example problems are presented here to illustrate the use of the general shear formula [Equation (13–1)] to compute the vertical shearing stress in a beam. The following procedure is typical of that used in solving such problems.

GUIDELINES FOR COMPUTING SHEARING STRESSES IN BEAMS ▼

The overall objective is to compute the shearing stress at any specified position on the beam at any specified axis within the cross section using the general shear formula,

$$\tau = \frac{VQ}{It}$$ (13–1)

1. Determine the vertical shearing force V at the section of interest. This may require preparation of the complete shearing force diagram using the procedures of Chapter 11.

2. Locate the centroid of the entire cross section and draw the neutral axis through the centroid.

3. Compute the moment of inertia of the section with respect to the neutral axis.

4. Identify the axis for which the shearing stress is to be computed and determine the thickness t at that axis. Include all parts of the section that are cut by the axis of interest when computing t.

5. Compute Q, the first moment, with respect to the neutral axis, of the partial area away from the axis of interest. Use the procedure developed in this section.

6. Compute the shearing stress using Equation (13–1).

Example Problem 13–4

Compute the shearing stress at the axis *a-a* for a beam with the rectangular cross section shown in Figure 13–6. The shearing force, *V*, on the section of interest is 1200 lb.

Solution **Objective:** Compute the shearing stress at the axis *a-a*.

Given: Cross section shape and dimensions in Figure 13–6. V = 1200 lb.

Analysis: Use the *guidelines for computing shearing stresses in beams.*

Results: **Step 1:** V = 1200 lb (given)

Step 2: For the rectangular shape, the centroid is at the midheight, as shown in Figure 13–6, coincident with axis a-a. $\bar{Y}$ = 5.00 in.

Step 3: $I = bh^3/12 = (2.0)(10.0)^3/12 = 166.7 \text{ in}^4$

Step 4: Thickness = t = 2.0 in at axis *a-a*.

Step 5: Normally we would compute $Q = A_p\bar{y}$ using the method shown earlier in this chapter. But the value of Q for the section in Figure 13–6 was computed in Example Problem 13–1. Use $Q = 25.0 \text{ in}^3$.

Step 6: Using Equation (13–1),

$$\tau = \frac{VQ}{It} = \frac{(1200 \text{ lb})(25.0 \text{ in}^3)}{(166.7 \text{ in}^4)(2.0 \text{ in})} = 90.0 \text{ psi}$$

Example Problem 13–5

Compute the shearing stress at the axes *a-a* and *b-b* for a beam with the T-shaped cross section shown in Figure 13–8. Axis *a-a* is at the very top of the vertical web, just below the flange. Axis *b-b* is at the very bottom of the flange. The shearing force, *V*, on the section of interest is 1200 lb.

Solution **Objective:** Compute the shearing stress at the axes *a-a* and *b-b*.

Given: Cross-section shape and dimensions in Figure 13–8. V = 1200 lb.

Analysis: Use the *guidelines for computing shearing stresses in beams.*

Results: For the axis *a-a*:

Step 1: V = 1200 lb (given)

Step 2: This particular T-shape was analyzed in Example Problem 13–3. Use $\bar{Y}$ = 6.86 in.

Step 3: We will use the methods of Chapter 6 to compute *I*. Let the web be part 1 and the flange be part 2. For each part, $I = bh^3/12$ and $d = \bar{Y} - \bar{y}$.

Part	I	A	d	Ad^2	$I + Ad^2$
1	64.00	12.0	2.86	98.15	162.15
2	5.33	16.0	2.14	73.27	78.60
				Total $I =$	240.75 in⁴

Step 4: Thickness = t = 1.5 in at axis *a-a* in the web.

Step 5: Normally we would compute $Q = A_p\bar{y}$ using the method shown earlier in this chapter. But the value of Q for the section in Figure 13–8 was computed in Example Problem 13–3. Use $Q = 34.2 \text{ in}^3$.

Step 6: Using Equation (13–1),

$$\tau = \frac{VQ}{It} = \frac{(1200 \text{ lb})(34.2 \text{ in}^3)}{(240.75 \text{ in}^4)(1.5 \text{ in})} = 114 \text{ psi}$$

For the axis *b-b*: Some of the data will be the same as at *a-a*.

Step 1: $V = 1200$ lb (given)

Step 2: Again, use $\bar{Y} = 6.86$ in.

Step 3: $I = 240.75$ in^4

Step 4: Thickness $= t = 8.0$ in at axis *b-b* in the flange.

Step 5: Again, use $Q = 34.2$ in^3. The value is the same as at axis *a-a* because both A_p and $\bar{y}$ are the same.

Step 6: Using Equation (13–1),

$$\tau = \frac{VQ}{It} = \frac{(1200\,\text{lb})(34.2\,\text{in}^3)}{(240.75\,\text{in}^4)(8.0\,\text{in})} = 21.3\,\text{psi}$$

Comment: Note the dramatic reduction in the value of the shearing stress when moving from the web to the flange.

13–4 DISTRIBUTION OF SHEARING STRESS IN BEAMS

Most applications require that the maximum shearing stress be determined to evaluate the acceptability of the stress relative to some criterion of design. For most sections used for beams, the maximum shearing stress occurs at the neutral axis, coincident with the centroidal axis, about which bending occurs. The following rule can be used to decide when to apply this observation.

> Provided that the thickness at the centroidal axis is not greater than at some other axis, the maximum shearing stress in the cross section of a beam occurs at the centroidal axis.

Thus, the computation of the shearing stress only at the centroidal axis would give the maximum shearing stress in the section, making the computations at other axes unnecessary.

The logic behind this rule can be seen by examining Equation (13–1), the general shear formula. To compute the shearing stress at any axis, the values of the shearing force V and the moment of inertia I are the same. Because the thickness, t, is in the denominator, the smallest thickness would tend to produce the largest shearing stress, as implied in the

statement of the rule. But the value of the first moment of the area Q also varies at different axes, decreasing as the axis of interest moves toward the outside of the section. Recall that Q is the product of the partial area A_p and the distance $\bar{y}$ to the centroid of A_p. For axes away from the centroidal axis, the area decreases at a faster rate than $\bar{y}$ increases, resulting in the value of Q decreasing. Thus, the maximum value of Q will be that for stress computed at the centroidal axis. It follows that the maximum shearing stress will always occur at the centroidal axis *unless the thickness at some other axis is smaller than that at the centroidal axis.*

The shapes shown in Figures 13–6, 13–7, and 13–8 are all examples that conform to the rule that the maximum shearing stress occurs at the neutral axis because each has its smallest thickness at the neutral axis. Figure 13–9 shows three examples where the rule *does not apply*. In each example, at some axes away from the neutral axis, the thickness is smaller than that at the neutral axis. In such cases, the maximum shearing stress *may* occur at some other axis. Example Problem 13–7 illustrates this observation by analyzing the triangular section.

The solid and hollow circular sections are important examples of where the maximum shearing stress does occur at the neutral axis even though the thickness decreases at other axes. It can be shown that the ratio Q/t continuously decreases for axes away from the neutral axis at the diameter.

The following example problems illustrate the shearing stress distribution in beams of different shapes. Note the comments at the end of each problem for some general conclusions.

(a)   (b)   (c)

FIGURE 13–9   Beam cross sections for which the maximum shearing stress may not occur at the centroidal axis, *c-c*.

Example Problem 13–6   Compute the distribution of shearing stress with position in the cross section for a beam with the rectangular shape shown in Figure 13–6. The actual dimensions are 2.0 in by 10.0 in. Plot the results. The shearing force, V, on the section of interest is 1200 lb.

Solution   **Objective:**   Compute the shearing stress at several axes and plot τ versus position.

Given: Cross-section shape and dimensions in Figure 13–6. $V = 1200$ lb.

Analysis: Use the *guidelines for computing shearing stresses in beams*. Because the shape is symmetrical with respect to the centroidal axis, we choose to compute the shearing stresses in the upper part at the axes *a-a*, *b-b*, *c-c*, and *d-d*, as shown in Figure 13–10. Then the values of stresses in the lower part at sections *b'-b'*, *c'-c'*, and *d'-d'* will be the same as the corresponding points above.

Results: **Step 1:** $V = 1200$ lb (given)

Step 2: For the rectangular shape, the centroid is at the midheight, as shown in Figure 13–6, coincident with axis *a-a*. $\bar{Y} = 5.00$ in.

Step 3: $I = bh^3/12 = (2.0)(10.0)^3/12 = 166.7$ in^4

Step 4: Thickness $= t = 2.0$ in at all axes.

(a) Location of axes of interest   Dimensions in inches   (b) Data for axis *a-a*

(c) Data for axis *b-b*   (d) Data for axis *c-c*   (e) Data for axis *d-d*

FIGURE 13–10   Data for Q calculations in Example Problem 13–6.

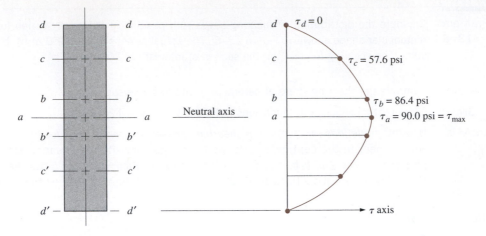

FIGURE 13–11   Distribution of shearing stress on rectangular section for Example Problem 13–6.

Step 5: We will compute $Q = A_p y$ for each axis using the method shown earlier in this chapter. Recall that the value of Q for this section at the centroidal axis was computed in Example Problem 13–1 where we found $Q = 25.0 \, in^3$. A similar calculation is summarized in the table following Step 6, using data from Figure 13–10.

Step 6: Using Equation (13–1), the calculation for shearing stress at the neutral axis a-a is shown here.

$$\tau = \frac{VQ}{It} = \frac{(1200 \, lb)(25.0 \, in^3)}{(166.7 \, in^4)(2.0 \, in)} = 90.0 \, psi$$

The calculation would be the same at the other axes with only the value of Q changing. See the following table.

Axis	V	I	t	A_p	y	$Q = A_p y$	$\tau = VQ/It$
a-a	1200	166.7	2.0	10.0	2.5	25.0	90.0 psi
b-b	1200	166.7	2.0	8.0	3.0	24.0	86.4 psi
c-c	1200	166.7	2.0	4.0	4.0	16.0	57.6 psi
d-d	1200	166.7	2.0	0.0	5.0	0.0	0.0 psi

The results of shearing stress versus position are shown in Figure 13–11 alongside the rectangular section itself.

LESSON 20

Comment: Note that the maximum shearing stress does occur at the neutral axis as predicted. The variation of shearing stress with position is parabolic, ending with zero stress at the top and bottom surfaces.

Example Problem 13–7   For the triangular beam cross section shown in Figure 13–12, compute the shearing stress that occurs at the axes *a* through *g*, each 50 mm apart. Plot the variation of stress with position on the section. The shearing force is 50 kN.

Solution   Objective: Compute the shearing stress at seven axes and plot τ versus position.

Given: Cross-section shape and dimensions in Figure 13–12. $V = 50$ kN.

Analysis: Use the *guidelines for computing shearing stresses in beams.*

Results: In the general shear formula, the values of V and I will be the same for all computations. V is given to be 50 kN and

$$I = \frac{bh^3}{36} = \frac{(300)(300)^3}{36} = 225 \times 10^6 \, mm^4$$

FIGURE 13–12 Triangular cross section for a beam for which the maximum shearing stress does not occur at the centroidal axis.

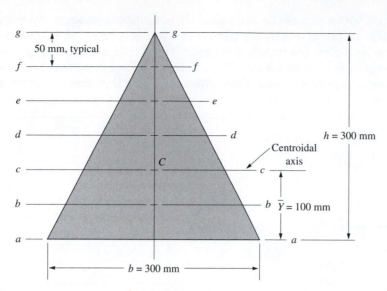

Table 13–1 shows the remaining computations. Obviously, the value for Q for axes a-a and g-g is zero because the area outside each axis is zero. Note that because of the unique shape of the given triangle, the thickness t at any axis is equal to the height of the triangle above the axis.

Figure 13–13 shows a plot of these stresses. The maximum shearing stress occurs at half the height of the section, and the stress at the centroid (at $h/3$) is lower. This illustrates the general statement made earlier that for sections whose minimum thickness does not occur at the centroidal axis, the maximum shearing stress may occur at some axis other than the centroidal axis.

TABLE 13–1

Axis	$A_p(\text{mm}^2)$	$\bar{y}(\text{mm})$	$Q = A_p\bar{y}(\text{mm}^3)$	t (mm)	τ(MPa)
a-a	0	100	0	300	0
b-b	13 750	75.8	1.042×10^6	250	0.92
c-c	20 000	66.7	1.333×10^6	200	1.48
d-d	11 250	100.0	1.125×10^6	150	1.67
e-e	5 000	133.3	0.667×10^6	100	1.48
f-f	1 250	166.7	0.208×10^6	50	0.92
g-g	0	200	0	0	0

FIGURE 13–13 Shearing stress distribution in the triangular cross section for Example Problem 13–7.

Comment: One further note can be made about the computations shown for the triangular section. For the axis b-b, the partial area A_p was taken as that area *below* b-b. The resulting section is the trapezoid between b-b and the bottom of the beam. For all other axes, the partial area A_p was taken as the triangular area *above* the axis. The area below the axis could have been used, but the computations would have been more difficult. When computing Q, it does not matter whether the area above or below the axis of interest is used for computing A_p and $\bar{y}$.

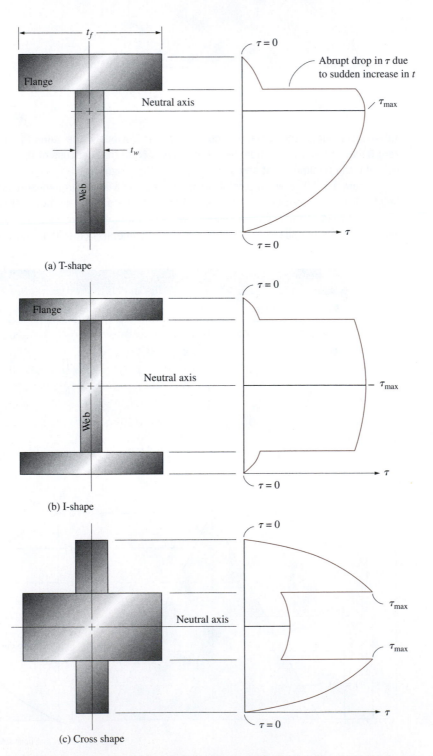

(a) T-shape

(b) I-shape

(c) Cross shape

FIGURE 13–14 Stress distribution for shapes with abrupt changes in thickness.

By reviewing the results of the example problems worked thus far in this chapter, the following conclusions can be drawn.

SUMMARY OF OBSERVATIONS ABOUT THE DISTRIBUTION OF SHEARING STRESS IN THE CROSS SECTION OF A BEAM ▼

1. The shearing stress at the outside of the section away from the centroidal axis is zero.

2. The maximum shearing stress in the cross section occurs at the centroidal axis provided that the thickness there is no greater than at some other axis.

3. Within a part of the cross section where the thickness is constant, the shearing stress varies in a curved fashion, decreasing as the distance from the centroidal axis increases. The curve is actually a part of a parabola.

4. At an axis where the thickness changes abruptly, as where the web of a tee or an I-shape joins the flange, the shearing stress also changes abruptly, being much smaller in the flange than in the thinner web. See Figure 13–14.

13–5 DEVELOPMENT OF THE GENERAL SHEAR FORMULA

This section presents the background information on the general shear formula. Figure 13–15 shows a beam carrying two transverse loads and the corresponding shearing force and bending moment diagrams to help you visualize certain relationships.

The *moment-area* principle of beam diagrams states that *the change in bending moment between two points on a beam is equal to the area under the shearing force curve between those two points.* For example, consider two points in segment *A-B* of the beam in Figure 13–15, marked x_1 and x_2, a small distance dx apart. The moment at x_1 is M_1 and the moment at x_2 is M_2. Then the moment-area rule states that

$$M_2 - M_1 = V(dx) = dM$$

This can also be stated as

$$V = \frac{dM}{dx} \qquad (13\text{--}4)$$

That is, the differential change in bending moment for a differential change in position on the beam is equal to the shearing force occurring at that position.

Equation (13–4) can also be developed by looking at a free body diagram of the small segment of the beam between x_1 and x_2, as shown in Figure 13–16(a). Since this is a cut section from the beam, the internal shearing forces and bending moments are shown acting on the cut faces. Because the beam itself is in equilibrium, this segment is also. Then the sum of moments about a point in the left face at *O* must be zero. This gives

$$\sum M_O = 0 = M_1 - M_2 + V(dx) = -dM + V(dx)$$

Or, as shown before,

$$V = \frac{dM}{dx}$$

Any *part* of the beam segment in Figure 13–16(a) must also be in equilibrium. The shaded portion isolated in Figure 13–16(b) is acted on by forces parallel to the axis of the beam. On the left side, F_1 is due to the bending stress acting at that section on the area. On the right side, F_2 is due to the bending stress acting at that section on the area. In general, the values of F_1 and F_2 will be different and there must be a third force acting on the bottom face of the shaded portion of the segment to maintain equilibrium. This is the shearing force, F_s, which causes the shearing stress in the beam. Figure 13–16(b) shows F_s acting on the area $t(dx)$. Then the shearing stress is

$$\tau = \frac{F_s}{t(dx)} \qquad (13\text{--}5)$$

By summing forces in the horizontal direction, we find

$$F_s = F_2 - F_1 \qquad (13\text{--}6)$$

We will now develop the equations for the forces F_1 and F_2. Each force is the product of the bending stress times the area over which it acts. But the bending stress varies with position in the cross section. From the flexure formula, the bending stress at any position y relative to the neutral axis is

$$\sigma = \frac{My}{I}$$

Then the total force acting on the shaded area of the left face of the beam segment is

$$F_1 = \int_A \sigma \, dA = \int_{y_a}^{y_b} \frac{M_1 y}{I} \, dA \qquad (13\text{--}7)$$

where dA is a small area within the shaded area. The values of M_1 and I are constant and can be taken outside the integral sign. Equation (13–7) then becomes

$$F_1 = \frac{M_1}{I} \int_{y_a}^{y_b} y \, dA \qquad (13\text{--}8)$$

Now the last part of Equation (13–8) corresponds to the definition of the centroid of the shaded area. That is,

$$\int_{y_a}^{y_b} y \, dA = \bar{y} A_p \qquad (13\text{--}9)$$

where A_p is the area of the shaded portion of the left face of the segment and $\bar{y}$ is the distance from the neutral axis to the centroid of A_p. This product of $\bar{y} A_p$ is called the first moment of the area Q in the general shear formula. Making this substitution in Equation (13–8) gives

$$F_1 = \frac{M_1}{I} \int_{y_a}^{y_b} y \, dA = \frac{M_1}{I} \bar{y} A_p = \frac{M_1 Q}{I} \qquad (13\text{--}10)$$

Similar reasoning can be used to develop the relationship for the force F_2 on the right face of the segment.

FIGURE 13–15   Beam diagrams used to develop general shear formula.

$$F_2 = \frac{M_2 Q}{I} \tag{13–11}$$

Substitutions can now be made for F_1 and F_2 in Equation (13–6) to complete the development of the shearing force.

$$F_s = F_2 - F_1 = \frac{M_2 Q}{I} - \frac{M_1 Q}{I} = \frac{Q}{I}(M_2 - M_1) \tag{13–12}$$

Earlier we defined $(M_2 - M_1) = dM$. Then

$$F_s = \frac{Q(dM)}{I} \tag{13–13}$$

Then, in Equation (13–5),

$$\tau = \frac{F_s}{t(dx)} = \frac{Q(dM)}{It(dx)}$$

But, from Equation (13–4), $V = dM/dx$. Then

$$\tau = \frac{VQ}{It}$$

This is the form of the general shear formula [Equation (13–1)] used in this chapter.

13–6 SPECIAL SHEAR FORMULAS

As demonstrated in several example problems, the general shear formula can be used to compute the shearing stress at any axis on any cross section of the beam. However, frequently it is desired to know only the *maximum shearing stress*. For many common shapes used for beams, it is possible to develop special simplified formulas that will give the maximum shearing stress quickly. The rectangle, circle, thin-walled hollow tube, and thin-webbed shapes can be analyzed this way. The formulas are developed in this section.

For all of these section shapes, the maximum shearing stress occurs at the neutral axis. The rectangle and thin-webbed shapes conform to the rule stated in Section 13–5 because the thickness at the neutral axis is no greater than at other axes in the section. The circle and the thin-walled tube do not conform to the rule. However, it can be shown that the ratio Q/t in the general shear formula decreases continuously as the axis of interest moves away from the neutral axis, resulting in the decrease in the shearing stress.

Rectangular Shape

Figure 13–17 shows a typical rectangular cross section having a thickness t and a height h. The three geometrical terms in the general shear formula can be expressed in terms of t and h.

$$I = \frac{th^3}{12}$$

$$t = t$$

(a)

(b)

FIGURE 13–16   Forces on a portion of a cut segment of a beam. (a) Free body diagram of beam segment. (b) Isolated portion of segment.

FIGURE 13–17 Rectangular shape.

$$Q = A_p\bar{y}\text{(for area above centroidal axis)}$$

$$Q = \frac{th}{2} \cdot \frac{h}{4} = \frac{th^2}{8}$$

Putting these terms in the general shear formula gives

$$\tau_{max} = \frac{VQ}{It} = V \cdot \frac{th^2}{8} \cdot \frac{12}{th^3} \cdot \frac{1}{t} = \frac{3}{2}\frac{V}{th}$$

But since th is the total area of the section,

Special Shear Formula for Rectangle ▼

$$\tau_{max} = \frac{3V}{2A} \qquad (13\text{–}14)$$

Equation (13–14) can be used to compute exactly the maximum shearing stress in a rectangular beam at its centroidal axis.

Note that $\tau = V/A$ represents the *average* shearing stress on the section. Thus, the maximum shearing stress on a rectangular cross section is 1.5 times higher than the average.

Example Problem 13–8 Compute the maximum shearing stress that would occur in the rectangular cross section of a beam like that shown in Figure 13–17. The shearing force is 1000 lb, $t = 2.0$ in, and $h = 8.0$ in.

Solution Using Equation (13–14) yields

$$\tau_{max} = \frac{3V}{2A} = \frac{3(1000\text{ lb})}{2(2\text{ in})(8\text{ in})} = 93.8\text{ psi}$$

Circular Shape

The special shear formula for the circular shape is developed in a similar manner to that used for the rectangular shape. Equations for Q, I, and t are written in terms of the primary size variable for the circular shape, its diameter. Then the general shear formula is simplified (refer to Figure 13–18).

$$t = D$$

$$I = \frac{\pi D^4}{64}$$

$$Q = A_p\bar{y} \quad \text{(for the semicircle above the centroid)}$$

$$Q = \frac{\pi D^2}{8} \cdot \frac{2D}{3\pi} = \frac{D^3}{12}$$

Then the maximum shearing stress is

$$\tau_{max} = \frac{VQ}{It} = V \times \frac{D^3}{12} \times \frac{64}{\pi D^4} \times \frac{1}{D} = \frac{64V}{12\pi D^2}$$

To refine the equation, factor out a 4 from the numerator and then note that the total area of the circular section is $A = \pi D^2/4$.

Special Shear Formula for Circle ▼

$$\tau_{max} = \frac{16(4)V}{12\pi D^2} = \frac{16V}{12A}$$

$$\tau_{max} = \frac{4V}{3A} \qquad (13\text{–}15)$$

This shows that the maximum shearing stress is 1.33 times higher than the average on the circular section.

FIGURE 13–18 Circular shape.

Example Problem
13–9

Compute the maximum shearing stress that would occur in a circular shaft, 50 mm in diameter, if it is subjected to a vertical shearing force of 110 kN.

Solution

Equation (13–15) will give the maximum shearing stress at the horizontal diameter of the shaft.

$$\tau_{max} = \frac{4V}{3A}$$

But

$$A = \frac{\pi D^2}{4} = \frac{\pi (50\ mm)^2}{4} = 1963\ mm^2$$

Then

$$\tau_{max} = \frac{4(110 \times 10^3\ N)}{3(1963\ mm^2)} = 74.7\ MPa$$

Hollow, Thin-Walled, Tubular Shape

Removing material from the center of a circular cross section tends to increase the local value of the shearing stress, especially near the diameter where the maximum shearing stress occurs. Although not giving a complete development here, it is observed that the maximum shearing stress in a thin-walled tube is approximately twice the average. That is,

Special Shear Formula for Thin-Walled Tube ▼

$$\tau_{max} \approx 2\frac{V}{A} \qquad (13\text{–}16)$$

where A is the total cross-sectional area of the tube.

Example Problem
13–10

Compute the approximate maximum shearing stress that would occur in a 3-in schedule 40 steel pipe if it is used as a beam and subjected to a shearing force of 6200 lb.

Solution

Equation (13–16) should be used. From Appendix A–12 we find that the cross-sectional area of the 3-in schedule 40 steel pipe is 2.228 in^2. Then an estimate of the maximum shearing stress in the pipe, occurring near the horizontal diameter, is

$$\tau_{max} \approx 2\frac{V}{A} = \frac{2(6200\ lb)}{2.228\ in^2} = 5566\ psi$$

Thin-Webbed Shapes

Structural shapes such as W- and S-beams have relatively thin webs. The distribution of shearing stress in such beams is typically like that shown in Figure 13–19. The maximum shearing stress is at the centroidal axis. It decreases slightly in the rest of the web and then drastically in the flanges. Thus, most of the resistance to the vertical shearing force is provided by the web. Also, the average shearing stress in the web would be just slightly smaller than the maximum stress. For these reasons, the *web shear formula* is often used to get a quick estimate of the shearing stress in thin-webbed shapes.

Web Shear Formula for Thin-Webbed Shapes ▼

$$\tau_{max} \approx \frac{V}{A_{web}} = \frac{V}{th} \qquad (13\text{–}17)$$

The thickness of the web is t. The simplest approach would be to use the full height of the beam for h. This would result in a shearing stress approximately 15% lower than the actual maximum shearing stress at the centroidal axis for typical beam shapes. Using just the web height between the flanges would produce a closer approximation of the maximum shearing stress, probably less than 10% lower than the actual

FIGURE 13–19   Distribution of shearing stress in a thin-webbed shape.

value. In problems using the web shear formula, we use the full height of the cross section unless otherwise stated.

In summary, for thin-webbed shapes, compute the shearing stress from the web shear formula using the full height of the beam for h and the actual thickness of the web for t. Then, to obtain a more accurate estimate of the maximum shearing stress, increase this value by about 15%.

Example Problem 13–11

Using the web shear formula, compute the shearing stress in a W12×16 beam if it is subjected to a shearing force of 25 000 lb.

Solution

In Appendix A–7 for W-beams, it is found that the web thickness is 0.220 in, and the overall depth (height) of the beam is 12.00 in. Then, using Equation (13–17), we have

$$\tau_{max} \approx \frac{V}{th} = \frac{25\,000 \text{ lb}}{(0.220 \text{ in})(12.00 \text{ in})} = 9470 \text{ psi}$$

13–7 DESIGN FOR SHEAR

The design shear stress depends greatly upon the material from which the beam is to be made and on the form of the member subjected to the shearing stress. A limited amount of data is presented in this book, and the reader is advised to check more complete references, such as those listed at the end of the chapter.

Steel

For shear stress in the webs of rolled steel beam shapes, the AISC generally recommends

$$\tau_d = 0.40s_y \qquad \textbf{(13–18)}$$

However, there are extensive discussions in References 3–5 for special cases of short beams; beams with unusually tall, thin webs; and beams with stiffeners applied either in the vertical or horizontal directions. Careful consideration of these factors is advised. See also References 11 and 13.

Aluminum

The Aluminum Association also provides extensive data for various conditions of loading and beam geometry. For example, Reference 1 gives actual data for allowable shear stress of the more popular aluminum alloys for several applications. It is not practical to summarize such data in this book.

As a general guideline, we will use the same design shear stress for ductile metals under static loads as listed in Chapter 9, Table 9–8, and Appendix A–21. That is, a design factor of $N = 2$ based on the yield strength of the material in shear, s_{ys}, is suggested. Also, an approximation for the value of s_{ys} is one-half of the yield strength in tension, s_y. In summary,

$$\tau_d = \frac{s_{ys}}{N} = \frac{0.5s_y}{N} = \frac{s_y}{2N} \qquad \textbf{(13–19)}$$

For $N = 2$,

$$\tau_d = \frac{s_y}{4} = 0.25s_y$$

Wood

For wood beams, data are given in Appendix A–19 for allowable horizontal shear stress. Note that the values are quite low, typically less than 100 psi (0.69 MPa). Shear failure is frequently the limiting factor for wood beams. See also References 9, 14, and 15.

Concrete

Shear strength of concrete is quite low compared with that of most metals. The American Concrete Institute (ACI) specifies that the ultimate shear strength is $2\sqrt{f_c'}$ where f_c' is the rated strength of the concrete, typically ranging from 2000 psi to 7000 psi. Then the shear strength ranges from 89 psi to 167 psi.

Furthermore, whenever a shear stress is created in concrete, a corresponding tensile stress results. As explained in Chapter 15, for example, when pure shear exists on an element, there is a tensile stress of equal magnitude produced on a plane 45 degrees from the original orientation of the element. With the extremely low tensile strength of concrete, tensile failures often occur in zones of high shear stress and they propagate at a 45 degree angle. To counteract this failure mode, the design of concrete beams always includes steel shear reinforcement, typically in the form of stirrups placed perpendicular to the longitudinal reinforcing bars required for bending strength. See Figure 13–20. Welded wire fabric or steel bars bent up at an angle of 30° to 60° are also sometimes used. See References 2 and 12 for design procedures to determine the size, layout, and spacing of the bars.

FIGURE 13–20

Masonry

The design of beams using masonry made from brick, stone, and concrete unit blocks must also consider shearing stresses. Similarly, load-bearing walls and shear walls subjected to bending loads from wind forces and forces transferred from floors, roofs, or supported beams must be designed to resist shear. Failure often occurs in the mortar joints along a diagonal in the wall, creating a noticeable zigzag pattern. See Reference 10.

Composite Materials

The design, testing, and analysis of structures made from composite materials are discussed in References 6–8. Composites made using laminated structures described in Chapter 8 often fail by the mechanism of *interlaminar shear,* where the internal shearing stresses cause the separation of layers within the structure. Reference 7 describes a testing method to evaluate a beam's tendency to fail in this manner. With other forms of composite materials, such as pultrusions, sheet molding compounds (SMC), those reinforced with long or short chopped fibers, or those employing woven fabrics or roving, failure can occur when the individual reinforcing fibers separate from the matrix. References 6 and 8 discuss methods of designing structural components to withstand shearing stresses.

13–8 SHEAR FLOW

Built-up sections used for beams, such as those shown in Figures 13–21 and 13–22, must be analyzed to determine the proper size and spacing of fasteners. The discussion in preceding sections showed that horizontal shearing forces exist at the planes joined by the nails, bolts, and rivets. Thus the fasteners are subjected to shear. Usually, the size and material of the fastener will permit the specification of an allowable shearing force on each. Then the beam must be analyzed to determine a suitable spacing for the fasteners that will ensure that all parts of the beam act together.

Dimensions in millimeters

FIGURE 13–22 Built-up beam shape for Example Problem 13–13.

The term *shear flow is* useful for analyzing built-up sections. Called q, the shear flow is found by multiplying the shearing stress at a section by the thickness at that section. That is,

$$q = \tau t \qquad (13\text{–}20)$$

But from the general shear formula,

$$\tau = \frac{VQ}{It}$$

Then

$$q = \tau t = \frac{VQ}{I} \qquad (13\text{–}21)$$

The units for q are *force per unit length,* such as N/m, N/mm, or lb/in. The shear flow is a measure of how much shearing force must be resisted at a particular section per unit length. Knowing the shearing force capacity of a fastener then allows the determination of a safe spacing for the fasteners.

For example, if a particular style of nail can safely withstand 150 lb of shearing force, we will define

$$F_{sd} = 150 \text{ lb}$$

Then, if in a particular place on a beam fabricated by nailing boards together with the shear flow computed to be $q = 28.5$ lb/in, the maximum spacing, s_{max} of the nail is

Dimensions in inches

FIGURE 13–21 Beam shape for Example Problem 13–12.

$$s_{max} = \frac{F_{sd}}{q} = \frac{150 \text{ lb}}{28.5 \text{ lb/in}} = 5.26 \text{ in} \qquad (13\text{–}22)$$

Example Problem 13–12

Determine the proper spacing of the nails used to secure the flange boards to the web of the built-up I-beam shown in Figure 13–21. All boards are standard 2×8 wood shapes. The nails to be used can safely resist 250 lb of shearing force each. The load on the beam is shown in Figure 13–15 with $P = 500$ lb.

Solution

Objective: Specify a suitable spacing for the nails.

Given: Loading in Figure 13–15. $P = 500$ lb; $F_{sd} = 250$ lb/nail.
Beam shape and dimensions in Figure 13–21.

Analysis: Use the *guidelines for specifying the spacing of fasteners*.

Results: The maximum shearing force on the beam is 500 lb, occurring between each support and the applied loads.

Step 1: The moment of inertia can be computed by subtracting the two open-space rectangles at the sides of the web from the full rectangle surrounding the I-shape.

$$I = \frac{7.25(10.25)^3}{12} - \frac{2(2.875)(7.25)^3}{12} = 468.0 \text{ in}^4$$

Step 2: At the place where the nails join the boards, Q is evaluated for the area of the top (or bottom) flange board.

$$Q = A_p\bar{y} = (1.5 \text{ in})(7.25 \text{ in})(4.375 \text{ in}) = 47.6 \text{ in}^3$$

Step 3: Then the shear flow is

$$q = \frac{VQ}{I} = \frac{(500 \text{ lb})(47.6 \text{ in}^3)}{468 \text{ in}^4} = 50.9 \text{ lb/in}$$

This means that 50.9 lb of force must be resisted along each inch of length of the beam at the point between the flange and the web boards.

Step 4: Since each nail can withstand 250 lb, the maximum spacing is

$$s_{max} = \frac{F_{sd}}{q} = \frac{250 \text{ lb}}{50.9 \text{ lb/in}} = 4.92 \text{ in}$$

Step 5: A spacing of $s = 4.5$ in would be reasonable.

The principle of shear flow also applies for sections like that shown in Figure 13–22, in which a beam section is fabricated by riveting square bars to a vertical web plate to form an I-shape. The shear flow occurs from the web plate to the flange bars. Thus, when evaluating the statical moment Q, the partial area, A_p, is taken to be the area of one of the flange bars.

Example Problem 13–13   A fabricated beam is made by riveting square aluminum bars to a vertical plate, as shown in Figure 13–22. The bars are 20 mm square. The plate is 6 mm thick and 200 mm high. The rivets can withstand 800 N of shearing force across one cross section. Determine the required spacing of the rivets if a shearing force of 5 kN is applied.

Solution   Objective: Specify a suitable spacing for the rivets.

Given: Shearing force = 5 kN. F_{sd} = 800 N/rivet.
Beam shape and dimensions in Figure 13–22.

Analysis: Use the *guidelines for specifying the spacing of fasteners.*

Results: **Step 1:** I is the moment of inertia of the entire cross section,

$$I = \frac{6(200)^3}{12} + 4\left[\frac{20^4}{12} + (20)(20)(90)^2\right]$$

$$I = 17.0 \times 10^6 \, \text{mm}^4$$

Step 2: Q is the product of $A_p\bar{y}$ for the area *outside* the section where the shear is to be calculated. In this case, the partial area A_p is the 20-mm square area *to the side* of the web. For the beam in Figure 13–22,

$$Q = A_p\bar{y} \text{ (for one square bar)}$$

$$Q = (20)(20)(90) \, \text{mm}^3 = 36\,000 \, \text{mm}^3$$

Step 3: Then for $V = 5$ kN,

$$q = \frac{VQ}{I} = \frac{(5 \times 10^3 \, \text{N})(36 \times 10^3 \, \text{mm}^3)}{17.0 \times 10^6 \, \text{mm}^4} = 10.6 \, \text{N/mm}$$

Thus, a shearing force of 10.6 N is to be resisted for each millimeter of length of the beam.

Step 4: Since each rivet can withstand 800 N of shearing force, the maximum spacing is

$$s_{max} = \frac{F_{sd}}{q} = \frac{800 \, \text{N}}{10.6 \, \text{N/mm}} = 75.5 \, \text{mm}$$

Step 5: Specify a spacing of $s = 75$ mm.

REFERENCES

1. Aluminum Association, *Aluminum Design Manual,* Washington, DC, 2005.

2. American Concrete Institute, *2006 Manual of Concrete Practice,* Farmington Hills, MI, American Concrete Institute, 2006.

3. American Institute of Steel Construction, *Specification for Structural Steel Buildings,* Chicago, IL, AISC, 2005.

4. American Institute of Steel Construction, *Manual of Steel Construction: Load and Resistance Factor Design,* 3rd ed., Chicago, IL, AISC, 2001.

5. American Institute of Steel Construction, *Manual of Steel Construction: Allowable Stress Design,* 9th ed., Chicago, IL, AISC, 1989.

6. ASM International, *ASM Handbook, Volume 21: Composites,* Materials Park, OH, 2001.

7. ASTM International, *Standard D2344—Standard Test Method for Short-Beam Strength of Polymer Matrix Composite Materials and Their Laminates,* ASTM International, West Conshohocken, PA, 2006.

8. ASTM International, *The Composite Materials Handbook—MIL 17,* ASTM International, West Conshohocken, PA, 2002.

9. Halperin, D. A. and T. G. Bible, *Principles of Timber Design for Architects and Builders,* New York, John Wiley & Sons, Inc., 1994.

10. The Masonry Society, *Masonry Designer's Guide,* 4th ed., Boulder, CO, The Masonry Society, 2003.

11. McCormac, J. C. and J. Nelson, *Structural Steel Design—LRFD Method,* 4th ed., Upper Saddle River, NJ, Prentice Hall, 2008.

12. Mehta, P. K. and P. J. M. Monteiro, *Concrete,* 3rd ed., New York, McGraw-Hill, 2006.

13. Speigel, L. and G. F. Limbrunner, *Applied Structural Steel Design,* 4th ed., Englewood Cliffs, NJ, Prentice Hall, 2002.

14. U.S. Department of Agriculture Forest Products Laboratory, *Wood Handbook—Wood as an Engineering Material,* Madison, WI, Forest Products Laboratory, 1999.

15. American Institute of Timber Construction, *Timber Construction Manual,* 5th ed., New York, John Wiley & Sons, Inc., 2004.

PROBLEMS

General Shear Formula

For Problems 13–1 through 13–20, compute the shearing stress at the horizontal neutral axis for a beam having the cross-sectional shape shown in the given figure for the given shearing force. Use the general shear formula.

13–1. Use a rectangular shape having a width of 50 mm and a height of 200 mm. $V = 7500$ N.

13–2. Use a rectangular shape having a width of 38 mm and a height of 180 mm. $V = 5000$ N.

13–3. Use a rectangular shape having a width of 1.5 in and a height of 7.25 in. $V = 12\,500$ lb.

13–4. Use a rectangular shape having a width of 3.5 in and a height of 11.25 in. $V = 20\,000$ lb.

13–5. Use a circular shape having a diameter of 50 mm. $V = 4500$ N.

13–6. Use a circular shape having a diameter of 38 mm. $V = 2500$ N.

13–7. Use a circular shape having a diameter of 2.00 in. $V = 7500$ lb.

13–8. Use a circular shape having a diameter of 0.63 in. $V = 850$ lb.

13–9. Use the shape shown in Figure P13–9. $V = 1500$ lb.

FIGURE P13–9

13–10. Use the shape shown in Figure P13–10. $V = 850$ lb.

FIGURE P13–10

13–11. Use the shape shown in Figure P13–11. $V = 850$ lb.

FIGURE P13–11

13–12. Use the shape shown in Figure P13–12. $V = 112$ kN.

FIGURE P13–12

13–13. Use the shape shown in Figure P13–13. $V = 71.2$ kN.

FIGURE P13–13

13–14. Use the shape shown in Figure P13–14. $V = 1780$ N.

FIGURE P13–14

13–15. Use the shape shown in Figure P13–15. $V = 675$ N.

FIGURE P13–15

13–16. Use the shape shown in Figure P13–16. $V = 2.5$ kN.

FIGURE P13–16

13–17. Use the shape shown in Figure P13–17. $V = 10.5$ kN.

FIGURE P13–17

13–18. Use the shape shown in Figure P13–18. $V = 1200$ lb.

FIGURE P13–18

13–19. Use the shape shown in Figure P13–19. $V = 775$ lb.

Dimensions in inches

FIGURE P13–19

13–20. Use the shape shown in Figure P13–20. $V = 2500$ lb.

FIGURE P13–20

For Problems 13–21 through 13–30, assume that the indicated shape is the cross section of a beam made from wood having an allowable shearing stress of 70 psi, which is that of No. 2 grade southern pine listed in Appendix A–19. Compute the maximum allowable shearing force for each shape. Use the general shear formula.

13–21. Use a standard 2 × 4 wooden beam with the long dimension vertical.

13–22. Use a standard 2 × 4 wooden beam with the long dimension horizontal.

13–23. Use a standard 2 × 12 wooden beam with the long dimension vertical.

13–24. Use a standard 2 × 12 wooden beam with the long dimension horizontal.

13–25. Use a standard 10 × 12 wooden beam with the long dimension vertical.

13–26. Use a standard 10 × 12 wooden beam with the long dimension horizontal.

13–27. Use the shape shown in Figure P13–27.

FIGURE P13–27

13–28. Use the shape shown in Figure P13–28.

FIGURE P13–28

13–29. Use the shape shown in Figure P13–29.

$\frac{1}{2}$-in plywood

24 in

2 × 4   2 × 4 $3\frac{1}{2}$ in

$\frac{1}{2}$ in

$1\frac{1}{2}$ in $1\frac{1}{2}$ in $\frac{1}{2}$ in

FIGURE P13–29

13–30. Use the shape shown in Figure P13–30.

$1\frac{1}{2}$ in

$11\frac{1}{4}$ in

2 × 12

2 × 12
2 × 12

$11\frac{1}{4}$ in

3 in

FIGURE P13–30

13–31. For a beam having the I-shape cross section shown in Figure P13–31, compute the shearing stress on horizontal axes 0.50 in apart from the bottom to the top. At the ends of the web where it joins the flanges, compute the stress in both the web and the flange. Use a shearing force of 500 lb. Then plot the results.

6 in

1 in

1 in

6 in

1 in

FIGURE P13–31

13–32. For a beam having the box-shape cross section shown in Figure P13–32, compute the shearing stress on horizontal axes 0.50 in apart from the bottom to the top. At the ends of the vertical sides where they join the flanges, compute the stress in both the web and the flange. Use a shearing force of 500 lb. Then plot the results.

6 in

1 in→ ←
typical

8 in

FIGURE P13–32

13–33. For a standard W14×43 steel beam, compute the shearing stress at the neutral axis when subjected to a shearing force of 33 500 lb. Use the general shear formula. Neglect the fillets at the intersection of the web with the flanges.

13–34. For the same conditions listed in Problem 13–33, compute the shearing stress at several axes and plot the variation of stress with position in the beam.

13–35. For a standard W14×43 steel beam, compute the shearing stress from the web shear formula when it carries a shearing force of 33 500 lb. Compare this value with that computed in Problem 13–33 and plot it on the graph produced for Problem 13–34.

13–36. For an Aluminum Association Standard I8×6.181 beam, compute the shearing stress at the neutral axis when subjected to a shearing force of 13 500 lb. Use the general shear formula. Neglect the fillets at the intersection of the web with the flanges.

13–37. For the same conditions listed in Problem 13–36, compute the shearing stress at several axes and plot the variation of stress with position in the beam.

13–38. For an aluminum I8×6.181 beam, compute the shearing stress from the web shear formula when the beam carries a shearing force of 13 500 lb. Compare this value with that computed in Problem 13–36 and plot it on the graph produced for Problem 13–37.

Use of Design Shear Stresses

Note: In problems calling for design stresses, use the following. For structural steel:

In bending: $\sigma_d = 0.66 s_y$

In shear: $\tau_d = 0.4 s_y$

For any other metal:

In bending: $\sigma_d = \dfrac{s_y}{N}$

In shear: $\tau_d = 0.5 \dfrac{s_y}{N}$

For wood:
Use allowable stresses in Appendix A–19.

13–39. The loading shown in Figure P13–39 is to be carried by a W12×16 steel beam. Compute the shearing stress using the web shear formula. Also compute the maximum bending stress. Then compare the stresses to the design stresses for ASTM A992 structural steel.

10 K   10 K

3 ft   8 ft   3 ft

FIGURE P13–39

13–40. Specify a suitable wide-flange beam to be made from ASTM A992 structural steel to carry the load shown in Figure P13–39 based on the design stress in bending. Then, for the beam selected, compute the shearing stress from the web shear formula and compare it with the design shear stress.

13–41. Specify a suitable wide-flange beam to be made from ASTM A992 structural steel to carry the load shown in Figure P13–41 based on the design stress in bending. Then, for the beam selected, compute the shearing stress from the web shear formula and compare it with the design shear stress.

FIGURE P13–41

13–42. Specify a suitable wide-flange beam to be made from ASTM A992 structural steel to carry the load shown in Figure P13–42 based on the design stress in bending. Then, for the beam selected, compute the shearing stress from the web shear formula and compare it with the design shear stress.

FIGURE P13–42

13–43. Specify a suitable standard steel pipe from Appendix A–12 to be made from ASTM A53, Grade B steel to carry the load shown in Figure P13–43 based on the design stress in bending with a design factor of 3. Then, for the pipe selected, compute the shearing stress from the special shear formula for hollow tubes and compute the resulting design factor from the design shear stress formula.

FIGURE P13–43

13–44. An Aluminum Association standard channel (Appendix A–10) is to be specified to carry the load shown in Figure P13–44 to produce a design factor of 4 in bending. The legs of the channel are to point down. The channel is made from 6061-T6 aluminum. For the channel selected, compute the maximum shearing stress.

FIGURE P13–44

13–45. A wooden joist in the floor of a building is to carry a uniformly distributed load of 200 lb/ft over a length of 12.0 ft. Specify a suitable standard wooden beam shape for the joist, made from No. 2 grade hemlock, to be safe in both bending and shear (see Appendices A–4 and A–19).

13–46. A wooden beam in an outdoor structure is to carry the load shown in Figure P13–46. If it is to be made from No. 3 grade Douglas fir, specify a suitable standard wooden beam to be safe in both bending and shear (see Appendices A–4 and A–19).

FIGURE P13–46

13–47. The box beam shown in Figure P13–47 is to be made from No. 1 grade southern pine. It is to be 14 ft long and carry two equal concentrated loads, each 3 ft from an end. The beam is simply supported at its ends. Specify the maximum allowable load for the beam to be safe in both bending and shear.

FIGURE P13–47

13–48. An aluminum I-beam, I229×12.44, carries the load shown in Figure P13–48. Compute the shearing stress in the beam using the web shear formula.

FIGURE P13–48

13–49. Compute the bending stress for the beam in Problem 13–48.

13–50. A 2 × 8 wooden floor joist in a home is simply supported, 12 ft long, and carries a uniformly distributed load of 80 lb/ft. Compute the shearing stress in the joist. Would it be safe if it is made from No. 2 grade southern pine?

13–51. A steel beam is made as a rectangle, 0.50 in wide by 4.00 in high.
 a. Compute the shearing stress in the beam if it carries the load shown in Figure P13–51.
 b. Compute the stress due to bending.
 c. Specify a suitable steel for the beam to produce a design factor of 3 for either bending or shear.

FIGURE P13–51

13–52. An aluminum beam is made as a rectangle, 16 mm wide by 60 mm high.
 a. Compute the shearing stress in the beam if it carries the load shown in Figure P13–52.
 b. Compute the stress due to bending.
 c. Specify a suitable aluminum for the beam to produce a design factor of 3 for either bending or shear.

FIGURE P13–52

13–53. It is planned to use a rectangular bar to carry the load shown in Figure P13–53. Its thickness is to be 12 mm and it is to be made from aluminum 6061-T6. Determine the required height of the rectangle to produce a design factor of 4 in bending based on yield strength. Then compute the shearing stress in the bar and the resulting design factor for shear.

FIGURE P13–53

13–54. A round shaft, 40 mm in diameter, carries the load shown in Figure P13–54.

FIGURE P13–54

 a. Compute the maximum shearing stress in the shaft.
 b. Compute the maximum stress due to bending.
 c. Specify a suitable steel for the shaft to produce a design factor of 4 based on yield strength in either shear or bending.

13–55. Compute the required diameter of a round bar to carry the load shown in Figure P13–55 while limiting the stress due to bending to 120 MPa. Then compute the resulting shearing stress in the bar and compare it to the bending stress.

FIGURE P13–55

13–56. Compute the maximum allowable vertical shearing force on a wood dowel having a diameter of 1.50 in, if the maximum allowable shearing stress is 70 psi.

13–57. A standard steel pipe is to be selected from Appendix A–12 to be used as a chinning bar in a gym. It is to be simply supported at the ends of its 36-in length. Men weighing up to 400 lb are expected to hang on the bar with either one or two hands at any place along its length. The pipe is to be made from AISI 1020 hot-rolled steel. Specify a suitable pipe to provide a design factor of 6 based on yield strength in either bending or shear.

13–58. A standard steel pipe is to be simply supported at its ends and carry a single concentrated load of 2800 lb at its center. The pipe is to be made from AISI 1020 hot-rolled steel. The minimum design factor is to be 4 based on yield strength for either bending or shear. Specify a suitable pipe size from Appendix A–12 if the length of the pipe is
 a. 1.5 in
 b. 3.0 in
 c. 4.5 in
 d. 6.0 in

Shear Flow Problems

13–59. The shape in Figure P13–59 is to be made by gluing the flat plate to the hat section. If the beam made from this section is subjected to a shearing force of 1200 lb, compute the shear flow at the joint. What must be the shearing strength of the glue in psi?

FIGURE P13–59

13–60. The shape in Figure P13–60 is to be made by using a metal-bonding glue between the S-beam and the web of the channel. Compute the shear flow at the joint and the required shearing strength of the glue for a shearing force of 2500 lb.

C12×25
steel

S12×50
steel

FIGURE P13–60

13–61. The shape in Figure P13–20 is to be fabricated by riveting the bottom plate to the angles and then welding the top plate to the angles. When used as a beam, there are four potential failure modes: bending stress, shearing stress in the angles, shear in the welds, and shear of the rivets. The shape is to be used as the seat of a bench carrying a uniformly distributed load over a span of 10.0 ft. Compute the maximum allowable distributed load for the following design limits.

a. The material of all components is aluminum 6061-T4, and a design factor of 4 is required for either bending or shear.

b. The allowable shear flow on each weld is 1800 lb/in.

c. The rivets are spaced 4.0 in apart along the entire length of the beam. Each rivet can withstand 600 lb of shear.

13–62. An alternative design for the bench described in Problem 13–61 is to use the built-up wood T-shape shown in Figure P13–30. The wood is to be No. 3 grade southern pine. One nail is to be driven into each vertical 2 × 12. Each nail can withstand 160 lb in shear, and the nails are spaced 6.0 in apart along the length of the beam. Compute the maximum allowable distributed load on the beam.

13–63. The shape shown in Figure P13–27 is glued together and the allowable shearing strength of the glue is 800 psi. The components are No. 2 grade Douglas fir. If the beam is to be simply supported and carry a single concentrated load at its center, compute the maximum allowable load. The length is 10 ft.

13–64. The I-section shown in Figure P13–27 is fabricated from three wooden boards by nailing through the top and bottom flanges into the web. Each nail can withstand 180 lb of shearing force. If the beam having this section carries a vertical shearing force of 300 lb, what spacing would be required between nails?

13–65. The built-up section shown in Figure P13–28 is nailed together by driving one nail through each side of the top and bottom boards into the $1\frac{1}{2}$-in-thick sides. If each nail can withstand 150 lb of shearing force, determine the required nail spacing when the beam carries a vertical shearing force of 600 lb.

13–66. The platform whose cross section is shown in Figure P13–29 is glued together. How much force per unit length of the platform must the glue withstand if it carries a vertical shearing force of 500 lb?

13–67. The built-up section shown in Figure P13–67 is fastened together by passing two $\frac{3}{8}$-in rivets through the top and bottom plates into the flanges of the beam. Each rivet will withstand 2650 lb in shear. Determine the required spacing of the rivets along the length of the beam if it carries a shearing force of 175 kN.

$\frac{1}{2} \times 8.0$ in plates (2)

W14×43
steel

Rivets

FIGURE P13–67

13–68. A fabricated beam having the cross section shown in Figure P13–60 carries a shearing force of 50 kN. The channel is riveted to the S-beam with two $\frac{1}{4}$-in-diameter rivets, which can withstand 1750 lb each in shear. Determine the required rivet spacing.

ADDITIONAL PRACTICE AND REVIEW PROBLEMS

Web Shear

For Problems 13–69 through 13–73, compute the shearing stress in the web of the given beam using the web shear formula. Then compare the result with the design shear stress defined in Section 13–7 to determine if the shearing stress is safe.

13–69. A W18×55 shape carries a vertical shearing force of 36.6 K. The material is ASTM A992 structural steel.

13–70. A W18×40 shape carries a vertical shearing force of 36.6 K. The material is ASTM A992 structural steel.

13–71. W14×26 shape carries a vertical shearing force of 10 000 lb. The material is ASTM A992 structural steel.

13–72. An aluminum I6×4.692 shape carries a vertical shearing force of 10 000 lb. The material is 6061-T6 aluminum.

13–73. W10×12 shape carries a vertical shearing force of 6750 lb. The material is ASTM A992 structural steel.

General Shear Formula and Special Shear Formulas

For Problems 13–74 through 13–79, compute the maximum shearing stress using the general shear formula or one of the special shear formulas. Then compare the result with the design shear stress defined in Section 13–7 to determine if the shearing stress is safe.

13–74. A standard rectangular steel tube, HSS6×2×$\frac{1}{4}$, carries a static vertical shearing force of 6750 lb. The material is ASTM A500 grade B structural steel.

13–75. A standard 2 × 8 wood beam made from No. 2 stress-rated southern pine is installed as a cantilever and carries a vertical shearing force of 480 lb.

13–76. The shape shown in Figure P13–29 is made from No. 2 Douglas fir and carries a shearing force of 750 lb.

13–77. A standard rectangular steel tube, HSS8×2×$\frac{1}{4}$, carries a vertical shearing force of 12 000 lb. The material is ASTM A500 grade B structural steel.

13–78. The shape shown in Figure P13–78 carries a vertical shearing force of 1800 lb. The materials are ASTM A992 structural steel for the W-beam and A500 grade B for the HSS sections.

13–79. A standard C10×6.136 aluminum channel is installed with the legs down and carries a vertical shearing force of 430 lb. The material is 6061-T6 aluminum.

Shear Flow

13–80. For the data of Problem 13–78, compute the shear flow at the interface between the W-shape and the tube.

13–81. The shape shown in Figure P13–29 is to carry a vertical shearing force of 500 lb. The plywood is to be nailed to the 2 × 4 members using nails that can carry a shearing force of 135 lb each. Specify a suitable spacing for the nails.

HSS4×2×1/4
Steel tube

W4×13

HSS4×2×1/4

FIGURE P13–78

DEFLECTION OF BEAMS

MAJOR TOPICS

14–1   The Big Picture—Deflection of Beams
14–2   The Need for Considering Beam Deflections
14–3   General Principles and Definitions of Terms
14–4   Beam Deflections Using the Formula Method
14–5   Comparison of the Manner of Support for Beams
14–6   Superposition Using Deflection Formulas
14–7   Successive Integration Method
14–8   Moment-Area Method

14–1 THE BIG PICTURE— DEFLECTION OF BEAMS

Discussion Map

- The proper performance of machine parts; the structural rigidity of buildings, vehicles, toys, and consumer products; and the tendency for a component to vibrate are all dependent on limiting deformations or deflections of beams to acceptable values. You will explore methods of analyzing beam deflections in this chapter.

- This topic builds on the competencies you developed in previous chapters.

- You discovered in Chapters 12 and 13 how to determine the normal stresses in beams due to bending moments and to the shearing stresses due to shearing forces that are created in beams.

- You also learned how to design beams to be sufficiently strong to ensure that they would not fail by fracture or yielding of the material of the beam.

- Another potential failure mode is the excessive deformation or deflection of the beam. Now you will explore ways that beams deflect under load and learn several ways for computing the predicted deflections.

- Two classes of beams will be considered:
  - *Statically determinate beams* (Figure 14–1)—those for which all unknown reaction forces and moments can be found using the classic equations of static equilibrium:
    - $\sum F = 0$ in any direction
    - $\sum M = 0$ about any point
  - *Statically indeterminate beams* (Figure 14–2)—those for which there are too many unknowns to solve by conventional methods of statics. Other analysis techniques are introduced here to enable these types of beams to be analyzed.

- You will also explore the effect of different types of support and loading conditions on the general flexibility or stiffness of beams so you can make sound judgments about which types to employ in your designs.

CHAPTER FOURTEEN

ACTIVITY   Chapter 14—Statically Determinate Beams

Acquire again the same materials you used for the activities in Chapters 6, 11, 12, and 13 when you explored the general behavior of beams, the shearing forces and bending moments developed in them, their cross-sectional shapes, bending stresses, and shearing stresses. There you used simple, flexible beam examples, a variety of supports, and some means of loading the beams.

Using Figure 14–1 as a guide, set up the various types of beam loading and support conditions described here.

1. Support the beam at its ends, place two concentrated loads near the middle, and observe the resulting shape of the beam. It deflects downward, forming a concave upward curved shape. This is the classical demonstration of positive bending [Figure 14–1 (a)].

2. Remove one load and move the other along the beam. Observe how the shape of the deflected beam changes [Figure 14–1(b)].

3. Move one of the supports in from the end of the beam to create an overhang. Then place one load near the end of the overhang. One thing you will need to do is reverse the direction of the support at the opposite end of the beam or else it will rotate up off the support. Then you should notice that the beam deflects into a convex or concave downward shape. This is an example of negative bending [Figure 14–1(c)].

W₁ and W₂ labels with "Simple supports" and "General shape of deflected beam"

(a) Simply supported beam with two equal loads

(b) Simply supported beam with one offset load

(c) Overhanging beam with one load at end of overhang

(d) Overhanging beam with two loads—both large

(e) Overhanging beam with two loads—$W_1 \gg W_2$

(f) Overhanging beam with two loads—$W_2 \gg W_1$

(g) Cantilever with one load

(h) Cantilever with two loads—both large

FIGURE 14–1 A variety of beam loading and support conditions.

4. Next, use the overhanging beam and place loads both at the end of the overhang and between the supports. Vary the magnitudes and placements of the two loads and see what kinds of shapes for the deflected beam you can produce [Figure 14–1(d), 14–1(e), 14–1(f)].

5. Now configure the beam as a cantilever by clamping it firmly to a table. Perhaps you can use your book and a firm hand to ensure that one end is held fixed. Place loads at various positions on the cantilever and in different directions and observe its shape. Also observe the relative flexibility of the cantilever depending on the shape of the beam, its length, and the placement of the load. You should see that a cantilever is somewhat more flexible than a similar beam of the same length supported at two points [Figure 14–1(g) and 14–1(h)].

ACTIVITY Chapter 14—Statically Indeterminate Beams

Set up demonstrations of the three types of statically indeterminate beams described here using the same materials from the activity for statically determinate beams but adding additional supports. Use Figure 14–2 as a guide.

1. A *continuous beam* is any beam with more than two supports. The name comes from the observation that the beam is continuous over several supports. Figure 14–2(a) shows one example. Use your beam-building materials to first set up a simply supported beam with supports on each end fairly widely spaced. Apply one or several loads with your hand. Note the flexibility of the beam. Now place a third support under the middle of the beam and apply the loads again. Do you sense the increased stiffness of the beam? Adding a fourth support will increase the stiffness even more.

   Where have you seen continuous beams? There are numerous examples in building structures and highway bridges. Look for them as you drive around the city or on the expressway. Bridges are typically supported on piers at each end and often employ intermediate supports such as in the median strip of a divided highway.

   Another example is in the basement of many ranch-style homes. Often a steel beam runs the length of the basement to support the joists for the floor

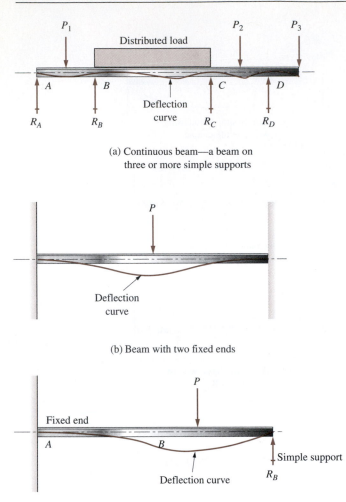

(a) Continuous beam—a beam on three or more simple supports

(b) Beam with two fixed ends

(c) Supported cantilever

FIGURE 14–2   Examples of statically indeterminate beams.

above. The beam rests on the basement walls and is typically supported at one or more places within the room by steel columns. These added columns make the beam into a continuous beam.

2. A *fixed-end beam* is one that is rigidly constrained at both ends to prevent rotation and to support the loads on the beam. See Figure 14–2(b) for a sketch. Notice that the deflected beam shape starts with a horizontal slope at each support because of the rigid end fixity there. This kind of beam is often used in the design of machine frames and building structures because they provide a very stiff, rigid assembly. The creation of the fixed type of end requires care. Figure 14–3 shows one way to fabricate this kind of beam by using an array of four bolts tightly fastening the beam to a strong, stiff support structure. You could also weld the beams to the supporting columns. Without due care, a condition between that of fixed ends and simple supports could result, and the beam will not perform as planned.

   Use your beam-building materials to simulate a fixed-end beam. Then load it and observe its relative stiffness compared with the other designs.

3. A *supported cantilever* is a beam with one fixed end and a second, simple support at the other as shown in Figure 14–2(c). Earlier we observed that a cantilever is one of the most flexible types of beams. The second support contributes a significant amount of additional rigidity. Figure 14–4 shows one way of constructing a supported cantilever.

   Make a supported cantilever yourself with your materials. Compare its stiffness with a simple cantilever.

FIGURE 14–3   Fixed-end beam.

FIGURE 14–4 Supported cantilever.

The analyses required for the statically indeterminate beams are somewhat more difficult than for statically determinate beams. This chapter provides several tools you can use to design and analyze such beams. Also included are some comparisons among the various types to help you make judgments about appropriate designs to use.

Methods of Analyzing Beam Deflections

In this chapter we present the principles on which the computation of the deflection of beams is based, along with four popular methods of deflection analysis: the *formula method,* the *superposition method,* the *successive integration method,* and the *moment-area method.*

Each method has its advantages and disadvantages, and your choice of which method to use depends on the nature of the problem.

- The formula method is the simplest, but it depends on the availability of a suitable formula to match the application.

- The superposition method, a modest extension of the formula method, dramatically expands the number of practical problems that can be handled without a significant increase in the complexity of your work.

- The moment-area method is fairly quick and simple, but it is typically used for computing the deflections of only one or a few points on the beam. Its use requires a high level of understanding of the principle of moments and the techniques of preparing bending moment diagrams.

- The successive integration method is perhaps the most general, and it can be used to solve virtually any combination of loading and support conditions for statically determinate beams. However, its use requires the ability to write the equations for the shearing force and bending moment diagrams and to derive equations for the slope and deflection of the beam using integral calculus. The successive integration method results in equations for the slope and deflection for the entire beam and enables the direct determination of the point of maximum deflection. Published formulas were developed using the successive integration or the moment-area method.

Several computer-assisted beam analysis programs are available to reduce the time and computation required to determine the deflection of beams. Although they can relieve the designer of much work, it is recommended that you understand the principles on which they are based before using them. See the list of Internet sites.

Objectives of This Chapter

After completing this chapter, you should be able to:

1. Understand the need for considering beam deflections.

2. Understand the development of the relationships between the manner of loading and support for a beam and the deflection of the beam.

3. Graphically show the relationships among the load, shearing force, bending moment, slope, and deflection curves for beams.

4. Define *statically determinate* and *statically indeterminate* as applied to beams.

5. Recognize statically indeterminate beams from given descriptions of the loading and support conditions.

6. Define *continuous beam.*

7. Define *supported cantilever.*

8. Define *fixed-end beam.*

9. Use standard formulas to compute the deflection of beams at selected points.

10. Use the principle of superposition along with standard formulas to solve problems of greater complexity.

11. Compare the relative strength and stiffness of beams having different support systems for given loadings.

12. Develop formulas for the deflection of beams for certain cases using the successive integration method.

13. Apply the method of successive integration to beams having a variety of loading and support conditions.

14. Use the moment-area method to solve for the slope and deflection for beams.

15. Write computer programs to assist in using the several methods of beam analysis described in this chapter.

The organization of the chapter allows selective coverage. In general, all the information necessary to use each method is included within that part of the chapter. An exception is that the understanding of the formula method is necessary before using the superposition method.

14–2 THE NEED FOR CONSIDERING BEAM DEFLECTIONS

The spindle of a lathe or drill press and the arbor of a milling machine carry cutting tools for machining metals. Deflection of the spindle or arbor would have an adverse effect on the accuracy that the machine could produce. The manner of loading and support of these machine elements indicates that they are beams, and the approach to computing their deflection will be discussed later in this chapter.

Precision measuring equipment must also be designed to be very rigid. Deflection caused by the application of measuring forces reduces the precision of the desired measurement.

Power-transmission shafts carrying gears must have sufficient rigidity to ensure that the gear teeth mesh properly. Excessive deflection of the shafts would tend to separate the mating gears, resulting in a movement away from the most desirable point of contact between the gear teeth. Noise generation, decreased power-transmitting capability, and increased wear would result. For straight spur gears, it is recommended that the movement between two gears not exceed 0.005 in (0.13 mm). This limit is the *sum* of the movement of the two shafts carrying the mating gears at the location of the gears.

The floors of buildings must have sufficient rigidity to carry expected loads. Occupants of the building should not notice floor deflections. Machines and other equipment require a stable floor support for proper operation. Beams carrying plastered ceilings must not deflect excessively so as not to crack the plaster. A limit of $\frac{1}{360}$ times the span of the beam carrying a ceiling is often used for deflection.

Frames of vehicles, metal-forming machines, automation devices, and process equipment must also possess sufficient rigidity to ensure satisfactory operation of the equipment carried by the frame. The bead of a lathe, the crown of a punch press, the structure of an automatic assembly device, and the frame of a truck are examples.

Vibration is caused by the forced oscillations of parts of a structure or machine. The tendency to vibrate at a certain frequency and the severity of the vibrations are functions of the flexibility of the parts. Of course, *flexibility* is a term used to describe how much a part deflects under load. Vibration problems can be solved by either *increasing* or *decreasing* the stiffness of the part, depending on the circumstances. In either case, an understanding of how to compute deflections of beams is important.

Recommended Deflection Limits

It is the designer's responsibility to specify the maximum allowable deflection for a beam in a machine, frame, or structure. Knowledge of the application should give guidance. In the absence of such guidance, the following limits are suggested in References 2 and 3:

General machine part: y_{max} = 0.0005 to 0.003 in/in or mm/mm of beam length

Moderate precision: y_{max} = 0.000 01 to 0.0005 in/in or mm/mm of beam length

High precision: y_{max} = 0.000 001 to 0.000 01 in/in or mm/mm of beam length

The numerical values given here could also be expressed as the ratio of the maximum deflection to the length of the beam. For example,

y_{max}/L = 0.0005 to 0.003 for a general machine part

y_{max}/L = 0.000 01 to 0.0005 for moderate precision

y_{max}/L = 0.000 001 to 0.000 01 for high precision

Then multiplying the number by the length of the beam would produce the recommended maximum deflection.

For example, consider a beam, 18 in (450 mm) long, for a general-purpose conveyor. The lower end of the range of recommended deflection is

$$y_{max} = \frac{0.0005 \text{ in}}{\text{in of length}} \times 18 \text{ in} = 0.009 \text{ in}$$

The same calculation applies to measurements in mm.

$$y_{max} = \frac{0.0005 \text{ mm}}{\text{mm of length}} \times 450 \text{ mm} = 0.225 \text{ mm}$$

If it does not adversely affect the operation of the system containing the conveyor, the deflection could be as high as

$$y_{max} = \frac{0.003 \text{ in}}{\text{in of length}} \times 18 \text{ in} = 0.054 \text{ in}$$

$$[\text{slightly less than } \tfrac{1}{16} \text{ in}]$$

Using the metric data gives

$$y_{max} = \frac{0.003 \text{ mm}}{\text{mm of length}} \times 450 \text{ mm} = 1.35 \text{ mm}$$

Now, if the conveyor is part of a high-speed transfer station for an automated assembly machine, these deflections are likely to be excessive because the parts on the conveyor must be accurately positioned at each station. Moving to the "moderate precision" data, the limits would be in the range of

$$y_{max} = \frac{0.000 \, 01 \text{ in}}{\text{in of length}} \times 18 \text{ in} = 0.000 \, 18 \text{ in to}$$

$$y_{max} = \frac{0.0005}{\text{in of length}} \times 18 \text{ in} = 0.009 \text{ in}$$

The corresponding limits using metric measurements are 0.0045 mm to 0.225 mm. These limits are readily achievable and should be applied to any piece of accurate machinery.

What if the beam were part of a precision inertial guidance system for a spacecraft? Here the "high precision" data should be applied. The range of deflection should be

$$y_{max} = \frac{0.000 \, 001 \text{ in}}{\text{in of length}} \times 18 \text{ in} = 0.000 \, 018 \text{ in to}$$

$$y_{max} = \frac{0.000 \, 01}{\text{in of length}} \times 18 \text{ in} = 0.00018 \text{ in}$$

The corresponding limits using metric measurements are 0.00045 mm to 0.0045 mm. These are, indeed, precision magnitudes for beam deflections.

14–3 GENERAL PRINCIPLES AND DEFINITIONS OF TERMS

To describe graphically the condition of a beam carrying a pattern of loading, five diagrams are used, as shown in Figure 14–5. The first three diagrams were used in earlier work. The *load diagram* is the free body diagram on which all external loads and support reactions are shown. From that, the *shearing force diagram* was developed, which enables the calculation of shearing stresses in the beam at any section. The *bending moment diagram* is a plot of the variation of bending moment with position on the beam, with the results used for computing the stress due to bending. The horizontal axis of these plots is the position on the beam, called x. It is typical to measure x relative to the left end of the beam, but any reference point can be used.

Deflection Diagram

The last two diagrams are related to the deformation of the beam under the influence of the loads. It is convenient to begin discussion with the last diagram, the *deflection diagram*, because this shows the shape of the deflected beam. Actually, it is the plot of the position of the neutral axis of the beam relative to its initial position. The initial position is taken to be the straight line between the two support points on the unloaded beam. The amount of deflection will be called y, with positive values measured upward. Typical beams carrying downward loads, such as the one shown in Figure 14–5, will result in downward (negative) deflections of the beam.

Slope Diagram

A line drawn tangent to the deflection curve at a point of interest would define the slope of the deflection curve at that

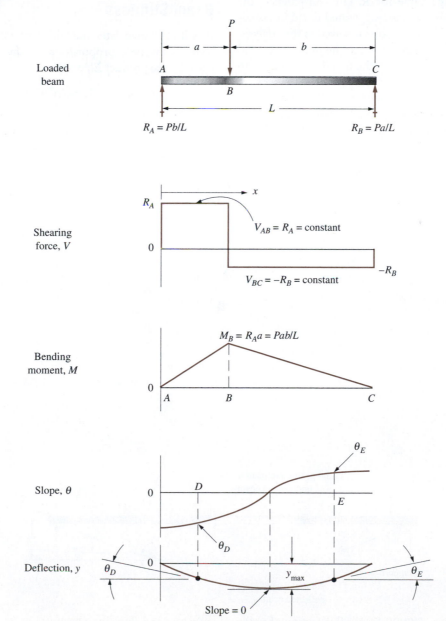

FIGURE 14–5 Five beam diagrams.

point. The slope is indicated as the angle θ, measured in radians, relative to the horizontal, as shown in Figure 14–5. The plot of the slope as a function of position on the beam is the *slope curve,* drawn below the bending moment curve and above the deflection curve. Note on the given beam that the slope of the left portion of the deflection curve is negative and the right portion has a positive slope. The point where the tangent line is itself horizontal is the point of zero slope and defines the location of the maximum deflection. This observation will be used in the discussion of the moment-area method and the successive integration method later in this chapter.

Radius of Curvature

Figure 14–6 shows the radius of curvature, R, at a particular point. For practical beams, the curvature is very slight, resulting in a very large value for R. For convenience, the shape of the deflection curve is exaggerated to aid in visualizing the principles and the variables involved in the analysis. Remember from analytic geometry that the radius of curvature at a point is perpendicular to the line drawn tangent to the curve at that point.

The relationship between slope and deflection is also illustrated in Figure 14–6. Over a small distance dx, the deflection changes by a small amount dy. A small portion of

the deflection curve itself completes the right triangle from which we can define

$$\tan \theta = \frac{dy}{dx} \tag{14–1}$$

The absolute value of θ will be very small because the curvature of the beam is very slight. We can then take advantage of the observation for small angles, $\tan \theta = \theta$. Then

$$\theta = \frac{dy}{dx} \tag{14–2}$$

It can then be concluded that

The slope of the deflection curve at a point is equal to the ratio of the change in the deflection to the change in position on the beam.

Beam Stiffness

It will be shown later that the amount of deflection for a beam is inversely proportional to the *beam stiffness,* indicated by the product EI, where

E = modulus of elasticity of the material of the beam

I = moment of inertia of the cross section of the beam with respect to the neutral axis

FIGURE 14–6 Illustration of radius of curvature and slope for a beam deflection curve.

It may help you to think of these terms as follows. The modulus of elasticity, *E*, is the *material stiffness*. The moment of inertia, *I*, is the *shape stiffness* for the cross section of the beam. As noted in Chapter 6, *I* is defined as the second moment of the area of the cross section. To minimize the deflection of a beam you should choose a material with the highest practical modulus of elasticity. And you should design the shape to have the largest practical moment of inertia. In general, the shape that is efficient for minimizing bending stresses, as discussed in Chapter 12, is also good for minimizing deflections of the beam.

14–4 BEAM DEFLECTIONS USING THE FORMULA METHOD

For many practical configurations of beam loading and support, formulas have been derived that allow the computation of the deflection at any point on the beam. The method of successive integration or the moment-area method, described later, may be used to develop the equations. Appendices A–23, A–24, and A–25 include many examples of formulas for beam deflections.

The deflection formulas are valid only for the cases where the cross section of the beam is uniform for its entire length. Example problems will demonstrate the application of the formulas.

Simply Supported Beams

Appendix A–23 includes ten different conditions of loading on beams that are simply supported, that is, beams having two and only two simple supports. Some are overhanging beams. We have shown earlier that such beams can be analyzed for the values of the reactions using the standard equations of equilibrium. Then the shearing force and bending moment diagrams can be developed using the methods from Chapter 11 from which the stress analysis of the beam can be completed, as discussed in Chapters 12 and 13. It must be understood that both the stress analysis and the deflection analysis of beams should be completed to assess the acceptability of a beam design.

The loading conditions in Appendix A–23 include a single concentrated load, two concentrated loads, a variety of distributed loads, and one case with a concentrated moment. The concentrated moment could be developed in the manner of the examples shown in Section 11–10. The thin phantom line in the diagrams is a sketch of the shape of the deflected beam, somewhat exaggerated. This can help you to visualize where critical points of deflection can be expected.

Note carefully the labeling for loads and dimensions in the diagrams of the beam deflection cases. It is essential that the real beam being analyzed match the general form of a given case and that you identify accurately the variables used in the formulas to the right of the diagrams. For most cases, formulas are given for the maximum deflection to be expected, deflections at the ends of overhangs, and deflections at points of applications of concentrated loads. Some cases include formulas for the deflection at any chosen point.

Take special notice of the general form of the deflection formulas. While some are more complex than others, the following general characteristics can be observed. Understanding these observations can help you make good decisions when designing beams.

1. Deflections are denoted by the variable *y* and are the change in the position of the neutral axis of the beam from its unloaded condition to the final, loaded condition, measured perpendicular to the original neutral axis.

2. Upward deflections are positive; downward deflections are negative.

3. The variable *x*, when used, denotes the horizontal position on the beam, measured from one of the supports. In some cases, a second position variable *v* is indicated, measured from the other support.

4. Deflections are proportional to the load applied to the beam.

5. Deflections are inversely proportional to the *stiffness* of the beam, defined as the product of *E*, the stiffness of the material from which the beam is made, and *I*, the moment of inertia of the cross section of the beam, also called *shape stiffness*.

6. Deflections are proportional to the *cube* of some critical length dimension, typically the span between the supports or the length of an overhang.

Cantilevers

Appendix A–24 includes four cases in which cantilever beams carry concentrated loads, distributed loads, or a concentrated moment. The maximum deflection obviously occurs at the free end of the beam. The fixed end constrains the beam against rotation at the support so that the deflection curve has a zero slope there.

Example Problem 14–1 Determine the maximum deflection of a simply supported beam carrying a hydraulic cylinder in a machine used to press bushings into a casting, as shown in Figure 14–7. The force exerted during the pressing operation is 15 kN. The beam is rectangular, 25 mm thick and 100 mm high, and made of steel.

Solution Objective: Compute the maximum deflection of the given beam.

Given: System in Figure 14–7. Load = *P* = 15 kN. Span = *L* = 1.60 m.
Beam cross section: 25 mm wide by 100 mm high. Steel beam.

FIGURE 14–7   Beam for Example Problem 14–1.

Analysis: The given beam can be considered to be a simply supported beam with a concentrated force applied in an upward direction at its center. Case *a* in Appendix A–23 applies.

Results: Using the formula from Appendix A–23–a, we find the maximum deflection to be

$$y = \frac{-PL^3}{48\,EI}$$

But *P* itself is negative, acting upward. From Appendix A–14, for steel, $E = 207$ GPa $= 207 \times 10^9$ N/m². For the rectangular beam,

$$I = \frac{(25)(100)^3}{12} = 2.083 \times 10^6 \, \text{mm}^4$$

Then

$$y = \frac{-PL^3}{48EI} = \frac{-(-15 \times 10^3 \, \text{N})(1.6 \, \text{m})^3}{48(207 \times 10^9 \, \text{N/m}^2)(2.083 \times 10^6 \, \text{mm}^4)} \times \frac{(10^3 \, \text{mm})^5}{\text{m}^5}$$

$$y = 2.97 \, \text{mm}$$

Comment: This is a relatively high deflection that could adversely affect the accuracy of the bushing placement operation. It would be useful to evaluate the ratio of the deflection to the span for this beam and compare it to the recommendations in Section 14–2.

$$y_{max}/L = 2.97 \, \text{mm}/1600 \, \text{mm} = 0.00186$$

This is within the range of acceptable deflections for a general machine part. However, to place the bushing accurately in the casting, greater precision is desired. A stiffer beam shape (one with a higher moment of inertia, *I*) should be considered. Alternatively, the support system could be modified to decrease the span between the supports, a desirable approach because the deflection is proportional to the cube of the length. Assuming that the overall operation of the system permits the redesign of the span to be one-half of the given span (0.80 m), the deflection would be only 0.37 mm, $\frac{1}{8}$ as much as the given design. Checking the ratio of the deflection to the span again gives

$$y_{max}/L = 0.37 \, \text{mm}/800 \, \text{mm} = 0.00046$$

This places the design within the more desirable moderate precision range.

  The stress in the beam should also be computed to assess the safety of the design. Figure 14–8 shows the load, shearing force, and bending moment diagrams for the original beam design from which we find the maximum bending moment in the beam to be $M = 6.00$ kN·m. The flexure formula can be used to compute the stress.

$$\sigma = \frac{Mc}{I} = \frac{(6.00 \, \text{kN·m})(50 \, \text{mm})}{2.083 \times 10^6 \, \text{mm}^4} \cdot \frac{10^3 \text{N}}{\text{kN}} \cdot \frac{10^3 \, \text{mm}}{\text{m}} = 144 \, \text{MPa}$$

FIGURE 14–8 Beam diagrams for Example Problem 14–1.

This is a relatively high stress level. To continue the analysis, note that the beam would be subjected to repeated bending stress. Therefore, the recommended design stress is

$$\sigma_d = s_u/8$$

Letting $\sigma_d = \sigma$, we can solve for the required ultimate strength.

$$s_u = 8\sigma = (8)(144\ \text{MPa}) = 1152\ \text{MPa}$$

Consulting Appendix A–14 for the properties of selected steels, we could specify AISI 4140 OQT 900 steel that has an ultimate strength of 1281 MPa. This is a fairly expensive, heat-treated steel. A redesign of the type discussed for limiting deflection would reduce the stress and permit the use of a lower-priced steel.

Example Problem 14–2

A round shaft, 45 mm in diameter, carries a 3500-N load, as shown in Figure 14–9. If the shaft is steel, compute the deflection at the load and at the point C, 100 mm from the right end of the shaft. Also compute the maximum deflection.

Solution Objective: Compute the deflection at points B and C and at the point of maximum deflection.

Given: Beam in Figure 14–9. Load = $P = 3500$ N
Beam is round shaft; $D = 45$ mm. Steel beam.

Analysis: Appendix A–23–b applies because the beam is simply supported and has a single concentrated load placed away from the middle of the beam.

FIGURE 14–9 Shaft for Example Problem 14–2.

Results: *At Point B:* The following formula applies.

$$y_B = \frac{-Pa^2b^2}{3EIL}$$

Note that the dimension *a* is the longer segment between the load and one support; *b* is the shorter. Express all data in N and mm for unit consistency.

Data:

$$L = 400 \, \text{mm}; \qquad a = 250 \, \text{mm}; \qquad b = 150 \, \text{mm}$$

$$I = \frac{\pi D^4}{64} = \frac{\pi (45 \, \text{mm})^4}{64} = 0.201 \times 10^6 \, \text{mm}^4$$

$$E = \frac{207 \times 10^9 \, \text{N}}{\text{m}^2} \times \frac{1 \, \text{m}^2}{(10^3 \, \text{mm})^2} = 207 \times 10^3 \, \text{N/mm}^2$$

Now y_B will be in mm.

$$y_B = \frac{-Pa^2b^2}{3EIL} = \frac{-(3500)(250)^2(150)^2}{3(207 \times 10^3)(0.201 \times 10^6)(400)} = -0.0985 \, \text{mm}$$

At Point C: Note that point *C* is in the longer segment. Previous data apply. Also, $x = 100$ mm from the right support to *C*.

$$y_C = \frac{-Pbx}{6EIL}(L^2 - b^2 - x^2)$$

$$y_C = \frac{-(3500)(150)(100)}{6(207 \times 10^3)(0.201 \times 10^6)(400)}(400^2 - 150^2 - 100^2)$$

$$y_C = -0.0670 \, \text{mm}$$

Maximum Deflection: Appendix A–23–b shows that the maximum deflection occurs in the longer segment of the beam at a distance x_1 from the support, where

$$x_1 = \sqrt{a(L + b)/3} = \sqrt{(250)(400 + 150)/3} = 214 \, \text{mm}$$

Then the deflection at that point is

$$y_{max} = \frac{-Pab(L + b)\sqrt{3a(L + b)}}{27(EIL)}$$

$$y_{max} = \frac{-(3500)(250)(150)(400 + 150)\sqrt{3(250)(400 + 150)}}{(27)(207 \times 10^3)(0.201 \times 10^6)(400)}$$

$$y_{max} = -0.103 \, \text{mm}$$

Comment: The results are shown in Figure 14–10.

FIGURE 14–10   Deflection curve for beam in Example Problem 14–2.

Formulas for Statically Indeterminate Beams

Appendix A–25 shows several examples of statically indeterminate beams along with formulas for computing the reactions at supports and the deflection at any point on the beams. These formulas can be applied directly in the manner demonstrated for statically determinate beams. Also shown are plots of the shearing force and bending moment diagrams along with the necessary formulas for computing the values at critical points.

The general characteristics of statically indeterminate beams are quite different from the statically determinate designs studied in earlier chapters. This is particularly true for the manner of computing the reaction moments and forces at supports, the distribution of the bending moment with respect to position on the beam, the magnitude of the deflection at various points on the beam, and the general shape of the deflection curve.

The formulas for reactions, shearing force, and bending moments given in Appendix A–25 were derived using the principles discussed in Chapter 11 with due consideration for the statically indeterminate nature of the loading and support conditions. In particular, the *theorem of three moments* from Section 11–12 was used for the continuous beams. The techniques of superposition shown later in this chapter were used to develop the equations for deflections.

As you review the formulas in Appendix A–25, note the following general characteristics.

GENERAL CHARACTERISTICS OF STATICALLY INDETERMINATE BEAMS ▼

Supported Cantilevers: (Cases *a* to *d* in Appendix A–25)

1. The fixed end provides a rigid support that resists any tendency for the beam to rotate. Thus, there is typically a significant bending moment there.
2. The second support is taken to be a simple support. If the simple support is at the free end of the beam, as in Cases *a*, *b*, and *c*, the bending moment there is zero.
3. If the supported cantilever has an overhang, as in Case *d*, the largest bending moment often occurs at the simple support. The shape of the bending moment curve is generally opposite that for the cases without an overhang.
4. There is a point of zero bending moment on a supported cantilever, generally near the fixed end.

Fixed-End Beams: (Cases *e*, *f*, and *g* in Appendix A–25)

1. The bending moments at the fixed ends are *not* zero and may be the maximum for the beam.
2. When loads are downward on a fixed-end beam, the bending moments at the ends are *negative*, indicating that the deflection curve near there is *concave downward*.
3. When loads are downward, the bending moments near the middle of fixed-end beams are *positive*, indicating that the deflection curve there is *concave upward*.

4. There are typically two points of zero bending moment in fixed-end beams.
5. The deflection curve for a fixed-end beam has a zero slope at the fixed ends because of the restraint provided there against rotation.
6. The means of fastening a fixed-end beam to its supports must be capable of resisting the bending moments and the shearing forces at these points. Chapter 18 on Connections should be consulted for techniques of designing and analyzing connections that must resist moments.

Continuous Beams: (Cases *h*, *i*, and *j* in Appendix A–25)

1. Points of maximum positive bending moment generally occur near the middle of spans between supports.
2. Points of maximum negative bending moment generally occur at the interior supports, and these are often the maximum bending moments.
3. Particularly for long continuous beams, it is often economically desirable to tailor the cross-sectional shape and dimensions to provide reinforcement over the sections that have the larger bending moments to accommodate the locally high values. An example would be to design the main beam shape to withstand the maximum positive bending moment between the supports and then to add reinforcing plates to the top or bottom of the beam near the supports to increase the moment of inertia and section modulus in the regions of high bending moment. Another approach would be to increase the depth of the beam near the supports. Highway overpasses and bridges over rivers often have these design features.
4. Where long continuous beams must be made in sections that are fastened together on site, it may be desirable to place the connection near a point of zero bending moment to simplify the design of the connection.

14–5 COMPARISON OF THE MANNER OF SUPPORT FOR BEAMS

In the following example problems we demonstrate the use of various formulas from Appendices A–23, A–24, and A–25 and also generate data with which to compare the performance of four different types of beam support to accomplish the same objective; that is, to support a given load at a given distance from one or two supports. The comparison is based on the magnitude of the bending stress and deflection in the four beams where each has the same material, shape, and size. The best-performing beam, then, is the one with the lower stress and the smaller deflection.

The parameters of the comparisons are the following:

1. The four types of beam support to be compared are
   a. Cantilever
   b. Simply supported beam
   c. Supported cantilever
   d. Fixed-end beam
2. Each beam is to support a single, static, concentrated load of 1200 lb.

3. The load is to be placed at a distance of 30 in from any support.

4. Each beam will be made from ASTM A36 structural steel for which we will use the following properties: $s_y = 36\,000$ psi; $E = 30 \times 10^6$ psi.

5. The maximum allowable bending stress will be based on the AISC standard,

$$\sigma_d = 0.66s_y = 0.66(36\,000\text{ psi}) = 23\,760\text{ psi}$$

6. The maximum allowable deflection will be based on a limit of $L/360$, where L is the length of the beam.

It can be shown that, of the four types of beams and support systems to be considered, the cantilever has the poorest performance with regard to stress and deflection. Therefore, we will start the process of comparison by specifying the beam

cross section shape and size that will ensure that the cantilever meets the requirements stated in items 5 and 6. Then the same shape and size will be used for the other three designs. Five example problems follow, generating the results for the desired comparison.

Example Problem 14–3: Design and analysis of the cantilever.

Example Problem 14–4: Analysis of the simply supported beam.

Example Problem 14–5: Analysis of the supported cantilever.

Example Problem 14–6: Analysis of the fixed-end beam.

Example Problem 14–7: Comparison of the four beam designs.

Example Problem 14–3 For the cantilever shown in Figure 14–11, specify the lightest standard steel channel shape that will limit the bending stress to 23 760 psi and limit the maximum deflection to $L/360$. The channel is to be positioned with the legs pointing downward and the flat side of the web on top to provide a surface for mounting the load. For the beam shape selected, compute the actual expected maximum stress and deflection.

Solution Objective: Design the cantilever and compute the resulting stress and deflection.

Given: Beam dimensions and loading as shown in Figure 14–11. Beam shape to be a standard steel channel with the legs pointing downward.

$$\sigma_d = 23\,760\text{ psi}; \qquad y_{max} = L/360 \quad \text{with } L = L_c = 30\text{ in}$$

FIGURE 14–11
Cantilever.

Analysis: Figure 14–11 contains the loading, shearing force, and bending moment diagrams from which we find that M_{max} = 36 000 lb·in. Then

$$\sigma = \sigma_d = M/S$$

Stress Analysis: For a safe stress, the required section modulus is

$$S = M/\sigma_d = (36000 \text{ lb·in})/(23760 \text{ lb/in}^2) = 1.515 \text{ in}^3$$

From Appendix A–6, considering the properties with respect to the Y-Y axis, we select C12×25 as the lightest suitable section. Its properties are

$$S = 1.88 \text{ in}^3; \quad I = 4.47 \text{ in}^4; \quad w = 25 \text{ lb/ft}$$

Deflection: The maximum allowable deflection is

$$y_{max} = -L/360 = -(30 \text{ in})/360 = -0.0833 \text{ in}$$

From Appendix A–24, the formula for the maximum deflection is

$$y_{max} = -PL^3/3EI \quad \text{at the end of the cantilever}$$

Then the required moment of inertia is

$$I = -PL^3/3Ey_{max}$$

$$I = \frac{-(1200 \text{ lb})(30 \text{ in})^3}{3(30 \times 10^6)(-0.0833 \text{ in})} = 4.32 \text{ in}^4$$

The previously selected beam shape is satisfactory for deflection.

Actual Bending Stress:

$$\sigma = M/S = (36000 \text{ lb·in})/(1.88 \text{ in}^3) = 19150 \text{ psi}$$

Actual Deflection:

$$y_{max} = \frac{-PL^3}{3EI} = \frac{-(1200 \text{ lb})(30 \text{ in})^3}{3(30 \times 10^6 \text{ lb/in}^2)(4.47 \text{ in}^4)} = -0.0805 \text{ in}$$

Results: *Summary of results:*
Beam shape: C12×25 standard steel channel

$$\sigma = 19150 \text{ psi}$$
$$y_{max} = -0.0805 \text{ in}$$

Comment: These results will be compared with the other designs in the following example problems.

Example Problem 14–4 The simply supported beam shown in Figure 14–12 is to be made from a standard steel channel, C12×25, with the legs pointing downward. Compute the actual expected maximum stress and deflection and compare them with the results of Example Problem 14–3 for the cantilever.

Solution Objective: Compute the maximum stress and deflection. Compare with the cantilever results.

Given: Beam dimensions and loading as shown in Figure 14–12. Beam shape is a standard steel channel, C12×25, with the legs pointing downward.
Beam properties: $S = 1.88 \text{ in}^3$; $I = 4.47 \text{ in}^4$

Analysis: From Figure 14–12 we find M_{max} = 18 000 lb·in. Then

Actual Bending Stress:

$$\sigma = M/S = (18000 \text{ lb·in})/(1.88 \text{ in}^3) = 9575 \text{ psi}$$

Actual Deflection:

$$y_{max} = \frac{-PL^3}{48EI} = \frac{-(1200 \text{ lb})(60 \text{ in})^3}{48(30 \times 10^6 \text{ lb/in}^2)(4.47 \text{ in}^4)} = -0.0403 \text{ in}$$

FIGURE 14–12 Simply supported beam.

Results: ***Comparison of Results with Cantilever:*** Using the subscript *1* for the cantilever and *2* for the simply supported beam,

$$\sigma_2/\sigma_1 = (9575 \text{ psi})/(19\,150 \text{ psi}) = 0.50$$

$$y_2/y_1 = (-0.0403 \text{ in})/(-0.0805 \text{ in}) = 0.50$$

Comment: These results will be compared with the other designs in Example Problem 14–7.

Example Problem 14–5

The supported cantilever shown in Figure 14–13 is to be made from a standard steel channel, C12×25, with the legs pointing downward. Compute the actual expected maximum stress and deflection and compare them with the results of Example Problem 14–3 for the cantilever.

FIGURE 14–13 Supported cantilever.

Solution Objective: Compute the maximum stress and deflection. Compare with the cantilever results.

Given: Beam dimensions and loading as shown in Figure 14–12. Beam shape is a standard steel channel, C12×25, with the legs pointing downward.
Beam properties: $S = 1.88 \text{ in}^3$; $I = 4.47 \text{ in}^4$

Analysis: From Figure 14–12 we find $M_{max} = 13\,500 \text{ lb} \cdot \text{in}$. Then

Actual Bending Stress:

$$\sigma = M/S = (13\,500 \text{ lb} \cdot \text{in})/(1.88 \text{ in}^3) = 7180 \text{ psi}$$

Actual Deflection: From Appendix A–25(a),

$$y_{max} = \frac{-PL^3}{107EI} = \frac{-(1200 \text{ lb})(60 \text{ in})^3}{107(30 \times 10^6 \text{ lb/in}^2)(4.47 \text{ in}^4)} = -0.0181 \text{ in}$$

Results: ***Comparison of Results with Cantilever:*** Using the subscript *1* for the cantilever and *3* for the supported cantilever,

$$\sigma_3/\sigma_1 = (7180 \text{ psi})/(19\,150 \text{ psi}) = 0.375$$
$$y_3/y_1 = (-0.0181 \text{ in})/(-0.0805 \text{ in}) = 0.225$$

Comment: These results will be compared with the other designs in Example Problem 14–7.

Example Problem 14–6

The fixed-end beam shown in Figure 14–14 is to be made from a standard steel channel, C12×25, with the legs pointing downward. Compute the actual expected maximum stress and deflection and compare them with the results of Example Problem 14–3 for the cantilever.

Solution Objective: Compute the maximum stress and deflection. Compare with the cantilever results.

Given: Beam dimensions and loading as shown in Figure 14–14. Beam shape is a standard steel channel, C12×25, with the legs pointing downward.
Beam properties: $S = 1.88 \text{ in}^3$; $I = 4.47 \text{ in}^4$

Analysis: From Figure 14–14 we find $M_{max} = 9000 \text{ lb} \cdot \text{in}$. Then

Actual Bending Stress:

$$\sigma = M/S = (9000 \text{ lb} \cdot \text{in})/(1.88 \text{ in}^3) = 4787 \text{ psi}$$

FIGURE 14–14 Fixed-end beam.

$$M_A = \frac{PL}{8} = 9000 \text{ lb} \cdot \text{in}$$

$$P = 1200 \text{ lb}$$

A $a = 30$ in   B $a = 30$ in   C

M_A

$$R_A = \frac{P}{2} = 600 \text{ lb}$$

y_{max}

$$R_C = \frac{P}{2} = 600 \text{ lb}$$

M_C

$$L = 60 \text{ in}$$

$$y_{max} = \frac{-PL^3}{192EI} = -0.0101 \text{ in}$$

Shearing force, V (lb)   600   0   −600

Bending moment, M (lb·in)   9000   0   −9000   −9000

Actual Deflection: From Appendix A–25(e),

$$y_{max} = \frac{-PL^3}{192EI} = \frac{-(1200 \text{ lb})(60 \text{ in})^3}{192(30 \times 10^6 \text{ lb/in}^2)(4.47 \text{ in}^4)} = -0.0101 \text{ in}$$

Results: **Comparison of Results with Cantilever:** Using the subscript *1* for the cantilever and *4* for the fixed-end beam,

$$\sigma_4/\sigma_1 = (4787 \text{ psi})/(19\,150 \text{ psi}) = 0.250$$

$$y_4/y_1 = (-0.0101 \text{ in})/(-0.0805 \text{ in}) = 0.125$$

Comment: These results will be compared with the other designs in Example Problem 14–7.

Example Problem 14–7   Compare the behavior of the four beams shown in Figures 14–11 through 14–14 with regard to shearing forces, bending moments, and maximum deflection. Use the results of Example Problems 14–3 through 14–6.

Solution   Objective: Compare the performance of the four beams.

Given: Beam designs shown in Figures 14–11 through 14–14.
Results of Example Problems 14–3 through 14–6.

Analysis: Figure 14–15 shows the comparison.

Results: In Figure 14–15, the graphs on the left show plots of the deflection, shearing force, and bending moment as a function of position on the beam for the four designs superimposed on each other. In each case, it is obvious that the cantilever produces the highest value by a significant margin. On the right are bar graphs of the relative performance where the values of deflection, shearing force, and bending moment for the cantilever are assigned a value of 1.0.

FIGURE 14–15
Comparison of performance of the four beam designs in Figures 14–11 through 14–14.

Comment: The results shown in Figure 14–15 indicate that it is desirable to provide two supports rather than one if that is practical. Also, providing two fixed ends for the beam is most desirable, producing only $\frac{1}{8}$ of the deflection, $\frac{1}{2}$ of the shearing force, and $\frac{1}{4}$ of the bending moment as compared with the cantilever for a given load, distance from the load to the support, beam shape, and size. Of course, it must be practical to provide the second support or the fixed ends.

14–6 SUPERPOSITION USING DEFLECTION FORMULAS

Formulas such as those used in the preceding section are available for a large number of cases of loading and support conditions. Obviously, these cases would allow the solution of many practical beam deflection problems. An even larger number of situations can be handled by the use of the *principle of superposition*.

If a particular loading and support pattern can be broken into components such that each component is like one of the cases for which a formula is available, then the total deflection at a point on the beam is equal to the sum of the deflections caused by each component. The deflection due to one component load is *superposed* on deflections due to the other loads, thus giving the name *superposition*.

An example of where superposition can be applied is shown in Figure 14–16. Represented in the figure is a roof beam carrying a uniformly distributed roof load of 800 lb/ft and also supporting a portion of a piece of process equipment that provides a concentrated load at the middle. Figure 14–17 shows how the loads are considered separately. Each component load produces a maximum deflection at the middle, point *B*. Therefore, the maximum total deflection will occur there also. Let the subscript *1* refer to the concentrated load case and the subscript *2* refer to the distributed load case. Then

$$y_1 = \frac{-PL^3}{48EI} \text{ (Appendix A–23, case a)}$$

$$y_2 = \frac{-5}{384}\frac{WL^3}{EI} \text{ (Appendix A–23, case d)}$$

The total deflection will be

$$y_T = y_1 + y_2$$

The terms *L*, *E*, and *I* will be the same for both cases. Be careful to keep units consistent.

$$L = 16\text{ ft} \times 12\text{ in/ft} = 192\text{ in}$$

$$E = 30 \times 10^6 \text{ psi for steel}$$

$$I = 103\text{ in}^4 \text{ for W12×16 beam}$$

To compute y_1, let $P = 2500$ lb.

$$y_1 = \frac{-2500(192)^3}{48(30 \times 10^6)(103)}\text{ in} = -0.119\text{ in}$$

To compute y_2, *W* is the total resultant of the distributed load.

$$W = (800\text{ lb/ft})(16\text{ ft}) = 12\,800\text{ lb}$$

Then

$$y_2 = \frac{-5(12\,800)(192)^3}{384(30 \times 10^6)(103)}\text{ in} = -0.382\text{ in}$$

and

$$y_T = y_1 + y_2 = -0.119\text{ in} - 0.382\text{ in} = -0.501\text{ in}$$

Since this is the total deflection, we could check to see if it meets the recommendation that the maximum deflection should be less than $\frac{1}{360}$ times the span of the beam. The span is *L*.

$$\frac{L}{360} = \frac{192\text{ in}}{360} = 0.533\text{ in}$$

The computed deflection was 0.501 in, which is satisfactory.

The superposition principle is valid for any place on the beam, not just at the loads. The following example problem illustrates this.

FIGURE 14–16   Roof beam.

FIGURE 14–17   Illustration of the superposition principle.

Example Problem 14–8

Figure 14–18 shows a shaft carrying two gears and simply supported at its ends by bearings. The mating gears above exert downward forces, which tend to separate the gears. Other forces acting horizontally are not considered in this analysis. Determine the total deflection at the gears B and C if the shaft is steel and has a diameter of 45 mm.

FIGURE 14–18 Shaft for Example Problem 14–8.

Solution

Objective: Compute the deflection at points B and C.

Given: Shaft $ABCD$ in Figure 14–18. Shaft is steel. $D = 45$ mm.
Load at $B = P_1 = 3.5$ kN $= 3500$ N. Load at $C = P_2 = 2.1$ kN $= 2100$ N.

Analysis: The two unsymmetrically placed, concentrated loads constitute a situation for which none of the given cases for beam deflection formulas is valid. However, it can be solved by using the formulas of Appendix A–23–b twice. Considering each load separately, the deflections at B and C can be found for each. The total, then, would be the sum of the component deflections. Figure 14–19 shows the logic from which we can say

$$y_B = y_{B1} + y_{B2}$$
$$y_C = y_{C1} + y_{C2}$$

where y_B = total deflection at B
   y_C = total deflection at C
   y_{B1} = deflection at B due to 3.5-kN load alone
   y_{C1} = deflection at C due to 3.5-kN load alone
   y_{B2} = deflection at B due to 2.1-kN load alone
   y_{C2} = deflection at C due to 2.1-kN load alone

FIGURE 14–19
Superposition logic for deflection of shaft in Figure 14–18.

For all the calculations, the values of E, I, and L will be needed. These are the same as those in Example Problem 14–2.

$$E = 207 \times 10^3 \, \text{N/mm}^2$$
$$I = 0.201 \times 10^6 \, \text{mm}^4$$
$$L = 400 \, \text{mm}$$

The product of EIL is present in all the formulas.

$$EIL = (207 \times 10^3)(0.201 \times 10^6)(400) = 16.64 \times 10^{12}$$

Now the individual component deflections will be computed. Note the values for the variables a, b, and x are different for each component load. See the data labeled in Figure 14–19.

Results: For component 1: y_{B1} is at the load. y_{C1} is in the longer segment.

$$y_{B1} = \frac{-P_1 a^2 b^2}{3EIL} = \frac{-(3.5 \times 10^3)(250)^2(150)^2}{3(16.64 \times 10^{12})} = -0.0985 \, \text{mm}$$

$$y_{C1} = \frac{-P_1 bx}{6EIL}(L^2 - b^2 - x^2)$$

$$y_{C1} = \frac{-(3.5 \times 10^3)(150)(100)}{6(16.64 \times 10^{12})}(400^2 - 150^2 - 100^2) = -0.0670 \, \text{mm}$$

For component 2, the load will be 2.1 kN at point C. Then y_{B2} is in the longer segment. y_{C2} is at the load.

$$y_{B2} = \frac{-P_2 bx}{6EIL}(L^2 - b^2 - x^2)$$

$$y_{B2} = \frac{-(2.1 \times 10^3)(100)(150)}{6(16.64 \times 10^{12})}(400^2 - 100^2 - 150^2) = -0.0402 \, \text{mm}$$

$$y_{C2} = \frac{-P_2 a^2 b^2}{3EIL} = \frac{-(2.1 \times 10^3)(300)^2(100)^2}{3(16.64 \times 10^{12})} = -0.0378 \, \text{mm}$$

Now, by superposition,

$$y_B = y_{B1} + y_{B2} = -0.0985 \, \text{mm} - 0.0402 \, \text{mm} = -0.1387 \, \text{mm}$$

$$y_C = y_{C1} + y_{C2} = -0.0670 \, \text{mm} - 0.0378 \, \text{mm} = -0.1048 \, \text{mm}$$

Comment: In Section 14–2 it was observed that a recommended limit for the movement of one gear relative to its mating gear is 0.13 mm. Thus, this shaft is too flexible since the deflection at B exceeds 0.13 mm without even considering the deflection of the mating shaft.

Superposition Applied to Statically Indeterminate Beams

Superposition can be applied to enable redundant reactions and moments in statically indeterminate beams to be determined as demonstrated here.

Consider first the supported cantilever shown in Figure 14–20. Because of the restraint at A and the simple support at B, the unknown reactions include

1. The vertical force R_B
2. The vertical force R_A
3. The restraining moment M_A

The conditions assumed for this beam are that the supports at A and B are absolutely rigid and at the same level, and that the connection at A does not allow any rotation of

the beam at that point. Conversely, the support at B will allow rotation and cannot resist moments.

If the support at B is removed, the beam would deflect downward, as shown in Figure 14–21(a), by an amount y_{B1} due to the load P. Now if the load is removed and the reaction

FIGURE 14–20   Supported cantilever.

FIGURE 14–21 Superposition applied to the supported cantilever.

force R_B is applied upward at B, the beam would deflect upward by an amount y_{B2}, as shown in Figure 14–21(b). In reality, of course, both forces are applied, and the deflection at B is *zero*. The principle of superposition would then provide the conclusion that

$$y_{B1} + y_{B2} = 0$$

This equation, along with the normal equations of static equilibrium, will permit the evaluation of all three unknowns, as demonstrated in the example problem that follows. It must be recognized that the principles of static equilibrium are still valid for statically indeterminate beams. However, they are not sufficient to allow a direct solution.

Example Problem 14–9   Determine the support reactions at A and B for the supported cantilever shown in Figure 14–20 if the load P is 2600 N and placed 1.20 m out from A. The total length of the beam is 1.80 m. Then draw the complete shearing force and bending moment diagrams and design the beam by specifying a configuration, a material, and the required dimensions of the beam. Use a design factor of 8 based on ultimate strength since the load will be repeated often.

Solution   Objective: Determine the support reactions, draw the shearing force and bending moment diagrams, and design the beam.

Given: Beam loading in Figure 14–20. P = 2600 N (repeated). L = 1.80 m.
a = 1.20 m. Use a design factor of N = 8 based on s_u.

Analysis: Use the superposition method.

Results: **Reaction at B, R_B**
We must first determine the reaction at B using superposition. Previously we observed,

$$y_{B1} + y_{B2} = 0$$

The equation for y_{B1} can be found from the beam deflection formulas in Appendix A–24. As suggested in Figure 14–21(a), the deflection at the end of a cantilever carrying an intermediate load is required. Then

$$y_{B1} = \frac{-Pa^2}{6EI}(3L - a)$$

where P = 2600 N
a = 1.20 m
L = 1.80 m

The values of E and I are still unknown, but we can express the deflection in terms of EI.

$$y_{B1} = \frac{(-2600\,\text{N})(1.20\,\text{m})^2}{6EI}[3(1.80\,\text{m}) - 1.20\,\text{m}]$$

$$y_{B1} = \frac{-2621\,\text{N}\cdot\text{m}^3}{EI}$$

Now looking at Figure 14–21(b), we need the deflection at the end of the cantilever due to a concentrated load there. Then

$$y_{B2} = \frac{PL^3}{3EI} = \frac{R_B(1.8\,\text{m})^3}{3EI} = \frac{R_B(1.944\,\text{m}^3)}{EI}$$

The sum of y_{B1} and y_{B2} is zero.

$$\frac{-2621 \text{ N} \cdot \text{m}^3}{EI} + \frac{R_B(1.944 \text{ m}^3)}{EI} = 0$$

The term EI can be canceled out, allowing the solution for R_B.

$$R_B = \frac{2621 \text{ N} \cdot \text{m}^3}{1.944 \text{ m}^3} = 1348 \text{ N}$$

The values of R_A and M_A can now be found using the equations of static equilibrium.

Reaction at A, R_A

$$\sum F = 0 \quad \text{(in the vertical direction)}$$
$$R_A + R_B - P = 0$$
$$R_A = P - R_B = 2600 \text{ N} - 1348 \text{ N} = 1252 \text{ N}$$

Bending Moment at A, M_A

Summing moments about point A gives

$$0 = M_A - 2600 \text{ N}(1.2 \text{ m}) + 1348 \text{ N}(1.8 \text{ m})$$
$$M_A = 693 \text{ N} \cdot \text{m}$$

The positive sign for the result indicates that the assumed sense of the reaction moment in Figure 14–20 is correct. However, this is a negative moment because it causes the beam to bend concave downward near the support A.

Shearing Force and Bending Moment Diagrams

The shearing force and bending moment diagrams can now be drawn, as shown in Figure 14–22, using conventional techniques. The maximum bending moment occurs at the load where $M = 809 \text{ N} \cdot \text{m}$.

Beam Design

The beam can now be designed. Let's assume that the actual installation is similar to that sketched in Figure 14–23, with the beam welded at its left end and resting on another beam at the right end. A rectangular bar would work well in this arrangement, and a ratio of $h = 3t$ will be assumed. A carbon steel such as AISI 1040, hot rolled, provides an ultimate strength of 621 MPa. Its percent

FIGURE 14–22
Shearing force and bending moment diagrams for the supported cantilever in Example Problem 14–9.

FIGURE 14–23
Physical implementation of a supported cantilever.

elongation, 25%, suggests good ductility, which will help it resist the repeated loads. The design should be based on bending stress.

$$\sigma = \frac{M}{S}$$

But let

$$\sigma = \sigma_d = \frac{s_u}{N} = \frac{621\,\text{MPa}}{8} = 77.6\,\text{MPa}$$

Then

$$S = \frac{M}{\sigma_d} = \frac{809\,\text{N} \cdot \text{m}}{77.6\,\text{N/mm}^2} \times \frac{10^3\,\text{mm}}{\text{m}} = 10\,425\,\text{mm}^3$$

For a rectangular bar,

$$S = \frac{th^2}{6} = \frac{t(3t)^2}{6} = \frac{9t^3}{6} = 1.5t^3$$

Then

$$1.5t^3 = 10\,425\,\text{mm}^3$$

$$t = 19.1\,\text{mm}$$

Let's use the preferred size of 20 mm for t. Then

$$h = 3t = 3(20\,\text{mm}) = 60\,\text{mm}$$

Comment: The final design can be summarized as a rectangular steel bar of AISI 1040, hot rolled, 20 mm thick and 60 mm high, welded to a rigid support at its left end and resting on a simple support at its right end. The maximum stress in the bar would be less than 77.6 MPa, providing a design factor of at least 8 based on ultimate strength.

The superposition method can be applied to any supported cantilever beam analysis for which the equation for the deflection due to the applied load can be found. Either the beam deflection formulas like those in the Appendix, the moment-area method, or the successive integration method developed later can be used.

Continuous beams can also be analyzed using superposition. Consider the beam on three supports shown in Figure 14–24. The three unknown support reactions make

FIGURE 14–24 Continuous beam.

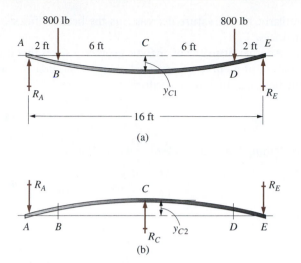

FIGURE 14–25   Superposition applied to a continuous beam.

the beam statically indeterminate. The "extra" reaction R_C can be found using the technique suggested in Figure 14–25. Removing the support at C would cause the deflection y_{C1} downward due to the two 800-lb loads. Case c of Appendix A–23 can be used to find y_{C1}. Then if the loads are imagined to be removed and the reaction R_C is replaced, the upward deflection y_{C2} would result. The formulas of Case a in Appendix A–23 can be used.

Here again, of course, the actual deflection at C is zero because of the unyielding support. Therefore,

$$y_{C1} + y_{C2} = 0$$

From this relationship, the value of R_C can be computed. The remaining reactions R_A and R_E can then be found in the

conventional manner, allowing the creation of the shearing force and bending moment diagrams.

14–7 SUCCESSIVE INTEGRATION METHOD

In this section we show the mathematical relationships among the moment, slope, and deflection curves from which you can solve for the actual equations for a given beam with a given loading and support condition.

Figure 14–26 shows a small segment of a beam in its initial straight shape and its deflected shape. The sides of the segment remain straight as the beam deflects, but they rotate with respect to a point at the neutral axis. This results in compression in the top portion of the segment and tension in the bottom portion, a fact used in the development of the flexure formula in Chapter 12.

The rotated sides of the segment intersect at the center of curvature and form the small angle $d\theta$. Note also the radius of curvature, R, measured from the center of curvature to the neutral axis. From the geometry shown in the figure,

$$\Delta s = R(d\theta) \qquad (14\text{–}3)$$

and

$$\delta = c(d\theta) \qquad (14\text{–}4)$$

where Δs is the length of the segment at the neutral axis

δ is the elongation of the bottom line of the segment that occurs as the beam deflects

c has the same meaning as in the flexure formula, the distance from the neutral axis to the outer fiber of the section

(a) Segment of a straight beam

(b) Segment of deflected beam

FIGURE 14–26   Relation between radius R and deformation δ.

Recall that the definition of the neutral axis states that no strain occurs there. Then the length Δs in the deflected beam segment equals the length Δx in the undeflected segment, and Equation (14–3) can be written

$$\Delta x = R(d\theta) \qquad (14\text{–}5)$$

Now both Equations (14–4) and (14–5) can be solved for $d\theta$.

$$d\theta = \frac{\delta}{c}$$

$$d\theta = \frac{\Delta x}{R}$$

Equating these values of $d\theta$ to each other gives

$$\frac{\delta}{c} = \frac{\Delta x}{R}$$

Another form of this equation is

$$\frac{c}{R} = \frac{\delta}{\Delta x}$$

The right side of this equation conforms to the definition of strain, ϵ. Then

$$\epsilon = \frac{c}{R} \qquad (14\text{–}6)$$

Earlier it was shown that

$$\epsilon = \frac{\sigma}{E}$$

where σ, the stress due to bending, can be computed from the flexure formula,

$$\sigma = \frac{Mc}{I}$$

Then

$$\epsilon = \frac{\sigma}{E} = \frac{Mc}{EI}$$

Combining this with Equation (14–6) gives

$$\frac{c}{R} = \frac{Mc}{EI}$$

Dividing both sides by c gives

$$\frac{1}{R} = \frac{M}{EI} \qquad (14\text{–}7)$$

Equation (14–7) is useful in developing the moment-area method for finding beam deflections. See Section 14–8.

In analytic geometry, the reciprocal of the radius of curvature, $1/R$, is defined as the *curvature* and denoted by κ, the lowercase Greek letter kappa. Then

$$\kappa = \frac{M}{EI} \qquad (14\text{–}8)$$

Equation (14–8) indicates that the curvature gets greater as the bending moment increases, which stands to reason.

Similarly, the curvature decreases as the beam stiffness, EI, increases.

Another principle of analytic geometry states that if the equation of a curve is expressed as $y = f(x)$, that is, y is a function of x, then the curvature is

$$\kappa = \frac{d^2y}{dx^2} \qquad (14\text{–}9)$$

Combining Equations (14–8) and (14–9) gives

$$\frac{M}{EI} = \frac{d^2y}{dx^2} \qquad (14\text{–}10)$$

or

$$M = EI\frac{d^2y}{dx^2} \qquad (14\text{–}11)$$

Equations (14–10) and (14–11) are useful in developing the successive integration method for finding beam deflections, described next. Remember that this relationship applies to beams that are initially straight and where the curvature after loading is very small.

Successive Integration Method— General Approach

This section presents a general approach that allows the determination of deflection at any point on the beam. The advantages of this approach are the following:

1. The result is a set of equations for deflection at all parts of the beam. Deflection at any point can then be found by substitution of the beam stiffness properties of E and I, and the position on the beam.

2. Data are easily obtained from which a plot of the shape of the deflection curve may be made.

3. The equations for the *slope* of the beam at any point are generated, as are deflections. This is important in some machinery applications such as shafts at bearings and shafts carrying gears. An excessive slope of the shaft would result in poor performance and reduced life of the bearings or gears.

4. The fundamental relationships among loads, manner of support, beam stiffness properties, slope, and deflections are emphasized in the solution procedure. The designer who understands these relationships can make more efficient designs.

5. The method requires the application of only simple mathematical concepts.

6. The point of maximum deflection can be found directly from the resulting equations.

The basis for the successive integration method has been developed in Sections 14–3 and 14–7. The five beam diagrams will be prepared, in a manner similar to that shown in Figure 14–5, to relate the loads, shearing forces, bending moments, slopes, and deflections over the entire length of the beam.

The load, shearing force, and bending moment diagrams can be drawn using the principles from Chapter 11. Then equations for the bending moment are derived for all segments of the bending moment diagram.

Equation (14–11) is then used to develop the slope and deflection equations from the moment equations by integrating twice with respect to the position on the beam, x, as follows.

$$M = EI\frac{d^2y}{dx^2} \qquad (14\text{–}11)$$

Now, integrating once with respect to x gives

$$\int M\,dx = EI\int \frac{d^2y}{dx^2}dx = EI\frac{dy}{dx} \qquad (14\text{–}12)$$

Earlier, in Section 14–3, Equation (14–2), we showed that $dy/dx = \theta$, the slope of the deflection curve. Then

$$\int M\,dx = EI\theta = \theta EI \qquad (14\text{–}13)$$

Equation (14–12) can be integrated again, giving

$$\int EI\theta\,dx = EI\int \frac{dy}{dx}dx = EIy = yEI \qquad (14\text{–}14)$$

After the final values for $EI\theta$ and EIy have been determined, they will be divided by the beam stiffness, EI, to obtain the values for slope, θ, and deflection, y.

The steps indicated by Equations (14–12) through (14–14) are to be completed for each segment of the beam for which the moment diagram is continuous. Also, because our objective is to obtain discrete equations for slope and deflection for particular beam-loading patterns, we will need to evaluate a constant of integration for each integration performed.

The development of the equations for the bending moment versus position is often accomplished by integrating the equations for the shearing force versus x, as shown in Chapter 11. This follows from the rule that the change in bending moment between two points on a beam is equal to the area under the shearing force curve between the same two points.

The step-by-step method used to find the deflection of beams using the general approach is as follows.

STEPS IN THE SUCCESSIVE INTEGRATION METHOD FOR BEAM DEFLECTIONS ▼

1. Determine the reactions at the supports for the beam.
2. Draw the shearing force and bending moment diagrams using the same procedures presented in Chapter 11, and identify the magnitudes at critical points.
3. Divide the beam into segments in which the shearing force diagram is continuous by naming points at the places where abrupt changes occur with the letters A, B, C, D, etc.

4. Write equations for the shearing force curve in each segment. In most cases, these will be equations of straight lines, that is, equations involving x to the first power. Sometimes, as for beams carrying concentrated loads, the equation will be simply of the form

$$V = \text{constant}$$

5. For each segment, perform the process,

$$M = \int V\,dx + C$$

To evaluate the constant of integration that ties the moment equation to the particular values already known for the moment diagram, insert known boundary conditions and solve for C.

6. For each segment, perform the process,

$$\theta EI = \int M\,dx + C$$

The constant of integration generated here cannot be evaluated directly right away. So each constant should be identified separately by a subscript such as C_1, C_2, C_3, etc. Then when they are evaluated (in step 9), they can be put in their proper places.

7. For each segment, perform the process,

$$y\,EI = \int \theta EI\,dx + C$$

Here again, the constants should be labeled with subscripts.

8. Establish *boundary conditions* for the slope and deflection diagrams. The same number of boundary conditions must be identified as there are unknown constants from steps 6 and 7. Boundary conditions express mathematically the special values of slope and deflection at certain points and the fact that both the slope curve and the deflection curve are continuous. Typical boundary conditions are:
   a. The deflection of the beam at each support is zero.
   b. The deflection of the beam at the end of one segment is equal to the deflection of the beam at the beginning of the next segment. This follows from the fact that the deflection curve is continuous; that is, it has no abrupt changes.
   c. The slope of the beam at the end of one segment is equal to the slope at the beginning of the next segment. The slope has no abrupt changes.
   d. For the special case of a cantilever, the slope of the beam at the support is also zero.

9. Combine all the boundary conditions to evaluate all the constants of integration. This typically involves the solution of a set of simultaneous equations in which the number of equations is equal to the number of unknown constants of integration. Equation-solving software or graphing calculators are very helpful for this step.

10. Substitute the constants of integration back into the slope and deflection equations, thus completing them. The value of the slope or deflection at any point can then be evaluated by simply placing the proper value of the position on the beam in the equation. Points of maximum deflection in any segment can also be found.

Example Problem 14–10 illustrates this method.

Example Problem 14–10

Figure 14–27 shows a beam used as a part of a special structure for a machine. The 20-K (20 000-lb) load at *A* and the 30-K (30 000-lb) load at *C* represent places where heavy equipment is supported. Between the two supports at *B* and *D*, the uniformly distributed load of 2 K/ft (2000 lb/ft) is due to stored bulk materials, which are in a bin supported by the beam. All loads are static. To maintain accuracy of the product produced by the machine, the maximum allowable deflection of the beam is 0.05 in. Specify an acceptable wide-flange steel beam and also check the stress in the beam.

FIGURE 14–27 Beam for Example Problem 14–10.

Solution

Objective: Specify a steel wide-flange beam shape to limit the deflection to 0.05 in. Check the stress in the selected beam to ensure safety.

Given: Beam loading in Figure 14–27.

Analysis: The beam will be analyzed to determine where the maximum deflection will occur. Then the required moment of inertia will be determined to limit the deflection to 0.05 in. A wide-flange beam that has the required moment of inertia will then be selected. The 10-step procedure discussed earlier will be used. The solution is shown in a programmed format. You should work through the problem yourself before looking at the given solution.

Results: **Steps 1 and 2:** Call for drawing the shearing force and bending moment diagrams. Do this now before checking the following result. Figure 14–28 shows the results. Now do step 3.

Step 3: Three segments are required: *AB*, *BC*, and *CD*. These are the segments over which the shearing force diagram is continuous. Now do step 4 to get the shearing force curve equations.

FIGURE 14–28 Load, shearing force, and bending moment diagrams for Example Problem 14–10.

Step 4: The results are

$$V_{AB} = -20 \qquad \text{(a)}$$
$$V_{BC} = -2x + 29 \qquad \text{(b)}$$
$$V_{CD} = -2x - 1 \qquad \text{(c)}$$

In the segments BC and CD, the shearing force curve is a straight line with a slope of -2 kip/ft, the same as the load. Any method of writing the equation of a straight line can be used to derive these equations.
Now do step 5 of the procedure.

Step 5: You should have the following for the moment equations. First,

$$M_{AB} = \int V_{AB}\,dx + C = \int -20\,dx + C = -20x + C$$

At $x = 0$, $M_{AB} = 0$. Therefore, $C = 0$ and

$$M_{AB} = -20x \qquad \text{(d)}$$

Next,

$$M_{BC} = \int V_{BC}\,dx + C = \int (-2x + 29)\,dx + C = -x^2 + 29x + C$$

At $x = 3$, $M_{BC} = -60$. Therefore, $C = -138$ and

$$M_{BC} = -x^2 + 29x - 138 \qquad \text{(e)}$$

Finally,

$$M_{CD} = \int V_{CD}\,dx + C = \int (-2x - 1)\,dx + C = -x^2 - x + C$$

At $x = 9$, $M_{CD} = 42$. Therefore, $C = 132$ and

$$M_{CD} = -x^2 - x + 132 \qquad \text{(f)}$$

Now do step 6 to get equations for θEI.

Step 6: By integrating the moment equations,

$$\theta_{AB} EI = \int M_{AB}\,dx + C = \int (-20x)\,dx + C$$
$$\theta_{AB} EI = -10x^2 + C_1 \qquad \text{(g)}$$
$$\theta_{BC} EI = \int M_{BC}\,dx + C = \int (-x^2 + 29x - 138)\,dx + C$$
$$\theta_{BC} EI = -x^3/3 + 14.5x^2 - 138x + C_2 \qquad \text{(h)}$$
$$\theta_{CD} EI = \int M_{CD}\,dx + C = \int (-x^2 - x + 132)\,dx + C$$
$$\theta_{CD} EI = -x^3/3 - x^2/2 + 132x + C_3 \qquad \text{(i)}$$

Now in step 7, integrate Equations (g), (h), and (i) to get the yEI equations.

Step 7: You should have

$$y_{AB} EI = \int \theta_{AB} EI\,dx + C$$
$$y_{AB} EI = -10x^3/3 + C_1 x + C_4 \qquad \text{(j)}$$
$$y_{BC} EI = \int \theta_{BC} EI\,dx + C$$
$$y_{BC} EI = -x^4/12 + 14.5x^3/3 - 69x^2 + C_2 x + C_5 \qquad \text{(k)}$$
$$yCD EI = \int \theta_{CD} EI\,dx + C$$
$$y_{CD} EI = -x^4/12 - x^3/6 + 66x^2 + C_3 x + C_6 \qquad \text{(l)}$$

Step 8 calls for identifying boundary conditions. Six are required since there are six unknown constants of integration in Equations (g) through (l). Write them now.

Step 8: Considering zero deflection points and the continuity of the slope and deflection curves, we can say

1. At $x = 3$, $y_{AB} EI = 0$ ⎫
2. At $x = 3$, $y_{BC} EI = 0$ ⎬ (zero deflection at supports)
3. At $x = 11$, $y_{CD} EI = 0$ ⎭
4. At $x = 9$, $y_{BC} EI = y_{CD} EI$ (continuous deflection curve at C)
5. At $x = 3$, $\theta_{AB} EI = \theta_{BC} EI$ ⎫
6. At $x = 9$, $\theta_{BC} EI = \theta_{CD} EI$ ⎬ (continuous slope curve at B and C)

We can now substitute these values of x into the proper equations and solve for C_1 through C_6. First make the substitutions and reduce the resulting equations to the form involving the constants.

For the previous six conditions listed, the following equations result:

1. $3C_1 + C_4$ $= 90$
2. $3C_2 + C_5$ $= 497.25$
3. $11C_3 + C_6$ $= -6544.08\overline{33} = -78\,529/12$
4. $9C_2 - 9C_3 + C_5 - C_6$ $= 7290$
5. $C_1 - C_2$ $= -202.5$
6. $C_2 - C_3$ $= 1215$

The value of the quantity on the right side of Equation 3 is expressed in excessively high precision. This is not often necessary but is being done for this example to eliminate the accumulation of round-off errors as the problem solution proceeds. There are many steps to the final solution, and inaccuracies at this stage can result in significant variation in the results that might frustrate you as you follow the solution. Note that writing the constant in Equation 3 as $-6544.08\overline{33}$ indicates that the 3's repeat to infinity. Thus, this is an inherently inaccurate representation of the number. Entering the number as the exact fraction ($-78\,529/12$) into an equation solver would eliminate the error. The use of a computer-based equation solver such as MATHCAD, TK Solver, MATLAB, or MAPLE facilitates the laborious calculations involved in the balance of the procedure. Many high-level calculators having graphing capability also contain simultaneous equation solvers.
   Now solve the six equations simultaneously for the values of C_1 through C_6.

Step 9: The results are

$$C_1 = 132.\overline{333} = 397/3 \qquad C_2 = 334.8\overline{33} = 4018/12$$
$$C_3 = -880.1\overline{66} = -5281/6 \qquad C_4 = -307 \text{ (exact)}$$
$$C_5 = -507.25 \text{ (exact)} \qquad C_6 = 3137.75 \text{ (exact)}$$

Step 10: The final equations for θ and y can now be written by substituting the constants into Equations (g) through (l). The results are shown next.

$$\theta_{AB} EI = -10x^2 + 132.\overline{333}$$
$$\theta_{BC} EI = -x^3/3 + 14.5x^2 - 138x + 334.8\overline{33}$$
$$\theta_{CD} EI = -x^3/3 - x^2/2 + 132x - 880.1\overline{66}$$
$$y_{AB} EI = -10x^3/3 + 132.\overline{333}x - 307$$
$$y_{BC} EI = -x^4/12 + 14.5x^3/3 - 69x^2 + 334.8\overline{33}x - 507.25$$
$$y_{CD} EI = -x^4/12 - x^3/6 + 66x^2 - 880.1\overline{66}x + 3137.75$$

Having the completed equations, we can now determine the point of maximum deflection, which is the primary objective of the analysis. Based on the loading, the probable shape of the deflected beam would be like that shown in Figure 14–29. Therefore, the maximum deflection could occur at point A at the end of the overhang, at a point to the right of B (upward), or at a point near the load at C (downward). It is probable that there are two points of zero slope at the points E and F, as shown in Figure 14–29. We would need to know where the slope equation $\theta_{BC}EI$ equals zero to determine where the maximum deflections occur.
   Notice that this is a third-degree equation. The use of a graphing calculator and an equation solver facilitates finding the points where $\theta_{BC}EI = 0$. Figure 14–30 shows an expanded plot of the BC segment of the beam showing that the zero points occur at $x = 3.836$ ft and at $x = 8.366$ ft.

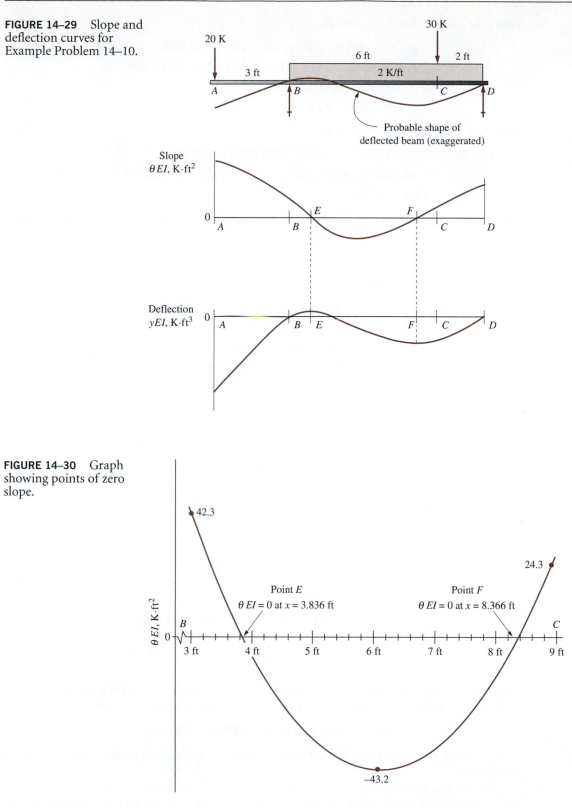

FIGURE 14–29 Slope and deflection curves for Example Problem 14–10.

FIGURE 14–30 Graph showing points of zero slope.

We can now determine the values for yEI at points A, E, and F to find out which is larger.

Point A. At $x = 0$ in segment AB,

$$y_{AB}EI = -10x^3/3 + 132.\overline{333}x - 307$$

$$y_A EI = -10(0.00)^3/3 + 132.\overline{333}(0.00) - 307$$

$$y_A EI = -307 \text{ K} \cdot \text{ft}^3$$

Point E. At $x = 3.836$ ft in segment BC,

$$y_{BC} EI = -x^4/12 + 14.5x^3/3 - 69x^2 + 334.8\overline{33}x - 507.25$$
$$y_E EI = -(3.836)^4/12 + 14.5(3.836)^3/3 - 69(3.836)^2$$
$$+334.8\overline{33}(3.836) - 507.25$$
$$y_E EI = +16.62\,\text{K} \cdot \text{ft}^3$$

Point F. At $x = 8.366$ ft in segment BC,

$$y_{BC} EI = -x^4/12 + 14.5x^3/3 - 69x^2 + 334.8\overline{33}x - 507.25$$
$$y_F EI = -(8.366)^4/12 + 14.5(8.366)^3/3 - 69(8.366)^2 +334.8\overline{33}(8.366) - 507.25$$
$$y_F EI = -113.5\,\text{K} \cdot \text{ft}^3$$

The largest value occurs at point A, so that is the critical point. We must choose a beam that limits the deflection at A to 0.05 in or less.

$$y_A EI = -307\,\text{K} \cdot \text{ft}^3$$

Let $y_A = -0.05$ in. Then the required I is

$$I = \frac{-307\,\text{K} \cdot \text{ft}^3}{Ey_A} \times \frac{1000\,\text{lb}}{\text{K}} \times \frac{(12\,\text{in})^3}{\text{ft}^3}$$
$$I = \frac{(-307)(1000)(1728)\,\text{lb} \cdot \text{in}^3}{(30 \times 10^6\,\text{lb/in}^2)(-0.05\,\text{in})} = 354\,\text{in}^4$$

Consult the table of wide-flange beams and select a suitable beam.

A W18×40 beam is the best choice from Appendix A–7 since it is the lightest beam that has a large enough value for I. For this beam, $I = 612\,\text{in}^4$, and the section modulus is $S = 68.4\,\text{in}^3$. The actual deflection will be somewhat less than the limiting value of 0.050 in because the actual moment of inertia is much greater than the required value. Because the deflection is directly proportional to the moment of inertia, we can find the actual deflection from

$$y_{max} = (0.050\,\text{in})(354\,\text{in}^4)/(612\,\text{in}^4) = 0.029\,\text{in}$$

Now compute the maximum bending stress and the web shear stress in the beam and check to see if they are safe.

In Figure 14–28, we find the maximum bending moment to be 60 K·ft. Then

$$\sigma = \frac{M}{S} = \frac{60\,\text{K} \cdot \text{ft}}{68.4\,\text{in}^3} \times \frac{1000\,\text{lb}}{\text{K}} \times \frac{12\,\text{in}}{\text{ft}} = 10\,526\,\text{psi}$$

We can specify that the beam be made from ASTM A992 structural steel having a yield strength of 50 000 psi. Using the AISC specifications for static loads, the allowable stress is

$$\sigma_d = 0.66\,s_y = (0.66)(50\,000\,\text{psi}) = 33\,000\,\text{psi}$$

This is much greater than the computed maximum stress in the beam. Therefore, the beam is safe for bending.

We can also check the shearing stress in the web. The maximum shearing force is 23 K or 23 000 lb at either points B or D. We use the web shear formula, $\tau = V/th$. From Appendix A–7 we find $t = 0.315$ in and $h = 17.90$ in. Then

$$\tau = V/th = (23\,000\,\text{lb})/(0.315\,\text{in})(17.90\,\text{in}) = 4079\,\text{psi}$$

The allowable web shearing stress is

$$\tau_d = 0.40\,s_y = (0.40)(50\,000\,\text{psi}) = 20\,000\,\text{psi}$$

Therefore, the beam is safe for web shear. We will also have to check the AISC specifications to determine the need for lateral bracing.

Comment: We have designed a beam to carry the load shown in Figure 14–27 that is safe for bending and shear and that will have a deflection no greater than 0.050 in. The beam is a W18×40 steel-rolled shape made from ASTM A992 structural steel. The maximum deflection occurs at point A at the left end of the overhang and has a value of 0.029 in. We also created complete equations for the shearing force, bending moment, slope, and deflection at all points on the beam.

14–8 MOMENT-AREA METHOD

The semigraphical procedure for finding beam deflections, called the *moment-area method*, is useful for problems in which a fairly complex loading pattern occurs or when the beam has a varying cross section along its length. Such cases are difficult to handle with the other methods presented in this chapter.

Shafts for mechanical drives are examples where the cross section varies throughout the length of the beam. Figure 14–31 shows a shaft designed to carry two gears where the changes in diameter provide shoulders against which to seat the gears and bearings to provide axial location. Note, also, that the bending moment decreases toward the ends of the shaft, allowing smaller sections to be safe with regard to bending stress.

In structural applications of beams, varying cross sections are often used to make more economical members. Larger sections having higher moments of inertia are used at sections where the bending moment is high, while decreased section sizes are used in places where the bending moment is lower. Figure 14–32 shows an example.

The moment-area method uses the quantity M/EI, the bending moment divided by the stiffness of the beam, to determine the deflection of the beam at selected points. Then it is convenient to prepare such a diagram as a part of the beam analysis procedure. If the beam has the same cross section over its entire length, the M/EI diagram looks similar to the familiar bending moment diagram except that its values have been divided by the quantity EI. However, if the moment of inertia of the cross section varies along the length of the beam, the shape of the M/EI diagram will be different. This is shown in Figure 14–33.

Recall Equation (14–10) from Section 14–7,

$$\frac{M}{EI} = \frac{d^2y}{dx^2} \qquad (14\text{–}10)$$

This formula relates the deflection of the beam, y, as a function of position, x, the bending moment, M, and the beam stiffness, EI. The right side of Equation (14–10) can be rewritten as

$$\frac{d^2y}{dx^2} = \frac{d}{dx}\left(\frac{dy}{dx}\right)$$

But note that dy/dx is defined as the slope of the deflection curve, θ; that is, $dy/dx = \theta$. Then

$$\frac{d^2y}{dx^2} = \frac{d\theta}{dx}$$

Equation (14–10) can then be written

$$\frac{M}{EI} = \frac{d\theta}{dx}$$

Solving for $d\theta$ gives

$$d\theta = \frac{M}{EI}\,dx \qquad (14\text{–}15)$$

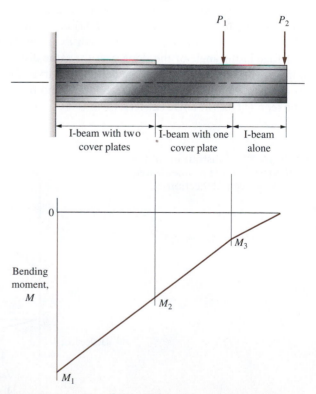

FIGURE 14–32 Cantilever with varying cross sections.

FIGURE 14–31 Shaft with varying cross sections.

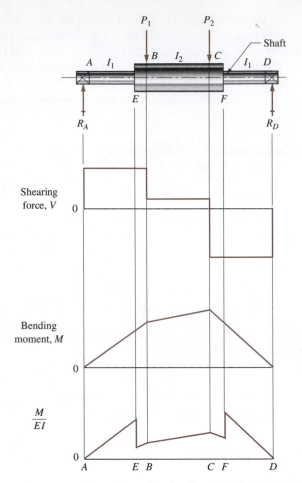

FIGURE 14–33 Illustration of *M/EI* diagrams for a beam with varying cross sections.

FIGURE 14–34 Principles of the moment-area method for beam deflections.

The interpretation of Equation (14–15) can be seen in Figure 14–34 in which the right side, $(M/EI)\ dx$, is the area under the *M/EI* diagram over a small length *dx*. Then $d\theta$ is the change in the angle of the slope over the same distance *dx*. If tangent lines are drawn to the deflection curve of the beam at the two points marking the beginning and the end of the segment *dx*, the angle between them is $d\theta$.

The change in angle $d\theta$ causes a change in the vertical position of a point at some distance *x* from the small element *dx*, as shown in Figure 14–34. Calling the change in vertical position *dt*, it can be found from

$$dt = x\ d\theta = \frac{M}{EI}x\ dx \qquad (14\text{–}16)$$

To determine the effect of the change in angle over a larger segment of the beam, Equations (14–15) and (14–16) must be integrated over the length of the segment. For example, over the segment A–B shown in Figure 14–35 from Equation (14–15),

$$\int_A^B d\theta = \theta_B - \theta_A = \int_A^B \frac{M}{EI}dx \qquad (14\text{–}17)$$

FIGURE 14–35 Illustrations of the two theorems of the moment-area method for beam deflections.

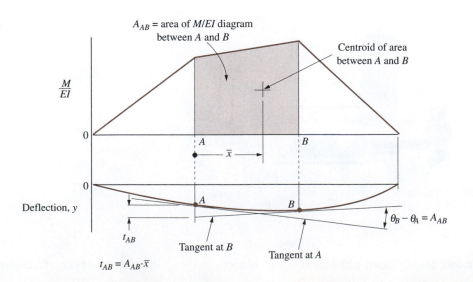

The last part of this equation is the area under the M/EI curve between A and B. This is equal to the change in the angle of the tangents at A and B, $\theta_B - \theta_A$.

From Equation (14–16),

$$\int_A^B dt = t_{AB} = \int_A^B \frac{M}{EI}x\,dx \qquad (14\text{–}18)$$

Here the term t_{AB} represents the tangential deviation of the point A from the tangent to point B, as shown in Figure 14–35. Also, the right side of Equation (14–18) is the *moment of the area of the M/EI diagram between points A and B*. In practice, the moment of the area is computed by multiplying the area under the M/EI curve by the distance to the centroid of the area, also shown in Figure 14–35.

Equations (14–17) and (14–18) form the basis for the two *theorems of the moment-area method for finding beam deflections*. They are the following:

THEOREM 1 ▼

The change in angle, in radians, between tangents drawn at two points A and B on the deflection curve for a beam, is equal to the area under the M/EI diagram between A and B.

THEOREM 2 ▼

The vertical deviation of point A on the deflection curve for a beam from the tangent through another point B on the curve is equal to the moment of the area under the M/EI curve with respect to point A.

Applications of the Moment-Area Method

In this section we show several examples of the use of the moment-area method for finding the deflection of beams. Procedures are developed for each class of beam according to the manner of loading and support. Considered are the following:

1. Cantilevers with a variety of loads
2. Symmetrically loaded simply supported beams
3. Beams with varying cross sections
4. Unsymmetrically loaded simply supported beams

Cantilevers

The definition of a cantilever includes the requirement that it is rigidly fixed to a support structure so that no rotation of the beam can occur at the support. Therefore, the tangent to the deflection curve at the support is always in line with the original position of the neutral axis of the beam in the unloaded state. If the beam is horizontal, as we usually picture it, this tangent is also horizontal.

The following procedure for finding the deflection of any point on a cantilever uses the two theorems developed in Section 14–10 along with the observation that the tangent to the deflection curve at the support is horizontal.

PROCEDURE FOR FINDING THE DEFLECTION OF A CANTILEVER— MOMENT-AREA METHOD ▼

1. Draw the load, shearing force, and bending moment diagrams.

2. Divide the bending moment values by the beam stiffness, EI, and draw the M/EI diagram. The dimension for the quantity M/EI is $(\text{length})^{-1}$; for example, $\text{m}^{-1}, \text{ft}^{-1}$, or in^{-1}.

3. Compute the area of the M/EI diagram and locate its centroid. If the diagram is not a simple shape, break it into parts and find the area and centroid for each part separately. If the deflection at the end of the cantilever is desired, the area of the entire M/EI diagram is used. If the deflection of some other point is desired, only the area between the support and the point of interest is used.

4. Use Theorem 2 to compute the vertical deviation of the point of interest from the tangent to the neutral axis of the beam at the support. Because the tangent is coincident with the original position of the neutral axis, the deviation thus found is the actual deflection of the beam at the point of interest. If all loads are in the same direction, the maximum deflection occurs at the end of the cantilever.

Example Problem 14–11   Use the moment-area method to determine the deflection at the end of the steel cantilever shown in Figure 14–36.

Solution   **Objective:** Compute the deflection at the end of the cantilever.

Given: Beam and loading in Figure 14–36.

Analysis: Use the *procedure for finding the deflection of a cantilever—moment-area method*.

Results: **Step 1:** Figure 14–36 shows the load, shearing force, and bending moment diagrams.

FIGURE 14–36 Load, shearing force, and bending moment diagrams for Example Problems 14–11 and 14–12.

Step 2: The stiffness is computed here:

$$E = 207\,\text{GPa} = 207 \times 10^9\,\text{N/m}^2$$

$$I = \frac{th^3}{12} = \frac{(0.02\,\text{m})(0.06\,\text{m})^3}{12} = 3.60 \times 10^{-7}\,\text{m}^4$$

$$EI = (207 \times 10^9\,\text{N/m}^2)(3.60 \times 10^{-7}\,\text{m}^4) = 7.45 \times 10^4\,\text{N} \cdot \text{m}^2$$

The M/EI diagram is drawn in Figure 14–37. Note that the only changes from the bending moment diagram are the units and the values because the beam has a constant stiffness along its length.

Step 3: The desired area is that of the entire triangular shape of the M/EI diagram, called A_{BA} here to indicate that it is used to compute the deflection of point B relative to A.

$$A_{BA} = (0.5)(24.15 \times 10^{-3}\,\text{m}^{-1})(1.20\,\text{m}) = 14.5 \times 10^{-3}\,\text{rad}$$

The centroid of this area is two-thirds of the distance from B to A, 0.80 m.

Step 4: To implement Theorem 2, we need to compute the moment of the area found in step 3. This is equal to t_{BA}, the vertical deviation of point B from the tangent drawn to the deflection curve at point A.

$$t_{BA} = A_{BA} \times \bar{x} = (14.5 \times 10^{-3}\,\text{rad})(0.80\,\text{m})$$

$$t_{BA} = y_B = 11.6 \times 10^{-3}\,\text{m} = 11.6\,\text{mm}$$

FIGURE 14–37 M/EI curve and deflection curve for Example Problem 14–11.

Because the tangent to point A is horizontal, t_{BA} is equal to the deflection of the beam at its end, point B.

Comment: This result is identical to that which would be found by applying the formula for Case a in Appendix A–24. The value of the moment-area method is much more evident when multiple loads are involved or if the cantilever has a varying cross section along its length.

Example Problem 14–12 For the same beam used in Example Problem 14–11 and shown in Figure 14–36, compute the deflection at a point 1.0 m from the support.

Solution Objective: Compute the deflection 1.0 m from the left end of the cantilever.

Given: Beam and loading in Figure 14–36.

Analysis: Use the *procedure for finding the deflection of a cantilever—moment-area method*. Steps 1 and 2 from Example Problem 14–11 are identical, resulting in the load, shearing force, bending moment, and M/EI diagrams shown in Figures 14–36 and 14–37. The solution procedure is continued at step 3.

Results: **Step 3:** This step changes because only the area of the M/EI diagram between the support and the point 1.0 m out on the beam is used, as shown in Figure 14–38. Calling the point of interest, point C, we need to compute t_{CA}, that is, the vertical deviation of point C relative to the tangent drawn to point A at the support. The required area is a trapezoid, and it is convenient to break it into a triangle and a rectangle and treat them as simple shapes. The calculation then takes the form

$$t_{CA} = A_{CA} \times \bar{x} = A_{CA1} \times x_1 + A_{CA2} \times x_2$$

where the areas and x-distances are shown in Figure 14–38. Note that the distances are measured *from the point C to the centroid of the component area.* Then

$$A_{CA1} \times x_1 = (0.004\,025\,\text{m}^{-1})(1.0\,\text{m})(0.50\,\text{m}) = 2.01 \times 10^{-3}\,\text{m}$$
$$A_{CA2} \times x_2 = (0.5)(0.020\,125\,\text{m}^{-1})(1.0\,\text{m})(0.667\,\text{m}) = 6.71 \times 10^{-3}\,\text{m}$$
$$t_{CA} = y_C = (2.01 + 6.71)(10^{-3})\,\text{m} = 8.72\,\text{mm}$$

As before, because the tangent to point A is horizontal, the vertical deviation, t_{CA}, is the true deflection of point C.

FIGURE 14–38 M/EI curve and deflection curve for Example Problem 14–12.

FIGURE 14–39   Load, shearing force, and bending moment diagrams for Example Problems 14–13 and 14–14.

Symmetrically Loaded Simply Supported Beams

This class of problems has the advantage that it is known that the maximum deflection occurs at the middle of the span of the beam. An example is shown in Figure 14–39, in which the beam carries two identical loads equally spaced from the supports. Of course, any loading for which the point of maximum deflection can be predicted can be solved by the procedure illustrated here.

PROCEDURE FOR FINDING THE DEFLECTION OF A SYMMETRICALLY LOADED SIMPLY SUPPORTED BEAM— MOMENT-AREA METHOD ▼

---

1. Draw the load, shearing force, and bending moment diagrams.

2. Divide the bending moment values by the beam stiffness, EI, and draw the M/EI diagram.

3. If the maximum deflection, at the middle of the span, is desired, use that part of the M/EI diagram between the middle and one of the supports; that is, half of the diagram.

4. Use Theorem 2 to compute the vertical deviation of the point at one of the supports from the tangent to the neutral axis of the beam at its middle. Because the tangent is horizontal and because the deflection at the support is actually zero, the deviation thus found is the actual deflection of the beam at its middle.

5. To determine the deflection at some other point on the same beam, use the area of the M/EI diagram between the middle and the point of interest. Use Theorem 2 to compute the vertical deviation of the point of interest from the point of maximum deflection at the middle of the beam. Then subtract this deviation from the maximum deflection found in step 4.

Example Problem 14–13   Determine the maximum deflection of the beam shown in Figure 14–39. The beam is an aluminum channel, C6×2.834, made from 6061-T6 aluminum, positioned with the legs downward.

Solution   Objective: Compute the maximum deflection of the beam.

Given: Beam and loading in Figure 14–39.

Analysis: Use the *procedure for finding the deflection of a symmetrically loaded simply supported beam— moment-area method*, steps 1–4. Because the loading pattern is symmetrical, the maximum deflection will occur at the middle of the beam.

Results: **Step 1:** The load, shearing force, and bending moment diagrams are shown in Figure 14–39, prepared in the traditional manner. The maximum bending moment is 7200 lb·in between B and C.

FIGURE 14–40 M/EI diagram and deflection curve for Example Problem 14–13.

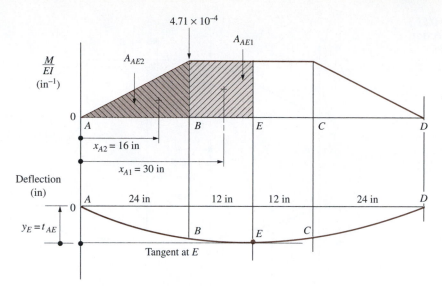

Step 2: The stiffness of the beam, EI, is found using data from the Appendixes. From Appendix A–18, E for 6061-T6 aluminum is 10×10^6 psi. From Appendix A–10, the moment of inertia of the channel, with respect to the Y-Y axis, is 1.53 in^4. Then

$$EI = (10 \times 10^6 \text{ lb/in}^2)(1.53 \text{ in}^4) = 1.53 \times 10^7 \text{ lb} \cdot \text{in}^2$$

Because the stiffness of the beam is uniform over its entire length, the M/EI diagram has the same shape as the bending moment diagram, but the values are different, as shown in Figure 14–40. The maximum value of M/EI is 4.71×10^{-4} in^{-1}.

Step 3: To determine the deflection at the middle of the beam, one-half of the M/EI diagram is used. For convenience, this is separated into a rectangle and a triangle, and the centroid of each part is shown.

Step 4: We need to find t_{AE}, the vertical deviation of point A from the tangent drawn to the deflection curve at point E, the middle of the beam. By Theorem 2,

$$t_{AE} = A_{AE1} \times x_{A1} + A_{AE2} \times x_{A2}$$

The symbols x_{A1} and x_{A2} indicate that the distances to the centroids of the areas must be measured *from point A*.

$$A_{AE1} \times x_{A1} = (4.71 \times 10^{-4} \text{ in}^{-1})(12 \text{ in})(30 \text{ in}) = 0.170 \text{ in}$$
$$A_{AE2} \times x_{A2} = (0.5)(4.71 \times 10^{-4} \text{ in}^{-1})(24 \text{ in})(16 \text{ in}) = 0.090 \text{ in}$$
$$t_{AE} = y_E = 0.170 + 0.090 = 0.260 \text{ in}$$

This is the vertical deviation of point A from the tangent to point E. Because the tangent is horizontal and because the actual deflection of point A is zero, this represents the true deflection of point E relative to the original position of the neutral axis of the beam.

Comment: This result is identical to that which would be found by applying the formula for Case c in Appendix A–23.

Example Problem 14–14

For the same beam used for Example Problem 14–13, shown in Figure 14–39, determine the deflection at point B under one of the loads.

Solution Objective: Compute the deflection at point B under one of the loads.

Given: Beam and loading in Figure 14–39.

Analysis: Use the *procedure for finding the deflection of a symmetrically loaded simply supported beam—moment-area method*, steps 1–5. Steps 1–4 from Example Problem 14–13 are identical, resulting in the load, shearing force, bending moment, M/EI, and deflection diagrams shown in Figures 14–39 and 14–40. The solution procedure is continued at step 5.

FIGURE 14–41 *M/EI* diagram and deflection curve for Example Problem 14–14.

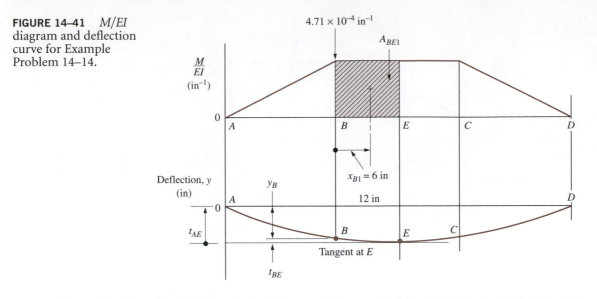

Results: **Step 5:** We can use the moment-area method to determine the vertical deviation, t_{BE}, of point B from the tangent to point E at the middle of the beam. Then subtracting that from the value of t_{AE} found in Example Problem 14–13 gives the true deflection of point B. Shown in Figure 14–41 are the data needed to compute t_{BE}.

$$t_{BE} = A_{BE1} \times x_{B1} = (4.71 \times 10^{-4} \text{ in}^{-1})(12 \text{ in})(6 \text{ in}) = 0.034 \text{ in}$$

Note that the distance x_{B1} must be measured from point B. Then the deflection of point B is

$$y_B = t_{AE} - t_{BE} = 0.260 - 0.034 = 0.226 \text{ in}$$

Beams with Varying Cross Sections

One of the major uses of the moment-area method is to find the deflection of a beam having a varying cross section along its length. Only one additional step is required as compared with beams having a uniform cross section, like those considered thus far.

An example of such a beam is illustrated in Figure 14–42. Note that it is a modification of the beam used in Example

FIGURE 14–42   Beam for Example Problem 14–15.

FIGURE 14–43 Beam diagrams for Problem 14–15.

Problems 14–13 and 14–14 and shown in Figure 14–39. Here we have added a rectangular plate, 0.25 in by 6.0 in, to the underside of the original channel over the middle 48 in of the length of the beam. The box shape would provide a significant increase in stiffness, thereby reducing the deflection of the beam. The stress in the beam would also be reduced.

The change in the procedure for analyzing the deflection of the beam is in the preparation of the M/EI diagram.

Figure 14–43 shows the load, shearing force, and bending moment diagrams as before. In the first and last 12 in of the M/EI diagram, the stiffness of the plain channel is used as before. For each segment over the middle 48 in, the stiffness of the box shape must be used. The M/EI diagram then includes the effect of the change in stiffness along the length of the beam. Example Problem 14–15 demonstrates this process.

**Example Problem
14–15**

Determine the deflection at the middle of the reinforced beam shown in Figure 14–42.

Solution Objective: Compute the deflection at the middle of the beam.

Given: Beam and loading in Figure 14–42. The beam shape is modified from that shown in Figure 14–39. The beam is a C6×2.834 aluminum channel with the legs down. The middle 48 in of the length of the beam has a 0.25-in plate welded to the bottom of the legs of the channel, forming a closed box section.

Analysis: Parts of the solution from Example Problem 14–13 are used. The M/EI diagram is adjusted to include the effect of the modified cross section of the beam.

Results: **Step 1:** The load, shearing force, and bending moment diagrams are prepared in the traditional manner, as shown in Figure 14–43.

Step 2: To prepare the M/EI diagram, two values for the stiffness, EI, are needed. The plain channel has the same value used in previous problems, 1.53×10^7 lb·in². For the box shape,

$$EI = (10 \times 10^6 \text{ lb/in}^2)(4.65 \text{ in}^4) = 4.65 \times 10^7 \text{ lb·in}^2$$

Then, at the point on the beam just before 12 in from A, where the bending moment is 3600 lb·in

$$\frac{M}{EI} = \frac{3600 \text{ lb·in}}{1.53 \times 10^7 \text{ lb·in}^2} = 2.35 \times 10^{-4} \text{ in}^{-1}$$

Just beyond 12 in from A,

$$\frac{M}{EI} = \frac{3600 \text{ lb·in}}{4.65 \times 10^7 \text{ lb·in}^2} = 7.74 \times 10^{-5} \text{ in}^{-1}$$

At point B, where $M = 7200$ lb·in and $EI = 4.65 \times 10^7$ lb·in²,

$$\frac{M}{EI} = \frac{7200 \text{ lb·in}}{4.65 \times 10^7 \text{ lb·in}^2} = 1.55 \times 10^{-4} \text{ in}^{-1}$$

These values establish the critical points on the M/EI diagram.

Step 3: The moment area for the left half of the M/EI diagram will be used to determine the value of t_{AE}, as done in Example Problem 14–13. For convenience, the total area is divided into four parts, as shown in Figure 14–43, with the locations of the centroids indicated relative to point A. The distances are

$$x_1 = (\tfrac{2}{3})(12 \text{ in}) = 8 \text{ in}$$
$$x_2 = (\tfrac{1}{2})(12 \text{ in}) + 12 \text{ in} = 18 \text{ in}$$
$$x_3 = (\tfrac{2}{3})(12 \text{ in}) + 12 \text{ in} = 20 \text{ in}$$
$$x_4 = (\tfrac{1}{2})(12 \text{ in}) + 24 \text{ in} = 30 \text{ in}$$

Step 4: We can now use Theorem 2 to compute the value of t_{AE}, the deviation of point A from the tangent to point E, by computing the moment of each of the four areas shown crosshatched in the M/EI diagram of Figure 14–43.

$$t_{AE} = A_1 x_1 + A_2 x_2 + A_3 x_3 + A_4 x_4$$
$$A_1 x_1 = (0.5)(2.35 \times 10^{-4} \text{ in}^{-1})(12 \text{ in})(8 \text{ in}) = 1.128 \times 10^{-2} \text{ in}$$
$$A_2 x_2 = (7.74 \times 10^{-5} \text{ in}^{-1})(12 \text{ in})(18 \text{ in}) = 1.672 \times 10^{-2} \text{ in}$$
$$A_3 x_3 = (0.5)(7.74 \times 10^{-5} \text{ in}^{-1})(12 \text{ in})(20 \text{ in}) = 9.293 \times 10^{-3} \text{ in}$$
$$A_4 x_4 = (1.55 \times 10^{-4} \text{ in}^{-1})(12 \text{ in})(30 \text{ in}) = 5.580 \times 10^{-2} \text{ in}$$

Then

$$t_{AE} = y_E = \sum(A_i x_i) = 9.309 \times 10^{-2} \text{ in} = 0.093 \text{ in}$$

Comment: As before, this value is equal to the deflection of point E at the middle of the beam. Comparing it to the deflection of 0.260 in found in Example Problem 14–14, the addition of the cover plate reduced the maximum deflection by approximately 64%.

Unsymmetrically Loaded Simply Supported Beams

The major difference between this type of beam and the ones considered earlier is that the point of maximum deflection is unknown. Special care is needed to describe the geometry of the M/EI diagram and the deflection curve for the beam.

The general procedure for finding the deflection at any point on the deflection curve for an unsymmetrically loaded simply supported beam is outlined in the following steps. Because of the myriad of different loading patterns, the specifics of applying this procedure may have to be adjusted for any given problem. You are advised to check the fundamental principles of the moment-area method as the solution of a problem is completed. The method will be demonstrated in Example Problem 14–16.

Example Problem 14–16

Determine the deflection of the beam shown in Figure 14–44 at its middle, 1.0 m from the supports. The beam is an American Standard steel beam, S3×5.7.

Solution

Objective: Compute the deflection at the middle of the beam.

Given: Beam and loading shown in Figure 14–44. Beam is steel.
Beam shape is an American Standard S3×5.7.

Analysis: Use the *procedure for finding the deflection of an unsymmetrically loaded simply supported beam—moment-area method*.

Results: **Step 1:** The load, shearing force, and bending moment diagrams are shown in Figure 14–44.

Step 2: The beam has a uniform stiffness along its entire length resulting in the shape of the *M/EI* diagram being the same as the bending moment diagram. The value of *M/EI* at point *B* can be computed by dividing the bending moment there (2.40 kN·m or 2400 N·m) by *EI*. We will use

FIGURE 14–44 Beam for Example Problem 14–16.

$E = 207$ GPa for steel. From Appendix A–8(SI) we find $I = 1.04 \times 10^6$ mm⁴. Converting this to the unit of m⁴ gives

$$I = 1.04 \times 10^6 \text{ mm}^4 \cdot \frac{\text{m}^4}{(10^3 \text{ mm})^4} = 1.04 \times 10^{-6} \text{ m}^4$$

Then the beam stiffness is

$$EI = (207 \times 10^9 \text{ N/m}^2)(1.04 \times 10^{-6} \text{ m}^4) = 2.153 \times 10^5 \text{ N} \cdot \text{m}^2$$

The value of M/EI at point B on the beam can now be computed.

$$\left(\frac{M}{EI}\right)_B = \frac{2400 \text{ N} \cdot \text{m}}{2.153 \times 10^5 \text{ N} \cdot \text{m}^2} = 0.01115 \text{ m}^{-1}$$

The M/EI diagram is drawn in Figure 14–44. It is desired to compute the deflection of the beam at its middle, labeled point D.

Step 3: Figure 14–44 shows an exaggerated sketch of the deflection curve for the beam. It is likely that the maximum deflection will occur very close to the middle of the beam where we are to determine the deflection, point D. Figure 14–45 shows the tangent to the deflection curve at point A at the left support and the vertical deviation of point C from this line. Note that point C is a known point on the deflection curve because the deflection there is zero. Now we can use Theorem 2 to compute t_{CA}. The entire M/EI diagram is used, broken into two triangles.

$$t_{CA} = A_{CA1}x_{C1} + A_{CA2}x_{C2}$$
$$A_{CA1}x_{C1} = (0.5)(0.01115 \text{ m}^{-1})(0.8 \text{ m})(0.533 \text{ m}) = 0.002379 \text{ m}$$
$$A_{CA2}x_{C2} = (0.5)(0.01115 \text{ m}^{-1})(1.2 \text{ m})(1.2 \text{ m}) = 0.008028 \text{ m}$$

FIGURE 14–45 Moment-area diagrams for Example Problem 14–16.

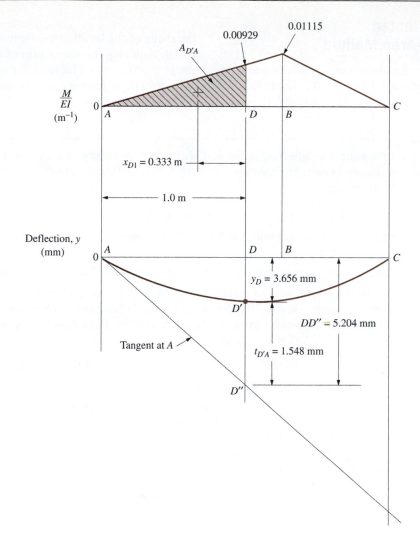

FIGURE 14–46   Moment-area diagrams for Example Problem 14–16.

Then

$$t_{CA} = 0.002379 + 0.008028 = 0.010407 \text{ m} = 10.407 \text{ mm}$$

Step 4: Use the principle of proportion to determine the distance DD'' from D to the tangent line.

$$\frac{t_{CA}}{CA} = \frac{DD''}{AD}$$

or

$$DD'' = t_{CA} \times \frac{AD}{CA} = (10.407 \text{ mm}) \times \frac{1.0 \text{ m}}{2.0 \text{ m}} = 5.204 \text{ mm}$$

Step 5: Compute the deviation, $t_{D'A}$, of point D' from the tangent line drawn to point A using Theorem 2. The part of the M/EI diagram between D and A is used as shown in Figure 14–46.

$$t_{D'A} = A_{D'A}x_{D1} = (0.5)(0.00929 \text{ m}^{-1})(1.0 \text{ m})(0.333 \text{ m}) = 0.001548 \text{ m}$$
$$t_{D'A} = 1.548 \text{ mm}$$

Step 6: From the geometry of the deflection diagram shown in Figure 14–46 the deflection at point D, y_D, is

$$y_D = DD' = DD'' - t_{D'A} = 5.204 - 1.548 = 3.656 \text{ mm}$$

Beams with Distributed Loads—Moment-Area Method

The general procedure for determining the deflection of beams carrying distributed loads is the same as that shown for beams carrying concentrated loads. However, the shape of the bending moment and M/EI curves is different, requiring the use of other formulas for computing the area and centroid location for use in the moment-area method. The following example shows the kind of differences to be expected.

Example Problem 14–17

Determine the deflection at the end of the cantilever carrying a uniformly distributed load shown in Figure 14–47. The beam is a $6 \times 2 \times \frac{1}{4}$ hollow rectangular steel tube with the 6.0-in dimension horizontal.

Solution Objective: Compute the deflection at the end of the cantilever.

Given: Beam and loading in Figure 14–47. Beam is a steel rectangular tube, $6 \times 2 \times \frac{1}{4}$, with 6.0-in dimension horizontal.

Analysis: The same basic procedure outlined for Example Problem 14–11 can be used.

Results: The solution starts with the preparation of the load, shearing force, and bending moment diagrams, shown in Figure 14–47. Then the M/EI curve will have the same shape as the bending moment curve because the stiffness of the beam is uniform. From Appendix A–9 we find $I = 2.21$ in^4. Then, using $E = 30 \times 10^6$ psi for steel,

$$EI = (30 \times 10^6 \text{ lb/in}^2)(2.21 \text{ in}^4) = 6.63 \times 10^7 \text{ lb} \cdot \text{in}^2$$

Now Figure 14–48 shows the M/EI curve along with the deflection curve for the beam. The horizontal line on the deflection diagram is the tangent drawn to the beam shape at point A, where the beam is fixed to the support. Then, at the right end of the beam, the deviation of the beam deflection curve from this tangent, t_{BA}, is equal to the deflection of the beam itself.

Using Theorem 2, the deviation t_{BA} is equal to the product of the area of the M/EI curve between B and A times the distance from point B to the centroid of the area. That is,

$$t_{BA} = A_{BA} \cdot x_B$$

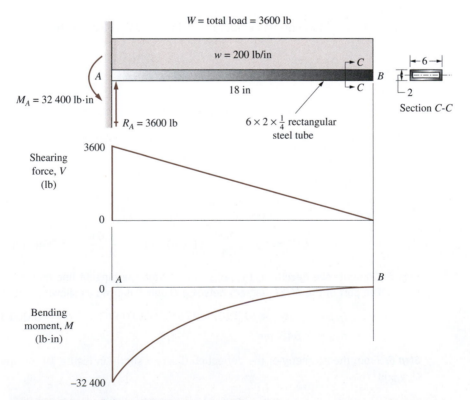

FIGURE 14–47 Load, shearing force, and bending moment diagrams for Example Problem 14–17.

FIGURE 14–48 M/EI and deflection curves for Example Problem 14–17.

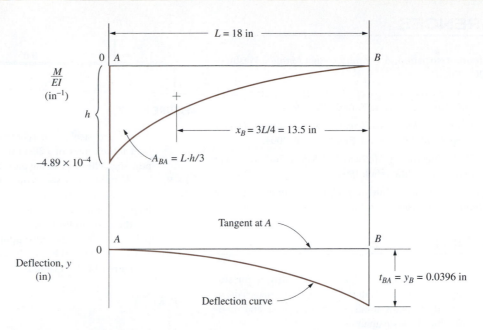

Recalling that the load, shearing force, and bending moment diagrams are related to each other in such a way that the higher curve is the derivative of the curve below it, the following conclusions can be drawn:

1. The shearing force curve is a first-degree curve (straight line with constant slope). Its equation is of the form

$$V = m \cdot x + b$$

where m is the slope of the line and b is its intercept with the vertical axis. The variable x is the position on the beam.

2. The bending moment curve is a second-degree curve, a parabola. The general equation of the curve is of the form

$$M = a \cdot x^2 + b$$

Appendix A–1 shows the relationships for computing the area and the location of the centroid for areas bounded by second-degree curves. For an area having the shape of the bending moment or M/EI curves,

$$\text{Area} = \frac{L \cdot h}{3}$$

$$x = \frac{L}{4}$$

where L = length of the base of the area
h = height of the area
x = distance from the side of the area to the centroid

Note that the corresponding distance from the vertex of the curve to the centroid is

$$x' = \frac{3L}{4}$$

Now, for the data shown in Figure 14–42,

$$A_{BA} = \frac{L \cdot h}{3} = \frac{(18 \text{ in})(4.89 \times 10^{-4} \text{ in}^{-1})}{3} = 2.932 \times 10^{-3}$$

$$x_B = \frac{3L}{4} = \frac{3(18 \text{ in})}{4} = 13.5 \text{ in}$$

Theorem 2 can now be used:

$$t_{BA} = A_{BA}x_B = (2.932 \times 10^{-3})(13.5 \text{ in}) = 0.0396 \text{ in}$$

This is equal to the deflection of the beam at its end, y_B.

REFERENCES

1. Aluminum Association, *Aluminum Design Manual*, Washington, DC, 2005.

2. Blodgett, Omer W., *Design of Weldments*, Cleveland, OH, James F. Lincoln Arc Welding Foundation, 1963.

3. Mott, Robert L., *Machine Elements in Mechanical Design*, 4th ed., Upper Saddle River, NJ, Prentice Hall, 2004.

4. Popov, E. P., *Engineering Mechanics of Solids*, 2nd ed. Upper Saddle River, NJ, Prentice Hall, 1999.

5. Young, W. C. and R. G. Budynas, *Roark's Formulas for Stress and Strain*, 7th ed., New York, McGraw-Hill, 2002.

INTERNET SITES

1. MDSolids www.mdsolids.com Educational software for students taking mechanics of materials, strength of materials, or mechanics of deformable solids; comprised of several modules, one of which is beam analysis.

2. Orand Systems www.orandsystems.com Producer of the Beam2D software for beam analysis.

3. Creative Engineering www.suverkrop.com/winbuild.html Structural analysis software with the facility to select the least weight beam for a given loading.

4. MITCalc www.mitcalc.com Calculation software for mechanical, industrial, and technical analysis, including beam analysis and many other technical topics.

5. Engineers Edge www.engineersedge.com/CalculatorsOnline.shtml A variety of online calculators, including *Beam Deflections and Stress*, *Section Properties of Selected Shapes*, *Pressure Vessels*, and other mechanical design phenomena.

6. Free Structural Software www.structural-engineering.fsnet.co.uk/free.htm Listings of many software packages, many of which are free and others that have a free trial period.

PROBLEMS

Formula Method—Statically Determinate Beams

14–1. A round shaft having a diameter of 32 mm is 700 mm long and carries a 3.0-kN load at its center. If the shaft is steel and simply supported at its ends, compute the deflection at the center.

14–2. For the shaft in Problem 14–1, compute the deflection if the shaft is 6061-T6 aluminum instead of steel.

14–3. For the shaft in Problem 14–1, compute the deflection if the ends are fixed against rotation instead of simply supported.

14–4. For the shaft in Problem 14–1, compute the deflection if the shaft is 350 mm long rather than 700 mm.

14–5. For the shaft in Problem 14–1, compute the deflection if the diameter is 25 mm instead of 32 mm.

14–6. For the shaft in Problem 14–1, compute the deflection if the load is placed 175 mm from the left support rather than in the center. Compute the deflection both at the load and at the center of the shaft.

14–7. A wide-flange steel beam, W12×16, carries the load shown in Figure P14–7. Compute the deflection at the loads and at the center of the beam.

FIGURE P14–7

14–8. A standard steel $1\frac{1}{2}$-in schedule 40 pipe carries a 650-lb load at the center of a 28-in span, simply supported. Compute the deflection of the pipe at the load.

14–9. An Aluminum Association Standard I-beam, I8×6.181, carries a uniformly distributed load of 1125 lb/ft over a 10-ft span. Compute the deflection at the center of the span.

14–10. For the beam in Problem 14–9, compute the deflection at a point 3.5 ft from the left support of the beam.

14–11. A wide-flange steel beam, W12×30, carries the load shown in Figure P14–11. Compute the deflection at the load.

FIGURE P14–11

14–12. For the beam in Problem 14–11, compute the deflection at the load if the left support is moved 2.0 ft toward the load.

14–13. For the beam in Problem 14–11, compute the maximum upward deflection and determine its location.

14–14. A 1-in schedule 40 steel pipe is used as a cantilever beam 8 in long to support a load of 120 lb at its end. Compute the deflection of the pipe at the end.

14–15. A round steel bar is to be used to carry a single concentrated load of 3.0 kN at the center of a 700-mm-long span on simple supports. Determine the required diameter of the bar if its deflection must not exceed 0.12 mm.

14–16. For the bar designed in Problem 14–15, compute the stress in the bar and specify a suitable steel to provide a design factor of 8 based on ultimate strength.

14–17. A flat strip of steel 0.100 in wide and 1.200 in long is clamped at one end and loaded at the other like a cantilever beam (as in Case *a* in Appendix A–24). What should be the thickness of the strip if it is to deflect 0.15 in under a load of 0.52 lb?

14–18. A wood joist in a commercial building is 14 ft 6 in long and carries a uniformly distributed load of 50 lb/ft. It is 1.50 in wide and 9.25 in high. If it is made of southern pine, compute the maximum deflection of the joist. Also compute the stress in the joist due to bending and horizontal shear, and compare them with the allowable stresses for No. 2 grade southern pine.

Formula Method—Statically Indeterminate Beams

For Problems 14–19 through 14–45, use the formulas in Appendix A–25 to complete any of the following according to the instructions given for a particular assignment.

a. Determine the reactions and draw the complete shearing force and bending moment diagrams. Report the maximum shearing force and the maximum bending moment and indicate where they occur.

b. Where deflection formulas are available, also compute the maximum deflection of the beam expressed in the form

$$y = C_d/EI$$

where EI is the beam stiffness, the product of the modulus of elasticity for the material of the beam and the moment of inertia of the cross section of the beam. The term C_d will then be the result of the computation of all other variables in the deflection equation for the particular beam support type, span length, and loading pattern.

c. Complete the design of the beam specifying a suitable material, and the shape and size for the cross section. The standard for design must include the specification that bending stresses and shearing stresses are safe for the given material. Unless otherwise specified by the assignment, consider all loads to be static.

d. Complete the design of the beam to limit the maximum deflection to some specified value as given by the assignment. In the absence of a specified limit, use $L/360$ as the maximum allowable deflection where L is the span between supports or the overall length of the beam. The design must specify a suitable material, and the shape and size for the cross section. This assignment may be linked to part b where the deflection was calculated in terms of the beam stiffness, EI. Then, for example, you may specify the material and its value for E, compute the limiting deflection, and solve for the required moment of inertia, I. The shape and size of the cross section can then be determined. Note that any design must also be shown to be safe with regard to bending stresses and shearing stresses as in part c.

14–19. Use A–25(a) with $P = 35$ kN, $L = 4.0$ m.
14–20. Use A–25(b) with $P = 35$ kN, $L = 4.0$ m, $a = 1.50$ m.
14–21. Use A–25(b) with $P = 35$ kN, $L = 4.0$ m, $a = 2.50$ m.
14–22. Use A–25(c) with $w = 400$ lb/ft, $L = 14.0$ ft.
14–23. Use A–25(c) with $w = 50$ lb/in, $L = 16.0$ in.
14–24. Use A–25(d) with $P = 350$ lb, $L = 10.8$ in, $a = 2.50$ in.
14–25. Use A–25(e) with $P = 35$ kN, $L = 4.0$ m.
14–26. Use A–25(f) with $P = 35$ kN, $L = 4.0$ m, $a = 1.50$ m.
14–27. Use A–25(f) with $P = 35$ kN, $L = 4.0$ m, $a = 2.50$ m.
14–28. Use A–25(g) with $w = 400$ lb/ft, $L = 14.0$ ft.
14–29. Use A–25(g) with $w = 50$ lb/in, $L = 16.0$ in.
14–30. Use A–25(h) with $w = 400$ lb/ft, $L = 7.0$ ft.
14–31. Use A–25(h) with $w = 50$ lb/in, $L = 8.0$ in.
14–32. Use A–25(i) with $w = 400$ lb/ft, $L = 56$ in.

14–33. Use A–25(i) with $w = 50$ lb/in, $L = 5.333$ in.
14–34. Use A–25(j) with $w = 400$ lb/ft, $L = 3.5$ ft.
14–35. Use A–25(j) with $w = 50$ lb/in, $L = 4.0$ in.
14–36. Use Figure P14–36.

FIGURE P14–36

14–37. Use Figure P14–37.

FIGURE P14–37

14–38. Use Figure P14–38.

FIGURE P14–38

14–39. Use A–25(d) with $P = 18$ kN, $L = 2.75$ m, $a = 1.40$ m.
14–40. Use A–25(f) with $P = 8500$ lb, $L = 109$ in, $a = 75$ in.
14–41. Use A–25(h) with $w = 4200$ lb/ft, $L = 16.0$ ft.
14–42. Use A–25(i) with $w = 50$ kN/m, $L = 3.60$ m.
14–43. Use A–25(j) with $w = 15$ lb/in, $L = 36$ in.
14–44. Use A–25(e) with $P = 140$ lb, $L = 54$ in.
14–45. Use A–25(b) with $P = 250$ N, $L = 55$ mm, $a = 15$ mm.

Comparisons of Beam Behavior

14–46. Compare the behavior of the four beams shown in Figure P14–46 with regard to shearing force, bending moment, and maximum deflection. In each case, the beam is designed to

FIGURE P14–46 Beams for Problem 14–46. (a) Simply supported beam. (b) Cantilever. (c) Supported cantilever. (d) Fixed-end beam.

support a uniformly distributed load across a given span. Complete the analysis in a manner similar to that used in Example Problems 14–3 through 14–7.

14–47. Compare Problems 14–22, 14–28, 14–30, 14–32, and 14–34 with regard to the maximum values of shearing force, bending moment, and deflection.

14–48. Compare Problems 14–23, 14–29, 14–31, 14–33, and 14–35 with regard to the maximum values of shearing force, bending moment, and deflection.

14–49. Specify a suitable design for a wooden beam, simply supported at its ends, to carry a uniformly distributed load of 120 lb/ft for a span of 24 ft. The beam must be safe for both bending and shear stresses when made from No. 2 grade southern pine. Then compute the maximum deflection for the beam you have designed.

14–50. Repeat Problem 14–49 except place an additional support at the middle of the beam, 12 ft from either end.

14–51. Repeat Problem 14–49 except use four supports 8 ft apart.

14–52. Compare the behavior of the three beams designed in Problems 14–49, 14–50, and 14–51 with regard to shearing force, bending moment, and maximum deflection. Complete the analysis in the manner used in Example Problems 14–3 through 14–7.

Superposition—Statically Determinate Beams

14–53. The load shown in Figure P14–53(a) is being carried by an extruded aluminum (6061-T6) beam having the shape shown in Figure P14–53(b). Compute the deflection of the beam at each load.

(a)   (b)

FIGURE P14–53

14–54. The loads shown in Figure P14–54(a) represent the feet of a motor on a machine frame. The frame member has the cross section shown in Figure P14–54(b), which has a moment of inertia of 16 956 mm^4. Compute the deflection at each load. The aluminum alloy 2014-T4 is used for the frame.

(a)   (b)

FIGURE P14–54

14–55. Compute the deflection at the middle of a steel W460×82 (W18×55) beam when it carries the load shown in Figure P14–55.

FIGURE P14–55

14–56. A 1-in schedule 40 steel pipe carries the two loads shown in Figure P14–56. Compute the deflection of the pipe at each load.

FIGURE P14–56

14–57. A cantilever beam carries two loads as shown in Figure P14–57. If a rectangular steel bar 20 mm wide by 80 mm high is used for the beam, compute the deflection at the end of the beam.

FIGURE P14–57

14–58. For the beam in Problem 14–57, compute the deflection if the bar is aluminum 2014-T4 rather than steel.

14–59. For the beam in Problem 14–57, compute the deflection if the bar is magnesium, ASTM AZ 63A-T6, instead of steel.

14–60. The load shown in Figure P14–60 is carried by a round steel bar having a diameter of 0.800 in. Compute the deflection of the bar at the right end.

FIGURE P14–60

14–61. Specify a standard wide-flange steel beam that would carry the loading shown in Figure P14–55 with a deflection at the middle of less than $\frac{1}{360}$ times the length of the beam.

14–62. It is planned to use an Aluminum Association channel with its legs down to carry the loads shown in Figure P14–62 so that the flat face can contact the load. The maximum allowable deflection is 0.080 in. Specify a suitable channel.

FIGURE P14–62

Superposition—Statically Indeterminate Beams

For Problems 14–63 through 14–70, use the superposition method to determine the reactions at all supports and draw the complete shearing force and bending moment diagrams. Indicate the maximum shearing force and bending moment for each beam.

14–63. Use Figure P14–36.
14–64. Use Figure P14–37.
14–65. Use Figure P14–38.
14–66. Use Figure P14–66.

FIGURE P14–66

14–67. Use Figure P14–67.

FIGURE P14–67

14–68. Use Figure P14–68.

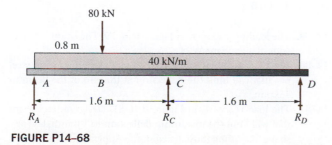

FIGURE P14–68

14–69. Use Figure P14–69.

FIGURE P14–69

14–70. Use Figure P14–70.

FIGURE P14–70

Design Problems—Stress and Deflection Limits

14–71. A plank from a wooden deck carries a load like that shown in Case j in Appendix A–25 with $w = 100$ lb/ft and $L = 24$ in. The plank is a standard 2×6 having a thickness of 1.50 in and a width of 5.50 in with the long dimension horizontal. Would the plank be safe if it is made from No. 2 grade southern pine?

14–72. Two designs for a diving board are proposed, as shown in Figure P14–72. Compare the designs with regard to shearing force, bending moment, and deflection. Express the deflection in terms of the beam stiffness, EI. Note that cases from Appendixes A–23 and A–25 can be used.

14–73. For each of the proposed diving board designs described in Problem 14–72 and shown in Figure P14–72, complete the design, specifying the material, the cross section, and the final dimensions. The board is to be 600 mm wide.

14–74. Figure P14–74 shows a roof beam over a loading dock of a factory building. Compute the reactions at the supports and draw the complete shearing force and bending moment diagrams.

14–75. For the roof beam described in Problem 14–74, complete its design, specifying the material, the cross section, and the final dimensions.

Successive Integration Method

For the following problems, 14–76 through 14–84, use the general approach outlined in Section 14–7 to determine the equations for the shape of the deflection curves of the beams. Unless otherwise noted, compute the maximum deflection of the beam from the equations and tell where it occurs.

14–76. The loading is shown in Figure P14–76. The beam is a rectangular steel bar, 1.0 in wide and 2.0 in high.

FIGURE P14–72   Proposals for diving boards for Problems 14–72 and 14–73. (a) Simple supports with overhang. (b) Supported cantilever with overhang.

FIGURE P14–74   Roof beam for a loading dock for Problems 14–74 and 14–75.

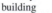

FIGURE P14–76

14–77. The loading is shown in Figure P14–77. The beam is a steel wide-flange shape W18×55.

FIGURE P14–77

14–78. The loading is shown in Figure P14–78. The beam is a $2\frac{1}{2}$-in schedule 40 steel pipe.

FIGURE P14–78

14–79. The loading is shown in Figure P14–79. The beam is a steel wide-flange shape W24×76.

14–80. The load is shown in Figure P14–80. Design a round steel bar that will limit the deflection at the end of the beam to 5.0 mm.

14–81. The load is shown in Figure P14–81. Design a steel beam that will limit the maximum deflection to 1.0 mm. Use any shape, including those listed in the Appendix.

FIGURE P14–79

FIGURE P14–80

FIGURE P14–81

14–82. The load is shown in Figure P14–82. Select an aluminum I-beam that will limit the stress to 120 MPa; then compute the maximum deflection in the beam.

FIGURE P14–82

14–83. A steel wide-flange beam W360×39 carries the loading shown in Figure P14–83. Compute the maximum deflection between the supports and at each end.

FIGURE P14–83

14–84. The loading shown in Figure P14–84 represents a steel shaft for a large machine. The loads are due to gears mounted on the shaft. Assuming that the entire shaft will be the same diameter, determine the required diameter to limit the deflection at any gear to 0.13 mm.

FIGURE P14–84

Moment-Area Method

Use the moment-area method to solve the following problems.

14–85. For the beam shown in Figure P14–76, compute the deflection at the load. The beam is a rectangular steel bar 1.0 in wide and 2.0 in high.

14–86. For the beam shown in Figure P14–76, compute the deflection at the middle, 8.0 in from either support. The beam is a rectangular steel bar 1.0 in wide and 2.0 in high.

14–87. For the beam shown in Figure P14–77, compute the deflection at the end. The beam is a steel wide-flange shape W18×55.

14–88. For the beam shown in Figure P14–78, compute the deflection at the end. The beam is a $2\frac{1}{2}$-in schedule 40 steel pipe (PIPE64STD).

14–89. For the beam shown in Figure P14–79, compute the deflection at the end. The beam is a steel wide-flange shape W24×76.

14–90. For the beam shown in Figure P14–80, compute the deflection at the end. The beam is a round aluminum bar 6061-T6 with a diameter of 100 mm.

14–91. For the beam shown in Figure P14–81, compute the deflection at the right end, point C. The beam is a square steel structural tube $2\times2\times\frac{1}{4}$ (HSS51×51×6.4).

14–92. For the beam shown in Figure P14–82, compute the deflection at point C. The beam is an aluminum I-beam I178×8.630 made from 6061-T6.

14–93. For the beam shown in Figure P14–83, compute the deflection at point A. The beam is a steel wide-flange shape W360×39.

14–94. Figure P14–94 shows a stepped round steel shaft carrying a single concentrated load at its middle. Compute the deflection under the load.

14–95. Figure P14–95 shows a composite beam made from steel structural tubing. Compute the deflection at the load.

14–96. Figure P14–96 shows a cantilever beam made from a steel wide-flange shape W18×55 with cover plates welded to the top and bottom of the beam over the first 6.0 ft. Compute the deflection at the end of the beam.

14–97. The top member for a gantry crane is made as shown in Figure P14–97. The two end pieces are $3\times3\times\frac{1}{4}$ steel structural tubing. The middle piece is $4\times4\times\frac{1}{2}$. They are combined for a distance of 10.0 in with the 3-in tube fitted tightly inside the 4-in tube and brazed together. Compute the deflection at the middle of the beam.

14–98. A crude diving board is made by nailing two wood planks together as shown in Figure P14–98. Compute the deflection at the end if the diver exerts a force of 1.80 kN at the end. The wood is No. 2 southern pine.

600 lb

3.0 in — 4.0 in — 4.0 in — 3.0 in

Bearings provide
simple support

A

B C D

E

R_A

R_E

1.40-in dia.

0.75-in dia.—both ends

FIGURE P14–94

4000 lb

60 in — 60 in

A B

A B E

A A B C D

24 in

A A B

B B

24 in

R_A

R_E

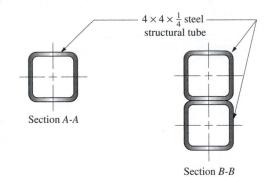

$4 \times 4 \times \frac{1}{4}$ steel
structural tube

Section A-A

Section B-B

FIGURE P14–95

10 ft

6 ft

20 K

A

B

A

B

$\frac{1}{2} \times 8$ plate

W18×55

$\frac{1}{2} \times 8$ plate

Section A-A

Section B-B

FIGURE P14–96

FIGURE P14–97   Gantry crane beam for Problem 14–97.

Section A-A   Section B-B

FIGURE P14–98   Diving board for Problem 14–98.

COMPUTER ASSIGNMENTS

1. Write a program or spreadsheet to evaluate the deflection of a simply supported beam carrying one concentrated load between the supports using the formulas given in Case b of Appendix A–23. The program should accept input for the length, position of the load, length of the overhang, if any, the beam stiffness values (E and I), and the point for which the deflection is to be computed.

Enhancements

a. Design the program so that it computes the deflection at a series of points that could permit the plotting of the complete deflection curve.

b. In addition to computing the deflection of the series of points, have the program plot the deflection curve on a plotter or printer.

c. Have the program compute the maximum deflection and the point at which it occurs.

2. Repeat Assignment 1 for any of the loading and support patterns shown in Appendix A–23.

3. Write a program similar to that of Assignment 1 for Case b of Appendix A–23, but have it accept two or more concentrated loads at any point on the beam and compute the deflection at specified points using the principle of superposition.

4. Combine two or more programs that solve for the deflection of beams for a given loading pattern so that superposition can be used to compute the deflection at any point due to the

combined loading. For example, combine Cases *b* and *d* of Appendix A–23 to handle any beam with a combination of concentrated loads and a complete uniformly distributed load. Or add Case *g* to include a distributed load over only part of the length of the beam.

5. Repeat Assignments 1 through 4 for the cantilevers from Appendix A–24.

6. Write a program or spreadsheet to solve for the critical shearing forces and bending moments for any of the statically indeterminate beam types in Appendix A–25.

Enhancements to Assignment 6

a. Use the graphics mode to plot the complete shearing force and bending moment diagrams for the beams.

b. Compute the required section modulus for the beam cross section to limit the stress due to bending a specified amount.

c. Include a table of section properties for steel wide-flange beams and search for suitable beam sizes to carry the load.

d. Assuming that the beam cross section will be a solid circular section, compute the required diameter.

e. Assuming that the beam cross section will be a rectangle with a given ratio of height to thickness, compute the required dimensions.

f. Assuming that the beam cross section will be a rectangle with a given height *or* thickness, compute the other required dimension.

g. Assuming that the beam is to be made from wood, in a rectangular shape, compute the required area of the beam cross section to limit the shearing stress to a specified value. Use the special shear formula for rectangular beams from Chapter 13.

h. Add the computation of the deflection at specified points on the beam using the formulas of Appendix A–25.

CHAPTER FIFTEEN

COMBINED STRESSES

MAJOR TOPICS

15–1 The Big Picture—Combined Stresses
15–2 The Stress Element
15–3 Stress Distribution Created by Basic Stresses
15–4 Creating the Initial Stress Element
15–5 Combined Normal Stresses
15–6 Combined Normal and Shear Stresses
15–7 Equations for Stresses in Any Direction
15–8 Maximum Stresses
15–9 Mohr's Circle for Stress
15–10 Stress Condition on Selected Planes
15–11 Special Case in Which Both Principal Stresses Have the Same Sign
15–12 Use of Strain-Gage Rosettes to Determine Principal Stresses

15–1 THE BIG PICTURE— COMBINED STRESSES

Discussion Map

- In the previous chapters the objective was to help you acquire the competencies to analyze and design load-carrying members that were subjected to a single type of stress. The following stresses were included:
  - Direct tension stress
  - Direct compression stress
  - Direct shear stress
  - Bearing stress
  - Torsional shear stress
  - Bending stress

  The associated strains and deformations were also considered for the individual cases.

- In Chapter 15 you will consider cases in which more than one type of stress exists in the member at the same time. First we discuss special combinations of stresses that occur frequently in machines, structures, and vehicles. We will consider the following:
  - Combined normal stresses produced by bending, direct tension, and compression
  - Combined normal and shearing stresses that may be produced by bending, axial loads, and torsion

- The successful analysis of combined stresses requires that you be able to visualize the stress distribution in the material from which the load-carrying member is made. The basic approach we use is to first visualize the stress distribution due to the *components* of the total load that can be analyzed with the same procedures that were developed in earlier chapters. We then combine these components at

specific points within the cross section of the member where the maximum combined stress is likely to occur.

- There are many other cases in which you will not know the direction of the maximum stresses. You will have to make whatever analysis you can from the given data and then use additional techniques to compute the maximum stress and determine the direction in which it acts.

- Other cases may require that you compute the stress in a particular direction that may differ from that for which you can conveniently compute the stress distribution.

ACTIVITY   Chapter 15—Combined Stresses

Look around you to find examples of load-carrying members that are subjected to combined stresses.

- Try to find beams in bending that also carry axial tensile or compressive stresses.
- Power transmission shafts carrying gears, belt sheaves, or chain sprockets typically transmit torque and also experience bending forces (Figure 7–2).
- Construction and agricultural machinery often contain members subjected to bending, axial, and torsional stresses simultaneously [Figure 1–1(d)].
- Crane booms that extend upward from a truck bed are subjected to combined bending and axial compression (Figure 15–1).
- Bicycle and motorcycle frames, cranks, seat supports, and other components carry complex arrays of forces, moments, and torques [Figure 1–1(b)].
- Automotive frames, suspension members, steering linkages, and drive train components experience multiple

FIGURE 15–1 *Source:* Pearson Education/PH College.

types of loading during different operating conditions such as smooth straight-line driving, turning, braking, acceleration, racing, and driving over rough roads or demanding off-road terrain [Figure 4–11(g)].

- Interior and body components of automobiles such as seat adjusters, window lifts, windshield wiper systems, hinges, and latches for doors, hoods, or trunks experience combined stresses.

- Aircraft and aerospace vehicles employ stressed skin construction with specially shaped members that distribute loads throughout the framework, producing lightweight, highly optimized structures [Figure 1–1(e)].

- Components for furniture such as chairs, tables, and cabinets for audiovisual equipment typically must withstand combined axial compression and bending.

- Highway signs extending over roadways that must not only support their own weight, but also must resist high wind forces, subject their support posts to combined bending, axial compression, and torsional shear stresses [Figure 4–11(c)].

- Frames and housings for industrial machinery, robots, engines, pumps, machine tools, measuring devices, and consumer products typically are subjected to complex stress fields that cannot be analyzed by the application of conventional stress analysis formulas (Figure 15–2).

- Structural beams in buildings and bridges which carry loads that are offset from their neutral axes experience both bending and torsional stresses (Figure 15–3).

In this chapter you will learn methods of analyzing these types of load-carrying members. First you will consider several commonly encountered special cases where combinations, adaptations, and extensions of stress analysis techniques already learned from this book can be applied. That is, combinations of direct axial tensile or compressive normal stresses, bending stresses, direct shearing stress, vertical shearing stresses in beams, and torsional shear stress will be considered. Then a more general method of combining stresses is developed and applied that is useful for more complex loading and component geometries.

In the general method, you will start with the known stress system at a particular point where any combination of normal stresses (tension or compression) and shearing stresses exist. The known stress element may be found from direct analysis, experimental stress analysis, or computer-aided finite element analysis. From the known stress element, you will learn how to compute the maximum and minimum principal (normal) stresses, the maximum shear stress, and the orientation of the stress element on which those stresses occur. For some stress analyses, it is useful to compute an additional combined stress, called the *von Mises* stress, from the principal stresses. The definition of the von Mises stress will be presented later.

These stresses can be compared with the inherent strength of the material from which the member is made to ensure that the member is safe and that it will perform to its expected requirements. The appropriate material strengths may include ultimate tensile strength, ultimate compressive

FIGURE 15–2 *Source:* Getty Images, Inc.—Taxi.

FIGURE 15–3 *Source:* Omni-Photo Communications, Inc.

strength, ultimate shear strength, tensile yield strength, compressive yield strength, yield strength in shear, or endurance (fatigue) strength.

Following courses of study, such as design of machine elements, finite element analysis, experimental stress analysis, aerospace structural design, structural steel design, design of concrete structures, or design of wood structures, extend these methods of stress analysis to even more complex cases. Furthermore, advanced courses in elasticity, plasticity, composite materials, material science, plates and shells, and fracture mechanics will extend your mastery of the wide range of applications you will encounter in your career.

Objectives of This Chapter

After completing this chapter, you should be able to:

1. Recognize cases for which combined stresses occur.

2. Represent the stress condition on a stress element.

3. Recognize the importance of visualizing the stress distribution over the cross section of a load-carrying member and considering the stress condition at a point.

4. Recognize the importance of free body diagrams of components of structures and mechanisms in the analysis of combined stresses.

5. Compute the combined normal stress resulting from the application of bending stress with either direct tensile or compressive stresses using the principle of superposition.

6. Evaluate the design factor for combined normal stress, including the properties of either isotropic or non-isotropic materials.

7. Optimize the shape and dimensions of a load-carrying member relative to the variation of stress in the member and its strength properties.

8. Analyze members subjected only to combined bending and torsion by computing the resulting maximum shear stress.

9. Use the maximum shear stress theory of failure properly.

10. Apply the equivalent torque technique to analyze members subjected to combined bending and torsion.

11. Consider the stress concentration factors when using the equivalent torque technique.

12. Understand the development of the equations for combined stresses, from which you can compute the following:

    a. The maximum and minimum principal stresses

    b. The orientation of the principal stress element

    c. The maximum shear stress on an element

    d. The orientation of the maximum shear stress element

    e. The normal stress that acts along with the maximum shear stress

    f. The normal and shear stress that occurs on the element oriented in any direction

13. Construct Mohr's circle for biaxial stress.

14. Interpret the information available from Mohr's circle for the stress condition at a point in any orientation.

15. Use the data from Mohr's circle to draw the principal stress element and the maximum shear stress element.

15–2 THE STRESS ELEMENT

In general, combined stress refers to cases in which two or more types of stress are exerted on a given point at the same time. The component stresses can be either *normal* (i.e., tension or compression) or *shear* stresses.

When a load-carrying member is subjected to two or more different kinds of stresses, the first task is to compute the stress due to each component. Then a decision is made about which point within the member has the highest *combination* of stresses, and the combined stress analysis is completed for that point. For some special cases, it is desired to know the stress condition at a given point regardless of whether or not it is the point of maximum stress. Examples would be near welds in a fabricated structure, along the grain of a wood member, or near a point of attachment of one member to another.

After the point of interest is identified, the stress condition at that point is determined from the classical stress analysis relationships presented in this book, if possible. At times, because of the complexity of the geometry of the member or the loading pattern, it is not possible to complete a reliable stress analysis entirely by computations. In such cases experimental stress analysis can be employed in which strain gages, photoelastic models, or strain-sensitive coatings give data experimentally. Also, with the aid of finite element stress analysis techniques, the stress condition can be determined using computer-based analysis.

Then, after using one of these methods, you should know the information required to construct the *stress element,* as shown in Figure 15–4. The element is assumed to be infinitesimally small and aligned with known directions on the member being analyzed. The complete element, as

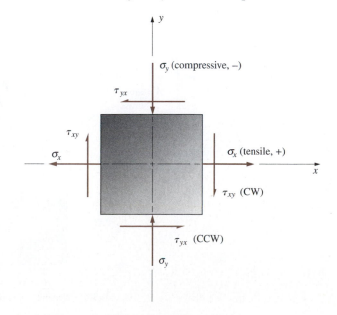

FIGURE 15–4   Complete stress element.

shown, could have a normal stress (either tensile or compressive) acting on each pair of faces in mutually perpendicular directions, usually named the x- and y-axes. As the name *normal stress* implies, these stresses act normal (perpendicular) to the faces. As shown, σ_x is aligned with the x-axis and is a tensile stress tending to pull the element apart. Recall that tensile stresses are considered positive. Then σ_y is shown to be compressive, tending to crush the element. Compressive stresses are considered negative.

In addition, there may be shear stresses acting along the sides of the element as if each side were being cut away from the adjacent material. Recall from the earlier discussion of shearing stresses that a set of four shears, all equal in magnitude, exist on any element in equilibrium. On any two opposite faces the shear stresses will act in opposite directions, thus creating a *couple* that tends to rotate the element. Then there must exist a pair of shear stresses on the adjacent faces producing an oppositely directed couple for the element to be in equilibrium. We will refer to each *pair* of shears using a double subscript notation. For example, τ_{xy} refers to the shear stress acting perpendicular to the x-axis and parallel to the y-axis. Conversely, τ_{yx} acts perpendicular to the y-axis and parallel to the x-axis. Rather than establish a convention for signs of the shear stresses, we will refer to them as *clockwise* (CW) or *counterclockwise* (CCW), according to how they tend to rotate the stress element.

15–3 STRESS DISTRIBUTION CREATED BY BASIC STRESSES

Figures 15–5, 15–6, and 15–7 show direct tension, direct compression, and bending stresses in which normal stresses were developed. It is important to know the distribution of stress within the member as shown in the figures. Also shown are stress elements subjected to these kinds of stresses. The following list contains the principal formulas for computing the value of these stresses.

Direct tension: (see Figure 15–5)	$\sigma = \dfrac{F}{A}$	Uniform over the area (Chapters 7, 9)
Direct compression: (see Figure 15–6)	$\sigma = \dfrac{-F}{A}$	Uniform over the area (Chapters 7, 9)
Stress due to bending: (see Figure 15–7)	$\sigma_{max} = \pm \dfrac{Mc}{I}$ $= \pm \dfrac{M}{S}$	Maximum stress at outer surfaces (Chapter 12)
	$\sigma = \pm \dfrac{My}{I}$	Bending stress at any point y (Chapter 12)

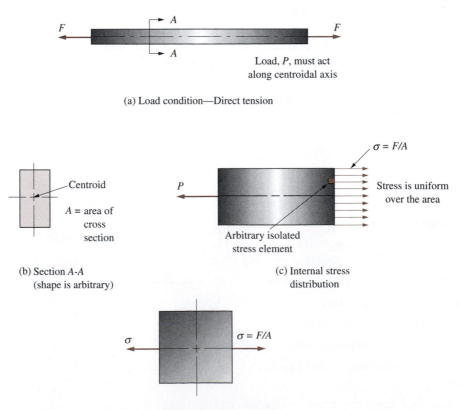

(a) Load condition—Direct tension

Load, P, must act along centroidal axis

(b) Section A-A (shape is arbitrary)

(c) Internal stress distribution

$\sigma = F/A$

Stress is uniform over the area

Arbitrary isolated stress element

Centroid

A = area of cross section

(d) Stress element—Normal tensile stress

$\sigma = F/A$

FIGURE 15–5 Distribution of normal stress for direct tension.

(a) Load condition—Direct compressive stress

Member must
be short

Load must act along
centroidal axis

(b) Section *A-A*
(shape is arbitrary)

Centroid

A = area of
cross
section

(c) Internal stress
distribution

$\sigma = -F/A$

Stress is uniform
over the area

Arbitrary isolated
stress element

(d) Stress element—Normal compressive stress

$\sigma = -F/A$

FIGURE 15–6   Distribution of normal stress for direct compression.

FIGURE 15–7   Distribution of
normal stress for bending.

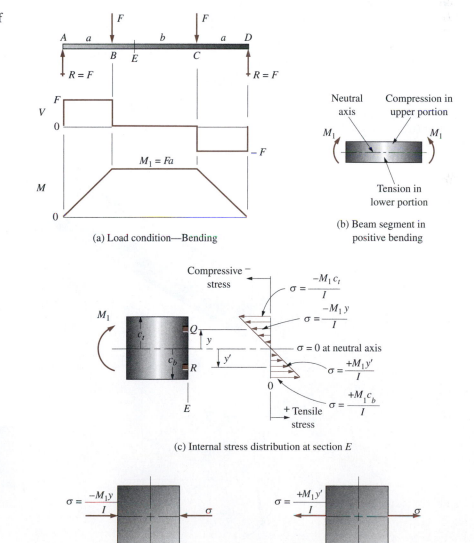

(a) Load condition—Bending

$M_1 = Fa$

(b) Beam segment in
positive bending

Neutral
axis

Compression in
upper portion

Tension in
lower portion

(c) Internal stress distribution at section *E*

Compressive
stress

$\sigma = \dfrac{-M_1 c_t}{I}$

$\sigma = \dfrac{-M_1 y}{I}$

$\sigma = 0$ at neutral axis

$\sigma = \dfrac{+M_1 y'}{I}$

$\sigma = \dfrac{+M_1 c_b}{I}$

+ Tensile
stress

(d) Stress element Q

$\sigma = \dfrac{-M_1 y}{I}$

(e) Stress element R

$\sigma = \dfrac{+M_1 y'}{I}$

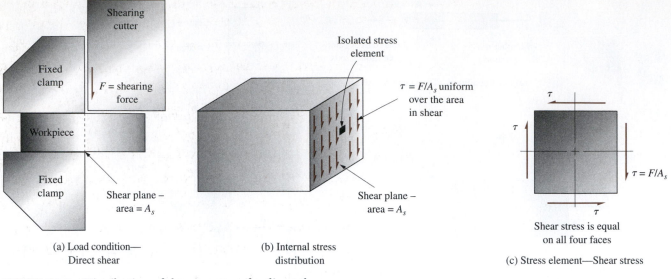

(a) Load condition—
Direct shear

(b) Internal stress
distribution

$\tau = F/A_s$ uniform
over the area
in shear

Shear stress is equal
on all four faces

(c) Stress element—Shear stress

FIGURE 15–8 Distribution of shearing stress for direct shear.

Figures 15–8, 15–9, and 15–10 show three cases in which shearing stresses are produced along with the stress distributions and the stress elements subjected to these types of stress. The following list contains the principal formulas used to compute shearing stresses.

Direct shear:
(see Figure 15–8)

$$\tau = \frac{F}{A_s}$$

Uniform over the area (Chapter 9)

Torsional shear:
(see Figure 15–9)

$$\tau_{max} = \frac{Tc}{J}$$ Maximum at outer surface (Chapter 10)

$$\tau = \frac{Tr}{J}$$ Torsional shear at any radius (Chapter 10)

Shearing stresses in beams: (see Figure 15–10)

$$\tau = \frac{VQ}{It}$$ (Chapter 13)

FIGURE 15–9 Distribution of torsional shear stress in a solid round shaft.

Stress element on surface of bar

(a) Load condition—Torsional shear on a circular bar

$$\tau_{max} = \frac{Tc}{J}$$

$$\tau = \frac{Tr}{J}$$

τ_{yx} (CW)

τ_{xy} (CCW)

$\tau_{yx} = \tau_{xy}$ in magnitude

(b) Stress distribution on a cross section

(c) Stress element—Torsional shear

(a) Load condition—
Beam in bending

(b) Free body diagram of segment
to left of Section 1-1

(c) Stress element

(d) Distribution of shearing
stress on Section 1-1

FIGURE 15–10   Distribution of shearing stress in a beam.

15–4 CREATING THE INITIAL STRESS ELEMENT

One major objective of this chapter is to develop relationships from which the *maximum principal stresses* and the *maximum shear stress* can be determined. Before this can be done, it is necessary to know the state of stress at a point of interest in some orientation. In this section we demonstrate the determination of the initial stress condition by direct calculation from the formulas for basic stresses.

Figure 15–11 shows an L-shaped lever attached to a rigid surface with a downward load P applied to its end. The shorter segment at the front of the lever is loaded as a cantilever beam as shown in Figure 15–12(a), with the moment at its left end resisted by the other segment of the lever.

Figure 15–12(b) shows the longer segment as a free body diagram. At the front, the force P and the torque T are the reactions to the force and moment at the left end of the shorter segment in (a). Then at the rear, there must be reactions M, P, and T to maintain equilibrium. Note carefully the directions of M, P, and T.

On the shorter segment in (a): P and M_1 act in the y–z plane

FIGURE 15–11   L-shaped lever illustrating combined stresses.

(b)

(a)

FIGURE 15–12   Free body diagrams of segments of an L-shaped lever. (a) Shorter segment of lever. (b) Longer segment of lever.

This part of the lever acts like a simple cantilever with the maximum bending moment at its left end and having a value of $M_1 = P \cdot a$.

On the longer segment in (b): P and M_2 act in the x–z plane

T acts in the y–z plane producing torsion about the x-axis

This part of the lever acts like a cantilever with added torsion. The magnitude of the torque, T, is equal to $M_1 = P \cdot a$, and is uniform along the entire length of the longer segment of the bar. The magnitude of M_2 has the value: $M_2 = P \cdot b$. It is largest at the far end where the bar is attached rigidly to the support plate.

FIGURE 15–13   Stress on element K from Figure 15–12 with data from Example Problem 15–1.

The bar is subjected to a combined stress condition with the following kinds of stresses:

Bending stress due to the bending moment

Torsional shear stress due to the torque

Shearing stress due to the vertical shearing force

We can conclude that one of the points where the stress is likely to be the highest is on the top of the long segment near the support. Called element K in Figure 15–12(b), it would see the maximum tensile bending stress and the maximum torsional shear stress. But the vertical shearing stress would be zero because it is at the outer surface away from the neutral axis.

Element K is then subjected to a combined stress, as shown in Figure 15–13. The tensile normal stress, σ_x, is directed parallel to the x-axis along the top surface of the bar. The applied torque tends to produce a shear stress τ_{xy} in the negative y-direction on the face toward the front of the bar and in the positive y-direction on the face toward the rear. Together they create a counterclockwise couple on the element. The element is completed by showing the shear stresses τ_{yx} producing a clockwise couple on the other faces.

Example Problem 15–1 includes illustrative calculations for the values of the stresses that are shown in Figure 15–13.

Example Problem 15–1   Figure 15–11 shows an L-shaped lever carrying a downward force at its end. Compute the stress condition that exists on a point on top of the lever near the support. Let $P = 1500$ N, $a = 150$ mm, $b = 300$ mm, and $D = 30$ mm. Show the stress condition on a stress element.

Solution   Objective: Compute the stress condition and draw the stress element.

Given: Geometry and loading in Figure 15–11. $P = 1500$ N
Dimensions: $a = 150$ mm, $b = 300$ mm, $D = 30$ mm.

Analysis: Figure 15–12 shows the lever separated into two free body diagrams. The point of interest is labeled "Element K" at the right end of the lever where it joins the fixed support. The element is subjected to bending stress due to the reaction moment at the support, M_2. It is also subjected to torsional shear stress due to the torque, T.

Results: Using the free body diagrams in Figure 15–12, we can show that

$$M_1 = T = P \cdot a = (1500 \text{ N})(150 \text{ mm}) = 225\,000 \text{ N} \cdot \text{mm}$$

$$M_2 = P \cdot b = (1500 \text{ N})(300 \text{ mm}) = 450\,000 \text{ N} \cdot \text{mm}$$

Then the bending stress on top of the bar, shown as element K in Figure 15–12(b), is

$$\sigma_x = \frac{M_2 c}{I} = \frac{M_2}{S}$$

The section modulus S is

$$S = \frac{\pi D^3}{32} = \frac{\pi (30 \text{ mm})^3}{32} = 2651 \text{ mm}^3$$

Then

$$\sigma_x = \frac{450\,000 \text{ N} \cdot \text{mm}}{2651 \text{ mm}^3} = 169.8 \text{ N/mm}^2 = 169.8 \text{ MPa}$$

The torsional shear stress is at its maximum value all around the outer surface of the bar, with the value of

$$\tau = \frac{T}{Z_p}$$

The polar section modulus Z_p is

$$Z_p = \frac{\pi D^3}{16} = \frac{\pi (30 \text{ mm})^3}{16} = 5301 \text{ mm}^3$$

Then

$$\tau = \frac{225\,000 \text{ N} \cdot \text{mm}}{5301 \text{ mm}^3} = 42.4 \text{ N/mm}^2 = 42.4 \text{ MPa}$$

Comment: The bending and shear stresses are shown on the stress element K in Figure 15–13. This is likely to be the point of maximum combined stress, which will be discussed later in this chapter. At a point on the side of the bar on the y-axis, a larger shearing stress would exist because the maximum torsional shear stress combines with the maximum vertical shear stress due to bending. But the bending stress there is zero. That element should also be analyzed.

15–5 COMBINED NORMAL STRESSES

This section examines one type of combined stress where the point of interest is subjected only to normal stress, that is, tension or compression. The following section addresses another special case of combined stress where combined normal and shear stresses act on the point of interest. Then the balance of the chapter develops and applies the more general case of combined stress where equations for the maximum normal stress, minimum normal stress, and maximum shear stress are computed for any combination of known stresses on a particular stress element. A semigraphical method called *Mohr's circle* is then used to facilitate the computations and to help you visualize the full range of possible stress conditions that can exist at a particular point in a load-carrying member.

The first combination to be considered is bending with direct tension or compression. In any combined stress problem, it is helpful to visualize the stress distribution caused by the various components of the total stress pattern. You should review Section 15–3 for summaries of the stress distribution

for bending and direct tension and compression. Notice that bending results in tensile and compressive stresses, as do both direct tension and direct compression. Since the same kinds of stresses are produced, a simple algebraic sum of the stresses produced at any point is all that is required to compute the resultant stress at that point. This process is called *superposition*.

GUIDELINES FOR SOLVING PROBLEMS WITH COMBINED NORMAL STRESSES ▽

These guidelines pertain to situations in which two or more loads or components of loads act in a manner that produces normal stresses (tension and/or compression) on the load-carrying member. In general, loads producing bending stress or direct tension or compression are to be included. The objective is to compute the maximum combined stress in the member. Let tensile stresses be positive (+) and compressive stresses be negative (−).

1. Draw the free body diagram for the load-carrying member and compute the magnitude of all applied forces.

2. For any force that acts at an angle to the neutral axis of the member, resolve the force into components perpendicular and parallel to the neutral axis.

3. Forces or components acting in a direction coincident with the neutral axis will produce either direct tension or direct compression with uniform stress distribution across the cross section. Compute these stresses.

4. Forces or components acting perpendicular to the neutral axis cause bending stresses. Using the methods of Chapter 11, determine the bending moments caused by these forces, either individually or in combination. Then, for the section subjected to the largest bending moment, compute the bending stress from the flexure formula, $\sigma = M/S$. The maximum stress will occur at the outermost fibers of the cross section. Note at which points the stress is tensile and which is compressive.

5. Forces or components acting parallel to the neutral axis but whose line of action is away from that axis also cause bending. The bending moment is the product of the force times the perpendicular distance from the neutral axis to the line of action of the force. Compute the bending stress due to such moments at any section where the combined stress may be the maximum.

6. Considering all normal stresses computed from steps 1–5, use superposition to combine them at any point within any cross section where the combined stress may be the maximum. Superposition is accomplished by algebraically summing all stresses acting at a point, taking care to note if each component stress is tensile (+) or compressive (−). It may be necessary to evaluate the stress condition at two or more points if it is not obvious where the maximum combined stress occurs. In general, the superposition process can be expressed as

$$\sigma_{comb} = \pm\frac{F}{A} \pm \frac{M}{S} \qquad (15\text{--}1)$$

where the term, $\pm F/A$, includes all direct tensile and compressive stresses acting at the point of interest and the term, $\pm M/S$, includes all bending stresses acting at that point. The sign for each stress must be logically determined from the load causing the individual stress.

7. The maximum combined stress on the member can then be compared with the design stress for the material from which the member is to be made to compute the resulting design factor and to assess the safety of the member. For isotropic materials, either tensile or compressive stress could cause failure, whichever is maximum. For nonisotropic materials having different strengths in tension and compression, it is necessary to compute the resulting design factor for both the tensile and the compressive stress to determine which is critical. Also, in general, it will be necessary to consider the stability of those parts of members subjected to compressive stresses by analyzing the tendency for buckling or local crippling of the member. See Chapter 16 for buckling of column-like compression members. The analysis for crippling and buckling of parts of members will require reference to other sources. See the references at the end of Chapters 12, 13, and 15.

An example of a member in which both bending and direct tensile stresses are developed is shown in Figure 15–14. The two horizontal beams support a 10 000-lb load by means of the cable assembly. The beams are rigidly attached to columns, so that they act as cantilever beams. The load at the end of each beam is equal to the tension in the cable. Figure 15–15 shows that the vertical component of the tension in each cable must be 5000 lb. That is,

$$F \cos 60° = 5000\,\text{lb}$$

FIGURE 15–14 Two cantilever beams subjected to combined bending and axial tensile stresses.

2.0 ft

30°   30°

$W = 10\,000\,\text{lb}$

5000 lb = $F \cos 60°$ 5000 lb = $F \cos 60°$

10 000 lb = F F = 10 000 lb

60° 60°

30°   30°

10 000 lb

FIGURE 15–15   Force analysis on cable system.

Aluminum I-beam
6I×4.03

2 ft

R_h

$F_h = F \cos 30° = 8660$ lb

30°

M

R_v

$F_v = F \sin 30° = 5000$ lb

F = 10 000 lb

FIGURE 15–16   Forces applied to each beam.

and the total tension in the cable is

$$F = \frac{5000 \text{ lb}}{\cos 60°} = 10\,000 \text{ lb}$$

This is the force applied to each beam, as shown in Figure 15–16.

Example Problem 15–2 completes the analysis of the combined stress condition in each of the horizontal cantilever beams in Figure 15–14.

Example Problem 15–2   For the system shown in Figure 15–14, determine the maximum combined tensile or compressive stress in each horizontal cantilever beam when a static load of 10 000 lb is suspended from the cable system between them. The beams are standard Aluminum Association I-beams, I6×4.03. Then, if the beams are to be made of 6061-T6 aluminum alloy, compute the resulting design factor.

Solution   Objective: Compute the maximum combined stress and the resulting design factor for the horizontal cantilever beams in Figure 15–14.

Given: Load weighs 10 000 lb. Force system detail shown in Figure 15–15. Tension in each cable is 10 000 lb. The 2.0-ft-long beams are aluminum I6×4.03 made from 6061-T6 aluminum alloy. From Appendix A–11, the properties of the beam are $A = 3.427 \text{ in}^2$ and $S = 7.33 \text{ in}^3$.

Analysis: The tension in the cable attached to the end of each cantilever beam will tend to cause direct tensile stress combined with bending stress in the beam. Use the *guidelines for solving problems with combined normal stresses* outlined in this section.

Results: **Step 1:** Figure 15–16 shows the free body diagram of one beam with the 10 000-lb force applied by the cable to the end of the beam. The reaction at the left end where the beam is rigidly attached to the column consists of a vertical reaction force, a horizontal reaction force, and a counterclockwise moment.

Step 2: Also shown in Figure 15–16 is the resolution of the 10 000-lb force into vertical and horizontal components where $F_v = 5000$ lb and $F_h = 8660$ lb.

Step 3: The horizontal force, F_h, acts in a direction that is coincident with the neutral axis of the beam. Therefore, it causes direct tensile stress with a magnitude of

$$\sigma_t = \frac{F_h}{A} = \frac{8660 \text{ lb}}{3.427 \text{ in}^2} = 2527 \text{ psi}$$

Step 4: The vertical force, F_v, causes bending in a downward direction such that the top of the beam is in tension and the bottom is in compression. The maximum bending moment will occur at the support at the left end, where

$$M = F_v(2.0 \text{ ft}) = (5000 \text{ lb})(2.0 \text{ ft})(12 \text{ in/ft}) = 120\,000 \text{ lb} \cdot \text{in}$$

Then the maximum bending stress caused by this moment is

$$\sigma_b = \frac{M}{S} = \frac{120\,000 \text{ lb} \cdot \text{in}}{7.33 \text{ in}^3} = 16\,371 \text{ psi}$$

$\sigma_b = +16\ 371$ psi

Neutral axis

$\sigma_b = -16\ 371$ psi

(a) Bending stress
distribution

$\sigma_t = 2527$ psi

(b) Direct tensile
stress distribution

$\sigma_{top} = \sigma_b + \sigma_t = 18\ 898$ psi

$\sigma_{bot} = \sigma_t - \sigma_b = -13\ 844$ psi

(c) Combined stress
distribution

FIGURE 15–17   Diagram of the superposition principle applied to the beams of Figure 15–14.

A stress of this magnitude occurs as a tensile stress at the top surface and as a compressive stress at the bottom surface of the beam at the support.

Step 5: This step does not apply to this problem because there is no horizontal force offset from the neutral axis.

Step 6: It can be reasoned that the maximum combined stress occurs at the top surface of the beam at the support because both the direct tensile stress computed in step 3 and the bending stress computed in step 4 are tensile at that point. Therefore, they will add together. Using superposition, as defined in Equation (15–1),

$$\sigma_{top} = 2527 \text{ psi} + 16\ 371 \text{ psi} = 18\ 898 \text{ psi tension}$$

For comparison, the combined stress at the bottom surface of the beam is

$$\sigma_{bot} = 2527 \text{ psi} - 16\ 371 \text{ psi} = -13\ 844 \text{ psi compression}$$

Figure 15–17 shows a set of diagrams that illustrate the process of superposition. Part (a) is the stress in the beam caused by bending. Part (b) shows the direct tensile stress due to F_h. Part (c) shows the combined stress distribution.

Step 7: Because the load is static, we can compute the design factor based on the yield strength of the 6061-T6 aluminum alloy, where $S_y = 40\ 000$ psi (Appendix A–18).

Then $N = s_y/\sigma_{top} = (40\ 000 \text{ psi})/(18\ 898 \text{ psi}) = 2.11$

Comment: This should be adequate for a purely static load. If there is uncertainty about the magnitude of the load or if there is a possibility of the load being applied with some shock or impact, a higher design factor would be preferred.

Example Problem 15–3   A picnic table in a park is made by supporting a circular top on a pipe that is rigidly held in concrete in the ground. Figure 15–18 shows the arrangement. Compute the maximum stress in the pipe if a person having a mass of 135 kg sits on the edge of the table. The pipe is made of an aluminum alloy and has an outside diameter of 170 mm and an inside diameter of 163 mm. If the aluminum is 6061-T4, compute the resulting design factor based on both yield strength and ultimate strength. Then comment on the suitability of the design.

Solution   Objective: Compute the maximum stress in the pipe in Figure 15–18 and the design factor based on both yield strength and ultimate strength.

Given: Loading and pipe dimensions shown in Figure 15–18. Load is the force due to a 135-kg mass on the edge of the table. Pipe is aluminum, 6061-T4; $s_y = 145$ MPa; $s_u = 241$ MPa.

Analysis: The pipe is subjected to combined bending and direct compression as illustrated in Figure 15–19, the free body diagram of the pipe. The effect of the load is to produce a downward force on the top of the pipe while exerting a moment in the clockwise direction. The moment is the product of the load times the radius of the table top. The reaction at the bottom of the pipe, supplied by the

FIGURE 15–18 Picnic table supported by a pipe.

FIGURE 15–19 Free body diagrams for tabletop and pipe for Example Problem 15–3.

(a) Free body diagram of tabletop

(b) Free body diagram of pipe

concrete, is an upward force combined with a counterclockwise moment. Use the *guidelines for solving problems with combined normal stresses*.

Results: **Step 1:** Figure 15–19 shows the free body diagram. The force is the gravitational attraction of the 135-kg mass.

$$F = m \cdot g = (135 \text{ kg})(9.81 \text{ m/s}^2) = 1324 \text{ N}$$

Step 2: No forces act at an angle to the axis of the pipe.

Step 3: Now the direct axial compressive stress in the pipe is

$$\sigma_a = -\frac{F}{A}$$

But

$$A = \frac{\pi(D_o^2 - D_i^2)}{4} = \frac{\pi(170^2 - 163^2)\,mm^2}{4} = 1831\ mm^2$$

Then

$$\sigma_a = -\frac{1324\ N}{1831\ mm^2} = -0.723\ MPa$$

This stress is a uniform, compressive stress across any cross section of the pipe.

Step 4: No forces act perpendicular to the axis of the pipe.

Step 5: Since the force acts at a distance of 1.1 m from the axis of the pipe, the moment is

$$M = F \cdot R = (1324\ N)(1.1\ m) = 1456\ N \cdot m$$

The bending stress computation requires the application of the flexure formula,

$$\sigma_b = \frac{Mc}{I}$$

where $c = \dfrac{D_o}{2} = \dfrac{170\ mm}{2} = 85\ mm$

$$I = \frac{\pi}{64}(D_o^4 - D_i^4) = \frac{\pi}{64}(170^4 - 163^4)mm^4$$
$$I = 6.347 \times 10^6\ mm^4$$

Then

$$\sigma_b = \frac{(1456\ N \cdot m)(85\ mm)}{6.347 \times 10^6\ mm^4} \times \frac{10^3\ mm}{m} = 19.5\ MPa$$

Step 6: The bending stress σ_b produces compressive stress on the right side of the pipe and tension on the left side. Since the direct compression stress adds to the bending stress on the right side, that is where the maximum stress would occur. The combined stress would then be, using Equation (15–1),

$$\sigma_c = -\sigma_a - \sigma_b = (-0.723 - 19.5)\,MPa = -20.22\ MPa$$

Step 7: The design factor based on yield strength is

$$N = \frac{s_y}{\sigma_c} = \frac{145\ MPa}{20.22\ MPa} = 7.17$$

The design factor based on ultimate strength is

$$N = \frac{s_u}{\sigma_c} = \frac{241\ MPa}{20.22\ MPa} = 11.9$$

Comment: These values should be acceptable for this application based on the normal tensile and compressive stresses. Table 9–2 in Chapter 9 suggests $N = 2$ based on yield strength for static loading and $N = 12$ based on ultimate strength for impact. If the person simply sits on the edge of the table, the loading would be considered to be static, but a person jumping on the edge would apply an impact load. The design factor of 11.9 is sufficiently close to the recommended value of 12. However, additional analysis should be performed to evaluate the tendency for the pipe to buckle as a column, as discussed in Chapter 16. Also, Reference 1 defines the procedures for evaluating the tendency for local buckling of a hollow circular pipe subjected to compression.

LESSON
21

15–6 COMBINED NORMAL AND SHEAR STRESSES

Rotating shafts in machines transmitting power represent good examples of members loaded in such a way as to produce combined bending and torsion. Figure 15–20 shows a shaft carrying two chain sprockets. Power is delivered to the shaft through the sprocket at C and transmitted down the shaft to the sprocket at B, which in turn delivers it to another shaft. Because it is transmitting power, the shaft between B and C is subject to a torque and torsional shear stress, as you learned in Chapter 10. In order for the sprockets to transmit torque, they must be pulled by one

FIGURE 15–20 Power transmission shafts.

side of the chain. At C, the back side of the chain must pull down with the force F_1 in order to drive the sprocket clockwise. Since the sprocket at B drives a mating sprocket, the front side of the chain would be in tension under the force F_2. The two forces, F_1 and F_2, acting downward cause bending of the shaft. Thus, the shaft must be analyzed for both torsional shear stress and bending stress. Then, since both stresses act at the same place on the shaft, their combined effect must be determined. The method of analysis to be used is called the *maximum shear stress theory of failure*, which is described next. Then example problems will be shown.

Maximum Shear Stress Theory of Failure

When the tensile or compressive stress caused by bending occurs at the same place that a shearing stress occurs, the two kinds of stress combine to produce a larger shearing stress. The maximum stress can be computed from

$$\tau_{max} = \sqrt{\left(\frac{\sigma}{2}\right)^2 + \tau^2} \qquad (15\text{–}2)$$

In Equation (15–2), σ refers to the magnitude of the tensile or compressive stress at the point, and τ is the shear stress at the same point. The result τ_{max} is the maximum shear stress at the point. The basis of Equation (15–2) is shown by using Mohr's circle later in this chapter.

The maximum shear stress theory of failure states that a member fails when the maximum shear stress exceeds the yield strength of the material in shear. This failure theory

shows good correlation with test results for ductile metals such as most steels.

Equivalent Torque

Equation (15–2) can be expressed in a simplified form for the particular case of a circular shaft subjected to bending and torsion. Evaluating the bending stress separately, the maximum tensile or compressive stress would be

$$\sigma = \frac{M}{S}$$

where $S = \dfrac{\pi D^3}{32}$ = section modulus

D = diameter of the shaft

M = bending moment on the section

The maximum stress due to bending occurs at the outside surface of the shaft, as shown in Figure 15–21.

Now consider the torsional shear stress separately. In Chapter 10 the torsional shear stress equation was derived as

$$\tau = \frac{T}{Z_p}$$

where $Z_p = \dfrac{\pi D^3}{16}$ = polar section modulus

T = torque on the section

FIGURE 15–21 Bending stress distribution in a round shaft.

FIGURE 15–22   Shear stress distribution in a round shaft.

The maximum shear stress occurs at the outer surface of the shaft all the way around the circumference, as shown in Figure 15–22.

Thus, the maximum tensile stress and the maximum torsional shear stress both occur at the same point in the shaft. Now let's use Equation (15–2) to get an expression for the combined stress in terms of the bending moment M, the torque T, and the shaft diameter D.

$$\tau_{max} = \sqrt{\left(\frac{\sigma}{2}\right)^2 + \tau^2}$$

$$\tau_{max} = \sqrt{\left(\frac{M}{2S}\right)^2 + \left(\frac{T}{Z_p}\right)^2} \qquad (15\text{–}3)$$

Notice that, from the definitions of S and Z_p given earlier,

$$2S = Z_p$$

Substituting this into Equation (15–3) gives

$$\tau_{max} = \sqrt{\left(\frac{M}{Z_p}\right)^2 + \left(\frac{T}{Z_p}\right)^2}$$

Factoring out Z_p yields

$$\tau_{max} = \frac{1}{Z_p}\sqrt{M^2 + T^2}$$

Sometimes the term $\sqrt{M^2 + T^2}$ is called the *equivalent torque* because it represents the amount of torque that would have to be applied to the shaft by itself to cause the equivalent magnitude of shear stress as the combination of bending and torsion. Calling the equivalent torque T_e,

$$T_e = \sqrt{M^2 + T^2} \qquad (15\text{–}4)$$

and

$$\tau_{max} = \frac{T_e}{Z_p} \qquad (15\text{–}5)$$

Equations (15–4) and (15–5) greatly simplify the calculation of the maximum shear stress in a circular shaft subjected to bending and torsion.

In the design of circular shafts subjected to bending and torsion, a design stress can be specified giving the

maximum allowable shear stress. This was done in Chapter 10.

$$\tau_d = \frac{s_{ys}}{N}$$

where s_{ys} is the yield strength of the material in shear. Since s_{ys} is seldom known, the approximate value found from $s_{ys} = s_y/2$ can be used. Then

$$\tau_d = \frac{s_y}{2N} \qquad (15\text{–}6)$$

where s_y is the yield strength in tension, as reported in most material property tables, such as those in Appendices A–14 through A–18. It is recommended that the value of the design factor be *no less than 4.* A rotating shaft subjected to bending is a good example of a repeated and reversed load. With each revolution of the shaft, a particular point on the surface is subjected to the maximum tensile and then the maximum compressive stress. Thus, fatigue is the expected mode of failure, and $N = 4$ or greater is recommended, based on yield strength.

Stress Concentrations

In shafts, stress concentrations are created by abrupt changes in geometry such as keyseats, shoulders, and grooves. See Appendix A–22 for values of stress concentration factors. The proper application of stress concentration factors to the equivalent torque Equations (15–4) and (15–5) should be considered carefully. If the value of K_t at a section of interest is equal for both bending and torsion, then it can be applied directly to Equation (15–5). That is,

$$\tau_{max} = \frac{T_e K_t}{Z_p} \qquad (15\text{–}7)$$

The form of Equation (15–7) can also be applied as a conservative calculation of τ_{max} by selecting K_t as the largest value for either torsion or bending.

To account for the proper K_t for both torsion and bending, Equation (15–4) can be modified as

$$T_e = \sqrt{(K_{tB}M)^2 + (K_{tT}T)^2} \qquad (15\text{–}8)$$

Then Equation (15–5) can be used directly to compute the maximum shear stress.

Example Problem 15–4

Specify a suitable material for the shaft shown in Figure 15–20. The shaft has a uniform diameter of 55 mm and rotates at 120 rpm while transmitting 3.75 kW of power. The chain sprockets at B and C are keyed to the shaft with sled-runner keyseats. Sprocket C receives the power, and sprocket B delivers it to another shaft. The bearings at A and D provide simple supports for the shaft.

Solution

Objective: Specify a suitable material for the shaft.

Given: Shaft and loading in Figure 15–20.
Power $= P = 3.75$ kW. Rotational speed $= n = 120$ rpm.
Shaft diameter $= D = 55$ mm.
Sled-runner keyseats at B and C.
Simple supports at A and D.

Analysis: The several steps to be used to solve this problem are outlined here.
1. The torque in the shaft will be computed for the known power and rotational speed from $T = P/n$, as developed in Chapter 10.
2. The tensions in the chains for sprockets B and C will be computed. These are the forces that produce bending in the shaft.
3. Considering the shaft to be a beam, the shearing force and bending moment diagrams will be drawn for it.
4. At the section where the maximum bending moment occurs, the equivalent torque T_e will be computed from Equation (15–4).
5. The polar section modulus Z_p and the stress concentration factor K_t will be determined.
6. The maximum shear stress will be computed from Equation (15–7).
7. The required yield strength of the shaft material will be computed by letting $\tau_{max} = \tau_d$ in Equation (15–6) and solving for s_y. Remember, let $N = 4$ or more.
8. A steel that has a sufficient yield strength will be selected from Appendix A–14.

Results: **Step 1:** The desirable unit for torque is N·m. Then it is most convenient to observe that the power unit of kilowatts is equivalent to the units of kN·m/s. Also, rotational speed must be expressed in rad/s.

$$n = \frac{120 \, \text{rev}}{\text{min}} \times \frac{2\pi \, \text{rad}}{\text{rev}} \times \frac{1 \, \text{min}}{60 \, \text{s}} = 12.57 \, \text{rad/s}$$

We can now compute torque.

$$T = \frac{P}{n} = \frac{3.75 \, \text{kN} \cdot \text{m}}{\text{s}} \times \frac{1}{12.57 \, \text{rad/s}} = 0.298 \, \text{kN} \cdot \text{m}$$

Step 2: The tensions in the chains are indicated in Figure 15–20 by the forces F_1 and F_2. For the shaft to be in equilibrium, the torque on both sprockets must be the same in magnitude but opposite in direction. On either sprocket the torque is the product of the chain force times the *radius* of the pulley. That is,

$$T = F_1 R_1 = F_2 R_2$$

The forces can now be computed.

$$F_1 = \frac{T}{R_1} = \frac{0.298 \, \text{kN} \cdot \text{m}}{75 \, \text{mm}} \times \frac{10^3 \, \text{mm}}{\text{m}} = 3.97 \, \text{kN}$$

$$F_2 = \frac{T}{R_2} = \frac{0.298 \, \text{kN} \cdot \text{m}}{50 \, \text{mm}} \times \frac{10^3 \, \text{mm}}{\text{m}} = 5.96 \, \text{kN}$$

Step 3: Figure 15–23 shows the complete shearing force and bending moment diagrams found by the methods of Chapter 11. The maximum bending moment is 1.06 kN·m at section B, where one of the sprockets is located.

Step 4: At section B, the torque in the shaft is 0.298 kN·m and the bending moment is 1.06 kN·m. Then

$$T_e = \sqrt{M^2 + T^2} = \sqrt{(1.06)^2 + (0.298)^2} = 1.10 \, \text{kN} \cdot \text{m}$$

Step 5: $Z_p = \dfrac{\pi D^3}{16} = \dfrac{\pi (55 \, \text{mm})^3}{16} = 32.67 \times 10^3 \, \text{mm}^3$

FIGURE 15-23
Shearing force and
bending moment dia-
grams for Example
Problem 15-4.

For the keyseat at section *B* securing the sprocket to the shaft, we will use $K_t = 1.6$, as reported in Figure 10-14 and Appendix A-22-11.

Step 6: $\tau_{max} = \dfrac{T_e K_t}{Z_p} = \dfrac{1.10 \times 10^3\, \text{N} \cdot \text{m}(1.6)}{32.67 \times 10^3\, \text{mm}^3} \times \dfrac{10^3\, \text{mm}}{\text{m}} = 53.9\, \text{MPa}$

Step 7: Let $\tau_{max} = \tau_d = \dfrac{s_y}{2N}$

Then $s_y = 2N\tau_{max} = (2)(4)(53.9\, \text{MPa}) = 431\, \text{MPa}$.

Step 8: Referring to Appendix A-14, we find that several alloys could be used. For example, AISI 1040, cold-drawn, has a yield strength of 490 MPa. Alloy AISI 1141 OQT 1300 has a yield strength of 469 MPa and also a very good ductility, as indicated by the 28% elongation. Either of these would be a reasonable choice.

15-7 EQUATIONS FOR STRESSES IN ANY DIRECTION

The *initial stress element* discussed in Section 15-4 was oriented in a convenient direction related to the member being analyzed. The methods of this section allow you to compute the stresses in any direction and to compute the maximum normal stresses and the maximum shear stress directly.

Figure 15-24 shows a stress element with orthogonal axes *u* and *v* superimposed on the initial element such that the *u*-axis is at an angle ϕ relative to the given *x*-axis. In general, there will be a normal stress σ_u and a shear stress τ_{uv} acting on the inclined surface *AC*. The following development will result in equations for those stresses.

Before going on, note that Figure 15-24(a) shows only two dimensions of an element that is really a three-dimensional cube. Part (b) of the figure shows the entire cube with each side having the dimension *h*.

Normal Stress in the *u*-Direction, σ_u

Visualize a wedge-shaped part of the initial element as shown in Figure 15-25. Note how the angle ϕ is located.

The side *AB* is the original left side of the initial element having the height *h*. The bottom of the wedge, side *BC*, is only a part of the bottom of the initial element where the length is determined by the angle ϕ itself.

$$BC = h \cdot \tan \phi$$

Also, the length of the sloped side of the wedge, *AC*, is

$$AC = \frac{h}{\cos \phi}$$

These lengths are important because now we are going to consider all the forces that act on the wedge. Because force is the product of stress times area, we need to know the area on which each stress acts. Starting with σ_x, it acts on the entire left face of the wedge having an area equal to h^2. Remember that the depth of the wedge perpendicular to the paper is also *h*. Then

$$\text{Force due to } \sigma_x = \sigma_x h^2$$

Using similar logic,

$$\text{Force due to } \sigma_y = \sigma_y h^2 \tan \phi$$

$$\text{Force due to } \tau_{xy} = \tau_{xy} h^2$$

$$\text{Force due to } \tau_{yx} = \tau_{yx} h^2 \tan \phi$$

(a)

(b)

FIGURE 15–24 Initial stress element with u and v axes included. (a) Stress element with inclined surface. (b) Three-dimensional element showing wedge.

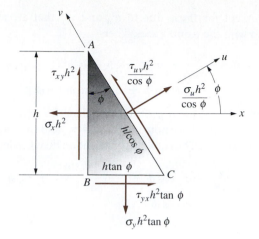

FIGURE 15–25 Free body diagram of wedge showing forces acting on each face.

The stresses acting on the inclined face of the wedge must also be considered:

$$\text{Force due to } \sigma_u = \frac{\sigma_u h^2}{\cos \phi}$$

$$\text{Force due to } \tau_{uv} = \frac{\tau_{uv} h^2}{\cos \phi}$$

Now using the principle of equilibrium, we can sum forces in the u-direction. From the resulting equation, we can solve for σ_u. The process is facilitated by resolving all forces into components perpendicular and parallel to the inclined face of the wedge. Figure 15–26 shows this for each

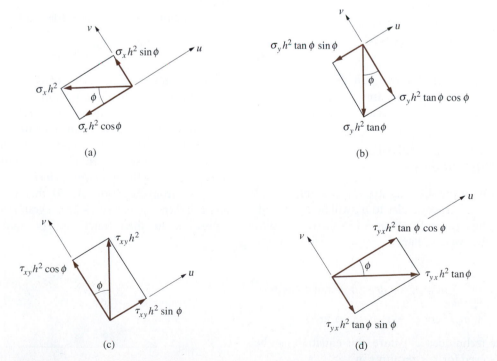

FIGURE 15–26 Resolution of forces along u and v directions. (a) Components of force due to σ_x. (b) Components of force due to σ_y. (c) Components of force due to τ_{xy}. (d) Components of force due to τ_{yx}.

force except for those due to σ_u and τ_{uv} that are already aligned with the u and v axes. Then

$$\sum F_u = 0 = \frac{\sigma_u h^2}{\cos \phi} - \sigma_x h^2 \cos \phi - \sigma_y h^2 \tan \phi \sin \phi$$
$$+ \tau_{xy} h^2 \sin \phi + \tau_{yx} h^2 \tan \phi \cos \phi$$

To begin solving for σ_u, all terms include h^2, which can be canceled. Also, we have noted that $\tau_{xy} = \tau_{yx}$ and we can let $\tan \phi = \sin \phi / \cos \phi$. The equilibrium equation then becomes

$$0 = \frac{\sigma_u}{\cos \phi} - \sigma_x \cos \phi - \frac{\sigma_y \sin \phi \sin \phi}{\cos \phi} + \tau_{xy} \sin \phi$$
$$+ \frac{\tau_{xy} \sin \phi \cos \phi}{\cos \phi}$$

Now multiply by $\cos \phi$ to obtain

$$0 = \sigma_u - \sigma_x \cos^2 \phi - \sigma_y \sin^2 \phi + \tau_{xy} \sin \phi \cos \phi$$
$$+ \tau_{xy} \sin \phi \cos \phi$$

Combine the last two terms and solve for σ_u.

$$\sigma_u = \sigma_x \cos^2 \phi + \sigma_y \sin^2 \phi - 2\tau_{xy} \sin \phi \cos \phi$$

This is a usable formula for computing σ_u, but a more convenient form can be obtained by using the following trigonometric identities:

$$\cos^2 \phi = \tfrac{1}{2} + \tfrac{1}{2} \cos 2\phi$$
$$\sin^2 \phi = \tfrac{1}{2} - \tfrac{1}{2} \cos 2\phi$$
$$\sin \phi \cos \phi = \tfrac{1}{2} \sin 2\phi$$

Making the substitutions gives

$$\sigma_u = \tfrac{1}{2}\sigma_x + \tfrac{1}{2}\sigma_x \cos 2\phi + \tfrac{1}{2}\sigma_y - \tfrac{1}{2}\sigma_y \cos 2\phi - \tau_{xy} \sin 2\phi$$

Combining terms, we obtain

Normal Stress in *u*-Direction ▼

$$\sigma_u = \tfrac{1}{2}(\sigma_x + \sigma_y) + \tfrac{1}{2}(\sigma_x - \sigma_y) \cos 2\phi - \tau_{xy} \sin 2\phi$$
$$(15\text{–}9)$$

Equation (15–9) can be used to compute the normal stress in any direction provided that the stress condition in some direction, indicated by the *x*- and *y*-axes, is known.

Shearing Stress, τ_{uv}, Acting Parallel to the Cut Plane

Now the equation for the shearing stress, τ_{uv}, acting parallel to the cut plane and perpendicular to σ_u will be developed. Again referring to Figures 15–25 and 15–26, we can sum the forces on the wedge-shaped element acting in the *v*-direction.

$$\sum F_v = 0 = \frac{\tau_{uv} h^2}{\cos \phi} + \sigma_x h^2 \sin \phi - \sigma_y h^2 \tan \phi \cos \phi$$
$$+ \tau_{xy} h^2 \cos \phi - \tau_{yx} h^2 \tan \phi \sin \phi$$

Using the same techniques as before, this equation can be simplified and solved for τ_{uv}, resulting in

Shearing Stress, τ_{uv} on Face of Element ▼

$$\tau_{uv} = -\tfrac{1}{2}(\sigma_x - \sigma_y) \sin 2\phi - \tau_{xy} \cos 2\phi \quad (15\text{–}10)$$

Equation (15–10) can be used to compute the shearing stress that acts on the face of the element at any angular orientation.

15–8 MAXIMUM STRESSES

In design and stress analysis, the maximum stresses are typically desired in order to ensure the safety of the load-carrying member. This section will develop equations for the *maximum principal stress, minimum principal stress,* and *maximum shear stress.* Both principal stresses are normal stresses, either tensile or compressive. The process includes using Equation (15–9) to find the angle ϕ at which the normal stress is maximum and using Equation (15–10) to find the angle at which the shear stress is maximum.

Principal Stresses   From the study of calculus, we know that the value of the angle ϕ at which the maximum or minimum normal stress occurs can be found by differentiating the function and setting the result equal to zero, then solving for ϕ. Differentiating Equation (15–9) gives

$$\frac{d\sigma_u}{d\phi} = 0 = 0 + \tfrac{1}{2}(\sigma_x - \sigma_y)(-\sin 2\phi)(2) - \tau_{xy} \cos 2\phi(2)$$

Dividing by $\cos 2\phi$ and simplifying gives

$$0 = -(\sigma_x - \sigma_y) \tan 2\phi - 2\tau_{xy}$$

Solving for $\tan 2\phi$ gives

$$\tan 2\phi = \frac{-2\tau_{xy}}{\sigma_x - \sigma_y} = \frac{-\tau_{xy}}{\tfrac{1}{2}(\sigma_x - \sigma_y)} \quad (15\text{–}11)$$

The angle ϕ is then

Angle Locating Maximum Principal Stress, σ_1 ▼

$$\phi = \frac{1}{2} \tan^{-1} \left[\frac{-\tau_{xy}}{\tfrac{1}{2}(\sigma_x - \sigma_y)} \right] \quad (15\text{–}12)$$

If we substitute the value of ϕ defined by Equations (15–11) and (15–12) into Equation (15–9), we can develop an equation for the maximum normal stress on the element. In addition, we can develop the equation for the minimum normal stress. These two stresses are called the *principal stresses* with σ_1 used to denote the *maximum principal stress* and σ_2 denoting the *minimum principal stress.*

Note from Equation (15–9) that we need values for $\sin 2\phi$ and $\cos 2\phi$. Figure 15–27 is a graphical aid to obtaining expressions for these functions. The right triangle has the

FIGURE 15–27   Development of $\sin 2\phi$ and $\cos 2\phi$ for principal stress formulas.

opposite and adjacent sides defined by the terms of the tangent function in Equation (15–11). The value of the hypotenuse is then found by the Pythagorean theorem as shown. It follows that

$$\tan 2\phi = \frac{-\tau_{xy}}{\frac{1}{2}(\sigma_x - \sigma_y)}$$

$$\sin 2\phi = \frac{-\tau_{xy}}{\sqrt{\left[\frac{1}{2}(\sigma_x - \sigma_y)\right]^2 + \tau_{xy}^2}}$$

$$\cos 2\phi = \frac{\frac{1}{2}(\sigma_x - \sigma_y)}{\sqrt{\left[\frac{1}{2}(\sigma_x - \sigma_y)\right]^2 + \tau_{xy}^2}}$$

Substituting into Equation (15–9) and simplifying gives

Maximum Principal Stress, σ_1 ▼

$$\sigma_{max} = \sigma_1 = \frac{1}{2}(\sigma_x + \sigma_y) + \sqrt{\left[\frac{1}{2}(\sigma_x - \sigma_y)\right]^2 + \tau_{xy}^2} \quad (15\text{–}13)$$

Because the square root has two possible values, + and −, we can also find the expression for the minimum principal stress, σ_2.

Minimum Principal Stress, σ_2 ▼

$$\sigma_{min} = \sigma_2 = \frac{1}{2}(\sigma_x + \sigma_y) - \sqrt{\left[\frac{1}{2}(\sigma_x - \sigma_y)\right]^2 + \tau_{xy}^2} \quad (15\text{–}14)$$

Conceivably, there would be a shear stress existing along with these normal stresses. It can be shown that by substituting the value of ϕ from Equation (15–12) into the shear stress Equation (15–10) that the result will be zero. In conclusion,

> On the element on which the principal stresses act, the shear stress is zero.

Maximum Shear Stress The same technique can be used to find the maximum shear stress by working with Equation (15–10). Differentiating with respect to ϕ and setting the result equal to zero gives

$$\frac{d\tau_{uv}}{d\phi} = 0 = -\frac{1}{2}(\sigma_x - \sigma_y)(\cos 2\phi)(2) - \tau_{xy}(-\sin 2\phi)(2)$$

Dividing by $\cos 2\phi$ and simplifying gives

$$\tan 2\phi = \frac{(\sigma_x - \sigma_y)}{2\tau_{xy}} = \frac{\frac{1}{2}(\sigma_x - \sigma_y)}{\tau_{xy}} \quad (15\text{–}15)$$

Solving for ϕ yields

Angle Locating Maximum Shear Stress, τ_{max} ▼

$$\phi = \frac{1}{2}\tan^{-1}\left[\frac{\frac{1}{2}(\sigma_x - \sigma_y)}{\tau_{xy}}\right] \quad (15\text{–}16)$$

Obviously, the value for ϕ from Equation (15–16) is different from that of Equation (15–12). In fact, we will see that the two values are *always 45 deg apart*.

Figure 15–28 shows the right triangle, from which we can find $\sin 2\phi$ and $\cos 2\phi$ as we did in Figure 15–27. The triangle is defined by Equation (15–15):

$$\tan 2\phi = \frac{\frac{1}{2}(\sigma_x - \sigma_y)}{\tau_{xy}}$$

$$\sin 2\phi = \frac{\frac{1}{2}(\sigma_x - \sigma_y)}{\sqrt{\left[\frac{1}{2}(\sigma_x - \sigma_y)\right]^2 + \tau_{xy}^2}}$$

$$\cos 2\phi = \frac{\tau_{xy}}{\sqrt{\left[\frac{1}{2}(\sigma_x - \sigma_y)\right]^2 + \tau_{xy}^2}}$$

Substituting these values into Equation (15–10) gives the maximum shear stress.

Maximum Shear Stress, τ_{max} ▼

$$\tau_{max} = \pm\sqrt{\left[\frac{1}{2}(\sigma_x - \sigma_y)\right]^2 + \tau_{xy}^2} \quad (15\text{–}17)$$

Here we should check to see if there is a normal stress existing on the element having the maximum shear stress. Substituting the value of ϕ from Equation (15–16) into the general normal stress Equation (15–9) gives

Normal Stress on Maximum Shear Stress Element ▼

$$\sigma_{avg} = \frac{1}{2}(\sigma_x + \sigma_y) \quad (15\text{–}18)$$

This is the formula for the *average* of the initial normal stresses, σ_x and σ_y. Thus, we can conclude:

> On the element on which the maximum shear stress occurs, there will also be a normal stress equal to the average of the initial normal stresses.

FIGURE 15–28 Development of $\sin 2\phi$ and $\cos 2\phi$ for maximum shear stress formula.

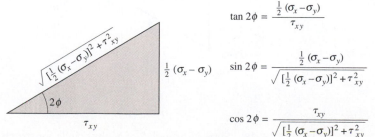

$$\tan 2\phi = \frac{\frac{1}{2}(\sigma_x - \sigma_y)}{\tau_{xy}}$$

$$\sin 2\phi = \frac{\frac{1}{2}(\sigma_x - \sigma_y)}{\sqrt{[\frac{1}{2}(\sigma_x - \sigma_y)]^2 + \tau_{xy}^2}}$$

$$\cos 2\phi = \frac{\tau_{xy}}{\sqrt{[\frac{1}{2}(\sigma_x - \sigma_y)]^2 + \tau_{xy}^2}}$$

15–9 MOHR'S CIRCLE FOR STRESS

The use of Equations (15–9) through (15–18) often presents difficulties because of the many possible combinations of signs for the terms σ_x, σ_y, τ_{xy}, and ϕ. Also, two roots of the square root and the fact that the inverse tangent function can result in angles in any of the four quadrants presents difficulties. Fortunately, there is a graphical aid, called *Mohr's circle*, available that can help to overcome these problems. The use of Mohr's circle should give you a better understanding of the general case of stress at a point.

It can be shown that the two equations, (15–9) and (15–10), for the normal and shear stress at a point in any direction can be combined and arranged in the form of the equation for a circle. First presented by Otto Mohr in 1895, the circle allows a rapid and exact computation of:

1. The maximum and minimum principal stresses [Equations (15–13) and (15–14)]

2. The maximum shear stress [Equation (15–17)]

3. The angles of orientation of the principal stress element and the maximum shear stress element [Equations (15–12) and (15–16)]

4. The normal stress that exists along with the maximum shear stress on the maximum shear stress element [Equation (15–18)]

5. The stress condition at any angular orientation of the stress element [Equations (15–9) and (15–10)]

The procedure listed next can be used to draw Mohr's circle. An example is illustrated in Figures 15–29 to 15–31 as the steps are discussed. Steps 1–7 are shown in Figure 15–29 with the initial stress element required for step 1 included. For this example we are using a generalized initial stress element with σ_x tensile (positive), σ_y compressive (negative), and τ_{xy} acting clockwise (CW). The circle will appear different for problems having stresses acting in different directions. No numerical data are given in this example problem, and the results are shown in symbol form to demonstrate the nature of the quantities that make up the complete Mohr's circle. Several problems with actual numerical data are shown later in this chapter.

Mohr's circle is drawn on a set of perpendicular axes with shear stress, τ, plotted vertically and normal stresses, σ, horizontally, as shown in Figure 15–29. The following convention is used in this book.

Sign Conventions

1. Positive normal stresses (tensile) are to the right.

2. Negative normal stresses (compressive) are to the left.

3. Shear stresses that tend to rotate the stress element clockwise (CW) are plotted upward on the τ-axis.

4. Shear stresses that tend to rotate the stress element counterclockwise (CCW) are plotted downward.

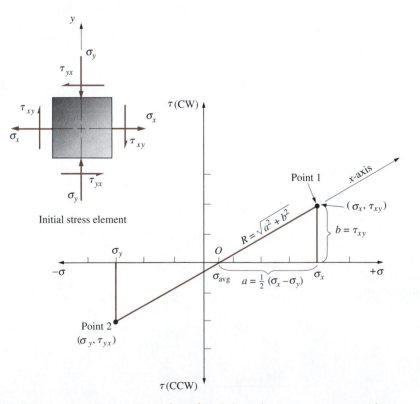

FIGURE 15–29 Steps 1 through 7 of Mohr's circle construction procedure.

FIGURE 15–30 Completed Mohr's circle.

FIGURE 15–31 General form of final results from Mohr's circle analysis.

(a) Initial stress element

(b) Principal stress element

(c) Maximum shear stress element

PROCEDURE FOR DRAWING MOHR'S CIRCLE ▼

1. Identify the stress condition at the point of interest and represent it as the *initial stress element* in the manner shown in Figure 15–29.

2. The combination σ_x and τ_{xy} is plotted as *point 1* on the $\sigma - \tau$ plane.

3. The combination σ_y and τ_{yx} is then plotted as *point 2*. Note that τ_{xy} and τ_{yx} always act in opposite directions. Therefore, one point will be plotted above the σ-axis and one will be below.

4. Draw a straight line between the two points.

5. This line crosses the σ-axis at the center of Mohr's circle, which is also the value of the *average normal stress* applied to the initial stress element. The location of the center can be observed from the data used to plot the points or computed from Equation (15–18), repeated here:

$$\sigma_{\text{avg}} = \tfrac{1}{2}(\sigma_x + \sigma_y)$$

For convenience, label the center O.

6. Identify the line from O through point 1 (σ_x, τ_{xy}) as the x-axis. This line corresponds to the original x-axis and is essential to correlating the data from Mohr's circle to the original x- and y-directions.

7. The points O, σ_x, and point 1 form an important right triangle because the distance from O to point 1, the hypotenuse of the triangle, is equal to the radius of the circle, R. Calling the other two sides a and b, the following calculations can be made:

$$a = \tfrac{1}{2}(\sigma_x - \sigma_y)$$

$$b = \tau_{xy}$$

$$R = \sqrt{a^2 + b^2} = \sqrt{\left[\tfrac{1}{2}(\sigma_x - \sigma_y)\right]^2 + \tau_{xy}^2}$$

Note that the equation for R is identical to Equation (15–17) for the maximum shear stress on the element. Thus,

The length of the radius of Mohr's circle is equal to the magnitude of the maximum shear stress.

Steps 8–11 are shown in Figure 15–30.

8. Draw the complete circle with the center at O and the radius R.

9. Draw the vertical diameter of the circle. The point at the top of the circle has the coordinates $(\sigma_{avg}, \tau_{max})$, where the shear stress has the clockwise (CW) direction. The point at the bottom of the circle represents $(\sigma_{avg}, \tau_{max})$ where the shear stress is counterclockwise (CCW).

10. Identify the points on the σ-axis at the ends of the horizontal diameter as σ_1 at the right (the maximum principal stress) and σ_2 at the left (the minimum principal stress). Note that the shear stress is zero at these points.

11. Determine the values for σ_1 and σ_2 from

$$\sigma_1 = \text{"}O\text{"} - R \qquad \textbf{(15–19)}$$

$$\sigma_2 = \text{"}O\text{"} - R \qquad \textbf{(15–20)}$$

where "O" represents the coordinate of the center of the circle, σ_{avg}, and R is the radius of the circle. Thus, Equations (15–19) and (15–20) are identical to Equations (15–13) and (15–14) for the principal stresses.

The following steps determine the angles of orientation of the principal stress element and maximum shear stress element. An important concept to remember is that *angles obtained from Mohr's circle are double the true angles*. The reason for this is that the equations on which it is based; Equations (15–9) and (15–10), are functions of 2ϕ.

12. The orientation of the principal stress element is determined by finding the angle *from the x-axis* to the "σ_1"-axis, labeled 2ϕ in Figure 15–30. From the data on the circle you can see that

$$2\phi = \tan^{-1}\left(\frac{b}{a}\right)$$

The argument of this inverse tangent function is the same as the absolute value of the argument shown in

Equation (15–12). Problems with signs for the resulting angle are avoided by noting the *direction from the x-axis to the "σ_1"-axis* on the circle, clockwise for the present example. Then the principal stress element is rotated *in the same direction* from the x-axis by an amount ϕ to locate the face on which the maximum principal stress σ_1 acts.

13. Draw the principal stress element in its proper orientation as determined from step 12 with the two principal stresses σ_1 and σ_2 shown [see Figure 15–31(a) and (b)].

14. The orientation of the maximum shear stress element is determined by finding the angle *from the x-axis to* the τ_{max}-axis, labeled $2\phi'$ in Figure 15–30. In the present example,

$$2\phi' = 90° - 2\phi$$

From trigonometry it can be shown that this is equivalent to finding the inverse tangent of a/b, the reciprocal of the argument used to find 2ϕ. Thus, it is a true evaluation of Equation (15–16), derived to find the angle of orientation of the element on which the maximum shear stress occurs.

Again problems with signs for the resulting angle are avoided by noting the *direction from the x-axis to the τ_{max}-axis* on the circle, counterclockwise for the present example. Then the maximum shear stress element is rotated *in the same direction* from the x-axis by an amount ϕ' to locate the face on which the clockwise maximum shear stress acts.

15. Draw the maximum shear stress element in its proper orientation as determined from step 14 with the shear stresses on all four faces and the average normal stress acting on each face [see Figure 15–31(c)].

As a whole, Figure 15–31 is the desired result from a Mohr's circle analysis. Shown are the initial stress element that establishes the x- and y-axes, the principal stress element drawn in proper rotation relative to the x-axis, and the maximum shear stress element also drawn in the proper rotation relative to the x-axis.

Example Problem 15–5 demonstrates this procedure with specific data for the stresses on the initial stress element.

Example Problem 15–5

It has been determined that a point in a load-carrying member is subjected to the following stress condition:

$$\sigma_x = 400\,\text{MPa} \qquad \sigma_y = -300\,\text{MPa} \qquad \tau_{xy} = 200\,\text{MPa (CW)}$$

Perform the following:
a. Draw the initial stress element.
b. Draw the complete Mohr's circle, labeling critical points.
c. Draw the complete principal stress element.
d. Draw the maximum shear stress element.

Solution

The 15–step *Procedure for drawing Mohr's circle* is used here to complete the problem. The numerical results from steps 1–12 are summarized here and shown in Figure 15–32.

Step 1: The initial stress element is shown at the upper left of Figure 15–32.

Step 2: Point 1 is plotted at $\sigma_x = 400$ MPa and $\tau_{xy} = 200$ MPa in quadrant 1.

Step 3: Point 2 is plotted at $\sigma_y = -300$ MPa and $\tau_{yx} = -200$ MPa in quadrant 3.

FIGURE 15–32
Complete Mohr's circle
for Example Problem
15–15.

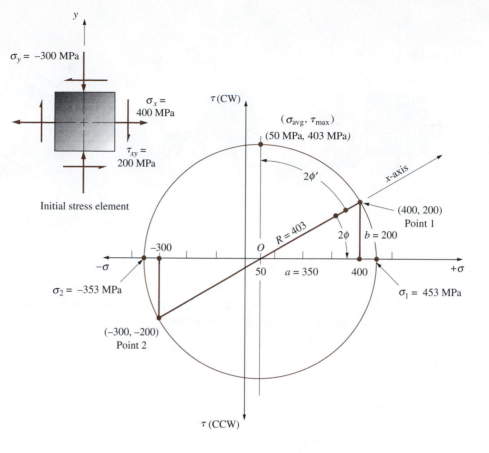

Step 4: The line from point 1 to point 2 has been drawn.

Step 5: The line from step 4 crosses the σ-axis at the average applied normal stress, called point O in Figure 15–32, computed from

$$\sigma_{avg} = \tfrac{1}{2}(\sigma_x + \sigma_y) = \tfrac{1}{2}[400 + (-300)] = 50 \text{ MPa}$$

Step 6: Point O is the center of the circle. The line from point O through point 1 is labeled as the x-axis to correspond with the x-axis on the initial stress element.

Step 7: The values of a, b, and R are found using the triangle formed by the lines from point O to point 1 to $\sigma_x = 400$ MPa and back to point O.
The lower side of the triangle,

$$a = \tfrac{1}{2}(\sigma_x - \sigma_y) = \tfrac{1}{2}[400 - (-300)] = 350 \text{ MPa}$$

The vertical side of the triangle, b, is completed from

$$b = \tau_{xy} = 200 \text{ MPa}$$

The radius of the circle, R, is completed from

$$R = \sqrt{a^2 + b^2} = \sqrt{(350)^2 + (200)^2} = 403 \text{ MPa}$$

Step 8: This is the drawing of the circle with point O as the center at $\sigma_{avg} = 50$ MPa and a radius of $R = 403$ MPa.

Step 9: The vertical diameter of the circle has been drawn through point O. The intersection of this line with the circle at the top indicates the value of $\tau_{max} = 403$ MPa, the same as the value of R.

Step 10: The maximum principal stress, σ_1, is at the right end of the horizontal diameter of the circle and the minimum principal stress, σ_2, is at the left.

Step 11: The values for σ_1 and σ_2 are

$$\sigma_1 = O + R = 50 + 403 = 453 \text{ MPa}$$
$$\sigma_2 = O - R = 50 - 403 = -353 \text{ MPa}$$

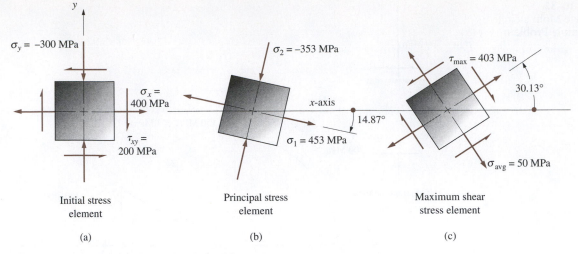

FIGURE 15–33 Results for Example Problem 15–5.

Step 12: The angle 2ϕ is shown on the circle as the angle from the x-axis to the σ_1-axis, a clockwise rotation. The value is computed from

$$2\phi = \tan^{-1}\left(\frac{b}{a}\right) = \tan^{-1}\left(\frac{200}{350}\right) = 29.74°$$

Note that 2ϕ is CW from the x-axis to σ_1 on the circle.

$$\phi = \frac{29.74°}{2} = 14.87°$$

Step 13: Using the results from steps 11 and 12, the principal stress element is drawn as shown in Figure 15–33(b). The element is rotated 14.87° CW from the original x-axis to the face on which the tensile stress $\sigma_1 = 453$ MPa acts. The compressive stress $\sigma_2 = -353$ MPa acts on the faces perpendicular to the σ_1 faces.

Step 14: The angle $2\phi'$ is shown in Figure 15–32 drawn from the x-axis CCW to the vertical diameter that locates τ_{max} at the top of the circle. Its value can be found in either of two ways. First, using Equation (15–16), observe that the numerator is the same as the value of a and the denominator is the same as the value of b from the construction of the circle. Then

$$2\phi' = \tan^{-1}(a/b) = \tan^{-1}(350/200) = 60.26° \text{ CCW}$$

Or, using the geometry of the circle, we can compute

$$2\phi' = 90° - 2\phi = 90° - 29.74° = 60.26° \text{ CCW}$$

Then the angle ϕ' is one-half of $2\phi'$.

$$\phi' = \frac{60.26°}{2} = 30.13°$$

Step 15: The maximum shear stress element is drawn in Figure 15–33(c), rotated 30.13° CCW from the original x-axis to the face on which the positive τ_{max} acts. The maximum shear stress of 403 MPa is shown on all four faces with vectors that create the two pairs of opposing couples characteristic of shear stresses on a stress element. Also shown is the tensile stress $\sigma_{avg} = 50$ MPa acting on all four faces of the element.

Summary of Results for
Example Problem 15–5
Mohr's Circle

Given: $\sigma_x = 400$ MPa $\qquad \sigma_y = -300$ MPa $\qquad \tau_{xy} = 200$ MPa CW

Results: Figures 15–32 and 15–33.

$\sigma_1 = 453$ MPa $\qquad \sigma_2 = -353$ MPa $\qquad \phi = 14.87°$ CW from x-axis

$\tau_{max} = 403$ MPa $\qquad \sigma_{avg} = 50$ MPa $\qquad \phi' = 30.13°$ CCW from x-axis

Comment: The x-axis is in the first quadrant.

Examples of the Use of Mohr's Circle

The data for Example Problem 15–5 in the preceding section and Example Problems 15–6 through 15–11 to follow have been selected to demonstrate a wide variety of results. A major variable is the quadrant in which the x-axis lies and the corresponding definition of the angles of rotation for the principal stress element and the maximum shear stress element.

Example Problems 15–9, 15–10, and 15–11 present the special cases of biaxial stress with no shear, uniaxial tension with no shear, and pure shear. These should help you understand the behavior of load-carrying members subjected to such stresses.

The solution for each Example Problem to follow is Mohr's circle itself along with the stress elements, properly labeled. For each problem, the objectives are to

a. Draw the initial stress element.

b. Draw the complete Mohr's circle, labeling critical points.

c. Draw the complete principal stress element.

d. Draw the complete shear stress element.

Example Problem 15–6
Mohr's Circle

Given: $\sigma_x = 60\text{ ksi}$ $\sigma_y = -40\text{ ksi}$ $\tau_{xy} = 30\text{ ksi CCW}$

Results: Figure 15–34.

$\sigma_1 = 68.3\text{ ksi}$ $\sigma_2 = -48.3\text{ ksi}$ $\phi = 15.48°\text{ CCW}$

$\tau_{max} = 58.3\text{ ksi}$ $\sigma_{avg} = 10.0\text{ ksi}$ $\phi' = 60.48°\text{ CCW}$

Comment: The x-axis is in the second quadrant.

FIGURE 15–34 Results for Example Problem 15–6. X-axis in the second quadrant.

**Example Problem
15–7
Mohr's Circle**

Given: $\sigma_x = -120\,\text{MPa}$ $\sigma_y = 180\,\text{MPa}$ $\tau_{xy} = 80\,\text{MPa CCW}$

Results: Figure 15–35.

$\sigma_1 = 200\,\text{MPa}$ $\sigma_2 = -140\,\text{MPa}$ $\phi = 75.96°\,\text{CCW}$

$\tau_{max} = 170\,\text{MPa}$ $\sigma_{avg} = 30\,\text{MPa}$ $\phi' = 59.04°\,\text{CW}$

Comment: The x-axis is in the third quadrant.

$\alpha = 28.07°$
$2\phi = 180 - \alpha = 151.93°$
$\phi = 75.96°\,\text{CCW}$
$2\phi' = 90° + \alpha = 118.07°$
$\phi' = 59.04°\,\text{CW}$

FIGURE 15–35 Results for Example Problem 15–7. X-axis in the third quadrant.

**Example Problem
15–8
Mohr's Circle**

Given: $\sigma_x = -30 \text{ ksi}$ $\sigma_y = 20 \text{ ksi}$ $\tau_{xy} = 40 \text{ ksi CW}$

Results: Figure 15–36.

$$\sigma_1 = 42.17 \text{ ksi} \qquad \sigma_2 = -52.17 \text{ ksi} \qquad \phi = 61.0° \text{ CW}$$
$$\tau_{max} = 47.17 \text{ ksi} \qquad \sigma_{avg} = -5.0 \text{ ksi} \qquad \phi' = 16.0° \text{ CW}$$

Comment: The x-axis is in the fourth quadrant.

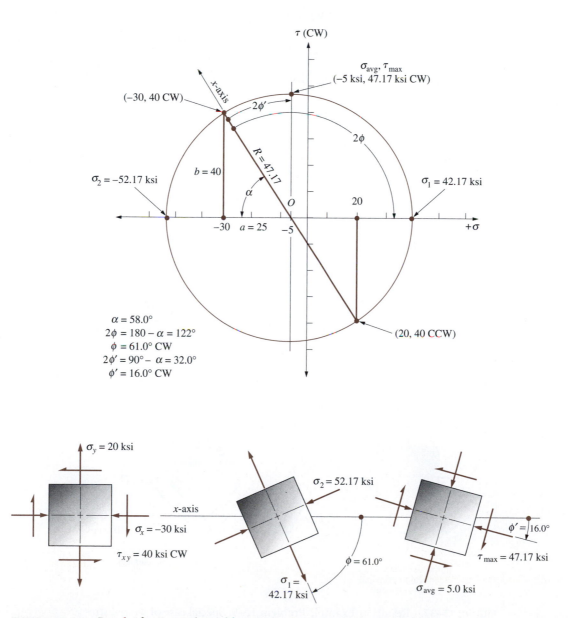

FIGURE 15–36   Results for Example Problem 15–8. X-axis in the fourth quadrant.

**Example Problem
15–9
Mohr's Circle**

Given: $\sigma_x = 220\,\text{MPa}$ $\sigma_y = -120\,\text{MPa}$ $\tau_{xy} = 0\,\text{MPa}$

Results: Figure 15–37.

$$\sigma_1 = 220\,\text{MPa} \qquad \sigma_2 = -120\,\text{MPa} \qquad \phi = 0°$$

$$\tau_{max} = 170\,\text{MPa} \qquad \sigma_{avg} = 50\,\text{MPa} \qquad \phi' = 45.0°\,\text{CCW}$$

Comment: Special case of biaxial stress with no shear on the given element.

FIGURE 15–37   Results of Example Problem 15–9. Special case of biaxial stress with no shear.

Example Problem
15–10
Mohr's Circle

Given: $\sigma_x = 40\,\text{ksi}$ $\sigma_y = 0\,\text{ksi}$ $\tau_{xy} = 0\,\text{ksi}$

Results: Figure 15–38.

$$\sigma_1 = 40\,\text{ksi} \qquad \sigma_2 = 0\,\text{ksi} \qquad \phi = 0°$$
$$\tau_{max} = 20\,\text{ksi} \qquad \sigma_{avg} = 20\,\text{ksi} \qquad \phi' = 45.0°\,\text{CCW}$$

Comment: Special case of uniaxial tension with no shear.

FIGURE 15–38   Results of Example Problem 15–10. Special case of uniaxial tension.

**Example Problem
15–11
Mohr's Circle**

Given: $\sigma_x = 0\,\text{ksi}$ $\sigma_y = 0\,\text{ksi}$ $\tau_{xy} = 40\,\text{ksi CW}$

Results: Figure 15–39.

$$\sigma_1 = 40\,\text{ksi} \qquad \sigma_2 = -40\,\text{ksi} \qquad \phi = 45°\,\text{CW}$$
$$\tau_{max} = 40\,\text{ksi} \qquad \sigma_{avg} = 0\,\text{ksi} \qquad \phi' = 0°$$

Comment: Special case of pure shear.

FIGURE 15–39   Results of Example Problem 15–11. Special case of pure shear.

Maximum Shear Stress Theory of Failure

One of the most widely used principles of design is the *maximum shear stress theory of failure,* which states:

> A ductile material can be expected to fail when the maximum shear stress to which the material is subjected exceeds the yield strength of the material in shear.

Of course, to apply this theory, it is necessary to be able to compute the magnitude of the maximum shear stress. If the member is subjected to pure shear, such as torsional shear stress, direct shear stress, or shearing stress in beams in bending, the maximum shear stress can be computed directly from formulas such as those developed in this book. But if a combined stress condition exists, the use of Equation (15–17) or Mohr's circle should be used to determine the maximum shear stress.

One special case of combined stress that occurs often is one in which a normal stress in only one direction is combined with a shear stress. For example, a round bar could be subjected to a direct axial tension while also being twisted. In many types of mechanical power transmissions, shafts are subjected to bending and torsion simultaneously. Certain fasteners may be subjected to tension combined with direct shear.

A simple formula can be developed for such cases using Mohr's circle or Equation (15–17). If only a normal stress in the x-direction, σ_x, combined with a shearing stress, τ_{xy} exists, the maximum shear stress is that given earlier as Equation (15–2).

$$\tau_{max} = \sqrt{(\sigma_x/2)^2 + \tau_{xy}^2} \qquad (15\text{–}2)$$

This formula can be developed from Equation (15–17) by letting $\sigma_y = 0$.

Example Problem 15–12   A solid circular bar is 45 mm in diameter and is subjected to an axial tensile force of 120 kN along with a torque of 1150 N·m. Compute the maximum shear stress in the bar.

Solution   Objective: Compute the maximum shear stress in the bar.

Given: Diameter = D = 45 mm.
Axial force = F = 120 kN = 120 000 N.
Torque = T = 1150 N·m = 1 150 000 N·mm.

Analysis: Use Equation (15–2) to compute τ_{max}.

Results: 1. First, the applied normal stress can be found using the direct stress formula.

$$\sigma = F/A$$
$$A = \pi D^2/4 = \pi(45\text{ mm})^2/4 = 1590\text{ mm}^2$$
$$\sigma = (120\,000\text{ N})/(1590\text{ mm}^2) = 75.5\text{ N/mm}^2 = 75.5\text{ MPa}$$

2. Next, the applied shear stress can be found from the torsional shear stress formula.

$$\tau = T/Z_p$$
$$Z_p = \pi D^3/16 = \pi(45\text{ mm})^3/16 = 17\,892\text{ mm}^3$$
$$\tau = (1\,150\,000\text{ N·mm})/(17\,892\text{ mm}^3) = 64.3\text{ N/mm}^2 = 64.3\text{ MPa}$$

3. Then using Equation (15–2) yields

$$\tau_{max} = \sqrt{\left(\frac{75.5\text{ MPa}}{2}\right)^2 + (64.3\text{ MPa})^2} = 74.6\text{ MPa}$$

Comment: This stress should be compared with the design shear stress.

15–10 STRESS CONDITION ON SELECTED PLANES

There are some cases in which it is desirable to know the stress condition on an element at some selected angle of orientation relative to the reference direction. Figures 15–40 and 15–41 show examples.

The wood block in Figure 15–40 shows that the grain of the wood is inclined at an angle of 30° CCW from the given x-axis. Because the wood is very weak in shear parallel to the grain, it is desirable to know the stresses in this direction.

Figure 15–41 shows a member fabricated by welding two components along a seam inclined at an angle relative to the given x-axis. The welding operation could produce a weaker material in the near vicinity of the weld, particularly if the component parts were made from steel that was heat-treated prior to welding. The same is true of many aluminum alloys. For such cases, the allowable stresses are somewhat lower along the weld line.

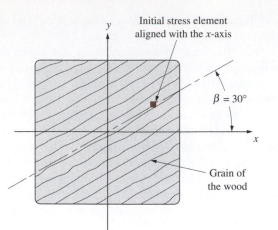

FIGURE 15–40   Cross section of a wood post with the grain running 30° with respect to the *x*-axis.

Additional examples include the following:

- A flat-sided tank made from welded steel plates for which some of the weld lines are inclined at some angle to the major axes along which loads are applied. The loads may be due to pressure in the tank, the weight of the structure, or accessory equipment mounted on the tank. In addition, there may be penetrations in the tank walls to install ports used to fill or drain the tank, for observation windows, or to install sensors. These elements may be welded in place at different angles, thus requiring the knowledge of stress levels along those angles. Many materials such as aluminum or heat-treated steel have significantly lower strength near welds in the heat-affected zone.

- Some large pipes are made from flat sheets of steel that are spiral wound into the desired tubular shape and then continuously welded along the seam. Knowing the stresses in line with the seam is important.

- Composite materials are inherently nonisotropic, being stronger in directions where the strong, stiff reinforcing fibers are laid and typically much weaker at other angles, depending on the manner of laying up the fibers. Analysis of stresses in several directions would be important. See Section 8–12 in Chapter 8.

- The environmental conditions to which a part is exposed during operation may also affect the material properties. For example, a furnace part may be subjected to local heating by radiant energy along a particular line. The

FIGURE 15–41   Flat bar welded along a 20° inclined seam.

strength of the heated material will be lower than that which remains cool, and thus it is desirable to know the stress condition along the angle of the heat-affected zone.

Mohr's circle can be used to determine the stress condition at specified angles of orientation of the stress element. The procedure is outlined here and demonstrated in Example Problem 15–13.

PROCEDURE FOR FINDING THE STRESS AT A SPECIFIED ANGLE ▼

Given: Stress condition on the given element aligned in the *x*- and *y*-directions.

Objective: Determine the normal and shear stresses on the element at a specified angle, β, relative to the given *x*-direction.

Step 1:   Draw the complete Mohr's circle for the element.

Step 2:   Identify the line representing the x-axis on the circle.

Step 3:   Measure the angle 2β from the x-axis and draw a line through the center of Mohr's circle extending to the two intersections with the circle. This line represents the axis aligned with the direction of interest.

Step 4:   Using the geometry of the circle, determine the coordinates (σ and τ) of the first point of intersection nearest to the x-axis. The σ-component is the normal stress acting on the element in the direction of β. The τ-component is the shear stress acting on the faces of the element. The coordinates of the second point represent the normal and shear stresses acting on the faces of the element of interest that are parallel to the β-axis.

Step 5:   Draw the element of interest showing the normal and shear stress acting on it.

Example Problem 15–13   For the flat bar in Figure 15–41 welded along a seam at an angle of 20° CCW from the *x*-axis, the stress element aligned parallel with the *x*- and *y*-axes is subjected to these stresses:

$$\sigma_x = 400\,\text{MPa} \qquad \sigma_y = -300\,\text{MPa} \qquad \tau_{xy} = 200\,\text{MPa CW}$$

Determine the stress condition on the element inclined at the 20° angle, aligned with the weld seam.

Solution   Objective: Draw the stress element aligned with the weld seam at 20° with the *x*-axis.

Given: Note that the given stress element is the same as that used in Example Problem 15–5. The basic Mohr's circle from that problem is shown in Figure 15–32 and it is reproduced in Figure 15–42 with additional construction described next.

FIGURE 15–42
Complete Mohr's circle
for Example Problem
15–13 showing stresses
on an element inclined
at 20° CCW from the
x-axis.

(a) Mohr's circle

(b) Given stress element   (c) Stress element at 20°

Analysis: Use the *procedure for finding the stress at a specified angle.*

Results: **Steps 1 and 2:** are shown on the original Mohr's circle.

Step 3: The desired axis is one inclined at 20° CCW from the x-axis. Recalling that angles in Mohr's circle are *double* the true angles, we can draw a line through the center of the circle at an angle of $2\beta = 40°$ CCW from the x-axis. The intersection of this line with the circle, labeled *A* in the figure, locates the point on the circle defining the stress condition of the desired element. The coordinates of this point, (σ_A, τ_A), give the normal and shear stresses acting on one set of faces of the desired stress element.

Step 4: Simple trigonometry and the basic geometry of the circle can be used to determine σ_A and τ_A by projecting lines vertically and horizontally from point *A* to the σ- and τ-axes, respectively. The total angle from the σ-axis to the axis through point *A*, called η (eta) in the figure, is the sum of 2ϕ and 2β. In Example Problem 15–5, we found $2\phi = 29.74°$. Then

$$\eta = 2\phi + 2\beta = 29.74° + 40° = 69.74°$$

A triangle has been identified in Figure 15–42 with sides labeled d, g, and R. From this triangle, we can compute

$$d = R\cos\eta = (403)\cos 69.74° = 140$$

$$g = R\sin\eta = (403)\sin 69.74° = 378$$

These values enable the computation of

$$\sigma_A = O + d = 50 + 140 = 190 \text{ MPa}$$

$$\tau_A = g = 378 \text{ MPa CW}$$

where O indicates the value of the normal stress at the center of Mohr's circle.

The stresses on the remaining set of faces of the desired stress element are the coordinates of the point A' located 180° from A on the circle and, therefore, 90° from the faces on which (σ_A, τ_A) act. Projecting vertically and horizontally from A' to the σ- and τ-axes locates $\sigma_{A'}$ and $\tau_{A'}$. By similar triangles we can say that $d' = d$ and $g' = g$. Then

$$\sigma_{A'} = O - d' = 50 - 140 = -90 \text{ MPa}$$

$$\tau_{A'} = g' = 378 \text{ MPa CCW}$$

Comment: Figure 15–42(c) shows the final element inclined at 20° from the x-axis. This is the stress condition experienced by the material along the weld line.

15–11 SPECIAL CASE IN WHICH BOTH PRINCIPAL STRESSES HAVE THE SAME SIGN

In preceding sections involving Mohr's circle, we used the convention that σ_1 is the maximum principal stress and σ_2 is the minimum principal stress. This is true for cases of plane stress (stresses applied in a single plane) if σ_1 and σ_2 have opposite signs, that is, when one is tensile and one is compressive. Also, in such cases, the shear stress found at the top of the circle (equal to the radius, R) is the true maximum shear stress on the element.

But special care must be exercised when Mohr's circle indicates that σ_1 and σ_2 have the same sign. Even though we are dealing with plane stress, the true stress element is three-dimensional and should be represented as a cube rather than a square, as shown in Figure 15–43. Faces 1, 2, 3, and 4 correspond to the sides of the square element and faces 5 and 6 are the "front" and "back." For plane stress the stresses on faces 5 and 6 are zero.

For the three-dimensional stress element, three principal stresses exist, called σ_1, σ_2, and σ_3, acting on the mutually perpendicular sides of the stress element. Convention calls for the following order:

$$\sigma_1 > \sigma_2 > \sigma_3$$

Thus, σ_3 is the true minimum principal stress and σ_1 is the true maximum principal stress. It can also be shown that the true maximum shear stress can be computed from

$$\tau_{max} = \tfrac{1}{2}(\sigma_1 - \sigma_3) \qquad (15\text{–}21)$$

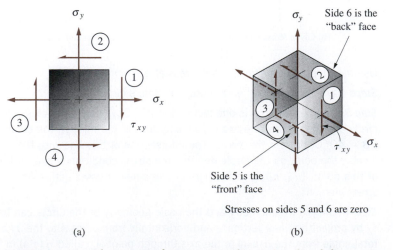

(a)

(b)

FIGURE 15–43   Plane stress shown as two-dimensional and three-dimensional stress elements. (a) Two-dimensional stress element. (b) Three-dimensional stress element.

$\sigma_y = 120$ MPa

$\sigma_x = 200$ MPa

$\tau_{xy} = 40$ MPa

(a) Initial stress element

τ (CW)

(160, 56.6)

(200, 40)

$\sigma_1 = 216.6$ MPa

$\sigma_2 = 103.4$ MPa

$R = 56.6$

σ

(120, −40)

(b) Mohr's circle

FIGURE 15–44 Mohr's circle for which σ_1 and σ_2 are both positive.

Figure 15–44 illustrates a case in which the three-dimensional stress element must be considered. The initial stress element, shown in part (a), carries the following stresses:

$$\sigma_x = 200 \text{ MPa} \qquad \sigma_y = 120 \text{ MPa} \qquad \tau_{xy} = 40 \text{ MPa CW}$$

Part (b) of the figure shows the conventional Mohr's circle, drawn according to the procedure outlined in Section 15–9. Note that both σ_1 and σ_2 are positive or tensile. Then, considering that the stress on the "front" and "back" faces is zero, this is the true minimum principal stress. We can then say that

$$\sigma_1 = 216.6 \text{ MPa}$$
$$\sigma_2 = 103.4 \text{ MPa}$$
$$\sigma_3 = 0 \text{ MPa}$$

From Equation (15–21), the true maximum shear stress is

$$\tau_{max} = \tfrac{1}{2}(\sigma_1 - \sigma_3) = \tfrac{1}{2}(216.6 - 0) = 108.3 \text{ MPa}$$

These concepts can be visualized graphically by creating a set of three Mohr's circles rather than only one. Figure 15–45 shows the circle obtained from the initial stress element, a second circle encompassing σ_1 and σ_3, and a third encompassing σ_2 and σ_3. Thus, each circle represents the

plane on which two of the three principal stresses act. The point at the top of each circle indicates the largest shear stress that would occur in that plane. Then the largest circle, drawn for σ_1 and σ_3, produces the true maximum shear stress, and its value is consistent with Equation (15–21).

Figure 15–46 illustrates another case in which the principal stresses from the initial stress element have the same sign, both negative in this case. The initial stresses are

$$\sigma_x = -50 \text{ MPa}$$
$$\sigma_y = -180 \text{ MPa}$$
$$\tau_{xy} = 30 \text{ MPa CCW}$$

Here, too, the supplementary circles must be drawn. But in this case, the zero stress on the "front" and "back" faces of the element becomes the *maximum* principal stress (σ_1). That is,

$$\sigma_1 = 0 \text{ MPa}$$
$$\sigma_2 = -43.4 \text{ MPa}$$
$$\sigma_3 = -186.6 \text{ MPa}$$

and the maximum shear stress is

$$\tau_{max} = \tfrac{1}{2}(\sigma_1 - \sigma_3) = \tfrac{1}{2}[0 - (-186.6)] = 93.3 \text{ MPa}$$

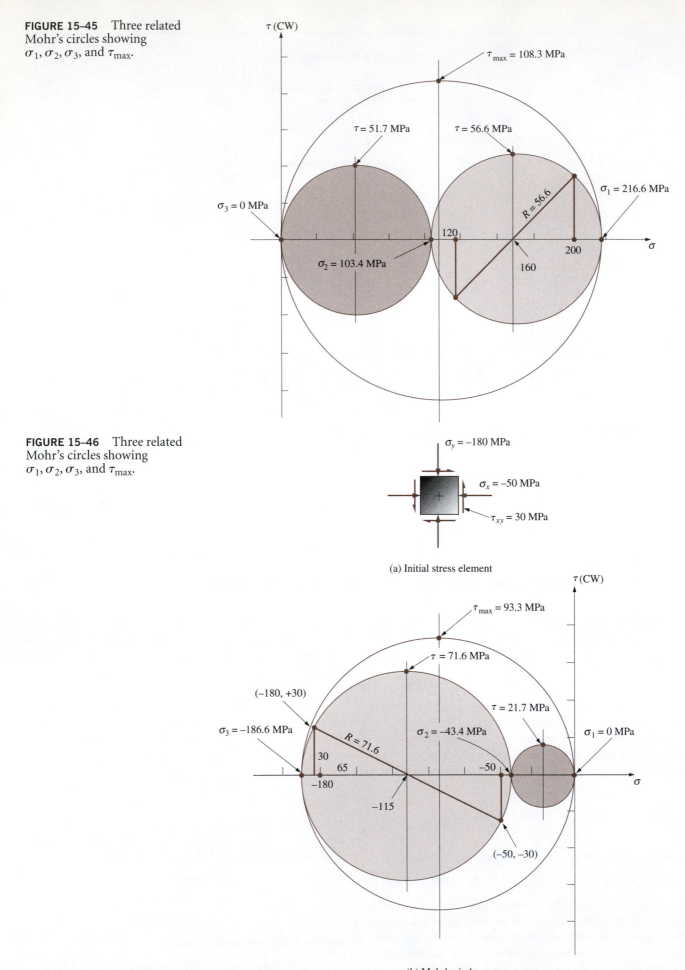

FIGURE 15–45 Three related Mohr's circles showing $\sigma_1, \sigma_2, \sigma_3$, and τ_{max}.

τ (CW)

$\tau_{max} = 108.3$ MPa

$\tau = 51.7$ MPa

$\tau = 56.6$ MPa

$\sigma_1 = 216.6$ MPa

$\sigma_3 = 0$ MPa

$R = 56.6$

120

200

σ

$\sigma_2 = 103.4$ MPa

160

FIGURE 15–46 Three related Mohr's circles showing $\sigma_1, \sigma_2, \sigma_3$, and τ_{max}.

$\sigma_y = -180$ MPa

$\sigma_x = -50$ MPa

$\tau_{xy} = 30$ MPa

(a) Initial stress element

τ (CW)

$\tau_{max} = 93.3$ MPa

$\tau = 71.6$ MPa

$\tau = 21.7$ MPa

$(-180, +30)$

$\sigma_2 = -43.4$ MPa

$\sigma_1 = 0$ MPa

$\sigma_3 = -186.6$ MPa

$R = 71.6$

30

65

-50

-180

σ

-115

$(-50, -30)$

(b) Mohr's circle

This summary pertains to the situation in which the Mohr's circle analysis of a stress element subjected to plane stress (stresses applied only in two dimensions) yields the result that both principal stresses (σ_1 and σ_2) are of the same sign; that is, both are tensile or both are compressive. In such cases, the following steps should be completed to provide a true picture of the stress condition on the three-dimensional element.

A. Draw the complete Mohr's circle for the plane stress condition, and identify the principal stresses σ_1 and σ_2.

B. If both principal stresses are tensile (positive) (see example in Figure 15–45):

1. Consider that the *zero stress* acting in the direction perpendicular to the initial stress element is the true minimum principal stress. Then it is necessary to define *three* principal stresses such that:

σ_1 = Maximum principal stress from the first Mohr's circle

σ_2 = Minimum principal stress from the first Mohr's circle

σ_3 = Zero (true minimum principal stress)

2. Draw a secondary Mohr's circle whose diameter extends from σ_1 to σ_3 on the σ-axis. The center of the circle will be at the average of σ_1 and σ_3, $(\sigma_1 + \sigma_3)/2$ But, because $\sigma_3 = 0$, the average is $\sigma_1/2$.

3. The maximum shear stress is at the top of the secondary circle and its value is also $\sigma_1/2$.

C. If both principal stresses are compressive (negative) (see example in Figure 15–46):

1. Consider that the *zero stress* acting in the direction perpendicular to the initial stress element is the true maximum principal stress. Then it is necessary to define *three* principal stresses such that

σ_1 = Zero (true maximum principal stress)

σ_2 = Maximum principal stress from the first Mohr's circle

σ_3 = Minimum principal stress from the first Mohr's circle

2. Draw a secondary Mohr's circle whose diameter extends from σ_1 to σ_3 on the σ-axis. The center of the circle will be at the average of σ_1 and σ_3, $(\sigma_1 + \sigma_3)/2$. But, because $\sigma_1 = 0$, the average is $\sigma_3/2$.

3. The maximum shear stress is at the top of the secondary circle, and its magnitude is also $\sigma_3/2$.

15–12 USE OF STRAIN-GAGE ROSETTES TO DETERMINE PRINCIPAL STRESSES

In the earlier sections of this chapter, data were given for the initial biaxial stress condition that existed on an element of a load-carrying member. Then the principal stresses and the maximum shear stress were calculated either with the equations developed in Sections 15–7 and 15–8 or by the use of Mohr's circle as discussed in Section 15–9. The stresses on the initial stress element may have been determined by direct calculation using the principles of stress analysis as demonstrated in Section 15–4.

An alternate method of determining the initial stress condition is to use experimental stress analysis techniques. This could be used to analyze, for example, the complex case for an automatic transmission for a car or the housing for an air compressor. Photoelastic techniques may first be used to determine where the regions of maximum stress are located. Then strain gages can be applied in those areas to measure more precisely the magnitudes of strains in particular directions. If known, the strain gages should be aligned as closely as possible with the directions of the principal stresses.

However, in cases of complex geometry and complex loading, the directions of principal stresses are not known. Then the use of a *strain-gage rosette* having an array of three gages with a precise geometric relationship among the gages is recommended. Figure 15–47 shows three different styles and others are available. Part (a) shows a 45-degree rosette, sometimes called a *rectangular rosette* or a 0°, 45°, 90° rosette to indicate the orientation of the three gages. Parts (b) and (c) show two styles of 60-degree rosettes, sometimes called *delta rosettes* or 0°, 60°, 120° rosettes. The stacked rosette in (c) is used where space is tight or where there are large gradients of stress. However, there are possible difficulties in mounting such gages because of their thickness and because the upper gages are displaced from the true surface where the strains are to be measured.

In general, the gage marked *number 1* is installed carefully aligned with some reference axis on the part to be tested. After mounting, the member is loaded and gage readings are taken from all three gages with their output typically called $\epsilon_1, \epsilon_2,$ and ϵ_3. From these measurements it

(a) 45-degree rosette

(b) 60-degree rosette

(c) Stacked rosette

FIGURE 15–47   Strain-gage rosettes.

Source: Measurements Group, Inc., Raleigh, NC.

can be shown that the maximum and minimum strains can be computed from the following equations. Note that different equations are used for the 45-degree rosette and for the 60-degree rosette because the angle between the gages enters into the calculations. Also, it is necessary to determine the orientation of the principal strain axes relative to the direction of gage number one. The angle from gage number one to the nearest principal strain is called β and is computed from the equation that follows the strain equations.

45° Rosette:

$$\epsilon_{max} = \frac{(\epsilon_1 + \epsilon_3)}{2} + \frac{\sqrt{(\epsilon_1 - \epsilon_2)^2 + (\epsilon_2 - \epsilon_3)^2}}{\sqrt{2}} \quad (15\text{–}22)$$

$$\epsilon_{min} = \frac{(\epsilon_1 + \epsilon_3)}{2} - \frac{\sqrt{(\epsilon_1 - \epsilon_2)^2 + (\epsilon_2 - \epsilon_3)^2}}{\sqrt{2}} \quad (15\text{–}23)$$

$$\beta = \frac{1}{2}\tan^{-1}\left[\frac{(\epsilon_2 - \epsilon_3) - (\epsilon_1 - \epsilon_2)}{(\epsilon_1 - \epsilon_3)}\right] \quad (15\text{–}24)$$

60° Rosette:

$$\epsilon_{max} = \frac{(\epsilon_1 + \epsilon_2 + \epsilon_3)}{3} + \frac{\sqrt{2}}{3}\sqrt{(\epsilon_1 - \epsilon_2)^2 + (\epsilon_2 - \epsilon_3)^2 + (\epsilon_1 - \epsilon_3)^2} \quad (15\text{–}25)$$

$$\epsilon_{min} = \frac{(\epsilon_1 + \epsilon_2 + \epsilon_3)}{3} - \frac{\sqrt{2}}{3}\sqrt{(\epsilon_1 - \epsilon_2)^2 + (\epsilon_2 - \epsilon_3)^2 + (\epsilon_1 - \epsilon_3)^2} \quad (15\text{–}26)$$

$$\beta = \frac{1}{2}\tan^{-1}\left[\frac{\sqrt{3}(\epsilon_2 - \epsilon_3)}{(\epsilon_1 - \epsilon_2) + (\epsilon_1 - \epsilon_3)}\right] \quad (15\text{–}27)$$

The measured strains are typically very small. For example, with using U.S. Customary units, the order of magnitude is usually under 5000×10^{-6} in/in. For convenience, some experimental stress analysts write this as 5000 $\mu\varepsilon$, read "5000 microstrains." Calculations in equations such as those in Equations (15–22) through (15–27) are then completed using only the whole number while recognizing that the actual complete value must be considered.

Principal Stresses Obtained from Principal Strains

The ultimate goal is to determine the principal stresses from these measurements of strain using the strain-gage rosette. To accomplish that goal we need to review some concepts and relationships discussed in Chapters 7–9.

We discussed the relationship between stress and strain by defining the *modulus of elasticity, E.* You should recall that, for *uniaxial stress,* such as direct tension or compression,

$$\sigma = E\epsilon$$

That is, stress is the product of the strain times the modulus of elasticity of the material.

Here we are considering the more general case of *biaxial stress,* where stresses in two directions occur simultaneously.

Now we must recall the fact that when a load-carrying member undergoes deformation in one direction, there also occurs deformation in the perpendicular directions. Figure 15–48 shows this concept as it was defined in Chapter 8. We use the term *Poisson's ratio, v,* to represent the ratio of the normal strain in the direction of the applied stress and the lateral strain in the perpendicular directions. Poisson's ratio is a property of the material, and typical values are given in Table 8–1.

Now, if we know the strain in one direction, say ϵ_x, the strain in the y-direction, perpendicular to x, is

$$\epsilon_y = v\epsilon_x$$

Under the influence of biaxial strain, the strain in each direction will affect the strain in the other direction. Thus, the calculation of the maximum stresses using the maximum strains is more complex than $\sigma = E\epsilon$. The development of the relationship is not shown here. The final result is

$$\sigma_{max} = \frac{E}{1 - v^2}(\epsilon_{max} + v\epsilon_{min}) \quad (15\text{–}28)$$

$$\sigma_{min} = \frac{E}{1 - v^2}(\epsilon_{min} + v\epsilon_{max}) \quad (15\text{–}29)$$

The maximum shearing strain in the plane of the original element, γ_{max}, can also be found from the absolute value of the difference between the maximum and minimum normal strains in the plane.

$$\gamma_{max} = |(\epsilon_{max} - \epsilon_{min})| \quad (15\text{–}30)$$

The units for γ are radians, considered to be dimensionless. Then the maximum shearing stress can be computed using the definition of G, the shear modulus of elasticity.

$$\tau_{max} = G\gamma_{max} \quad (15\text{–}31)$$

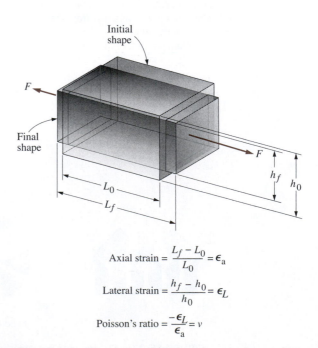

Axial strain $= \dfrac{L_f - L_0}{L_0} = \epsilon_a$

Lateral strain $= \dfrac{h_f - h_0}{h_0} = \epsilon_L$

Poisson's ratio $= \dfrac{-\epsilon_L}{\epsilon_a} = v$

FIGURE 15–48 Illustration of Poisson's ratio for an element in tension.

The units for τ will be the same as those for G. An alternate form for G was shown in Chapter 8. Then Equation 15–31 can also be written

$$\tau_{max} = \frac{E\gamma_{max}}{2(1 + \nu)} \quad \textbf{(15–32)}$$

The same precautions apply when the maximum and minimum principal stresses have the same sign as discussed in Section 15–11.

PROCEDURE FOR ANALYZING THE DATA FROM A STRAIN-GAGE ROSETTE ▼

In summary, strain-gage rosettes can be used to determine the maximum and minimum principal stresses and the maximum shearing stress by the following procedure:

1. Determine the area where maximum stresses are likely to occur. You may use judgment, photoelastic stress analysis, or finite element analysis.

2. Apply a strain-gage rosette to the area of the load-carrying member where measurements are to be taken,
being careful to align gage number 1 with a known axis on the part.

3. Operate the equipment, applying loads representative of the largest expected in service.

4. Record the strain readings from each gage giving values for ϵ_1, ϵ_2, and ϵ_3.

5. Depending on the type of rosette, use the equations shown as (15–22) to (15–24) or (15–25) to (15–27) to determine the maximum and minimum strains at the point of interest and the angle of orientation of these strains relative to the alignment of gage number 1.

6. Use Equations (15–28) and (15–29) to compute the maximum and minimum principal stresses. The directions of the maximum principal stresses are the same as the directions of the principal strains.

7. Compute the maximum shearing strain from Equation (15–30).

8. Compute the maximum shearing stress from Equation (15–31) or (15–32).

9. Check to see if the maximum and minimum principal stresses have the same sign. If so, complete the additional analysis discussed in Section 15–11.

Example Problem 15–14

As part of the product development process for a new pump, the designer has instrumented a critical area of the housing with a rectangular strain-gage rosette like that shown in Figure 15–47(a). Gage 1 is aligned with the horizontal centerline of the pump inlet passage. During the test under high-capacity operating conditions, the following readings were taken for the strains from the three arms of the gage: $\epsilon_1 = 950 \times 10^{-6}$ in/in, $\epsilon_2 = -375 \times 10^{-6}$ in/in, $\epsilon_3 = 525 \times 10^{-6}$ in/in. The material for the pump housing is aluminum 2014-T6. Compute the maximum principal stress, the minimum principal stress, and the maximum shear stress at the location where the rosette is mounted.

Solution   Objective: Compute σ_{max}, σ_{min}, and τ_{max}.

Given: Strain readings from a 0°, 45°, 90° strain-gage rosette.

$\epsilon_1 = 950 \times 10^{-6}$ in/in, $\epsilon_2 = -375 \times 10^{-6}$ in/in, $\epsilon_3 = 525 \times 10^{-6}$ in/in.

Aluminum 2014-T6: $E = 10.6 \times 10^6$ psi, $s_y = 60\,000$ psi (Appendix A–18) $\nu = 0.33$ (Table 8–1)

Analysis: Use the *procedure for analyzing the data from a strain-gage rosette.*

Results: **Steps 1–4:** Have already been completed.

Step 5: In using Equations (15–22) to (15–24), we will show only the whole number part of the strain values. The unit is then microstrains, $\mu\epsilon$.
The maximum principal strain is

$$\epsilon_{max} = \frac{(\epsilon_1 + \epsilon_3)}{2} + \frac{\sqrt{(\epsilon_1 - \epsilon_2)^2 + (\epsilon_2 - \epsilon_3)^2}}{\sqrt{2}}$$

$$\epsilon_{max} = \frac{(950 + 525)}{2} + \frac{\sqrt{[(950 - (-375)]^2 + (-375 - 525)^2}}{\sqrt{2}} = 1870 \; \mu\epsilon$$

The minimum principal strain is

$$\epsilon_{min} = \frac{(\epsilon_1 + \epsilon_3)}{2} - \frac{\sqrt{(\epsilon_1 - \epsilon_2)^2 + (\epsilon_2 - \epsilon_3)^2}}{\sqrt{2}}$$

$$\epsilon_{min} = \frac{(950 + 525)}{2} - \frac{\sqrt{[(950 - (-375)]^2 + (-375 - 525)^2}}{\sqrt{2}} = -395 \; \mu\epsilon$$

The angle from gage number 1 to the nearest principal strain axis is

$$\beta = \frac{1}{2}\tan^{-1}\left[\frac{(\epsilon_2 - \epsilon_3) - (\epsilon_1 - \epsilon_2)}{(\epsilon_1 - \epsilon_3)}\right]$$

$$\beta = \frac{1}{2}\tan^{-1}\left[\frac{(-375 - 525) - [950 - (-375)]}{(950 - 525)}\right] = -39.6°$$

Step 6: The maximum principal stress is found from Equation (15–28).

$$\sigma_{max} = \frac{E}{1 - \nu^2}(\epsilon_{max} + \nu\epsilon_{min})$$

$$\sigma_{max} = \frac{10.6 \times 10^6 \text{ psi}}{1 - (0.33)^2}[1870 + 0.33(-395)](10^{-6}) = 20\,694 \text{ psi}$$

The minimum principal stress is found from Equation (15–29).

$$\sigma_{min} = \frac{E}{1 - \nu^2}(\epsilon_{min} + \nu\epsilon_{max})$$

$$\sigma_{min} = \frac{10.6 \times 10^6 \text{ psi}}{1 - (0.33)^2}[-395 + 0.33(1870)](10^{-6}) = 2642 \text{ psi}$$

Step 7: The maximum shearing strain is found from Equation (15–30).

$$\gamma_{max} = |(\epsilon_{max} - \epsilon_{min})| = |1870 - (-395)| = 2265 \times 10^{-6} \text{ rad}$$

Step 8: The maximum shearing stress in the plane of the initial element is found from Equation (15–32).

$$\tau_{max} = \frac{E\gamma_{max}}{2(1 + \nu)} = \frac{(10.6 \times 10^6 \text{ psi})(2265 \times 10^{-6})}{2(1 + 0.33)} = 9026 \text{ psi}$$

Step 9: We note that both the maximum principal stress and the minimum principal stress are positive or tensile. Therefore, we must draw a supplemental Mohr's circle to determine the true maximum shearing strain. See Figure 15–49. The circle is simply drawn by plotting both principal stresses on the horizontal axis and drawing the circle to include both. The supplementary circle is drawn through σ_{max} and the origin of the axes because that represents a zero stress perpendicular to the plane of the initial stress element. The true maximum shearing stress is equal to the radius of this circle, found from

$$\tau_{max} = (\sigma_{max} - 0)/2 = (20\,694 \text{ psi})/2 = 10\,347 \text{ psi}$$

FIGURE 15–49   Mohr's circles for Example Problem 15–14.

Summary: The final results are

$\sigma_{max} = 20\,694$ psi $\sigma_{min} = 2642$ psi in the plane of the initial element

True $\sigma_{min} = 0$ perpendicular to the plane of the initial element

$\tau_{max} = 10\,347$ psi

Spreadsheet Solution for Calculations for Strain-Gage Rosettes

The calculations for stress and strain for rosettes is cumbersome, time-consuming, and prone to error. It is very desirable to use a spreadsheet, computer algebra routine, or programmable calculator to perform the calculations. Figure 15–50 shows a spreadsheet that accomplishes this objective. Here are some of its features.

1. The nine-step *procedure for analyzing the data from a strain-gage rosette* is implemented.

FIGURE 15–50 Spreadsheet with calculations for strain-gage rosettes.

SPREADSHEET FOR COMPUTING PRINCIPAL STRAINS AND STRESSES FROM STRAIN-GAGE ROSETTE OUTPUT DATA

Refer to Section 15–12 for method.
Input data in shaded elements

Material Properties	U.S. Customary Units		SI Metric Units
Modulus of Elasticity	10.6×10^6 psi		73.1×10^9 Pa
Poisson's Ratio	0.33		0.33

Rectangular [0, 45, 90 degree] Rosette Data [Use Equations 15–22 to 15–24]

	U.S. Customary Units		SI Metric Units
Strain from Gage 1	950×10^{-6} in/in		950×10^{-6} m/m
Strain from Gage 2	-375×10^{-6} in/in		-375×10^{-6} m/m
Strain from Gage 3	525×10^{-6} in/in		525×10^{-6} m/m
	Results:		Results:
Max Principal Strain	1870×10^{-6} in/in		1870×10^{-6} m/m
Min Principal Strain	-395×10^{-6} in/in		-395×10^{-6} m/m

Angle, β, from the axis of gage 1 to the nearer principal axis

Angle β	–39.6 degrees		–39.6 degrees

Max Principal Stress	20695 psi		143 MPa
Min Principal Stress	2641 psi		18.2 MPa

Max Shear Strain	2265×10^{-6} radians [Dimensionless]		2265×10^{-6} radians
Max Shear Stress	9027 psi [in plane of initial element]		62.3 MPa

Only in the case that Max and Min Principal Stresses have the same sign

True Max Shear Stress	10347 psi		71.4 MPa

[Assuming stress = 0 perpendicular to plane of initial element]

Delta [0, 60, 120 degree] Rosette Data [Use Equations 15–25 to 15–27]

	U.S. Customary Units		SI Metric Units
Strain from Gage 1	1250×10^{-6} in/in		1250×10^{-6} m/m
Strain from Gage 2	-235×10^{-6} in/in		-235×10^{-6} m/m
Strain from Gage 3	645×10^{-6} in/in		645×10^{-6} m/m
	Results:		Results:
Max Principal Strain	1416×10^{-6} in/in		1416×10^{-6} m/m
Min Principal Strain	-309×10^{-6} in/in		-309×10^{-6} m/m

Angle, β, from the axis of gage 1 to the nearer principal axis

Angle β	–18.1 degrees		–18.1 degrees

Max Principal Stress	15626 psi		108 MPa
Min Principal Stress	1882 psi		13.0 MPa

Max Shear Strain	1725×10^{-6} radians [Dimensionless]		1725×10^{-6} radians
Max Shear Stress	6872 psi [in plane of initial element]		47.4 MPa

Only in the case that Max and Min Principal Stresses have the same sign

True Max Shear Stress	7813 psi		53.9 MPa

[Assuming stress = 0 perpendicular to plane of initial element]

2. It assumes that the stress condition is biaxial and that the stress perpendicular to the plane of the measured strains is zero.

3. It includes sections for both rectangular and delta-style rosettes.

4. It allows the use of either U.S. Customary units or SI metric units.

5. Input data required are highlighted by shading the appropriate cells.

6. Enter material property data for modulus of elasticity, E, and Poisson's ratio, ν. For E, enter only the significant digits as shown. It is assumed that E is in millions of psi or GPa (10^9 Pa).

7. Enter strain data as whole numbers of microstrains. The spreadsheet uses this value times 10^{-6} in/in or 10^{-6} m/m. Of course, the strain data are actually dimensionless.

8. The first set of data shown in the upper left of the spreadsheet is identical to the data from Example Problem 15–14.

9. Take care in interpreting the results for the maximum shear stress. Two values are calculated. One is the maximum shear stress in the plane of the initial stress element. This is the true maximum shear stress only if the maximum and minimum principal stresses have different signs. If they have the same signs, the alternate calculation labeled "True Max Shear Stress" is used. This recognizes that there is a larger shear stress in a plane other than the plane of the initial element. A supplementary Mohr's circle is drawn as discussed in Section 15–11 and as demonstrated in Example Problem 15–14 and Figure 15–49.

REFERENCES

1. Aluminum Association, *Aluminum Design Manual*, Washington, DC, 2005.

2. American Institute of Steel Construction, *Steel Construction Manual*, 13th ed., Chicago, 2005.

3. Budynas, R. G. and J. K. Nisbett, *Shigley's Mechanical Engineering Design*, 8th ed., New York, McGraw-Hill, 2007.

4. Dally, James W. and W. F. Riley, *Experimental Stress Analysis*, 4th ed., Knoxville, TN, College House Enterprises, 2005.

5. Doyle, James F., *Modern Experimental Stress Analysis*, New York, John Wiley & Sons, Inc., 2004.

6. Doyle, James F. and J. W. Phillips, eds., *Manual on Experimental Stress Analysis*, 5th ed., Westport, CT, Society for Experimental Stress Analysis, 1989.

7. Mott, Robert L., *Machine Elements in Mechanical Design*, 4th ed., Upper Saddle River, NJ, Prentice Hall, 2004.

8. Young, W. C. and R. D. Cook, *Advanced Mechanics of Materials*, 2nd ed., Upper Saddle River, NJ, Prentice Hall, 1999.

INTERNET SITES

1. Vishay Micro-Measurements www.vishay.com Through its Measurement Group, Vishay produces strain gages and associated equipment, photoelastic stress measurement devices and materials, test and measurement systems, and accessories. From the complete listing of products, select *strain gages, PhotoStress® Plus*, or *Test and measurements*.

2. Vishay—Experimental Stress Analysis Notebook www.vishay.com/company/brands/measurements-group/schools/nbindex.htm A collection of 30 issues of the Notebook that covers details of strain-gage technology and many practical applications and student projects that use strain gages.

3. HBM www.hbm.com A member of the Spectris plc instrumentation and electronic controls group, HBM produces strain gages and a variety of sensors, transducers, and electronic instrumentation systems.

4. Society for Experimental Mechanics (SEM) http://sem.org SEM promotes research and application of engineering measurement using experimental methods for testing materials and the forces that affect them. Interest areas are materials, modeling and analysis, strain measurement, and structural testing.

5. Stress Photonics, Inc. www.stressphotonics.com Providers of full-field stress and strain measurement systems and real time nondestructive evaluation (NDE) and inspection systems, including the Grey-field Polariscope Photoelastic Strain Measurement System and the DeltaTherm Thermoelastic Stress Measurement System.

6. Omega.com www.omega.com/literature/transactions/volume3/strain.html An online introduction to strain-gage technology, sensor designs, measuring circuits, applications, and installation.

7. Geokon, Incorporated www.geokon.com Designers and manufacturers of geotechnical instrumentation for the construction industry, including embedment strain gages, weldable strain gages, rebar strain meters, joint meters, strand meters for cables, and many other devices. Select *Product Information*.

PROBLEMS

Combined Normal Stresses

15–1. A $2\frac{1}{2}$-in schedule 40 steel pipe is used as a support for a basketball backboard, as shown in Figure P15–1. It is securely fixed into the ground. Compute the stress that would be developed in the pipe if a 230-lb player hung on the base of the rim of the basket.

15–2. The bracket shown in Figure P15–2 has a rectangular cross section 18 mm wide by 75 mm high. It is securely attached to the wall. Compute the maximum stress in the bracket.

15–3. The beam shown in Figure P15–3 carries a 6000-lb load attached to a bracket below the beam. Compute the stress at points M and N, where it is attached to the column.

15–4. For the beam shown in Figure P15–3, compute the stress at points M and N if the 6000-lb load acts vertically downward instead of at an angle.

15–5. For the beam shown in Figure P15–3, compute the stress at points M and N if the 6000-lb load acts back toward the column at an angle of 40 deg below the horizontal instead of as shown.

FIGURE P15–1   Basketball backboard for Problem 15–1.

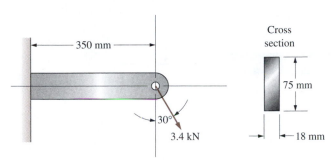

FIGURE P15–2   Bracket for Problem 15–2.

FIGURE P15–3   Beam for Problems 15–3, 15–4, and 15–5.

15–6. Compute the maximum stress in the top portion of the coping saw frame shown in Figure P15–6 if the tension in the blade is 125 N.

FIGURE P15–6   Coping saw frame for Problem 15–6.

15–7. Compute the maximum stress in the crane beam shown in Figure P15–7 when a load of 12 kN is applied at the middle of the beam.

FIGURE P15–7   Crane beam for Problem 15–7.

15–8. Figure P15–8 shows a metal-cutting hacksaw. Its frame is made of hollow tubing having an outside diameter of 12 mm and a wall thickness of 1.0 mm. The blade is pulled taut by the wing nut so that a tensile force of 160 N is applied to the blade. Compute the maximum stress in the top section of the tubular frame.

FIGURE P15–8   Hacksaw frame for Problem 15–8.

15–9. The C-clamp in Figure P15–9 is made of cast zinc, ZA12. Determine the allowable clamping force that the clamp can exert if it is desired to have a design factor of 4 based on ultimate strength in either tension or compression.

FIGURE P15–9 C-clamp for Problem 15–9.

26 mm

3 mm

3 mm 14 mm

8 mm

Section *A-A*

FIGURE P15–10 C-clamp for Problem 15–10.

42 mm

6 mm

3 mm

3 mm

26 mm

3 mm

12 mm

Section *A-A*

15–10. The C-clamp shown in Figure P15–10 is made of cast malleable iron, ASTM A220 grade 45008. Determine the allowable clamping force that the clamp can exert if it is desired to have a design factor of 4 based on ultimate strength in either tension or compression.

15–11. A tool used for compressing a coil spring to allow its installation in a car is shown in Figure P15–11. A force of 1200 N is applied near the ends of the extended lugs, as shown. Compute the maximum tensile stress in the threaded rod. Assume a stress concentration factor of 3.0 at the thread root for both tensile and bending stresses. Then, using a design factor of 2 based on yield strength, specify a suitable material for the rod.

15–12. Figure P15–12 shows a portion of the steering mechanism for a car. The steering arm has the detailed design shown. Notice that the arm has a constant thickness of 10 mm so

Coil spring

Tool

M16 × 2.00 threaded rod
(Root dia. = 13.2 mm)

44 mm

Spring forces

1200 N   1200 N

FIGURE P15–11 Tool for compressing springs for Problem 15–11.

FIGURE P15–12   Steering arm
for Problem 15–12.

225 N

225 N

60°

60°

Steering
arm

A

14 mm

30 mm

375 mm

B

25 mm

Resisting
moment

Cross section

A   *B*

14 mm   25 mm

10 mm

that all cross sections between the end lugs are rectangular. The 225-N force is applied to the arm at a 60-deg angle. Compute the stress in the arm at sections *A* and *B*. Then, if the arm is made of ductile iron, ASTM A536, grade 80-55-6, compute the minimum design factor at these two points based on ultimate strength.

15–13. An S3×5.7 American Standard I-beam is subjected to the forces shown in Figure P15–13. The 4600-lb force acts di-

rectly in line with the axis of the beam. The 500-lb downward force at *A* produces the reactions shown at the supports *B* and *C*. Compute the maximum tensile and compressive stresses in the beam.

15–14. The horizontal crane boom shown in Figure P15–14 is made of a hollow rectangular steel tube. Compute the stress in the boom just to the left of point *B* when a mass of 1000 kg is supported at the end.

FIGURE P15–13   Beam for
Problem 15–13.

S3 × 5.7 beam

A   *B*   *C*

R_C

4600 lb   4600 lb

6 ft   6 ft

R_B

500 lb

FIGURE P15–14   Crane boom
for Problem 15–14.

Cross section
of boom
HSS 152×51×6.4

152 mm

6.4 mm
nominal

51 mm

A   0.6 m

45°

B

1.0 m

Boom

C

Load

15–15. For the crane boom shown in Figure P15–14, compute the load that could be supported if a design factor of 3 based on yield strength is desired. The boom is made of AISI 1040 hot-rolled steel. Analyze only sections where the full box section carries the load, assuming that sections at connections are adequately reinforced.

15–16. The member *EF* in the truss shown in Figure P15–16 carries an axial tensile load of 54 000 lb in addition to the two 1200-lb loads shown. It is planned to use two steel angles for the member, arranged back to back. Specify a suitable size for the angles if they are to be made of ASTM A36 structural steel with an allowable stress of 0.6 times the yield strength.

FIGURE P15–16 Truss for Problem 15–16.

For Problems 15–17 through 15–20

Specify a suitable size for the horizontal portion of the member that will maintain the combined tensile or compressive stress to 6000 psi if the data are in the U.S. Customary unit system and 42 MPa if the SI metric system is used. The cross section is to be square.

15–17. Use Figure P15–17.

FIGURE P15–17

15–18. Use Figure P15–18.

FIGURE P15–18

15–19. Use Figure P15–19.

FIGURE P15–19

15–20. Use Figure P15–20.

FIGURE P15–20

Combined Normal and Shear Stresses

15–21. A solid circular bar is 40 mm in diameter and is subjected to an axial tensile force of 150 kN along with a torque of 500 N · m. Compute the maximum shear stress in the bar.

15–22. A solid circular bar has a diameter of 2.25 in and is subjected to an axial tensile force of 47 000 lb along with a torque of 8500 lb · in. Compute the maximum shear stress in the bar.

15–23. A short solid circular bar has a diameter of 4.00 in and is subjected to an axial compressive force of 40 000 lb along with a torque of 25 000 lb · in. Compute the maximum shear stress in the bar.

15–24. A short hollow circular post is made from a 12-in schedule 40 steel pipe and it carries an axial compressive load of 250 000 lb along with a torque of 180 000 lb · in. Compute the maximum shear stress in the post.

15–25. A short hollow circular support is made from a 3-in schedule 40 steel pipe and carries an axial compressive load of 25 000 lb along with a torque of 15 500 lb · in. Compute the maximum shear stress in the support.

15–26. A television antenna is mounted on a hollow aluminum tube as sketched in Figure P15–26. During installation, a force of 20 lb is applied to the end of the antenna as shown. Calculate the torsional shear stress in the tube and the stress due to bending. Consider the tube to be simply supported against bending at the clamps, but assume that rotation is not permitted. If the tube is made of 6061-T6 aluminum, would it be safe under this load? The tube has an outside diameter of 1.50 in and a wall thickness of $\frac{1}{16}$ in.

FIGURE P15–26 Antenna mounting tube for Problem 15–26.

15–27. Figure P15–27 shows a crank to which a force *F* of 1200 N is applied. Compute the maximum stress in the circular portion of the crank.

15–28. A standard steel pipe is to be used to support a bar carrying four loads, as shown in Figure P15–28. Specify a suitable pipe that would keep the maximum shear stress to 8000 psi.

FIGURE P15–27   Crank for Problem 15–27.

FIGURE P15–28   Bracket for Problem 15–28.

Rotating Shafts—Combined Torsional Shear and Bending Stresses

15–29. A circular shaft carries the load shown in Figure P15–29. The shaft carries a torque of 1500 N·m between sections *B* and *C*. Compute the maximum shear stress in the shaft near section *B*.

FIGURE P15–30   Shaft for Problem 15–30.

FIGURE P15–29   Shaft for Problem 15–29.

15–30. A circular shaft carries the load shown in Figure P15–30. The shaft carries a torque of 4500 N·m between sections *B* and *C*. Compute the maximum shear stress in the shaft near section *B*.

15–31. A 1.0-in-diameter solid round shaft will carry 25 hp while rotating at 1150 rpm in the arrangement shown in Figure P15–31. The total bending loads at the gears *A* and *C* are

FIGURE P15–31   Shaft for Problem 15–31.

shown, along with the reactions at the bearings B and D. Use a design factor of 6 for the maximum shear stress theory of failure and determine a suitable steel for the shaft.

15–32. Three pulleys are mounted on a rotating shaft, as shown in Figure P15–32. Belt tensions are as shown in the end view. All power is received by the shaft through pulley C from below. Pulleys A and E deliver power to mating pulleys above.

   a. Compute the torque at all points in the shaft.
   b. Compute the bending stress and the torsional shear stress in the shaft at the point where the maximum bending moment occurs if the shaft diameter is 1.75 in.
   c. Compute the maximum shear stress at the point used in step (b). Then specify a suitable material for the shaft.

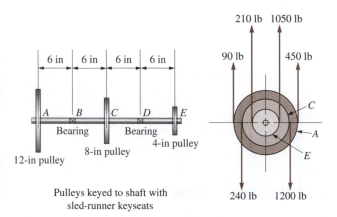

FIGURE P15–32   Shaft for Problem 15–32.

15–33. The vertical shaft shown in Figure P15–33 carries two belt pulleys. The tensile forces in the belts under operating conditions are shown. Also, the shaft carries an axial compressive load of 6.2 kN. Considering torsion, bending, and axial compressive stresses, compute the maximum shear stress using Equation (15–2).

FIGURE P15–33   Shaft for Problem 15–33.

15–34. For the shaft in Problem 15–33, specify a suitable steel that would provide a design factor of 4 based on yield strength in shear.

Combined Axial Tension and Direct Shear Stresses

15–35. A machine screw has Number 8-32 UNC American Standard threads (see Appendix A–3). The screw is subjected to an axial tensile force that produces a direct tensile stress in the threads of 15 000 psi based on the tensile stress area. There is a section under the head without threads that has a diameter equal to the major diameter of the threads. This section is also subjected to a direct shearing force of 120 lb. Compute the maximum shear stress in this section.

15–36. Repeat Problem 15–35 except the screw threads are $\frac{1}{4}$-20 UNC American Standard and the shearing force is 775 lb.

15–37. Repeat Problem 15–35 except the screw threads are No. 4-48 UNF American Standard and the shearing force is 50 lb.

15–38. Repeat Problem 15–35 except the screw threads are $1\frac{1}{4}$-12 UNF and the shearing force is 2500 lb.

15–39. A machine screw has metric threads with a major diameter of 16 mm and a pitch of 2.0 mm (see Appendix A–3). The screw is subjected to an axial force that produces a direct tensile stress in the threads of 120 MPa based on the tensile stress area. There is a section under the head without threads that has a diameter equal to the major diameter of the threads. This section also is subjected to a direct shearing force of 8.0 kN. Compute the maximum shear stress in this section.

15–40. A machine screw has metric threads with a major diameter of 48 mm and a pitch of 5.0 mm (see Appendix A–3). The screw is subjected to an axial tensile force that produces a direct tensile stress in the threads of 120 MPa based on the tensile stress area. There is a section under the head without threads that has a diameter equal to the major diameter of the threads. This section also is subjected to a direct shearing force of 80 kN. Compute the maximum shear stress in this section.

Combined Bending and Vertical Shear Stresses

15–41. A rectangular bar is used as a beam carrying a concentrated load of 5500 lb at the middle of its 60-in length. The cross section is 2.00 in wide and 6.00 in high with the 6.00-in dimension oriented vertically. Compute the maximum shear stress that occurs in the bar near the load at the following points within the section:

   a. At the bottom of the bar.
   b. At the top of the bar.
   c. At the neutral axis.
   d. At a point 1.0 in above the bottom of the bar.
   e. At a point 2.0 in above the bottom of the bar.

15–42. Repeat Problem 15–41 except the beam is an aluminum I-beam, I6×4.692.

15–43. Repeat Problem 15–41 except the load is a uniformly distributed load of 100 lb/in over the entire length. Consider cross sections near the middle of the beam, near the supports, and 15 in from the left support.

Noncircular Sections—Combined Normal and Torsional Shear Stresses

15–44. A square bar is 25 mm on a side and carries an axial tensile load of 75 kN along with a torque of 245 N · m. Compute the maximum shear stress in the bar. (*Note:* Refer to Section 10–11 and Figure 10–27.)

250 mm

85 N ←

113 mm

→ A

→ A

11.0 mm

X ——— X

1.5 mm
typical

7.0 mm

4.5 mm

$I_X = 128$ mm^4
Area = 33.0 mm^2

Section A-A

FIGURE P15–48

15–45. A rectangular bar is 30 mm by 50 mm in cross section and is subjected to an axial tensile force of 175 kN along with a torque of 525 N · m. Compute the maximum shear stress in the bar. (*Note:* Refer to Section 10–11 and Figure 10–27.)

15–46. A bar has a cross section in the form of an equilateral triangle, 50 mm on a side. It carries an axial tensile force of 115 kN along with a torque of 775 N · m. Compute the maximum shear stress in the bar. (*Note:* Refer to Section 10–11 and Figure 10–27.)

15–47. A link in a large mechanism is made from a square structural tube HSS3×3×$\frac{1}{4}$ (see Appendix A–9). It is originally designed to carry an axial tensile load that produces a design factor of 3 based on the yield strength of ASTM A500 cold-formed structural steel, shaped, grade C.
  a. Determine this load and the maximum shear stress that results in the tube under this load.
  b. In operation, a torque of 950 lb · ft is experienced by the tube in addition to the axial load. Compute the maximum shear stress under this combined loading and compute the resulting design factor based on the yield strength of the steel in shear. (See Section 10–11 and Figure 10–27.)

ADDITIONAL REVIEW AND PRACTICE PROBLEMS

15–48. Compute the maximum tensile and compressive stresses in the horizontal portion of the bracket shown in Figure P15–48.

15–49. Repeat Problem 15–48 with the load applied downward and to the left at an angle of 35° to the horizontal.

15–50. Repeat Problem 15–48 with the load applied to the right rather than to the left.

15–51. Repeat Problem 15–48 with the load applied upward and to the right at an angle of 65° to the horizontal.

15–52. The bracket shown in Figure P15–52 is made from structural steel tubing, HSS3×3×$\frac{1}{4}$. It is attached rigidly to the

15 in

6 in

20°

1800 lb

FIGURE P15–52

ceiling. Compute the maximum tensile and compressive stresses in the bracket and state where they occur.

15–53. Figure P15–53 shows a small crane. The load is 34.0 kN. The connection at each point A to F is a 30-mm-diameter pin in a clevis. Determine the maximum normal stresses in the horizontal and vertical members of the crane.

15–54. Figure P15–54 shows a rotating shaft carrying transverse loads. The distributed load between the bearings is provided by a large flat pulley. The load at the end of the overhang is applied by a chain drive. The shaft transmits 30.5 hp while rotating at 320 rpm. Compute the maximum shear stress in the shaft and specify a suitable material.

15–55. The cantilevered bar shown in Figure P15–55 carries an offset load as shown. Determine the stress condition that occurs on elements *M* and *N*.

FIGURE P15–53

Boom

Section A-A
Dimensions in millimeters

FIGURE P15–54

650 lb

32 lb/in

14.0 in — 5.0 in
(a) Loading

1.50-in dia.
1.00-in dia.

r = 0.03 in
at all steps

1.25-in dia.
at bearings
(b) Shaft details

FIGURE P15–55

Rigid
support

M
10

N
14

28

22

450 N

92

Dimensions in
millimeters

20   8

Mohr's Circle

A. For Problems 15–56 to 15–83, determine the principal stresses and the maximum shear stress using Mohr's circle. The following data sets give the stresses on the initial stress element. Perform the following operations.

  a. Draw the complete Mohr's circle, labeling critical points including σ_1, σ_2, τ_{max}, and σ_{avg}.

  b. On Mohr's circle, indicate which line represents the x-axis on the initial stress element.

  c. On Mohr's circle, indicate the angles from the line representing the x-axis to the σ_1-axis and the τ_{max}-axis.

  d. Draw the principal stress element and the maximum shear stress element in their proper orientation relative to the initial stress element.

Problem	σ_x	σ_y	τ_{xy}
15–56.	300 MPa	−100 MPa	80 MPa CW
15–57.	250 MPa	−50 MPa	40 MPa CW
15–58.	80 MPa	−10 MPa	60 MPa CW
15–59.	150 MPa	10 MPa	100 MPa CW
15–60.	20 ksi	−5 ksi	10 ksi CCW
15–61.	38 ksi	−25 ksi	18 ksi CCW
15–62.	55 ksi	15 ksi	40 ksi CCW
15–63.	32 ksi	−50 ksi	20 ksi CCW
15–64.	−900 kPa	600 kPa	350 kPa CCW
15–65.	−580 kPa	130 kPa	75 kPa CCW
15–66.	−840 kPa	−35 kPa	650 kPa CCW
15–67.	−325 kPa	50 kPa	110 kPa CCW
15–68.	−1800 psi	300 psi	800 psi CW
15–69.	−6500 psi	1500 psi	1200 psi CW
15–70.	−4250 psi	3250 psi	2800 psi CW
15–71.	−150 psi	8600 psi	80 psi CW
15–72.	260 MPa	0 MPa	190 MPa CCW
15–73.	1450 kPa	0 kPa	830 kPa CW
15–74.	22 ksi	0 ksi	6.8 ksi CW
15–75.	6750 psi	0 psi	3120 psi CCW
15–76.	0 ksi	−28 ksi	12 ksi CW
15–77.	0 MPa	440 MPa	215 MPa CW
15–78.	0 MPa	260 MPa	140 MPa CCW
15–79.	0 kPa	−1560 kPa	810 kPa CCW
15–80.	225 MPa	−85 MPa	0 MPa
15–81.	6250 psi	−875 psi	0 psi
15–82.	775 kPa	−145 kPa	0 kPa
15–83.	38.6 ksi	−13.4 ksi	0 ksi

B. For Problems 15–84 to 15–95 that result in the principal stresses from the initial Mohr's circle having the same sign, use the procedures from Section 15–11 to draw supplementary circles and find the following:

  a. The three principal stresses: σ_1, σ_2, and σ_3.

  b. The true maximum shear stress.

Problem	σ_x	σ_y	τ_{xy}
15–84.	300 MPa	100 MPa	80 MPa CW
15–85.	250 MPa	150 MPa	40 MPa CW
15–86.	180 MPa	110 MPa	60 MPa CW
15–87.	150 MPa	80 MPa	30 MPa CW
15–88.	30 ksi	15 ksi	10 ksi CCW

Problem	σ_x	σ_y	τ_{xy}
15–89.	38 ksi	25 ksi	8 ksi CCW
15–90.	55 ksi	15 ksi	5 ksi CCW
15–91.	32 ksi	50 ksi	20 ksi CCW
15–92.	−840 kPa	−335 kPa	120 kPa CCW
15–93.	−325 kPa	−50 kPa	60 kPa CCW
15–94.	−1800 psi	−300 psi	80 psi CW
15–95.	−6500 psi	−2500 psi	1200 psi CW

C. For Problems 15–96 to 15–105, use the data from the indicated problem for the initial stress element to draw Mohr's circle. Then determine the stress condition on the element at the specified angle of rotation from the given x-axis. Draw the rotated element in its proper relation to the initial stress element and indicate the normal and shear stresses acting on it.

Problem	Problem for the initial stress data	Angle of rotation from the x-axis
15–96.	15–56	30 deg CCW
15–97.	15–56	30 deg CW
15–98.	15–59	70 deg CCW
15–99.	15–61	20 deg CW
15–100.	15–63	50 deg CCW
15–101.	15–65	45 deg CW
15–102.	15–68	10 deg CCW
15–103.	15–70	25 deg CW
15–104.	15–71	80 deg CW
15–105.	15–73	65 deg CW

Computing Maximum Shear Stress

D. For the following problems, use Equation (15–2) to compute the magnitude of the maximum shear stress for the data from the indicated problem.

15–106. Use data from Problem 15–72.
15–107. Use data from Problem 15–73.
15–108. Use data from Problem 15–74.
15–109. Use data from Problem 15–75.

Strain-Gage Rosettes—Rectangular

A. The following data are obtained from tests in which a rectangular [0°, 45°, and 90°] strain-gage rosette was applied to a product made from the given material. Locate material property data from the Appendix and from Table 8–1. Use the procedure outlined in Section 15–12 to compute the following:

  a. Maximum principal strain

  b. Minimum principal strain

  c. Angle of orientation of the principal strains from the axis for gage 1

  d. Maximum principal stress

  e. Minimum principal stress

  f. Maximum shearing strain

  g. Maximum shearing stress

15–110. $\epsilon_1 = 1480\ \mu\epsilon$, $\epsilon_2 = 165\ \mu\epsilon$, $\epsilon_3 = 428\ \mu\epsilon$. Aluminum 6061-T6. SI metric units.

15–111. $\epsilon_1 = 853\ \mu\epsilon$, $\epsilon_2 = 406\ \mu\epsilon$, $\epsilon_3 = 641\ \mu\epsilon$. Aluminum 7075-T6. SI metric units.

15–112. $\epsilon_1 = 389\ \mu\epsilon$, $\epsilon_2 = 737\ \mu\epsilon$, $\epsilon_3 = -290\ \mu\epsilon$. AISI 1040 cold-drawn steel. SI metric units.

15–113. $\epsilon_1 = 925\ \mu\epsilon$, $\epsilon_2 = -631\ \mu\epsilon$, $\epsilon_3 = 552\ \mu\epsilon$. AISI 4140 OQT 900 steel. SI metric units.

15–114. $\epsilon_1 = 169\ \mu\epsilon$, $\epsilon_2 = -266\ \mu\epsilon$, $\epsilon_3 = 543\ \mu\epsilon$. Copper C14500 hard. U.S. Customary units.

15–115. $\epsilon_1 = 775\ \mu\epsilon$, $\epsilon_2 = 369\ \mu\epsilon$, $\epsilon_3 = -318\ \mu\epsilon$. Titanium Ti-6A1-4V, aged. U.S. Customary units.

15–116. $\epsilon_1 = 389\ \mu\epsilon$, $\epsilon_2 = 737\ \mu\epsilon$, $\epsilon_3 = -290\ \mu\epsilon$. Ductile iron, ASTM A536, 80-55-6. U.S. Customary units.

15–117. $\epsilon_1 = 1532\ \mu\epsilon$, $\epsilon_2 = -228\ \mu\epsilon$, $\epsilon_3 = 893\ \mu\epsilon$. AISI 501 OQT 1000 Stainless steel. U.S. Customary units.

Strain-Gage Rosettes—Delta

Repeat the problems in section E, but the data are taken from a delta [0°, 60°, and 120°] strain-gage rosette.

15–118. Use data from Problem 15–110.
15–119. Use data from Problem 15–111.
15–120. Use data from Problem 15–112.
15–121. Use data from Problem 15–113.
15–122. Use data from Problem 15–114.
15–123. Use data from Problem 15–115.
15–124. Use data from Problem 15–116.
15–125. Use data from Problem 15–117.

COMPUTER ASSIGNMENTS

1. Write a program for a computer, spreadsheet, or a programmable calculator to aid in the construction of Mohr's circle. Input the initial stresses, σ_x, σ_y, and τ_{xy}. Have the program compute the radius of the circle, the maximum and minimum principal stresses, the maximum shear stress, and the average stress. Use the program in conjunction with freehand sketching of the circle for the data for Problems 15–56 through 15–79.

2. Enhance the program in Assignment 1 by computing the angle of orientation of the principal stress element and the angle of orientation of the maximum shear stress element.

3. Enhance the program in Assignment 1 by computing the normal and shear stresses on the element rotated at any specified angle relative to the original x-axis.

4. Enhance the program in Assignment 1 by causing it to detect if the principal stresses from the initial Mohr's circle are of the same sign; and in such cases, print out the three principal stresses in the proper order, σ_1, σ_2, σ_3. Also have the program compute the true maximum shear stress from Equation (15–21).

5. Produce a spreadsheet similar to that shown in Figure 15–50 to analyze strain-gage rosette data for either a rectangular or delta-style rosette. The results are to include the items identified in the problem set E.

COLUMNS

MAJOR TOPICS

16–1 The Big Picture—Columns

16–2 Slenderness Ratio

16–3 Transition Slenderness Ratio

16–4 The Euler Formula for Long Columns

16–5 The J. B. Johnson Formula for Short Columns

16–6 Summary—Buckling Formulas

16–7 Design Factors for Columns and Allowable Load

16–8 Summary—Method of Analyzing Columns

16–9 Column Analysis Spreadsheet

16–10 Efficient Shapes for Column Cross Sections

16–11 Specifications of the AISC

16–12 Specifications of the Aluminum Association

16–13 Noncentrally Loaded Columns

16–1 THE BIG PICTURE— COLUMNS

Discussion Map

A column is a relatively long, slender member loaded in compression.

The failure mode for a column is called *buckling*, a common term for the condition of *elastic instability*, when the load on an initially straight column causes it to bend significantly. If the load is increased a small amount from the buckling load, the column would collapse suddenly—a very dangerous situation.

- How do we determine when a compression member is long and slender?

- How do we determine the magnitude of the load at which buckling would occur?

- What kind of cross-sectional shapes are preferred for columns?

- What influence does the manner of holding the ends of a column have on the buckling load?

- What industry standards apply to columns?

- These and other details about column analysis and design are presented in this chapter.

Activity

Let's explore this concept of a column. Find some examples of things around you that may fit that definition. Then describe each of them, giving its length, the shape and dimensions of its cross section, and the material from which it is made. Here are a few examples to get you started.

A meter stick: It is typically made from wood or aluminum. Obviously its length is 1.00 m. The cross section is typically about 30 mm wide and 4 mm thick. It even looks long and slender.

A steel 6-in (152-mm) rule: You might have used one of these in a class where you operated metal-cutting machine tools or wood lathes. Again its length is obviously 6.00 in (152 mm). It is made from a flat strip of steel, 0.75 in wide and 0.020 in thick (19 mm × 0.50 mm). Even though it is much shorter than the meter stick, it is thinner, that is, more slender. Both length and slenderness matter.

A wood dowel that you might get at a hardware store. Perhaps it is 36 in long and has a diameter of $\frac{3}{8}$ in (914 mm × 9.5 mm).

What Examples Have You Found?

Now let's try to load one of these items with a direct axial compressive load. This means that the line of action of the load is in line with the long axis of the column. Just support it on a table or the floor and push down on it with your hand. Try to push straight down and not sideways at all, but don't hold it tight with your fingers. Be careful not to push too hard and break it!

What Happened?

Here we describe the behavior of the wood meter stick. Loading it slowly, we find that it can support a very small load while remaining straight. But without much exertion,

Force

X   Y

Y   X

Buckling occurs
about the thinner
Y-Y axes

Base

FIGURE 16–1   Illustration of buckling of a meter stick.

we can cause the stick to bend noticeably. This phenomenon is called *buckling*. Be careful! With just a modest increase in load after buckling occurs, the stick would break easily. Notice that the meter stick buckles about the thin dimension of its cross section. Figure 16–1 shows a sketch of what happened. Figure 16–2(a) shows a side view. You probably would have predicted that based on your own experience. Later we will quantify why that happens.

Now let's change the procedure a bit. It appears that the meter stick tends to bow out near the middle, say at the 0.50-m point. What if we provide some lateral support on the sides of the stick at that point? Try it yourself if you have a meter stick. Place your fingers lightly on either side to restrain the tendency for it to bow outward. Now push down on it again as you did before.

What Happened?

Now we can apply a much higher load to the stick without it buckling. But there is a point where the load is high enough to see a quite different form of buckling. The lower half and the upper half of the stick buckle with one going one way and one going the other. In fact, it looks as if the stick takes the form of a complete sine wave. We will discuss this observation later.

Let's change the procedure again. Grab both ends of the meter stick with a firm grip and try as hard as you can to keep the stick from rotating while simultaneously applying an axial load that will cause buckling.

What Happened?

First, you should notice that it takes a much higher force to cause buckling. Also, you should notice that the shape of the buckled stick is different from that when you did not restrain the ends. Two factors are working here. Grabbing the stick with your fists effectively shortens the column by about 90 mm (3.5 inches) on each end. Because the column is shorter, it takes a higher load to cause buckling. But also your effort at keeping the ends from rotating caused the buckled shape to be similar to that shown in Figure 16–2(b). You essentially *fixed the ends*. We will discuss this phenomenon later also.

Are All of the Columns You Found Perfectly Straight?

Probably not. Of those tested here, most tended to buckle in a particular direction because they were initially crooked. Pushing down on them had the additional effect of bending the crooked section even more in the same direction. We will explore that phenomenon later also.

The examples discussed here are all items that were not meant to carry axial compressive loads. They do serve well as demonstrators for buckling.

FIGURE 16–2   Comparison of shapes for buckled columns.

F

Force guided
to act axially

Initially
straight
column

Shape of buckled
column

Both ends may
rotate

(a) Buckled shape of column
with unrestrained ends

F

Initially
straight
column

Top fixed from
rotation and
guided

Shape of buckled
column

Fixed end

(b) Buckled shape of column
with fixed ends

What examples of columns can you find that are more substantial and that were designed to be sufficiently strong and stable to withstand sizable axial compressive loads?

You might not be able to carry all of these examples into a classroom, laboratory, or office, but here are a few.

- The vertical columns of a steel-framed building: The lower columns on a multistory building must be strong and stiff enough to hold up all the weight above them. Even in a one-story building, they must hold the roof structure and, possibly, a snow load on top of that.

- The steel posts holding up the beam across the length of the basement of a home: The beam supports the joists from the floor above and all the weight of the furniture and the people there. The posts carry that load to the basement floor or the foundation. The posts are likely made from steel pipe or round hollow structural sections (HSS). That is an efficient shape for a column as we will discuss later in this chapter.

- The cylinder rod of a hydraulic actuator: You may have seen these on a piece of construction equipment, agricultural machinery, or in an industrial automation system. Some of these cylinder rods push with great force and they must be designed not to buckle as they extend out from the cylinder.

What Others Have You Found?

Now summarize the observations so far.

- We have demonstrated that a long, slender member tends to buckle when subjected to an axial compressive load. But when does a member become long and slender? We define the term *slenderness ratio* later to quantify that. It is a function of the length of the column, how its ends are held, and the shape and size of the cross section of the column.

- We demonstrated that a column can take some magnitude of axial load before buckling begins. Then the onset of buckling is quite sudden. At what load will it buckle? We show several methods of predicting that in this chapter.

- We demonstrated that the way the ends of the column are held affects the buckling load. We elaborate on that later using the term *end fixity*.

- The columns we found were possibly made from different materials, such as steel, aluminum, wood, or plastic. What effect does the material have on its tendency to buckle? We show in this chapter that the material's modulus of elasticity, *E*, has a major effect on the tendency for a long column to buckle. For shorter columns, the yield strength is also a factor.

- Some of the columns we found were initially crooked. It seemed that they buckled at a lower load than the straight ones and always in the direction of the initial crookedness. Can we quantify that? Yes, additional approaches to analyze crooked columns and those that have the load applied off the axis (called *eccentrically loaded columns*) will be presented.

Objectives of This Chapter

After completing this chapter, you should be able to:

1. Define *column*.
2. Differentiate between a column and a short compression member.
3. Describe the phenomenon of *buckling*, also called *elastic instability*.
4. Define *radius of gyration* for the cross section of a column and be able to compute its magnitude.
5. Understand that a column is expected to buckle about the axis for which the radius of gyration is the minimum.
6. Define *end-fixity factor, K*, and specify the appropriate value depending on the manner of supporting the ends of a column.
7. Define *effective length, L_e, slenderness ratio*, and *transition slenderness ratio* (also called the *column constant, C_c*) and compute their values.
8. Use the values for the slenderness ratio and the column constant to determine when a column is *long* or *short*.
9. Use the *Euler formula* for computing the critical buckling load for long columns and the *J. B. Johnson formula* for short columns.
10. Apply a design factor to the critical buckling load to determine the *allowable load* on a column.
11. Recognize efficient shapes for column cross sections.
12. Design columns to safely carry given axial compression loads.
13. Apply the specifications of the American Institute of Steel Construction (AISC) and the Aluminum Association to the analysis of columns.
14. Analyze columns that are initially crooked to determine their critical buckling load.
15. Analyze columns for which the applied load acts eccentric to the axis of the column.

16–2 SLENDERNESS RATIO

A column has been described as a relatively long, slender member loaded in compression. This description is stated in relative terms and is not very useful for analysis.

The measure of the slenderness of a column must take into account the length, the cross-sectional shape and dimensions of the column, and the manner of attaching the ends of the column to the structures that supply loads and reactions to the column. The commonly used measure of slenderness is the *slenderness ratio*, defined as

Slenderness Ratio ▼

$$SR = \frac{KL}{r} = \frac{L_e}{r} \qquad (16\text{–}1)$$

where   *L* = *actual length* of the column between points of support or lateral restraint
   K = *end-fixity factor*

L_e = *effective length,* taking into account the manner of attaching the ends (note that $L_e = KL$)

r = *smallest radius of gyration* of the cross section of the column

Each of these terms is discussed here.

Actual Length, *L*

For a simple column having the load applied at one end and the reaction provided at the other, the actual length is, obviously, the length between its ends. But for components of structures loaded in compression where a means of restraining the member laterally to prevent buckling is provided, the actual length is taken between points of restraint. Each part is then considered to be a separate column.

End-Fixity Factor, *K*

The end-fixity factor is a measure of the degree to which each end of the column is restrained against rotation. Three classic types of end connections are typically considered: the pinned end, the fixed end, and the free end. Figure 16–3

shows these end types in several combinations with the corresponding values of K. Note that two values of K are given. One is the theoretical value and the other is the one typically used in practical situations, recognizing that it is difficult to achieve the truly fixed end, as discussed later. See also Section 16–6.

Figure 16–3(a) shows a commercially available demonstration device that illustrates the relative stiffness and resistance to buckling of four end-fixity conditions. The sizes of the stacks of weights on the upper end of the first three indicate the approximate load at which buckling is initiated. The very small load on the right end column (the fixed-free condition) does not represent the true buckling load because the column itself is unstable and bends to one side easily if there is any degree of misalignment of the load from the axis of the column.

Pinned ends of columns are essentially unrestrained against rotation. When a column with two pinned ends buckles, it assumes the shape of a smooth curve between its ends, as shown in Figure 16–3(b). This is the basic case of column buckling, and the value of $K = 1.0$ is applied to columns having two pinned ends. An ideal implementation

FIGURE 16–3 Values of K for effective length, $L_e = KL$, for four different end fixities.

Source: P.A. Hilton Ltd. Hi-Tech, Hampshire, England.

(a) Commercially available demonstrator for end fixity

	Pinned-pinned	Fixed-pinned	Fixed-fixed	Fixed-free
Theoretical values	$K = 1.0$	$K = 0.7$	$K = 0.5$	$K = 2.0$
Practical values	$K = 1.0$	$K = 0.8$	$K = 0.65$	$K = 2.10$
	(b)	(c)	(d)	(e)

of the pinned end is the frictionless spherical ball joint that would permit the column to rotate in any direction about any axis. For a cylindrical pin joint, free rotation is permitted about the centerline of the pin, but some restraint is provided in the plane perpendicular to the centerline. Care must be exercised in applying end-fixity factors to cylindrical pins for this reason. It is assumed that the pinned end is guided in some way so that the line of action of the axial load remains unchanged.

The combination of one fixed end and one pinned end is shown in Figure 16–3(c). Notice that the buckled shape approaches the fixed end with a zero slope while the pinned end rotates freely. The theoretical value of $K = 0.7$ applies to such an end fixity while $K = 0.80$ is recommended for practical use.

Fixed ends provide theoretically perfect restraint against rotation of the column at its ends. As the column tends to buckle, the deflected shape of the axis of the column must approach the fixed end with a zero slope, as illustrated in Figure 16–3(d). The buckled shape bows out in the middle but exhibits two points of inflection to reverse the direction of curvature near the ends. The theoretical value of the end-fixity factor is $K = 0.5$, indicating that the column acts as if it were only one-half as long as it really is. Columns with fixed ends are much stiffer than pinned-end columns and can, therefore, take higher loads before buckling. It should be understood that it is very difficult to provide perfectly fixed ends for a column. It requires that the connection to the column is rigid and stiff and that the structure to which the loads are transferred is also rigid and stiff. For this reason, the higher value of $K = 0.65$ is recommended for practical use.

A free end for a column is unrestrained against rotation and also against translation. Because it can move in any direction, this is the worst case for column end fixity. The only practical way of using a column with a free end is to have the opposite end fixed, as illustrated in Figure 16–3(e). Such a column is sometimes referred to as the *flagpole* case because the fixed end is similar to the flagpole inserted deeply into a tight-fitting socket while the other end is free to move in any direction. Called the fixed-free end condition, the theoretical value of K is 2.0. A practical value is $K = 2.10$.

Effective Length, L_e

Effective length combines the actual length with the end-fixity factor; $L_e = KL$. In the problems in this book we use the recommended practical values for the end-fixity factor, as shown in Figure 16–3. In summary, the following relationships will be used to compute the effective length:

Effective Length ▼

1. Pinned-end columns: $L_e = KL = 1.0(L) = L$
2. Fixed-pinned columns: $L_e = KL = 0.80(L)$
3. Fixed-end columns: $L_e = KL = 0.65(L)$
4. Fixed-free columns: $L_e = KL = 2.10(L)$

Radius of Gyration, r

The measure of slenderness of the cross section of the column is its radius of gyration, r, defined as

Radius of Gyration ▼

$$r = \sqrt{\frac{I}{A}} \qquad (16\text{–}2)$$

where I = moment of inertia of the cross section of the column with respect to one of the principal axes
   A = area of the cross section

Because both I and A are geometrical properties of the cross section, the radius of gyration, r, is also. Formulas for computing r for several common shapes are given in Appendix A–1. Also, r is listed with other properties for some of the standard shapes in the Appendix. For those for which r is not listed, the values of I and A are available and Equation (16–2) can be used to compute r very simply. Section 6–10 in Chapter 6 includes additional discussion of the radius of gyration with examples and practice problems.

Note that the value of the radius of gyration, r, is dependent on the axis about which it is to be computed. In most cases, it is required that you determine the axis for which the *radius of gyration is the smallest,* because that is the axis about which the column would likely buckle. Consider, for example, a column made from a rectangular section whose width is much greater than its thickness, as sketched in Figure 16–1. The meter stick demonstrates that when loaded in axial compression with little or no restraint at the ends, the column will always buckle with respect to the axis through the thinnest dimension.

Refer to Figure 16–4 to illustrate this point. Shown there are sketches of the thin rectangular cross section of the meter stick sketched in Figure 16–1. Part (a) shows the shape with respect to the centroidal axis Y-Y. The thickness of the rectangle is t and the width is h. Then the moment of inertia of the rectangle with respect to the Y-Y axis is

$$I_Y = ht^3/12$$

The area is simply

$$A = th$$

Now, using Equation (16–2) we can obtain a relationship for the radius of gyration, r_Y.

$$r_Y = \sqrt{\frac{I_Y}{A}} = \sqrt{\frac{ht^3/12}{th}} = \sqrt{\frac{t^2}{12}} = \frac{t}{\sqrt{12}} = 0.289t$$

Similarly, using Figure 16–4(b), we can obtain an equation for r_X.

$$I_X = th^3/12$$
$$A = th$$

$$r_X = \sqrt{\frac{I_X}{A}} = \sqrt{\frac{th^3/12}{th}} = \sqrt{\frac{h^2}{12}} = \frac{h}{\sqrt{12}} = 0.289h$$

FIGURE 16–4 Radius of gyration of the cross section of a thin, rectangular column.

Column cross section

Critical buckling axis
For axis Y-Y: $r = 0.289t$

(a) Radius of gyration for Y-Y axis

For axis X-X: $r = 0.289h$

(b) Radius of gyration for X-X axis

Observe that because $h > t$, $r_X > r_Y$ and then r_Y is the smallest radius of gyration for the section.

For the wide-flange beams (Appendix A–7) and American Standard beams (Appendix A–8), the minimum value of r is that computed with respect to the Y-Y axis; that is,

$$r_{\min} = r_Y = \sqrt{\frac{I_Y}{A}}$$

Similarly, for rectangular hollow structural sections (HSS) (Appendix A–9), the minimum radius of gyration is that with respect to the Y-Y axis. Values for r are listed in the table.

For structural steel angles, called L-shapes, neither the X-X nor the Y-Y axis provides the minimum radius of gyration. As illustrated in Appendix A–5, $r_{\min}$ is computed with respect to the Z-Z axis, with the values listed in the table.

For symmetrical sections, the value of r is the same with respect to any principal axis. Such shapes are the solid or hollow circular section and the solid or hollow square section.

SUMMARY OF THE METHOD FOR COMPUTING THE SLENDERNESS RATIO ▼

1. Determine the actual length of the column, L, between endpoints or between points of lateral restraint.

2. Determine the end-fixity factor from the manner of support of the ends and by reference to Figure 16–3.

3. Compute the effective length, $L_e = KL$.

4. Compute the *smallest* radius of gyration for the cross section of the column.

5. Compute the slenderness ratio from

$$SR = \frac{L_e}{r_{\min}}$$

16–3 TRANSITION SLENDERNESS RATIO

When is a column considered long? The answer to this question requires the determination of the *transition slenderness ratio*, or column constant C_c.

Transition Slenderness Ratio ▼

$$C_c = \sqrt{\frac{2\pi^2 E}{s_y}} \qquad (16\text{–}3)$$

The following rules are applied to determine if a given column is long or short.

If the actual effective slenderness ratio L_e/r is greater than C_c, then the column is long, and the Euler formula, defined in the next section, should be used to analyze the column.

If the actual ratio L_e/r is less than C_c, then the column is short. In these cases, either the J. B. Johnson formula, special codes, or the direct compressive stress formula should be used, as discussed in later sections.

Where a column is being analyzed to determine the load it will carry, the value of C_c and the actual ratio L_e/r should be computed first to determine which method of analysis should be used. Notice that C_c depends on the material properties of yield strength s_y and modulus of elasticity E. In working with steel, E is usually taken to be 207 GPa (30×10^6 psi). Using this value and assuming a range of values for yield strength, we obtain the values for C_c shown in Figure 16–5. Be aware that the value of E for structural steels is typically taken to be 200 GPa (29×10^6 psi), shifting the curve downward very slightly.

For aluminum, E is approximately 69 GPa (10×10^6 psi). The corresponding values for C_c are shown in Figure 16–6.

FIGURE 16–5 Transition slenderness ratio C_c versus yield strength for steel.

FIGURE 16–6 Transition slenderness ratio C_c versus yield strength for aluminum.

16–4 THE EULER FORMULA FOR LONG COLUMNS

For long columns having an effective slenderness ratio greater than the transition value C_c, the Euler formula can be used to predict the critical load at which the column would be expected to buckle. The formula is

Euler Formula for Long Columns ▽

$$P_{cr} = \frac{\pi^2 E A}{(L_e/r)^2} \qquad (16\text{--}4)$$

where A is the cross-sectional area of the column. An alternative form can be expressed in terms of the moment of inertia by noting that $r^2 = I/A$. Then the formula becomes

Euler Formula for Long Columns ▼

$$P_{cr} = \frac{\pi^2 EI}{L_e^2} \qquad (16-5)$$

16–5 THE J. B. JOHNSON FORMULA FOR SHORT COLUMNS

If a column has an actual effective slenderness ratio L_e/r less than the transition value C_c, the Euler formula predicts an unreasonably high critical load. One formula recommended for machine design applications in the range of L_e/r less than C_c is the J. B. Johnson formula.

J. B. Johnson Formula for Short Columns ▼

$$P_{cr} = As_y\left[1 - \frac{s_y(L_e/r)^2}{4\pi^2 E}\right] \qquad (16-6)$$

This is one form of a set of equations called parabolic formulas, and it agrees well with the performance of steel columns in typical machinery.

The Johnson formula gives the same result as the Euler formula for the critical load at the transition slenderness ratio C_c. Then for very short columns, the critical load approaches that which would be predicted from the direct compressive stress equation, $\sigma = P/A$. Therefore, it could be said that the Johnson formula applies best to columns of intermediate length.

16–6 SUMMARY—BUCKLING FORMULAS

This section summarizes the basis for the Euler and Johnson formulas for column analysis and elaborates on the behavior of columns of varying materials and slenderness ratios.

The buckling phenomenon is not a failure of the material from which a column is made. Rather, it is a failure of the column as a whole to maintain its shape. This type of failure is called *elastic instability*.

Recall the exercises you did during the Big Picture section of this chapter. As you loaded a long, slender column, such as the meter stick, you observed that the column buckled at some moderate load. If the load was removed after observing the occurrence of buckling, you should notice that the column was not damaged. There was no yielding or fracture of the material.

To design a safe column, you must ensure that it remains elastically stable. The basis of both the Euler and Johnson formulas is developed from the field of stress analysis known as *elasticity*. Reference 6 is a useful source for this development.

The principle of elastic stability states that a column is stable if it maintains its straight shape as the load is increased. There is a level of load, though, at which the column is unable to maintain its shape. Then it buckles. We call the load at which buckling occurs the *critical buckling load*, P_{cr}. Obviously, as the designer of the column, you must ensure that the actual load applied to the column is well below P_{cr}.

When the axial load on a column is below the critical buckling load, it has sufficient stiffness to resist the tendency to deviate from the straight-line orientation of its neutral axis. Even when the load is slightly away from the axis, the column is able to maintain its shape. We can visualize this by referring to Figure 16–2. Part (a) shows a straight, pinned-end column carrying an axial compressive load. It also shows, in an exaggerated manner, that at the point of incipient buckling, the column takes the shape of one-half of a sine wave. Any deviation of the neutral axis from a straight line introduces bending in the column. When the applied force is less than P_{cr}, the stiffness of the column is sufficient to resist this deformation and to maintain the straightness of the column. It acts like a kind of spring, sometimes called a *leaf spring*, to return the column to its straight form whenever there is a tendency to deviate. The mathematical techniques of differential equations were used by the Swiss mathematician Leonhard Euler (1707–1783) in 1757 to analyze this condition and to produce what is now called the *Euler formula* in his honor [Equation (16–4)].

Recognizing that the deformed, buckled, pinned-end column takes the shape of one-half of a sine wave can also help us to visualize why different end-fixity conditions affect the magnitude of the critical buckling load. Refer to Figure 16–3 that shows the buckled shape of four columns with different end fixity. The explanation of the theoretical value for K, the end-fixity factor, is given here.

a. Part (b) is the pinned-end design, and the entire length conforms to the sine wave form after buckling starts. This is the reason that $L_e = L$.

b. Part (d) shows the fixed-fixed design, and the sine wave form occurs only over the middle half of the column. Therefore, the theoretical effective length is $L_e = 0.5L$.

c. Part (e) shows the fixed-free design, and the deformed shape of the column is only one-fourth of a sine wave. To complete the one-half sine wave that characterizes the buckled shape, the curved line for the deformed neutral axis would have to be extended an equal distance *below* the bottom, fixed end of the column. Therefore, the effective length is $L_e = 2.0 L$.

d. Part (c) shows the fixed-pinned design. You should logically see that this design is somewhat midway between those shown in parts (b) and (d). Indeed, the sine wave form occurs in approximately the upper two-thirds of the column. Therefore, the theoretical effective length is $L_e = 0.7 L$.

Again, because it is very difficult to produce a perfectly fixed end for a column, the design values for the end-fixity

FIGURE 16–7 Average column stress vs. slenderness ratio.

Average Column Stress vs. Slenderness Ratio

factor, K, are used in practical problems such as those in this book.

Comparison of the Euler and Johnson Formulas

We have noted that the Euler formula is only valid for columns having slenderness ratios greater than the transition slenderness ratio, C_c. When the actual L_e/r is less than C_c, the Johnson formula is recommended. Then, for very small values of L_e/r, the result from the Johnson formula approaches the load at which the material of the column would fail by yielding under direct axial compression.

Another way to examine these concepts is to divide both the Euler [Equation (16–4)] and Johnson [(Equation (16–6)] formulas by the cross-sectional area, A. The left side of the formulas then have the form P/A and represent an average stress on the column cross section.

Figure 16–7 illustrates these concepts in graphical form. The vertical axis plots the average stress in the column when the axial load is equal to the critical buckling load. The horizontal axis is the slenderness ratio, L_e/r. The material properties of modulus of elasticity, E, and yield strength, s_y, affect the critical buckling load and so the graph presents a family of curves representative of different materials. For a given material, the right-most part is for large values of $L_e/r > C_c$. In this region, the behavior of columns made from any alloy of the material conforms to the Euler formula. At $L_e/r = C_c$, the Euler and Johnson formulas are tangent. Then, for $L_e/r < C_c$, the Johnson formula applies. Finally, for very short columns, the average stress approaches the yield strength of the material representing failure by direct compression without any buckling effect.

Laterally Braced Columns

Recall the exercise described in the Big Picture section. When the long, slender meter stick was held lightly at midlength, the stick could withstand a much higher load before buckling. The lateral bracing effectively divides the column into two separate columns, one-half as long as the full column. The critical buckling load is then increased dramatically. When buckling does occur, each part of the column deforms in the shape of the one-half sine wave as shown in Figure 16–8. The whole column then takes the shape of a full sine wave. An example problem shown later illustrates this result.

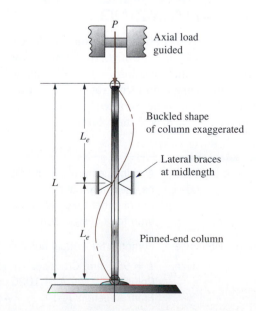

FIGURE 16–8 Laterally braced column.

16–7 DESIGN FACTORS FOR COLUMNS AND ALLOWABLE LOAD

Because the mode of failure for a column is buckling rather than yielding or ultimate failure of the material, the methods used before for computing design stress do not apply to columns.

Instead, an *allowable load* is computed by dividing the critical buckling load from either the Euler formula [Equation (16–4)] or the Johnson formula [Equation (16–6)] by a design factor, *N*. That is,

Allowable Load on a Column ▽

$$P_a = \frac{P_{cr}}{N} \qquad (16\text{–}7)$$

where P_a = allowable, safe load
P_{cr} = critical buckling load
N = design factor

The selection of the design factor is the responsibility of the designer unless the project comes under a code. Factors to be considered in the selection of a design factor are similar to those used for design factors applied to stresses. A common factor used in mechanical design is $N = 3.0$, chosen because of the uncertainty of material properties, end fixity, straightness of the column, or the possibility that the load will be applied with some eccentricity rather than along the axis of the column. Larger factors are sometimes used for critical situations and for very long columns.

In building construction, where the design is governed by the specifications of the American Institute of Steel Construction, AISC, a factor of 1.92 is recommended for long columns. The Aluminum Association calls for $N = 1.95$ for long columns. See Sections 16–12 and 16–13.

16–8 SUMMARY—METHOD OF ANALYZING COLUMNS

The purpose of this section is to summarize the concepts presented in Sections 16–3 through 16–7 into a procedure that can be used to analyze columns. It can be applied to a straight column having a uniform cross section throughout its length and for which the compression load is applied in line with the centroidal axis of the column.

METHOD OF ANALYZING COLUMNS ▽

To start, it is assumed that the following factors are known:

1. The actual length, *L*
2. The manner of connecting the column to its supports
3. The shape of the cross section of the column and its dimensions
4. The material from which the column is made.

Then the procedure is

1. Determine the end-fixity factor, *K*, by comparing the manner of connection of the column to its supports with the information in Figure 16–3.
2. Compute the effective length, $L_e = KL$.
3. Compute the minimum value of the radius of gyration of the cross section from $r_{min} = \sqrt{I_{min}/A}$; or determine r_{min} from tables of data.
4. Compute the maximum slenderness ratio from

$$SR_{max} = \frac{L_e}{r_{min}}$$

5. Using the modulus of elasticity, *E*, and the yield strength, s_y, for the material, compute the column constant,

$$C_c = \sqrt{\frac{2\pi^2 E}{s_y}}$$

6. Compare the value of *SR* with C_c.
   a. If $SR > C_c$, the column is long. Use the Euler formula to compute the critical buckling load,

   $$P_{cr} = \frac{\pi^2 EA}{(SR)^2} \qquad (16\text{–}4)$$

   b. If $SR < C_c$, the column is short. Use the Johnson formula to compute the critical buckling load,

   $$P_{cr} = As_y\left[1 - \frac{s_y(SR)^2}{4\pi^2 E}\right] \qquad (16\text{–}6)$$

7. Specify the design factor, *N*.
8. Compute the allowable load, P_a,

$$P_a = \frac{P_{cr}}{N}$$

Example Problem 16–1   A round compression member with both ends pinned and made of AISI 1020 cold-drawn steel is to be used in a machine. Its diameter is 25 mm, and its length is 950 mm. What maximum load can the member take before buckling would be expected? Also compute the allowable load on the column for a design factor of $N = 3$.

Solution   Objective: Compute the critical buckling load for the column and the allowable load for a design factor of $N = 3$.

Given: $L = 950$ mm. Cross section is circular; $D = 25$ mm. Pinned ends.
Column is steel; AISI 1020 cold-drawn.
From Appendix A–14: $s_y = 441$ MPa; $E = 207$ GPa $= 207 \times 10^9$ N/m^2

Analysis: Use the *Method of Analyzing Columns*.

Results: **Step 1:** Determine the end-fixity factor. For the pinned-end column, $K = 1.0$.

Step 2: Compute the effective length.

$$L_e = KL = 1.0(L) = 950 \text{ mm}$$

Step 3: Compute the smallest value of the radius of gyration. From Appendix A–1, for any axis of a circular cross section, $r = D/4$. Then

$$r = \frac{D}{4} = \frac{25 \text{ mm}}{4} = 6.25 \text{ mm}$$

Step 4: Compute the slenderness ratio, $SR = L_e/r$.

$$SR = \frac{L_e}{r} = \frac{950 \text{ mm}}{6.25 \text{ mm}} = 152$$

Step 5: Compute the column constant, C_c.

$$C_c = \sqrt{\frac{2\pi^2 E}{s_y}} = \sqrt{\frac{2\pi^2(207 \times 10^9 \text{ N/m}^2)}{441 \times 10^6 \text{ N/m}^2}} = 96.3$$

Step 6: Compare C_c with SR and decide if column is long or short. Then use the appropriate column formula to compute the critical buckling load. Since SR is greater than C_c, Euler's formula applies.

$$P_{cr} = \frac{\pi^2 EA}{(SR)^2}$$

The area is

$$A = \frac{\pi D^2}{4} = \frac{\pi(25 \text{ mm})^2}{4} = 491 \text{ mm}^2$$

Then

$$P_{cr} = \frac{\pi^2(207 \times 10^9 \text{ N/m}^2)(491 \text{ mm}^2)}{(152)^2} \times \frac{1 \text{ m}^2}{(10^3 \text{ mm})^2} = 43.4 \text{ kN}$$

Step 7: A design factor of $N = 3$ is specified.

Step 8: The allowable load, P_a, is

$$P_a = \frac{P_{cr}}{N} = \frac{43.4 \text{ kN}}{3} = 14.5 \text{ kN}$$

Example Problem 16–2 Determine the critical load on a steel column having a square cross section 12 mm on a side with a length of 300 mm. The column is to be made of AISI 1040, hot-rolled. It will be rigidly welded to a firm support at one end and connected by a pin joint at the other. Also compute the allowable load on the column for a design factor of $N = 3$.

Solution Objective: Compute the critical buckling load for the column and the allowable load for a design factor of $N = 3$.

Given: $L = 300$ mm. Cross section is square; each side is $b = 12$ mm.
One pinned end; one fixed end. Column is steel; AISI 1040 hot-rolled.
From Appendix A–14: $s_y = 414$ MPa; $E = 207$ GPa $= 207 \times 10^9$ N/m^2

Analysis: Use the *Method of Analyzing Columns*.

Results: **Step 1:** Determine the end-fixity factor. For the fixed-pinned-end column, $K = 0.80$ is a practical value (Figure 16–3).

Step 2: Compute the effective length.

$$L_e = KL = 0.80(L) = 0.80(300 \text{ mm}) = 240 \text{ mm}$$

Step 3: Compute the smallest value of the radius of gyration. From Appendix A–1, for a square cross section, $r = b/\sqrt{12}$. Then

$$r = \frac{b}{\sqrt{12}} = \frac{12 \text{ mm}}{\sqrt{12}} = 3.46 \text{ mm}$$

Step 4: Compute the slenderness ratio, $SR = L_e/r$.

$$SR = \frac{L_e}{r} = \frac{KL}{r} = \frac{(0.8)(300\,\text{mm})}{3.46\,\text{mm}} = 69.4$$

Step 5: Normally, we would compute the value of the column constant, C_c. But, in this case, let's use Figure 16–5. For a steel having a yield strength of 414 MPa, $C_c = 96$, approximately.

Step 6: Compare C_c with SR and decide if the column is long or short. Then use the appropriate column formula to compute the critical buckling load. Since SR is less than C_c, the Johnson formula, Equation (16–6), should be used.

$$P_{cr} = As_y\left[1 - \frac{s_y(SR)^2}{4\pi^2 E}\right]$$

The area of the square cross section is

$$A = b^2 = (12\,\text{mm})^2 = 144\,\text{mm}^2$$

Then

$$P_{cr} = (144\,\text{mm}^2)\left(\frac{414\,\text{N}}{\text{mm}^2}\right)\left[1 - \frac{(414 \times 10^6\,\text{N/m}^2)(69.4)^2}{4\pi^2(207 \times 10^9\,\text{N/m}^2)}\right]$$

$$P_{cr} = 45.1\,\text{kN}$$

Step 7: A design factor of $N = 3$ is specified.

Step 8: The allowable load, P_a, is

$$P_a = \frac{P_{cr}}{N} = \frac{45.1\,\text{kN}}{3} = 15.0\,\text{kN}$$

Example Problem 16–3   Refer to the results of Example Problem 16–1 in which a 25-mm-diameter steel column with a length of 950 mm was analyzed to determine the critical buckling load. The column was made from AISI 1020 cold-drawn steel. It was found that $P_{cr} = 43.4\,\text{kN}$. Now consider a redesign of the structure for which this column is a part. It is decided to provide lateral bracing in all directions for the column at its midlength. Determine the critical buckling load for the redesigned column.

Solution   Objective: Compute the critical buckling load for the braced column.

Given: Data from Example Problem 16–1. Circular cross section, $D = 25\,\text{mm}$. $L = 950\,\text{mm}$. Pinned ends. AISI 1020 CD steel; $E = 207\,\text{GPa}$; $s_y = 441\,\text{MPa}$. Column braced at 450 mm from each end.

Analysis: Use the *Method of Analyzing Columns*.

Results: **Step 1:** End-fixity factor, $K = 1.0$ for pinned ends.

Step 2: Effective length: The unbraced length is now 450 mm. Then $L_e = KL = 1.0(450\,\text{mm}) = 450\,\text{mm}$

Step 3: Radius of gyration $= r = 6.25\,\text{mm}$ [from Example Problem 16–1]

Step 4: Slenderness ratio $= SR = L_e/r = (450\,\text{mm})/(6.25\,\text{mm}) = 72$

Step 5: Column Constant $= C_c = 96.2$ [from Example Problem 16–1]

Step 6: Because $SR < C_c$, use the Johnson formula. $A = 491\,\text{mm}^2$

$$P_{cr} = As_y\left[1 - \frac{s_y(SR)^2}{4\pi^2 E}\right]$$

$$P_{cr} = (491\,\text{mm}^2)\left(\frac{441\,\text{N}}{\text{mm}^2}\right)\left[1 - \frac{(441 \times 10^6\,\text{N/m}^2)(72)^2}{4\pi^2(207 \times 10^9\,\text{N/m}^2)}\right] = 156\,\text{kN}$$

Comment: The critical buckling load has been increased from 43.4 kN to 156 kN, over $3\frac{1}{2}$ times. That is a significant improvement. The column behaves as if it were only one-half as long.

16–9 COLUMN ANALYSIS SPREADSHEET

Completing the process described in Section 16–8 using a calculator, pencil, and paper is tedious. A spreadsheet automates the calculations after you have entered the pertinent data for the particular column to be analyzed. Figure 16–9 shows the output of a spreadsheet used to solve Example Problem 16–1. The layout of the spreadsheet could be done in many ways, and you are encouraged to develop your own style. The following comments describe the features of the given spreadsheet.

1. At the top of the sheet, instructions to the user are given for entering data and for units. This sheet is for SI metric units only. A different sheet would be used if U.S. Customary data were to be used.

2. On the left side of the sheet are listed the various data that must be provided by the user to run the calculations. On the right are listed the output values. Formulas for computing L_e, C_c, KL/r, and allowable load are written directly into the cell where the computed values show. The output data for the message "Column is: *long*" and the critical buckling load are produced by *functions* set up within *macros* written in Visual Basic and placed on a separate sheet of the spreadsheet. Figure 16–10 shows the two macros used. The first (*LorS*) carries out the decision process to test whether the column is long or short as indicated by comparison of its slenderness ratio with the column constant. The

second (*Pcr*) computes the critical buckling load using either the Euler formula or the J. B. Johnson formula, depending on the result of the *LorS* macro. These functions are called by statements in the cells where "long" and the computed value of the critical buckling load (43.42 kN) are located.

3. Having such a spreadsheet can enable you to analyze several design options quickly. For example, the given problem statement indicated that the ends were pinned, resulting in an end-fixity value of $K = 1$. What would happen if both ends were fixed? Simply changing the value of that one cell to $K = 0.65$ would cause the entire sheet to be recalculated, and the revised value of critical buckling load would be available almost instantly. The result is that $P_{cr} = 102.76$ kN, an increase of 2.37 times the original value. With that kind of improvement, you, the designer, might be inclined to change the design to produce fixed ends.

16–10 EFFICIENT SHAPES FOR COLUMN CROSS SECTIONS

When designing a column to carry a specified load, the designer has the responsibility for selecting the general shape of the column cross section and then to determine the required dimensions. The following principles may aid in the initial selection of the cross-sectional shape.

FIGURE 16–9 Spreadsheet for column analysis with data from Example Problem 16–1 [SI data].

FIGURE 16–10   Macros used in the column analysis spreadsheet.

```
'LorS Macro
'Determines if column is long or short.
Function LorS(SR,CC)
    If SR > CC Then
        LorS = "long"
    Else
        LorS = "short"
    End If
End Function

'Critical Load Macro
'Uses Euler formula for long columns
'Uses Johnson formula for short columns
Function Pcr(LorS, SR, E, A, Sy)
Const Pi = 3.1415926
    If LorS = "long" Then
        Pcr = Pi ^ 2 * E * A / SR ^ 2
        'Euler Equation; Eq. (16-4)
    Else
        Pcr = A * Sy(1—(Sy * SR ^ 2 / (4 * Pi ^ 2 * E)))
        'Johnson Equation; Eq. (16-7)
    End If
End Function
```

An efficient shape is one that uses a small amount of material to perform a given function. For columns, the goal is to maximize the radius of gyration in order to reduce the slenderness ratio. Note also that because $r = \sqrt{I/A}$, maximizing the moment of inertia for a given area has the same effect.

When discussing moment of inertia in Chapters 6 and 12, it was noted that it is desirable to put as much of the area of the cross section as far away from the centroid as possible.

For beams (discussed in Chapter 12) there was usually only one important axis, the axis about which the bending occurred. For columns, buckling can, in general, occur in any direction. Therefore, it is desirable to have uniform properties with respect to any axis. The hollow circular section, commonly called a pipe, then makes a very efficient shape for a column. Closely approximating that is the hollow square tube. Fabricated sections made from standard structural sections can also be used, as shown in Figure 16–11.

FIGURE 16–11   Examples of efficient column shapes. (a) Hollow circular section, pipe. (b) Hollow square tube. (c) Box section made from wood beams. (d) Equal-leg angles with plates. (e) Aluminum channels with plates. (f) Two equal-leg angles.

(a) W12×16 beam shape

Area = 4.71 in^2
I_x = 103 in^4
I_y = 2.82 in^4
r_x = 4.68 in
r_y = 0.77 in
r_x/r_y = 6.08

(b) W12×65 column section

Area = 19.1 in^2
I_x = 533 in^4
I_y = 174 in^4
r_x = 5.28 in
r_y = 3.02 in
r_x/r_y = 1.75
r_y more nearly equal to r_x

FIGURE 16–12 Comparison of a wide-flange beam shape with a column section.

Building columns are often made from special wide-flange shapes called *column sections*. They have relatively wide, thick flanges as compared with the shapes typically selected for beams. This makes the moment of inertia with respect to the *Y-Y* axis more nearly equal to that for the *X-X* axis. The result is that the radii of gyration for the two axes are more nearly equal also. Figure 16–12 shows a comparison of two 12-in wide-flange shapes: one a column section and one a typical beam shape. Note that the smaller radius of gyration should be used in computing the slenderness ratio.

16–11 SPECIFICATIONS OF THE AISC

Columns are essential elements of many structures. The design and analysis of columns in construction applications are governed by the specifications of the AISC, the American Institute of Steel Construction (Reference 2), briefly summarized here for column sections that are loaded through their centroidal axes and that do not exhibit local buckling of thin extended flanges. The method involves the following variables. Note that symbols used here are similar to those used in earlier sections of this chapter and are not necessarily the same as those in the AISC manual.

Transition slenderness ratio = $SR_t = 4.71\sqrt{E/s_y}$ **(16–8)**

(SR_t is approximately 6% larger than C_c)

Elastic critical buckling stress = $s_e = \pi^2 E/(SR)^2$ **(16–9)**

(sometimes called *Euler stress*)

Flexural buckling stress = s_{cr} whose value is dependent on the actual *SR*

If $SR \leq SR_t$, then the column is short and

$s_{cr} = [0.658^d]s_y$ and the exponent $d = s_y/s_e$ **(16–10)**

If $SR > SR_t$, then the column is long and

$s_{cr} = 0.877\, s_e$ **(16–11)**

Then the *nominal buckling strength* = $P_n = s_{cr} A_g$ **(16–12)**

(A_g is the column's gross area)

Finally, the *allowable compressive strength* = $P_a = P_n/1.67$ **(16–13)**

It should be noted that these formulas are simpler and somewhat different from those reported in previous editions of the AISC manual. However, the resulting values of P_a are within approximately 2% of the previous results.

Figure 16–13 shows a graph of the *flexural buckling stress, s_{cr},* versus the actual slenderness ratio of a column. Data are for ASTM A992 structural steel, the most common steel for W-beams and column sections. Note how Equations (16–10) and (16–11) are tangent at the transition slenderness ratio value and how the buckling stress approaches the yield strength for very short columns. The AISC recommends that the largest usable slenderness ratio be 200.

FIGURE 16–13 Flexure buckling stress vs. slenderness ratio—AISC method.

Example Problem 16–4

Compute the allowable compressive strength, P_a, for a column made from steel rectangular structural tubing, HSS102×51×6.4. The material is ASTM A500, grade B, structural steel. The column length is 3050 mm and its ends are pinned.

Solution

We will use Equations (16–8) to (16–13) with $E = 200\,GPa = 200\,000\,MPa$ and $s_y = 290\,MPa$. The tube is expected to buckle about the Y-Y axis so $r_{min} = r_y = 19.8\,mm$ [Appendix A–9(SI)] and $A_g = 1570\,mm^2$.

$$\text{Actual slenderness ratio} = SR = KL/r = 1.00(3050\,mm/19.8\,mm) = 154$$
$$\text{Transition slenderness ratio} = SR_t = 4.71\sqrt{E/S_y} = 4.71\sqrt{(200\,000/290)} = 123.7$$
$$\text{Elastic critical buckling stress} = s_e = \pi^2\,E/(SR)^2 = \pi^2\,(200\,000\,MPa)/(154)^2$$
$$s_e = 83.23\,MPa$$

Because $SR > SR_t$, the column is long and we use Equation (16–11):

$$s_{cr} = 0.877s_e = 0.877(83.23\,MPa) = 73.0\,MPa$$

Now, using Equation (16–12),
The *nominal buckling strength* = $P_n = s_{cr}A_g = (73.0\,N/mm^2)(1570\,mm^2) = 114.6\,kN$
Using Equation (16–13),
The *allowable compressive strength* = $P_a = P_n/1.67 = 114.6\,kN/1.67 = 68.6\,kN$

Example Problem 16–5

Compute the allowable compressive strength, P_a, for the column described in Example Problem 16–4 except it will be installed with both ends fixed instead of pinned.

Solution

Actual slenderness ratio = $SR = KL/r = (0.65)(3050\,mm/19.8\,mm) = 100$
From Example Problem 16–4, *transition slenderness ratio* = $SR_t = 123.7$
Elastic critical buckling stress = $s_e = \pi^2 E/(SR)^2 = \pi^2\,(200\,000\,MPa)/(100)^2$

$$s_e = 197.4\,MPa$$

Then the column is short and Equation (16–10) is applied next.
The exponent $d = s_y/s_e = 290\,MPa/197.4\,MPa = 1.469$ and

$$s_{cr} = [0.658^d]\,s_y = [0.658^{1.469}](290\,MPa) = 156.8\,MPa$$

The *nominal buckling strength* = $P_n = s_{cr}A_g = (156.8\,N/mm^2)(1570\,mm^2) = 246.2\,kN$
Using Equation (16–13),

The *allowable compressive strength* = P_a = $P_n/1.67$ = 246.2 kN/1.67 = 147.4 kN

Comment: The resulting allowable compressive strength for the fixed-end column is 2.15 times greater than the column design having pinned ends.

16–12 SPECIFICATIONS OF THE ALUMINUM ASSOCIATION

The Aluminum Association publication, *Aluminum Design Manual* (see Reference 1), defines allowable stresses for columns for each of several aluminum alloys and their heat treatments. Three different equations are given for short, intermediate, and long columns defined in relation to slenderness limits. The equations are of the form

$$\frac{P_a}{A} = \frac{s_y}{FS} \qquad \text{(short columns)} \qquad (16\text{--}14)$$

$$\frac{P_a}{A} = \frac{B_c - D_c(L_e/r)}{FS} \qquad \text{(intermediate columns)} \qquad (16\text{--}15)$$

$$\frac{P_a}{A} = \frac{\pi^2 E}{FS(L_e/r)^2} \qquad \text{(long columns)} \qquad (16\text{--}16)$$

In all three cases, it is recommended that $FS = 1.95$ for buildings and similar structures. The short column analysis assumes that buckling will not occur and that safety is dependent on the yield strength of the material. Equation (16–16) for long columns is the Euler formula with a factor of safety applied. The intermediate column formula [Equation (16–15)] depends on buckling constants B_c and D_c, which are functions of the yield strength of the aluminum alloy and the modulus of elasticity. The division between intermediate and long columns is similar to the C_c used previously in this chapter.

Following are the specific equations for the alloy 6061-T6 used in building structures in the forms of sheet, plate, extrusions, structural shapes, rod, bar, tube, and pipe.

Short columns and intermediate columns: $0 < L_e/r < 66$

$$\frac{P_a}{A} = \left(20.2 - 0.126\frac{L_e}{r}\right) \text{ksi} \qquad (16\text{--}17a)$$

$$\frac{P_a}{A} = \left(139 - 0.869\frac{L_e}{r}\right) \text{MPa} \qquad (16\text{--}17b)$$

Long columns: $L_e/r > 66$

$$\frac{P_a}{A} = \frac{51\,000}{(L_e/r)^2} \text{ksi} \qquad (16\text{--}18a)$$

$$\frac{P_a}{A} = \frac{352\,000}{(L_e/r)^2} \text{MPa} \qquad (16\text{--}18b)$$

See Reference 1 for column design stresses for other aluminum alloys.

16–13 NONCENTRALLY LOADED COLUMNS

All of the analysis methods discussed in this chapter so far have been limited to loadings in which the compressive loads on the columns act in-line with the centroidal axis of the column cross section. Also, it is assumed that the column axis is perfectly straight prior to the application of the loads. We have used the term *straight centrally loaded column* to describe such a case.

Many real columns violate these assumptions to some degree. Figure 16–14 shows two such conditions. If a column is initially *crooked*, the applied compressive force on the column

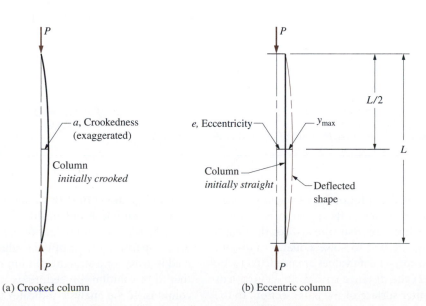

(a) Crooked column   (b) Eccentric column

FIGURE 16–14   Illustration of crooked and eccentric columns.

tends to cause bending in the column in addition to buckling, and failure would occur at a lower load than that predicted from the equations used so far in this chapter. An *eccentrically loaded column* is one in which there is some purposeful offsetting of the line of action of the compressive load from the centroidal axis of the column. Here, again, there is some bending stress produced in addition to the axial compressive stress that tends to cause buckling.

Crooked Columns

The crooked column formula allows an initial crookedness, a, to be considered (see References 4, 5, and 6):

Crooked Column Formula ▼

$$P_a^2 - \frac{1}{N}\left[s_yA + \left(1 + \frac{ac}{r^2}\right)P_{cr}\right]P_a + \frac{s_yAP_{cr}}{N^2} = 0 \quad (16\text{--}19)$$

where c = distance from the neutral axis of the cross section about which bending occurs to its outer edge

P_{cr} is defined to be the critical load found from the *Euler formula.*

Although this formula may become increasingly inaccurate for shorter columns, it is not appropriate to switch to the Johnson formula as it is for straight columns.

The crooked column formula is a quadratic with respect to the allowable load P_a. Evaluating all constant terms in Equation (16–19) produces an equation of the form

$$P_a^2 + C_1 P_a + C_2 = 0$$

Then, from the solution for a quadratic equation,

$$P_a = 0.5\left[-C_1 - \sqrt{C_1^2 - 4C_2}\right]$$

The smaller of the two possible solutions is selected.

Example Problem 16–6   A column has both ends pinned and has a length of 32 in. It has a circular cross section with a diameter of 0.75 in and an initial crookedness of 0.125 in. The material is AISI 1040 hot-rolled steel. Compute the allowable load for a design factor of 3.

Solution   Objective: Specify the allowable load for the column.

Given: Solid circular cross section: $D = 0.75$ in; $L = 32$ in; use $N = 3$.
Both ends are pinned. Initial crookedness = $a = 0.125$ in.
Material: AISI 1040 hot-rolled steel.

Analysis: Use Equation (16–19). First evaluate C_1 and C_2. Then solve the quadratic equation for P_a.

Results:
$$s_y = 60\,000 \text{ psi}$$
$$A = \pi D^2/4 = (\pi)(0.75)^2/4 = 0.442 \text{ in}^2$$
$$r = D/4 = 0.75/4 = 0.188 \text{ in}$$
$$c = D/2 = 0.75/2 = 0.375 \text{ in}$$
$$KL/r = [(1.0)(32)]/0.188 = 171$$
$$P_{cr} = \frac{\pi^2 EA}{(KL/r)^2} = \frac{\pi^2(30\,000\,000)(0.442)}{(171)^2} = 4476 \text{ lb}$$
$$C_1 = \frac{-1}{N}\left[s_yA + \left(1 + \frac{ac}{r^2}\right)P_{cr}\right] = -12\,311$$
$$C_2 = \frac{s_yAP_{cr}}{N^2} = 1.319 \times 10^7$$

The quadratic is, therefore,

$$P_a^2 - 12\,311P_a + 1.319 \times 10^7 = 0$$

Comment: From this, $P_a = 1186$ lb is the allowable load.

Figure 16–15 shows the solution of Example Problem 16–6 using a spreadsheet. Whereas its appearance is similar to that of the earlier column analysis spreadsheet, the details follow the calculations needed to solve Equation (16–19). On the lower left, two special data values are needed: (1) the crookedness *a* and (2) the distance *c* from the neutral axis for buckling to the outer surface of the cross section. In the middle of the right part are listed some intermediate values

used in Equation (16–19): C_1 and C_2 as defined in the solution to Example Problem 16–6. The result, the allowable load, P_a, is at the lower right of the spreadsheet. Above that, for comparison, the computed value of the critical buckling load is given for a straight column of the same design. Note that this solution procedure is most accurate for long columns. If the analysis indicates that the column is *short* rather than *long*, the designer should take note of how short

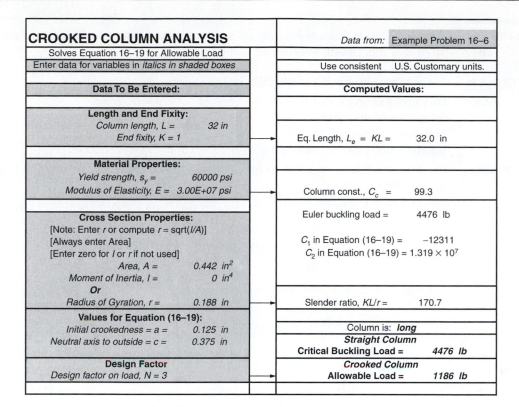

FIGURE 16–15 Spreadsheet for analysis of crooked columns—U.S. Customary units.

it is by comparing the slenderness ratio, KL/r, with the column constant, C_c. If the column is quite short, the designer should not rely on the accuracy of the result from Equation (16–19).

Eccentrically Loaded Columns

An *eccentric load* is one that is applied away from the centroidal axis of the cross section of the column, as shown in Figure 16–14(b). Such a load exerts bending in addition to the column action that results in the deflected shape shown in the figure. The maximum stress in the deflected column occurs in the outermost fibers of the cross section at the mid-length of the column where the maximum deflection, y_{max}, occurs. Let's denote the stress at this point as $\sigma_{L/2}$. Then, for any applied load, P,

Secant Formula for Eccentrically Loaded Columns ▼

$$\sigma_{L/2} = \frac{P}{A}\left[1 + \frac{ec}{r^2}\sec\left(\frac{KL}{2r}\sqrt{\frac{P}{AE}}\right)\right] \quad (16\text{–}20)$$

(See Reference 6.) Note that this stress is *not* directly proportional to the load. When evaluating the secant in this formula, note that its argument in the parentheses is in *radians*. Also, because most calculators do not have the secant function, recall that the secant is equal to 1/cosine.

For design purposes, we would like to specify a design factor, N, that can be applied to the *failure load* similar to that defined for straight, centrally loaded columns. However, in this case, failure is predicted when the maximum stress in the column exceeds the yield strength of the

material. Let's now define a new term, P_y, to be the load applied to the eccentrically loaded column when the maximum stress is equal to the yield strength. Equation (16–20) then becomes

$$s_y = \frac{P_y}{A}\left[1 + \frac{ec}{r^2}\sec\left(\frac{KL}{2r}\sqrt{\frac{P_y}{AE}}\right)\right]$$

Now, if we define the *allowable load* to be

$$P_a = P_y/N$$

or

$$P_y = NP_a$$

this equation becomes

Design Equation for Eccentrically Loaded Columns ▼

Required $$s_y = \frac{NP_a}{A}\left[1 + \frac{ec}{r^2}\sec\left(\frac{KL}{2r}\sqrt{\frac{NP_a}{AE}}\right)\right] \quad (16\text{–}21)$$

This equation cannot be solved for either N or P_a. Therefore, an iterative solution is required, as will be demonstrated in Example Problem 16–7.

Another critical factor may be the amount of deflection of the axis of the column due to the eccentric load:

Maximum Deflection in an Eccentrically Loaded Column ▼

$$y_{max} = e\left[\sec\left(\frac{KL}{R}\sqrt{\frac{P}{AE}}\right) - 1\right] \quad (16\text{–}22)$$

Note that the argument of the secant is the same as that used in Equation (16–20).

Example Problem
16–7

For the column of Example Problem 16–6, compute the maximum stress and deflection if a load of 1075 lb is applied with an eccentricity of 0.75 in.

Solution   Objective: Compute the stress and the deflection for the eccentrically loaded column.

Given: Data from Example Problem 16–6, but eccentricity = e = 0.75 in.
Solid circular cross section: D = 0.75 in; L = 32 in.
Both ends are pinned; KL = 32 in; r = 0.188 in; c = $D/2$ = 0.375 in.
Material: AISI 1040 hot-rolled steel; E = 30 × 10^6 psi, s_y = 60 000 psi

Analysis: Use Equation (16–20) to compute maximum stress. Then use Equation (16–22) to compute maximum deflection.

Results: All terms have been evaluated before. Then the maximum stress is found from Equation (16–20):

$$\sigma_{L/2} = \frac{1075}{0.422}\left[1 + \frac{(0.75)(0.375)}{(0.188)^2}\sec\left(\frac{32}{2(0.188)}\sqrt{\frac{1075}{(0.422)(30 \times 10^6)}}\right)\right]$$

$$\sigma_{L/2} = 29\,300\text{ psi}$$

The maximum deflection is found from Equation (16–22):

$$y_{max} = 0.75\left[\sec\left(\frac{32}{2(0.188)}\sqrt{\frac{1075}{(0.442)(30 \times 10^6)}}\right) - 1\right] = 0.293\text{ in}$$

Comments: The maximum stress is 29 300 psi at the midlength of the column. The deflection there is 0.293 in.

Example Problem
16–8

The stress in the column found in Example Problem 16–7 seems high for the AISI 1040 hot-rolled steel. Redesign the column to achieve a design factor of at least 3. Use only the preferred sizes listed in Appendix A–2.

Solution   Objective: Redesign the eccentrically loaded column of Example Problem 16–7 to reduce the stress and achieve a design factor of at least 3.

Given: Data from Example Problems 16–6 and 16–7.

Analysis: Use a larger diameter. Use Equation (16–21) to compute the required strength. Then compare that with the strength of AISI 1040 hot-rolled steel. Iterate until the stress is satisfactory.

Results: Appendix 3 gives the value for the yield strength of AISI 1040 HR to be 60 000 psi. If we choose to retain the same material, the cross-sectional dimensions of the column must be increased to decrease the stress. Equation (16–21) can be used to evaluate a design alternative.

The objective is to find suitable values for A, c, and r for the cross section such that P_a = 1075 lb; N = 3; L_e = 32 in; e = 0.75 in; and the value of the entire right side of the equation is less than 60 000 psi. The original design had a circular cross section with a diameter of 0.75 in. Let's try increasing the diameter to D = 1.00 in. Then

$$A = \pi D^2/4 = \pi(1.00\text{ in})^2/4 = 0.785\text{ in}^2$$
$$r = D/4 = (1.00\text{ in})/4 = 0.250\text{ in}$$
$$r^2 = (0.250\text{ in})^2 = 0.0625\text{ in}^2$$
$$c = D/2 = (1.00\text{ in})/2 = 0.50\text{ in}$$

Now let's call the right side of Equation (16–21) s_y'. Then

$$s_y' = \frac{3(1075)}{0.785}\left[1 + \frac{(0.75)(0.50)}{(0.0625)}\sec\left(\frac{32}{2(0.250)}\sqrt{\frac{(3)(1075)}{(0.785)(30 \times 10^6)}}\right)\right]$$

$$s_y' = 37\,740\text{ psi} = \text{required value of } s_y$$

This is well below the value of s_y = 60 000 psi for the given steel, and it gives the desired design factor of 3.0 or greater. Trying the only smaller preferred size from Appendix A–2 (7/8 in = 0.875 in), the required s_y is 65 825 psi, and that is too high. Therefore, specify D = 1.00 in.

Now we can evaluate the expected maximum deflection with the new design using Equation (16–22):

$$y_{max} = 0.75\left[\sec\left(\frac{32}{2(0.250)}\sqrt{\frac{1075}{(0.785(30\times10^6))}}\right) - 1\right]$$

$$y_{max} = 0.076 \text{ in}$$

Comments: The diameter of 1.00 in is satisfactory. The maximum deflection for the column is 0.076 in.

Figure 16–16 shows the solution of the eccentric column problem of Example Problem 16–8 using a spreadsheet to evaluate Equations (16–21) and (16–22). It is a design aid that facilitates the iteration required to determine an acceptable geometry for a column to carry a specified load with a desired design factor. Note that the data are in U.S. Customary units. At the lower left of the spreadsheet, data required for Equations (16–21) and (16–22) are entered by the designer, along with the other data discussed for earlier column

analysis spreadsheets. The "FINAL RESULTS" at the lower right show the computed value of the required yield strength of the material for the column and compare it with the given value entered by the designer near the upper left. The designer must ensure that the actual value is greater than the computed value by trying different values for the diameter. The last part of the right side of the spreadsheet gives the computed maximum deflection of the column that occurs at its midlength.

FIGURE 16–16   Spreadsheet for analysis of eccentric columns—U.S. customary units.

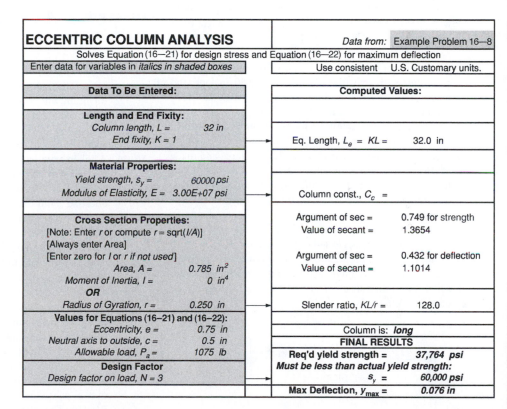

REFERENCES

1. Aluminum Association, Aluminum Design Manual, Washington, DC, 2005.

2. American Institute of Steel Construction, *Steel Construction Manual*, 9th ed., Chicago, IL 2005.

3. Mott, Robert L., *Machine Elements in Mechanical Design*, 4th ed., Upper Saddle River, NJ, Prentice Hall, 2004.

4. Spotts, M. F., T. E. Shoup, and L. E. Hornberger. *Design of Machine Elements*, 8th ed. Upper Saddle River, NJ, Prentice Hall, 2004.

5. Timoshenko, S. *Strength of Materials*, Vol. 2, 2nd ed., New York, Van Nostrand Reinhold, 1941.

6. Timoshenko, S. and J. M. Gere, *Theory of Elastic Stability*, 2nd ed., New York, McGraw-Hill, 1961.

PROBLEMS

16–1.   Determine the critical load for a pinned-end column made of a circular bar of AISI 1020 hot-rolled steel. The diameter of the bar is 20 mm, and its length is 800 mm.

16–2.   Repeat Problem 16–1 with the length of 350 mm.

16–3. Repeat Problem 16–1 with the bar made of 6061-T6 aluminum instead of steel.

16–4. Repeat Problem 16–1 with the column ends fixed instead of pinned.

16–5. Repeat Problem 16–1 with a square steel bar with the same cross-sectional area as the circular bar.

16–6. For a PIPE 25STD (1-in schedule 40) steel pipe used as a column, determine the critical load if it is 2.05 m long. The material is similar to AISI 1020 hot-rolled steel. Compute the critical load for each of the four end conditions described in Figure 16–3.

16–7. A rectangular steel bar has cross-sectional dimensions of 12 mm by 25 mm and is 210 mm long. Assuming that the bar has pinned ends and is made from AISI 1141 OQT 1300 steel, compute the critical load when the bar is subjected to an axial compressive load.

16–8. Compute the allowable load on a column with fixed ends if it is 5.45 m long and made from an S150×18.6 (S6×12.5) beam. The material is ASTM A36 steel. Use the AISC formula.

16–9. A raised platform is 20 ft by 40 ft in area and is being designed for 75 pounds per square foot uniform loading. It is proposed that standard 3-in schedule 40 steel pipe be used as columns to support the platform 8 ft above the ground with the base fixed and the top free. How many columns would be required if a design factor of 3.0 is desired? Use $s_y = 30\,000$ psi.

16–10. An aluminum I-beam, I254×12.87 (I10×8.646), is used as a column with two pinned ends. It is 2.80 m long and made of 6061-T6 aluminum. Using Equations (16–17b) and (16–18b), compute the allowable load on the column.

16–11. Compute the allowable load for the column described in Problem 16–10 if the length is only 1.40 m.

16–12. A column is a W8×10 steel beam, 12.50 ft long, and made of ASTM A992 steel. Its ends are attached in such a way that L_e is approximately 0.80L. Using the AISC formulas, determine the allowable load on the column.

16–13. A built-up column is made of four angles, as shown in Figure P16–13. The angles are held together with lacing bars, which can be neglected in the analysis of geometrical properties. Using the standard Johnson or Euler equations with $L_e = L$ and a design factor of 3.0, compute the allowable load on the column if it is 18.4 ft long. The angles are of ASTM A36 steel.

16–14. Compute the allowable load on a built-up column having the cross section shown in Figure P16–14. Use $L_e = L$ and 6061-T6 aluminum. The column is 10.5 ft long. Use the Aluminum Association formulas.

Aluminum
C6 × 2.834

6 in

FIGURE P16–14 Built-up column cross section for Problem 16–14.

16–15. Figure P16–15 shows a beam supported at its ends by pin-joints. The inclined bar at the top supports the right end of the beam, but also places an axial compressive force in the beam. Would a standard S150×18.6 (S6×12.5) beam be satisfactory in this application if it carries 1320 kg at its end? The beam is made from ASTM A36 steel.

16–16. A link in a mechanism which is 8.40 in long, has a rectangular cross section $\frac{1}{4}$ in × $\frac{1}{8}$ in, and is subjected to a compressive load of 50 lb. If the link has pinned ends, is it safe from buckling? Cold-drawn AISI 1040 steel is used in the link.

16–17. A piston rod on a shock absorber is 12 mm in diameter and has a maximum length of 190 mm outside the shock absorber body. The rod is made of AISI 1040 CD steel. Consider one end to be pinned and the other to be fixed. What axial compressive load on the rod would be one-third of the critical buckling load?

16–18. A stabilizing rod in an automobile suspension system is a round bar loaded in compression. It is subjected to 1375 lb of axial load and supported at its ends by pin-type connections, 28.5 in apart. Would a 0.800-in diameter bar of AISI 1020 hot-rolled steel be satisfactory for this application?

16–19. A structure is being designed to support a large hopper over a plastic extruding machine, as sketched in Figure P16–19. The hopper is to be carried by four columns which share the load equally. The structure is cross-braced by rods. It is proposed that the columns be made from standard 2-in schedule 40 pipe. They will be fixed at the floor. Because of the cross-bracing, the top of each column is guided so that it behaves as if it were rounded or pinned. The material for the pipe is AISI 1020 steel, hot rolled. The hopper is designed to hold 20 000 lb of plastic powder. Are the proposed columns adequate for this load?

16–20. Discuss how the column design in Problem 16–19 would be affected if a careless forklift driver runs into the cross braces and breaks them.

16–21. The assembly shown in Figure P16–21 is used to test parts by pulling on them repeatedly with the hydraulic cylinder.

12 in

L3 × 3 × $\frac{1}{4}$

12 in

FIGURE P16–13 Built-up column cross section for Problem 16–13.

FIGURE P16–15   Structure for Problem 16–15.

1.00 m

2.40 m

1320 kg

FIGURE P16–19   Hopper for Problems 16–19 and 16–20.

Hopper

Cross braces

Column

14 ft

FIGURE P16–21   Test fixture for Problem 16–21.

36 in

18 in

Beam

Hydraulic cylinder

Column

Column length
L = 40 in

Column

Column

Tension link

Note: Cylinder pulls up on tension link and down on beam with a force of 3000 lb.

FIGURE P16–22 Solid waste compactor for Problems 16–22, 16–23, 16–24, and 16–25.

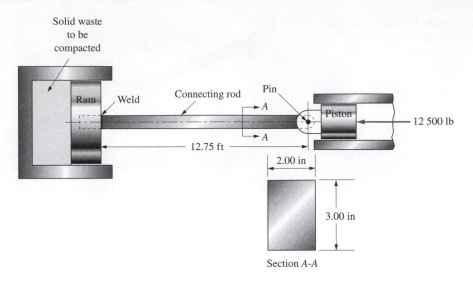

Section *A-A*

A maximum force of 3000 lb can be exerted by the cylinder. The parts of the assembly of concern here are the columns. It is proposed that the two columns be made of $1\frac{1}{4}$-in square bars of aluminum alloy 6061-T6. The columns are fixed at the bottom and free at the top. Determine the acceptability of the proposal.

16–22. Figure P16–22 shows the proposed design for a hydraulic press used to compact solid waste. A piston at the right is capable of exerting a force of 12 500 lb through the connecting rod to the ram. The rod is straight and centrally loaded. It is made from AISI 1040 WQT 1100 steel. Compute the resulting design factor for this design.

16–23. For the conditions described in Problem 16–22, specify the required diameter of the connecting rod if it is made as a solid circular cross section. Use a design factor of 4.0.

16–24. For the conditions described in Problem 16–22, specify a suitable standard steel pipe for use as the connecting rod. Use a design factor of 4.0. The pipe is to be made from ASTM A501 structural steel.

16–25. For the conditions described in Problem 16–22, specify a suitable standard I-beam for use as the connecting rod. Use a design factor of 4.0. The I-beam is to be made from aluminum alloy 6061-T6. The connection at the piston is as shown in Figure P16–25.

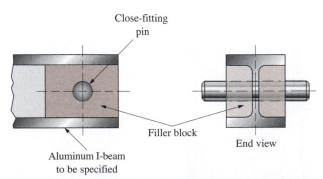

FIGURE P16–25 End connection for I-beam for Problem 16–25.

16–26. A hollow square tube, HSS3×3×$\frac{1}{4}$, made from ASTM A500 steel, grade B, is used as a building column having a length of 16.5 ft. Using $L_e = 0.80L$, compute the allowable load on the column for a design factor of 3.0.

16–27. A hollow rectangular tube, HSS4×2×$\frac{1}{4}$, made from ASTM A500 steel, grade B, is used as a building column having a length of 16.5 ft. Using $L_e = 0.80L$, compute the allowable load on the column for a design factor of 3.0.

16–28. A column is made by welding two standard steel angles, 3×3×$\frac{1}{4}$, into the form shown in Figure 16–11(f). The angles are made from ASTM A36 structural steel. If the column has a length of 16.5 ft and $L_e = 0.8L$, compute the allowable load on the column for a design factor of 3.0.

16–29. A rectangular steel bar, made of AISI 1020 hot-rolled steel, is used as a safety brace to hold the ram of a large punch press while dies are installed in the press. The bar has cross-sectional dimensions of 60 mm by 40 mm. Its length is 750 mm and its ends are welded to heavy flat plates that rest on the flat bed of the press and the flat underside of the ram. Specify a safe load that could be applied to the brace.

16–30. It is planned to use an aluminum channel, C102× 2.586(C4×1.738), as a column having a length of 4.25 m. The ends can be considered to be pinned. The aluminum is alloy 6061-T4. Compute the allowable load on the column for a design factor of 4.0.

16–31. In an attempt to improve the load-carrying capacity of the column described in Problem 16–30, alloy 6061-T6 is proposed in place of the 6061-T4 to take advantage of its higher strength. Evaluate the effect of this proposed change on the allowable load.

16–32. Compute the allowable load on the steel W12×65 column section shown in Figure 16–12(b) if it is 22.5 ft long, made from ASTM A992 steel, and installed such that $L_e = 0.8L$. Use the AISC code.

ADDITIONAL PRACTICE AND REVIEW PROBLEMS

16–33. A hollow rectangular steel tube, HSS3×2×$\frac{1}{4}$, is used to support an axial compressive load. The tube is 13.6 ft long and rigidly fixed at each end. It is made from ASTM A501, hot-formed structural steel. Compute the allowable load on the column for a design factor of 3.0.

16–34. Repeat Problem 16–33 if the tube is laterally braced at a point 80 in from its lower end.

16–35. Repeat Problem 16–33 if the tube is pin-connected at its top.

16–36. Repeat Problem 16–33 if the tube is HSS3×3×¼.

16–37. A hollow rectangular steel tube, HSS102×51×6.4 (4× 2×1/4), has pinned ends, is 2.65 m long and carries an axial compressive load of 75.0 kN. Compute the design factor that results from this design. Use ASTM A501 steel.

16–38. Redesign the column described in Problem 16–37 to result in a design factor of no less that 3.0. Verify that your re-design is satisfactory.

16–39. Figure P16–39 shows a truss. Specify a suitable design for each member loaded in compression that will achieve a minimum design factor of 3.0.

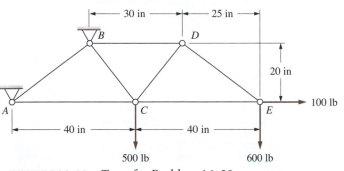

FIGURE P16–39   Truss for Problem 16–39.

16–40. For the truss in Figure P16–40, specify a suitable design for each member loaded in compression that will result in a minimum design factor of 2.5.

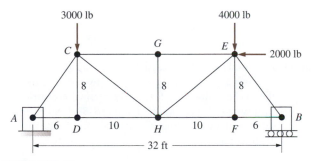

FIGURE P16–40

16–41. For the truss in Figure P16–41, specify a suitable design for each member loaded in compression that will result in a minimum design factor of 2.5.

16–42. A sling, sketched in Figure P16–42, is to carry a load of 18 000 lb. Design the spreader.

FIGURE P16–42

16–43. Repeat Problem 16–42 if the angle shown is changed from 30° to 15°.

Crooked Columns: Find P_a for $N = 3$

16–44. Repeat Problem 16–1 if the column has an initial crooked-ness of 4.0 mm.

16–45. Repeat Problem 16–7 if the column has an initial crooked-ness of 1.60 mm.

16–46. Repeat Problem 16–10 if the column has an initial crookedness of 14.0 mm.

16–47. Repeat Problem 16–12 if the column has an initial crookedness of 0.75 in.

16–48. Repeat Problem 16–33 if the column has an initial crookedness of 1.25 in.

16–49. Repeat Problem 16–37 if the column has an initial crookedness of 32 mm.

Eccentrically Loaded Columns

16–50. An aluminum (6061-T4) column is 42 in long and has a square cross section, 1.25 in on a side. If it carries a compressive load of 1250 lb, applied with an eccentricity of 0.60 in, compute the maximum stress in the column and the maximum deflection.

16–51. A steel (AISI 1020 hot-rolled) column is 3.2 m long and is made from a standard PIPE 75STD (3-in schedule 40) steel pipe (see Appendix A–12). If a compressive load of 30.5 kN is applied with an eccentricity of 150 mm, compute the maxi-mum stress in the column and the maximum deflection.

16–52. A connecting link in a mechanism is 14.75 in long and has a square cross section 0.250 in on a side. It is made from annealed AISI 301 stainless steel. Use $E = 28\,000\,000$ psi. If it carries a compressive load of 45 lb with an eccentricity of 0.30 in, compute the maximum stress and the maxi-mum deflection.

FIGURE P16–41

FIGURE P16–54

Section A-A

16–53. A hollow square steel tube, 40 in long, is proposed for use as a prop to hold up the ram of a punch press during installation of new dies. The ram weighs 75 000 lb. The prop is made from HSS4×4×$\frac{1}{4}$ structural tubing. It is made from steel similar to structural steel, ASTM A500, grade C. If the load applied by the ram could have an eccentricity of 0.50 in would the prop be safe?

16–54. Compute the maximum stress and deflection that can be expected in the steel machine member carrying an eccentric load as shown in Figure P16–54. The load P is 1000 lb. If a design factor of 3 is desired, specify a suitable steel.

16–55. An axial load of 4000 lb is applied to a C5×9 steel channel that is 112 in long and made from ASTM A36 structural steel. The line of action of the load acts at the mid-depth of the web and the mid-width of the flanges. The ends are pinned. Would the channel be suitable if a design factor of 3.0 is desirable?

16–56. Figure P16–56 shows a HSS4×4×$\frac{1}{2}$ steel column made from ASTM A500, grade B, structural steel. To accommodate

a special mounting restriction, the load is applied eccentrically as shown. Determine the amount of load that the column can safely support. The top end of the column is supported laterally by the structure.

16–57. The device shown in Figure P16–57 is subjected to opposing forces F. Determine the maximum allowable load to achieve a design factor of 3. The device is made from aluminum 6061-T6.

Section A-A

FIGURE P16–57

16–58. A hydraulic cylinder is capable of exerting a force of 5200 N to move a heavy casting along a conveyor. The design of the pusher causes the load to be applied eccentrically to the piston rod as shown in Figure P16–58. Is the piston rod safe under this loading if it is made from AISI501 OQT 1000 stainless steel?

FIGURE P16–56

Section A-A

FIGURE P16–58

16–59. A standard 2-in schedule 40 steel pipe is proposed to be used to support the roof of a porch during renovation. Its length is 13.0 ft. The pipe is made from ASTM A501 structural steel.

   a. Determine the safe load on the pipe to achieve a design factor of 3 if the pipe is straight.

   b. Determine the safe load if the pipe has an initial crookedness of 1.25 in.

COMPUTER ASSIGNMENTS

1. Write a computer program or spreadsheet to analyze proposed column designs using the procedure outlined in Section 16–8. Have all the essential design data for material, end fixity, length, and cross-sectional properties input by the user. Have the program output the critical load and the allowable load for a given design factor.

Enhancements to Assignment 1

   a. Include a table of data for standard schedule 40 steel pipe for use by the program to determine the cross-sectional properties for a specified pipe size.

   b. Design the program to handle columns made from solid circular cross sections and compute the cross-sectional properties for a given diameter.

   c. Add a table of data for standard structural steel square tubing for use by the program to determine the cross-sectional properties for a specified size.

   d. Have the program use the specifications of the AISC as stated in Section 16–11 for computing the allowable load and factor of safety for steel columns.

   e. Have the program use the specifications of the Aluminum Association as stated in Section 16–12 for computing the allowable load for columns made from 6061-T6.

2. Write a program to design a column with a solid circular cross section to carry a given load with a given design factor. Note that the program will have to check to see that the correct method of analysis is used, either the Euler formula for long columns or the Johnson formula for short columns, after an initial assumption is made.

3. Write a program to design a column with a solid square cross section to carry a given load with a given design factor.

4. Write a program to select a suitable schedule 40 steel pipe to carry a given load with a given design factor. The program could be designed to search through a table of data for standard pipe sections from the smallest to the largest until a suitable pipe was found. For each trial section, the allowable load could be computed using the Euler or Johnson formula, as required, and compared with the design load.

5. Create a spreadsheet to analyze a crooked column as described in Section 16–13.

6. Create a spreadsheet to analyze an eccentrically loaded column as described in Section 16–13.

PRESSURE VESSELS

MAJOR TOPICS

17–1 The Big Picture—Pressure Vessels

17–2 Distinction Between Thin-Walled and Thick-Walled Pressure Vessels

17–3 Thin-Walled Spheres

17–4 Thin-Walled Cylinders

17–5 Thick-Walled Cylinders and Spheres

17–6 Analysis and Design Procedures for Pressure Vessels

17–7 Spreadsheet Aid for Analyzing Thick-Walled Spheres and Cylinders

17–8 Shearing Stress in Cylinders and Spheres

17–9 Other Design Considerations for Pressure Vessels

17–10 Composite Pressure Vessels

17–1 THE BIG PICTURE— PRESSURE VESSELS

Discussion Map

A pressure vessel is a container designed to hold liquids or gases under internal pressure.

- Typical pressure vessels are either spherical or cylindrical in shape.

- It is important for you to understand how a pressure vessel can fail so you can design it to be safe under a specific applied pressure.

- Failure occurs when the internal pressure causes an excessively high tensile stress in the walls of the vessel.

- The size of the vessel and the thickness of its wall are the primary variables that affect the level of stress.

- Two different types of analysis are discussed in this chapter, one for thin-walled vessels and one for thick-walled vessels. We quantify the difference between these two classifications of vessels later.

- Now let's explore further this concept of a pressure vessel.

Activity Find some examples of things that you are familiar with that may fit that definition. Then describe each of them, giving its function, a general description of its shape, the material it is made from, its length, and the overall dimensions. Here are a few examples to get you started. See Figure 17–1.

- **A Tank to Hold the Pressurized Propane Fuel for a Household Gas Grill:** It is typically a cylinder with domed ends, made from steel, with a diameter of about 300 mm (12 in). The length of the straight, cylindrical part is about 225 mm (9 in) and the overall length is about 350 mm (14 in). The

cylindrical or spherical shape gives the tank a sturdy, stiff feel, but when struck lightly by something hard, it seems that the thickness of the steel making up its walls is relatively thin. There is a penetration in the top of the domed end for a valve to enable you to draw out some propane to cook the perfect steak, chicken, hamburger, or fish.

- **A Compressed Air Tank:** You may have seen examples of these in an automobile service station, on a construction site, or in an industrial plant. They are connected to a compressor that delivers air at approximately 400 to 700 kPa (60 to 100 psi) to the tank where a relatively large volume is stored until needed to add air to a tire, to blow debris from the floor, or to raise a car on the lift. While they vary greatly in size and capacity, they are typically of a similar shape as the propane tank, being mostly cylindrical with domed ends. The tank incorporates a pressure switch that senses tank pressure and cuts the compressor off when the tank reaches the desired pressure setting. This is important because the tank is designed to hold safely a specified maximum pressure. The rupture of a compressed air tank is very dangerous. In series with the discharge line from the tank valve, there is usually a pressure regulator to limit the pressure at the working end of the line.

- **A Spherical High-Pressure Storage Tank:** Have you ever seen one of these in a photograph or video of a spacecraft? In this chapter we will see that the spherical shape is an optimum shape for a pressure vessel. They are used when there is a need to store gases or liquids at high pressure to be able to use the thinnest, lightest material. Light, strong materials such as titanium or advanced composites are often used in such applications.

Continue your own search for examples of pressure vessels. Some you may be able to find in person and, if not, the

FIGURE 17–1 Examples of pressure vessels.

Source: (c) Photo Researchers, Inc.

20# vertical

(a) Typical household propane cylinder

(b) General-purpose air compressor with pressure tank

(c) High-pressure spherical tank

Internet can help you find quite a variety. Check out the sites listed at the end of this chapter by going first to the group labeled *Pressure Vessel Manufacturers*. These are companies that specialize in the design and fabrication of pressure vessels for such applications as

- Medical sterilization systems
- Chemical processing plants
- Food production facilities
- Compressed air tanks
- Petrochemical and pharmaceutical manufacturing and many others

Another interesting Internet search you could try is to investigate the several ways in which hydrogen can be stored for use in fuel cells for vehicle power or in electrical generation systems. When using hydrogen in gaseous form, it is important to store it at high pressure so that only a moderate volume is required to hold sufficient fuel to operate a vehicle for approximately 300 miles (480 km) or an electrical power system for several hours. Practical designs for cylindrical tanks have been proposed that are maintained at 3000 to 10 000 psi (20.7 MPa to 69.0 MPa) so that a vehicle hydrogen fuel tank would take approximately the same space as a typical gasoline fuel tank. Check out the designs by Quantum Technologies, Inc., at www.qtww.com. Other methods of storing hydrogen are being studied that allow operation at lower pressures.

What Examples Have You Found?

In this chapter you will learn how the internal pressure causes stresses in the walls of the pressure vessel. This will enable you to design safe vessels yourself and to recognize the importance of protecting and maintaining existing tanks to keep them safe. Most of the pressure vessels you will encounter in your career are likely to be of the *thin-walled* type. But some, particularly those carrying high-pressure fluids, may be in the range of *thick-walled* designs. In this chapter we quantify the definition of when a vessel is considered to have a thin and thick wall. This is an important distinction because the method of analysis is very different for these two classes of vessels.

Objectives of This Chapter

After completing this chapter, you should be able to:

1. Determine whether a pressure vessel should be classified as *thin-walled* or *thick-walled*.

2. Describe *hoop stress* as it is applied to spheres carrying an internal pressure and apply the hoop stress formula to compute the maximum stress in the wall of a thin-walled sphere.

3. Describe *hoop stress* as it is applied to cylinders carrying an internal pressure and apply the hoop stress formula to compute the maximum stress in the wall of a thin-walled cylinder.

4. Describe *longitudinal stress* as it is applied to cylinders carrying internal pressure and apply the longitudinal stress formula to compute the stress in the wall of a thin-walled cylinder that acts in the direction parallel to the axis of the cylinder.

5. Identify the hoop stress, longitudinal stress, and radial stress developed in the wall of a thick-walled sphere or cylinder due to internal pressure and apply the formulas for computing the maximum values of the stresses.

17–2 DISTINCTION BETWEEN THIN-WALLED AND THICK-WALLED PRESSURE VESSELS

In general, the magnitude of the stress in the wall of a pressure vessel varies as a function of position within the wall. A precise analysis enables the computation of the stress at any point. The formulas for making such a computation will be shown in a later section.

However, when the wall thickness of the pressure vessel is small, the assumption that the stress is uniform throughout the wall results in very little error. Also, this assumption permits the development of relatively simple formulas for stress.

First you need to understand the basic geometry of a pressure vessel and to define some terms. Figure 17–2 shows the definition of key diameters, radii, and the wall thickness for cylinders and spheres. Most are obvious from the figure. The mean radius, R_m, is defined as the average of the outside radius and the inside radius. That is,

Mean Radius ▼

$$R_m = \frac{R_o + R_i}{2} \qquad (17–1)$$

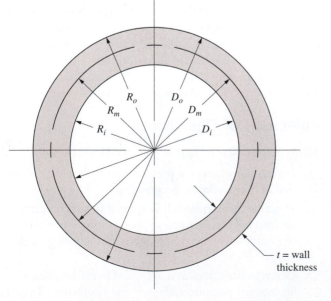

FIGURE 17–2 Definition of key diameters, radii, and wall thickness for cylinders and spheres.

We can also define the *mean diameter* as

Mean Diameter ▼

$$D_m = \frac{D_o + D_i}{2} \qquad (17–2)$$

Other useful formulas are

$$R_i = R_o - t \qquad D_i = D_o - 2t$$

$$R_m = R_o - \frac{t}{2} \qquad D_m = D_o - t$$

$$R_m = R_i + \frac{t}{2} \qquad D_m = D_i + t$$

The criterion for determining when a pressure vessel can be considered thin-walled is as follows:

If the ratio of the mean radius of the vessel to its wall thickness is 10 or greater, the stress is very nearly uniform and it can be assumed that all the material of the wall shares equally to resist the applied forces. Such pressure vessels are called thin-walled vessels.

Then a pressure vessel is considered to be thin if

$$\frac{R_m}{t} \ge 10 \qquad (17–3)$$

where t is the wall thickness of the vessel.

Because the diameter is twice the radius, the criterion for a vessel to be considered thin-walled is also written as

$$\frac{D_m}{t} \ge 20 \qquad (17–4)$$

Obviously, if the vessel does not satisfy the criteria listed in Equations (17–3) and (17–4), it is considered to be thick-walled.

The next two sections are devoted to the analysis of *thin-walled* spheres and cylinders. Then Section 17–5 is concerned with *thick-walled* spheres and cylinders.

17–3 THIN-WALLED SPHERES

In analyzing a spherical pressure vessel, the objective is to determine the stress in the wall of the vessel to ensure safety. Because of the symmetry of a sphere, a convenient free body for use in the analysis is one-half of the sphere, as shown in Figure 17–3. The internal pressure of the liquid or gas contained in the sphere acts perpendicular to the walls, uniformly over all the interior surface. Because the sphere was cut through a diameter, the forces in the walls all act horizontally. Therefore, only the horizontal component of the forces due to the fluid pressure needs to be considered in determining the magnitude of the force in the walls. If a pressure p acts on an area A, the force exerted on the area is

$$F = pA \qquad (17–5)$$

FIGURE 17–3 Free body diagram for a sphere carrying internal pressure.

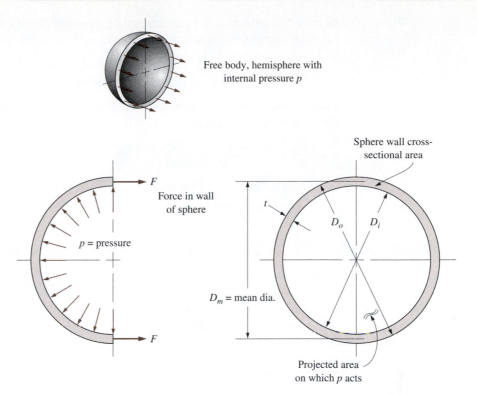

Free body, hemisphere with internal pressure p

Force in wall of sphere

F

p = pressure

F

D_m = mean dia.

Sphere wall cross-sectional area

t

D_o D_i

Projected area on which p acts

Taking all of the force acting on the entire inside of the sphere and finding the horizontal component, we find the resultant force in the horizontal direction to be

$$F_R = pA_p \qquad (17\text{–}6)$$

where A_p is the *projected area* of the sphere on the plane through the diameter. Therefore,

$$A_p = \frac{\pi D_m^2}{4} \qquad (17\text{–}7)$$

Because of the equilibrium of the horizontal forces on the free body, the forces in the walls must also equal F_R, as computed in Equation (17–6). These tensile forces acting on the cross-sectional area of the walls of the sphere cause tensile stresses to be developed. That is,

$$\sigma = \frac{F_R}{A_w} \qquad (17\text{–}8)$$

where A_w is the area of the annular ring cut to create the free body, as shown in Figure 17–2. The actual area is

$$A_w = \frac{\pi}{4}(D_o^2 - D_i^2) \qquad (17\text{–}9)$$

However, for thin-walled spheres with a wall thickness t, less than about $\frac{1}{10}$ of the radius of the sphere, the wall area can be closely approximated as

$$A_w = \pi D_m t \qquad (17\text{–}10)$$

This is the area of a rectangular strip having a thickness t and a length equal to the mean circumference of the sphere, πD_m.

Equations (17–6) and (17–8) can be combined to yield an equation for stress,

$$\sigma = \frac{F_R}{A_w} = \frac{pA_p}{A_w} \qquad (17\text{–}11)$$

Expressing A_p and A_w in terms of D_m and t from Equations (17–7) and (17–10) gives

Stress in a Thin-Walled Sphere ▽

$$\sigma = \frac{p(\pi D_m^2/4)}{\pi D_m t} = \frac{pD_m}{4t} \qquad 17\text{–}12$$

This is the expression for the stress in the wall of a thin-walled sphere subjected to internal pressure. Very little error (less than 5%) will result from using either the outside or inside diameter in place of the mean diameter.

Example Problem 17–1   Compute the stress in the wall of a sphere having an inside diameter of 300 mm and a wall thickness of 1.50 mm when carrying nitrogen gas at 3500 kPa internal pressure.

Solution   Objective: Compute the stress in the wall of the sphere.

Given: $p = 3500$ kPa; $D_i = 300$ mm; $t = 1.50$ mm.

Analysis: We must first determine if the sphere can be considered to be thin-walled by computing the ratio of the mean diameter to the wall thickness.

$$D_m = D_i + t = 300\,\text{mm} + 1.50\,\text{mm} = 301.5\,\text{mm}$$

$$D_m/t = 301.5\,\text{mm}/1.50\,\text{mm} = 201$$

Because this is far greater than the lower limit of 20, the sphere is thin. Then Equation (17–12) should be used to compute the stress.

Results: $$\sigma = \frac{pD_m}{4t} = \frac{(3500 \times 10^3\,\text{Pa})(301.5\,\text{mm})}{4(1.50\,\text{mm})}$$

$$\sigma = 175.9 \times 10^6\,\text{Pa} = 175.9\,\text{MPa}$$

17–4 THIN-WALLED CYLINDERS

Cylinders are frequently used for pressure vessels, for example, as storage tanks, hydraulic and pneumatic actuators, and for piping of fluids under pressure. The stresses in the walls of cylinders are similar to those found for spheres, although the maximum value is greater.

Two separate analyses are shown here. In one case, the tendency for the internal pressure to pull the cylinder apart in a direction parallel to its axis is found. This is called *longitudinal stress*. Next, a ring around the cylinder is analyzed to determine the stress tending to pull the ring apart. This is called *hoop stress*, or *tangential stress*.

Longitudinal Stress

Figure 17–4 shows a part of a cylinder, which is subjected to an internal pressure, cut perpendicular to its axis to create a free body. Assuming that the end of the cylinder is closed, the pressure acting on the circular area of the end would produce a resultant force of

$$F_R = pA = p\left(\frac{\pi D_m^2}{4}\right) \qquad \text{(17–13)}$$

This force must be resisted by the force in the walls of the cylinder, which, in turn, creates a tensile stress in the walls. The stress is

$$\sigma = \frac{F_R}{A_w} \qquad \text{(17–14)}$$

Assuming that the walls are thin, as we did for spheres,

$$A_w = \pi D_m t \qquad \text{(17–15)}$$

where t is the wall thickness.

FIGURE 17–4   Free body diagram of a cylinder carrying internal pressure showing longitudinal stress.

Free body, closed-end cylinder with internal pressure p

(a)

Internal pressure p

Force in wall of cylinder

Cylinder wall cross-sectional area

(b)

FIGURE 17–5 Free body diagram of a cylinder carrying internal pressure showing hoop stress.

Ring of any length L with internal pressure p

(a)

F

Tangential forces in cylinder wall

p = pressure

D_m

L

Area on which F acts

t

t

F

(b)

Now combining Equations (17–13), (17–14), and (17–15),

Longitudinal Stress in a Thin-Walled Cylinder ▼

$$\sigma = \frac{F_R}{A_w} = \frac{p(\pi D_m^2/4)}{\pi D_m t} = \frac{pD_m}{4t} \quad (17\text{–}16)$$

This is the stress in the wall of the cylinder in a direction parallel to the axis, called the longitudinal stress. Notice that it is of the same magnitude as that found for the wall of a sphere. But this is not the maximum stress, as shown next.

Hoop Stress

The presence of the tangential or hoop stress can be visualized by isolating a ring from the cylinder, as shown in Figure 17–5. The internal pressure pushes outward evenly all around the ring. The ring must develop a tensile stress in a direction tangential to the circumference of the ring to resist the tendency of the pressure to burst the ring. The magnitude of the stress can be determined by using half of the ring as a free body, as shown in Figure 17–5(b).

The resultant of the forces due to the internal pressure must be determined in the horizontal direction and balanced with the forces in the walls of the ring. Using the same reasoning as we did for the analysis of the sphere, we find that the resultant force is the product of the pressure and

the *projected area* of the ring. For a ring with a mean diameter D_m and a length L,

$$F_R = pA_p = p(D_m L) \quad (17\text{–}17)$$

The tensile stress in the wall of the cylinder is equal to the resisting force divided by the cross-sectional area of the wall. Again assuming that the wall is thin, the wall area is

$$A_w = 2tL \quad (17\text{–}18)$$

Then the stress is

$$\sigma = \frac{F_R}{A_w} = \frac{F_R}{2tL} \quad (17\text{–}19)$$

Combining Equations (17–17) and (17–19) gives

Hoop Stress in a Thin-Walled Cylinder ▼

$$\sigma = \frac{F_R}{A_w} = \frac{pD_m L}{2tL} = \frac{pD_m}{2t} \quad (17\text{–}20)$$

This is the equation for the hoop stress in a thin cylinder subjected to internal pressure. Notice that the magnitude of the hoop stress is *twice* that of the longitudinal stress. Also, the hoop stress is twice that of the stress in a spherical container of the same diameter carrying the same pressure.

Example Problem 17–2 A cylindrical tank holding oxygen at 2000 kPa pressure has an outside diameter of 450 mm and a wall thickness of 10 mm. Compute the hoop stress and the longitudinal stress in the wall of the cylinder.

Solution Objective: Compute the hoop stress and the longitudinal stress in the wall of the cylinder.

Given: $p = 2000$ kPa; $D_o = 450$ mm; $t = 10$ mm.

Analysis: We must first determine if the cylinder can be considered to be thin-walled by computing the ratio of the mean diameter to the wall thickness.

$$D_m = D_o - t = 450\,\text{mm} - 10\,\text{mm} = 440\,\text{mm}$$

$$D_m/t = 440\,\text{mm}/10\,\text{mm} = 44$$

Because this is far greater than the lower limit of 20, the cylinder is thin. Then Equation (17–20) should be used to compute the hoop stress and Equation (17–16) should be used to compute the longitudinal stress. The hoop stress is computed first.

Results: $\sigma = \dfrac{pD_m}{2t} = \dfrac{(2000 \times 10^3\,\text{Pa})(440\,\text{mm})}{2(10\,\text{mm})} = 44.0\,\text{MPa}$

The longitudinal stress, from Equation (17–12), is

$$\sigma = \frac{pD_m}{4t} = 22.0\,\text{MPa}$$

Example Problem 17–3

Determine the pressure required to burst a standard 8-in schedule 40 steel pipe if the ultimate tensile strength of the steel is 40 000 psi.

Solution Objective: Compute the pressure required to burst the steel pipe.

Given: Ultimate tensile strength of steel = s_u = 40 000 psi
Pipe is a standard 8-in schedule 40 steel pipe.
The dimensions of the pipe are found in Appendix A–12 to be

$$\text{Outside diameter} = 8.625\,\text{in} = D_o$$

$$\text{Inside diameter} = 7.981\,\text{in} = D_i$$

$$\text{Wall thickness} = 0.322\,\text{in} = t$$

Analysis: We should first check to determine if the pipe should be called a thin-walled cylinder by computing the ratio of the mean diameter to the wall thickness.

$$D_m = \text{mean diameter} = \frac{D_o + D_i}{2} = 8.303\,\text{in}$$

$$\frac{D_m}{t} = \frac{8.303\,\text{in}}{0.322\,\text{in}} = 25.8$$

Since this ratio is greater than 20, the thin-wall equations can be used. The hoop stress is the maximum stress and should be used to compute the bursting pressure.

Results: Use Equation (17–20).

$$\sigma = \frac{pD_m}{2t} \qquad\qquad \textbf{(17–20)}$$

Letting σ = 40 000 psi and using the mean diameter gives the bursting pressure to be

$$p = \frac{2t\sigma}{D_m} = \frac{(2)(0.322\,\text{in})(40\,000\,\text{lb/in}^2)}{8.303\,\text{in}} = 3102\,\text{psi}$$

Comment: A design factor of 6 or greater is usually applied to the bursting pressure to get an allowable operating pressure. This cylinder would be limited to approximately 500 psi of internal pressure.

17–5 THICK-WALLED CYLINDERS AND SPHERES

The formulas in the preceding sections for thin-walled cylinders and spheres were derived under the assumption that the stress is uniform throughout the wall of the container.

As stated, if the ratio of the diameter of the container to the wall thickness is greater than 20, this assumption is reasonably correct. Conversely, if the ratio is less than 20, the walls are considered to be thick, and a different analysis technique is required.

σ_1 = hoop stress σ_3 = radial stress $\sigma_1 = \sigma_2$ = tangential stress
σ_2 = longitudinal stress

FIGURE 17–6   Notation for stresses in thick-walled cylinders and spheres.

The detailed derivation of the thick-wall formulas will not be given here because of their complexity. See Reference 12. But the application of the formulas will be shown.

For a thick-walled cylinder, Figure 17–6 shows the notation to be used. The geometry is characterized by the inner radius a, the outer radius b, and any radial position between a and b, called r. The *hoop stress* is called σ_1; the *longitudinal stress* is σ_2. These have the same meaning as they did for thin-walled vessels, except now they will have varying magnitudes at different positions in the wall. In addition to hoop and longitudinal stresses, a *radial stress* σ_3 is created in a thick-walled vessel. As the name implies, the radial stress acts along a radius of the cylinder or sphere. It is a compressive stress and varies from a magnitude of zero at the outer surface to a maximum at the inner surface, where it is equal to the internal pressure.

Table 17–1 shows a summary of the formulas needed to compute the three stresses in the walls of thick-walled cylinders and spheres subjected to internal pressure. The terms *longitudinal stress* and *hoop stress* do not apply to spheres.

Instead, we refer to the *tangential stress*, which is the same in all directions around the sphere. Then

$$\text{Tangential stress} = \sigma_1 = \sigma_2$$

17–6 ANALYSIS AND DESIGN PROCEDURES FOR PRESSURE VESSELS

Presented here is a summary of the principles discussed in this chapter related to the stress analysis of both thin-walled and thick-walled spheres and cylinders. The summary is given in the form of general procedures for analyzing and designing pressure vessels.

For design stresses, it is advised that Sections 9–3 to 9–6 be reviewed. It will be assumed here that the failure of a pressure vessel subjected to internal pressure is due to the tensile stresses occurring tangentially in the walls of the vessel. Design stresses should take into account the material from which the vessel is made, the operating environment, and whether the pressure is constant or varying in a cyclic fashion.

TABLE 17–1 Stresses in Thick-Walled Cylinders and Spheres[*]		
	Stress at position r	**Maximum stress**
Thick-walled cylinder		
Hoop (tangential)	$\sigma_1 = \dfrac{pa^2(b^2 + r^2)}{r^2(b^2 - a^2)}$	$\sigma_1 = \dfrac{p(b^2 + a^2)}{b^2 - a^2}$ (at inner surface)
Longitudinal	$\sigma_2 = \dfrac{pa^2}{b^2 - a^2}$	$\sigma_2 = \dfrac{pa^2}{b^2 - a^2}$ (uniform throughout wall)
Radial	$\sigma_3 = \dfrac{-pa^2(b^2 - r^2)}{r^2(b^2 - a^2)}$	$\sigma_3 = -p$ (at inner surface)
Thick-walled sphere		
Tangential	$\sigma_1 = \sigma_2 = \dfrac{pa^3(b^3 + 2r^3)}{2r^3(b^3 - a^3)}$	$\sigma_1 = \sigma_2 = \dfrac{p(b^3 + 2a^3)}{2(b^3 - a^3)}$ (at inner surface)
Radial	$\sigma_3 = \dfrac{-pa^3(b^3 - r^3)}{r^3(b^3 - a^3)}$	$\sigma_3 = -p$ (at inner surface)

[*]Symbols used here are as follows: a = inner radius; b = outer radius; r = any radius between a and b; p = internal pressure, uniform in all directions. Stresses are tensile when positive, compressive when negative.

See also Section 17–7 for a discussion of other modes of failure in vessels having penetrations, structural supports, reinforcing rings, and other features that differ from the simple cylindrical or spherical shape.

Design Stresses

For steady pressure, the design stress can be based on the yield strength of the material

$$\sigma_d = s_y/N$$

The choice of the design factor, N, is often dictated by code because of the danger created when a pressure vessel fails. This is particularly true for vessels containing gases or steam under pressure because failures produce violent expulsion of the gas as the high level of stored energy is released. In the absence of a code, we will use $N = 4$ as a minimum and larger values should be used for critical applications or where uncertainty exists in the operating conditions or material properties. Another suggested guideline is to limit the pressure in a vessel to no more than $\frac{1}{6}$ of the pressure that would be predicted to burst it. This effectively calls for a design stress related to the ultimate tensile strength of the material of

$$\sigma_d = s_u/N = s_u/6$$

For cycling pressure, base the design stress on the ultimate strength,

$$\sigma_d = s_u/N$$

Use $N = 8$ as a minimum to produce a design stress related to the fatigue strength of the material.

A. PROCEDURE FOR ANALYZING PRESSURE VESSELS ▼

Given: Internal pressure in the vessel, p.
Material from which the vessel is made. Ductile metal assumed.
Outside diameter, D_o; inside diameter, D_i; and wall thickness, t, for the vessel.
Objective: Determine the maximum stress in the vessel and check the safety of that stress level with regard to the design stress in the material from which the vessel is made.

1. Compute the mean diameter, D_m, for the vessel from Equation (17–2), $D_m = (D_o + D_i)/2$.
2. Compute the ratio of the mean diameter to the wall thickness for the vessel, D_m/t.
3. If $D_m/t \geq 20$, the vessel can be considered to be thin-walled. Use Equation (17–12) for spheres or Equation (17–20) for cylinders to compute the maximum tangential stress in the walls of the vessel.

$$\sigma = pD_m/4t \quad \text{for spheres} \qquad \textbf{(17–12)}$$

$$\sigma = pD_m/2t \quad \text{for cylinders} \qquad \textbf{(17–20)}$$

4. If $D_m/t < 20$, the vessel must be considered to be thick-walled. Use equations from Table 17–1 to compute

the maximum tangential or hoop stress in the walls of the vessel.

$$\sigma = \frac{p(b^3 + 2a^3)}{2(b^3 - a^3)} \quad \text{for spheres}$$

$$\sigma = \frac{p(b^2 + a^2)}{b^2 - a^2} \quad \text{for cylinders}$$

5. Compute the design stress for the material from which the vessel is made.
6. The actual maximum stress must be less than the design stress for safety.

B. PROCEDURE FOR DESIGNING PRESSURE VESSELS FOR A GIVEN MATERIAL ▼

Given: Internal pressure in the vessel, p.
Material from which the vessel is made. Ductile metal assumed.
Nominal internal diameter of the vessel based on the desired volumetric capacity.
Objective: Specify the outside diameter, D_o; inside diameter, D_i; and wall thickness, t, for the vessel to ensure the safety of the vessel with regard to the design stress in the material from which the vessel is made.

1. Use the given diameter as an estimate of the mean diameter, D_m, for the vessel.
2. Assume at first that the vessel will be thin-walled and that the maximum stress can be computed from Equation (17–12) for a sphere or Equation (17–20) for a cylinder. This assumption will be checked later.
3. Compute the design stress for the material from which the vessel is to be made.
4. In the appropriate stress equation, substitute the design stress for the maximum stress and solve for the minimum required wall thickness, t.
5. Specify convenient values for t, D_i, and D_o, based on available material thicknesses. Appendix A–2 may also be used to specify preferred basic sizes.
6. Compute the actual mean diameter for the vessel using the specified dimensions.
7. Compute the ratio of the mean diameter to the wall thickness for the vessel, D_m/t.
8. If $D_m/t \geq 20$, the vessel is thin-walled as assumed and the design is finished.
9. If $D_m/t < 20$, the vessel must be considered to be thick-walled. Use equations from Table 17–1 to compute the maximum tangential or hoop stress in the walls of the vessel and compare that stress with the design stress. If the actual stress is less than the design stress, the design is satisfactory. If the actual maximum stress is greater than the design stress, increase the wall thickness and recompute the resulting stress. Continue this process until a satisfactory stress level and convenient dimensions for the vessel are obtained. Equation-solving computer software, spreadsheets, or a graphing calculator with equation-solving capability may facilitate this process.

C. PROCEDURE FOR SPECIFYING A DUCTILE METAL FOR A PRESSURE VESSEL OF A GIVEN SIZE ▼

Given: Internal pressure in the vessel, p.
Outside diameter, D_o; inside diameter, D_i; and wall thickness, t, for the vessel.
Objective: Specify a suitable ductile metal from which the vessel is to be made.

1. Compute the mean diameter, D_m, for the vessel from Equation (17–2): $D_m = (D_o + D_i)/2$.
2. Compute the ratio of the mean diameter to the wall thickness for the vessel, D_m/t.
3. If $D_m/t \geq 20$, the vessel can be considered to be thin-walled. Use Equation (17–12) for spheres or Equation (17–20) for cylinders to compute the maximum tangential stress in the walls of the vessel.

$$\sigma = pD_m/4t \quad \text{for spheres} \qquad \text{(7–12)}$$
$$\sigma = pD_m/2t \quad \text{for cylinders} \qquad \text{(7–20)}$$

4. If $D_m/t < 20$, the vessel must be considered to be thick-walled. Use equations from Table 17–1 to compute the maximum tangential or hoop stress in the walls of the vessel.

$$\sigma = \frac{p(b^3 + 2a^3)}{2(b^3 - a^3)} \quad \text{for spheres}$$
$$\sigma = \frac{p(b^2 + a^2)}{b^2 - a^2} \quad \text{for cylinders}$$

5. Specify a suitable equation for the design stress from the discussion earlier in this section.
6. Let the design stress equal the computed maximum stress from step 3 or 4. Then solve for the appropriate material strength, either s_y or s_u from the design stress equation.
7. Specify a suitable material that has a strength greater than the minimum required value.

Example Problem 17–4   Compute the magnitude of the maximum longitudinal, hoop, and radial stresses in a cylinder carrying helium at a steady pressure of 10 000 psi. The outside diameter is 8.00 in and the inside diameter is 6.40 in. Specify a suitable material for the cylinder.

Solution   Objective: Compute the maximum stresses and specify a material.

Given: Pressure $= p = 10\,000$ psi. $D_o = 8.00$ in. $D_i = 6.40$ in.

Analysis: Use Procedure C from this section.

Results: **Step 1:** $D_m = (D_o + D_i)/2 = (8.00 + 6.40)/2 = 7.20$ in

Step 2: $t = (D_o - D_i)/2 = (8.00 - 6.40)/2 = 0.80$ in
$D_m/t = 7.20/0.80 = 9.00$

Step 3: This step does not apply. Cylinder is thick.

Step 4: Use equations from Table 17–1.

$$a = D_i/2 = 6.40/2 = 3.20 \text{ in}$$
$$b = D_o/2 = 8.00/2 = 4.00 \text{ in}$$
$$\sigma_1 = \frac{p(b^2 + a^2)}{b^2 - a^2} = \frac{(10\,000\,\text{psi})(4.00^2 + 3.20^2)\,\text{in}^2}{(4.00^2 - 3.20^2)\,\text{in}^2} = 45\,560 \text{ psi hoop}$$
$$\sigma_2 = \frac{pa^2}{b^2 - a^2} = \frac{(10\,000\,\text{psi})(3.20\,\text{in})^2}{(4.00^2 - 3.20^2)\,\text{in}^2} = 17\,780 \text{ psi longitudinal}$$
$$\sigma_3 = -p = -10\,000 \text{ psi radial}$$

All three stresses are a maximum at the inner surface of the cylinder.

Step 5: Let the design stress $= \sigma_d = s_y/4$.

Step 6: The maximum stress is the hoop stress, $\sigma_1 = 45\,560$ psi. Then the required yield strength for the material is

$$s_y = N(\sigma_2) = 4(45\,560\,\text{psi}) = 182\,200 \text{ psi} = 182 \text{ ksi}$$

Step 7: From Appendix A–14, we can specify AISI 4140 OQT 700 steel that has a yield strength of 212 ksi.

Example Problem 17–5   Compute the magnitude of the maximum tangential and radial stresses in a sphere carrying helium at a steady pressure of 10 000 psi. The outside diameter is 8.00 in and the inside diameter is 6.40 in. Specify a suitable material for the cylinder.

Solution Objective: Compute the maximum stresses and specify a material.

Given: Pressure $= p = 10\,000$ psi. $D_o = 8.00$ in. $D_i = 6.40$ in.

Analysis: Use Procedure C from this section. These data are the same as those used in Example Problem 17–4. Some values will be carried forward.

Results: **Steps 1, 2, 3:** Sphere is thick-walled.

Step 4: Use equations from Table 17–1. $a = 3.20$ in. $b = 4.00$ in.

$$\sigma_1 = \sigma_2 = \frac{p(b^3 + 2a^3)}{2(b^3 - a^3)} = \frac{(10\,000 \text{ psi})[4.00^3 + 2(3.20)^3] \text{ in}^3}{2(4.00^3 - 3.20^3) \text{ in}^3}$$

$$\sigma_1 = \sigma_2 = 20\,740 \text{ psi tangential}$$

$$\sigma_3 = -p = -10\,000 \text{ psi radial}$$

Each of these stresses is a maximum at the inner surface.

Steps 5, 6: For a maximum stress of 20 740 psi, the required yield strength for the material is

$$s_y = N(\sigma_2) = 4(20\,740 \text{ psi}) = 82\,960 \text{ psi} = 83 \text{ ksi}$$

Step 7: From Appendix A–14, we can specify AISI 4140 OQT 1300 steel that has a yield strength of 101 ksi. Others could be used.

Comment: The maximum stress in the sphere is less than half that in the cylinder of the same size, allowing a material with a much lower strength to be used. Alternatively, it would be possible to design the sphere with the same material but with a smaller wall thickness.

Example Problem 17–6

A cylindrical vessel has an outside diameter of 400 mm and an inside diameter of 300 mm. For an internal pressure of 20.1 MPa, compute the hoop stress σ_1 at the inner and outer surfaces and at points within the wall at intervals of 10 mm. Plot a graph of σ_1 versus the radial position in the wall.

Solution Objective: Compute the hoop stress at specified positions in the wall of the cylinder.

Given: Pressure $= p = 20.1$ MPa. $D_o = 400$ mm. $D_i = 300$ mm.
Use 10 mm increments for radius within the wall from the outside surface to the inside surface.

Analysis: Use steps 1–4 from Procedure A from this section.

Results: **Step 1:** $D_m = (D_o + D_i)/2 = (400 + 300)/2 = 350$ mm

Step 2: $t = (D_o - D_i)/2 = (400 - 300)/2 = 50$ mm
$D_m/t = 350/50 = 7.00 < 20$; thick-walled cylinder

Step 3: This step does not apply.

Step 4: Use the equation for tangential stress from Table 17–1.

$$\sigma_1 = \frac{pa^2(b^2 + r^2)}{r^2(b^2 - a^2)}$$

$$a = D_i/2 = 300/2 = 150 \text{ mm}$$

$$b = D_o/2 = 400/2 = 200 \text{ mm}$$

The results are shown in tabular form below.

r (mm)	σ_2 (MPa)	
200	51.7	(Minimum at outer surface)
190	54.5	
180	57.7	
170	61.6	
160	66.2	
150	71.8	(Maximum at inner surface)

FIGURE 17–7
Variation of tangential
stress in the wall of the
thick-walled cylinder in
Example Problem 17–6.

Comment: Figure 17–7 shows the graph of tangential stress versus position in the wall. The graph illustrates clearly that the assumption of uniform stress in the wall of a thick-walled cylinder would *not* be valid.

Example Problem 17–7

Design a cylinder to be made from aged titanium Ti-6Al-4V to carry compressed natural gas at 7500 psi. The internal diameter must be 24.00 in to provide the necessary volume. The design stress is to be $\frac{1}{6}$ of the ultimate strength of the titanium.

Solution Objective: Design the cylinder.

Given: Pressure $= p = 7500$ psi. $D_i = 24.0$ in.
Titanium Ti-6Al-4V; $s_u = 170$ ksi (Appendix A–15)

Analysis: Use Procedure B from this section.

Results: **Step 1:** Let $D_m = 24.00$ in.

Step 2: Assume thin-walled cylinder.

Step 3: Design stress,

$$\sigma_d = s_u/6 = (170\,000\text{ psi})/6 = 28\,333\text{ psi}$$

Step 4: Use the Equation (17–20) to compute nominal value for t.

$$t = \frac{pD_m}{2\sigma_d} = \frac{(7500\text{ psi})(24.0\text{ in})}{2(28\,333\text{ psi})} = 3.18\text{ in}$$

Step 5: Trial #1: $D_i = 24.00$; $t = 3.50$ in; $D_o = D_i + 2t = 31.00$ in

Step 6: $D_m = D_i + t = 24.00 + 3.50 = 27.50$ in

Step 7: $D_m/t = 27.50/3.50 = 7.86 < 20$; thick-walled.

Step 8: This step does not apply.

Step 9: Use the equation for σ_1 from Table 17–1.

$$a = D_i/2 = 24.00/2 = 12.00\text{ in}$$
$$b = D_o/2 = 31.00/2 = 15.50\text{ in}$$
$$\sigma_1 = \frac{p(b^2 + a^2)}{(b^2 - a^2)} = \frac{(7500\text{ psi})(15.50^2 + 12.00^2)}{(15.50^2 - 12.00^2)}$$
$$\sigma_1 = 29\,940\text{ psi}$$

Slightly high. Repeat steps 5 and 9.

Step 5: Increase $t = 3.75$ in; $D_o = D_i + 2t = 31.50$ in; thick-walled.

Step 9: Use the equation for σ_1 from Table 17–1.

$$a = D_i/2 = 24.00/2 = 12.00 \text{ in}$$
$$b = D_o/2 = 31.50/2 = 15.75 \text{ in}$$

Then $\sigma_1 = 28\,250$ psi. This is less than the design stress. OK.

Comment: The wall thickness is quite large, which would result in a heavy cylinder. Consider using a sphere and higher strength material for the vessel. A composite sphere may produce a lighter design.

17–7 SPREADSHEET AID FOR ANALYZING THICK-WALLED SPHERES AND CYLINDERS

The calculations called for in analyzing thick-walled spheres and cylinders can be tedious. The equations shown in Table 17–1 require numerous calculations. The use of spreadsheets, computer programs, programmable calculators, or computer algebra systems can simplify your task significantly.

Figure 17–8 shows an example spreadsheet that analyzes thick-walled spheres and cylinders for the maximum tangential, longitudinal, and radial stresses. Only the few data values in the shaded areas need to be entered by the user.

If the vessel is a sphere, the results are shown near the middle of the sheet. If the vessel is a cylinder, the maximum stresses are shown just below the middle of the sheet.

Also included in the lower part of the spreadsheet is the calculation of the tangential stress in a cylinder as a function of the radial position within the wall of the cylinder. This requires that the user specify the values of the radii for which the calculations are to be done.

The sample data included in the spreadsheet are the same as used in Example Problem 17–6 for a thick-walled cylinder and a graph of the distribution of tangential stress in the wall of the cylinder is shown in Figure 17–7.

Enhancements of this spreadsheet could be made to enable it to be used for designing pressure vessels using Procedures A, B, and C outlined earlier in this chapter. You are encouraged to produce such design aids.

STRESSES IN THICK-WALLED CYLINDERS AND SPHERES

DATA REQUIRED:

Pressure = p =	20100 kPa		Wall thickness = t =	50 mm
$R_i = a$ =	150 mm		Mean Dia = D_m =	350 mm
$R_o = b$ =	200 mm		Ratio: D_m/t =	7.0
			If Ratio < 20, vessel is thick	

Analysis of a Sphere

Max Tangential Stress	32.05 MPa	At inner surface
Max Radial Stress	–20.10 MPa	At inner surface

Analysis of a Cylinder

Max Tangential Stress	71.79 MPa	At inner surface
Max Longitudinal Stress	25.84 MPa	Uniform through wall
Max Radial Stress	–20.10 MPa	At inner surface

Stress vs. Radius - Tangential Stress Only

Radius	Stress	
150	71.79 MPa	At inner surface
160	66.22 MPa	
170	61.61 MPa	
180	57.75 MPa	
190	54.48 MPa	
200	51.69 MPa	At outer surface

FIGURE 17–8   Spreadsheet for analyzing stresses in thick-walled spheres and cylinders.

17–8 SHEARING STRESS IN CYLINDERS AND SPHERES

We have seen that internal pressure in a pressure vessel causes normal tensile stresses to be created in the walls of the cylinder or sphere. If these are the only externally generated stresses on the vessel, they will be the principal stresses on any planar surface element. No externally applied shearing stresses are created in the plane of the surface. This assumes that there is no torsion applied to the vessel and there are no external supports on or penetrations into the wall of the vessel in the area of interest.

Figure 17–9 shows the stress element at the surface of a cylinder along with the Mohr's circle for that element. It shows that there will be shearing stresses at angles other than 0° or 90° with respect to the x-axis where the longitudinal stress acts. The maximum shearing stress in the plane of the element is $\sigma_2/2$ and it occurs on an angle at 45° to the maximum principal stress axis.

Because both of the principal stresses are tensile, this is a classic case of both principal stresses having the same sign. It was observed in Section 15–11, during the discussion of Mohr's circle, that in such cases there is a need to consider the combination of the maximum principal stress with the stress perpendicular to the surface stress element to determine the true maximum shear stress on the element. The stress perpendicular to the plane of the cylinder, σ_3, is zero at the surface. Then, as we show in Figure 17–9, a supplementary Mohr's circle is drawn with σ_1 and σ_3 on the horizontal diameter. For this combination, the true maximum shearing stress is $\sigma_1/2$, acting in the x–z plane, at 45° to the maximum principal stress axis.

Pressure vessels are typically made from flat plates rolled into a cylindrical shape and welded along the seam. The design of the welds and the strength of the material in the vicinity of the welds are critical design parameters. The seam is often longitudinal for convenience of the rolling operation. The stress on the seam would be the tangential stress, which is the maximum principal stress.

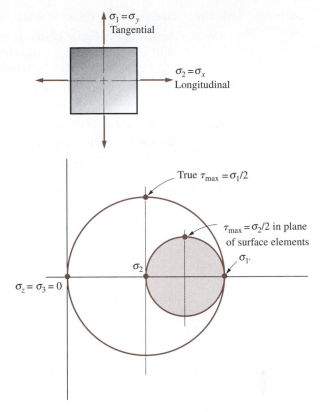

FIGURE 17–9 Principal stresses and shearing stresses in a pressurized cylinder.

Another method of making the cylindrical form is to roll the flat sheet along a spiral path as sketched in Figure 17–10. Then the entire spiral seam is welded. This is particularly attractive when making relatively long, large-diameter tanks or pipes. In this case, you will need to know the stress condition on an element in line with the spiral seam. Mohr's circle can also be used for that task as demonstrated in Example Problem 17–8.

FIGURE 17–10 Tank for Example Problem 17–8.

Example Problem 17–8

Figure 17–10 shows a tank that is to be made by rolling flat sheets of AISI 1040 cold-drawn steel into the spiral shape shown, where the spiral makes an angle of 65° with the horizontal axis of the tank. The design internal pressure is 1750 kPa. The inside diameter has been specified to be 900 mm to create the desired capacity in the tank. (a) Specify a suitable thickness for the steel sheet to provide a design factor of 4 based on yield strength or to produce a design factor of 6 based on ultimate strength. (b) For the final design of the tank, determine the stress condition on an element aligned with the weld.

Solution   Objective: Specify the tank thickness and determine the stress on the weld.

Given: Tank design in Figure 17–10. $D_i = 900$ mm. $p = 1750$ kPa.
AISI 1040 CD steel, $s_y = 565$ MPa, $s_u = 669$ MPa.
$\sigma_d = s_y/4$ or $\sigma_d = s_u/6$, whichever is smaller.

Analysis: Use Procedure B from Section 17– 6 to design the tank.
Then use Mohr's circle to determine the stress on the weld.

Results: **Steps 1 and 2:** First assume $D_m = 900$ mm. Use Equation (17–20).

Step 3: Compute two design stresses and specify the smaller value.

$$\sigma_d = s_y/4 = (565\,\text{MPa})/4 = 141.3\,\text{MPa}$$
$$\sigma_d = s_u/6 = (669\,\text{MPa})/6 = 111.5\,\text{MPa}$$

Step 4: Solve for minimum required wall thickness, t. Let $\sigma_{max} = \sigma_d$.

$$\sigma_{max} = (pD_m)/(2t) \qquad \text{Equation (17–20)}$$
$$t = \frac{pD_m}{2\sigma_{max}} = \frac{(1750 \times 10^3\,\text{Pa})(900\,\text{mm})}{2(111.5 \times 10^6\,\text{Pa})} = 7.06\,\text{mm}$$

Step 5: Specify $t = 8.0$ mm. Then

$$D_i = 900\,\text{mm}$$
$$D_o = D_i + 2t = 900 + 2(8.0) = 916\,\text{mm}$$

Step 6: $D_m = D_i + t = 900 + 8.0 = 908$ mm

Step 7: $D_m/t = (908\,\text{mm})/(8.0\,\text{mm}) = 113.5$

Step 8: $D_m/t > 20$. The cylinder is very thin.

Mohr's Circle Analysis. See Figure 17–11.
The actual maximum principal stress, σ_1, is

$$\sigma_1 = (pD_m)/(2t)$$
$$\sigma_1 = \frac{(1750 \times 10^3\,\text{Pa})(908\,\text{mm})}{2(8.0\,\text{mm})} = 99.31 \times 10^6\,\text{Pa} = 99.31\,\text{MPa}$$

The actual minimum principal stress, σ_2, is for this special case

$$\sigma_2 = (pD_m)/(4t) = \sigma_1/2 = 49.66\,\text{MPa}$$

The stress element in the plane of the tank is shown in Figure 17–11 with $\sigma_2 = \sigma_x$ and $\sigma_1 = \sigma_y$. No shear stress exists on the element in this orientation.

Mohr's circle is also shown in Figure 17–11. We need to find the stress on an element that is rotated 25° CCW from the x-axis.

FIGURE 17–11   Stress elements and Mohr's circle for Example Problem 17–8.

$\sigma_1 = 99.31$ MPa
$\sigma_2 = 49.66$ MPa
$\sigma_{avg} = 74.48$ MPa
$\sigma_w = 58.52$ MPa
$\sigma_v = 90.44$ MPa
$\tau_{wv} = 19.02$ MPa

In the Mohr's circle, the x-axis is the line from the center of the circle to σ_2. We then rotate CCW by an angle of $2(25°) = 50°$ to find the point on the circle that represents the stress condition along the weld. That point is labeled A on the circle, and it has the coordinates of σ_w, τ_{wv}. Here we are using the w subscript to indicate the direction perpendicular to the weld line and the v subscript to indicate the direction along the weld line. Then t_{wv} is the shearing stress perpendicular to the w-direction and parallel to the v-direction.

Using the geometry of the circle we find

$$\sigma_w = \sigma_{avg} - R\cos(50°)$$

where R is the radius of the circle. We find R from

$$R = (\sigma_1 - \sigma_2)/2 = (99.31 - 49.66)/2 = 24.83\,\text{MPa}$$

Then

$$\sigma_w = \sigma_{avg} - R(\cos 50°) = 74.48 - (24.83)(\cos 50°) = 58.52\,\text{MPa}$$

Also,

$$\tau_{wv} = R\sin(50°) = (24.83)(\sin 50°) = 19.02\,\text{MPa}$$

Finally, we find the normal stress parallel to the direction of the weld. That is found 180° from point A, called point B on the circle. Using similar logic,

$$\sigma_v = \sigma_{avg} + R(\cos 50°) = 74.48 + (24.83)(\cos 50°) = 90.44\,\text{MPa}$$

The final stress element oriented on the weld line is shown in Figure 17–11.

Summary: The final design can be summarized as follows.

The tank is made from AISI 1040 CD steel.
$D_i = 900\,\text{mm}.\,D_o = 916\,\text{mm}.\,t = 8.0\,\text{mm}.$
Normal stress perpendicular to the weld line is $\sigma_w = 58.52\,\text{MPa}.$
Normal stress parallel to the weld line is $\sigma_v = 90.44\,\text{MPa}.$
Shear stress parallel to the weld line is $\tau_{w_v} = 19.02\,\text{MPa}.$

17–9 OTHER DESIGN CONSIDERATIONS FOR PRESSURE VESSELS

Pressure vessel design and analysis techniques presented thus far have related only to the basic stress analysis of ideal cylinders and spheres without consideration of penetrations or other changes in geometry. Of course, most practical pressure vessels incorporate several types of features that cause the vessel to differ from the ideal shape. Figure 17–12 shows a computer-generated drawing of such a vessel. It has two cylindrical sections with different diameters, a conical transition section between them, spherical domed ends of two diameters, and three penetrations intersecting the shell with different geometries.

External loads are often applied that create stresses that combine with the stress due to internal pressure. For example:

- A spherical or cylindrical pressure vessel would typically have one or more ports used to fill or empty the vessel. The ports would often be welded into the vessel, causing a discontinuity in the geometry, as well as modifying the material properties in the vicinity of the weld.

- Some pressure vessels used for chemical reactions or other material processing applications contain view

FIGURE 17–12   Computer-generated drawing of a pressure vessel having several geometry changes.

ports for observation of the process. The view ports may contain flanges to hold the transparent window.

- Cylindrical vessels are often made with domed or hemispherical ends to provide a more optimum design to resist the internal pressure. But, because the tangential stress in the spherical end is less than that in the cylinder, special attention should be paid to the design at the intersection of the ends with the straight cylindrical portion.

- Large cylinders may have reinforcing bands or ribs applied to the inside or outside to stiffen the vessel structurally.

- Large cylinders and spheres may experience large stresses due to the actual weight of the vessel and its contents that combine with the stresses produced by the internal pressure. For example: a relatively long cylindrical tank laid horizontally and supported near its ends is subjected to bending stresses; a cylindrical tank positioned with its axis vertical is subjected to axial compressive stress.

- Large cylinders and spheres must be fitted with supports that transmit the weight of the vessel and its contents to a floor or the earth. Special stress conditions exist in the vicinity of such supports.

- Pressure vessels used in ground transportation equipment often experience dynamic loads caused by stopping, starting, product movement within the vessel, and vibrations caused by uneven roadways.

- Pressure vessels on aircraft and spacecraft are subjected to high acceleration forces during landings, takeoffs, launches, and rapid maneuvers.

- Joints between sections of pressure vessels made from two or more pieces often contain geometric discontinuities requiring special analysis techniques and careful fabrication.

Sources of Additional Information on Pressure Vessels

Analysis techniques for the conditions listed previously are not covered in this book. The lists of references and Internet sites at the end of this chapter offer a wide variety of standards, guidelines, and computational aids to ensure safe and economical design of pressure vessels. The discussions that follow offer annotations about the references and Internet sites listed at the end of this chapter.

1. The primary reference for pressure vessel design standards in the United States is Reference 2, the *ASME Boiler and Pressure Vessel Code.* The most current revision should be used. Its major sections are the following:

   a. I Power Boilers

   b. II Materials

   c. III Rules for Construction of Nuclear Facility Components

   d. IV Heating Boilers

   e. V Nondestructive Examination

   f. VI Recommended Rules for the Care and Operation of Heating Boilers

   g. VII Recommended Guidelines for the Care of Power Boilers

   h. VIII Pressure Vessels

   i. IX Welding and Brazing Qualifications

   j. X Fiber-Reinforced Plastic Pressure Vessels

   k. XI Rules for Inservice Inspection of Nuclear Power Plant Components

   l. XII Rules for Construction and Continued Service of Transport Tanks

   Numerous seminars, short courses, and specialized training programs based on the BPV Code are available, and many firms provide certified design, testing, and construction services in applying the code. Section VIII, *Pressure Vessels,* is the most pertinent to this book. See Internet site 1 for more information. Strict adherence to these standards is essential for ensuring protection of life and property.

2. Piping systems themselves perform as pressure vessels in conjunction with tanks, boilers, heat exchangers, and other special devices under pressure. Specialized requirements for materials, components (such as flanges, fittings, valves, and so forth), design, fabrication, assembly, erection, examination, inspection, and testing of piping are included in the following standards.

   a. Reference 3, ASME standard B31.1, provides standards for piping systems for electrical power generation stations, industrial plants, and central heating plants that typically operate at high temperatures and moderate to high pressures.

   b. Reference 4, ASME standard B31.3, covers piping typically found in petroleum refineries; chemical, pharmaceutical, textile, paper, semiconductor, and related processing plants; and terminals where shipping of the liquid or gaseous products occurs.

3. References 1, 5–8, 10, 11, and 14 are manuals and guidebooks that expand on the ASME standards and provide examples of the applications of provisions from the standards.

4. Reference 9 is part of a large, comprehensive manual for engineers, contractors, and vendors that design or provide equipment for use in the Lawrence Livermore National Laboratories, one of the major U.S. government research laboratories. The document on *Pressure Vessel and System Design* is the most pertinent part of the manual for this book.

   a. Section 4.0 discusses design controls for pressure vessels including materials selection, materials specifications (carbon and alloy steels, stainless steels, and titanium), design considerations, maximum allowable working pressure (MAWP), maximum operating pressure (MOP), test pressures, and calculation guide for thin- and thick-walled vessels and end closures.

b. Section 5.3 describes requirements for gas-pressure containment vessels that will contain toxic, radioactive, corrosive, or flammable materials.

c. Section 6.3 provides guidelines and standards for pipe and tube support.

d. Section 6.6 describes pressure relief devices that limit the pressure level in a pressure vessel or piping system.

The manual can be viewed and downloaded from Internet site 3.

5. References 12 and 13 provide more background, derivation, and extension of the analytical methods and formulas used in this book for stress analysis and design of pressure vessels and piping systems and their components.

6. While the ASME code is most often used in the United States, Internet site 2 describes a corresponding standard used in Great Britain and some other parts of Europe. Standards for other parts of the world exist and should be consulted when providing equipment for those areas.

7. Internet sites 4–9 connect to several providers of computer software packages for pressure vessel design that perform the complex computations required to design and analyze pressure vessels and their attachments. Most provide three-dimensional CAD capability, the ability to export the design drawings into other CAD systems, and complete analyses of the shell, stiffening rings, various styles of heads, nozzles, ports, flanges, and joints between elements. All stress analysis methods are correlated with the ASME codes and standards and consider material properties, design temperatures for vessel operation, welds, testing requirements, and documentation forms that must be completed and filed with the product designs.

8. Internet sites 10–14 connect to a small sample of geographically dispersed companies in the United States that offer a diversified array of products and services in the design and fabrication of pressure vessels, tanks, and piping systems. These sites may be useful to students and other users of this book who have limited experience with the details of such systems to help them visualize their complexity.

17–10 COMPOSITE PRESSURE VESSELS

The applications and examples presented in this chapter have emphasized the use of metals for the structural walls of pressure vessels. Other materials, particularly composite materials and reinforced plastics, are often used as well. The special characteristics of these materials must be understood when applying them to pressure vessels.

High-strength composite materials are well suited for use in fabricating pressure vessels. The fact that the primary stresses are either tangential (hoop) or longitudinal leads the designer of composite vessels to call for alignment of the composite fibers in the direction of the maximum stresses. Circumferential wrapping of a prepreg tape around a liner made from metal or plastic offers significant weight savings as compared with a design using only metal or plastic. To resist longitudinal stresses due to internal pressure along with other external forces, some tanks are wrapped in a helical fashion in addition to the circumferential wrapping. The thickness and direction of plies can be tailored to the specific set of loads expected in a particular application.

Materials selected for composite pressure vessels include E-glass/epoxy, structural glass/epoxy, and carbon/epoxy. Cost is a major factor in material specification.

Care must be exercised to ensure that the composite material bonds well and fits the geometry of any liner used in the vessel. Particular attention is needed in the domed ends of pressure cylinders and in the locations of ports. Ports are typically placed at the top or bottom at the poles of the domed ends in such a manner that the composite fibers are continuous. Placing ports in the sides of a tank would interrupt the integrity of the filament windings. Also, the geometry of the shape of the tank is often tailored to produce gradually varying stresses at joints between the cylindrical part and the domed ends. The thickness of composite plies is also varied to match the expected stresses.

The primary applications for composite pressure vessels include those in which light weight is a major design goal. The air supply tank for self-contained breathing apparatus (SCBA) used by firefighters is a good example because the lighter tank allows more mobility and less fatigue. Weight reductions in space and aeronautical applications allow greater payloads or higher performance of aerospace vehicles.

The development of ground vehicles using compressed natural gas (CNG) or hydrogen calls for the production of lightweight, compact cylinders to carry the fuel. Demonstration units incorporating fuel storage tanks made from advanced composites are in current use in buses, commercial fleet vehicles, automobiles, utility vehicles, and even power units for portable computers, remotely operated sensing systems, aerospace applications, and manufacturing equipment. Weight savings can be significant.

REFERENCES

1. Aluminum Association, *Aluminum Design Manual, Specifications for Aluminum Structures,* Washington, D.C., Aluminum Association, 2005.

2. American Society of Mechanical Engineers (ASME), *ASME Boiler and Pressure Vessel Code,* Fairfield, NJ, ASME, 2007.

3. American Society of Mechanical Engineers (ASME), *ASME Standard B31.1—2007 Power Piping,* Fairfield, NJ, ASME, 2007.

4. American Society of Mechanical Engineers (ASME), *ASME Standard B31.3—2006 Process Piping,* Fairfield, NJ, ASME, 2006.

5. Ball, B. E. and W. J. Carter, *CASTI Guidebook to ASME Section VIII Div. 1 Pressure Vessels,* 4th ed., Edmonton, Alberta, Canada, CASTI Publishing, 2005.

6. Chattopadhyay, S., *Pressure Vessels: Design and Practice*, Boca Raton, FL, CRC Press, 2004.

7. Ellenberger, P., R. Chuse, and B. E. Carson, *Pressure Vessels*, New York, McGraw-Hill, 2004.

8. Farr, J. R. and M. H. Jawad, *Guidebook for the Design of ASME Section VIII Pressure Vessels*, 3rd ed., Fairfield, NJ, American Society of Mechanical Engineers, 2006.

9. Lawrence Livermore National Laboratory (LLNL), *Environment, Safety and Health Manual*, Volume II, Part 18: Pressure/Noise/Hazardous Atmospheres, Document 18.2 *Pressure Vessel and System Design*, Livermore, CA, LLNL, 2005. (See Internet site 3.)

10. Moss, D. R., *Pressure Vessel Design Manual*, 3rd ed., New York, Elsevier, 2003.

11. Rao, K. R. (ed.), *Companion Guide to ASME Boiler & Pressure Vessel Code*, Volume 1 (2001), Volume 2 (2006), Volume 3 (2006), New York, American Society of Mechanical Engineers.

12. Young, W. C. and R. D. Cook, *Advanced Mechanics of Materials*, 2nd ed., Upper Saddle River, NJ, Prentice Hall, 1999.

13. Young, W. C. and R. G. Budynas, *Roark's Formulas for Stress and Strain*, 6th ed., New York, McGraw-Hill, 2002.

14. Zeman, F. L., F. Rsuscher, and S. Schindler, *Pressure Vessel Design: The Direct Route*, New York, Elsevier, 2006.

INTERNET SITES

General Information on Pressure Vessels

1. American Society of Mechanical Engineers www.asme.org/ Codes/Publications/BPVC/ A listing of the ASME resources available related to the *ASME Boiler & Pressure Vessel Code*.

2. British Standards Institution www.bsi-global.com/PSS/ About/index.xalter This page is the overall description of the BSI with links to sources of information about standards issued by BSI. Most relevant to this chapter is Standard BS EN 13445, *Unfired Pressure Vessels*, a widely used standard in Europe.

3. Lawrence Livermore National Laboratory (LLNL) www.llnl. gov/es_and_h/hsm/doc_18.02/doc18-02.html Part of LLNL's Environment, Safety and Health Manual, Volume II, Part 18: Pressure/Noise/Hazardous Atmospheres, Document 18.2 *Pressure Vessel and System Design*, 2005. This document is a good overview of a rigorous design process for pressure vessels, even though it is focused on LLNL's mission. Sections 4, 5, and 6 are most relevant to design considerations.

Computer Software for Pressure Vessel Design

4. Computer Engineering, Inc. www.computereng.com/products/ advanced_pressure_vessel/ This software performs calculations in accordance with the ASME Section VIII of the ASME Boiler & Pressure Vessel Code, including the basic vessel, nozzles, heads, and accessories.

5. Algor—Center for Mechanical Design Technology www.algor .com/products/PVDesi1475/default.asp Producer of the PV/Designer software that facilitates the efficient design of pressure vessels in compliance with the ASME Boiler & Pressure Vessel Code.

6. COADE Engineering Software http://coade.com Producer of the PVElite software for pressure vessel design and the design of tall process towers in compliance with the ASME Boiler & Pressure Vessel Code. The companion CODECALC software assists in the design of heat exchangers, flanges, nozzles, and piping according to the ANSI B31.3 code.

7. Heat Transfer Consultants, Inc. www.htcsoftware.com Producer of the PVX-2007 Pressure Vessel & Heat Exchanger software.

8. Codeware, Inc. www.codeware.com Producer of the COMPRESS software for the design of pressure vessels in compliance with the ASME Boiler & Pressure Vessel Code.

9. Chempute Software www.chempute.com Supplier of software for the chemical, mechanical, and other engineering disciplines for chemical processing, power generation, and oil refining industries, including several packages dealing with pressure vessels and piping systems.

Pressure Vessel Manufacturers

The following list is a sampling of companies that offer a wide variety of services in the design and fabrication of pressure vessels, tanks, and piping systems.

10. PMF Industries, Inc. www.pmfind.com Producer of stainless steel pressure vessels for medical sterilizing and other sanitary applications; located in Williamsport, PA.

11. West Metal Works, Inc. www.westmetalworks.com Producer of pressure vessels according to the ASME code along with other containers and special fabrications; located in Buffalo, NY.

12. Bay Tank & Manufacturing Co., Inc. www.baytankfab.com Producer of ASME code pressure vessels, reactors, columns, storage tanks, stacks, rotary kilns, scrubbers, and other products for petrochemical, pharmaceutical, power generation, and other industries; located in Panama City, FL.

13. Enerfab www.enerfab.com Producer of ASME pressure vessels, columns, reactors, fermenters, piping systems, and complete processing systems using a wide variety of materials; based in Cincinnati, OH, with operations in many other locations.

14. Roy E. Hanson, Jr. Manufacturing www.hansontank.com Hanson Tank manufacturers ASME pressure vessels, water storage tanks, air receivers, propane tanks, and a wide variety of other products; located in Los Angeles, CA.

PROBLEMS

17–1. Compute the stress in a sphere having an outside diameter of 200 mm and an inside diameter of 184 mm if an internal pressure of 19.2 MPa is applied.

17–2. A large, spherical storage tank for a compressed gas in a chemical plant is 10.5 m in diameter and is made of AISI 1040 hot-rolled steel plate, 12 mm thick. What internal pressure could the tank withstand if a design factor of 4.0 based on yield strength is desired?

17–3. Titanium 6A1-4V is to be used to make a spherical tank having an outside diameter of 1200 mm. The working pressure in the tank is to be 4.20 MPa. Determine the required thickness of the tank wall if a design factor of 4.0 based on yield strength is desired.

17–4. If the tank of Problem 17–3 was made of aluminum 2014-T6 sheet instead of titanium, compute the required wall thickness. Which design would weigh less?

17–5. Compute the hoop stress in the walls of a 10-in schedule 40 steel pipe if it carries water at 150 psi.

17–6. A pneumatic cylinder has a bore of 80 mm and a wall thickness of 3.5 mm. Compute the hoop stress in the cylinder wall if an internal pressure of 2.85 MPa is applied.

17–7. A cylinder for carrying acetylene has a diameter of 300 mm and will hold the acetylene at 1.7 MPa. If a design factor of 4 is desired based on yield strength, compute the required wall thickness for the tank. Use AISI 1040 cold-drawn steel.

17–8. The companion oxygen cylinder for the acetylene discussed in Problem 17–7 carries oxygen at 15.2 MPa. Its diameter is 250 mm. Compute the required wall thickness using the same design criteria.

17–9. A propane tank for a recreation vehicle is made of AISI 1040 hot-rolled steel, 2.20 mm thick. The tank diameter is 450 mm. Determine what design factor would result based on yield strength if propane at 750 kPa is put into the tank.

17–10. The supply tank for propane at the distributor is a cylinder having a diameter of 1800 mm. If it is desired to have a design factor of 4 based on yield strength using AISI 1040 hot-rolled steel, compute the required thickness of the tank walls when the internal pressure is 750 kPa.

17–11. Oxygen on a spacecraft is carried at a pressure of 70.0 MPa in order to minimize the volume required. The spherical vessel has an outside diameter of 250 mm and a wall thickness of 18 mm. Compute the maximum tangential and radial stresses in the sphere.

17–12. Compute the maximum longitudinal, hoop, and radial stresses in the wall of a standard $\frac{1}{2}$-in schedule 40 steel pipe when carrying an internal pressure of 1.72 MPa (250 psi).

17–13. The barrel of a large field-artillery piece has a bore of 220 mm and an outside diameter of 300 mm. Compute the magnitude of the hoop stress in the barrel at points 10 mm apart from the inside to the outside surfaces. The internal pressure is 50 MPa.

17–14. A $1\frac{1}{2}$-in schedule 40 steel pipe has a mean radius less than 10 times the wall thickness and thus should be classified as a thick-walled cylinder. Compute what maximum stresses would result from both the thin-walled and the thick-walled formulas due to an internal pressure of 10.0 MPa.

17–15. A cylinder has an outside diameter of 50 mm and an inside diameter of 30 mm. Compute the maximum tangential stress in the wall of the cylinder due to an internal pressure of 7.0 MPa.

17–16. For the cylinder of Problem 17–15, compute the tangential stress in the wall at increments of 2.0 mm from the inside to the outside. Then plot the results for stress versus radius.

17–17. For the cylinder of Problem 17–15, compute the radial stress in the wall at increments of 2.0 mm from the inside to the outside. Then plot the results for stress versus radius.

17–18. For the cylinder of Problem 17–15, compute the tangential stress that would have been predicted if thin-walled theory were used instead of thick-walled theory. Compare the result with the stress found in Problem 17–15.

17–19. A sphere is made from stainless steel, AISI 501 OQT 1000. Its outside diameter is 500 mm and the wall thickness is 40 mm. Compute the maximum pressure that could be placed in the sphere if the maximum stress is to be one-fourth of the yield strength of the steel.

17–20. A sphere has an outside diameter of 500 mm and an inside diameter of 420 mm. Compute the tangential stress in the wall at increments of 5.0 mm from the inside to the outside. Then plot the results. Use a pressure of 100 MPa.

17–21. A sphere has an outside diameter of 500 mm and an inside diameter of 420 mm. Compute the radial stress in the wall at increments of 5.0 mm from the inside to the outside. Then plot the results. Use a pressure of 100 MPa.

17–22. To visualize the importance of using the thick-walled formulas for computing stresses in the walls of a cylinder, compute the maximum predicted tangential stress in the wall of a cylinder from both the thin-walled and thick-walled formulas for the following conditions. The outside diameter for all designs is to be 400 mm. The wall thickness is to vary from 5.0 mm to 85.0 mm in 10.0 mm increments. Use a pressure of 10.0 MPa. Then compute the ratio of D_m/t and plot percent difference between the stress from the thick-walled and thin-walled theory versus that ratio. Note the increase in the percent difference as the value of D_m/t decreases, that is, as t increases.

17–23. A sphere has an outside diameter of 400 mm and an inside diameter of 325 mm. Compute the variation of the tangential stress from the inside to the outside in increments of 7.5 mm. Use a pressure of 10.0 MPa.

17–24. A sphere has an outside diameter of 400 mm and an inside diameter of 325 mm. Compute the variation of the radial stress from the inside to the outside in increments of 7.5 mm. Use a pressure of 10.0 MPa.

17–25. Appendix A–12 lists the dimensions of American National Standard schedule 40 steel pipe. Which of the pipe sizes should be classified as thick-walled and which can be considered thin-walled?

17–26. Design a cylindrical pressure vessel to carry compressed air for a self-contained breathing apparatus for use by firefighters when operating in smoke-filled buildings. The minimum inside diameter is to be 6.00 in and the length of the cylindrical portion of the tank is to be 15.0 in. It must withstand a service pressure of 450 psi. Use a design stress of $s_u/8$ to account for a large number of pressurization cycles. Also, check the final design for its ability to withstand a maximum pressure of 900 psi by computing the design factor based on yield strength. The tank is to be made from aluminum alloy 6061-T6. Compute the weight of just the cylindrical portion.

17–27. Repeat Problem 17–26 but use titanium Ti-6A1-4V.

17–28. Repeat Problem 17–26 but use stainless steel 17-4PH H900.

17–29. For any of the designs for the SCBA air cylinder for Problems 17–26, 17–27, or 17–28, make a sketch of the complete tank by placing hemispherical heads on each end. Show a port at one end for attaching the pressure regulator and discharge device. Assuming that the wall thickness of the heads is the same as the wall thickness of the cylindrical portion, compute the approximate weight of the complete tank.

17–30. Repeat Problem 17–26 but use the graphite/epoxy composite material listed in Table 8–13 in Chapter 8 that has a tensile strength of 278 ksi. Check the final design by computing the design factor with respect to tensile strength against the maximum pressure of 900 psi. The tank will be lined with a thin polymeric shell and wrapped fully with the unidirectional composite in a circumferential pattern to resist the hoop stress in the cylinder. Neglect the contribution of the liner in the stress analysis and in the weight calculation. (Note that the cylindrical shell would also likely require some plies of the composite to be placed in a helical fashion to resist the longitudinal stress and to permit the formation of the dome-shaped ends. Therefore, the final weight will be somewhat higher than computed for just the circumferentially wrapped part.)

17–31. Design a spherical tank to carry oxygen at a pressure of 3000 psi with an internal diameter of 18.0 in. Use stainless steel AISI 501 OQT 1000 and a design factor of 6 based on the ultimate strength. Compute the weight of the tank.

17–32. Repeat Problem 17–31 but use aluminum alloy 7075-T6.

17–33. Repeat Problem 17–31 but use titanium alloy Ti-6A1-4V.

17–34. Design a cylindrical tank for compressed natural gas at a pressure of 4.20 MPa. The minimum inside diameter is to be 450 mm. Use aluminum alloy 6061-T6 and a design factor of 8 based on the ultimate strength.

17–35. Design a cylindrical tank for compressed air that will be used to provide remote service for truck tire repair. The air pressure will be 300 psi. The minimum internal diameter for the tank is to be 24 in. Use AISI 1040 cold-drawn steel and a design factor of 8 based on the ultimate strength. Check the final design for a maximum pressure of 600 psi by computing the design factor based on yield strength.

COMPUTER ASSIGNMENTS

1. Write a program or spreadsheet to compute the tangential stress in the wall of a thin-walled sphere. Include the computation of the mean diameter and the ratio of mean diameter to thickness to verify that it is thin-walled.

2. Write a program or spreadsheet to compute the tangential stress in the wall of a thin-walled cylinder. Include the computation of the mean diameter and the ratio of mean diameter to thickness to verify that it is thin-walled.

3. Write a program or spreadsheet to compute the longitudinal stress in the wall of a thin-walled cylinder. Include the computation of the mean diameter and the ratio of mean diameter to thickness to verify that it is thin-walled.

4. Combine the programs or spreadsheets of Assignments 2 and 3.

5. Combine the programs or spreadsheets of Assignments 1, 2, and 3, and let the user specify whether the vessel is a cylinder or a sphere.

6. Rewrite the programs or spreadsheets of Assignments 1, 2, and 5 so that the objective is to compute the required wall thickness for the pressure vessel to produce a given maximum stress for a given internal pressure.

7. Write a program or spreadsheet to compute the maximum longitudinal, hoop, and radial stress in the wall of a thick-walled cylinder using the formulas from Table 17–1.

8. Write a program or spreadsheet to compute the tangential stress at any radius within the wall of a thick-walled cylinder using the formulas from Table 17–1.

9. Write a program or spreadsheet to compute the radial stress at any radius within the wall of a thick-walled cylinder using the formulas from Table 17–1.

10. Write a program or spreadsheet to compute the tangential stress at any radius within the wall of a thick-walled sphere using the formulas from Table 17–1.

11. Write a program or spreadsheet to compute the radial stress at any radius within the wall of a thick-walled sphere using the formulas from Table 17–1.

12. Combine the programs or spreadsheets of Assignments 8 through 11.

13. Write a program or spreadsheet to compute the tangential stress distribution within the wall of a thick-walled cylinder using the formulas from Table 17–1. Start at the inside radius and specify a number of increments between the inside and the outside.

14. Write a program or spreadsheet to compute the radial stress distribution within the wall of a thick-walled cylinder using the formulas from Table 17–1. Start at the inside radius and specify a number of increments between the inside and the outside.

15. Write a program or spreadsheet to compute the tangential stress distribution within the wall of a thick-walled sphere using the formulas from Table 17–1. Start at the inside radius and specify a number of increments between the inside and the outside.

16. Write a program or spreadsheet to compute the radial stress distribution within the wall of a thick-walled sphere using the formulas from Table 17–1. Start at the inside radius and specify a number of increments between the inside and the outside.

17. Write a program or spreadsheet to perform the computations of the type called for in Problem 17–22.

18. Write a program or spreadsheet to perform the computations of the type called for in Problem 17–22, except do it for a sphere.

19. Write a program or spreadsheet to compute the maximum tangential stress in any standard schedule 40 pipe for a given internal pressure. Include a table of data for the dimensions of the pipe sizes listed in Appendix A–12. Include a check to see if the pipe is thick-walled or thin-walled.

CONNECTIONS

MAJOR TOPICS

18–1 The Big Picture—Connections

18–2 Modes of Failure for Bolted Joints

18–3 Design of Bolted Connections

18–4 Riveted Joints

18–5 Eccentrically Loaded Riveted and Bolted Joints

18–6 Welded Joints with Concentric Loads

18–1 THE BIG PICTURE— CONNECTIONS

Discussion Map

- Load-carrying members that make up structures and machines must act *together* to perform their desired functions. After completing the design or analysis of the primary members, it is necessary to specify suitable connections between them. As their name implies, connections provide the linkage between members.

- Structures and mechanical devices rely on the connections between load-carrying elements to maintain the integrity of the assemblies. The connections provide the path by which loads are transferred from one element to another.

- Three common types of connections are riveting, welding, and bolting. Figure 18–1 shows a bulk storage hopper supported by rectangular straps from a tee beam. During fabrication of the hopper, the support tabs were welded to the outside of the side walls. The tabs contain a pattern of holes, allowing the straps to be bolted on at the assembly site. Prior to installation of the tee beam, the straps were riveted to the web.

- The load due to the weight of the hopper and the materials in it must be transferred from the hopper walls into the tabs through the welds. Then the bolts transfer the load to the straps, which act as tension members. Finally, the rivets transfer the load into the tee.

Activity Find some examples in structures, machines, vehicles, or consumer products where fasteners play a major role. Try to find at least ten examples that include rivets, welds, or bolted joints. In addition to the hopper shown in Figure 18–1, here are a few examples to get you started.

- The bolts that hold the alternator, air conditioning compressor, and other accessories to the engine of your car. Each of these fasteners is critical to maintaining the location of the device with respect to the engine block and to the mating parts. The engine components themselves are fastened together with a variety of bolts that are very highly stressed and that must be tightened with a calibrated torque wrench. The bolts provide primarily a clamping function to ensure that the mating parts do not separate or move during loading. They are usually loaded in some combination of tension and shear.

- The rivets that hold the steps of a stepladder to the side rails. These critical fasteners ensure that the force transferred from your body weight to the steps is transferred safely into the side rails and thus to the ground. The other functional parts of the ladder are also typically fastened with tubular rivets that are inserted through holes in both parts and then upset with a tool that spreads the head against a solid surface. Rivets usually carry loads in shear.

- The welds that hold the various parts of a bicycle frame together. The frame members are usually tubular steel or some lighter metal that must be welded at the joints to obtain an integral, stiff, strong structure. The welds are subjected to combinations of bending, torsion, and direct shear as the cyclist executes maneuvers. Notice the detail differences in the frame among the touring, racing, and mountain bike styles.

- Structural fasteners in building construction. Find a construction site where a steel-framed building is being erected and look specifically at how the columns, beams, and other structural members are connected using bolting or welding. Examine the type of fasteners used, the arrangement of bolts in multiple-bolt joints, weld line patterns, and the design of the connected members. Figure 18–2 is a photograph of a sculpture that illustrates many of the types of structural joints used for instruction of students in civil engineering and technology, construction, and architecture. Notice

FIGURE 18–1   Illustration of three types of joints—riveted, bolted, and welded.

the variety of angle clips, added plates, and modifications to the main structural members that facilitate fastening.

Now, what other examples are on your list? Discuss them with your colleagues and the instructor and find out what different examples they have found.

Search the Internet

How much do you currently know about connections, fasteners, welding, and joining? This chapter will provide useful information on the functions of connections, their modes of failure, riveted connections, bolted connections, and welded joints. Before getting into the technical details, check out some of the Internet sites listed at the end of this chapter so you can get a good overview of

- The technology of connection design and fabrication
- The standards that guide connection design
- Software that provides computer-assisted design and analysis of bolted joints
- A sampling of manufacturers of many types of fasteners used in building construction, automotive, aerospace, consumer products, agricultural, industrial, furniture, and other applications
- Welding technology and the design of welded joints

Objectives of This Chapter

The primary objective of this chapter is to provide data and methods of analysis for the safe design of riveted joints, bolted joints, and welded joints.

After completing this chapter, you should be able to:

1. Describe the typical geometry of riveted and bolted joints.
2. Identify the probable modes of failure for a joint.
3. Recognize typical styles of rivets.
4. Identify when a fastener is in single shear or double shear.
5. Analyze a riveted or bolted joint for shearing force capacity.
6. Analyze a riveted or bolted joint for tensile force capacity.
7. Analyze a riveted or bolted joint for bearing capacity.
8. Use the allowable stresses for steel structural connections as published by the American Institute of Steel Construction (AISC).
9. Describe the difference between a friction-type connection and a bearing-type connection and complete the appropriate analysis.
10. Use the allowable stresses for aluminum structural connections as published by the Aluminum Association.

FIGURE 18–2   Sculpture illustrating numerous methods of connecting structural members.
Source: University of Dayton School of Engineering, Dayton, OH.

11. Analyze both symmetrically loaded joints and eccentrically loaded joints.

12. Analyze welded joints with concentric loads.

18–2 MODES OF FAILURE FOR BOLTED JOINTS

Figure 18–3 shows simple lap joints in which two plates are connected with two bolts and nuts. The function of the joint is to clamp the plates together and to transfer a load applied to one plate through the bolt to the other plate. Follow the load path and visualize the types of stress created:

1. From plate 1, the load passes to the side surfaces of the bolts.

2. The bolts bear on the holes, tending to crush the plate material [Figure 18–3(a)].

3. The load passes through the bolt and into plate 2, producing bearing stress on the holes.

4. The opposing forces acting on plates 1 and 2 tend to cut (shear) the bolt at the interface between the two plates [Figure 18–3(b)].

5. The tensile forces on plates 1 and 2 tend to tear the material across the section with minimum area to resist the tensile force. That occurs through the section where the bolt holes are located [Figure 18–3(c)].

6. As the bolts bear on the side surfaces of the holes in the plates, there is a tendency to tear out the material from the bolt to the edge of the material in either plate [Figure 18–3(d)].

Properly designed connections should have an edge distance from the centerline of the bolt to the edge of the plate being joined of at least three times the diameter of the bolt. The edge distance is measured in the direction of the bearing pressure. If this recommendation is heeded, then end tearout should not occur. Although this should be checked with analysis, it will be assumed to apply to example problems in this chapter. Thus, shear, bearing, and tensile failure modes only will be considered in evaluating joint strength.

Material crushed by rivets

t_1

w

t_2

(a) Bearing failure

(b) Shear of rivets

(c) Tensile failure

(d) End tearout

FIGURE 18–3 Types of failure of bolted connections.

18–3 DESIGN OF BOLTED CONNECTIONS

Two types of bolted connections are used in building construction:

- **Bearing-Type Connections:** It is assumed that the joined plates are *not* clamped tightly enough to cause frictional forces between the plates to transmit loads. Therefore, the bolts do bear on the holes, and bearing failure must be investigated. Both shear failure and tensile failure could also occur.

- **Slip-Critical Connections:** High clamping forces are produced in this type of joint to prevent slipping, and careful fabrication techniques are required to ensure that friction forces do share in the transmission of connection forces.

The design of slip-critical connections involves many variations and steps. Users are advised to see References 2 and 3 for details. Only bearing-type connections are discussed in this chapter.

Table 18–1 shows sample data for allowable stresses on bolted connections for structural steel, listing three types of

bolts: ASTM A307, ASTM A325, and ASTM A490, with progressively increased strengths. Note the difference in allowable shearing stress, depending on whether or not threads are included in the shear plane. Figure 18–4 shows a standard hex-head bolt with a portion of its length threaded. It is preferred to design the joint so that the full-diameter part of the bolt is in the shear plane.

The methods of analyzing shear, bearing, and tensile failure modes are described next.

Shear Failure

The bolt body is assumed to be in direct shear when a tensile load is applied to a joint, provided that the line of action of the load passes through the centroid of the pattern of bolts. It is also assumed that the total applied load is shared equally among all the bolts. The capacity of a joint with regard to shear of the bolts is

$$F_s = \tau_a A_s \qquad (18–1)$$

where F_s = capacity of the joint in shear
τ_a = allowable shear stress in bolts
A_s = area in shear

TABLE 18–1	Allowable Stresses for Steel Structural Connections for Bearing-Type Connections[*]					
	Allowable shear stress				Allowable tensile stress	
	No threads in shear plane		Threads in shear plane			
Bolts	ksi	MPa	ksi	MPa	ksi	MPa
ASTM A307	12.0	82.5	12.0	82.5	22.5	155
ASTM A325	30.0	207	24.0	165	45.0	310
ASTM A490	37.5	260	30.0	207	56.5	390
Connected members			Allowable bearing stress		Allowable tensile stress[‡]	
All alloys			$1.20s_u$		$0.6s_y$	

[*]AISC specifications.

[‡]See Appendix A–16 for structural steels.

(a) Bolt (b) Bolted joint

FIGURE 18–4 Bolted joint.

The area in shear is dependent on the number of cross sections of bolts available to resist shear. Calling this number N_s,

$$A_s = \frac{N_s \pi D^2}{4} \qquad (18\text{–}2)$$

where D is the bolt diameter.

To determine N_s, it must be observed whether *single shear* or *double shear* exists in the joint. Figure 18–3 shows an example of single shear. Only one cross section of each bolt resists the applied load. Then N_s is equal to the number of bolts in the joint. The straps used to support the hopper in Figure 18–1 place the rivets and bolts in double shear. Two cross sections of each bolt resist the applied load. Then N_s is twice the number of bolts in the joint.

Bearing Failure

When a cylindrical bolt bears against the wall of a hole in the plate, a nonuniform pressure exists between them. As a simplification of the actual stress distribution, it is assumed that the area in bearing, A_b, is the rectangular area found by multiplying the plate thickness t by the diameter of the bolt D. This can be considered to be the *projected area* of the bolt hole. Then the bearing capacity of a joint is

$$F_b = \sigma_{ba} A_b \qquad (18\text{–}3)$$

where F_b = capacity of the joint in bearing
σ_{ba} = allowable bearing stress
A_b = bearing area = $N_b D t$ $\qquad$ (18–4)
N_b = number of bearing surfaces
t = plate thickness

The allowable bearing stress is typically based on the yield strength of the connected material because the fastener usually has higher strength. This should be checked.

Tensile Failure

A direct tensile force applied through the centroid of the bolt pattern would produce a tensile stress. Then the capacity of the joint in tension would be

$$F_t = \sigma_{ta} A_t \qquad (18\text{–}5)$$

where F_t = capacity of the joint in tension
σ_{ta} = allowable stress in tension
A_t = net tensile area

The evaluation of A_t requires the subtraction of the diameter of all the holes from the width of the plates being joined. Then

$$A_t = (w - N D_H) t \qquad (18\text{–}6)$$

where w = width of the plate
D_H = hole diameter (in structures use
$D_H = D + \frac{1}{16}$ in or $D + 1.6$ mm)
N = number of holes at the section of interest
t = thickness of the plate

Example Problem 18–1

For the single lap bearing-type joint shown in Figure 18–3, determine the allowable load on the joint if the two plates are $\frac{1}{4}$ in thick by 2 in wide and joined by two steel bolts, $\frac{3}{8}$ in diameter, ASTM A490. The plates are ASTM A36 structural steel.

Solution Objective: Compute the allowable load on the joint.

Given: Plate thickness = t = 0.25 in; plate width = w = 2.00 in
Plates are ASTM A36 structural steel. s_u = 58 ksi, s_y = 36 ksi
Bolts: Diameter = D = 0.375 in; ASTM A490
Bearing type connection; no threads in the shear plane

Analysis: Possible failure in shear, bearing, and tension will be investigated. The lowest of the three values is the limiting load on the joint.

Results: **Shear Failure**

$$F_s = \tau_a A_s \tag{18-1}$$

$$\tau_a = 37\,500 \text{ psi} \qquad (\text{Table } 18-1)$$

$$A_s = \frac{N_s \pi D^2}{4} = \frac{2\pi(0.375 \text{ in})^2}{4} = 0.221 \text{ in}^2 \tag{18-2}$$

Then

$$F_s = (37\,500 \text{ lb/in}^2)(0.221 \text{ in}^2) = 8288 \text{ lb}$$

Bearing Failure

$$F_b = \sigma_{ba} A_b \tag{18-3}$$

$$\sigma_{ba} = 1.20(58\,000 \text{ psi}) = 69\,600 \text{ psi}$$

$$A_b = N_b D t = (2)(0.375 \text{ in})(0.25 \text{ in}) = 0.188 \text{ in}^2 \tag{18-4}$$

Then

$$F_b = (69\,600 \text{ lb/in}^2)(0.188 \text{ in}^2) = 13\,050 \text{ lb}$$

Tensile Failure

$$F_t = \sigma_{ta} A_t \tag{18-5}$$

$$\sigma_{ta} = 0.6(36\,000 \text{ psi}) = 21\,600 \text{ psi}$$

$$A_t = (W - ND_H)t = [2.0 \text{ in} - 2(0.375 + 0.063) \text{ in}](0.25 \text{ in}) = 0.281 \text{ in}^2 \tag{18-6}$$

Then

$$F_t = (21\,600 \text{ lb/in}^2)(0.281 \text{ in}^2) = 6070 \text{ lb}$$

Comment: Now the capacity in tension is the lowest, so the capacity of the joint is 6070 lb.

18–4 RIVETED JOINTS

Riveted joints are similar to those shown in Figure 18–3 except that the fastener takes one of the forms shown in Figure 18–5. The cylindrical body of the rivet is inserted into the holes in the members to be connected. With the preformed head held firmly from one side of the joint, the opposite end is pressed or hammered to form a head on the opposite side, clamping the members together. The holes for the rivets are typically quite close in size to the body diameter and the upsetting action during installation causes the body to swell, filling the hole and preventing relative motion between the connected members and the rivets. The blind rivet is unique in that the body part is inserted into the holes from one side and a special tool pulls on the mandrel, causing the head on the opposite side to spread out and grip the connected members. After forming, the mandrel snaps off and is discarded. The advantages of this technique are that it requires access to only one side of the connected members and the complete installation is quite rapid.

FIGURE 18–5   Examples of rivet styles.

Button head   Brazier head   Flat head

Flat countersunk head   Oval countersunk head   Countersunk, semitubular   Oval semitubular   Blind rivet

TABLE 18–2   Allowable Shear Force Capacity of Typical Rivets [Force in lb and (N)]

Body diameter [in (mm)]	Material			
	Aluminum	Carbon steel	Stainless steel	MONEL®
3/32 (2.38)	70 (311)	130 (578)	230 (1023)	200 (890)
1/8 (3.18)	120 (534)	260 (1156)	420 (1868)	350 (1557)
5/32 (3.97)	190 (845)	370 (1646)	650 (2891)	550 (2446)
3/16 (4.76)	260 (1156)	540 (2402)	950 (4226)	800 (3558)
1/4 (6.35)	460 (2046)	700 (3114)	1700 (7562)	1400 (6227)

The basic analysis methods for riveted joints are similar to those described for bolted joints. Shear, bearing, and tensile failure modes must be analyzed. Equations (18–1) to (18–6) can be applied as illustrated in Example Problem 18–1. However, for tensile failure of the connected members, the hole diameter is taken to be equal to the body diameter of the rivet.

Table 18–2 shows representative data for the allowable shear force capacity of rivets made from aluminum, carbon steel, stainless steel, and MONEL® (a nickel alloy with excellent corrosion resistance often used in marine and chemical plant applications; MONEL is a registered trademark of Special Metals Corporation). These data will be used in this book. However, strength data from specific suppliers should be sought for critical designs. See Internet site 13. Table 18–3 shows additional data for aluminum rivets, bolts, and connected members.

TABLE 18–3   Allowable Stresses for Aluminum Structural Connections for Building-Type Structures

Rivets			
Alloy and temper before driving*		Allowable shear stress	
		ksi	MPa
1100-H14		4.1	28
2017-T4		14	96
6053-T61		8.5	58
6061-T6		10.5	72

Bolts					
	Allowable shears stress[†]			Allowable tensile stress[†]	
Alloy and temper	ksi	MPa		ksi	MPa
2024-T4	16	110		26	179
6061-T6	11	76		18	124
7075-T73	18	124		29	200

Connected members		
Alloy and temper	Allowable bearing stress	
	ksi	MPa
1100-H14	16	110
2014-T6	62	427
3003-H14	21	145
6061-T6	39	269
6063-T6	31	214

Source: Aluminum Association, *Aluminum Design Manual.* Washington, D.C., 2000.

*All cold-driven.

[†]Stresses are based on the area corresponding to the nominal diameter of the bolt unless the threads are in the shear plane. Then the shear area is based on the root diameter.

18-5 ECCENTRICALLY LOADED RIVETED AND BOLTED JOINTS

Previously considered joints were restricted to cases in which the line of action of the load on the joint passed through the centroid of the pattern of rivets or bolts. In such cases, the applied load is divided equally among all the fasteners. When the load does not pass through the centroid of the fastener pattern, it is called an *eccentrically loaded joint,* and a nonuniform distribution of forces occurs in the fasteners.

In eccentrically loaded joints, the effect of the moment or couple on the fastener must be considered. Figure 18–6 shows a bracket attached to the side of a column and used to support an electric motor. The net downward force exerted by the weight of the motor and the belt tension act at a distance *a* from the center of the column flange. Then the total force system acting on the bolts of the bracket consists of the direct shearing force P plus the forces produced by the moment $P \times a$. Each of these components can be considered separately and then added together by using the principle of superposition.

Figure 18–7(a) shows that for the direct shearing force P, each bolt is assumed to carry an equal share of the load, just as in concentrically loaded joints. But in part (b) of the figure, because of the moment, each bolt is subjected to a force acting perpendicular to a radial line from the centroid to the bolt pattern. It is assumed that the magnitude of the force in a bolt due to the moment load is proportional to its distance r from the centroid. This magnitude is

$$R_i = \frac{Mr_i}{\sum r^2} \qquad (18\text{–}7)$$

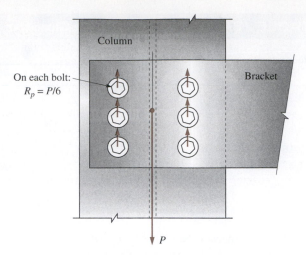

On each bolt:
$R_p = P/6$

(a) Forces resisting P, the shearing force

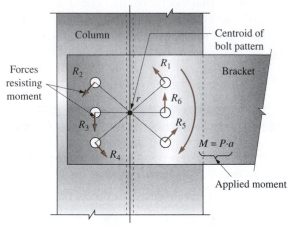

(b) Forces resisting the moment

FIGURE 18–7   Loads on bolts that are eccentrically loaded.

where R_i = shearing force in bolt i due to the moment M
r_i = radial distance to bolt i from the centroid to the bolt pattern
$\sum r^2$ = sum of the radial distances to *all* bolts in the pattern squared

If it is more convenient to work with horizontal and vertical components of forces, they can be found from

$$R_{ix} = \frac{My_i}{\sum r^2} = \frac{My_i}{\sum (x^2 + y^2)} \qquad (18\text{–}8)$$

$$R_{iy} = \frac{Mx_i}{\sum r^2} = \frac{Mx_i}{\sum (x^2 + y^2)} \qquad (18\text{–}9)$$

where y_i = vertical distance to bolt i from centroid
x_i = horizontal distance to bolt i from centroid
$\sum (x^2 + y^2)$ = sum of horizontal and vertical distances squared for all bolts in the pattern

Finally, all horizontal forces are summed and all vertical forces are summed for any particular bolt. Then the resultant of the horizontal and vertical forces is determined.

FIGURE 18–6   Eccentric load on a bolted joint.

Example Problem 18–2

In Figure 18–6, the net downward force P is 26.4 kN on each side plate of the bracket. The distance a is 0.75 m. Determine the required size of ASTM A325 bolts to secure the bracket. Consider the connection to be of the bearing type with no threads in the shear plane.

Solution **Objective:** Specify the size of the bolts in the joint.

Given: Load $= P = 26.4$ kN downward. Moment arm $= a = 0.75$ m.
Bolt pattern in Figure 18–6. Bolts: ASTM A325
Bearing-type connection.

Analysis: To determine the shearing force in each bolt to carry the direct vertical shearing force of $P = 26.4$ kN, each of the six bolts will be assumed to carry an equal share of the load. Then Equations (18–8) and (18–9) will be used to compute the forces in the most highly stressed bolt to resist the moment load, where

$$M = P \times a$$

The resulting forces will be combined vectorially to determine the resultant load on the most highly stressed bolt. Then the required size of that bolt will be computed based on the allowable shearing stress for the ASTM A325 bolts.

Results: *Direct Shearing Force*
The total downward shearing force is divided among six bolts. Then the load on each bolt, called R_p, is

$$R_p = \frac{P}{6} = \frac{26.4\ \text{kN}}{6} = 4.4\ \text{kN}$$

Figure 18–7(a) shows this to be an upward reaction force on each bolt.

Forces Resisting the Moment. In Equations (18–8) and (18–9), the following term is required:

$$\sum(x^2 + y^2) = 6(100\ \text{mm})^2 + 4(75\ \text{mm})^2 = 82\,500\ \text{mm}^2$$

The moment on the joint is

$$M = P \times a = 26.4\ \text{kN}\,(0.75\ \text{m}) = 19.8\ \text{kN} \cdot \text{m}$$

Starting first with bolt 1 at the upper right (see Figure 18–8),

$$R_{1x} = \frac{My_1}{\sum(x^2 + y^2)} = \frac{19.8\ \text{kN} \cdot \text{m}(75\ \text{mm})}{82\,500\ \text{mm}^2} \times \frac{10^3\ \text{mm}}{\text{m}}$$

$$R_{1x} = 18.0\ \text{kN} \leftarrow (\text{acts toward the left})$$

$$R_{1y} = \frac{Mx_1}{\sum(x^2 + y^2)} = \frac{(19.8\ \text{kN} \cdot \text{m})(100\ \text{mm})}{82\,500\ \text{mm}^2} \times \frac{10^3\ \text{mm}}{\text{m}}$$

$$R_{1y} = 24.0\ \text{kN} \uparrow (\text{acts upward})$$

FIGURE 18–8 Forces on each bolt.

Now the resultant of these forces can be found. In the vertical direction, R_p and R_{1y} both act upward.

$$R_p + R_{1y} = 4.4\,\text{kN} + 24.0\,\text{kN} = 28.4\,\text{kN}$$

Only R_{1x} acts in the horizontal direction. Calling the total resultant force on bolt 1, R_{t1}.

$$R_{t1} = \sqrt{28.4^2 + 18.0^2} = 33.6\,\text{kN}$$

Investigating the other five bolts in a similar manner would show that bolt 1 is the most highly stressed. Then its diameter will be determined to limit the shear stress to 207 MPa (30.0 ksi) for the ASTM A325 bolts, as listed in Table 18–1.

$$\tau = \frac{R_{t1}}{A}$$

$$A = \frac{R_{t1}}{\tau_a} = \frac{33.6\,\text{kN}}{207\,\text{N/mm}^2} = 162\,\text{mm}^2 = \frac{\pi D^2}{4}$$

$$D = \sqrt{\frac{4A}{\pi}} = \sqrt{\frac{4(162)\,\text{mm}^2}{\pi}} = 14.4\,\text{mm}$$

The nearest preferred metric size is 16 mm. If conventional inch units are to be specified,

$$D = 14.4\,\text{mm} \times \frac{1\,\text{in}}{25.4\,\text{mm}} = 0.567\,\text{in}$$

The nearest standard size is $\frac{5}{8}$ in (0.625 in).

Comment: Specify either $D = 16$ mm or $D = \frac{5}{8}$-in bolts.

18–6 WELDED JOINTS WITH CONCENTRIC LOADS

Welding is a joining process in which heat is applied to cause two pieces of metal to become metallurgically bonded. The heat may be applied by a gas flame, an electric arc, a laser beam or by a combination of electric resistance heating and pressure.

Types of welds include butt, groove, fillet, and spot welds (as shown in Figure 18–9), and others. Groove and fillet welds are most frequently used in structural connections since they are readily adaptable to the shapes and plates that make up the structures. Spot welds are used for joining relatively light-gage steel sheets and cold-formed shapes.

The variables involved in designing welded joints are the shape and size of the weld, the choice of filler metal, the length of weld, and the position of the weld relative to the applied load.

Fillet welds are assumed to have a slope of 45 degrees between the two surfaces joined, as shown in Figure 18–9(b). The size of the weld is denoted as the height of one side of the triangular-shaped fillet. Typical sizes range from $\frac{1}{8}$ in to $\frac{1}{2}$ in in steps of $\frac{1}{16}$ in. The stress developed in fillet welds is assumed to be *shear stress* regardless of the direction of application of the load. The maximum shear stress would occur at the throat of the fillet (see Figure 18–9), where the thickness is 0.707 times the nominal size of the weld. Then the shearing stress in the weld due to a load P is

$$\tau = \frac{P}{Lt} \qquad (18\text{–}10)$$

where L = the length of weld

t = the thickness at the throat

Equation (18–10) is used *only* for concentrically loaded members. This requires that the line of action of the force on the welds passes through the centroid of the weld pattern. Eccentricity of the load produces a moment, in addition to the direct shearing force, which must be resisted by the weld metal. References 2, 3, 6, 11, and 15 at the end of this chapter contain pertinent information with regard to eccentrically loaded welded joints.

In electric arc welding, used mostly for structural connections, a filler rod is normally used to add metal to the welded zone. As the two parts to be joined are heated to a molten state, filler metal is added, which combines with the base metal. On cooling, the resulting weld metal is normally stronger than the original base metal. Therefore, a properly designed and made welded joint should fail in the base metal rather than in the weld.

In structural welding, the electrodes are given a code beginning with an E and followed by two or three digits, such as E60, E80, or E100. The number denotes the ultimate tensile strength in ksi of the weld metal in the rod. Thus, an E80 rod would have a tensile strength of 80 000 psi. Other digits may be added to the code number to denote special properties. The allowable shear stress for fillet welds using electrodes is 0.3 times the tensile strength of the electrode according to AISC. Table 18–4 lists some common electrodes and their allowable stresses.

Aluminum products are welded using either the inert gas-shielded arc process or the resistance welding process. For the inert gas-shielded arc process, filler alloys are specified by the Aluminum Association for joining particular base metal alloys, as indicated in Table 18–5. The allowable shear stresses for such welds are also listed. It should be noted that the heat of welding lowers the properties of most aluminum alloys within 1.0 in (25 mm) of the weld, and allowance for this must be made in the design of welded assemblies.

See Reference 4 for additional data and application considerations for welded joints.

FIGURE 18–9 Types of welds.

(a) Butt weld with single bevel groove

(b) Fillet weld

(c) Spot weld

TABLE 18–4 Properties of Welding Electrodes for Steel

| Electrode type | Minimum tensile strength | | Allowable shear stress | | Typical metals joined |
	ksi	MPa	ksi	MPa	
E60	60	414	18	124	A36, A500
E70	70	483	21	145	A572 Gr. 50
					A913 Gr. 50, A992
E80	80	552	24	165	A913 Gr. 65

TABLE 18–5 Allowable Shear Stresses in Fillet Welds in Aluminum Building-Type Structures

| Base metal | Filler alloy | | | | | | | |
| | 1100 | | 4043 | | 5356 | | 5556 | |
	ksi	MPa	ksi	MPa	ksi	MPa	ksi	MPa
1100	3.2	22	4.8	33	—	—	—	—
3003	3.2	22	5.0	34	—	—	—	—
6061	—	—	5.0	34	7.0	48	8.5	59
6063	—	—	5.0	34	6.5	45	6.5	45

Example Problem
18–3

A lap joint is made by placing two $\frac{3}{8}$-in fillet welds across the full width of two $\frac{1}{2}$-in ASTM A36 steel plates, as shown in Figure 18–10. The shielded metal-arc method is used, using an E60 electrode. Compute the allowable load, *P*, that can be applied to the joint.

FIGURE 18–10 Welded lap joint.

Solution

Objective: Compute the allowable load, *P*, on the joint.

Given: Joint design in Figure 18–10. Plates are ASTM A36 steel.
E60 electrode used in shielded metal-arc method.

Analysis: The load is assumed to be equally distributed on all parts of the weld, so Equation (18–10) can be used with *L* = 8.0 in.

$$\tau = \frac{P}{Lt}$$

Let τ equal the allowable stress of 18 ksi, listed in Table 18–4. The thickness *t* is

$$t = 0.707(\tfrac{3}{8}\,\text{in}) = 0.265\,\text{in}$$

Now we can solve for *P*.

$$P = \tau_a Lt = (18\,000\,\text{lb/in}^2)\,(8.0\,\text{in})\,(0.265\,\text{in}) = 38\,200\,\text{lb}$$

REFERENCES

1. Aluminum Association, *Aluminum Design Manual,* Washington, DC, 2005.

2. American Institute of Steel Construction, *Specification for Structural Steel Buildings,* Chicago, IL, AISC, 2005.

3. American Institute of Steel Construction, *Allowable Stress Design Specification for Structural Joints Using ASTM A325 or A490 Bolts,* New York, American Institute of Steel Construction, 2001.

4. American Welding Society, *Standard AWS D1.1/D1.1M Structural Welding Code Steel,* Miami, FL, American Welding Society, 2006.

5. Bickford, J. H. *An Introduction to the Design and Behavior of Bolted Joints,* 4th ed., New York, Marcel Dekker, 2008.

6. Blodgett, O. W., *Design of Weldments,* Cleveland, OH, James F. Lincoln Arc Welding Foundation, 1963.

7. Brockenbrough, R. L. and F. S. Merritt, *Structural Steel Designer's Handbook,* New York, McGraw-Hill, 2005.

8. Industrial Fasteners Institute, *Fastener Standards,* 7th ed., Cleveland, OH, Industrial Fasteners Institute, 2003.

9. Kissell, J. R. and R. Ferry, *Aluminum Structures: A Guide to Their Specifications and Design,* New York, John Wiley & Sons, Inc., 2002.

10. McCormac, J. C. and J. Nelson, *Structural Steel Design—LRFD Method,* 4th ed., Upper Saddle River, NJ, Prentice Hall, 2008.

11. Mott, R. L., *Machine Elements in Mechanical Design,* 4th ed., Upper Saddle River, NJ, Prentice Hall, 2004.

12. Oberg E., F. D. Jones, and H. L. Horton, *Machinery's Handbook,* 28th ed., New York, Industrial Press, 2008.

13. Parmley, R. O., *Standard Handbook of Fastening and Joining,* 3rd ed., New York, McGraw-Hill, 1997.

14. Society of Automotive Engineers, *SAE Fastener Standards Manual—2005 Edition,* Warrendale, PA, SAE International, 2005.

15. Tamboli, A. R., *Handbook of Structural Steel Connection Design and Details,* New York, McGraw-Hill, 2000.

INTERNET SITES

Standards and Professional Associations

1. Aluminum Association www.aluminum.org The association of the aluminum industry. Source of information and documents on aluminum used for fasteners and information on the design of connections for aluminum structures.

2. American Institute of Steel Construction www.aisc.org Publisher of the *Specifications for Structural Steel Buildings* that contains extensive data on connections and procedures for designing and fabricating connections for steel buildings.

3. ASTM International www.astm.org Formerly known as the American Society for Testing and Materials. Developer of numerous standards for fasteners used in the building construction industry.

4. Industrial Fasteners Institute www.industrial-fasteners.org An association of manufacturers and suppliers of bolts, nuts, screws, rivets, and special formed parts and the materials and equipment to make them. The site includes standards, reference publications, and educational opportunities.

5. Research Council on Structural Connections (RCSC) www.boltcouncil.org An organization that stimulates and supports research on structural connections and prepares and publishes standards.

6. SAE International www.sae.org The Society of Automotive Engineers, the engineering society for advancing mobility on land or sea, in air or space. Publisher of the *SAE Handbook* and the *SAE Fastener Manual,* each containing useful information about the use of fasteners and their properties.

7. NASA Ames Research Center (ARC) www.windtunnels.arc.nasa.gov/strucjo.html Part of a larger site for ARC that lists standards for structural joints for any structure or equipment supplied for use in their wind tunnels. The list is a good overview of the design and fabrication considerations important to successful structural joints.

Software for Joint Design

8. Bolt Science www.boltscience.com Producer of software for the analysis of bolted joints. The site includes an informative *Tutorial on the Basics of Bolted Joints.*

9. Sensor Products, Inc. www.sensorproducts.com Developer of the BoltFAST software for bolted joints that includes joint analysis, thread analysis, and tightening torque.

Fastener Manufacturers

10. St. Louis Screw & Bolt Company www.stlouisscrewbolt.com Manufacturer of bolts, nuts, and washers to ASTM standards for the construction industry.

11. Nucor Fastener Division www.nucor-fastener.com Manufacturer of hex-head cap screws in SAE, ASTM, and metric grades; hex nuts; and structural bolts, nuts, and washers.

12. Nylok Corporation www.nvlok.com Manufacturer of Nylok® self-locking fasteners for automotive, aerospace, consumer products, agricultural, industrial, furniture, and many other applications.

13. The Fastener Group www.fastenergroup.com A supplier of bolts, cap screws, nuts, rivets, and numerous other types of fasteners for general industry uses.

14. SPS Technologies, Inc. www.spstech.com/unbrako Manufacturer of engineered fasteners under the Unbrako®, Flexloc®, and Durlok® brands, including socket-head cap screws, locknuts, and vibration resistant nuts and bolts for industrial machinery, automotive, and aerospace applications. The site includes catalogs and engineering data.

Welding Technology and Standards

15. American Welding Society www.aws.org Publisher of the Standard AWS D1.1/D1.1M *Structural Welding Code Steel* and many other publications related to the design of welded structures, the process of welding, and the welding industry.

16. James F. Lincoln Foundation www.jflf.org An organization that promotes education and training in welding technology. The site includes much information about welding processes, joint design, and guides to welded steel construction.

17. Miller Electric Company www.millerwelds.com A manufacturer of welding equipment and accessories. The site includes a Training/Education section with information about welding processes.

18. Hobart Institute of Welding Technology www.welding.org An educational organization that provides instruction in the performance of welding techniques. The site includes welding tips and a glossary of welding terms.

PROBLEMS

18–1. Determine the allowable loads on the joints shown in Figure P18–1. The fasteners are all carbon steel rivets having the shearing force capacity shown in Table 18–2. The plates are ASTM A36 structural steel.

18–2. Determine the allowable loads on the joints shown in Figure P18–2. The fasteners are all stainless steel rivets having the shearing force capacity shown in Table 18–2. The plates are AISI 430 stainless steel in the full hard condition.

18–3. Determine the allowable loads on the joints shown in Figure P18–1 if the fasteners are all ASTM A307 steel bolts providing a bearing-type connection. The plates are ASTM A242 HSLA corrosion-resistant structural steel.

18–4. Determine the allowable loads on the joints shown in Figure P18–2 if the fasteners are all ASTM A325 high-strength steel bolts providing a bearing-type connection. The plates are ASTM A514 quenched and tempered alloy structural steel.

(a)   (b)

(c)   (d)

FIGURE P18–1 Joints for Problems 18–1 and 18–3.

18–5. Determine the required diameter of the bolts used to attach the cantilever beam to the column as shown in Figure P18–5. Use ASTM A325 bolts. The steel plate is made from ASTM A36 structural steel and the column is made from ASTM A992 structural steel.

18–6. Design the connection of the channel to the column for the bracket shown in Figure P18–6. The channel is made from ASTM A36 structural steel and the column is made

from ASTM A992 structural steel. Specify the bolt material, the number of bolts, the bolt pattern (location and spacing), and the size of the bolts. Use the data from Table 18–1.

18–7. For the connection shown in Figure P18–1(a), assume that, instead of the two rivets, the two plates were welded across the ends of the 3-in-wide plates using $\frac{5}{16}$-in welds. The plates are ASTM A36 steel and the electric arc welding

(a)

(b)

(c)

(d)

FIGURE P18–2   Joints for Problems 18–2 and 18–4.

technique is used with E60 electrodes. Determine the allowable load on the connection.

18–8.   Determine the allowable load on the joint shown in Figure P18–2(c) if $\frac{1}{4}$-in welds using E70 electrodes were placed along both ends of both cover plates. The plates are ASTM A572 grade 50 steel.

18–9.   Design the joint at the top of the straps in Figure 18–1 if the total load in the hopper is 15.0 megagrams (Mg). The beam is a WT12×34 made of ASTM A36 steel and has a

web thickness of 10.6 mm. The clear vertical height of the web is about 275 mm. Use steel rivets and specify the pattern, number of rivets, rivet diameter, rivet material, strap material, and strap dimensions. Use data from Table 18–2.

18–10.   Design the joint at the bottom of the straps in Figure 18–1 if the total load in the hopper is 15.0 Mg. Use steel bolts and a bearing-type connection. Specify the pattern, number of bolts, bolt diameter, bolt material, strap material, and strap dimensions. You may want to coordinate the

FIGURE P18–5 Connection for Problem 18–5.

FIGURE P18–6 Connection for Problem 18–6.

strap design with the results of Problem 18–9. The design of the tab in Problem 18–11 is also affected by the design of the bolted joint.

18–11. Design the tab to be welded to the hopper for connection to the support straps, as shown in Figure 18–1. The hopper

load is 15.0 Mg. The material from which the hopper is made is ASTM A36 steel. Specify the width and thickness of the tab and the design of the welded joint. You may want to coordinate the tab design with the bolted connection called for in Problem 18–10.

LIST OF APPENDIXES

A–1   Properties of Areas

A–2   Preferred Basic Sizes

A–3   Screw Threads

A–4   Properties of Standard Wood Beams

A–5   Properties of Steel Angles (L-Shapes)

A–6   Properties of American Standard Steel Channels (C-Shapes)

A–7   Properties of Steel Wide-Flange Shapes (W-Shapes)

A–8   Properties of American Standard Steel Beams (S-Shapes)

A–9   Properties of Steel Structural Tubing (HSS-Shapes)

A–10   Properties of Aluminum Association Standard Channels

A–11   Properties of Aluminum Association Standard I-Beam Shapes

A–12   Properties of Steel Pipe

A–13   Properties of Steel Mechanical Tubing

A–14   Typical Properties of Carbon and Alloy Steels

A–15   Typical Properties of Stainless Steels and Nonferrous Metals

A–16   Properties of Structural Steels

A–17   Typical Properties of Cast Iron

A–18   Typical Properties of Aluminum Alloys

A–19   Typical Properties of Wood

A–20   Typical Properties of Selected Plastics

A–21   Design Stress Guidelines

A–22   Stress Concentration Factors

A–23   Beam Deflection Formulas for Simply Supported Beams

A–24   Beam Deflection Formulas for Cantilevers

A–25   Beam Diagrams and Deflection Formulas for Statically Indeterminate Beams

A–26   Conversion Factors

A–1 Properties of areas.*

*Symbols used are:

A = area $\qquad r$ = radius of gyration = $\sqrt{I/A}$
I = moment of inertia $\qquad J$ = polar moment of inertia
S = section modulus $\qquad Z_p$ = polar section modulus

Circle

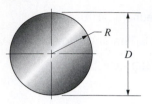

$$A = \frac{\pi D^2}{4} = \pi R^2 \qquad r = \frac{D}{4} = \frac{R}{2}$$

$$I = \frac{\pi D^2}{64} \qquad\qquad J = \frac{\pi D^4}{32}$$

$$S = \frac{\pi D^3}{32} \qquad\qquad Z_p = \frac{\pi D^3}{16}$$

Circumference = $\pi D = 2\pi R$

Hollow circle (tube)

$$A = \frac{\pi(D^2 - d^2)}{4} \qquad r = \frac{\sqrt{D^2 + d^2}}{4}$$

$$I = \frac{\pi(D^4 - d^4)}{64} \qquad J = \frac{\pi(D^4 - d^4)}{32}$$

$$S = \frac{\pi(D^4 - d^4)}{32D} \qquad Z_p = \frac{\pi(D^4 - d^4)}{16D}$$

Square

$$A = s^2$$

$$I_x = \frac{s^4}{12} \qquad\qquad r_x = \frac{s}{\sqrt{12}}$$

$$S_x = \frac{s^3}{6}$$

Rectangle

$$A = bh$$

$$I_x = \frac{bh^3}{12} \qquad S_x = \frac{bh^2}{6} \qquad r_x = \frac{h}{\sqrt{12}}$$

$$I_y = \frac{hb^3}{12} \qquad S_y = \frac{hb^2}{6} \qquad r_y = \frac{b}{\sqrt{12}}$$

Triangle

$$A = \frac{bh}{2}$$

$$I_x = \frac{bh^3}{36} \qquad r_x = \frac{h}{\sqrt{18}}$$

$$S_x = \frac{bh^2}{24}$$

Semicircle

$c = 0.288D$

$\bar{y} = 2D/3\pi = 0.212D$

$$A = \frac{\pi D^2}{8}$$

$I_x = 0.00686D^4$	$S_x = 0.0238D^3$	$r_x = 0.132D$
$I_y = 0.0245D^4$	$S_y = 0.0491D^3$	$r_y = D/4 = R/2$

Regular hexagon

$\bar{y} = 0.500h$

$c = 0.577h$

$$A = 0.866h^2$$

$I_x = 0.0601h^4$	$S_x = 0.120h^3$	$r_x = 0.264h$
$I_y = 0.0601h^4$	$S_y = 0.104h^3$	$r_y = 0.264h$

Area under a second-degree curve

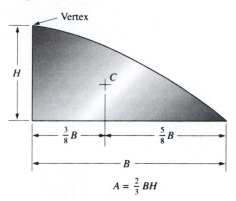

Vertex

$\frac{3}{8}B$ $\frac{5}{8}B$

B

$A = \frac{2}{3}BH$

Area over a second-degree curve

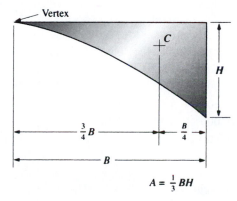

Vertex

$\frac{3}{4}B$ $\frac{B}{4}$

B

$A = \frac{1}{3}BH$

Area under a third-degree curve

$0.4B$ $0.6B$

B

$A = \frac{3}{4}BH$

Area over a third-degree curve

$0.8B$ $0.2B$

B

$A = \frac{1}{4}BH$

A–2 Preferred Basic Sizes

Fractional (in)				Decimal (in)			Metric (mm) First	Second	First	Second	First	Second
$\frac{1}{64}$	0.015 625	5	5.000	0.010	2.00	8.50	1		10		100	
$\frac{1}{32}$	0.031 25	$5\frac{1}{4}$	5.250	0.012	2.20	9.00		1.1		11		110
$\frac{1}{16}$	0.062 5	$5\frac{1}{2}$	5.500	0.016	2.40	9.50	1.2		12		120	
$\frac{3}{32}$	0.093 75	$5\frac{3}{4}$	5.750	0.020	2.60	10.00		1.4		14		140
$\frac{1}{8}$	0.125 0	6	6.000	0.025	2.80	10.50	1.6		16		160	
$\frac{5}{32}$	0.156 25	$6\frac{1}{2}$	6.500	0.032	3.00	11.00		1.8		18		180
$\frac{3}{16}$	0.187 5	7	7.000	0.040	3.20	11.50	2		20		200	
$\frac{1}{4}$	0.250 0	$7\frac{1}{2}$	7.500	0.05	3.40	12.00		2.2		22		220
$\frac{5}{16}$	0.312 5	8	8.000	0.06	3.60	12.50	2.5		25		250	
$\frac{3}{8}$	0.375 0	$8\frac{1}{2}$	8.500	0.08	3.80	13.00		2.8		28		280
$\frac{7}{16}$	0.437 5	9	9.000	0.10	4.00	13.50	3		30		300	
$\frac{1}{2}$	0.500 0	$9\frac{1}{2}$	9.500	0.12	4.20	14.00		3.5		35		350
$\frac{9}{16}$	0.562 5	10	10.000	0.16	4.40	14.50	4		40		400	
$\frac{5}{8}$	0.625 0	$10\frac{1}{2}$	10.500	0.20	4.60	15.00		4.5		45		450
$\frac{11}{16}$	0.687 5	11	11.000	0.24	4.80	15.50	5		50		500	
$\frac{3}{4}$	0.750 0	$11\frac{1}{2}$	11.500	0.30	5.00	16.00		5.5		55		550
$\frac{7}{8}$	0.875 0	12	12.000	0.40	5.20	16.50	6		60		600	
1	1.000	$12\frac{1}{2}$	12.500	0.50	5.40	17.00		7		70		700
$1\frac{1}{4}$	1.250	13	13.000	0.60	5.60	17.50	8		80		800	
$1\frac{1}{2}$	1.500	$13\frac{1}{2}$	13.500	0.80	5.80	18.00		9		90		900
$1\frac{3}{4}$	1.750	14	14.000	1.00	6.00	18.50					1000	
2	2.000	$14\frac{1}{2}$	14.500	1.20	6.50	19.00						
$2\frac{1}{4}$	2.250	15	15.000	1.40	7.00	19.50						
$2\frac{1}{2}$	2.500	$15\frac{1}{2}$	15.500	1.60	7.50	20.00						
$2\frac{3}{4}$	2.750	16	16.000	1.80	8.00							
3	3.000	$16\frac{1}{2}$	16.500									
$3\frac{1}{4}$	3.250	17	17.000									
$3\frac{1}{2}$	3.500	$17\frac{1}{2}$	17.500									
$3\frac{3}{4}$	3.750	18	18.000									
4	4.000	$18\frac{1}{2}$	18.500									
$4\frac{1}{4}$	4.250	19	19.000									
$4\frac{1}{2}$	4.500	$19\frac{1}{2}$	19.500									
$4\frac{3}{4}$	4.750	20	20.000									

A–3 Screw Threads

(a) American Standard thread dimensions, numbered sizes

Size	Basic major diameter, D (in)	Coarse threads: UNC		Fine threads: UNF	
		Threads per inch, n	Tensile stress area (in^2)	Threads per inch, n	Tensile stress area (in^2)
0	0.060 0	—	—	80	0.001 80
1	0.073 0	64	0.002 63	72	0.002 78
2	0.086 0	56	0.003 70	64	0.003 94
3	0.099 0	48	0.004 87	56	0.005 23
4	0.112 0	40	0.006 04	48	0.006 61
5	0.125 0	40	0.007 96	44	0.008 30
6	0.138 0	32	0.009 09	40	0.010 15
8	0.164 0	32	0.014 0	36	0.014 74
10	0.190 0	24	0.017 5	32	0.020 0
12	0.216 0	24	0.024 2	28	0.025 8

(b) American Standard thread dimensions, fractional sizes

Size	Basic major diameter, D (in)	Coarse threads: UNC		Fine threads: UNF	
		Threads per inch, n	Tensile stress area (in^2)	Threads per inch, n	Tensile stress area (in^2)
$\frac{1}{4}$	0.250 0	20	0.031 8	28	0.036 4
$\frac{3}{16}$	0.312 5	18	0.052 4	24	0.058 0
$\frac{3}{8}$	0.375 0	16	0.077 5	24	0.087 8
$\frac{7}{16}$	0.437 5	14	0.106 3	20	0.118 7
$\frac{1}{2}$	0.500 0	13	0.141 9	20	0.159 9
$\frac{9}{16}$	0.562 5	12	0.182	18	0.203
$\frac{5}{8}$	0.625 0	11	0.226	18	0.256
$\frac{3}{4}$	0.750 0	10	0.334	16	0.373
$\frac{7}{8}$	0.875 0	9	0.462	14	0.509
1	1.000	8	0.606	12	0.663
$1\frac{1}{8}$	1.125	7	0.763	12	0.856
$1\frac{1}{4}$	1.250	7	0.969	12	1.073
$1\frac{3}{8}$	1.375	6	1.155	12	1.315
$1\frac{1}{2}$	1.500	6	1.405	12	1.581
$1\frac{3}{4}$	1.750	5	1.90	—	—
2	2.000	$4\frac{1}{2}$	2.50	—	—

(Continued)

A–3 (*Continued*)

(c) Metric thread dimensions

Basic major diameter, D (mm)	Coarse threads		Fine threads	
	Pitch (mm)	Tensile stress area (mm^2)	Pitch (mm)	Tensile stress area (mm^2)
1	0.25	0.460	—	—
1.6	0.35	1.27	0.20	1.57
2	0.4	2.07	0.25	2.45
2.5	0.45	3.39	0.35	3.70
3	0.5	5.03	0.35	5.61
4	0.7	8.78	0.5	9.79
5	0.8	14.2	0.5	16.1
6	1	20.1	0.75	22.0
8	1.25	36.6	1	39.2
10	1.5	58.0	1.25	61.2
12	1.75	84.3	1.25	92.1
16	2	157	1.5	167
20	2.5	245	1.5	272
24	3	353	2	384
30	3.5	561	2	621
36	4	817	3	865
42	4.5	1121	—	—
48	5	1473	—	—

A–4 Properties of Standard Wood Beams

Nominal size	Actual size		Area of section		Moment of inertia, I_x		Section modulus, S_x	
	in	mm	in^2	mm^2	in^4	mm^4	in^3	mm^3
2×4	1.5×3.5	38×89	5.25	3.39×10^3	5.36	2.23×10^6	3.06	50.1×10^3
2×6	1.5×5.5	38×140	8.25	5.32×10^3	20.8	8.66×10^6	7.56	124×10^3
2×8	1.5×7.25	38×184	10.87	7.01×10^3	47.6	19.8×10^6	13.14	215×10^3
2×10	1.5×9.25	38×235	13.87	8.95×10^3	98.9	41.2×10^6	21.4	351×10^3
2×12	1.5×11.25	38×286	16.87	10.88×10^3	178	74.1×10^6	31.6	518×10^3
4×4	3.5×3.5	89×89	12.25	7.90×10^3	12.51	5.21×10^6	7.15	117×10^3
4×6	3.5×5.5	89×140	19.25	12.42×10^3	48.5	20.2×10^6	17.65	289×10^3
4×8	3.5×7.25	89×184	25.4	16.39×10^3	111.1	46.2×10^6	30.7	503×10^3
4×10	3.5×9.25	89×235	32.4	20.90×10^3	231	96.1×10^6	49.9	818×10^3
4×12	3.5×11.25	89×286	39.4	25.42×10^3	415	172×10^6	73.9	1211×10^3
6×6	5.5×5.5	140×140	30.3	19.55×10^3	76.3	31.8×10^6	27.7	454×10^3
6×8	5.5×7.5	140×191	41.3	26.65×10^3	193	80.3×10^6	51.6	846×10^3
6×10	5.5×9.5	140×241	52.3	33.74×10^3	393	164×10^6	82.7	1355×10^3
6×12	5.5×11.5	140×292	63.3	40.84×10^3	697	290×10^6	121	1983×10^3
8×8	7.5×7.5	191×191	56.3	36.32×10^3	264	110×10^6	70.3	1152×10^3
8×10	7.5×9.5	191×241	71.3	46.00×10^3	536	223×10^6	113	1852×10^3
8×12	7.5×11.5	191×292	86.3	55.68×10^3	951	396×10^6	165	2704×10^3
10×10	9.5×9.5	241×241	90.3	58.26×10^3	679	283×10^6	143	2343×10^3
10×12	9.5×11.5	241×292	109.3	70.52×10^3	1204	501×10^6	209	3425×10^3
12×12	11.5×11.5	292×292	132.3	85.35×10^3	1458	607×10^6	253	4146×10^3

A–5 Properties of Steel Angles (L-Shapes) U.S. Customary Units

Ref.	Shape (in) (in) (in)	Weight per foot (lb/ft)	Area, A (in²)	Axis X-X			Axis Y-Y			Axis Z-Z	
				I_x (in⁴)	S_x (in³)	y (in)	I_y (in⁴)	S_y (in³)	x (in)	r (in)	α (deg.)
a	L 8 × 8 × 1	51.3	15.1	89.1	15.8	2.36	89.1	15.8	2.36	1.56	45.0
b	L 8 × 8 × $\frac{1}{2}$	26.7	7.84	48.8	8.36	2.17	48.8	8.36	2.17	1.59	45.0
c	L 8 × 4 × 1	37.6	11.10	69.7	14.00	3.03	11.6	3.94	1.04	0.844	13.9
d	L 8 × 4 × $\frac{1}{2}$	19.7	5.80	38.6	7.48	2.84	6.75	2.15	0.854	0.863	14.9
e	L 6 × 6 × $\frac{3}{4}$	28.8	8.46	28.1	6.64	1.77	28.1	6.64	1.77	1.17	45.0
f	L 6 × 6 × $\frac{3}{8}$	14.9	4.38	15.4	3.51	1.62	15.4	3.51	1.62	1.19	45.0
g	L 6 × 4 × $\frac{3}{4}$	23.5	6.90	24.4	6.23	2.08	8.63	2.95	1.08	0.857	23.2
h	L 6 × 4 × $\frac{3}{8}$	12.2	3.58	13.4	3.30	1.94	4.84	1.58	0.940	0.871	24.1
i	L 4 × 4 × $\frac{1}{2}$	12.7	3.75	5.52	1.96	1.18	5.52	1.96	1.18	0.776	45.0
j	L 4 × 4 × $\frac{1}{4}$	6.58	1.93	3.00	1.03	1.08	3.00	1.03	1.08	0.783	45.0
k	L 4 × 3 × $\frac{1}{2}$	11.1	3.25	5.02	1.87	1.32	2.40	1.10	0.822	0.633	28.5
l	L 4 × 3 × $\frac{1}{4}$	5.75	1.69	2.75	0.988	1.22	1.33	0.585	0.725	0.639	29.2
m	L 3 × 3 × $\frac{1}{2}$	9.35	2.75	2.20	1.06	0.929	2.20	1.06	0.929	0.580	45.0
n	L 3 × 3 × $\frac{1}{4}$	4.89	1.44	1.23	0.569	0.836	1.23	0.569	0.836	0.585	45.0
o	L 3 × 2 × $\frac{1}{2}$	7.70	2.26	1.92	1.00	1.08	0.667	0.470	0.580	0.425	22.4
p	L 3 × 2 × $\frac{1}{4}$	4.09	1.20	1.09	0.541	0.980	0.390	0.258	0.487	0.431	23.6
q	L 2 × 2 × $\frac{3}{8}$	4.65	1.37	0.476	0.348	0.632	0.476	0.348	0.632	0.386	45.0
r	L 2 × 2 × $\frac{1}{4}$	3.21	0.944	0.346	0.244	0.586	0.346	0.244	0.586	0.387	45.0
s	L 2 × 2 × $\frac{1}{8}$	1.67	0.491	0.189	0.129	0.534	0.189	0.129	0.534	0.391	45.0

Section properties

Centroid

A–5 (SI) Properties of Steel Angles (L-Shapes) SI Units

Ref.	Shape (mm) (mm)	Mass per m (kg/m)	Weight per m (N/m)	Area, A (mm²)	Section properties Axis X-X I_x (mm⁴)	Axis X-X S_x (mm³)	Axis X-X y (mm)	Axis Y-Y I_y (mm⁴)	Axis Y-Y S_y (mm³)	Axis Y-Y x (mm)	Axis Z-Z r (mm)	Axis Z-Z α (deg.)
a	L 203 × 203 × 25.4	76.3	749	9740	3.71E+07	2.59E+05	59.9	3.71E+07	2.59E+05	59.9	39.6	45.0
b	L 203 × 203 × 12.7	39.7	390	5060	2.03E+07	1.37E+05	55.1	2.03E+07	1.37E+05	55.1	40.4	45.0
c	L 203 × 102 × 25.4	55.9	549	7160	2.90E+07	2.29E+05	77.0	4.83E+06	6.46E+04	26.4	21.4	13.9
d	L 203 × 102 × 12.7	29.3	288	3740	1.61E+07	1.23E+05	72.1	2.81E+06	3.52E+04	21.7	21.9	14.9
e	L 152 × 152 × 19	42.9	420	5460	1.17E+07	1.09E+05	45.0	1.17E+07	1.09E+05	45.0	29.7	45.0
f	L 152 × 152 × 9.5	22.2	217	2830	6.41E+06	5.75E+04	41.1	6.41E+06	5.75E+04	41.1	30.2	45.0
g	L 152 × 102 × 19	35.0	343	4450	1.02E+07	1.02E+05	52.8	3.59E+06	4.84E+04	27.4	21.8	23.2
h	L 152 × 102 × 9.5	18.2	178	2310	5.58E+06	5.41E+04	49.3	2.01E+06	2.59E+04	23.9	22.1	24.1
i	L 102 × 102 × 12.7	18.9	185	2420	2.30E+06	3.21E+04	30.0	2.30E+06	3.21E+04	30.0	19.7	45.0
j	L 102 × 102 × 6.4	9.79	96.0	1250	1.25E+06	1.69E+04	27.4	1.25E+06	1.69E+04	27.4	19.9	45.0
k	L 102 × 76 × 12.7	16.5	162	2100	2.09E+06	3.06E+04	33.5	9.99E+05	1.80E+04	20.9	16.1	28.5
l	L 102 × 76 × 6.4	8.56	83.9	1090	1.14E+06	1.62E+04	31.0	5.54E+05	9.59E+03	18.4	16.2	29.2
m	L 76 × 76 × 12.7	13.9	136	1770	9.16E+05	1.74E+04	23.6	9.16E+05	1.74E+04	23.6	14.7	45.0
n	L 76 × 76 × 6.4	7.28	71.4	929	5.12E+05	9.33E+03	21.2	5.12E+05	9.33E+03	21.2	14.9	45.0
o	L 76 × 51 × 12.7	11.5	112	1460	7.99E+05	1.64E+04	27.4	2.78E+05	7.70E+03	14.7	10.8	22.4
p	L 76 × 51 × 6.4	6.09	59.7	774	4.54E+05	8.87E+03	24.9	1.62E+05	4.23E+03	12.4	10.9	23.6
q	L 51 × 51 × 9.5	6.92	67.9	884	1.98E+05	5.70E+03	16.1	1.98E+05	5.70E+03	16.1	9.80	45.0
r	L 51 × 51 × 6.4	4.78	46.9	609	1.44E+05	4.00E+03	14.9	1.44E+05	4.00E+03	14.9	9.83	45.0
s	L 51 × 51 × 3.2	2.48	24.4	317	7.87E+04	2.11E+03	13.6	7.87E+04	2.11E+03	13.6	9.93	45.0

A–6 Properties of American Standard Steel Channels (C-Shapes) U.S. Customary Units

	Shape		Area, A (in²)	Depth, d (in)	Web thickness, t_w (in)	Flange			Section properties					
						Width, b_f (in)	Thickness, t_f average (in)		Axis X-X			Axis Y-Y		
Ref.	(in)	(lb/ft)							I_x (in⁴)	S_x (in³)		I_y (in⁴)	S_y (in³)	x (in)
a	C 15 × 50		14.7	15.0	0.716	3.72	0.650		404	53.8		11.0	3.77	0.799
b	C 15 × 40		11.8	15.0	0.520	3.52	0.650		348	46.5		9.17	3.34	0.778
c	C 12 × 30		8.82	12.0	0.510	3.17	0.501		162	27.0		5.12	2.05	0.674
d	C 12 × 25		7.35	12.0	0.387	3.05	0.501		144	24.0		4.45	1.87	0.674
e	C 10 × 30		8.82	10.0	0.673	3.03	0.436		103	20.7		3.93	1.65	0.649
f	C 10 × 20		5.88	10.0	0.379	2.74	0.436		78.9	15.8		2.80	1.31	0.606
g	C 9 × 20		5.88	9.0	0.448	2.65	0.413		60.9	13.5		2.41	1.17	0.583
h	C 9 × 15		4.41	9.0	0.285	2.49	0.413		51.0	11.3		1.91	1.01	0.586
i	C 8 × 18.75		5.51	8.0	0.487	2.53	0.390		43.9	11.0		1.97	1.01	0.565
j	C 8 × 11.5		3.38	8.0	0.220	2.26	0.390		32.5	8.14		1.31	0.775	0.572
k	C 7 × 14.75		4.33	7.0	0.419	2.30	0.366		27.2	7.78		1.37	0.772	0.532
l	C 7 × 9.8		2.87	7.0	0.210	2.09	0.366		21.2	6.07		0.957	0.617	0.541
m	C 6 × 13		3.83	6.0	0.437	2.16	0.343		17.3	5.78		1.05	0.638	0.514
n	C 6 × 8.2		2.40	6.0	0.200	1.92	0.343		13.1	4.35		0.687	0.488	0.512
o	C 5 × 9		2.64	5.0	0.325	1.89	0.320		8.89	3.56		0.624	0.444	0.478
p	C 5 × 6.7		1.97	5.0	0.190	1.75	0.320		7.48	2.99		0.470	0.372	0.484
q	C 4 × 7.25		2.13	4.0	0.321	1.72	0.296		4.58	2.29		0.425	0.337	0.459
r	C 4 × 5.4		1.59	4.0	0.184	1.58	0.296		3.85	1.92		0.312	0.277	0.457
s	C 3 × 6		1.76	3.0	0.356	1.60	0.273		2.07	1.38		0.300	0.263	0.455
t	C 3 × 4.1		1.21	3.0	0.170	1.41	0.273		1.65	1.10		0.191	0.196	0.437

A–6 (SI) Properties of American Standard Steel Channels (C-Shapes) SI Units

Ref.	Shape (mm) (kg/m)	Wt/m (KN/m)	Area, A (mm²)	Depth, d (mm)	Web thickness, t_w (mm)	Flange Width, b_f (mm)	Flange Thickness, t_f Average (mm)	Axis X-X I_x (mm⁴)	Axis X-X S_x (mm³)	Axis Y-Y I_y (mm⁴)	Axis Y-Y S_y (mm³)	x (mm)
a	C 380 × 74	0.730	9480	381	18.2	94.4	16.5	1.68E+08	8.82E+05	4.58E+06	6.18E+04	20.3
b	C 380 × 60	0.584	7610	381	13.2	89.4	16.5	1.45E+08	7.62E+05	3.82E+06	5.47E+04	19.8
c	C 300 × 45	0.438	5690	305	13.0	80.5	12.7	6.74E+07	4.43E+05	2.13E+06	3.36E+04	17.1
d	C 300 × 37	0.365	4740	305	9.83	77.4	12.7	5.99E+07	3.93E+05	1.85E+06	3.06E+04	17.1
e	C 250 × 45	0.438	5690	254	17.1	77.0	11.1	4.29E+07	3.93E+05	1.64E+06	2.70E+04	16.5
f	C 250 × 30	0.292	3790	254	9.63	69.6	11.1	3.28E+07	2.59E+05	1.17E+06	2.15E+04	15.4
g	C 230 × 30	0.292	3790	229	11.4	67.3	10.5	2.53E+07	2.21E+05	1.00E+06	1.92E+04	14.8
h	C 230 × 22	0.219	2850	229	7.24	63.1	10.5	2.12E+07	1.85E+05	7.95E+05	1.66E+04	14.9
i	C 200 × 27.9	0.274	3560	203	12.4	64.2	9.91	1.83E+07	1.80E+05	8.20E+05	1.66E+04	14.4
j	C 200 × 17.1	0.168	2180	203	5.59	57.4	9.91	1.35E+07	1.33E+05	5.45E+05	1.27E+04	14.5
k	C 180 × 22	0.215	2790	178	10.6	58.4	9.30	1.13E+07	1.28E+05	5.70E+05	1.27E+04	13.5
l	C 180 × 14.6	0.143	1850	178	5.33	53.1	9.30	8.82E+06	9.95E+04	3.98E+05	1.01E+04	13.7
m	C 150 × 19.3	0.190	2470	152	11.1	54.8	8.71	7.20E+06	9.47E+04	4.37E+05	1.05E+04	13.1
n	C 150 × 12.2	0.120	1550	152	5.08	48.8	8.71	5.45E+06	7.13E+04	2.86E+05	8.00E+03	13.0
o	C 130 × 13	0.128	1700	127	8.26	47.9	8.13	3.70E+06	5.83E+04	2.60E+05	7.28E+03	12.1
p	C 130 × 10.4	0.102	1270	127	4.83	44.5	8.13	3.11E+06	4.90E+04	1.96E+05	6.10E+03	12.3
q	C 100 × 10.8	0.106	1370	102	8.15	43.7	7.52	1.91E+06	3.75E+04	1.77E+05	5.52E+03	11.7
r	C 100 × 8	0.0788	1020	102	4.67	40.2	7.52	1.60E+06	3.15E+04	1.30E+05	4.54E+03	11.6
s	C 80 × 8.9	0.0876	1140	76.2	9.04	40.5	6.93	8.62E+05	2.26E+04	1.25E+05	4.31E+03	11.6
t	C 80 × 6.1	0.0598	777	76.2	4.32	35.8	6.93	6.87E+05	1.80E+04	7.95E+04	3.21E+03	11.1

A–7 Properties of Steel Wide-Flange Shapes (W-Shapes) U.S. Customary Units

Ref.	Shape (in) (lb/ft)	Area, A (in²)	Depth, d (in)	Web thickness, t_w (in)	Flange Width, b_f (in)	Flange Thickness, t_f (in)	Axis X-X I_x (in⁴)	Axis X-X S_x (in³)	Axis Y-Y I_y (in⁴)	Axis Y-Y S_y (in³)
a	W 30 × 173	51.0	30.4	0.655	15.0	1.070	8230	541	598	79.8
b	W 30 × 108	31.7	29.8	0.545	10.5	0.760	4470	299	146	27.9
c	W 27 × 146	43.1	27.4	0.605	14.0	0.975	5660	414	443	63.5
d	W 27 × 102	30.0	27.1	0.515	10.0	0.830	3620	267	139	27.8
e	W 24 × 76	22.4	23.9	0.440	8.99	0.680	2100	176	82.5	18.4
f	W 24 × 68	20.1	23.7	0.415	8.97	0.585	1830	154	70.4	15.7
g	W 21 × 73	21.5	21.2	0.455	8.30	0.740	1600	151	70.6	17.0
h	W 21 × 57	16.7	21.1	0.405	6.56	0.650	1170	111	30.6	9.35
i	W 18 × 55	16.2	18.1	0.390	7.53	0.630	890	98.3	44.9	11.9
j	W 18 × 40	11.8	17.9	0.315	6.02	0.525	612	68.4	19.1	6.35
k	W 14 × 43	12.6	13.7	0.305	8.00	0.530	428	62.7	45.2	11.3
l	W 14 × 26	7.69	13.9	0.255	5.03	0.420	245	35.3	8.91	3.54
m	W 12 × 65	19.1	12.1	0.390	12.0	0.605	533	87.9	174	29.1
n	W 12 × 30	8.79	12.3	0.260	6.52	0.440	238	38.6	20.3	6.24
o	W 12 × 16	4.71	12.0	0.220	3.99	0.265	103	17.1	2.82	1.41
p	W 10 × 60	17.60	10.2	0.420	10.1	0.680	341	66.7	116	23.0
q	W 10 × 30	8.84	10.5	0.300	5.81	0.510	170	32.4	16.7	5.75
r	W 10 × 12	3.54	9.87	0.190	3.96	0.210	53.8	10.9	2.18	1.10
s	W 8 × 40	11.70	8.25	0.360	8.07	0.560	146	35.5	49.1	12.2
t	W 8 × 21	6.16	8.28	0.250	5.27	0.400	75.3	18.2	9.77	3.71
u	W 8 × 10	2.96	7.89	0.170	3.94	0.205	30.8	7.81	2.09	1.06
v	W 6 × 15	4.43	5.99	0.230	5.99	0.260	29.1	9.72	9.32	3.11
w	W 6 × 12	3.55	6.03	0.230	4.00	0.280	22.1	7.31	2.99	1.50
x	W 5 × 19	5.54	5.15	0.270	5.03	0.430	26.2	10.2	9.13	3.63
y	W 5 × 16	4.68	5.01	0.240	5.00	0.360	21.3	8.51	7.51	3.00
z	W 4 × 13	3.83	4.16	0.280	4.06	0.345	11.3	5.46	3.86	1.90

Ref.	Shape (mm)	(kg/m)	Wt/m (KN/M)	Area, A (mm²)	Depth, d (mm)	Web thickness, t_w (mm)	Flange Width, b_f (mm)	Flange Thickness, t_f (mm)	Axis X-X I_x (mm⁴)	Axis X-X S_x (mm³)	Axis Y-Y I_y (mm⁴)	Axis Y-Y S_y (mm³)
a	W 760 × 257		2.525	32 900	772	16.6	381	27.2	3.43E+09	8.87E+06	2.49E+08	1.31E+06
b	W 760 × 161		1.576	20 500	757	13.8	267	19.3	1.86E+09	4.90E+06	6.08E+07	4.57E+05
c	W 690 × 217		2.131	27 800	696	15.4	356	24.8	2.36E+09	6.79E+06	1.84E+08	1.04E+06
d	W 690 × 152		1.489	19 400	688	13.1	254	21.1	1.51E+09	4.38E+06	5.79E+07	4.56E+05
e	W 610 × 113		1.109	14 500	607	11.2	228	17.3	8.74E+08	2.88E+06	3.43E+07	3.02E+05
f	W 610 × 101		0.993	13 000	602	10.5	228	14.9	7.62E+08	2.52E+06	2.93E+07	2.57E+05
g	W 530 × 109		1.066	13 900	538	11.6	211	18.8	6.66E+08	2.47E+06	2.94E+07	2.79E+05
h	W 530 × 85		0.832	10 800	536	10.3	167	16.5	4.87E+08	1.82E+06	1.27E+07	1.53E+05
i	W 460 × 82		0.803	10 500	460	9.91	191	16.0	3.70E+08	1.61E+06	1.87E+07	1.95E+05
j	W 460 × 60		0.584	7610	455	8.00	153	13.3	2.55E+08	1.12E+06	7.95E+06	1.04E+05
k	W 360 × 64		0.628	8130	348	7.75	203	13.5	1.78E+08	1.03E+06	1.88E+07	1.85E+05
l	W 360 × 39		0.380	4960	353	6.48	128	10.7	1.02E+08	5.79E+05	3.71E+06	5.80E+04
m	W 310 × 97		0.949	12 300	307	9.91	305	15.4	2.22E+08	1.44E+06	7.24E+07	4.77E+05
n	W 310 × 44.5		0.437	5670	312	6.60	166	11.2	9.91E+07	6.33E+05	8.45E+06	1.02E+05
o	W 310 × 23.8		0.233	3040	305	5.59	101	6.73	4.29E+07	2.80E+05	1.17E+06	2.31E+04
p	W 250 × 89		0.876	11 400	259	10.7	257	17.3	1.42E+08	1.09E+06	4.83E+07	3.77E+05
q	W 250 × 44.8		0.439	5700	267	7.62	148	13	7.08E+07	5.31E+05	6.95E+06	9.42E+04
r	W 250 × 17.9		0.176	2280	251	4.83	101	5.33	2.24E+07	1.79E+05	9.07E+05	1.80E+04
s	W 200 × 59		0.579	7550	210	9.14	205	14.2	6.08E+07	5.82E+05	2.04E+07	2.00E+05
t	W 200 × 31.3		0.307	3970	210	6.35	134	10.2	3.13E+07	2.98E+05	4.07E+06	6.08E+04
u	W 200 × 15		0.146	1910	200	4.32	100	5.21	1.28E+07	1.28E+05	8.70E+05	1.74E+04
v	W 150 × 22.5		0.221	2860	152	5.84	152	6.60	1.21E+07	1.59E+05	3.88E+06	5.10E+04
w	W 150 × 18		0.175	2290	153	5.84	102	7.11	9.20E+06	1.20E+05	1.24E+06	2.46E+04
x	W 130 × 28.1		0.276	3570	131	6.86	128	10.9	1.09E+07	1.67E+05	3.80E+06	5.95E+04
y	W 130 × 23.8		0.233	3020	127	6.10	127	9.14	8.87E+06	1.39E+05	3.13E+06	4.92E+04
z	W 100 × 19.3		0.189	2470	106	7.11	103	8.76	4.70E+06	8.95E+04	1.61E+06	3.11E+04

A–8 Properties of American Standard Steel Beams (S-Shapes) U.S. Customary Units

Ref.	Shape (in) (lb/ft)	Area, A (in²)	Depth, d (in)	Web thickness, t_w (in)	Flange Width, b_f (in)	Flange Thickness, t_f (in)	Axis X-X I_x (in⁴)	Axis X-X S_x (in³)	Axis Y-Y I_y (in⁴)	Axis Y-Y S_y (in³)
a	S 24 × 121	35.5	24.5	0.800	8.05	1.090	3160	258	83.0	20.6
b	S 24 × 90	26.5	24.0	0.625	7.13	0.870	2250	187	44.7	12.5
c	S 20 × 96	28.2	20.3	0.800	7.20	0.920	1670	165	49.9	13.9
d	S 20 × 75	22.0	20.0	0.635	6.39	0.795	1280	128	29.5	9.25
e	S 20 × 66	19.4	20.0	0.505	6.26	0.795	1190	119	27.5	8.78
f	S 18 × 70	20.5	18.0	0.711	6.25	0.691	923	103	24.0	7.69
g	S 18 × 54.7	16.0	18.0	0.461	6.00	0.691	801	89.0	20.7	6.91
h	S 15 × 50	14.7	15.0	0.550	5.64	0.622	485	64.7	15.6	5.53
i	S 15 × 42.9	12.6	15.0	0.411	5.50	0.622	446	59.4	14.3	5.19
j	S 12 × 50	14.6	12.0	0.687	5.48	0.659	303	50.6	15.6	5.69
k	S 12 × 35	10.2	12.0	0.426	5.08	0.544	228	38.1	9.84	3.88
l	S 10 × 35	10.3	10.0	0.594	4.94	0.491	147	29.4	8.30	3.36
m	S 10 × 25.4	7.45	10.0	0.311	4.66	0.491	123	24.6	6.73	2.89
n	S 8 × 23	6.76	8.00	0.441	4.17	0.425	64.7	16.2	4.27	2.05
o	S 8 × 18.4	5.40	8.00	0.271	4.00	0.425	57.5	14.4	3.69	1.84
p	S 6 × 17.25	5.06	6.00	0.465	3.57	0.359	26.2	8.74	2.29	1.28
q	S 6 × 12.5	3.66	6.00	0.232	3.33	0.359	22.0	7.34	1.80	1.08
r	S 5 × 10	2.93	5.00	0.214	3.00	0.326	12.3	4.90	1.19	0.795
s	S 4 × 9.5	2.79	4.00	0.326	2.80	0.293	6.8	3.38	0.887	0.635
t	S 4 × 7.7	2.26	4.00	0.193	2.66	0.293	6.05	3.03	0.748	0.562
u	S 3 × 7.5	2.20	3.00	0.349	2.51	0.260	2.91	1.94	0.578	0.461
v	S 3 × 5.7	1.66	3.00	0.170	2.33	0.260	2.50	1.67	0.447	0.383

A–8 (SI) Properties of American Standard Steel Beams (S-Shapes) SI Units

Ref.	Shape (mm) (kg/m)	Wt/m (KN/m)	Area, A (mm²)	Depth, d (mm)	Web thickness, t_w (mm)	Flange Width, b_f (mm)	Flange Thickness, t_f (mm)	Axis X-X I_x (mm⁴)	Axis X-X S_x (mm³)	Axis Y-Y I_y (mm⁴)	Axis Y-Y S_y (mm³)
a	S 610 × 180	1.766	22 900	622	20.3	204	27.7	1.32E+09	4.23E+06	3.45E+07	3.38E+05
b	S 610 × 134	1.314	17 100	610	15.9	181	22.1	9.36E+08	3.06E+06	1.86E+07	2.05E+05
c	S 510 × 143	1.401	18 200	516	20.3	183	23.4	6.95E+08	2.70E+06	2.08E+07	2.28E+05
d	S 510 × 112	1.095	14 200	508	16.1	162	20.2	5.33E+08	2.10E+06	1.23E+07	1.52E+05
e	S 510 × 98.2	0.963	12 500	508	12.8	159	20.2	4.95E+08	1.95E+06	1.14E+07	1.44E+05
f	S 460 × 104	1.022	13 200	457	18.1	159	17.6	3.84E+08	1.69E+06	9.99E+06	1.26E+05
g	S 460 × 81.4	0.799	10 300	457	11.7	152	17.6	3.33E+08	1.46E+06	8.62E+06	1.13E+05
h	S 380 × 74	0.730	9480	381	14.0	143	15.8	2.02E+08	1.06E+06	6.49E+06	9.06E+04
i	S 380 × 64	0.626	8130	381	10.4	140	15.8	1.86E+08	9.74E+05	5.95E+06	8.51E+04
j	S 300 × 74	0.730	9420	305	17.4	139	16.7	1.26E+08	8.29E+05	6.49E+06	9.33E+04
k	S 300 × 52	0.511	6580	305	10.8	129	13.8	9.49E+07	6.24E+05	4.10E+06	6.36E+04
l	S 250 × 52	0.511	6650	254	15.1	125	12.5	6.12E+07	4.82E+05	3.45E+06	5.51E+04
m	S 250 × 37.8	0.371	4810	254	7.9	118	12.5	5.12E+07	4.03E+05	2.80E+06	4.74E+04
n	S 200 × 34	0.336	4360	203	11.2	106	10.8	2.69E+07	2.66E+05	1.78E+06	3.36E+04
o	S 200 × 27.4	0.269	3480	203	6.9	102	10.8	2.39E+07	2.36E+05	1.54E+06	3.02E+04
p	S 150 × 25.7	0.252	3260	152	11.8	90.7	9.1	1.09E+07	1.43E+05	9.53E+05	2.10E+04
q	S 150 × 18.6	0.182	2360	152	5.9	84.6	9.1	9.16E+06	1.20E+05	7.49E+05	1.77E+04
r	S 130 × 15	0.146	1890	127	5.4	76.2	8.3	5.12E+06	8.03E+04	4.95E+05	1.30E+04
s	S 100 × 14.1	0.138	1800	102	8.3	71.1	7.4	2.81E+06	5.54E+04	3.69E+05	1.04E+04
t	S 100 × 11.5	0.113	1460	102	4.9	67.6	7.4	2.52E+06	4.97E+04	3.11E+05	9.21E+03
u	S 80 × 11.2	0.110	1420	76.2	8.9	63.8	6.6	1.21E+06	3.18E+04	2.41E+05	7.56E+03
v	S 80 × 8.5	0.083	1070	76.2	4.3	59.2	6.6	1.04E+06	2.74E+04	1.86E+05	6.28E+03

A–9 Properties of Steel Structural Tubing (HSS-Shapes) U.S. Customary Units

Ref.	Shape (in) (in) (in)	Design wall thickness t_w (in)	Weight per foot (lb/ft)	Area, A (in²)	Axis X-X			Axis Y-Y			Torsional constants	
					I_x (in⁴)	S_x (in³)	r_x (in)	I_y (in⁴)	S_y (in³)	r_y (in)	J (in⁴)	C (in³)
a	HSS $8 \times 8 \times \frac{1}{2}$	0.465	48.7	13.5	125	31.2	3.04	125	31.2	3.04	204	52.4
b	HSS $8 \times 8 \times \frac{1}{4}$	0.233	25.8	7.10	70.7	17.7	3.15	70.7	17.7	3.15	111	28.1
c	HSS $8 \times 4 \times \frac{1}{2}$	0.465	35.1	9.74	71.8	17.9	2.71	23.6	11.8	1.56	61.1	24.4
d	HSS $8 \times 4 \times \frac{1}{4}$	0.233	19.0	5.24	42.5	10.6	2.85	14.4	7.21	1.66	35.3	13.6
e	HSS $8 \times 2 \times \frac{1}{4}$	0.233	15.6	4.30	28.5	7.12	2.57	2.94	2.94	0.827	9.36	6.35
f	HSS $6 \times 6 \times \frac{1}{2}$	0.465	35.1	9.74	48.3	16.1	2.23	48.3	16.1	2.23	81.1	28.1
g	HSS $6 \times 6 \times \frac{1}{4}$	0.233	19.0	5.24	28.6	9.54	2.34	28.6	9.54	2.34	45.6	15.4
h	HSS $6 \times 4 \times \frac{1}{4}$	0.233	15.6	4.30	20.9	6.96	2.20	11.1	5.56	1.61	23.6	10.1
i	HSS $6 \times 2 \times \frac{1}{4}$	0.233	12.2	3.37	13.1	4.37	1.97	2.21	2.21	0.810	6.55	4.70
j	HSS $4 \times 4 \times \frac{1}{2}$	0.465	21.5	6.02	11.9	5.97	1.41	11.9	5.97	1.41	21.0	11.2
k	HSS $4 \times 4 \times \frac{1}{4}$	0.233	12.2	3.37	7.80	3.90	1.52	7.80	3.90	1.52	12.8	6.56
l	HSS $4 \times 2 \times \frac{1}{4}$	0.233	8.78	2.44	4.49	2.25	1.36	1.48	1.48	0.779	3.82	3.05
m	HSS $3 \times 3 \times \frac{1}{4}$	0.233	8.78	2.44	3.02	2.01	1.11	3.02	2.01	1.11	5.08	3.52
n	HSS $3 \times 2 \times \frac{1}{4}$	0.233	7.08	1.97	2.13	1.42	1.04	1.11	1.11	0.751	2.52	2.23
o	HSS $2 \times 2 \times \frac{1}{4}$	0.233	5.38	1.51	0.747	0.747	0.704	0.747	0.747	0.704	1.31	1.41

Section properties

A–9 (SI) Properties of Steel Structural Tubing (HSS-Shapes) SI Units

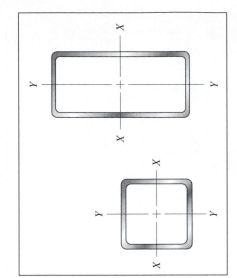

Ref.	Shape (mm) (mm)	Design wall thickness, t_w (mm)	Mass per m (kg/m)	Weight per m (N/m)	Area A (mm²)	Axis X-X I_x (mm⁴)	S_x (mm³)	r_x (mm)	Axis Y-Y I_y (mm⁴)	S_y (mm³)	r_y (mm)	Torsional constants J (mm⁴)	C (mm³)
a	HSS 203 × 203 × 12.7	11.8	72.5	711	8710	5.20E+07	5.11E+05	77.2	5.20E+07	5.11E+05	77.2	8.49E+07	8.59E+05
b	HSS 203 × 203 × 6.4	5.92	38.4	377	4580	2.94E+07	2.90E+05	80.0	2.94E+07	2.90E+05	80.0	4.62E+07	4.61E+05
c	HSS 203 × 102 × 12.7	11.8	52.2	512	6280	2.99E+07	2.93E+05	68.8	9.82E+06	1.93E+05	39.6	2.54E+07	4.00E+05
d	HSS 203 × 102 × 6.4	5.92	28.3	277	3380	1.77E+07	1.74E+05	72.4	5.99E+06	1.18E+05	42.2	1.47E+07	2.23E+05
e	HSS 203 × 51 × 6.4	5.92	23.2	228	2770	1.19E+07	1.17E+05	65.3	1.22E+06	4.82E+04	21.0	3.90E+06	1.04E+05
f	HSS 152 × 152 × 12.7	11.8	52.2	512	6280	2.01E+07	2.64E+05	56.6	2.01E+07	2.64E+05	56.6	3.38E+07	4.61E+05
g	HSS 152 × 152 × 6.4	5.92	28.3	277	3380	1.19E+07	1.56E+05	59.4	1.19E+07	1.56E+05	59.4	1.90E+07	2.52E+05
h	HSS 152 × 102 × 6.4	5.92	23.2	228	2770	8.70E+06	1.14E+05	55.9	4.62E+06	9.11E+04	40.9	9.82E+06	1.66E+05
i	HSS 152 × 51 × 6.4	5.92	18.2	178	2170	5.45E+06	7.16E+04	50.0	9.20E+05	3.62E+04	20.6	2.73E+06	7.70E+04
j	HSS 102 × 102 × 12.7	11.8	32.0	314	3880	4.95E+06	9.78E+04	35.8	4.95E+06	9.78E+04	35.8	8.74E+06	1.84E+05
k	HSS 102 × 102 × 6.4	5.92	18.2	178	2170	3.25E+06	6.39E+04	38.6	3.25E+06	6.39E+04	38.6	5.33E+06	1.08E+05
l	HSS 102 × 51 × 6.4	5.92	13.1	128	1570	1.87E+06	3.69E+04	34.5	6.16E+05	2.43E+04	19.8	1.59E+06	5.00E+04
m	HSS 76 × 76 × 6.4	5.92	13.1	128	1570	1.26E+06	3.29E+04	28.2	1.26E+06	3.29E+04	28.2	2.11E+06	5.77E+04
n	HSS 76 × 51 × 6.4	5.92	10.5	103	1271	8.87E+05	2.33E+04	26.4	4.62E+05	1.82E+04	19.1	1.05E+06	3.65E+04
o	HSS 51 × 51 × 6.4	5.92	8.01	78.5	974	3.11E+05	1.22E+04	17.9	3.11E+05	1.22E+04	17.9	5.45E+05	2.31E+04

Section properties

A–10 Properties of Aluminum Association Standard Channels U.S. Customary Units

Ref.	Shape (in)	(lb/ft)	Depth, A (in)	Width, B (in)	Area (in²)	Flange thickness, t_1 (in)	Web thickness, t (in)	Axis X-X I_x (in⁴)	S_x (in³)	r_x (in)	Axis Y-Y I_y (in⁴)	S_y (in³)	r_y (in)	x (in)
a	C 2 × 0.577		2.00	1.00	0.491	0.13	0.13	0.288	0.288	0.766	0.045	0.064	0.303	0.298
b	C 2 × 1.071		2.00	1.25	0.911	0.26	0.17	0.546	0.546	0.774	0.139	0.178	0.391	0.471
c	C 3 × 1.135		3.00	1.50	0.965	0.20	0.13	1.41	0.94	1.21	0.22	0.22	0.47	0.49
d	C 3 × 1.597		3.00	1.75	1.358	0.26	0.17	1.97	1.31	1.20	0.42	0.37	0.55	0.62
e	C 4 × 1.738		4.00	2.00	1.478	0.23	0.15	3.91	1.95	1.63	0.60	0.45	0.64	0.65
f	C 4 × 2.331		4.00	2.25	1.982	0.29	0.19	5.21	2.60	1.62	1.02	0.69	0.72	0.78
g	C 5 × 2.212		5.00	2.25	1.881	0.26	0.15	7.88	3.15	2.05	0.98	0.64	0.72	0.73
h	C 5 × 3.089		5.00	2.75	2.627	0.32	0.19	11.14	4.45	2.06	2.05	1.14	0.88	0.95
i	C 6 × 2.834		6.00	2.50	2.410	0.29	0.17	14.35	4.78	2.44	1.53	0.90	0.80	0.79
j	C 6 × 4.030		6.00	3.25	3.427	0.35	0.21	21.04	7.01	2.48	3.76	1.76	1.05	1.12
k	C 7 × 3.205		7.00	2.75	2.725	0.29	0.17	22.09	6.31	2.85	2.10	1.10	0.88	0.84
l	C 7 × 4.715		7.00	3.50	4.009	0.38	0.21	33.79	9.65	2.90	5.13	2.23	1.13	1.20
m	C 8 × 4.147		8.00	3.00	3.526	0.35	0.19	37.40	9.35	3.26	3.25	1.57	0.96	0.93
n	C 8 × 5.789		8.00	3.75	4.923	0.41	0.25	52.69	13.17	3.27	7.13	2.82	1.20	1.22
o	C 9 × 4.983		9.00	3.25	4.237	0.35	0.23	54.41	12.09	3.58	4.40	1.89	1.02	0.93
p	C 9 × 6.970		9.00	4.00	5.927	0.44	0.29	78.31	17.40	3.63	9.61	3.49	1.27	1.25
q	C 10 × 6.136		10.00	3.50	5.218	0.41	0.25	83.22	16.64	3.99	6.33	2.56	1.10	1.02
r	C 10 × 8.360		10.00	4.25	7.109	0.50	0.31	116.15	23.23	4.04	13.02	4.47	1.35	1.34
s	C 12 × 8.274		12.00	4.00	7.036	0.47	0.29	159.76	26.63	4.77	11.03	3.86	1.25	1.14
t	C 12 × 11.822		12.00	5.00	10.053	0.62	0.35	239.69	39.95	4.88	25.74	7.60	1.60	1.61

A–10 (SI) Properties of Aluminum Association Standard Channels SI Units

Ref.	Shape (mm)	(kg/m)	Wt/m (N/m)	Depth, A (mm)	Width, B (mm)	Area (mm²)	Flange thickness, t_1 (mm)	Web thickness, t (mm)	Axis X-X I_x (mm⁴)	S_x (mm³)	r_x (mm)	Axis Y-Y I_y (mm⁴)	S_y (mm³)	r_y (mm)	x (mm)
a	C 51 × 0.859	8.42		51	25	317	3.3	3.3	1.20E+05	4.72E+03	19.46	1.87E+04	1.05E+03	7.70	7.57
b	C 51 × 1.594	15.63		51	32	588	6.6	4.3	2.27E+05	8.95E+03	19.66	5.79E+04	2.92E+03	9.93	12.0
c	C 76 × 1.689	16.57		76	38	623	5.1	3.3	5.87E+05	1.54E+04	30.73	9.16E+04	3.61E+03	11.94	12.4
d	C 76 × 2.376	23.31		76	44	876	6.6	4.3	8.20E+05	2.15E+04	30.48	1.75E+05	6.06E+03	13.97	15.7
e	C 102 × 2.586	25.37		102	51	954	5.8	3.8	1.63E+06	3.20E+04	41.40	2.50E+05	7.38E+03	16.26	16.5
f	C 102 × 3.468	34.03		102	57	1279	7.4	4.8	2.17E+06	4.26E+04	41.15	4.25E+05	1.13E+04	18.29	19.8
g	C 127 × 3.291	32.29		127	57	1214	6.6	3.8	3.28E+06	5.16E+04	52.07	4.08E+05	1.05E+04	18.29	18.5
h	C 127 × 4.596	45.09		127	70	1695	8.1	4.8	4.64E+06	7.29E+04	52.32	8.53E+05	1.87E+04	22.35	24.1
i	C 152 × 4.217	41.37		152	64	1555	7.4	4.3	5.97E+06	7.83E+04	61.98	6.37E+05	1.48E+04	20.32	20.1
j	C 152 × 6.00	58.8		152	83	2211	8.9	5.3	8.76E+06	1.15E+05	62.99	1.56E+06	2.88E+04	26.67	28.4
k	C 178 × 4.77	46.8		178	70	1758	7.4	4.3	9.19E+06	1.03E+05	72.39	8.74E+05	1.80E+04	22.35	21.3
l	C 178 × 7.02	68.8		178	89	2587	9.7	5.3	1.41E+07	1.58E+05	73.7	2.14E+06	3.65E+04	28.70	30.5
m	C 203 × 6.17	60.5		203	76	2275	8.9	4.8	1.56E+07	1.53E+05	82.8	1.35E+06	2.57E+04	24.38	23.6
n	C 203 × 8.61	84.5		203	95	3176	10.4	6.4	2.19E+07	2.16E+05	83.1	2.97E+06	4.62E+04	30.48	31.0
o	C 229 × 7.41	72.7		229	83	2734	8.9	5.8	2.26E+07	1.98E+05	90.9	1.83E+06	3.10E+04	25.91	23.6
p	C 229 × 10.37	101.7		229	102	3824	11.2	7.4	3.26E+07	2.85E+05	92.2	4.00E+06	5.72E+04	32.26	31.8
q	C 254 × 9.13	89.6		254	89	3367	10.4	6.4	3.46E+07	2.73E+05	101.3	2.63E+06	4.20E+04	27.94	25.9
r	C 254 × 12.44	122.0		254	108	4587	12.7	7.9	4.83E+07	3.81E+05	102.6	5.42E+06	7.33E+04	34.29	34.0
s	C 305 × 12.31	120.8		305	102	4540	11.9	7.4	6.65E+07	4.36E+05	121.2	4.59E+06	6.33E+04	31.75	29.0
t	C 305 × 17.59	172.6		305	127	6486	15.7	8.9	9.98E+07	6.55E+05	124.0	1.07E+07	1.25E+05	40.64	40.9

Section properties

Ref.	Shape (in)	(lb/ft)	Depth, A (in)	Width, B (in)	Area (in²)	Flange thickness, t_1 (in)	Web thickness, t (in)	Axis X-X I_x (in⁴)	S_x (in³)	r_x (in)	Axis Y-Y I_y (in⁴)	S_y (in³)	r_y (in)
a	I 3 × 1.637	3.00	2.50	1.392	0.20	0.13	2.24	1.49	1.27	0.52	0.42	0.61	
b	I 3 × 2.030	3.00	2.50	1.726	0.26	0.15	2.71	1.81	1.25	0.68	0.54	0.63	
c	I 4 × 2.311	4.00	3.00	1.965	0.23	0.15	5.62	2.81	1.69	1.04	0.69	0.73	
d	I 4 × 2.793	4.00	3.00	2.375	0.29	0.17	6.71	3.36	1.68	1.31	0.87	0.74	
e	I 5 × 3.700	5.00	3.50	3.146	0.32	0.19	13.94	5.58	2.11	2.29	1.31	0.85	
f	I 6 × 4.030	6.00	4.00	3.427	0.29	0.19	21.99	7.33	2.53	3.10	1.55	0.95	
g	I 6 × 4.692	6.00	4.00	3.990	0.35	0.21	25.50	8.50	2.53	3.74	1.87	0.97	
h	I 7 × 5.800	7.00	4.50	4.932	0.38	0.23	42.89	12.25	2.95	5.78	2.57	1.08	
i	I 8 × 6.181	8.00	5.00	5.256	0.35	0.23	59.69	14.92	3.37	7.30	2.92	1.18	
j	I 8 × 7.023	8.00	5.00	5.972	0.41	0.25	67.78	16.94	3.37	8.55	3.42	1.20	
k	I 9 × 8.361	9.00	5.50	7.110	0.44	0.27	102.02	22.67	3.79	12.22	4.44	1.31	
l	I 10 × 8.646	10.00	6.00	7.352	0.41	0.25	132.09	26.42	4.24	14.78	4.93	1.42	
m	I 10 × 10.286	10.00	6.00	8.747	0.50	0.29	155.79	31.16	4.22	18.03	6.01	1.44	
n	I 12 × 11.672	12.00	7.00	9.925	0.47	0.29	255.57	42.60	5.07	26.90	7.69	1.65	
o	I 12 × 14.292	12.00	7.00	12.153	0.62	0.31	317.33	52.89	5.11	35.48	10.14	1.71	

A–11 (SI) Properties of Aluminum Association Standard I-Beam Shapes SI Units

Ref.	Shape (mm) × (kg/m)	Wt/m (N/m)	Depth, A (mm)	Width, B (mm)	Area (mm²)	Flange thickness, t_1 (mm)	Web thickness, t (mm)	Axis X-X I_x (mm⁴)	Axis X-X S_x (mm³)	Axis X-X r_x (mm)	Axis Y-Y I_y (mm⁴)	Axis Y-Y S_y (mm³)	Axis Y-Y r_y (mm)
a	I 76 × 2.436	23.90	76	64	898	5.1	3.3	9.32E+05	2.44E+04	32.26	2.16E+05	6.88E+03	15.49
b	I 76 × 3.021	29.63	76	64	1114	6.6	3.8	1.13E+06	2.97E+04	31.75	2.83E+05	8.85E+03	16.00
c	I 102 × 3.439	33.73	102	76	1268	5.8	3.8	2.34E+06	4.61E+04	42.93	4.33E+05	1.13E+04	18.54
d	I 102 × 4.156	40.77	102	76	1532	7.4	4.3	2.79E+06	5.51E+04	42.67	5.45E+05	1.43E+04	18.80
e	I 127 × 5.506	54.01	127	89	2030	8.1	4.8	5.80E+06	9.15E+04	53.59	9.53E+05	2.15E+04	21.59
f	I 152 × 5.997	58.83	152	102	2211	7.4	4.8	9.15E+06	1.20E+05	64.26	1.29E+06	2.54E+04	24.13
g	I 152 × 6.982	68.49	152	102	2574	8.9	5.3	1.06E+07	1.39E+05	64.26	1.56E+06	3.06E+04	24.64
h	I 178 × 8.630	84.66	178	114	3182	9.7	5.8	1.79E+07	2.01E+05	74.93	2.41E+06	4.21E+04	27.43
i	I 203 × 9.197	90.22	203	127	3391	8.9	5.8	2.48E+07	2.45E+05	85.60	3.04E+06	4.79E+04	29.97
j	I 203 × 10.45	102.5	203	127	3853	10.4	6.4	2.82E+07	2.78E+05	85.60	3.56E+06	5.61E+04	30.48
k	I 229 × 12.44	122.0	229	140	4587	11.2	6.9	4.25E+07	3.72E+05	96.27	5.09E+06	7.28E+04	33.27
l	I 254 × 12.87	126.2	254	152	4744	10.4	6.4	5.50E+07	4.33E+05	107.7	6.15E+06	8.08E+04	36.07
m	I 254 × 15.31	150.1	254	152	5644	12.7	7.4	6.48E+07	5.11E+05	107.2	7.50E+06	9.85E+04	36.58
n	I 305 × 44.50	436.5	305	178	6404	11.9	7.4	1.06E+08	6.98E+05	128.8	1.12E+07	1.26E+05	41.91
o	I 305 × 23.80	233.5	305	178	7841	15.7	7.9	1.32E+08	8.67E+05	129.8	1.48E+07	1.66E+05	43.43

A–12 Properties of Steel Pipe U.S. Customary Units

Ref.	Nominal size (in)	Outside diameter (in)	Inside diameter (in)	Wall thickness, t_w (in)	Area, A (in^2)	Section properties I (in^4)	S (in^3)	r (in)	Torsional constants J (in^4)	Z_p (in^3)
	Schedule 40 pipe									
a	1/8 in	0.405	0.269	0.068	0.072	1.06E-03	5.25E-03	0.122	2.13E-03	1.05E-02
b	1/4 in	0.540	0.364	0.088	0.125	3.31E-03	1.23E-02	0.163	6.62E-03	2.45E-02
c	3/8 in	0.675	0.493	0.091	0.167	7.29E-03	2.16E-02	0.209	1.46E-02	4.32E-02
d	PIPE 1/2 STD	0.840	0.622	0.109	0.250	1.71E-02	4.07E-02	0.261	3.42E-02	8.14E-02
e	PIPE 3/4 STD	1.050	0.824	0.113	0.333	3.70E-02	7.05E-02	0.334	7.41E-02	0.1411
f	PIPE 1 STD	1.315	1.049	0.133	0.494	8.73E-02	0.1328	0.421	0.1747	0.2657
g	PIPE 1-1/4 STD	1.660	1.380	0.140	0.669	0.1947	0.2346	0.540	0.3894	0.4692
h	PIPE 1-1/2 STD	1.900	1.610	0.145	0.799	0.3099	0.3262	0.623	0.6198	0.6524
i	PIPE 2 STD	2.375	2.067	0.154	1.075	0.6657	0.5606	0.787	1.331	1.121
j	PIPE 2-1/2 STD	2.875	2.469	0.203	1.704	1.530	1.064	0.947	3.059	2.128
k	PIPE 3 STD	3.500	3.068	0.216	2.228	3.017	1.724	1.164	6.034	3.448
l	PIPE 3-1/2 STD	4.000	3.548	0.226	2.680	4.788	2.394	1.337	9.575	4.788
m	PIPE 4 STD	4.500	4.026	0.237	3.174	7.233	3.214	1.510	14.47	6.429
n	PIPE 5 STD	5.563	5.047	0.258	4.300	15.16	5.451	1.878	30.32	10.90
o	PIPE 6 STD	6.625	6.065	0.280	5.581	28.14	8.496	2.245	56.28	16.99
p	PIPE 8 STD	8.625	7.981	0.322	8.399	72.49	16.81	2.938	145.0	33.62
q	PIPE 10 STD	10.750	10.020	0.365	11.908	160.7	29.90	3.674	321.5	59.81
r	12 in	12.750	11.938	0.406	15.745	300.2	47.09	4.367	600.4	94.18
s	16 in	16.000	15.000	0.500	24.347	731.9	91.49	5.483	1464	183.0
t	18 in	18.000	16.876	0.562	30.788	1171	130.2	6.168	2343	260.3

Note: All values shown are for standard schedule 40 steel pipe.
Rows d–q conform to AISC standards for dimensions of standard weight pipe. Rows a–c and r–t do not.
Many other sizes of round hollow structural sections (HSS) are available. See AISC Manual.

A–12 (SI) Properties of Steel Pipe SI Units

					Section properties				Torsional constants	
Ref.	Nominal size (mm)	Outside diameter (mm)	Inside diameter (mm)	Wall thickness, t_w (mm)	Area, A (mm^2)	I (mm^4)	S (mm^3)	r (mm)	J (mm^4)	Z_p (mm^3)
	Schedule 40 pipe									
a	3.2 mm	10.29	6.83	1.73	46.45	442.7	86.07	3.087	885.4	172.1
b	6.4 mm	13.72	9.25	2.24	80.62	1379	201.0	4.135	2757	402.1
c	9.5 mm	17.15	12.52	2.31	107.7	3035	354.0	5.308	6069	708.0
d	PIPE 13 STD	21.34	15.80	2.77	161.5	7114	666.9	6.637	14 228	1334
e	PIPE 19 STD	26.67	20.93	2.87	214.6	15416	1156	8.475	30 831	2312
f	PIPE 25 STD	33.40	26.64	3.38	318.6	36355	2177	10.68	72 710	4354
g	PIPE 32 STD	42.16	35.05	3.56	431.3	81044	3844	13.71	1.62E+05	7688
h	PIPE 38 STD	48.26	40.89	3.68	515.8	1.29E+05	5346	15.81	2.58E+05	10 691
i	PIPE 51 STD	60.33	52.50	3.91	693.2	2.77E+05	9187	19.99	5.54E+05	18 374
j	PIPE 64 STD	73.03	62.71	5.16	1099	6.37E+05	17436	24.06	1.27E+06	34 873
k	PIPE 75 STD	88.90	77.93	5.49	1438	1.26E+06	28253	29.55	2.51E+06	56 506
l	PIPE 89 STD	101.6	90.12	5.74	1729	1.99E+06	39228	33.95	3.99E+06	78 457
m	PIPE 102 STD	114.3	102.3	6.02	2048	3.01E+06	52676	38.34	6.02E+06	1.05E+05
n	PIPE 127 STD	141.3	128.2	6.55	2774	6.31E+06	89327	47.70	1.26E+07	1.79E+05
o	PIPE 152 STD	168.3	154.1	7.11	3601	1.17E+07	1.39E+05	57.04	2.34E+07	2.78E+05
p	PIPE 203 STD	219.1	202.7	8.18	5419	3.02E+07	2.75E+05	74.62	6.03E+07	5.51E+05
q	PIPE 254 STD	273.1	254.5	9.27	7683	6.69E+07	4.90E+05	93.32	1.34E+08	9.80E+05
r	305 mm	323.9	303.2	10.31	10 158	1.25E+08	7.72E+05	110.9	2.50E+08	1.54E+06
s	406 mm	406.4	381.0	12.70	15 708	3.05E+08	1.50E+06	139.3	6.09E+08	3.00E+06
t	457 mm	457.2	428.7	14.27	19 863	4.88E+08	2.13E+06	156.7	9.75E+08	4.27E+06

Note: All values shown are for standard schedule 40 steel pipe, converted to SI units.
Rows d–q conform to AISC standards for dimensions of standard weight pipe. Rows a–c and r–t do not.
Many other sizes of round hollow structural sections (HSS) are available. See AISC Manual.

A–13 Properties of Steel Mechanical Tubing U.S. Customary Units

Ref.	Nominal size OD (in)	Wall gage	Outside diameter (in)	Inside diameter (in)	Wall thickness, t_w (in)	Area, A (in²)	Section properties I (in⁴)	S (in³)	r (in)	Torsional constants J (in⁴)	Z_p (in³)
a	$\frac{1}{2}$	17	0.500	0.384	0.058	0.081	0.002 00	0.008 00	0.158	0.004 00	0.016 0
b	$\frac{1}{2}$	14	0.500	0.334	0.083	0.109	0.002 46	0.009 83	0.150	0.004 91	0.019 7
c	1	16	1.000	0.870	0.065	0.191	0.021 0	0.041 9	0.331	0.041 9	0.083 9
d	1	10	1.000	0.732	0.134	0.365	0.035 0	0.070 0	0.310	0.070 0	0.140
e	$1\frac{1}{2}$	16	1.500	1.370	0.065	0.293	0.075 6	0.101	0.508	0.151	0.202
f	$1\frac{1}{2}$	10	1.500	1.232	0.134	0.575	0.135	0.181	0.485	0.271	0.361
g	2	16	2.000	1.870	0.065	0.395	0.185	0.185	0.685	0.370	0.370
h	2	10	2.000	1.732	0.134	0.786	0.344	0.344	0.661	0.687	0.687
i	$2\frac{1}{2}$	10	2.500	2.232	0.134	0.996	0.699	0.559	0.838	1.398	1.119
j	$2\frac{1}{2}$	5	2.500	2.060	0.220	1.576	1.034	0.827	0.810	2.067	1.654
k	3	10	3.000	2.732	0.134	1.207	1.241	0.828	1.014	2.483	1.655
l	3	5	3.000	2.560	0.220	1.921	1.868	1.245	0.986	3.736	2.490
m	$3\frac{1}{2}$	10	3.500	3.232	0.134	1.417	2.010	1.149	1.191	4.020	2.297
n	$3\frac{1}{2}$	5	3.500	3.060	0.220	2.267	3.062	1.750	1.162	6.125	3.500
o	4	5	4.000	3.560	0.220	2.613	4.682	2.341	1.339	9.364	4.682
p	4	1	4.000	3.400	0.300	3.487	6.007	3.003	1.312	12.013	6.007
q	$4\frac{1}{2}$	5	4.500	4.060	0.220	2.958	6.791	3.018	1.515	13.583	6.037
r	$4\frac{1}{2}$	1	4.500	3.900	0.300	3.958	8.773	3.899	1.489	17.546	7.798
s	5	5	5.000	4.560	0.220	3.304	9.456	3.782	1.692	18.911	7.564
t	5	1	5.000	4.400	0.300	4.430	12.281	4.912	1.665	24.562	9.825

A–13 (SI) Properties of Steel Mechanical Tubing SI Units

	Nominal size		Outside diameter (mm)	Inside diameter (mm)	Wall thickness, t_w (mm)	Area, A (mm²)	Section properties				Torsional constants	
Ref.	OD (mm)	Wall gage					I (mm⁴)	S (mm³)	r (mm)		J (mm⁴)	Z_p (mm³)
a	12.7	17	12.70	9.754	1.473	51.96	833	131.1	4.00		1665	262.3
b	12.7	14	12.70	8.484	2.108	70.15	1023	161.1	3.82		2045	322.1
c	25.4	16	25.40	22.098	1.651	123.2	8726	687.1	8.42		1.75E+04	1374
d	25.4	10	25.40	18.593	3.404	235.2	1.46E+04	1147	7.87		2.91E+04	2294
e	38.1	16	38.10	34.798	1.651	189.1	3.15E+04	1651	12.9		6.29E+04	3303
f	38.1	10	38.10	31.293	3.404	371.0	5.64E+04	2959	12.3		1.13E+05	5918
g	50.8	16	50.80	47.498	1.651	254.9	7.71E+04	3034	17.4		1.54E+05	6068
h	50.8	10	50.80	43.993	3.404	506.8	1.43E+05	5632	16.8		2.86E+05	1.13E+04
i	63.5	10	63.50	56.693	3.404	642.6	2.91E+05	9166	21.3		5.82E+05	1.83E+04
j	63.5	5	63.50	52.324	5.588	1017	4.30E+05	1.35E+04	20.6		8.60E+05	2.71E+04
k	76.2	10	76.20	69.393	3.404	778.4	5.17E+05	1.36E+04	25.8		1.03E+06	2.71E+04
l	76.2	5	76.20	65.024	5.588	1240	7.77E+05	2.04E+04	25.0		1.55E+06	4.08E+04
m	88.9	10	88.90	82.093	3.404	914.2	8.37E+05	1.88E+04	30.3		1.67E+06	3.76E+04
n	88.9	5	88.90	77.724	5.588	1463	1.27E+06	2.87E+04	29.5		2.55E+06	5.74E+04
o	101.6	5	101.60	90.424	5.588	1686	1.95E+06	3.84E+04	34.0		3.90E+06	7.67E+04
p	101.6	1	101.60	86.360	7.620	2250	2.50E+06	4.92E+04	33.3		5.00E+06	9.84E+04
q	114.3	5	114.30	103.124	5.588	1908	2.83E+06	4.95E+04	38.5		5.65E+06	9.89E+04
r	114.3	1	114.30	99.060	7.620	2554	3.65E+06	6.39E+04	37.8		7.30E+06	1.28E+05
s	127.0	5	127.00	115.824	5.588	2131	3.94E+06	6.20E+04	43.0		7.87E+06	1.24E+05
t	127.0	1	127.00	111.760	7.620	2858	5.11E+06	8.05E+04	42.3		1.02E+07	1.61E+05

A–14 Typical Properties of Carbon and Alloy Steels*

Material AISI no.	Condition[†]	Ultimate strength, s_u		Yield strength, s_y		Percent elongation
		ksi	MPa	ksi	MPa	
1020	Annealed	57	393	43	296	36
1020	Hot-rolled	65	448	48	331	36
1020	Cold-drawn	75	517	64	441	20
1040	Annealed	75	517	51	352	30
1040	Hot-rolled	90	621	60	414	25
1040	Cold-drawn	97	669	82	565	16
1040	WQT 700	127	876	93	641	19
1040	WQT 900	118	814	90	621	22
1040	WQT 1100	107	738	80	552	24
1040	WQT 1300	87	600	63	434	32
1080	Annealed	89	614	54	372	25
1080	OQT 700	189	1303	141	972	12
1080	OQT 900	179	1234	129	889	13
1080	OQT 1100	145	1000	103	710	17
1080	OQT 1300	117	807	70	483	23
1141	Annealed	87	600	51	352	26
1141	Cold-drawn	112	772	95	655	14
1141	OQT 700	193	1331	172	1186	9
1141	OQT 900	146	1007	129	889	15
1141	OQT 1100	116	800	97	669	20
1141	OQT 1300	94	648	68	469	28
4140	Annealed	95	655	60	414	26
4140	OQT 700	231	1593	212	1462	12
4140	OQT 900	187	1289	173	1193	15
4140	OQT 1100	147	1014	131	903	18
4140	OQT 1300	118	814	101	696	23
5160	Annealed	105	724	40	276	17
5160	OQT 700	263	1813	238	1641	9
5160	OQT 900	196	1351	179	1234	12
5160	OQT 1100	149	1027	132	910	17
5160	OQT 1300	115	793	103	710	23

*Other properties approximately the same for all carbon and alloy steels:

Modulus of elasticity in tension = 30 000 000 psi (207 GPa)

Modulus of elasticity in shear = 11 500 000 psi (80 GPa)

Density = 0.283 lb_m/in^3 (7680 kg/m^3)

[†]OQT means oil-quenched and tempered. WQT means water-quenched and tempered.

A–15 Typical Properties of Stainless Steels and Nonferrous Metals

Material and condition		Ultimate strength, s_u		Yield strength, s_y		Percent elongation	Density		Modulus of elasticity, E	
		ksi	MPa	ksi	MPa		lb/in³†	kg/m³	psi	GPa
Stainless steels										
AISI 301 annealed		110	758	40	276	60	0.290	8030	28×10^6	193
AISI 301 full hard		185	1280	140	965	8	0.290	8030	28×10^6	193
AISI 430 annealed		75	517	40	276	30	0.280	7750	29×10^6	200
AISI 430 full hard		90	621	80	552	15	0.280	7750	29×10^6	200
AISI 501 annealed		70	483	30	207	28	0.280	7750	29×10^6	200
AISI 5010QT 1000		175	1210	135	931	15	0.280	7750	29×10^6	200
17-4PH H900		210	1450	185	1280	14	0.281	7780	28.5×10^6	197
PH 13-8 Mo H1000		215	1480	205	1410	13	0.279	7720	29.4×10^6	203
Copper and its alloys										
C14500 copper,	soft	32	221	10	69	50	0.323	8940	17×10^6	117
	hard	48	331	44	303	20				
C17200 Beryllium copper,	soft	72	496	20	138	20	0.298	8250	19×10^6	131
	hard	195	1344	145	1000	4				
C36000 brass,	soft	44	305	18	124	20	0.308	8530	16×10^6	110
	hard	70	480	35	240	4				
C54400 bronze,	hard	68	469	57	393	20	0.318	8800	17×10^6	117
Magnesium—cast										
ASTM AZ 63A-T6		40	276	19	131	5	0.066	1830	6.5×10^6	45
Zinc—cast-ZA 12		58	400	47	324	5	0.218	6030	12×10^6	83

(*Continued*)

Titanium and its alloys

Material								
Pure alpha Ti-65A Wrought	65	448	55	379	18	0.163	4515	15×10^6 / 103
Alpha alloy Ti-0.2Pd Wrought	50	345	40	276	20	0.163	4515	14.9×10^6 / 103
Beta alloy Ti-3Al-13V-11Cr Aged	185	1280	175	1210	6	0.176	4875	16.0×10^6 / 110
Alpha-beta alloy Ti-6Al-4V Aged	170	1170	155	1070	8	0.160	4432	16.5×10^6 / 114

Nickel-based alloys

Material								
N06600—annealed								
70°F (21°C)	93	640	37	255	45	0.304	8420	30×10^6 / 207
800°F (427°C)	89	614	30	207	49			
1200°F (649°C)	65	448	27	186	39			
N06110—40% cold-worked								
70°F (21°C)	175	1205	150	1034	18	0.302	8330	30×10^6 / 207
500°F (260°C)			130	896	18			
800°F (427°C)			120	827	18			
N04400—annealed [At 70°F (21°C)]								
Annealed	80	550	30	207	50	0.318	8800	26×10^6 / 181
Cold-drawn	100	690	75	517	30			

†This can be used as specific weight or mass density in lb_m/in^3.

A–16 Properties of Structural Steels

Material ASTM no. and products	Ultimate strength, s_u*		Yield strength, s_y*		Percent elongation in 2 in
	ksi	MPa	ksi	MPa	
A36—Carbon steel: shapes, plates, and bars	58	400	36	248	21
A 53—grade B-pipe	60	414	35	240	—
A242—HSLA corrosion resistant: shapes, plates, and bars					
$\leq \frac{3}{4}$ in thick	70	483	50	345	21
$\frac{3}{4}$ to $1\frac{1}{2}$ in thick	67	462	46	317	21
$1\frac{1}{2}$ to 4 in thick	63	434	42	290	21
A500—Cold-formed structural tubing					
Round, grade B	58	400	42	290	23
Round grade C	62	427	46	317	21
Shaped, grade B	58	400	46	317	23
Shaped, grade C	62	427	50	345	21
A501—Hot-formed structural tubing, round or shaped	58	400	36	248	23
A514—Quenched and tempered alloy steel: plate					
$\leq 2\frac{1}{2}$ in thick	110	758	100	690	18
$2\frac{1}{2}$ to 6 in thick	100	690	90	620	16
A572—HSLA columbium-vanadium steel: shapes, plates and bars					
Grade 42	60	414	42	290	24
Grade 50	65	448	50	345	21
Grade 60	75	517	60	414	18
Grade 65	80	552	65	448	17
A913—HSLA, grade 65: shapes	80	552	65	448	17
A992—HSLA: W-Shapes only	65	448	50	345	21

*Minimum values; may range higher.

HSLA-High strength low-alloy

The American Institute of Steel Construction specifies $E = 29 \times 10^6$ psi (200 GPa) for structural steel.

A–17 Typical Properties of Cast Iron*

Material type and grade	s_u[†] ksi	s_u[†] MPa	s_{uc}[‡] ksi	s_{uc}[‡] MPa	s_{us} ksi	s_{us} MPa	s_{yt}[‡] ksi	s_{yt}[‡] MPa	Modulus of elasticity, E[‡] psi	Modulus of elasticity, E[‡] GPa	Percent elongation
Gray iron ASTM A48											
Grade 20	20	138	80	552	32	221	—	—	12.2×10^6	84	<1
Grade 40	40	276	140	965	57	393	—	—	19.4×10^6	134	<0.8
Grade 60	55	379	170	1170	72	496	—	—	21.5×10^6	148	<0.5
Ductile iron ASTM A536											
60-40-18	60	414	—	—	57	393	40	276	24×10^6	165	18
80-55- 6	80	552	—	—	73	503	55	379	24×10^6	165	6
100-70- 3	100	690	—	—	—	—	70	483	24×10^6	165	3
120-90- 2	120	827	180	1240	—	—	90	621	23×10^6	159	2
Austempered ductile iron (ADI)											
Grade 1	125	862	—	—	—	—	80	552	24×10^6	165	10
Grade 2	150	1034	—	—	—	—	100	690	24×10^6	165	7
Grade 3	175	1207	—	—	—	—	125	862	24×10^6	165	4
Grade 4	200	1379	—	—	—	—	155	1069	24×10^6	165	1
Malleable iron ASTM A220											
45 008	65	448	240	1650	49	338	45	310	26×10^6	170	8
60 004	80	552	240	1650	65	448	60	414	27×10^6	186	4
80 002	95	655	240	1650	75	517	80	552	27×10^6	186	2

*The density of cast iron ranges from 0.25 to 0.27 lb_m/in^3 (6920 to 7480 kg/m^3).

[†]Minimum values; may range higher.

[‡]Approximate values; may range higher or lower by about 15%.

A–18 Typical Properties of Aluminum Alloys*

Alloy and temper	Ultimate strength, s_u		Yield strength, s_y		Percent elongation	Shear strength, s_u	
	ksi	MPa	ksi	MPa		ksi	MPa
1100-H12	16	110	15	103	25	10	69
1100-H18	24	165	22	152	15	13	90
2014-0	27	186	14	97	18	18	124
2014-T4	62	427	42	290	20	38	262
2014-T6	70	483	60	414	13	42	290
3003-0	16	110	6	41	40	11	76
3003-H12	19	131	18	124	20	12	83
3003-H18	29	200	27	186	10	16	110
5154-0	35	241	17	117	27	22	152
5154-H32	39	269	30	207	15	22	152
5154-H38	48	331	39	269	10	28	193
6061-0	18	124	8	55	30	12	83
6061-T4	35	241	21	145	25	24	165
6061-T6	45	310	40	276	17	30	207
7075-0	33	228	15	103	16	22	152
7075-T6	83	572	73	503	11	48	331
Casting alloys (permanent mold castings)							
204.0-T4	48	331	29	200	8	—	—
206.0-T6	65	445	59	405	6	—	—
356.0-T6	41	283	30	207	10	—	—

*Modulus of elasticity E for most aluminum alloys, including 1100, 3003, 6061, and 6063, is 10×10^6 psi (69.0 GPa). For 2014, $E = 10.6 \times 10^6$ psi (73.1 GPa). For 5154, $E = 10.2 \times 10^6$ psi (70.3 GPa). For 7075, $E = 10.4 \times 10^6$ psi (71.7 GPa). Density of most aluminum alloys is approximately 0.10 lb_m/in^3 (2770 kg/m^3).

A-19 Typical Properties of Wood

	Allowable stress										Modulus of elasticity	
	Bending		Tension parallel to grain		Horizontal shear		Compression					
							Perpendicular to grain		Parallel to grain			
Type and grade	psi	MPa	psi	MPa	psi	MPa	psi	MPa	psi	MPa	ksi	GPa
Douglas fir—2 to 4 in thick, 6 in and wider												
No. 1	1750	12.1	1050	7.2	95	0.66	385	2.65	1250	8.62	1800	12.4
No. 2	1450	10.0	850	5.9	95	0.66	385	2.65	1000	6.90	1700	11.7
No. 3	800	5.5	475	3.3	95	0.66	385	2.65	600	4.14	1500	10.3
Hemlock—2 to 4 in thick, 6 in and wider												
No. 1	1400	9.6	825	5.7	75	0.52	245	1.69	1000	6.90	1500	10.3
No. 2	1150	7.9	675	4.7	75	0.52	245	1.69	800	5.52	1400	9.7
No. 3	625	4.3	375	2.6	75	0.52	245	1.69	500	3.45	1200	8.3
Southern pine—$2\frac{1}{2}$ to 4 in thick, 6 in and wider												
No. 1	1400	9.6	825	5.7	80	0.55	270	1.86	850	5.86	1600	11.0
No. 2	1000	6.9	575	4.0	70	0.48	230	1.59	550	3.79	1300	9.0
No. 3	650	4.5	375	2.6	70	0.48	230	1.59	400	2.76	1300	9.0

A–20 Typical Properties of Selected Plastics

Material	Type	Tensile strength (ksi)	Tensile strength (MPa)	Tensile modulus (ksi)	Tensile modulus (MPa)	Flexural strength (ksi)	Flexural strength (MPa)	Flexural modulus (ksi)	Flexural modulus (MPa)	Impact strength IZOD (ft·lb/in of notch)
Nylon 66	Dry	21.0	146	1200	8700	32.0	221	1100	7900	
30% Glass	50% R.H.	15.0	102	800	5500					
ABS	Medium-impact	6.0	41	360	2480	11.5	79	310	2140	4.0
	High-impact	5.0	34	250	1720	8.0	55	260	1790	7.0
Polycarbonate	General-purpose	9.0	62	340	2340	11.0	76	300	2070	12.0
Acrylic	Standard	10.5	72	430	2960	16.0	110	460	3170	0.4
	High-impact	5.4	37	220	1520	7.0	48	230	1590	1.2
PVC	Rigid	6.0	41	350	2410			300	2070	0.4–20.0 (varies widely)
Polyimide	25% graphite powder filler	5.7	39			12.8	88	900	6210	0.25
	Glass-fiber filler	27.0	186			50.0	345	3250	22 400	17.0
	Laminate	50.0	345			70.0	483	4000	27 580	13.0
Acetal	Copolymer	8.0	55	410	2830	13.0	90	375	2590	1.3
Polyurethane	Elastomer	5.0	34	100	690	0.6	4			No break
Phenolic	General	6.5	45	1100	7580	9.0	62	1100	7580	0.3
Polyester with glass-fiber mat reinforcement (approx. 30% glass by weight)										
	Lay-up, contact mold	9.0	62			16.0	110	800	5520	
	Cold press molded	12.0	83			22.0	152	1300	8960	
	Compression molded	25.0	172			10.0	69	1300	8960	

A–21 Design Stress Guidelines

Direct Normal Stresses—General machine and structural design

Manner of loading	Ductile materials (% Elongation >5%)	Brittle materials (% Elongation <5%)
Static loads	$\sigma_d = s_y/2$	$\sigma_d = s_u/6$
Repeated loads	$\sigma_d = s_u/8$	$\sigma_d = s_u/10$
Impact or shock	$\sigma_d = s_u/12$	$\sigma_d = s_u/15$

Direct Normal Stresses—Static loads on steel members of building-like structures

AISC Code: $\sigma_d = s_y/1.67 = 0.60\, s_y$ or $\sigma_d = s_u/2.00 = 0.50\, s_u$
Whichever is lower.

Direct Normal Stresses—Static loads on aluminum members of building-like structures

Aluminum Association: $\sigma_d = s_y/1.65 = 0.61\, s_y$ or $\sigma_d = s_u/1.95 = 0.51\, s_u$
Whichever is lower.

Design Shear Stresses—For direct shear and for torsional shear stresses

Based on maximum shear stress theory of failure:

$$\tau_d = s_{ys}/N = 0.5\, s_y/N = s_y/2N$$

Manner of loading	Design factor	Design shear stress
Static loads	Use $N = 2$	$\tau_d = s_y/4$
Repeated loads	Use $N = 4$	$\tau_d = s_y/8$
Shock or impact	Use $N = 6$	$\tau_d = s_y/12$

Estimates for the Ultimate Strength in Shear

Formula	Material
$s_{us} = 0.65\, s_u$	Aluminum alloys
$s_{us} = 0.82\, s_u$	Steel—Plain carbon and alloy
$s_{us} = 0.90\, s_u$	Malleable iron and copper alloys
$s_{us} = 1.30\, s_u$	Gray cast iron

Allowable Bearing Stress

Steel—Flat surfaces or the projected area of pins in reamed, drilled, or bored holes:

$$\sigma_{bd} = 0.90 s_y$$

Allowable Bearing Load, R_a—Steel roller on flat steel plate

U.S. Customary Units	SI Metric Units
$R_a = (s_y - 13)\,(0.03dL)$	$R_a = (s_y - 90)\,(3.0 \times 10^{-5} dL)$

Where: R_a = allowable bearing load in kips or kN
  s_y = yield strength of steel in ksi or MPa
  d = roller diameter in inches or mm
  L = length of roller in inches or mm

(Continued)

A–21 (Continued)

Allowable bearing stresses on masonry and soils for use in this book

Material	Allowable bearing stress, σ_{bd}	
	psi	MPa
Sandstone and limestone	400	2.76
Brick in cement mortar	250	1.72
Solid hard rock	350	2.41
Shale or medium rock	140	0.96
Soft rock	70	0.48
Hard clay or compact gravel	55	0.38
Soft clay or loose sand	15	0.10

Concrete: $\sigma_{bd} = Kf'_c = (0.34\sqrt{A_2/A_1})f'_c$ (But maximum $\sigma_{bd} = 0.68 f'_c$)

Where f'_c = rated strength of concrete

A_1 = bearing area

A_2 = full area of the support

$A_2/A_1 = 1.0$ $A_2/A_1 > 1.0$

(Continued)

A–21 Design Stress Guidelines (*Continued*)

Design Bending Stresses—General machine and structural design

Manner of loading	Ductile materials (% Elongation >5%)	Brittle materials (% elongation <5%)
Static loads	$\sigma_d = s_y/2$	$\sigma_d = s_u/6$
Repeated loads	$\sigma_d = s_u/8$	$\sigma_d = s_u/10$
Impact or shock	$\sigma_d = s_u/12$	$\sigma_d = s_u/15$

Design Bending Stresses—AISC specifications for structural steel carrying static loads on building-like structures

$$\sigma_d = s_y/1.5 = 0.66\, s_y$$

Design Bending Stresses—Aluminum Association specifications for aluminum carrying static loads on building-like structures

$$\sigma_d = s_y/1.65 = 0.61\, s_y \qquad \text{or} \qquad \sigma_d = s_u/1.95 = 0.51\, s_u$$

Whichever is lower.

Design Shear Stresses for Beams in Bending
Rolled structural steel beam shapes—allowable web shear stress (AISC)

$$\tau_d = 0.40\, s_y$$

General ductile materials carrying static loads: Based on yield strength of the material in shear with design factor, $N = 2$:

$$\tau_d = s_{ys}/N = 0.5\, s_y/N = s_y/2N = s_y/2(2) = s_y/4 = 0.25\, s_y$$

A–22 Stress Concentration Factors

A–22–1 Axially loaded grooved round bar in tension

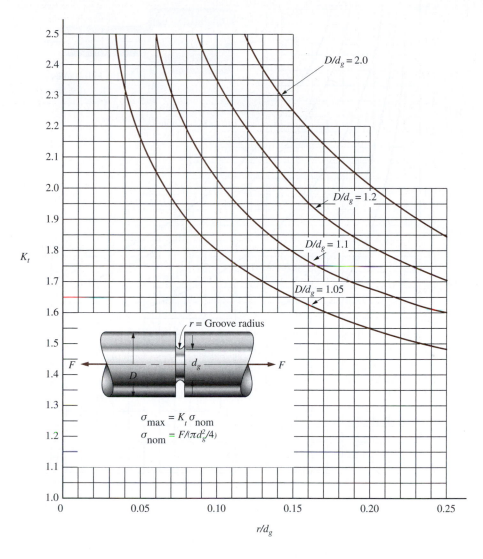

$$\sigma_{max} = K_t \sigma_{nom}$$
$$\sigma_{nom} = F/(\pi d_g^2/4)$$

A–22–2 Axially loaded stepped round bar in tension

A–22–3 Axially loaded stepped flat plate in tension

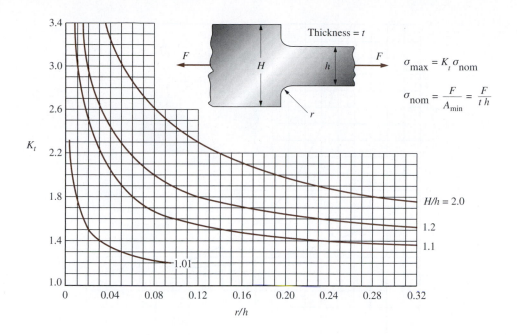

A–22–4 Flat plate with central hole in tension and bending

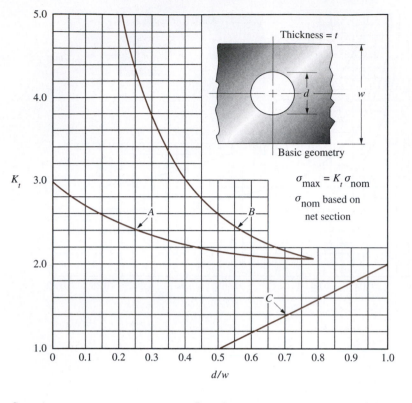

Thickness = t

Basic geometry

$\sigma_{max} = K_t \sigma_{nom}$

σ_{nom} based on net section

K_t

d/w

Curve A

Direct tension on plate

$$\sigma_{nom} = \frac{F}{A_{net}} = \frac{F}{(w-d)t}$$

F F = total load

Curve B

Tension load applied through a pin in the hole

$$\sigma_{nom} = \frac{F}{A_{net}} = \frac{F}{(w-d)t}$$

F

Curve C

Bending in the plane of the plate

$$\sigma_{nom} = \frac{Mc}{I_{net}} = \frac{6Mw}{(w^3-d^3)t}$$

M M

Note: $K_t = 1.0$ for $d/w < 0.5$

A–22–5 Round bar with transverse hole in tension, bending, and torsion

Basic geometry

Note: K_{t_g} is based on the nominal stress in a round bar without a hole (gross section).

$$\sigma_{max}=K_{t_g}\,\sigma_{gross} \qquad \tau_{max}=K_{t_g}\,\tau_{gross}$$

Curve A	Curve B	Curve C
Tension	Bending	Torsion

$$\sigma_{gross}=\frac{F}{A}=\frac{F}{\pi D^2/4} \qquad \sigma_{gross}=\frac{M}{S}=\frac{M}{\pi D^3/32} \qquad \tau_{gross}=\frac{T}{Z_p}=\frac{T}{\pi D^3/16}$$

A–22–6 Grooved round bar in torsion

A–22–7 Stepped round bar in torsion

$$\tau_{max} = K_t \, \tau_{nom}$$
$$\tau_{nom} = T/(\pi d_g^3/16)$$

A–22–8 Grooved round bar in bending

$\sigma_{max} = K_t\, \sigma_{nom}$

$\sigma_{nom} = M/(\pi d_g^3/32)$

A–22–9 Stepped round bar in bending

$\sigma_{max} = K_t\,\sigma_{nom}$
$\sigma_{nom} = M/(\pi d^3/32)$

r = Fillet radius

A–22–10 Stepped flat plate in bending

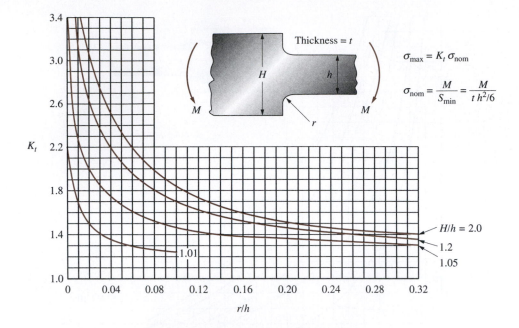

A–22–11 Shafts with Keyseats—Bending and Torsion	
Type of keyseat	K_t*
Sled-runner	1.6
Profile	2.0

$*K_t$ is to be applied to the stress computed for the full nominal diameter of the shaft where the keyseat is located.

A–23 Beam Deflection Formulas for Simply Supported Beams

(a)

$$y_B = y_{max} = \frac{-PL^3}{48EI} \quad \text{at center}$$

Between A and B:

$$y = \frac{-Px}{48EI}(3L^2 - 4x^2)$$

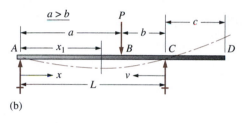

(b)

$$y_{max} = \frac{Pab(L + b)\sqrt{3a(L + b)}}{27EIL}$$

$$\text{at } x_1 = \sqrt{a(L + b)/3}$$

$$y_B = \frac{-Pa^2b^2}{3EIL} \quad \text{at load}$$

Between A and B (the longer segment):

$$y = \frac{-Pbx}{6EIL}(L^2 - b^2 - x^2)$$

Between B and C (the shorter segment):

$$y = \frac{-Pav}{6EIL}(L^2 - v^2 - a^2)$$

At end of overhang at D:

$$y_D = \frac{Pabc}{6EIL}(L + a)$$

(c)

$$y_E = y_{max} = \frac{-Pa}{24EI}(3L^2 - 4a^2) \quad \text{at center}$$

$$y_B = y_C = \frac{-Pa^2}{6EI}(3L - 4a) \quad \text{at loads}$$

Between A and B:

$$y = \frac{-Px}{6EI}(3aL - 3a^2 - x^2)$$

Between B and C:

$$y = \frac{-Pa}{6EI}(3Lx - 3x^2 - a^2)$$

A–23 (*Continued*)

(d)

$$y_B = y_{max} = \frac{-5wL^4}{384EI} = \frac{-5WL^3}{384EI} \text{ at center}$$

Between A and B:

$$y = \frac{-wx}{24EI}(L^3 - 2Lx^2 + x^3)$$

At D at end:

$$y_D = \frac{wL^3 a}{24EI}$$

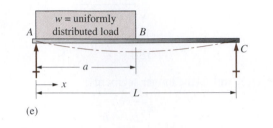

(e)

Between A and B:

$$y = \frac{-wx}{24EIL}[a^2(2L-a)^2 - 2ax^2(2L-a) + Lx^3]$$

Between B and C:

$$y = \frac{-wa^2(L-x)}{24EIL}(4Lx - 2x^2 - a^2)$$

M_B = concentrated moment at B

Between A and B:

$$y = \frac{-M_B}{6EI}\left[\left(6a - \frac{3a^2}{L} - 2L\right)x - \frac{x^3}{L}\right]$$

Between B and C:

$$y = \frac{M_B}{6EI}\left[3a^2 + 3x^2 - \frac{x^3}{L} - \left(2L + \frac{3a^2}{L}\right)x\right]$$

(f)

(g)

At C at end of overhang:

$$y_C = \frac{-Pa^2}{3EI}(L + a)$$

At D, maximum upward deflection:

$$y_D = 0.06415\frac{PaL^2}{EI}$$

A–23 (*Continued*)

(h)

At *C* at center:

$$y = \frac{-W(L - 2a)^3}{384EI}\left[\frac{5}{L}(L - 2a) - \frac{24}{L}\left(\frac{a^2}{L - 2a}\right)\right]$$

At *A* and *E* at ends:

$$y = \frac{-W(L - 2a)^3 a}{24EIL}\left[-1 + 6\left(\frac{a}{L - 2a}\right)^2 + 3\left(\frac{a}{L - 2a}\right)^3\right]$$

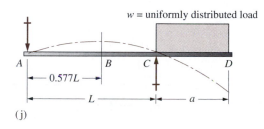

(i)

At *C* at center:

$$y = \frac{PL^2 a}{8EI}$$

At *A* and E at ends at loads:

$$y = \frac{-Pa^2}{3EI}\left(a + \frac{3}{2}L\right)$$

At *B*:

$$y = 0.03208\,\frac{wa^2 L^2}{EI}$$

At *D* at end:

$$y = \frac{-wa^3}{24EI}(4L + 3a)$$

(j)

A–24 Beam Deflection Formulas for Cantilevers

(a)

At B at end:

$$y_B = y_{max} = \frac{-PL^3}{3EI}$$

Between A and B:

$$y = \frac{-Px^2}{6EI}(3L - x)$$

At B at load:

$$y_B = \frac{-Pa^3}{3EI}$$

At C at end:

$$y_C = y_{max} = \frac{-Pa^2}{6EI}(3L - a)$$

Between A and B:

$$y = \frac{-Px^2}{6EI}(3a - x)$$

Between B and C:

$$y = \frac{-Pa^2}{6EI}(3x - a)$$

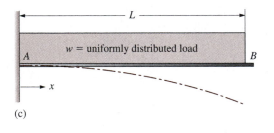

(b)

W = total load = wL

At B at end:

$$y_B = y_{max} = \frac{-WL^3}{8EI}$$

Between A and B:

$$y = \frac{-Wx^2}{24EIL}[2L^2 + (2L - x)^2]$$

(c)

M_B = concentrated moment at end

At B at end:

$$y_B = y_{max} = \frac{-M_B L^2}{2EI}$$

Between A and B:

$$y = \frac{-M_B x^2}{2EI}$$

(d)

A–25 Beam Diagrams and Deflection Formulas for Statically Indeterminate Beams

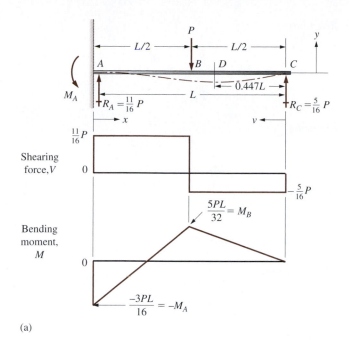

Deflections
At B at load:

$$y_B = \frac{-7}{768}\frac{PL^3}{EI}$$

y_{max} is at $v = 0.447L$ at D:

$$y_D = y_{max} = \frac{-PL^3}{107EI}$$

Between A and B:

$$y = \frac{-Px^2}{96EI}(9L - 11x)$$

Between B and C:

$$y = \frac{-Pv}{96EI}(3L^2 - 5v^2)$$

(a)

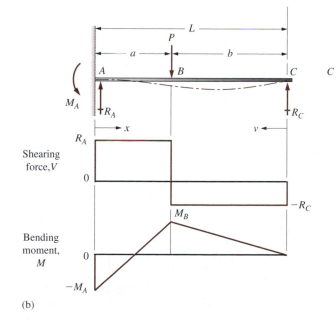

(b)

Reactions

$$R_A = \frac{Pb}{2L^3}(3L^2 - b^2)$$

$$R_C = \frac{Pa^2}{2L^3}(b + 2L)$$

Moments

$$M_A = \frac{-Pab}{2L^2}(b + L)$$

$$M_B = \frac{Pa^2b}{2L^3}(b + 2L)$$

Deflections
At B at load:

$$y_B = \frac{-Pa^3b^2}{12EIL^3}(3L + b)$$

Between A and B:

$$y = \frac{-Px^2b}{12EIL^3}(3C_1 - C_2x)$$

$$C_1 = aL(L + b); C_2 = (L + a)(L + b) + aL$$

Between B and C:

$$y = \frac{-Pa^2v}{12EIL^3}[3L^2b - v^2(3L - a)]$$

A–25 (*Continued*)

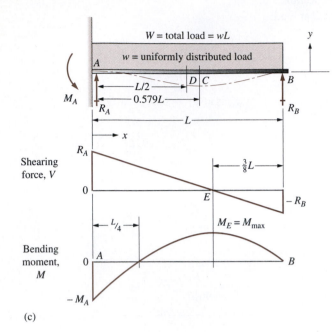

(c)

Reactions

$$R_A = \frac{5}{8} W$$

$$R_B = \frac{3}{8} W$$

Moments

$$M_A = -0.125\,WL$$

$$M_E = 0.0703\,WL$$

Deflections
At C at $x = 0.579L$:

$$y_C = y_{max} = \frac{-WL^3}{185EI}$$

At D at center:

$$y_D = \frac{-WL^3}{192EI}$$

Between A and B:

$$y = \frac{-Wx^2(L - x)}{48EIL}(3L - 2x)$$

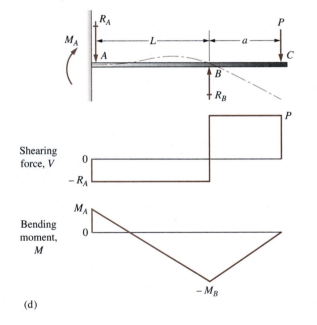

(d)

Reactions

$$R_A = \frac{-3Pa}{2L}$$

$$R_B = P\left(1 + \frac{3a}{2L}\right)$$

Moments

$$M_A = \frac{Pa}{2}$$

$$M_B = -Pa$$

Deflection
At C at end:

$$y_C = \frac{-PL^3}{EI}\left(\frac{a^2}{4L^2} + \frac{a^3}{3L^3}\right)$$

A–25 (*Continued*)

(e)

Moments

$$M_A = M_B = M_C = \frac{PL}{8}$$

Deflections
At *B* at center:

$$y_B = y_{max} = \frac{-PL^3}{192EI}$$

Between *A* and *B*:

$$y = \frac{-Px^2}{48EI}(3L - 4x)$$

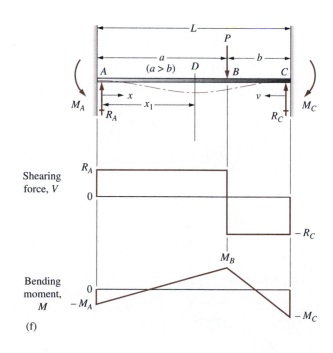

(f)

Reactions

$$R_A = \frac{Pb^2}{L^3}(3a + b)$$

$$R_C = \frac{Pa^2}{L^3}(3b + a)$$

Moments

$$M_A = \frac{-Pab^2}{L^2}$$

$$M_B = \frac{2Pa^2b^2}{L^3}$$

$$M_C = \frac{-Pa^2b}{L^2}$$

Deflections
At *B* at load:

$$y_B = \frac{-Pa^3b^3}{3EIL^3}$$

At *D* at $x_1 = \frac{2aL}{3a + b}$

$$y_D = y_{max} = \frac{-2Pa^3b^2}{3EI(3a + b)^2}$$

Between *A* and *B* (longer segment):

$$y = \frac{-Px^2b^2}{6EIL^3}[2a(L - x) + L(a - x)]$$

Between *B* and *C* (shorter segment):

$$y = \frac{-Pv^2a^2}{6EIL^3}[2b(L - v) + L(b - v)]$$

A–25 (*Continued*)

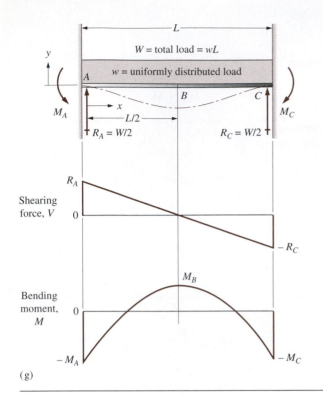

(g)

Moments

$$M_A = M_C = \frac{-WL}{12}$$

$$M_B = \frac{WL}{24}$$

Deflections
At *B* at center:

$$y_B = y_{max} = \frac{-WL^3}{384EI}$$

Between *A* and *C*:

$$y = \frac{-wx^2}{24EI}(L - x)^2$$

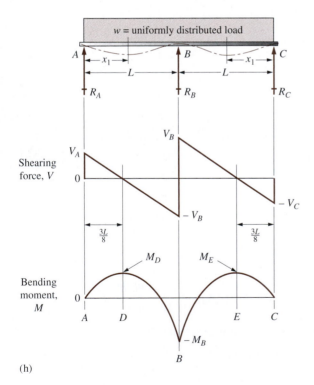

(h)

Reactions

$$R_A = R_C = \frac{3wL}{8}$$

$$R_B = 1.25wL$$

Shearing forces

$$V_A = V_C = R_A = R_C = \frac{3wL}{8}$$

$$V_B = \frac{5wL}{8}$$

Moments

$$M_D = M_E = 0.0703wL^2$$
$$M_B = -0.125wL^2$$

Deflections
At $x_1 = 0.4215L$ from *A* or *C*:

$$y_{max} = \frac{-wL^4}{185EI}$$

Between *A* and *B*:

$$y = \frac{-w}{48EI}(L^3x - 3Lx^3 + 2x^4)$$

A–25 (*Continued*)

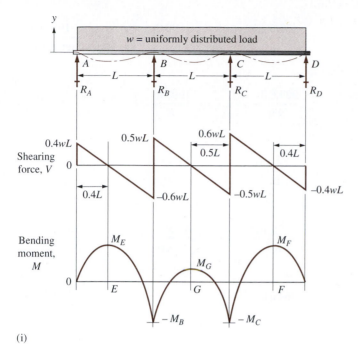

Reactions

$$R_A = R_D = 0.4wL$$
$$R_B = R_C = 1.10wL$$

Moments

$$M_E = M_F = 0.08wL^2$$
$$M_B = M_C = -0.10wL^2 = M_{max}$$
$$M_G = 0.025wL^2$$

(i)

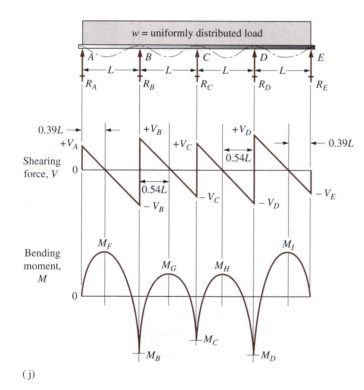

Reactions

$$R_A = R_E = 0.393wL$$
$$R_B = R_D = 1.143wL$$
$$R_C = 0.928wL$$

Shearing forces

$$V_A = +0.393wL$$
$$-V_B = -0.607wL$$
$$+V_B = +0.536wL$$
$$-V_C = -0.464wL$$
$$+V_C = +0.464wL$$
$$-V_D = -0.536wL$$
$$+V_D = +0.607wL$$
$$-V_E = -0.393wL$$

Moments

$$M_B = M_D = -0.1071wL^2 = M_{max}$$
$$M_F = M_I = 0.0772wL^2$$
$$M_C = -0.0714wL^2$$
$$M_G = M_H = 0.0364wL^2$$

(j)

A–26 Conversion Factors

Mass   Standard SI unit: Kilogram (kg). Equivalent unit: $N \cdot s^2/m$.

$\dfrac{14.59 \text{ kg}}{\text{slug}}$	$\dfrac{32.174 \text{ lb}_m}{\text{slug}}$	$\dfrac{2.205 \text{ lb}_m}{\text{kg}}$	$\dfrac{453.6 \text{ grams}}{\text{lb}_m}$	$\dfrac{2000 \text{ lb}_m}{\text{ton}_m}$	$\dfrac{1000 \text{ kg}}{\text{metric ton}_m}$

Force   Standard SI unit: Newton (N). Equivalent unit: $kg \cdot m/s^2$.

$\dfrac{4.448 \text{ N}}{\text{lb}_f}$	$\dfrac{10^5 \text{ dynes}}{\text{N}}$	$\dfrac{4.448 \times 10^5 \text{ dynes}}{\text{lb}_f}$	$\dfrac{224.8 \text{ lb}_f}{\text{kN}}$	$\dfrac{1000 \text{ lb}}{K}$

Length

$\dfrac{3.281 \text{ ft}}{\text{m}}$	$\dfrac{39.37 \text{ in}}{\text{m}}$	$\dfrac{12 \text{ in}}{\text{ft}}$	$\dfrac{25.4 \text{ mm}}{\text{in}}$	$\dfrac{1.609 \text{ km}}{\text{mi}}$	$\dfrac{5280 \text{ ft}}{\text{mi}}$

Area

$\dfrac{144 \text{ in}^2}{\text{ft}^2}$	$\dfrac{10.76 \text{ ft}^2}{\text{m}^2}$	$\dfrac{645.2 \text{ mm}^2}{\text{in}^2}$	$\dfrac{10^6 \text{ mm}^2}{\text{m}^2}$	$\dfrac{43\,560 \text{ ft}^2}{\text{acre}}$	$\dfrac{10^4 \text{ m}^2}{\text{hectare}}$

Volume

$\dfrac{1728 \text{ in}^3}{\text{ft}^3}$	$\dfrac{231 \text{ in}^3}{\text{gal}}$	$\dfrac{7.48 \text{ gal}}{\text{ft}^3}$	$\dfrac{264 \text{ gal}}{\text{m}^3}$	$\dfrac{3.785 \text{ L}}{\text{gal}}$	$\dfrac{35.3 \text{ ft}^3}{\text{m}^3}$

Section Modulus

$\dfrac{1.639 \times 10^4 \text{ mm}^3}{\text{in}^3}$	$\dfrac{10^9 \text{ mm}^3}{\text{m}^3}$

Moment of Inertia or Second Moment of an Area

$\dfrac{4.162 \times 10^5 \text{ mm}^4}{\text{in}^4}$	$\dfrac{10^{12} \text{ mm}^4}{\text{m}^4}$

Density (Mass/Unit Volume)

$\dfrac{515.4 \text{ kg/m}^3}{\text{slug/ft}^3}$	$\dfrac{1000 \text{ kg/m}^3}{\text{gram/cm}^3}$	$\dfrac{32.17 \text{ lb}_m/\text{ft}^3}{\text{slug/ft}^3}$	$\dfrac{16.018 \text{ kg/m}^3}{\text{lb}_m/\text{ft}^3}$

Specific Weight (Weight/Unit Volume)

$\dfrac{157.1 \text{ N/m}^3}{\text{lb}_f/\text{ft}^3}$	$\dfrac{1728 \text{ lb/ft}^3}{\text{lb/in}^3}$

Bending Moment or Torque

$\dfrac{8.851 \text{ lb} \cdot \text{in}}{\text{N} \cdot \text{m}}$	$\dfrac{1.356 \text{ N} \cdot \text{m}}{\text{lb} \cdot \text{ft}}$

Pressure, Stress, or Loading   Standard SI unit: Pascal (Pa). Equivalent units: N/m^2 or $kg/m \cdot s^2$.

$\dfrac{144 \text{ lb/ft}^2}{\text{lb/in}^2}$	$\dfrac{47.88 \text{ Pa}}{\text{lb/ft}^2}$	$\dfrac{6895 \text{ Pa}}{\text{lb/in}^2}$	$\dfrac{1 \text{ Pa}}{\text{N/m}^2}$	$\dfrac{6.895 \text{ MPa}}{\text{ksi}}$

Energy   Standard SI unit: Joule (J). Equivalent units: $N \cdot m$ or $kg \cdot m^2/s^2$.

$\dfrac{1.356 \text{ J}}{\text{lb} \cdot \text{ft}}$	$\dfrac{1.0 \text{ J}}{\text{N} \cdot \text{m}}$	$\dfrac{8.85 \text{ lb} \cdot \text{in}}{\text{J}}$	$\dfrac{1.055 \text{ kJ}}{\text{Btu}}$	$\dfrac{3.600 \text{ kJ}}{\text{W} \cdot \text{hr}}$	$\dfrac{778 \text{ ft} \cdot \text{lb}}{\text{Btu}}$

Power   Standard SI unit: Watt (W). Equivalent unit: J/s or $N \cdot m/s$.

$\dfrac{745.7 \text{ W}}{\text{hp}}$	$\dfrac{1.0 \text{ W}}{\text{N} \cdot \text{m/s}}$	$\dfrac{550 \text{ lb} \cdot \text{ft/s}}{\text{hp}}$	$\dfrac{1.356 \text{ W}}{\text{lb} \cdot \text{ft/s}}$	$\dfrac{3.412 \text{ Btu/hr}}{\text{W}}$	$\dfrac{1.341 \text{ hp}}{\text{kW}}$

General approach to application of conversion factors: Arrange the conversion factor from the table in such a manner that, when multiplied by the given quantity, the original units cancel out, leaving the desired units. See examples on next page.

Example 1. Convert a stress of 36 ksi to MPa.

$$\sigma = 36 \, \text{ksi} \times \frac{6.895 \, \text{MPa}}{\text{ksi}} = 248 \, \text{MPa}$$

Example 2. Convert a stress of 1272 MPa to ksi.

$$\sigma = 1272 \, \text{MPa} \times \frac{1.0 \, \text{ksi}}{6.895 \, \text{MPa}} = 184 \, \text{ksi}$$

ANSWERS TO SELECTED PROBLEMS

Chapter 1 Basic Concepts in Statics and Strength of Materials

Definitions and Unit Systems

1–1. Length: foot (ft); meter (m)

1–2. Time: second (s); second (s)

1–3. Force: pound (lb); newton (N)

1–4. Temperature: degree Fahrenheit (°F); Kelvin (K)

1–5. Angle: degree (deg); radian (rad)

1–6. Length: millimeter; kilometer

1–7. Time: minute (min); hour (h)

1–8. Force: $kg \cdot m/s^2$

1–9. Temperature: degree Celsius (°C)

1–10. Angle: degree (deg)

1–11. Length: inch (in); yard (yd); mile (mi)

1–12. Time: minute (min); hour (h)

1–13. Force: kip (K) [K = 1000 lb]

1–14. Temperature: degree Celsius (°C); degree Rankine (°R)

1–15. Angle: radian (rad)

1–17. 45.2×10^3 m or 45.2 km

1–19. 48.0×10^{-3} m or 48.0 mm

1–21. 338×10^3 N or 338 kN

1–23. 5.33×10^3 N or 5.33 kN

1–25. 29.8×10^6 N or 29.8 MN

1–26. Mass refers to the amount of substance in a body.

1–27. kilogram (kg) or $N \cdot s^2/m$

1–28. slug or $lb \cdot s^2/ft$

1–29. Weight is the force of gravitational pull on a body.

1–30. newton (N) or $kg \cdot m/s^2$

1–31. pound (lb); kip (K) [kip = 1000 lb]

1–32. Density is the amount of mass per unit volume of a material.

1–33. kg/m^3

1–34. $slugs/ft^3$

1–35. Specific weight is the amount of weight per unit volume of a material.

1–36. N/m^3

1–37. lb/ft^3

1–38. Pressure is the amount of force applied to a given unit of area.

1–39. N/m^2, Pa, kPa, MPa

1–40. lb/in^2, psi, lb/ft^2, psf, K/in^2, ksi, K/ft^2, ksf

Mass-Weight Conversions

1–41. 17.7 kN

1–43. 3.81 kPa

1–47. 3970 lb

1–49. 79.6 lb/ft^2

1–51. 85.4 slugs

Unit Conversions

1–55. 149 MPa

1–57. 183 rad/s

1–59. 2.03 mm

1–61. 0.200 in^2; 129 mm^2

Moments, Torques, and Couples

1–63. 795 $N \cdot m$ CCW

1–65. 3.1 $kN \cdot m$ CCW

1–67. 4.189 $kN \cdot m$ CCW

1–69. 7.378 $kN \cdot m$ CCW

1–71. 4800 $lb \cdot ft$ CW

1–73. $M_{A1} = (1200 \, lb \cdot in)$ CW
$M_{A2} = (480 \, lb \cdot in)$ CCW

1–75. 150 lb upward

1–77. $T = 0; M = 43.5 \, N \cdot m$

1–79. 189 $lb \cdot in$ CW

1–81. 642 N

1–83. 261 $N \cdot m$

1–85. 5368 $N \cdot m$ CW

Chapter 2 Forces and Free Body Diagrams

Questions 1–28 relate to basic principles for which the answers are in the text.

Components of Forces

2–29. $F_x = 679 \, N \rightarrow$; $F_y = 392 \, N \uparrow$

2–31. $F_x = 63.0 \, lb \rightarrow$; $F_y = 25.5 \, lb \uparrow$

2–33. $F_x = 325 \, lb \rightarrow$; $F_y = -1213 \, lb \downarrow$

2–35. $F_x = -351 \, kN \leftarrow$; $F_y = 870 \, kN \uparrow$

2–37. $F_x = -538 \, N \leftarrow$; $F_y = 114 \, N \uparrow$

Distributed Loads on Surfaces

2–41. 131 904 lb

2–43. On joist: $w = 2198.4$ lb/ft; on edge beam: $w = 1099.2$ lb/ft

2–45. On joist: $w = 16.55$ lb/ft; on edge beam: $w = 8.275$ lb/ft

2–47. Weight $= 286$ N; weight per unit length $= 170.2$ N/m

2–49. Weight $= 90\ 950$ lb
  A. Unit loading $= 1098$ lb/ft^2;
  B. Unit loading varies linearly from 1098 lb/ft^2 to 1830 lb/ft^2 from left to right.
  C. Unit loading $= 1830$ lb/ft^2
  D. Unit loading varies linearly from 1830 lb/ft^2 to 1098 lb/ft^2 from left to right.

Forces—Resultants

2–51. $R = 10.60$ kN at the angle of 38.6° to the left from the negative y-axis.

2–55. $R = 14\ 238$ lb at the angle of 22.1° to the left from the negative y-axis.

2–57. $R = 46.7$ kN at the angle of 44.5° to the right from the negative x-axis.

Chapter 3   Static Equilibrium

Static Equilibrium of Equipment and Structures

3–1. 71.6 lb on each right foot
  111.4 on each left foot

3–3. 88.8 lb

3–7. 458 N

3–8. 78 N on each rear wheel
  12.19 kN on each front wheel

3–13. 1.864 m forward of the rear axle

3–15. 13.49 kN on trailer axle
  8.958 kN on rear truck axle; 4.214 kN gain
  11.604 kN on front truck axle; 3.968 kN gain

3–16. Center of mass: 6.19 m aft of the nose; 0.01 m forward of the net lift force vector

Reactions on Beams—Weight of Beam Neglected

[R_L = Left reaction force; R_R = Right reaction force]
[Reactions act vertically upward unless otherwise noted.]

3–19. $R_L = 11.43$ K; $R_R = 4.57$ K

3–21. $R_L = 585.7$ lb; $R_R = 514.3$ lb

3–23. $R_L = 1189$ N; $R_R = 1451$ N

3–25. $R_L = 1485$ lb; $R_R = 1745$ lb

3–27. $R_L = 687$ N; $R_R = 113$ N

3–29. $R_L = 37.4$ kN$\downarrow$; $R_R = 38.3$ kN$\uparrow$

3–31. Cantilever: $R_S = 160$ lb at support
  Moment $= M_S = 1580$ lb · in CCW

3–33. Cantilever: $R_S = 45$ K at support
  Moment $= M_S = 320$ K · ft CCW

3–35. $R_L = R_R = 12$ K

3–37. $R_L = 16.7$ K; $R_R = 23.3$ K

3–39. $R_L = 42$ kN; $R_R = 50$ kN

3–41. $R_L = 4.0$ K; $R_R = 20.0$ K

3–43. Cantilever: $R_S = 600$ N at support
  Moment $= M_S = 200$ N · m CCW

3–45. $R_L = 305$ lb; $R_R = 155$ lb

3–47. $R_L = 66.5$ kN; $R_R = 35.9$ kN

3–49. $R_L = 34.5$ K; $R_R = 72.5$ K

3–51. Cantilever: $R_S = 230$ kN at support
  Moment $= M_S = 430$ kN · m CCW

3–53. $R_L = 135$ lb; $R_R = 135$ lb

3–55. Cantilever: $R_S = 10.1$ kN at support
  Moment $= M_S = 8.06$ kN · m CCW

Reactions on Beams—Weight of Beam Considered

3–57. $R_L = 587.5$ lb; $R_R = 516.1$ lb

3–59. $R_L = 631.8$ N; $R_R = 381.9$ N

Forces on Supports for Structural Components and Mechanical Devices

3–61. $R_{Lx} = 0$; $R_{Ly} = 282.4$ N$\downarrow$;
  $R_{Ry} = 282.4$ N$\uparrow$; $R_{Rx} = 800$ N $\leftarrow$

3–63. $R_{Ly} = 162.2$ N$\downarrow$; $R_{Lx} = 1200$ N $\leftarrow$;
  $R_{Ry} = 162.2$ N$\uparrow$; $R_{Rx} = 0$

3–65. $R_{Lx} = 0$; $R_{Ly} = 165.5$ N$\uparrow$; $R_{Ry} = 18.4$ N$\uparrow$;
  $R_{Rx} = 183.9$ N $\rightarrow$

3–67. $R_{Ly} = 4.55$ N$\uparrow$; $R_{Lx} = 22.7$ N $\leftarrow$;
  $R_{Ry} = 131.8$ N$\downarrow$; $R_{Lx} = 0$

Forces on Supports for Structures

3–69. $R_{Ay} = 10\ 261$ lb$\uparrow$; $R_{Ax} = 3562$ lb $\rightarrow$; $R_{Fy} = 9130$ lb$\uparrow$;
  $R_{Fx} = 0$

3–71. $R_{Ay} = 25.23$ kN$\downarrow$; $R_{Ax} = 11.3$ kN $\rightarrow$;
  $R_{By} = 57.53$ kN$\uparrow$; $R_{Bx} = 0$

3–73. $R_{Ay} = 1400$ lb$\uparrow$; $R_{Ax} = 4687.5$ N $\leftarrow$; $R_{Dy} = 0$;
  $R_{Dx} = 4487.5$ N $\rightarrow$

3–75. $F_S = 1276$ N; $R_A = 436$ N; directions perpendicular to beam in directions shown in the figure.

3–77. $R_{Ay} = 294.6$ lb$\uparrow$; $R_{Ax} = 409$ lb $\rightarrow$; $R_{Dy} = 5.43$ lb$\uparrow$;
  $R_{Dx} = 0$

Rotational Equilibrium

3–79. $R_{Ay} = 114.3$ lb$\uparrow$; $R_{Ax} = 0$; $R_{Cy} = 285.7$ lb$\uparrow$;
  $R_{Cx} = 0$; $T_{AB} = 6400$ lb · in CW viewed from A;
  $T_{BC} = 0$

3–81. $R_{Ay} = 2.432$ kN$\uparrow$; $R_{Ax} = 0$; $R_{By} = 1.182$ kN$\downarrow$;
  $R_{Bx} = 0$; $T = 375$ kN · mm CW from load to A only, viewed toward A; $T_{AB} = 0$

3–83. $F_2 = 187.5$ N$\downarrow$; (O is pivot) $R_{Ox} = 450$ N $\leftarrow$;
  $R_{Oy} = 187.5$ N$\uparrow$

Pulley Systems

3–87. Operator pull $= 262.5$ N; 3.50 m of rope

Chapter 4   Forces in Frames, Structures, and Trusses

Note: Complete solutions for Chapter 4 problems include sketches of each member of the frame, structure, or truss depicted as a free body diagram with all forces indicated in magnitude and direction. Only certain force magnitudes are listed here.

Pin-Connected Frames

4–1. Member AB is a two-force member in tension with 20.0 kN of force. On beam CBD: $C_y = 7.5$ kN$\downarrow$; $B_y = 20.0$ kN$\uparrow$; $D_y = 12.5$ kN$\downarrow$

4–3. Member BC is a two-force member in compression with 54.5 kN of force. On beam ABD: $A_y = 20.6$ kN$\downarrow$; $A_x = 25.67$ kN $\leftarrow$; $B_y = 48.13$ kN$\uparrow$; $B_x = 25.67$ kN $\rightarrow$, $D_y = 27.5$ kN$\downarrow$

4–5. Forces given for one support rod and one set of supports. Loading is symmetrical. Angled bar AC is a two-force member in tension with 53.7 kN of force. Shelf is BC with reaction at B: $B_x = 41.1$ kN $\rightarrow$; $B_y = 55.0$ kN$\uparrow$; at C: $C_x = 41.1$ kN $\leftarrow$; $C_y = 34.5$ kN$\uparrow$

4–7. Member BC is a two-force member in tension with 20.76 kN of force. On beam AB: $A_y = 1.00$ kN$\uparrow$; $A_x = 17.60$ kN $\rightarrow$; $B_y = 11.00$ kN$\uparrow$; $B_x = 17.60$ kN $\leftarrow$

4–9. Member BD is a two-force member in compression with 20.16 kN of force. On beam ABC: $A_y = 8.91$ kN$\downarrow$; $A_x = 14.26$ kN $\leftarrow$; $B_y = 14.26$ kN$\uparrow$; $B_x = 14.26$ kN $\rightarrow$; $C_y = 5.346$ kN$\downarrow$

4–11. Member BD is a two-force member in compression with 4326 lb of force. On beam ABC: $A_y = 2400$ lb$\downarrow$; $A_x = 2911$ lb $\leftarrow$; $B_y = 3200$ lb$\uparrow$; $B_x = 2911$ lb $\rightarrow$; $C_y = 800$ lb$\downarrow$

4–13. Cable AB: $AB = 73.2$ kN tension; Cable AE: $AE = 110.8$ kN tension
Boom DEF: $D_y = 34.0$ kN$\downarrow$; $D_x = 0$; $E_x = 102.2$ kN $\rightarrow$; $E_y = 42.95$ kN$\uparrow$; $F_x = 102.2$ kN $\leftarrow$; $F_y = 8.95$ kN$\downarrow$
Post AFC: $A_x = 74.9$ kN $\leftarrow$; $A_y = 110.9$ kN$\downarrow$; $F_x = 102.2$ kN $\rightarrow$; $F_y = 8.95$ kN$\uparrow$; $C_x = 27.2$ kN $\leftarrow$; $C_y = 102$ kN$\uparrow$

4–15. Each rod: 4611 N tension; each angled cable: 9494 N tension; cable from crane: 9221 N tension; spreader bar: Let A, B, C, D be attachment points for rods and cables; $A_y = D_y = 4600$ N$\downarrow$; $A_x = D_x = 0$; $B_x = 8299$ N $\rightarrow$; $B_y = 4611$ N$\uparrow$; $C_x = 8299$ N $\leftarrow$; $C_y = 4611$ N$\uparrow$

4–17. Cable forces: $AB = 34.8$ kN; $BC = 31.6$ kN; $BD = 14.7$ kN

4–19. Cable forces: $AB = 37.5$ kN; $BC = 28.4$ kN; $BD = 41.2$ kN load

4–21. Each link sees 618 lb axial compression; screw sees 1236 lb tension.

4–23. Force on each roller is 46 147 lb.

4–25. Force on wire = 122 N; force on pin = 177 N

4–27. 1214 N

4–29. a. 21 802 lb;
   b. 21 540 lb

Trusses

4–31. $AB = 1818$ lb (C); $AD = 1286$ lb (T); $BD = 1286$ lb (T)
$BC = 1286$ lb (C); $CE = 1500$ lb(T); $CD = 155$ lb (C)
$DE = 1416$ lb (T); $EF = 1416$ lb(T); $CF = 2000$ lb (C)

4–33. $AD = 10.5$ kN(C); $AB = 9.09$ kN(T); $BC = 9.09$ kN(T)
$BD = 10.5$ kN (T); $CD = 10.5$ kN (C)

4–35. $AB = 11\,667$ lb (C); $AD = 4000$ lb (T); $BD = 0$
$BC = 6500$ lb (C); $CE = 3333$ lb (C); $DE = 4000$ lb (T)
$BE = 4167$ lb (T); $EF = 6500$ lb (T); $CF = 10\,833$ lb (C)

4–37. $AB = 2465$ lb (T); $AC = 1925$ lb (C); $BD = 1200$ lb (T)
$BC = 1375$ lb (T); $CE = 650$ lb (C); $CD = 750$ lb (C)
$DE = 960.5$ lb (T)

4–39. $AB = 1688$ N (T); $AD = 600$ N (T); $BD = 1113$ N (C)
$DE = 550$ N (C); $EF = 550$ N (C); $EG = 200$ N (T)
$BF = 960.5$ N (T); $BE = BC = CF = CG = 0$

4–43. $AB = 1600$ lb (C); $AC = 1789$ lb (T); $BC = 1200$ lb (T)
$BE = 1600$ lb (C); $CD = 5600$ lb (T); $CE = 4472$ lb (C)
$DE = 2000$ lb (T)

4–45. $AB = 1600$ lb (C); $AD = 2000$ lb (T); $BD = 667$ lb (C)
$DE = 2133$ lb (T); $BC = 1067$ lb (C); $CE = 1333$ lb (T)
$BE = 1333$ lb (C)

Frames

Note: Paired problems listed: $4-49 + 4-53$; $4-51 + 4-55$

4–49. $F_x = 3.21$ kN; $F_y = 7.48$ kN; $F_z = 11.8$ kN

4–53. $\cos\theta_x = 0.2242$; $\theta_x = 77.0°$
$\cos\theta_y = 0.5232$; $\theta_y = 58.5°$
$\cos\theta_z = 0.8222$; $\theta_z = 34.7°$

4–51. $F_x = -7965$ lb; $F_y = 3983$ lb; $F_z = 8629$ lb

4–55. $\cos\theta_x = -0.6423$; $\theta_x = 130.0°$
$\cos\theta_y = 0.3212$; $\theta_y = 71.3°$
$\cos\theta_z = 0.6959$; $\theta_z = 45.9°$

4–57. a. $F_x = 450$ lb; $F_y = 620$ lb; $F_z = 900$ lb
   b. $\cos\theta_x = 0.3807$; $\theta_x = 67.6°$
   $\cos\theta_y = 0.5245$; $\theta_y = 58.4°$
   $\cos\theta_z = 0.7614$; $\theta_z = 40.4°$

4–59. $\theta_y = 71.5°$; $F_x = -881$ lb; $F_y = 746$ lb; $F_z = 2049$ lb

4–61. $F_1 = 740$ N; $F_2 = 786$ N; $F_3 = 324$ N

4–63. 57.2 lb

4–65. $BC = BD = 3642$ lb (T); $AB = 5390$ lb (C)

4–67. $AB = 600.5$ lb; $AC = 899.7$ lb; $AD = 629.3$ lb

4–69. $BC_x = 22.52$ kN; $BC_y = 6.03$ kN; $BC_z = 61.74$ kN
$\cos\theta_x = 0.3412$; $\cos\theta_y = 0.09143$; $\cos\theta_z = 0.9355$

Chapter 5   Frictional Forces

Measuring and Using Coefficient of Friction

5–1. 0.41

5–3. 0.57

5–5. 222 lb

5–7. 31.2 lb

5–9. 454 N

5–11. 11.1 kN

5–13. 9.9 kN

5–15. 0.36

5–17. 0.53

5–19. 38.0°

5–21. 43.2°

Rolling Friction

5–23. 3.94 lb

5–25. 118 N

5–27. 897 N

5–29. 241 lb

Tipping and Sliding

5–31. 0.277

5–33. 34.3 in

5–35. $F = 34.29$ lb; no tipping

5–37. 70.8 kg

5–39. Crate will slide down ramp; $F_f = 260$ lb

5–41. $h_{max} = 1250$ mm

5–43. Ladder will not slide.

5–45. 30.7°

5–47. 12.0°

5–49. $h_{min} = 21.8$ mm

5–51. Both cartons will slide down ramp.

5–53. $F_{max} = 189$ N

5–55. 15.17 kN

5–57. Wedge will not slide.

Brakes, Belts, and Rope Friction

5–59. 238 N

5–61. 2.55 lb

5–63. 811 lb

5–65. 117 lb

5–67. $T_1 = 39.4$ N · m; $T_2 = 88.6$ N · m

Chapter 6   Centroids and Moments of Inertia

Note: The following answers refer to Figures P6–1 through P6–39. The first number is the distance to the centroid from the bottom of the section unless noted otherwise. The second number is the moment of inertia with respect to the horizontal centroidal axis.

P6–1. 0.663 in; 0.3156 in^4

P6–3. 4.00 in; 184 in^4

P6–5. 35.0 mm; 2.66×10^5 mm^4

P6–7. 20.0 mm; 7.29×10^4 mm^4

P6–9. 20.0 mm; 1.35×10^5 mm^4

P6–11. 21.81 mm; 1.86×10^5 mm^4

P6–13. 23.33 mm; 1.41×10^5 mm^4

P6–15. 1.068 in; 0.3572 in^4

P6–17. 125 mm; 6.73×10^7 mm^4

P6–19. 0.9305 in; 1.2506 in^4

P6–21. 4.25 in; 151.4 in^4

P6–23. 2.25 in; 107.2 in^4

P6–25. 7.35 in; 831.5 in^4

P6–27. 7.40 in; 423.5 in^4

P6–29. 3.50 in; 89.26 in^4

P6–31. 3.00 in from center of either pipe; 17.87 in^4

P6–33. 2.717 in; 46.76 in^4

P6–35. 2.609 in; 58.25 in^4

P6–37. 3.50 in; 16.95 in^4

P6–39. 3.361 in; 44.34 in^4

P6–41. 8.172 in; 357.5 in^4

P6–43. 8.20 in; 376.1 in^4

P6–45. 5.023 in; 100.2 in^4

P6–47. 14.641 mm; 41 647 mm^4

Radius of Gyration—Horizontal Axis

6–49. 3.04 in

6–51. 70.35 mm

6–53. 8.78 mm

6–55. 15.67 mm

6–57. 6.11 mm

6–59. 0.72 in

6–61. 68.78 mm

6–63. 2.33 in

6–65. 4.02 in

Radius of Gyration—Vertical Axis

6–67. 1.424 in

6–69. 34.91 mm

6–71. 0.212 in

6–73. 0.809 in

6–75. 8.483 in

6–77. 2.065 in

6–79. 1.783 in

6–81. 0.732 in

Metric Versions of Previous Problems

*Figure numbers listed with **M** suffix.*
First number is distance to centroid from the bottom of section; second number is moment of inertia with respect to the centroidal axis.

P6–21M. 108 mm; 6.31×10^7 mm^4

P6–23M. 57.20 mm; 4.47×10^7 mm^4

P6–25M. 186 mm; 3.44×10^8 mm^4

P6–27M. 186.7 mm; 1.75×10^8 mm^4

P6–29M. 87 mm; 3.38×10^7 mm^4

P6–31M. 75 mm from center of either pipe; 7.11×10^6 mm^4

P6–33M. 68.07 mm; 1.84×10^7 mm^4

P6–35M. 85.39 mm; 2.36×10^7 mm^4

P6–37M. 889.25 mm; 7.08×10^6 mm^4

P6–39M. 85.16 mm; 1.83×10^7 mm^4

P6–41M. 207.3 mm; 1.49×10^8 mm^4

P6–43M. 207.9 mm; 1.56×10^8 mm^4

P6–45M. 127.5 mm; 4.18×10^7 mm^4

Chapter 7   Basic Concepts in Strength of Materials

7–14. 48.63 MPa

7–16. 40.7 MPa

7–18. 5375 psi

7–20. 79.8 MPa

7–22. 803 psi

7–24. $\sigma_{AB} = 107.4$ MPa
$\sigma_{BC} = 75.2$ MPa
$\sigma_{BD} = 131.1$ MPa

7–26. $\sigma_{AB} = 167$ MPa tension
$\sigma_{BC} = 77.8$ MPa tension
$\sigma_{CD} = 122$ MPa tension

7–28. $\sigma_{AB} = 20\,471$ psi tension
$\sigma_{BC} = 3129$ psi tension

7–30. Forces: $AD = CD = 10.5$ kN
$AB = BC = 9.09$ kN
Stresses: $\sigma_{AB} = \sigma_{BC} = 25.3$ MPa tension
$\sigma_{BD} = 17.5$ MPa tension
$\sigma_{AD} = \sigma_{CD} = 21.0$ MPa compression

7–32. 50.0 MPa

7–34. 11 791 psi

7–36. 146 MPa

7–38. 151 MPa

7–40. 81.1 MPa

7–42. 24.7 MPa

7–44. Pin: $\tau = 50\,930$ psi
Collar: $\tau = 38\,800$ psi

7–46. 183 MPa

7–48. 73.9 MPa

7–50. 22.6 MPa

Chapter 8   Design Properties of Materials

8–15. 1020 HR

8–19. 16.4 lb

8–21. Magnesium

8–29. $s_{ut} = 40$ ksi; $s_{uc} = 140$ Ksi

8–31. Bending: $\sigma_d = 1450$ psi
Tension: $\sigma_d = 850$ psi
Compression: $\sigma_d = 1000$ psi parallel to grain
Compression: $\sigma_d = 385$ psi perpendicular to grain
Shear: $\tau_d = 95$ psi

Problems 8–67 to 8–77: Data approximated from Figure P8–66

8–67.
a. $s_y = 173$ ksi—Yield point
b. $s_u = 187$ ksi
c. $s_p = 162$ ksi
d. $s_{el} = 168$ ksi
e. $E = 29.0 \times 10^6$ psi
f. 15% elongation
g. Ductile
h. Steel
i. AISI 4140 OQT 900

8–69.
a. $s_y = 49$ ksi—Yield point
b. $s_u = 65$ ksi
c. $s_p = 46$ ksi
d. $s_{el} = 48$ ksi
e. $E = 26.5 \times 10^6$ psi
f. 36% elongation
g. Ductile
h. Steel
i. AISI 1020 CD

8–71.
a. $s_y = 53$ ksi—offset
b. $s_u = 59$ ksi
c. $s_p = 31$ ksi
d. $s_{el} = 42$ ksi
e. $E = 12.0 \times 10^6$ psi
f. 5.0% elongation
g. Borderline brittle/ductile
h. Zinc
i. Cast ZA-12

8–73.
a. $s_y = 19$ ksi—offset
b. $s_u = 40$ ksi
c. $s_p = 14$ ksi
d. $s_{el} = 17$ ksi
e. $E = 6 \times 10^6$ psi
f. 5% elongation
g. Borderline brittle/ductile
h. Magnesium
i. ASTM AZ 63A-T6

8–75.
a. $s_y = 40$ ksi—offset
b. $s_u = 45$ ksi
c. $s_p = 30$ ksi
d. $s_{el} = 35$ ksi
e. $E = 10.0 \times 10^6$ psi
f. 17% elongation
g. Ductile
h. Aluminum
i. 6061-T6

8–77.
a. $s_y = 80$ ksi—offset
b. $s_u = 95$ ksi
c. $s_p = 55$ ksi
d. $s_{el} = 68$ ksi
e. $E = 26 \times 10^6$ psi
f. 2.0% elongation
g. Brittle, but does yield
h. Malleable iron
i. ASTM A220 Grade 80002

Chapter 9   Direct Stress, Deformation, and Design

9–1. Required s_y = 216 MPa
9–3. Required s_u = 86 000 psi
9–5. No. Tensile stress too high.
9–7. d_{min} = 0.824 in
9–9. d_{min} = 12.4 mm
9–11. Required σ_d > 803 psi
9–13. 16.7 kN
9–15. d_{min} = 0.412 in
9–17. For sides B and H: B_{min} = 22.2 mm; H_{min} = 44.4 mm
9–19. Required s_y = 360 MPa
9–21. Required s_u = 400 MPa

Elastic Deformation

9–23. 0.041 in
9–25. Force = 2357 lb
       σ = 3655 psi
9–27. a. and b. d_{min} = 10.63 mm; mass = 0.430 kg
       c. d_{min} = 18.4 mm; mass = 0.465 kg
9–29. a. 0.857 mm
       b. 0.488 mm
9–31. Elongation = 0.0040 in
       Compression = 0.00045 in
9–33. Force = 3214 lb; unsafe
9–35. δ = 0.016 mm
       σ = 27.9 MPa
9–37. 0.804 mm
9–39. 4.42 mm shorter
9–41. a. δ = 0.276 in; σ = 37 300 psi (close to s_y)
       b. σ = 62 200 psi—greater than s_u. Wire will break.
9–43. Force = 6737 lb
       δ = 0.055 in
9–45. Mass = 132 kg
       σ = 183 MPa
9–47. 0.806 in
9–49. 180 MPa
9–51. a. 0.459 mm
       b. 213 MPa
9–53. 693 psi
9–55. 234.8°C
9–57. Brass: δ = 6.46 mm
       Stainless steel: δ = 3.51 mm
9–59. 38.7 MPa
9–61. σ = 37 500 psi compression. Bar should fail in compression and/or buckle.
9–63. 0.157 in
9–65. 154 MPa

9–67. σ_c = 17.1 MPa; σ_s = 109 MPa
9–69. 13.8 in
9–71. d_{min} = 6.20 mm
9–73. σ_s = 427 MPa; σ_a = 49.4 MPa

Additional Practice Problems

9–75. F_{max} = 12.56 kN; δ = 0.751 mm
9–77. a. t = 110.6°C
       b. σ = 50.4 MPa
       c. Safe for compression. Check buckling.
9–79. δ = 0.341 mm. But stress is near yield strength.
9–81. For AB: δ = 15.2 mm
       For BC: δ = 15.2 mm

Problems 9–83 to 9–89:

	Length (in)	Strain (in/in)	Force (lb)
9–83.	2.0143	0.00714	42 840
9–85.	2.0137	0.00687	26 100
9–87.	2.0071	0.00353	8460
9–89.	2.0116	0.00581	27 900

Problems 9–91 to 9–99:

	Length (mm)	Strain (mm/mm)	Force (kN)
9–91.	50.716	0.01431	8.30
9–93.	50.486	0.00972	5.50
9–95.	50.148	0.00297	4.50
9–97.	50.455	0.00909	137.9
9–99.	50.083	0.00167	110.3

Stress Concentrations—Direct Axial Stresses

9–101. 20 140 psi
9–103. 224 MPa
9–105. 239 MPa
9–107. 52 800 psi
9–109. 33 127 psi
9–111. 63.7 MPa
9–113. 833 MPa
9–115. 34 240 psi
9–117. 34 020 psi
9–119. 531 MPa
9–121. 62.1 MPa
9–123. 15.9 MPa
9–125. 1905 N

Bearing Stress

9–127. A. 5869 psi,   C. 181 psi
        B. 80.2 psi,   D. 20.1 psi
9–129. a. 1610 psi,
        b. 311 psi

9–131. **a.** Pin/tube interface, $\sigma_b = 106\,700$ psi—Very high
b. Collar/tube interface, $\sigma_b = 31\,950$ psi

9–133. **a.** On middle part: $\sigma_b = 28.3$ MPa
b. On outer parts: $\sigma_b = 21.25$ MPa

9–135. **a.** Steel plate: $\sigma_b = 2.25$ MPa—OK
b. Top of concrete: $\sigma_b = 0.563$ MPa—OK
c. Soil: $\sigma_b = 22.5$ kPa—OK

9–137. $R_a = 33.1$ kips

9–139. **a.** $R_a = 27.6$ kips
b. $R_a = 39.6$ kips

9–141. $R_a = 284$ kN

9–143. Required $A_b = 3.49$ in^2

9–145. 8633 lb

Direct Shear Stress

9–147. $a_{min} = 0.438$ in

9–149. $\tau = 151$ MPa; Required $s_y = 1208$ MPa

9–151. 185 kN

9–153. 256 kN

9–155. 18 300 lb

9–157. 62 650 lb

9–159. 119 500 lb

9–161. 119 700 lb

9–163. Pin A: $D_{A_{min}} = 13.8$ mm
Pins B and C: $D_{min} = 17.7$ mm

9–165. Lifting force = 2184 lb; $\tau_{axle} = 6275$ psi (double shear)

9–167. $D_{min} = 0.206$ in

9–169. 35 020 lb

9–171. 58.06 kN

9–173. 110.9 kN

9–175. 137.4 kN

Problems with More Than One Kind of Direct Stress and Design Problems

9–177. $D_{min} = 0.808$ in based on shear

9–179. **a.** $F_{allow} = 6200$ lb shear
b. $F_{allow} = 35\,340$ lb bearing
c. $F_{allow} = 2832$ lb tension—governs design

9–181. **a.** $F_{allow} = 3148$ lb shear—governs design
b. $F_{allow} = 9375$ lb tension
c. $F_{allow} = 10\,969$ lb bearing

9–183. **a.** $F_{allow} = 6626$ lb tension—governs design
b. $F_{allow} = 10\,713$ lb shear
c. $F_{allow} = 81\,843$ lb bearing

Chapter 10   Torsional Shear Stress and Torsional Deformation

10–1. 178 MPa

10–3. 4042 psi

10–5. 83.8 MPa

10–7. $\tau = 6716$ psi; safe

10–9. $\tau = 5190$ psi; required $s_y = 62\,300$ psi

10–11. $\tau = 65.5$ MPa; $\theta = 0.0378$ rad; required $s_y = 262$ MPa

10–13. $D_i = 49.0$ mm; $D_o = 61.3$ mm

10–15. $d_{min} = 0.512$ in

10–17. Power = 0.0686 hp; $\tau = 8488$ psi; required $s_y = 67\,900$ psi

10–19. $D_i = 12.09$ in; $D_o = 15.11$ in

10–21. 1.96 N · m

10–23. 0.1509 rad

10–25. 0.267 rad

10–27. 0.0756 rad

10–29. 0.278 rad

10–31. $\theta_{AB} = 0.0636$ rad; $\theta_{AC} = 0.0976$ rad

10–33. $\tau = 9.06$ MPa; $\theta = 0.0046$ rad

10–35. 49.0 MPa

10–37. 1370 lb · in

10–39. 2902 lb · in

10–41. 0.083 rad

10–43. 0.112 rad

10–45. 0.0667 rad

10–47. 1.82 MPa

10–49. 0.00363 rad

10–51. 0.0042 rad

10–53. 78 150 lb · in

10–55. 144 800 lb · in

10–57. $\tau_{tube}/\tau_{pipe} = 1.100$
$\theta_{tube}/\theta_{pipe} = 1.266$

Additional Practice Problems

10–58. $\tau_{max} = 123$ MPa

10–59. Required $s_y = 493$ MPa

10–61. Required $s_y = 1480$ MPa

10–63. $D_{min} = 0.419$ in; Specify $D = 0.50$ in

10–65. $\theta = 1.85$ degrees

10–67. $\tau = 211$ MPa at A at keyseat $= \tau_{max}$
$\tau = 171$ MPa at right of A at shoulder
$\tau = 55.7$ MPa at right of bearing seat
$\tau = 39.6$ MPa at retaining ring groove to left of B
$\tau = 26.4$ MPa at B at keyseat
$\tau = 14.7$ MPa at right of B at shoulder
$\tau = 10.7$ MPa at step from 50 mm to 30 mm
$\tau = 10.3$ MPa at left of bearing seat
$\tau = 22.5$ MPa at left of C at shoulder
$\tau = 30.1$ MPa at C at keyseat

10–69. $L_{min} = 1.088$ m

10–70. $D_o = 21.79$ mm; $D_i = 14.53$ mm

10–71. $\tau = 408.8$ MPa

Chapter 11   Shearing Forces and Bending Moments in Beams

The following answers refer to Figures P11–1 through P11–84 and P11–93 through P11–110. For reactions, R_1 is the left; R_2 is the right. V and M refer to the maximum absolute values for shearing force and bending moment, respectively. The complete solutions require the construction of the complete shearing force and bending moment diagrams.

P11–1. $R_1 = R_2 = 325$ lb
$V = 325$ lb
$M = 4550$ lb $\cdot$ in

P11–3. $R_1 = 11.43$ K; $R_2 = 4.57$ K
$V = 11.43$ K
$M = 45.7$ K $\cdot$ ft

P11–5. $R_1 = 575$ N; $R_2 = 325$ N
$V = 575$ N
$M = 195$ N $\cdot$ m

P11–7. $R_1 = 46.36$ kN; $R_2 = 23.64$ kN
$V = 46.36$ kN
$M = 71.54$ kN $\cdot$ m

P11–9. $R_1 = 1557$ lb; $R_2 = 1743$ lb
$V = 1557$ lb
$M = 6228$ lb $\cdot$ in

P11–11. $R_1 = 7.5$ K; $R_2 = 37.5$ K
$V = 20$ K
$M = 60$ K $\cdot$ ft

P11–13. $R_1 = R_2 = 250$ N
$V = 850$ N
$M = 362.5$ N $\cdot$ m

P11–15. $R_1 = 37.4$ kN (down); $R_2 = 38.3$ kN (up)
$V = 24.9$ kN
$M = 50$ kN $\cdot$ m

P11–17. $R = 120$ lb
$V = 120$ lb
$M = 960$ lb $\cdot$ in

P11–19. $R = 24$ K
$V = 24$ K
$M = 168$ K $\cdot$ ft

P11–21. $R = 1800$ N
$V = 1800$ N
$M = 1020$ N $\cdot$ m

P11–23. $R = 120$ kN
$V = 120$ kN
$M = 240$ kN $\cdot$ m

P11–25. $R_1 = R_2 = 180$ lb
$V = 180$ lb
$M = 810$ lb $\cdot$ in

P11–27. $R_1 = 240$ lb; $R_2 = 120$ lb
$V = 240$ lb
$M = 640$ lb $\cdot$ in

P11–29. $R_1 = 99.2$ N; $R_2 = 65.8$ N
$V = 99.2$ N
$M = 9.9$ N $\cdot$ m

P11–31. $R_1 = 42$ kN; $R_2 = 50$ kN
$V = 50$ kN
$M = 152.2$ kN $\cdot$ m

P11–33. $R_1 = R_2 = 440$ lb
$V = 240$ lb
$M = 360$ lb $\cdot$ in

P11–35. $R_1 = 1456$ N; $R_2 = 644$ N
$V = 956$ N
$M = 125$ N $\cdot$ m

P11–37. $R_1 = 35.3$ N; $R_2 = 92.3$ N
$V = 52.2$ N
$M = 4.0$ N $\cdot$ m

P11–39. $R = 360$ lb
$V = 360$ lb
$M = 1620$ lb $\cdot$ in

P11–41. $R = 600$ N
$V = 600$ N
$M = 200$ N $\cdot$ m

P11–43. $R_1 = R_2 = 330$ lb
$V = 330$ lb
$M = 4200$ lb $\cdot$ in

P11–45. $R_1 = 36.6$ K; $R_2 = 30.4$ K
$V = 36.6$ K
$M = 183.2$ K $\cdot$ ft

P11–47. $R_1 = R_2 = 450$ N
$V = 450$ N
$M = 172.5$ N $\cdot$ m

P11–49. $R_1 = 180$ kN; $R_2 = 190$ kN
$V = 190$ kN
$M = 630$ kN $\cdot$ m

P11–51. $R_1 = 636$ lb; $R_2 = 1344$ lb
$V = 804$ lb
$M = 2528$ lb $\cdot$ in

P11–53. $R_1 = 4950$ N; $R_2 = 3100$ N
$V = 2950$ N
$M = 3350$ N $\cdot$ m

P11–55. $R = 236$ lb
$V = 236$ lb
$M = 1504$ lb $\cdot$ in

P11–57. $R = 1130$ N
$V = 1130$ N
$M = 709$ N $\cdot$ m

P11–59. $R = 230$ kN
$V = 230$ kN
$M = 430$ kN $\cdot$ m

P11–61. $R = 1400$ lb
$V = 1500$ lb
$M = 99\,000$ lb $\cdot$ in

P11–63. $R = 1250$ N
$V = 1250$ N
$M = 1450$ N $\cdot$ m

P11–65. $R_1 = 1333$ lb; $R_2 = 2667$ lb
$V = 2667$ lb
$M = 5132$ lb $\cdot$ ft

P11–67. $R_1 = R_2 = 75 \text{ N}$
$V = 75 \text{ N}$
$M = 15 \text{ N} \cdot \text{m}$

P11–69. $R_1 = 8.60 \text{ kN}; R_2 = 12.2 \text{ kN}$
$V = 12.2 \text{ kN}$
$M = 9.30 \text{ kN} \cdot \text{m}$

P11–71. $R_1 = R_2 = 5400 \text{ lb}$
$V = 5400 \text{ lb}$
$M = 19\,800 \text{ lb} \cdot \text{ft}$

P11–73. $R = 10.08 \text{ kN}$
$V = 10.08 \text{ kN}$
$M = 8.064 \text{ kN} \cdot \text{m}$

P11–75. $R = 7875 \text{ lb}$
$V = 7875 \text{ lb}$
$M = 21\,063 \text{ lb} \cdot \text{ft}$

For Problems P11–77 through P11–83, results are shown for the main horizontal section only.

P11–77. $R_1 = R_2 = 282 \text{ N}$
$V = 282 \text{ N}$
$M = 120 \text{ N} \cdot \text{m}$

P11–79. $R_1 = R_2 = 162 \text{ N}$
$V = 162 \text{ N}$
$M = 42.2 \text{ N} \cdot \text{m}$

P11–81. $R_1 = 165.4 \text{ N}; R_2 = 18.4 \text{ N}$
$V = 165.4 \text{ N}$
$M = 16.54 \text{ N} \cdot \text{m}$

P11–83. $R_1 = 4.35 \text{ N}; R_2 = 131.35 \text{ N}$
$V = 127 \text{ N}$
$M = 6.35 \text{ N} \cdot \text{m}$

Additional Practice Problems

Shearing Force and Bending Moment Using Free Body Diagram Method

11–85. $V = -7.36 \text{ kN}; M = 29.96 \text{ kN} \cdot \text{m}$

11–87. $V = 1000 \text{ N}; M = -170 \text{ N} \cdot \text{m}$

11–89. $V = -1250 \text{ N}; M = 2400 \text{ N} \cdot \text{m}$

11–91. $V = -4.20 \text{ kN}; M = 8.20 \text{ kN} \cdot \text{m}$

For the format of answers for Problems 11–93 to 11–109, see note before P11–1.

11–93. $R_1 = 1200 \text{ lb}; R_2 = 1200 \text{ lb}$
$V = 1200 \text{ lb}$
$M = 54\,000 \text{ lb} \cdot \text{in}$

11–95. $R_1 = 11\,300 \text{ lb}; R_2 = 1250 \text{ lb}$
$V = 6750 \text{ lb}$
$M = 202\,500 \text{ lb} \cdot \text{in}$

11–97. $R_1 = 66.25 \text{ kN}; R_2 = 128.75 \text{ kN}$
$V = 78.75 \text{ kN}$
$M = 132.5 \text{ kN} \cdot \text{m}$

11–99. $R_1 = 3100 \text{ lb}; R_2 = 3100 \text{ lb}$
$V = 3100 \text{ lb}$
$M = 141\,000 \text{ lb} \cdot \text{in}$

11–101. $R_1 = 495 \text{ lb}; R_2 = 1405 \text{ lb}$
$V = 805 \text{ lb}$
$M = 2450.25 \text{ lb} \cdot \text{in}$

11–103. $R_1 = 27.0 \text{ kN}; R_2 = 9.0 \text{ kN}$
$V = 15.0 \text{ kN}$
$M = 18.0 \text{ kN} \cdot \text{m}$

11–105. $R_1 = 6200 \text{ N}; R_2 = 36\,800 \text{ N}$
$V = 18\,800 \text{ N}$
$M = 3150 \text{ N} \cdot \text{m}$

11–107. $R_1 = 920 \text{ lb}; R_2 = 520 \text{ lb}$
$V = 920 \text{ lb}$
$M = 6720 \text{ lb} \cdot \text{in}$

11–109. $R = 350 \text{ N}$
$V = 350 \text{ N}$
$M = 122.5 \text{ N} \cdot \text{m}$

Problems with Given Shearing Force Diagrams

Load and bending moment diagrams to be found.

11–111. Simple beam with three concentrated loads at B, C, and D; reactions at A and E:
$R_A = 35 \text{ kN}, R_E = 45 \text{ kN}; F_B = 26 \text{ kN}$,
$F_C = 30 \text{ kN}, F_D = 24 \text{ kN}$
$M_A = 0 \text{ kN} \cdot \text{m}, M_B = 52.5 \text{ kN} \cdot \text{m}$,
$M_C = 66.0 \text{ kN} \cdot \text{m (max)}, M_D = 45.0 \text{ kN} \cdot \text{m}$,
$M_E = 0 \text{ kN} \cdot \text{m}$

11–113. Cantilever, supported at A, with two concentrated loads at B and C:
$R_A = 80 \text{ lb}, M_A = 500 \text{ lb} \cdot \text{ft CCW}$;
$F_B = 60 \text{ lb}, F_C = 20 \text{ lb}$
$M_A = -500 \text{ lb} \cdot \text{ft}, M_B = -100 \text{ lb} \cdot \text{ft}$

11–115. Simple beam with partial uniformly distributed load from B to C; reactions at A and C:
$R_A = 4050 \text{ lb}, R_C = 6750 \text{ lb}$; distributed load from B to C; $w = 1200 \text{ lb/ft}$
$M_A = 0 \text{ lb} \cdot \text{ft}, M_B = 12\,150 \text{ lb} \cdot \text{ft}$,
$M_C = 0 \text{ lb} \cdot \text{ft}$,
$M_D = 18\,984.4 \text{ lb} \cdot \text{ft (max at } x = 6.375 \text{ ft from } A)$

Problems with Given Bending Moment Diagrams

Load and shearing force diagrams to be found.

11–117. Simple beam with two concentrated loads at B and C; reactions at A and D:
$R_A = 100 \text{ lb}, R_D = 150 \text{ lb}; F_B = 75 \text{ lb}$,
$F_C = 175 \text{ lb}$
$V_{A-B} = 100 \text{ lb}, V_{B-C} = 25 \text{ lb}$,
$V_{C-D} = -150 \text{ lb}$

11–119. Simple beam with two concentrated loads at A and C; reactions at B and D:
$R_B = 19.17 \text{ kN}, R_D = 10.0 \text{ kN}$,
$F_A = 6.67 \text{ kN}, F_C = 22.5 \text{ kN}$
$V_{A-B} = -6.67 \text{ kN}, V_{B-C} = 19.17 \text{ kN}$,
$V_{C-D} = -10.0 \text{ kN}$

11–121. Simple beam with uniformly distributed load over entire length; reactions at A and B:
Uniformly distributed load $= w = 900$ lb/ft;
$R_A = 5400$ lb, $R_B = 5400$ lb
$V_A = 5400$ lb, $V_B = -5400$ lb,
$V_C = 0$ lb at 6.0 ft from A is point where maximum bending moment occurs

Continuous Beams—Theorem of Three Moments

11–124. $R_A = R_E = 371$ lb at ends of beam
$R_C = 858$ lb at middle support
$V_{max} = 429$ lb from B to D between two 800 lb loads
$M_{max} = 1113$ lb · ft under each load

11–126. $R_A = 56.5$ kN, $R_C = 135$ kN at middle support
$R_D = 16.5$ kN at right end
$V_{max} = -87.5$ kN at C
$M_{max} = 32.4$ kN · m at 80 kN load

11–128. $R_A = 8.90$ kN, $R_C = 34.1$ kN at middle support
$R_D = 6.00$ kN at right end
$V_{max} = 18.0$ kN at C
$M_{max} = 8.9$ kN · m at B at 25 kN load

11–130. $R_A = 21.1$ kN, $R_C = 101.8$ kN at middle support
$R_E = 37.1$ kN
$V_{max} = 62.9$ kN at C
$M_{max} = 31.7$ kN · m at B under 60 kN load

Chapter 12 Stress Due to Bending

12–1. 94.4 MPa

12–3. **a.** 20 620 psi
b. 41 240 psi

12–5. 21 050 psi

12–7. $\sigma_t = 6882$ psi
$\sigma_c = 12\,970$ psi

12–9. 5794 psi (39.9 MPa)

12–11. 13 963 psi

12–13. Required $S = 2636$ mm^3; for $h/b = 3.0$;
$b = 12.1$ mm; $h = 36.3$ mm
Convenient dimensions from Appendix A–2:
$b = 12$ mm; $h = 40$ mm; $S = 3200$ mm^3;
$A = 480$ mm^2; $h/b = 3.33$
$b = 14$ mm; $h = 35$ mm; $S = 2858$ mm^3;
$A = 490$ mm^2; $h/b = 2.5$

12–15. Required $s_u = 290$ MPa; possible material—6061-T6

12–17. Required $S = 88.2$ in^3; W20×66 steel beam

12–19. Required $s_u = 10.6$ ksi; OK for 6061-T4

12–21. Required $s_y = 284$ MPa; OK for 2014-T4

12–23. Required $S = 1.35$ in^3; 3-in schedule 40 steel pipe

12–24. Required $S = 5.84$ in^3; $6 \times 4 \times \frac{1}{4}$ or $8 \times 2 \times \frac{1}{4}$ steel tube

12–25. Required $S = 6.72$ in^3; 6I×4.030 aluminum beam

12–26. Required $S = 5.38$ in^3; W8×10 steel beam

12–27. Required $S = 7.47$ in^3; no suitable channel

12–28. Required $S = 7.47$ in^3; 6-in schedule 40 steel pipe

12–31. Required $S = 9.76$ in^3; W10×12 steel beam or required $S = 1.60 \times 10^5$ mm^3; W250×17.9

12–33. Required $S = 16.6$ in^3; W12×16 steel beam

12–35. Required $S = 1.89 \times 10^5$ mm^3; W310×223.8 steel beam

12–37. Required $S = 7.47 \times 10^5$ mm^3; W460×60 steel beam

12–39. Required $S = 14.5$ in^3; W12×16 steel beam

12–41. Required $S = 6.37 \times 10^3$ mm^3; W200×15 steel beam

12–43. Required $S = 16.6$ in^3; S10×25.4 steel beam

12–45. Required $S = 1.89 \times 10^5$ mm^3; S200×27.4 steel beam

12–47. Required $S = 7.47 \times 10^5$ mm^3; S380×64 steel beam

12–49. Required $S = 14.5$ in^3; S8×23 steel beam

12–51. Required $S = 6.37 \times 10^3$ mm^3; S80×8.5 steel beam

12–53. Required $S = 13.85$ in^3; W12×16 steel beam

12–55. Required $S = 1.58 \times 10^5$ mm^3; W250×17.9 steel beam

12–57. Required $S = 6.23 \times 10^5$ mm^3; W310×44.5 steel beam

12–59. Required $S = 12.12$ in^3; W12×16 steel beam

12–61. Required $S = 5.31 \times 10^3$ mm^3; W200×15 steel beam

12–63. Required $S = 18.8$ in^3; 2 × 10 wood beam

12–65. 2 × 8 wood beam

12–68. Required $S = 11.1$ in^3; 2 × 8 wood beam

12–69. Required $S = 25.0$ in^3; 2 × 12 wood beam

12–70. Required $S = 2.79$ in^3; 2 × 4 wood beam

12–71. Required $S = 12.5$ in^3; 2 × 8 wood beam

12–75. 10 × 12 wood beam; No. 2 southern pine W8×10 steel beam; ASTM A992 steel

12–77. $\sigma_d = 4.3$ MPa
At A: $\sigma = 3.81$ MPa; OK
At B: $\sigma = 5.16$ MPa; unsafe
At C: $\sigma = 4.62$ MPa; unsafe

12–79. 4.86 N/mm

12–81. 1094 N

12–83. 676 lb

12–85. 102 lb

12–87. 6.77 lb/in

12–89. 3.80 ft from wall to joint; 4-in pipe is safe at wall

12–91. 398 MPa at C

12–93. At fulcrum, $\sigma = 8000$ psi
At bottom hole, $\sigma = 5067$ psi
At next hole, $\sigma = 3800$ psi
At next hole, $\sigma = 2534$ psi
At next hole, $\sigma = 1267$ psi

12–94. At fulcrum, $\sigma = 8000$ psi
At bottom hole, $\sigma = 10\,000$ psi
At next hole, $\sigma = 7506$ psi
At next hole, $\sigma = 5004$ psi
At next hole, $\sigma = 2500$ psi

12–95. a. With pivot in end hole as shown:
At fulcrum, $\sigma = 8000$ psi
At bottom hole, $\sigma = 8064$ psi
For pivot in any other hole, maximum stress is
at fulcrum. For pivot in:
Hole 2: $\sigma = 6800$ psi
Hole 3: $\sigma = 5600$ psi
Hole 4: $\sigma = 4400$ psi
Hole 5: $\sigma = 3200$ psi

12–97. 109 MPa

12–98. 149 MPa

12–99. Required $s_u = 1195$ MPa
AISI 4140 OQT 900 (others possible)

12–100. 2513 N

12–101. 1622 N

12–103. Impossible

12–105. Yes. $d_{max} = 37.2$ mm

12–106. 118 MPa at first step (L_3)

12–107. Required $s_u = 946$ MPa
AISI 1141 OQT 900 (others possible)

12–109. $L_{1max} = 206$ mm
$L_{2max} = 83.4$ mm
$L_{3max} = 24.7$ mm

12–111.

x (mm)	σ (MPa)
0	0
40	52.1
80	76.5
120	87.9
160	92.6
200	93.8
240	112.5

12–113. $h_1 = 22$ mm; $h_2 = 22$ mm

12–115. Required $S = 16.36$ in^3 W12×16

12–117. For composite beam, $w = 4.18$ K/ft
For S-beam alone; $w = 3.58$ K/ft

12–118. 11.5 mm

12–120. 0.805 in

12–122. 25.5 mm

12–124. 46 mm from center

12–126. $e = 12.9$ mm

12–127. 4.94 lb/in

12–129. 822 N

12–131. 625 N

12–133. 21.0 kN

12–134. 6.94 kN/m

12–135. 9.69 kN/m

12–136. 48.0 kN

12–137. 126 kN

12–139. $\sigma_d = 46.0$ MPa; $\sigma_{max} = 68.4$ MPa; unsafe

12–141. $\sigma_{max} = 47.0$ MPa at step 50 mm from R_1

12–143. Required $S = 66.7$ in^3; W18×40

12–145. Required $S = 3.25$ in^3; $4 \times 4 \times \frac{1}{4}$ or $6 \times 2 \times \frac{1}{4}$
Or: Required $S = 5.33 \times 10^4$ mm^3;
HSS102×102 × 6.4 or HSS152×51 × 6.4

12–147. Required $S = 7.45$ in^3; $8 \times 2 \times \frac{1}{4}$

12–148. Allowable load $= 3963$ lb (tension)

12–150. Required $S = 6.14$ in^3; W8×10;
Total weight $= 83$ lb

Chapter 13   Shearing Stresses in Beams

13–1. 1.125 MPa

13–3. 1724 psi

13–5. 3.05 MPa

13–7. 3180 psi

13–9. 1661 psi

13–11. 69.3 psi

13–13. 7.46 MPa

13–15. 2.79 MPa

13–17. 10.3 MPa

13–19. 2342 psi

13–21. 245 lb

13–23. 788 lb

13–25. 5098 lb

13–27. 661 lb

13–29. 787 lb

13–31. Let $y =$ distance from bottom of the I-shape:

y(in)	τ (psi)
0.0	0.0
0.5	5.65
1.0(−)	10.54
1.0(+)	63.25
1.5	67.39
2.0	70.78
2.5	73.42
3.0	75.30
3.5	76.43
4.0	76.81

13–33. 8733 psi

13–35. From web shear formula:
$\tau = 8017$ psi
Approximately 8% low compared with τ_{max} from
13–33.

13–37. Let $y =$ distance from bottom of the I-shape:

y (in)	τ (psi)
0.0	0
0.175	155
0.35($-$)	303
0.35($+$)	6582
1.0	7073
2.0	7638
3.0	7976
4.0	8089

13–39. $\tau = 3788$ psi;
$\tau_d = 20\,000$ psi; OK
$\sigma = 21\,053$ psi;
$\sigma_d = 33\,000$ psi; safe

13–41. W18×40; $\tau = 7892$ psi;
$\tau_d = 20\,000$ psi; OK

13–43. $1\frac{1}{4}$ in schedule 40 pipe;
$\tau = 2404$ psi;
$\tau_d = 8000$ psi; OK

13–45. 4 × 10 beam

13–47. 1256 lb

13–49. 116.3 MPa

13–51. **a.** $\tau = 1125$ psi
b. $\sigma = 6750$ psi
c. $s_y = 20\,250$ psi; any steel

13–53. 35.4 mm; $\tau = 1.59$ MPa;
$N = 86.7$ for shear

13–55. 24.5 mm; $\tau = 1.28$ MPa;
$\sigma_d = 120$ MPa

13–57. Required $S = 0.45$ in^3; 2-in schedule 40 pipe

13–59. $q = 736$ lb/in; $\tau = 526$ psi

13–61. 433 lb/ft based on strength of rivets

13–63. 1722 lb based on bending

13–65. Maximum spacing $= 4.36$ in

13–67. Maximum spacing $= 3.94$ in

Web Shear

13–69. $\tau = 5185$ psi; $\tau_d = 20\,000$ psi; safe

13–71. $\tau = 2821$ psi; $\tau_d = 20\,000$ psi; safe

13–73. $\tau = 3599$ psi; $\tau_d = 20\,000$ psi; safe

General Shear and Special Shear Formulas

13–75. $\tau = 66.2$ psi; $\tau_d = 70$ psi; safe

13–77. $\tau = 4623$ psi; $\tau_d = 11\,500$ psi; safe

13–79. $\tau = 209$ psi; $\tau_d = 10\,000$ psi; safe

Shear Flow

13–81. $q = 112$ lb/in; spacing $= 2.41$ in maximum

Chapter 14 Deflection of Beams Formula Method—Statically Determinate Beams

14–1. -2.01 mm

14–3. -0.503 mm

14–5. -5.40 mm

14–7. At loads: -0.251 in
At center: -0.385 in

14–9. -0.424 in

14–11. -0.271 in

14–13. $+0.093$-in deflection at $x = 69.2$ in

14–15. $D = 64.8$ mm

14–17. $t = 0.020$ in

Formula Method—Statically Indeterminate Beams

For Problems 14–19 through 14–45, the following values are reported:

Reactions at all supports
Shearing forces at critical points
Bending moments at critical points
Maximum deflection or deflection at selected points in the form $y = C_d/EI$

When you specify the beam material, shape, and dimensions, you can compute the beam stiffness, EI, and use that to calculate the deflection. Deflections will be in the given unit of length when E and I are in the same units of force and length given in the answers. For example, in Problem 14–19, length is in m, force is in N. Then deflection is in m when E is in N/m^2 (Pa) and I is in m^4.

14–19. $R_A = V_A = 24\,063$ N, $R_C = -V_C = 10\,938$ N
$M_A = -26\,250$ N·m, $M_B = 21\,875$ N·m,
$M_C = 0$ N·m
$y_{max} = (-20\,934)/EI$ at 1.79 m from C

14–21. $R_A = V_A = 18\,760$ N, $R_C = -V_C = 16\,235$ N
$M_A = -22\,560$ N·m, $M_B = 24\,353$ N·m,
$M_C = 0$ N·m
$y_B = (-21\,629)/EI$ at load

14–23. $R_A = V_A = 500$ lb, $R_B = -V_B = 300$ lb
$M_A = -1600$ lb·in, M_E(max) $= 900$ lb·in at
$x = 10.0$ in, $M_B = 0$ lb·in
$y_{max} = (-17\,712)/EI$ at $x = 9.264$ in

14–25. $R_A = V_A = 17\,500$ N, $R_C = -V_C = 17\,500$ N
$M_A = -17\,500$ N·m, $M_B = 17\,500$ N·m,
$M_C = -17\,500$ N·m
$y_{max} = (-11\,667)/EI$ at B at center

14–27. $R_A = V_A = 11\,074$ N, $R_C = -V_C = 23\,926$ N
$M_A = -12\,305$ N·m, $M_B = 15\,381$ N·m,
$M_C = -20\,508$ N·m
$y_{max} = (-10\,127)/EI$ at B at center

14–29. $R_A = V_A = 400$ lb, $R_B = -V_B = 400$ lb
$M_A = -1067$ lb·in, $M_B = 533$ lb·in,
$M_C = -1067$ lb·in
$y_{max} = (-8533)/EI$ at center

14–31. $R_A = V_A = 150$ lb, $R_B = 500$ lb,
$R_C = -V_C = 150$ lb
$M_A = 0$ lb·in, $M_B = -400$ lb·in, $M_C = 0$ lb·in
$M_D = M_E = 225$ lb·in at $x = 3.00$ in from A and C
$y_{max} = (-1107)/EI$ at $x = 3.372$ in from A or C

14–33. $R_A = R_D = 106.7$ lb, $R_B = R_C = 293.3$ lb
$V_A = -V_D = 106.7$ lb, $-V_B = V_C = 160$ lb
$M_A = M_D = 0$ lb·in, $M_B = M_C = -142.2$ lb·in,
$M_G = 35.6$ lb·in
$M_E = M_F = 113.8$ lb·in at $x = 2.13$ in from A and D

14–35. $R_A = R_E = 78.6$ lb, $R_B = R_D = 228.6$ lb,
$R_C = 185.6$ lb
$V_A = -V_E = 78.6$ lb, $-V_B = V_D = 121.4$ lb,
$V_C = 92.8$ lb
$M_A = M_E = 0$ lb·in, $M_B = M_D = -85.7$ lb·in,
$M_C = -57.1$ lb·in
$M_F = M_I = 61.8$ lb·in at $x = 1.56$ in from A and E
$M_G = M_H = 29.1$ lb·in at $x = 2.16$ in from B and D

14–37. $R_A = V_A = 22\,500$ N, $R_B = V_B = 13\,500$ N
$M_A = -8100$ N·m, $M_E = 4556$ N·m at 0.675 in
from B
$y_{max} = -1135/EI$ at 1.042 in from A

14–39. $R_A = V_A = -13\,750$ N, $R_B = 31\,750$ N
$V_B = -18\,000$ N
$M_A = 12\,600$ N·m, $M_B = -25\,200$ N·m, $M_C = 0$
$y_C = -40\,719/EI$ at overhang at right end

14–41. $R_A = R_C = V_A = -V_C = 25\,200$ lb, $R_B = 8400$ lb
$V_B = 42\,000$ lb
$M_A = M_C = 0$, $M_B = -134\,400$ lb·ft
$M_D = M_E = 75\,587$ lb·ft at $x = 6.0$ ft from A or C
$y_{max} = (-1.488 \times 10^6 \text{ lb·ft}^3)/EI$ at 6.74 ft from A
or C

14–43. $R_A = R_E = V_E = -V_E = 212$ lb,
$R_B = R_D = 617$ lb, $R_C = 501$ lb
$V_B = V_D = 328$ lb, $V_C = 251$ lb
$M_A = M_E = 0$, $M_C = -1388$ lb·in
$M_B = M_D = -2082$ lb·in
$M_F = M_I = 1501$ lb·in at 14.04 in from A or E
$M_G = M_H = 708$ lb·in at 19.44 in from B or D

14–45. $R_A = V_A = 224.6$ N, $R_C = -V_C = 25.4$ N
$M_A = -2.355$ N·m, $M_B = 1.014$ N·m, $M_C = 0$
$y_B = (-1.386 \times 10^{-4})/EI$ at load

Comparisons of Beam Behavior

14–46. Comparison of four beam designs to carry a uniformly distributed load:

	V_{max}	V/V_1	M_{max}	M/M_1	y_{max}	y/y_1
a.	3600 N	1.0	3600 N·m	1.0	$-6000/EI$	1.0
b.	7200 N	2.0	$-14\,400$ N·m	4.0	$-57\,600/EI$	9.6
c.	4500 N	1.25	-3600 N·m	1.0	$-2490/EI$	0.415
d.	3600 N	1.0	-2400 N·m	0.67	$-1200/EI$	0.20

14–48. Comparison of results of five problems:

		V_{max}	V/V_1	M_{max}	M/M_1	y_{max}	y/y_1
1.	14–23	500 lb	1.0	-1600 lb·in	1.0	$-17\,712/EI$	1.0
2.	14–29	400 lb	0.80	-1067 lb·in	0.667	$-8533/EI$	0.482
3.	14–31	250 lb	0.50	-400 lb·in	0.250	$-1107/EI$	0.0625
4.	14–33	160 lb	0.320	-142 lb·in	0.089	—	—
5.	14–35	121 lb	0.243	-85.7 lb·in	0.054	—	—

14–52. Comparison of results of three problems:

		V_{max}	V/V_1	M_{max}	M/M_1	y_{max}	y/y_1	A	A/A_1
1.	14–49	1440 lb	1.0	103 680 lb·in	1.0	$-896 \times 10^6/EI$	1.0	63.3 in^2	1.0
2.	14–50	900 lb	0.625	25 920 lb·in	0.25	$-372 \times 10^6/EI$	0.415	25.4 in^2	0.401
3.	14–51	576 lb	0.40	9 216 lb·in	0.089	—	—	13.87 in^2	0.172

Superposition—Statically Determinate Beams

14–53. $y_B = -1.291$ mm at 840 N load
$y_C = -3.055$ mm at 600 N load
$y_D = -1.353$ mm at 1200 N load

14–56. $y_B = -0.0140$ in at 85 lb load
$y_C = -0.0262$ in at 75 lb load at end

14–57. -0.869 mm

14–59. -3.997 mm

14–60. −0.0498 in

14–62. C5 × 2.212 Aluminum channel

Superposition—Statically Indeterminate Beams

14–64. $R_A = V_A = 22\,500\,\text{N}, R_B = V_B = 13\,500\,\text{N}$
$M_A = -8100\,\text{N} \cdot \text{m}, M_E = 4556\,\text{N} \cdot \text{m}$ at 0.675 in from B
$y_{max} = -1135/EI$ at 1.042 in from A

14–66. $R_A = R_E = 371$ lb at ends of beam
$R_C = 858$ lb at middle support
$V_{max} = 429$ lb from B to D between two 800-lb loads
$M_{max} = 1113$ lb · ft under each load

14–68. $R_A = 56.5\,\text{kN}, R_C = 135\,\text{kN}$ at middle support
$R_D = 16.5\,\text{kN}$ at right end
$V_{max} = -87.5\,\text{kN}$ at C
$M_{max} = 32.4\,\text{kN} \cdot \text{m}$ at 80-kN load

14–70. $R_A = 4009$ lb at fixed end
$R_D = 1991$ lb at right support
$V_{max} = 4009$ lb at A
$M_{max} = -8490$ lb · ft at A

Design Problems—Stress and Deflection Limits

14–72. Comparison of two designs in Figure P14–72:
   a. $V = 2250\,\text{N}, M = -4500\,\text{N} \cdot \text{m}$,
   $y = -22\,500/EI$
   b. $V = 3375\,\text{N}, M = -4500\,\text{N} \cdot \text{m}$,
   $y = -20\,250/EI = 0.90\,y_a$
   Little difference in the performance of the two designs.

14–74. $R_A = V_A = 32.75\,\text{kN}, R_B = -V_B = 19.65\,\text{kN}$
$M_A = -42.9\,\text{kN} \cdot \text{m}, M_E = 24.1\,\text{kN} \cdot \text{m}$

Successive Integration Method

14–76. $y = -0.0078$ in at $x = 8.56$ in

14–78. $y = -3.79$ mm

14–80. $D = 69.2$ mm

14–82. I178×8.630 (I7×5.800) aluminum beam
$y = -5.37$ mm at $x = 1.01$ m

14–84. $D = 109$ mm

Moment-Area Method

14–86. −0.0078 in

14–88. −3.79 mm

14–90. −3.445 mm

14–92. −5.13 mm

14–94. −0.01138 in

14–96. −0.257 in

14–98. −71.7 mm

Chapter 15 Combined Stresses

15–1. −10 510 psi

15–3. $\sigma_N = 9480$ psi;
$\sigma_M = -7530$ psi

15–5. $\sigma_N = 13\,980$ psi;
$\sigma_M = -15\,931$ psi

15–7. −64.3 MPa

15–9. 415 N

15–11. $\sigma = 724$ MPa; required $s_y = 1448$ MPa;
AISI 4140 OQT 700

15–13. At B: $\sigma = 24\,328$ psi tension on top of beam
$\sigma = -18\,785$ psi compression on bottom of beam

15–15. Load = 9081 N;
Mass = 926 kg

15–17. 26 mm

15–19. 18.7 mm

15–21. 71.6 MPa

15–23. 2548 psi

15–25. 7189 psi

15–27. 51.6 MPa

15–29. 982 MPa

15–31. $\tau = 9923$ psi, $s_y = 119$ ksi

15–33. 67.1 MPa

15–35. 7548 psi

15–37. 7149 psi

15–39. 61.5 MPa

15–41. a. 3438 psi
   b. 3438 psi
   c. 344 psi
   d. 2300 psi
   e. 1186 psi

15–43.

	Middle of beam	Near supports	15 in from A
a.	1875 psi	0 psi	1406 psi
b.	1875 psi	0 psi	1406 psi
c.	0 psi	375 psi	188 psi
d.	1250 psi	208 psi	943 psi
e.	625 psi	333 psi	498 psi

15–45. 75.3 MPa

15–47. a. $P = 40\,667$ lb;
   $\tau_{max} = 8333$ psi
   b. $\tau_{max} = 8925$ psi; $N = 2.80$

Additional Practice Problems

15–49. $\sigma_{max} = 397$ MPa tensile; $\sigma_{max} = 709$ MPa compressive

15–51. $\sigma_{max} = 821$ MPa tensile; $\sigma_{max} = 458$ MPa compressive

15–55. At M: $\sigma_{max} = 2.18$ MPa tensile
At N: $\sigma_{max} = 3.54$ MPa compressive

Note: The complete solutions for Problems 15–56 through 15–82 require the construction of the complete Mohr's circle and the drawing of the principal stress element and the maximum shear stress element. Listed below are the significant numerical results.

Prob. No.	σ_1	σ_2	$\phi(\deg)$	τ_{max}	σ_{avg}	$\phi'(\deg)$
15–56.	315.4 MPa	−115.4 MPa	10.9 CW	215.4 MPa	100.0 MPa	34.1 CCW
15–58.	110.0 MPa	−40.0 MPa	26.6 CW	75.0 MPa	35.0 MPa	18.4 CCW
15–60.	23.5 ksi	−8.5 ksi	19.3 CCW	16.0 ksi	7.5 ksi	64.3 CCW
15–62.	79.7 ksi	−9.7 ksi	31.7 CCW	44.7 ksi	35.0 ksi	76.7 CCW
15–64.	677.6 kPa	−977.6 kPa	77.5 CCW	827.6 kPa	−150.0 kPa	57.5 CW
15–66.	327.0 kPa	−1202.0 kPa	60.9 CCW	764.5 kPa	−437.5 kPa	74.1 CW
15–68.	570.0 psi	−2070.0 psi	71.3 CW	1320.0 psi	−750.0 psi	26.3 CW
15–70.	4180.0 psi	−5180.0 psi	71.6 CW	4680.0 psi	−500.0 psi	26.6 CW
15–72.	360.2 MPa	−100.2 MPa	27.8 CCW	230.2 MPa	130.0 MPa	72.8 CCW
15–74.	23.9 ksi	−1.9 ksi	15.9 CW	12.9 ksi	11.0 ksi	29.1 CCW
15–76.	4.4 ksi	−32.4 ksi	20.3 CW	18.4 ksi	−14.0 ksi	24.7 CCW
15–78.	321.0 MPa	−61.6 MPa	64.4 CCW	191.0 MPa	130.0 MPa	68.6 CW
15–80.	225.0 MPa	−85.0 MPa	0.0	155.0 MPa	70.0 MPa	45.0 CCW
15–82.	775.0 kPa	−145.0 kPa	0.0	460.0 kPa	315.0 kPa	45.0 CCW

For Problems 15–84 through 15–94, Mohr's circle from the given data results in both principal stresses having the same sign. For this class of problems, the supplementary circle is drawn using the procedures discussed in Section 15–11 of the text. The results include three principal stresses where $\sigma_1 > \sigma_2 > \sigma_3$. Also, the maximum shear stress is found from the radius of the circle containing σ_1 and σ_3 and is equal to $\frac{1}{2}\sigma_1$ or $\frac{1}{2}\sigma_3$, whichever has the greatest magnitude. Angles of rotation of the resulting elements are not requested.

Prob. No.	σ_1	σ_2	σ_3	τ_{max}
15–84.	328.1 MPa	71.9 MPa	0.0 MPa	164.0 MPa
15–86.	214.5 MPa	75.5 MPa	0.0 MPa	107.2 MPa
15–88.	35.0 ksi	10.0 ksi	0.0 ksi	17.5 ksi
15–90.	55.6 ksi	14.4 ksi	0.0 ksi	27.8 ksi
15–92.	0.0 kPa	−307.9 kPa	−867.1 kPa	433.5 kPa
15–94.	0.0 psi	−295.7 psi	−1804.3 psi	902.1 psi

For Problems 15–96 through 15–104, Mohr's circles from earlier problems are used to find the stress condition on the element at some specified angle of rotation. The listed results include the two normal stresses and the shear stress on the specified element.

Prob. No.	σ_A	$\sigma_{A'}$	τ_A
15–96.	130.7 MPa	69.3 MPa	213.2 MPa CW
15–98.	−37.9 MPa	197.9 MPa	31.6 MPa CCW
15–100.	3.6 ksi	−21.6 ksi	43.9 ksi CW
15–102.	−2010.3 psi	510.3 psi	392.6 psi CW
15–104.	8363.5 psi	86.5 psi	1421.2 psi CW

15–106. $\tau_{max} = 230.2$ MPa

15–108. $\tau_{max} = 12.9$ ksi

Strain-Gage Rosettes

The following table lists answers for parts (a) through (g) for Problems 15–110 through 15–124.

	(a)	(b)	(c)	(d)	(e)	(f)	(g)
15–110.	1902×10^{-6}	6.0×10^{-6}	$-28.2°$	147 MPa	49.1 MPa	1897 rad	49.2 MPa
15–112.	816×10^{-6}	-717×10^{-6}	$31.9°$	138 MPa	-109 MPa	1534 rad	123 MPa
15–114.	1006×10^{-6}	-294×10^{-6}	$36.6°$	17.3 ksi	0.731 ksi	1299 rad	8.30 ksi
15–116.	816×10^{-6}	-717×10^{-6}	$31.9°$	16.1 ksi	-12.9 ksi	1534 rad	14.5 ksi
15–118.	1494×10^{-6}	-112×10^{-6}	$-5.4°$	113 MPa	29.5 MPa	1607 rad	41.7 MPa
15–120.	882×10^{-6}	-324×10^{-6}	$39.7°$	178 MPa	-15.5 MPa	1206 rad	96.8 MPa
15–122.	616×10^{-6}	-319×10^{-6}	$-43.8°$	9.75 ksi	2.20 ksi	935 rad	5.98 ksi
15–124.	882×10^{-6}	-324×10^{-6}	$39.7°$	20.6 ksi	-2.24 ksi	1206 rad	11.4 ksi

Chapter 16   Columns

16–1. 25.1 kN

16–3. 8.35 kN

16–5. 26.2 kN

16–7. 111 kN

16–9. $P_a = 7318$ lb/column; use 9 columns

16–11. 499 kN

16–13. 65 300 lb

16–15. Axial force = 31.1 kN; critical load = 256.7 kN; $N = 8.25$ OK

16–17. 15.1 kN

16–19. Critical load = 10 914 lb; actual load = 5000 lb; $N = 2.18$; low

16–21. Critical load = 2849 lb; actual load = 1500 lb; $N = 1.90$; low

16–23. 2.68 in

16–25. Required $I = 5.02$ in^4; I7×5.800

16–27. 5649 lb

16–29. 245 kN

16–31. No improvement

Additional Practice Problems

16–33. 9420 lb

16–35. 6219 lb

16–37. $N = 2.27$

16–39. Design problem. Example solution. Compression members: ASTM A36 Structural steel
Round bar; nearest 1/16 in
$AC = 1925$ lb; 40 in length; 15/16 in round bar;
$P_{all} = 2253$ lb
$CD = 750$ lb: 25 in length; 9/16 in round bar;
$P_{all} = 750$ lb
$DE = 650$ lb; 40 in length; 7/16 in round bar;
$P_{all} = 654$ lb

16–41. Design problem. Example solution.
Compression members: Aluminum 6061–T6 Square bar; nearest preferred mm size (App. 2)
$DE = 2300$ N; 250 mm length; 9.0 mm side;
$P_{all} = 2383$ N
$BD = 2597$ N; 297 mm length; 11.0 mm side;
$P_{all} = 3767$ N
$EF = 2300$ N; 200 mm length; 8.0 mm side;
$P_{all} = 2324$ N
$CF = 800$ N; 160 mm length; 6.0 mm side;
$P_{all} = 1149$ N
$FG = 550$ N; 150 mm length; 5.0 mm side;
$P_{all} = 630$ N

16–43. Design problem. Example solution. Lightest steel tube.
Compressive force = 33 588 lb
HSS4×4×$\frac{1}{4}$ Steel tube; ASTM A501 Structural steel;
$P_{all} = 42\,441$ lb.

16–45. $P_a = 20.64$ kN for $N = 3$

16–47. $P_a = 8143$ lb for $N = 3$

16–49. $N = 1.068$—Low

16–51. $\sigma_{max} = 211$ MPa (Unsafe), $y_{max} = 25.8$ mm (High)

16–53. Not safe. Required $s_y = 103$ ksi for $N = 3$.

16–55. Not safe. Required $s_y = 104$ ksi for $N = 3$.

16–57. $F_a = 675$ lb for eccentric loading in the plane of the drawing
$F_a = 164$ lb for buckling about the narrow thickness

16–59. **a.** $P_a = 2610$ lb for straight pipe
b. $P_a = 1676$ lb for crooked pipe

Chapter 17   Pressure Vessels

17–1. 115 MPa

17–3. 4.70 mm minimum

17–5. 2134 psi

17–7. 1.80 mm

17–9. $N = 3.80$

17–11. $\sigma_2 = 212$ MPa; $\sigma_3 = -70.0$ MPa

17–13.

Radius (mm)	σ_2 (MPa)
110	166 inner surface
120	149
130	136
140	125
150	116 outer surface

17–15. 14.88 MPa

17–17.

Radius (mm)	σ_3 (MPa)
15	−7.00 inner surface
17	−4.58
19	−2.88
21	−1.64
23	−0.71
25	0.00 outer surface

17–19. 86.8 MPa

17–21.

Radius (mm)	σ_3 (MPa)
210	−100.0 inner surface
215	−83.3
220	−68.0
225	−54.1
230	−41.4
235	−29.7
240	−19.0
245	−9.1
250	0.00 outer surface

17–23.

Radius (mm)	σ_2 (MPa)
162.5	22.35 inner surface
170	20.99
177.5	19.84
185	18.88
192.5	18.06
200	17.35 outer surface

72–25. Sizes $\frac{1}{8}$ through 4 are thick-walled. Sizes 5 through 18 are thin-walled.

Chapter 18   Connections

18–1.   **a.** $F_s = 1400$ lb (Limit); $F_b = 13\,050$ lb; $F_t = 20\,250$ lb
     c. $F_s = 2160$ lb (Limit); $F_b = 9788$ lb; $F_t = 21\,263$ lb

18–2.   **a.** $F_s = 3800$ lb (Limit); $F_b = 40\,500$ lb; $F_t = 87\,000$ lb
     c. $F_s = 7600$ lb (Limit); $F_b = 40\,500$ lb; $F_t = 87\,000$ lb

18–3.   **a.** $F_s = 1178$ lb (Limit); $F_b = 15\,750$ lb; $F_t = 26\,708$ lb
     c. $F_s = 1325$ lb (Limit); $F_b = 11\,813$ lb; $F_t = 28\,114$ lb

18–4.   **a.** $F_s = 3313$ lb (Limit); $F_b = 49\,500$ lb; $F_t = 112\,485$ lb
     c. $F_s = 6627$ lb (Limit); $F_b = 49\,500$ lb; $F_t = 104\,970$ lb

18–5. $\frac{3}{4}$-in bolts

18–7.   23 860 lb on welds

Note: Problems 18–9 to 18–11 are design problems for which there are no unique answers.

INDEX

A

AASHTO, 49–52
Acceleration due to gravity, 17
Air bearings, 150
Allowable stress, 265
Allowable stress design (ASD), 271
Alloy steel, 234, 680
Aluminum Association, 24, 233, 266, 270, 607, 645, 672–675, 688, 690
Aluminum Association standard channels and I-beams, 20, 24, 672–675
Aluminum alloys, 239, 685
    cast, 239, 240, 685
    coefficient of thermal expansion, 278
    design bearing stress, 292
    designation system, 239
    series, 239
    tempers, 239
American Institute of Steel Construction, 24, 266, 270, 418, 642, 690
American Iron and Steel Institute, 24, 233
American National Standards Institute, 270
ASTM International (American Society for Testing and Materials), 24, 237
American Society of Civil Engineers, 48–51
American Society of Mechanical Engineers, 270, 634
American Society of State Highway and Transportation Officials, 49–52
American Standard beams (S-shapes), 20, 22, 668
Angle of friction, 151
Angle of twist, 338
Angles, structural steel, 20, 21, 662, 663
Anisotropic materials, 241, 251, 433
Annealing, 235
Areas, properties of, 656, 657
Austempered ductile iron, 238, 684
Austempering, 238
Axial deformation, 273, 274

B

Basic sizes, preferred, 264, 658
Beam deflection formulas, 701–709
Beam diagrams, mathematical analysis, 387, 393
Beam loading patterns, 359–362
Beam stiffness, 488
Beam support types, 362–363, 483–485
Beam types, 363, 393, 483–485
Beam, definition, 355, 359
    stress distribution, 413
Beams, analysis, 415–417
    comparison of support types, 359, 483–485, 493
    continuous, 363, 393, 483, 493
    deflection (See Deflection of beams)
    design examples, 417–423
    design procedures, 419–419
    failure modes, 408–409
    flexure formula, 409–411, 414
    fixed-end, 363
    made from anisotropic materials, 433
    made from composite materials, 434
    made from thin materials, 432
    preferred shapes, 431
    reactions, 364–367
    shearing stresses, 454–474 (See Shearing stress in beams)
    statically indeterminate beams, 363, 393, 483–484, 493, 501, 705–709
    stress concentrations, 424, 694, 695, 698–700
Bearing friction, 150
Bearing stress, 287
    contact stress, 289
    design, 290–294, 690
    in pin joints, 288
Belt drives, 150, 161
Bending moment diagrams, area rule, 370–371
    rules for drawing, 371, 378, 381
Bending moments in beams, 358, 368, 376, 381
Bending stress, 409, 410
    comparison of support types, 493
    stress distribution, 413
Beryllium copper, 240
Block and tackle, 79
Bolted connections, 642, 646
Brakes and clutches, 150, 159–161
Brass, 240, 681
Bridge standards, 48
Brittle material, 226
Bronze, 240, 681
    coefficient of thermal expansion, 278
Buckling, columns, 197, 591–611 (See Columns)
Building codes, 48–51

C

Cables, forces, 32
Calculations, 16
Canada, Institute for Research in Construction, 49
Canada, National Research Council, 49
Cantilever, 363, 381, 489, 515, 704–706
Capstan, 161–162
Carbon steel, 234, 680
Cast iron, 237, 684
    austempered ductile iron, 238, 684
    coefficient of thermal expansion, 278
    ductile iron, 237, 684
    gray iron, 237, 684
    malleable iron, 238, 684
    properties, 684
    white, 238
Center of mass (center of gravity), 31, 171
Centroid of an area, 66, 69, 171–174

Centroidal axis (Neutral axis), 174, 175, 413
Channels, aluminum, 20, 24, 372, 673
    structural steel, 20, 22, 664, 665
Charpy test, 230
Codes and standards, 48, 265, 270, 418
Coefficient of thermal expansion, 233, 278, 279
Column constant, 596
Column section, commercially available steel, 605
Columns, 591–611
    AISC specifications, 605
    allowable load, 600
    Aluminum Association specifications, 607
    analysis method, 600
    buckling, 197
    critical buckling load, 597
    crooked, 608
    eccentric, 609
    efficient shapes for, 603
    Euler formula, 597
    J. B. Johnson formula, 598
    laterally braced, 599
    non-centrally loaded, 607
    spreadsheet for analysis, 603, 609, 611
Combined stress, 537–580
Combined stresses, normal, 545–550
    normal and shear, 550–554
    Strain-Gage Rosette, 575–579
Composite compression members in construction, 285
Composite materials, beams, 434
Composites, 245–255
    advantages of, 248
    beams, 434
    coefficient of thermal expansion, 279
    comparison with metals, 249, 250
    filament winding, 253
    filler materials, 247, 248
    laminated construction, 251, 252, 256
    limitations, 251
    matrix materials, 247
    modulus of elasticity, 248, 254
    predicting properties, 253
    prepreg, 253
    pressure vessels, 635
    processing, 252
    products-examples, 245, 246
    rule of mixtures, 254
    sheet molding compound, 253
    specific modulus, 248, 257
    specific stiffness, 248, 257
    specific strength, 248, 257
    volume fraction, 253
Compound beam, 364
Compressive stress, 202, 224
Concentrated loads, 359, 367
Concentrated moments, 362, 383–387
Concrete, 242, 689
    bearing strength, 294, 689
    coefficient of thermal expansion, 278
    density, 243,
    modulus of elasticity, 243
    strength, rated, 243
Connections, 639–650
    allowable stresses, 642, 648, 649
    bolted, 642

bolted, eccentrically loaded, 646
    modes of failure, 641
    riveted, 644, 645
    types, 639, 642
    welded, 648, 649
Contact stress, 289
Continuous beams, 363, 364, 393, 493, 483, 708, 709
Conversion factors, 710
Copper, 240, 681
    coefficient of thermal expansion, 233
Copper Development Association, 233
Cosines, law of, 39
Couples, 7
Cranes, 95, 100–105
Creep, 230
Cylinders, thick-walled, 620, 624–625
    thin-walled, 620

D
Deflection formulas, beams, 489–505, 701–709
Deflection limits, 486
Deflection of beams, 418, 482–528
    definition of terms, 487
    deflection diagram, 487
    formula method, 489–505, 701–709
    methods of analyzing, 485
    moment-area method, 513–526
        beams with varying cross section, 520
        cantilever, 515–518, 526
        distributed loads, 526
        simply supported beams, 518
        theorems, 515
        unsymmetrically loaded, 522
    need for, 485
    radius of curvature, 488
    slope diagram, 487
    successive integration method, 505–513
    superposition, 499
Deformation, 204, 262, 273
    axial, 273
    normal, 274
    thermal, 278
    torsional, 337
    unit deformation, 204
Density, 18, 204, 243
Department of Defense, 232, 266
Derrick, 95
Design bearing stress, 290, 294, 689
Design factor, 265–267, 328
Design factor, column loading, 600
Design factor, direct tensile stresses, 265
    guidelines, 267
    torsional shear, 328
Design for direct stresses, 264
Design stress, 204, 265, 267, 270, 328, 418, 419, 471, 625–627, 688
    bending, 418, 419, 688
    methods of computing, 270
    normal, 265
    pressure vessels, 625–627
    shear, 204, 328, 471, 688
    torsional shear, 328
Design stress guidelines, 267, 688
Direct normal stress, 201

Direct shear stress, 204, 296
Distributed loads, 34–36, 69–73, 360–362, 376, 382
Double shear, 205
Drop weight tester, 231
Ductile iron, 237, 684
Ductile material, 226
Ductility, 225

E

Effective length, 594, 595
Elastic deformation in tension, 273
Elastic limit, 223
Elastic torsional deformation, 337, 345
Elongation, 225
End-fixity factor, 594
Endurance strength, 230
English Gravitational Unit System, 16
Equivalent torque, 551–552
Euler formula for columns, 597
Experimental stress analysis, 209, 286, 413, 424, 575–579

F

Factor of safety, 265, 328 (See also Design factor)
Failure modes, 198–200
Fasteners, spacing, shear flow, 473
Fatigue strength, 230
Filler materials for composites, 247, 248
Finite element analysis, 211
Fixed support, 44
Flexural center, 428, 429
Flexural modulus, 227
Flexural strength, 227
Flexure formula, 409, 410
    conditions on use, 411
    derivation, 414
Forces, 3, 17, 30–42
    cables, 32
    components, 36
    compressive, 31
    distributed, 34
    reactions, 31
    resultants, 36–42
    ropes, 32
    shearing, 31
    two-force member, 96
    tensile, 31
    weight, 17, 32
Fracture, 196
Frames, 93–106
Free body diagrams, 30–48
    creating, 44–48
    parts of beams, 368, 376, 383–387
Friction, 146–162
    air bearings, 150
    analysis, 150–162
    angle of friction, 154
    bearing friction, 150
    belt drives, 150, 161
    brakes and clutches, 150, 159–162
    capstans, 161–162
    coefficient of friction, 147
    definition, 146
    hydrostatic bearings, 150

rolling friction, 149–150
table of coefficient of friction, 148, 150
testing for coefficient of friction, 152–154
tipping or sliding, 154–157
rope friction, 161
wedges, 157–159

G

Gage length, 225
General shear formula, 457
    development of, 467
    guidelines for use of, 460
Glass, coefficient of thermal expansion, 278
Gray iron, 237, 684

H

Hardness, 227
    tables, 228–229
Heat treatment, 235
Hollow structured shapes (HSS), 20, 22, 670–671
Hooke's law, 224
House, wood frame construction, 10
Hydrostatic friction, 150

I

I-beams, 20–24, 666, 667, 674, 675
    aluminum, 20, 24, 674, 675
Impact energy, 229
Institute for Research in Construction-Canada, 49
International Code Council (ICC), 48
IZOD test, 230

J

J. B. Johnson formula for columns, 598
J, Polar moment of inertia, 324, 326, 327, 345

K

Keys, 207
Keyseats, 334, 700

L

Laminated composite construction, 257
Law of cosines, 39
Law of sines, 39
Length, effective, for columns, 594, 695
LRFD, Load and resistance factor design, 271, 418
Loads,
    bridge, 51
    concentrated, 359, 367
    distributed, uniformly, 34–36, 68–71, 360–362, 376
    distributed, varying, 36, 71–73, 362, 382
    highway, 51
    live, 49
    load path, 8
    load per unit area, 19
    roof, 49–50
    seismic, 51
    snow, 49–51
    table of, 50
    Wind, 51

M

Magnesium, 240, 681
    coefficient of thermal expansion, 278, 279
Malleable iron, 238, 684
Margin of safety, 265
Masonry, design bearing stress, 293, 294, 689
Mass, 17
Mass, force, and weight relationship, 17
Materials, anisotropic, 251, 433
    classifications, 256
    composites, 245–255 (See Composites)
    composites, beams, 434
    concrete, 242
    design properties, 222
    plastics, 86
    wood, 82
Materials selection, 255–257
Matrix materials for composites, 247
Maximum shear stress, 551, 557
Maximum shear stress theory of failure, 551, 569
Mean diameter for pressure vessels, 620
Mechanical advantage, 77
Mechanisms, 96
Mechanical devices, forces on, 12
Members made of more than one material, 283
Metals, 233–241
    brittle, 226
    classifications, 223
    ductile, 226
    proportional limit, 223
    stiffness, 224
    strength, 222–224
    stress-strain curve, 223
    tensile strength and ultimate tensile strength, 223
    yield strength and yield point, 223
Metric Unit System, 16
Metrication of structural shapes, 21
Military standards, 266
Modulus, flexural, 227
Modulus of elasticity, 224, 243
Modulus of elasticity in shear, 227, 338
Mohr's circle for stress, 558–568
    examples, 560–568
    procedure for drawing, 559
    special case when both principal stresses have same
      sign, 572
Moment of inertia of an area,
    definition, 174, 178
    for composite shapes, 179–182
    for shapes with rectangular parts, 182
    for structural shapes, 179
    general procedure for computing, 177
    parallel axis theorem, 176
Moments,
    bending moments in beams, 358, 368, 376, 381
    concentrated, 362, 383, 387
    definition, 5

N

National Research Council Canada, 49
Newton, the unit of force, 17
Newton's law of gravitation, 17

N

Nickel-based alloys, 241, 682
Noncircular sections, torsion, 344
    table, 345
Nonmetals in design, 241–255
Normal stress, 201
Normal stress element, 209
Normalizing, 235

O

Offset—for yield strength, 223
Overhanging beam, 363

P

Parallel axis theorem, 176
Parallelogram rule, 40
Percent elongation, 225
Photoelastic stress analysis, 209–210, 286, 413, 424
Pin joints, bearing stress in, 288
Pipe, steel, 20, 23, 676–677
Plastics, 244, 687
    coefficient of thermal expansion, 229
    selection, 244–245
Poisson's ratio, 226
    Table, 227
Polar moment of inertia, 324, 326–327, 345
Polar section modulus, 328
Power, 322–323
Power transmission systems, 12
Precipitation hardening steel, 236
Preferred basic sizes, 246, 658
Pressure, 19
Pressure vessels, 618–635
    ASME boiler & pressure vessel code, 634
    composites, 635
    cylinders, thin-walled, 622
    design considerations, 633
    design procedure, 625–627
    hoop stress, 623
    longitudinal stress, 622
    mean diameter, 620
    shearing stresses in, 631
    spheres, thin-walled, 620–622
    spreadsheet aid for analyzing, 630
    thick-walled, 620, 624–625
    thin-walled, 620–622
    wall thickness, thin/thick, 620
Principal stresses, 556
Problem solving, 15
Proportional limit, 223
Pulley systems, 77
Punching operation, 202

Q

Quenching, 235

R

Radius of gyration, 184, 595
Reactions at supports, 31, 43
    beams, 68–73, 364–367
Resistance factor, 271
Resultants, 36–42
Retaining rings, 334–335

Ring groove in tension, 287
Rivets, 644–645
Rope friction, 161
Rollers, bearing, 291
Rolling friction, 149–150
Rosettes, strain gage, 575–579
Rotational equilibrium, 75
Rotational speed, 322
Rule of mixtures, 254

S

S-shapes, structural steel, 20, 22, 668–669
Safety factor, 265
    (See also, Design factor)
Scissors-lift table, 105–107
Screw threads, 264, 659–660
Section modulus, 186, 419
    torsional, 328
Shafts with flats, 345
Shafts, solid and hollow, comparison, 330
Shapes, structural, 19–23, 662–677
Shear, double, 205
Shear, single, 205
Shear center, 428–429
Shear failure of rivets, 473, 644–645
Shear flow, 472
Shear modulus of elasticity, 227, 338
Shear pin, 297
Shear stress, design, 296, 328
    direct, 204
    maximum, 551, 557
    maximum, theory of failure, 551, 569
    ultimate strength in shear, 296
    yield strength in shear, 296
Shear stress element, 209
Shearing force, definition, 358, 367
Shearing force diagrams, 358, 371, 378
    for distributed loads, 376
    guidelines for drawing, 371, 378, 381
Shearing forces for cantilevers, 381
Shearing forces in beams, 356, 358, 371, 376
Shearing strain, 227, 332
Shearing stress in beams, 454
    circular shapes, 469
    design, 471
    distribution of, 462
    first moment of the area, 457–458
    general shear formula, 457
    development of, 467
    guidelines, 460
    hollow thin-walled tubular shapes, 470
    importance, 456
    rectangular shapes, 458, 463, 468
    special shear formulas, 468–471
    statical moment, 457–458
    stress distribution, 462
    stressed skin structures, 457
    thin-webbed, 457, 470
    tubular shapes, 470
    visualization, 454–455
    wooden, 456, 471
SI Metric Unit System, 16
Simple beam, 363, 489

Simply supported beams, formulas, 489, 701–703
Sines, law of, 39
Single shear, 205
Sizes, preferred basic, 264, 658
Slenderness ratio, 593
    transition, 596
Slug, the unit of mass, 17
Society of Automotive Engineers, 233
Soils, design bearing stress, 294, 689
Special shear formulas, 468–471
Specific modulus, 248
Specific stiffness, 248
Specific strength, 248
Specific weight, 18
Spheres, thick-walled, 620, 624–625
    thin-walled, 620
Stability, 197, 412, 591, 598
Stainless steel, 236, 681
    coefficient of thermal expansion, 278
Standard structural shapes, 19–23, 662–677
Static equilibrium, 64–79
    equations of, 65
    examples, 67–76
    pulley systems, 77
    rotational, 75
    structures, 73
Statical moment, 457
Statically indeterminate beams, 393, 705–709
Statically indeterminate members, axial, 283
Steel, 233, 680–683
    alloy, 234, 680
    annealing, 235
    carbon, 234, 680
    carbon and alloy, properties, 680–683
    coefficient of thermal expansion, 278
    cold drawn, 234
    cold rolled, 234
    conditions for, 234
    designation system, 233
    hot rolled, 234
    normalizing, 235
    precipitation hardening, 236
    quenching, 235
    stainless, 236, 681
    stress relief annealing, 236
    structural, 237–238, 683
    tempering, 235
    weathering, 237
Stiffness, 224
    beam, 488
    torsional, 37, 345
Strain, definition, 204
    shearing, 227, 332
Strain gaging, 210
    Rosettes, 575–579
Strength, ultimate tensile, 222
    yield, 223
Stress concentration factors, charts, 691–700
    in beams, 424
    in combined stress, 552
    normal stresses, 285–287
    torsion, 332
Stress condition on selected planes, 554, 569

Stress element, normal stress, 209
    general, 539, 543
    shear stress, 209, 227
Stress, apparent, 222
    bearing, 287
    bending, 409–419
        distribution, 413
        derivation, 414
        design, 417–419
    compressive, 202, 224
    contact, 289
    definition, 201
    design for bending, 417, 419
    hoop, pressure vessels, 623
    longitudinal, pressure vessels, 622
    normal, 201
    shear, direct, 204
    tensile, 202
    thermal, 281
    units, 201
Stress-relief annealing, 235
Stress-strain curve, 223
Stresses, allowable on connections, 642, 648–649
    combined, normal, 545–550
    combined, normal and shear, 550, 554
    principal, 556
Structures, 8, 93–130
    forces in, 93
    frames, pin-connected, 97–106
    three-dimensional, 93, 123–130
    trusses, 93, 107–123
Structural shapes, 19–23, 662–677
    Metrication, 21
    designations, 20
    torsional properties, 346
St. Venant principal, 424
Superposition, for statically indeterminate beams, 501
Support types, 43–44, 359, 362–363, 483–485

T

Tempering, 235, 239
Tensile stress, direct, 201–202
Theorem of three moments, 393, 396
Thermal expansion, 278
Thermal stress, 281
Thermoplastic plastic, 244
Thermosetting plastic, 244
Three-dimensional structures, 93, 123–130
    direction angles, 124
    direction cosines, 124
    force vectors, 124
    moments in, 126
    resultants, 124
    summary, 129–130
Three moment theorem, 393–396
Tipping and sliding, friction, 154–157
Titanium, 241, 682
    coefficient of thermal expansion, 278
Torque, 6, 318, 321
    equivalent, 551–552
Torque sensors, 323
Torsional deformation, 332, 345

Torsional shear stress, 323–344
    circular sections, 324, 326
    design, 328
    formula, 324
    formula, derivation, 325
    hollow circular bars, 326–327, 330
    noncircular sections, 344
    stress concentrations, 332, 695–697
Torsional stiffness, 337, 345
    recommended, 337
    torsional properties of structural shapes, 346
Toughness, 229
Truck loads, 52
Trusses, 93, 107–122
    arrangements in structures, 112–113
    force analysis, 113–123
    method of joints, 114–120
    method of sections, 120–123
    statically determinate, conditions, 113–114
    types, 110–112
    uses for, 107–110
Tubing, mechanical, 20, 22, 24, 678–679
Tubing, structural, 22, 670–671
Twisting, 320, 337
Two-force member, 96
Tup, for drop testing, 230

U

U.S. Customary Unit System, 16
Ultimate strength in shear, 296
Ultimate tensile strength, 222
Unified Numbering System, 233
Unit systems, 16
Units, stress, 19, 201

V

Vector triangle, 39
Volume fraction, 253

W

Weathering steel, 237
Wedges, 157–159
Weight, 17, 31
Welded joints, 648–649
Welding electrodes, 649
Welds, types of, 649
Wheatstone bridge, 210, 212
White iron, 238
Wide-flange shapes (W-shapes), 20, 22, 666–667
Wood, 241–242
    properties, 242, 686
Wood beam shapes, 19, 661
Wood frame house construction, 10
Working stress, 265

Y

Yield point, 223
Yield strength, 223–224
Yield strength in shear, 296

Z

Zinc, 240, 681